THIRD EDITION

BACTERIAL PATHOGENESIS

A Molecular Approach

THIRD EDITION

BACTERIAL PATHOGENESIS

A Molecular Approach

BRENDA A. WILSON
Department of Microbiology, University of Illinois,
Urbana-Champaign, Illinois

ABIGAIL A. SALYERS
Department of Microbiology, University of Illinois,
Urbana-Champaign, Illinois

DIXIE D. WHITT
Department of Microbiology, University of Illinois,
Urbana-Champaign, Illinois

MALCOLM E. WINKLER
Department of Biology, Indiana University,
Bloomington, Indiana

ASM
PRESS
WASHINGTON, DC

Address editorial correspondence to ASM Press, 1752 N St. NW, Washington, DC 20036-2904, USA

Send orders to ASM Press, P.O. Box 605, Herndon, VA 20172, USA
Phone: 800-546-2416; 703-661-1593
Fax: 703-661-1501
E-mail: books@asmusa.org
Online: estore.asm.org

Library of Congress Cataloging-in-Publication Data

Bacterial pathogenesis : a molecular approach / Brenda A. Wilson ... [et al.].—3rd ed.
    p. ; cm.
  Rev. ed. of: Bacterial pathogenesis : a molecular approach / Abigail A. Salyers and Dixie D. Whitt. 2nd ed. c2002.
  Includes bibliographical references and index.
  ISBN 978-1-55581-418-2 (softcover : alk. paper) 1. Bacterial diseases—Pathogenesis. 2. Molecular microbiology. I. Wilson, Brenda A. II. Salyers, Abigail A. Bacterial pathogenesis.
  [DNLM: 1. Bacteria—pathogenicity. 2. Bacterial Infections—etiology. 3. Bacterial Infections—prevention & control. 4. Host-Parasite Interactions. 5. Virulence. QZ 65]
  QR201.B34S24 2011
  616.9′207—dc22
                                                                2010029309

Current printing (last digit)
10 9 8 7 6 5 4 3 2 1

Cover and interior design: Susan Brown Schmidler

To Stanley Falkow and Charles Yanofsky
in recognition of their immense contributions to
bacterial pathogenesis and molecular genetics

Contents

Preface

Since the second edition of *Bacterial Pathogenesis* was written, the field has changed and matured considerably. These changes have necessitated an extensive rewriting of the textbook. One of these changes is evident from the front cover: there are two new authors, Brenda Wilson and Malcolm Winkler. When Abigail Salyers and Dixie Whitt first wrote *Bacterial Pathogenesis*, it could be argued that there was an advantage to having authors who were not in the center of pathogenesis research. An outside perspective of the emerging new discipline had the merit of not being committed to any particular perspective. Now, as the field has begun to take new form and importance, this advantage is no longer so clear. The new authors are at the center of pathogenesis research and thus bring a more immediate, modern perspective to the subject. The rewriting of the textbook reflects their expertise and insights.

Previous editions of this book have tended to take an organism-based approach, with separate chapters devoted to individual bacterial pathogens. Such a format can be defended on the basis that each bacterial pathogen has its own personality, but it obscures the emergence of underlying similarities among pathogens and their mechanisms of action that have been emerging over the past several years. The third edition provides a more accurate representation of the way in which scientists now view the field. This shift in emphasis should help instructors who are teaching a one-semester course because it focuses attention on core principles that are better adapted to a single semester.

More emphasis is given to a topic that has received increasing attention during the past several years: the importance of the normal microbial populations of the human body in health and disease. Although these populations are normally beneficial or neutral, some components of these populations, called opportunistic pathogens, can cause disease and are responsible for more infections today than classical pathogens such as the bacterium that causes plague. The reality that these opportunistic pathogens can and do cause serious diseases has raised questions about some of the original concepts of virulence factors. It may be that the ultimate virulence factor for many disease-causing organisms is not a single protein toxin or a surface adhesion molecule but rather the ability to survive if the organism manages to enter tissue or the bloodstream of the host.

An important addition to this book is the inclusion at the end of many chapters of problems based on research examples. In previous editions, there were study questions at the end of each chapter that were designed to challenge students to probe more deeply into issues covered in the chapter. Now, there are also problems that take actual data generated in research laboratories and challenge students to interpret the data and to then think of new experiments that could be performed. Most textbooks tend to present material as something to memorize and regurgitate on tests. There is little or no indication of what it is like to do the experimental research that develops into the material included in a textbook. The addition of actual research problems to the discussion questions provides students with insight into this continuing research process. We are grateful to all of our colleagues for helpful discussions and insights that have contributed to the enhancement of the educational experience for the students through this means. In particular, we thank Shelley Haydel, David Nunn, Richard Tapping, James Slauch, Joseph Barbieri, and the late Roderick MacLeod for sharing their expertise and a few of their favorite pathogenesis problems.

One aspect of this textbook is preserved. This is a textbook that is written for students, not for professors. A past criticism of this text made by some is that it occasionally lapses into moments of unseemly levity or that it is not suitably "serious." In our opinion, "professional" is not synonymous with dull. Research is a messy, sometimes humorous, enterprise. Why hide this from students? Why should we deny the excitement and fun of discovery in the name of making our discipline seem "professional"? We remain unrepentant renegades in this regard.

A student-friendly aspect of this text, which is not usually mentioned in prefaces, is the continuing attempt to keep the cost of the textbook to a minimum. One way to do this is to publish the book in paperback. Another way of keeping the cost of the textbook down is the plain format of the figures. We were advised initially that this plain format was a mistake because students want lots of color and intricate details. In fact, surprisingly, one of the most common compliments that we have received on the first two editions was about the simplicity of the figures. Apparently, students appreciate any help they can get that enables them to focus on the important aspects of the material.

A final important feature of this edition of the textbook is that it is published by ASM Press. In our experience as authors, this publisher is unparalleled for its ability to get thorough, and in some cases painful, reviews of every chapter from research scientists and teachers. Whereas most publishers are interested in hearing how great a chapter is, ASM Press actually wants to know what is wrong with the text material, and they make sure that the authors take all criticisms seriously and make suitable modifications. The authors in turn welcome any suggestions or corrections to the material presented that might enhance the learning experience of the students. We want to make a special acknowledgment of the contribution of Senior Editor Greg Payne, the editor who has been most intimately involved in the development of this textbook and in keeping us on track. We also give special thanks to Production Editor Susan Birch whose attention to detail has served us well through all three editions of this textbook. Cathy Balogh deserves thanks for her assistance as well. All of the ASM Press staff has been helpful and conscientious.

Textbooks are notorious for being ossified entities that rarely change except in minor details. Our textbook breaks free from this image. This textbook is a living thing that changes in response to the constantly changing landscape of research in the exciting field of bacterial pathogenesis. The field of pathogenesis is changing at such a rapid rate that there should probably be a new edition every year. The fact that this has not been, and is unlikely to be the case in the future is a reflection of the fact that the authors are all active research scientists and teachers, not professional textbook writers. Fortunately, the increased Internet access to research publications makes it possible for students themselves to become active participants in the virtual upgrading of the content of this textbook. This textbook should be considered as a template onto which emerging research findings can be applied.

1

The Power of Bacteria

I t is unwise to underestimate a potential adversary that has had a 3-billion-year evolutionary head start.

Why Are Bacteria Once Again in the Public Health Spotlight?

Antibiotics were first introduced into widespread clinical use in the 1950s. The term given them at the time, "miracle drugs," exemplifies the euphoria felt by physicians and the public when this new therapy became available. They came at a time when the medical community was gaining greater control over infectious diseases than ever before. In clinics and hospitals, hygienic practices, such as hand washing and disinfectant use, were reducing the risk of disease transmission. In the community, improved nutrition made people better able to resist infections, while less crowded conditions and a clean water supply had reduced disease transmission. Vaccines gave protection against some much-feared diseases. Nonetheless, bacterial infections, such as pneumonia, tuberculosis, and syphilis, continued to take a heavy toll, and infectious diseases were still a leading cause of death. Antibiotics appeared to be the superweapon that would give humans the final decisive victory over bacteria.

In the early euphoria over the success of antibiotics, scientists and policy makers alike concluded that bacterial infections were no longer a threat and turned their attention to other problems, such as cancer, heart disease, and viral infections. For the next 3 decades, bacteria were of interest mainly as tractable model systems for studying physiology and genetics and as a source of tools for the new molecular biology and genetic engineering techniques that were revolutionizing all of biology. Confidence that bacterial diseases were completely under control was also bolstered by the fact that there was a glut of new antibiotics on the market.

There is a story still circulating about a comment allegedly made by the U.S. Surgeon General William H. Stewart while testifying before Congress in the late 1960s. He is supposed to have stated

1

confidently that it was "time to close the book on infectious disease," meaning primarily bacterial diseases, because these diseases could now be treated easily with antibiotics. He further went on to state, so the story goes, that it was time to move on to other illnesses, such as cancer and heart disease. Apparently, there is no evidence that this statement was ever made, at least by the U.S. Surgeon General, but it expressed a perspective that was widely shared in the medical and research community at the time.

Unnoticed by all but a few, the first cracks soon began to appear in the protective shield against bacterial disease. Antibiotics were no longer the highly profitable products they had once been, especially compared to heart medications or tranquilizers, which had to be taken daily for long periods of time. Also, new antibiotics were becoming harder to discover and more expensive to develop. One pharmaceutical company after another quietly cut back or dismantled its antibiotic discovery programs. For a while, the cracks appeared not to matter. There were enough new antibiotics that still worked on the bacteria that had become resistant to the old standbys, like penicillin. Warnings from scientists that bacteria were rapidly becoming even more resistant to antibiotics, especially the newer ones, were largely ignored.

During the late 1980s, however, scientists and health officials began to notice an alarming increase in bacterial infections. By 1995, infectious diseases became one of the top five causes of death in the United States. Even with the AIDS epidemic in full swing, most infectious disease deaths were caused by bacterial diseases, such as pneumonia and bacterial bloodstream infections (sepsis). Why was the incidence of bacterial pneumonia and sepsis increasing? For one thing, the population was aging, and older people are more likely to contract these diseases. For another, modern medicine had created an increasingly large population of patients whose immune systems had been temporarily disrupted due to cancer chemotherapy or immunosuppressive therapy following organ transplants.

Another development that caught many in the medical community by surprise was the appearance of new diseases that were dubbed **emerging infectious diseases.** In the past, scientists had assumed that any microorganism capable of causing disease would surely have done so by now, given the millions of years humans had occupied the planet. This view overlooked two important facts. First, bacteria can change their genetic makeup very rapidly to take advantage of new opportunities. Members of some bacterial populations are hypermutable, making it possible for them to try many genetic combinations in

seeking the one that is most appropriate for the current environment the bacterium is experiencing. Also, bacteria can acquire genes that confer new virulence traits or resistance to antibiotics from other bacteria through a phenomenon known as **horizontal gene transfer.** Second, changing human practices, such as increased global travel, the widespread use of air-conditioning, and the creation of crowded intensive-care wards in big hospitals, brought susceptible people into contact with microorganisms that had not previously had the opportunity to cause human infections.

A new category of disease-causing bacteria was recognized: **opportunistic pathogens.** These bacteria normally were unable to cause disease in healthy people but could infect and cause disease in people whose defenses were compromised in some way. In fact, many of the opportunists were normally found in or on the human body and had thus been assumed to be innocuous. Others were bacteria commonly found in soil. During the early antibiotic era, these soil bacteria were thought to be beneficial to humans because scientists were finding that many of them were producers of antibiotics. However, these bacteria were suddenly being seen as the only bacterium isolated from the blood, lungs, or wounds of seriously ill patients. They also tended to exhibit a troubling not-so-friendly characteristic. Because of the antibiotics present in their natural environment, they were often resistant to a variety of antibiotics, a fact that made opportunistic infections by these bacteria hard to treat.

Scientists and physicians were reluctantly beginning to realize that a decisive human victory over bacteria had not occurred and that the problem was not just that known pathogens were changing to be more resistant to antibiotics or more able to cause disease. New pathogens were emerging. The infectious disease picture was changing in a way that made it increasingly difficult to predict new patterns of bacterial disease.

Bacteria, an Ancient Life Form

The forgoing brief account of how bacterial diseases have come back into prominence as a health problem explored the recent past. However, to understand fully why no one should have been surprised by this development and why bacteria are such formidable opponents, it is necessary to take a closer look at the long history of bacteria, a history that explains their impressive ability to adapt to new conditions.

Today, we realize that Earth is a microbial planet. Bacteria were probably the first form of life to appear on Earth, about 3.5 to 4 billion years ago (Figure 1–1). Bacteria and another type of prokaryote, the archaea, ruled the world undisputed for at least a billion years before the first eukaryotes appeared. During this period, they created the global geochemical cycles that made the Earth habitable for larger life forms. Bacteria and archaea are master recyclers. Bacteria put the first molecular oxygen in the Earth's atmosphere, creating the ozone layer, which protected the Earth's surface from the killing radiation that had formerly bombarded it. Life on the Earth's surface was thereby made possible. By adding molecular oxygen to the atmosphere, bacteria also created conditions that permitted the later evolution of oxygen-utilizing creatures, including us.

In the course of their long history, bacteria developed a variety of metabolic capabilities that allowed them to survive under an impressive range of conditions. There are bacteria that can obtain energy by oxidizing sulfides, by reducing sulfate, by oxidizing ammonia, by reducing nitrate, and by oxidizing methane, to name only a few of the vast number of metabolic types represented in the bacterial world. Bacteria also learned how to maximize the plasticity of their genomes, constantly acquiring new DNA and mutating or rearranging existing genes. In this way, they evolved new capabilities that enabled them to colonize the many niches the Earth provided. So far, no part of the Earth has been found to be free of bacteria. They can be found in arctic ice, in the deep subsurface of landmasses, on the surfaces and in the depths of the oceans, and in boiling hot springs. The genetic plasticity that made the evolution of such metabolically diverse organisms possible stands them in good stead today as they face new challenges and opportunities.

Bacteria became specialists in metabolic diversity. In fact, acquisition of bacteria or archaea as coinhabitants **(endosymbionts)** enabled eukaryotes to expand their metabolic diversity. Cells of what later became plants, for example, acquired the ability to photosynthesize by acquiring photosynthetic bacteria as endosymbionts. Some plants also recruited for their root cells bacterial endosymbionts that could fix atmospheric nitrogen into nitrogenous compounds that the plant could use as fertilizer.

Bacteria were also quick to take advantage of the warm, wet environment offered by the intestinal tracts of animals and humans, and in turn, many of the bacteria help to provide nutrients for the animal or human host. There is no point in human history, except for the brief time the fetus spends in the uterus, when the human body is not heavily colonized by high concentrations of bacteria, especially on the skin, in the mouth, in the intestinal tract, and in the vaginal tract. These bacteria are highly adapted to the conditions they encounter in and on the human body, and their constant presence puts them in a position to take advantage of any breach in the defenses that protect the interior tissues and the bloodstream from bacterial invasion.

About a billion years after bacteria first came into existence on Earth, the first **eukaryotes,** the single-celled **protozoa,** appeared. Although some eukaryotic microbes, such as algae, have a photosynthetic lifestyle, many others, especially amoebas and other protozoa, live by feeding on bacteria and archaea. There is an interesting aspect of protozoal grazing that is seldom mentioned but that has attracted some attention recently. Protozoa have properties that are remarkably similar to those of human phagocytic cells, cells that form an important part of the defenses of the human body. Some of these human phagocytic cells function mainly to engulf and kill bacteria in blood and tissues. Others engulf and then break down bacteria and present segments of their proteins to the cells of the immune system.

Bacteria capable of causing disease are often able to do so because they have developed strategies for evading phagocytosis (engulfment) or for surviving inside phagocytic cells. The evolution of such strategies could have begun soon after the appearance of the first protozoa, well in advance of the appearance of animals and humans. Some of the toxic proteins that disease-causing bacteria use to kill human cells could have evolved originally to allow the bacteria to evade or survive predation by their protozoal adversaries. Today, scientists are finding that some bacterial pathogens that are harmful to humans normally live inside amoebas in nature. If this view of bacterial evolution is correct, then there are likely to be more unidentified disease-causing bacteria in nature than we thought.

When animals and humans finally appeared on the evolutionary scene, bacteria immediately took advan-

Figure 1–1 Overview of microbial evolution. Microorganisms appeared 3.5 to 4 billion years ago and changed Earth so that eukaryotes could evolve.

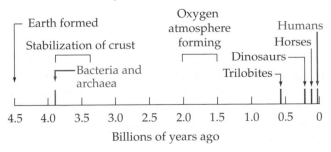

Billions of years ago

tage of them as rich niches in which to grow. To a bacterium used to the vagaries of the external environment, where the temperature and the availability of water and nutrients can vary widely (and unpredictably), a warm-blooded animal whose temperature is stably maintained and which spends most of its time searching for food and water must be as close as it gets to bacterial heaven. Given this, it should not surprise us that the bodies of humans and animals carry dense bacterial loads, especially in the mouth, intestinal tract, and vaginal tract. Small wonder that the human or animal body is often referred to in the scientific literature as the **"host."**

Scientists who study the evolution of insects, animals, and humans have almost completely ignored the selective pressure exerted by the long-term presence of large and diverse populations of bacteria. As will become evident below, the effects of microbial pressure can be seen clearly in the design of the human skin, eyes, lungs, intestinal tract, and vaginal tract, and in particular the immune system.

Pressing Current Infectious Disease Issues

Having made this digression into ancient history, let us now return to the present and examine some of the burning public health issues that have brought bacterial infections once again to the forefront. These include emerging infectious diseases, increasing problems with large outbreaks of **food-borne** and **water-borne infections, hospital-acquired (nosocomial) infections, bioterrorism, antibiotic resistance, microbiota shift diseases, pathogen evolution,** and **disease transmission.**

Emerging and Reemerging Infectious Diseases

The emergence of apparently new bacterial diseases and the reemergence of old diseases that were thought to be under control (at least in developed countries) were an unpleasant shock to the health care community. Emerging and reemerging infectious diseases illustrate an important principle: disease patterns change, both because bacteria change and because changing human activities can create new opportunities for bacteria to cause disease.

Not all diseases are truly emerging in the sense of being completely new to the human population. In some cases, the disease symptoms have been around for a long time as a known disease, but the bacterial cause has only recently been identified. A good example of this phenomenon is gastric ulcers, which are now known to be caused by a bacterium called *Helicobacter pylori*. This bacterium was missed previously because the methods for cultivating and identifying it were not yet commonly used and because many medical researchers were convinced that no bacterium could colonize the human stomach. Such diseases are "emerging" only in the sense of awareness by the public, not in the minds of the scientists who study them.

Old diseases can return if the conditions change to favor their reemergence. Tuberculosis made a comeback in developed countries, such as the United States and Europe, during the 1980s and 1990s as a result of the dismantling of the preventive infrastructure that had been developed in the 1950s to contain the spread of this insidious disease, plus the presence of unprecedented large populations of highly susceptible populations in crowded settings, such as prisons and homeless shelters. The reemergence of tuberculosis was further complicated by the emergence of diseases, such as AIDS, that suppress the immune system. Another factor was the development of resistance to the traditional antituberculosis drugs, which had not been updated or improved since their original introduction because no one thought tuberculosis would return.

Food-Borne and Water-Borne Infections

Many food-borne and water-borne infections fit the category of emerging or reemerging infectious diseases. We treat this subject as a separate category because of its unique impact on the public perception of disease risks. Ironically, as the food and water supplies have become cleaner, the public's concern about their integrity has become greater rather than less. A review of news articles from the past few years makes this quite evident. The integrity of the food and water supplies is a nonnegotiable issue as far as the public is concerned.

From the 1960s through the 1980s, the public's main concern about the food supply focused on pesticide residues and other chemical adulterants that might cause cancer. This problem has been largely solved by more stringent regulations, which limit the amount of pesticide residues and other harmful chemicals that might be found in food sold for human consumption. More recently, concern has arisen about another hazard that had been around all along but had not been perceived as a threat: **food-borne bacterial diseases.** The catalyst for this abrupt swing in public concern was *Escherichia coli* O157:H7, a type of *E. coli* that can cause kidney failure and death, especially in children. This problem first attracted attention when an outbreak of disease was caused by undercooked, contaminated hamburger dispensed by a fast-food chain. Cases occurred in many western and northwestern states before the source of the outbreak was identified and the contaminated meat was recalled. Since then, there have been numerous cases of *E. coli* O157:H7 infections spread by undercooked meat, radish sprouts, and even apple juice. The apple juice incident, where only one death was involved, nearly bankrupted the company that had produced the contaminated juice, which had not been pasteurized. The lesson that juice was not exempt from contamination was learned very quickly by the industry, and it is now rare to see unpasteurized juices in supermarkets.

Earlier in the 20th century, before the advent of centralized food processing and distribution, food-borne disease outbreaks tended to be confined to church so-cials, family gatherings, or business-sponsored employee picnics (Box 1–1). As the food industry became more centralized, however, a different pattern of food-borne disease emerged: the multistate (or even multicountry) outbreak of food-borne disease derived from a single source. In the case of the *E. coli* O157:H7 outbreak mentioned above, a single processing plant was the source. Contaminated radish sprouts caused another outbreak of *E. coli* O157:H7 infection in Japanese school children. The seeds used by the Japanese sprouting companies came from a single source in the northwestern United States, where the initial contamination event probably occurred. In 2006, *E. coli* O157:H7-contaminated spinach grown in California and used by consumers in spinach salads was spread throughout the United States. More re-

BOX 1–1   Terrorism Hits Oregon Salad Bars

In 1984, a large outbreak of salmonellosis involving at least 750 people occurred in The Dalles, the county seat of Wasco County in Oregon. At the time, a religious commune was at odds with long-term local inhabitants over land use restrictions placed on the commune in an attempt by the townspeople to eliminate it. Members of the commune felt that the outcome of an upcoming election was critical to their future ability to grow. In an apparent attempt to disrupt the election, commune members planned to cause an outbreak of salmonellosis that would keep people home from the polls. The actual outbreak was a trial run to determine the best way to create the most havoc. At least 10 restaurants were involved, with the salad bar being the main site of intentional contamination. Contamination attempts were also made at local grocery stores, but the most effective source of disease was the restaurant salad bars. Unfortunately for commune members, their trial run was too successful and attracted the attention of the Public Health Department and the police.

Even so, it took nearly a year to trace the epidemic source and to suspect intentional contamination. Such events are fortunately quite rare, and thus, intentional contamination was not even considered at first as a possible explanation for the outbreak. Careful questioning of the victims led investigators to deduce that salad bars had been the source of the outbreak. This in itself was somewhat unusual, because large outbreaks of salmonellosis are usually associated with meat, milk, or eggs. Such infections acquired from vegetables, while not unheard of, are relatively uncommon. Meanwhile, interrogation of commune members by police and FBI agents revealed that the commune had been the source of the outbreak. The commune had its own laboratory, where the strain of *Salmonella enterica* serovar Typhimurium was grown and prepared for inoculation of the salad bars. Commune members had apparently gotten the strain by ordering it from the American Type Culture Collection, a widely respected repository of bacterial strains that distributes strains to scientific laboratories for a modest fee. Nearly 2 years after the outbreak, two commune members were sentenced to 1 to 2 years in prison for conspiring to tamper with consumer products. An earlier episode of product tampering involving the introduction of cyanide into Tylenol capsules had been responsible for a rash of antitampering legislation. The antitampering laws were used to prosecute the commune members.

Source: T. J. Torok, R. V. Tauxe, R. P. Wise, J. R. Livengood, R. Sokolow, S. Mauvais, K. S. Birkness, M. R. Skeels, J. M. Horan, and L. R. Foster. 1997. A large community outbreak of salmonellosis caused by intentional contamination of restaurant salad bars. *JAMA* **278**:389–395.

cently, in 2009, *E. coli* O157:H7 showed up again, this time in refrigerated cookie dough sold throughout the United States. The nature of food-borne outbreaks has changed considerably from the days of the church social-type outbreaks, and the effect on the public perception of food safety has been dramatic.

Outbreaks of water-borne disease continue to be fairly localized in occurrence. Most of the water-borne outbreaks that have made the news lately have been caused by protozoal parasites, such as *Cryptosporidium parvum,* but the same conditions that allow protozoal contamination of water—aging pipes and water treatment plants or mammal and bird fecal material in water reservoirs—could easily spread bacterial water-borne diseases in the future. The media rapidly picked up on this, as was seen on a cover of *Time* magazine (3 August 1998) showing *E. coli* cascading out of a kitchen tap. The event that stimulated this media response was an epidemic that occurred among 61 residents of Alpine, WY (total population, 470), who drank tap water on one particular weekend. The spring that was the source of the town's tap water had probably been contaminated by wild animals.

An often-overlooked aspect of water-borne infectious microorganisms is that contaminated water can produce a contaminated food product if that water is used to wash the food. In most cases, water used to wash the dirt off fruits and vegetables prior to shipping is not tap water quality and can be what is referred to as "gray water," water processed to remove the worst contamination but not microbiologically clean. In some food-borne outbreaks, the contamination could actually have come from the water used to wash the food. At first, vegetarians felt safe because food-borne diseases were so often spread by meat. However, most people concerned with ensuring food safety privately consider foods that are usually consumed raw, such as fruits and vegetables, a potentially more serious threat to public health. If meat or other food is properly cooked, the contamination problem is solved, but even careful washing of raw fruits or vegetables is not always sufficient to render contaminated raw food safe.

Modern Medicine as a Source of New Diseases

Modern medicine has made impressive breakthroughs in therapies for many human diseases. Surgeons now routinely transplant new organs into patients whose own organs are failing. Cancer chemotherapy is becoming more and more effective. This progress has had a cost, however. Transplant patients and patients receiving cancer chemotherapy

have suppressed immune systems due to the medications they are taking. This immunosuppression is temporary in the case of cancer patients and ends when the chemotherapy is finished, but transplant patients usually take their immunosuppressive therapy for life. Also, there are a large number of individuals suffering from infectious diseases, such as AIDS, that are immunosuppressive. Not surprisingly, these immunocompromised patients can become infected with bacteria never before suspected of being able to cause human disease.

Other bacteria cause disease, not because the patient is immunocompromised, but because a physical barrier against bacterial invasion is bypassed. For example, accidental perforation of the colon during surgery releases bacteria into tissue and blood. Patients with certain types of respiratory infections may have ventilator tubes inserted to keep their airways open. This may allow bacteria to bypass some of the defenses of the respiratory tract and directly enter the lungs.

These bacteria that are normally not able to cause disease but can do so if they have opportunities have been called "opportunistic pathogens," because some defense of the body that normally keeps them at bay has to be breached in order for them to have the opportunity to cause infection. For an interesting recent example of an opportunist, see Box 1–2. The term "opportunist" makes such bacteria seem somehow less dangerous than "real" disease-causing bacteria. Do not be fooled by the seemingly innocuous nature of the opportunists. In most developed countries, a person is far more likely to die from an opportunistic infection than from the epidemic diseases that serve as the public's mental image of infectious diseases.

Another way in which modern medicine has affected the infectious disease picture is by increasing the human life span. The increasing number of elderly people, whose immune defenses are beginning to decline and who are more likely to be receiving therapies that undermine the defenses of the body, provides an expanding population of individuals highly susceptible to diseases. Put these elderly people in crowded conditions, such as those experienced in nursing homes, and an even greater opportunity is created for infectious diseases to spread.

Postsurgical and Other Wound Infections

Most recent studies of wound infections have focused on the infections that can be a serious complication of surgery (**postsurgical infections**). In the preantibiotic era, infections were a major complication of surgery. However skillful the surgeon, an infection could kill

BOX 1–2 Enterprising Bacteria Always on the Alert for New Infection Opportunities

An example of how bacteria can rapidly act to take advantage of new opportunities is provided by an outbreak of pneumonia in an intensive-care ward. Many of the patients were very ill and were on respirators to support breathing. The type of respirator being used required that a tube be inserted deep into the airway of the lungs, an ideal conduit to carry bacteria into the lung. After a number of cases of respirator-associated lung infections, hospital personnel had learned to be very careful not to contaminate the respirator itself or to allow bacterial contaminants to enter the air being forced into the lung.

No one, however, thought about mouthwash. Since patients on respirators are often unable to attend to their own dental hygiene, hospital staff workers use mouthwash to clean and freshen the mouth every day. The cause of the lung infections was identified as *Burkholderia* (formerly *Pseudomonas*) *cepacia*.

Although this species has caused some infections in people with cystic fibrosis, it is generally considered to be a relatively innocuous soil bacterium. In fact, it is used as a biocontrol agent to degrade herbicides, such as 2,4,5-trichlorophenoxyacetic acid. *B. cepacia* is ubiquitous in soil and water and apparently managed to contaminate many lots of the mouthwash, which did not contain alcohol and thus contained nothing to discourage bacterial growth. In effect, the hospital workers taking care of the respirator patients were inoculating their teeth and gums daily with a contaminated solution, which placed the bacteria in an ideal location to gain access to the lungs.

Source: Centers for Disease Control and Prevention. 1998. Nosocomial *Burkholderia cepacia* infection and colonization associated with intrinsically contaminated mouthwash—Arizona 1998. *MMWR Morb. Mortal. Wkly. Rep.* **47:**926–928.

the recipient of the most successful surgery. This may have been the origin of the grim old joke that the surgery was a success, but the patient died. Antibiotics changed all this and made routine surgery possible because antibiotics eliminated any bacteria that managed to penetrate the barrier of the surgical scrub and other hygienic procedures.

The first shadow in this rosy picture appeared when surgeons and other health care workers began to take patient survival for granted and to become more lax in time-consuming hygienic practices. To make matters worse, hospitals tried to save money by cutting funds for nurses and janitors, people who had been responsible for the cleanliness that used to characterize hospitals in developed countries. Unfortunately, the bacteria that often cause postsurgical problems tend to be resistant to antibiotics. This has consequences for hospitals, as well as for patients. For patients, of course, the serious consequence is damage to major organs or even death. For hospitals and insurance companies, the consequence is much higher costs for patients who contract postsurgical infections.

The state of Pennsylvania made history not long ago by publishing postsurgical-infection data from its hospitals. Until this unprecedented move to transparency, infection rates in hospitals were secrets guarded almost as fiercely as top secret data in the CIA files. The reason is easy to understand. No hospital wants potential users of its facilities, especially people getting elective surgery, to identify the hospital as a place where people go in healthy and come out dead. The Pennsylvania figures confirmed what everyone in the infectious disease community already knew: patients who contracted a postsurgical infection, especially one caused by antibiotic-resistant bacteria, cost over four times more to treat than people who did not contract an infection. Unfortunately, this type of statistic has attracted a lot more attention than the suffering of the patients involved, but the good news is that something may now be done that will help protect the people who go into hospitals.

Recently, attention has once again focused on another old problem: war-related infections. Accounts of the antibiotic revolution often point out that due to a combination of antibiotics and improved surgical interventions, wound infections that once killed soldiers more often than the trauma of the wounds themselves are now readily treated. World War I was the last war in which infectious diseases—not just wound infections, but also diarrhea and pneumonia—were the main cause of soldiers' deaths.

An ominous development has been the appearance of a soil bacterium, *Acinetobacter baumannii,* as a wound infection problem in soldiers. *A. baumannii* was not a stranger to microbiologists when it made its appearance as a problem for soldiers wounded in the Iraq war. There had been a few outbreaks in intensive-care wards, but it was only in military hospitals during the Iraq war that *A. baumannii* began to attract real attention. *A. baumannii*'s claim to fame is that it was one of the first bacteria to be called "pan-resistant" because it is resistant to almost all antibiotics. It had an illustrious precursor: *Pseudomonas aeruginosa,* another soil bacterium that is resistant to many antibiotics. *P. aeruginosa* has long been known as an infectious disease problem in burn patients and in patients with cystic fibrosis. *A. baumannii* is, if anything, worse than *P. aeruginosa* on the resistance front. A number of other soil bacteria and bacteria normally found in or on the human body also seem to be resistant to multiple antibiotics. Unfortunately, these ubiquitous bacteria are common causes of hospital-acquired infections.

A new understanding of antibiotic resistance is that the physiological state of a bacterium can be as important as its complement of resistance genes. Many bacteria can form **biofilms,** multilayer assemblages of bacteria that are held together by a sticky polysaccharide matrix. Biofilms are found in many places in nature, most notably in areas such as streams, where the fast flow of water makes it necessary for bacteria to resort to biofilm formation to stay in a particular site. In hospital patients, biofilms seem to form very readily on plastic or metal implants and are very difficult to eliminate with antibiotics because bacteria in the biofilm become much less susceptible to antibiotics, although the reasons for this are not yet clear. Frequently, a patient with a biofilm-contaminated implant has to submit to a second surgery that removes the implant so that antibiotics can be used to eliminate any remaining bacteria prior to another try at inserting a new implant.

Bioterrorism

No discussion of current infectious disease issues would be complete without a mention of bioterrorism. Germ warfare—using infectious microorganisms as weapons—is an old idea that has, fortunately, never worked very well. The nature of germ warfare has been changing in recent years. In the past, the purpose of germ warfare was to kill or incapacitate large numbers of soldiers. Recently, the aim of terrorists has changed. Now, the goal is to frighten the general population. A small number of deaths are sufficient to achieve this goal. Among bacteria, *Bacillus anthracis,* the cause of anthrax, has been identified as a particularly useful weapon by bioterrorists. *B. anthracis* is a spore former and is thus easier to store and to "weaponize" than a fragile organism such as *Yersinia pestis,* the cause of bubonic plague. The U.S. Army was worried enough about possible anthrax attacks to begin to administer the anthrax vaccine to soldiers going to Iraq and Afghanistan. This sparked a controversy, because the efficacy and safety of the current anthrax vaccines remain controversial. Unfortunately, however, the anthrax attacks through the U.S. postal system in late 2001 only served to fuel the fear and the realization that bioterrorism is a reality that we must now address.

Another bacterial choice of the bioterrorists is *Clostridium botulinum,* the bacterium that produces botulinum neurotoxin. Producing botulinum neurotoxin in your garage is inadvisable and can be extremely hazardous to your health, but this toxin is produced commercially (as Botox) for use in a variety of medical and cosmetic applications ranging from correcting facial tics and crossed eyes to eliminating wrinkles. Thus, it is conceivable that terrorists could hijack commercially produced Botox from factories that produce it. Whether emptying a load of toxin into a city's water supply would actually result in any deaths is not clear, due to dilution of the toxin and breakdown of purified toxin in the environment, but it is better to err on the side of caution. The most recent concern, so far only theoretical, is that botulinum toxin might be introduced into milk, juice, or soft drinks during processing.

A New Respect for Prevention

Major changes have been occurring in the approach to controlling infectious diseases, changes that hold great promise for the future. Traditionally, medical establishments in developed countries have opted for a treatment-based approach to controlling infectious diseases. Granted, vaccinations were given to prevent some diseases, and doctors used antibiotics prophylactically to prevent others, such as postsurgical infections or infections in cancer chemotherapy patients, but the most common approach to infectious diseases was to wait for an infected person to seek medical help before intervening in the disease process. This approach has been criticized for being expensive and for allowing diseases to gain a foothold in the body before action—a delay that in some cases results in long-term damage to the patient even if the treatment is successful in eliminating the infecting bacterium from the body.

Treatment-based approaches have also become much less effective as increasingly resistant bacteria make it more difficult to make the right initial choices about which antibiotic to use. Physicians' response has been to use more advanced broad-spectrum antibiotics to treat infections that might be treatable with less expensive antibiotics. Physicians have been advised repeatedly to use the front-line antibiotics first if they must but to send samples to the microbiological laboratory for antibiotic resistance evaluation and then to adjust the therapy if laboratory results indicate that this is appropriate. Once physicians get used to using the front-line antibiotics, they are loath to abandon a strategy that is working for a particular patient. The result is increased selective pressure for development of resistance to front-line antibiotics.

A far preferable approach to controlling a disease is preventing it in the first place. This approach has been successful in ensuring the safety of food and water. Now, more and more public health officials, hospital managers, and executives of health management organizations are rediscovering that prevention is far more effective—and far less expensive—than treatment. Prevention is suddenly center stage again. However, for a preventive approach to work, it is first necessary to have information about disease patterns and early-warning systems that signal some new disease trend. Led by the Centers for Disease Control and Prevention (CDC), a variety of **surveillance programs** have been implemented to monitor the appearance of new diseases, the increased incidence of existing diseases, and the occurrence of antibiotic-resistant bacteria.

The CDC has been monitoring a subset of infectious diseases for years, but the list of diseases covered was far from exhaustive. Only recently have some major infectious diseases, such as chlamydial infections, been added to the list of reportable diseases. A problem the CDC has had to cope with is that reporting of diseases is voluntary on the part of state public health departments. Overworked and underfunded state health departments have sometimes, understandably, given reporting of diseases a low priority. The CDC and the National Institutes of Health are fighting to alert government agencies to the importance of having consistently funded monitoring programs.

Surveillance: an Early-Warning System

An interesting example of a relatively new surveillance program is **Foodnet,** a program that tries to count all cases of salmonellosis and *Campylobacter* food-borne disease in selected states and to estimate from these data the incidence of the diseases nationwide. Prior to the introduction of Foodnet, the CDC had abundant information about large outbreaks of food-borne disease but had no idea how many isolated cases of food-borne disease occur. Attempts are also being made in several areas to monitor antibiotic-resistant pathogens.

Monitoring disease prevalence is only the first step. Next must come effective action. There are encouraging signs that such programs are beginning to be implemented. A program for prevention of food-borne diseases, the **hazard analysis and critical control point (HACCP)** program, has been implemented by the Food and Drug Administration. Previously, companies waited to test foods for microbiological safety until the final step in food processing. Because it can take days to weeks for microbiology test results to be obtained and since many products must be shipped immediately for reasons of shelf life and economy, this approach allowed shipments of contaminated food to leave the processing plant and reach many distribution points before the results of the tests were known. HACCP programs monitor the food at control points where contamination is likely to occur. This approach not only lessens the likelihood that contaminated foods will be shipped, but identifies contamination problems early so that they can be rectified. At first, the food industry was leery of the HACCP approach, viewing it as a needless and potentially expensive government intrusion. The food industry has now become more enthusiastic about HACCP programs after seeing how expensive and injurious to the reputation of a company a large recall of contaminated products can be. A good HACCP program not only protects the public from disease, but protects the company from recalls and lawsuits. There is still grumbling about specifics, but the HACCP concept seems to be catching on.

Making Hospitals Safe for Patients

Fears concerning increasingly antibiotic-resistant bacteria have led to a number of changes in the way hospitals handle infectious disease problems. At one time, a surgeon could infect numerous patients and never be held accountable, or even be informed of such incidents, due to lack of communication between physicians who cared for patients during the post-surgical period and the surgeons who performed the operation. Previously, hospitals that had an infection control program often delegated the job of infection control officer to someone of low status who had no authority to take on physicians whose postsurgical infection record was poor or whose drug-prescribing practices were open to criticism. Now, this job is being taken much more seriously and is usually held by an infectious disease physician. Communication lines are also being improved.

Managed health care organizations, alarmed at the costs associated with antibiotic-resistant bacterial infections, have started cracking down on physician abuse and overuse of antibiotics. Ironically, cost-cutting measures by the health care plan bean counters who had pressured physicians earlier to reduce drastically the number of laboratory tests they ordered actually contributed to the resistance problem by encouraging physicians to use the strongest drugs available regardless of need. Thus, although there is still a certain amount of confusion and sending of contradictory signals between managers and physicians, at least a beginning has been made.

A somewhat less successful effort, so far, has been the campaign within hospitals to persuade health care workers to wash their hands after each patient and to implement other supposedly standard precautions, such as changing gloves when moving from one patient to another. Because many outbreaks of hospital-acquired infections are likely spread by the hands of health care workers, preventing transmission by this route has become an important priority. Unfortunately, health care workers have gotten out of the habit of washing their hands and instead rely on antibiotics. The high-stress atmosphere of a modern hospital, in which fewer health care workers are expected to treat more and sicker patients, has made it difficult to perform hygienic procedures by the book. Now that antibiotics are becoming less effective, hygienic practices have become more important than ever. A recent trend, the use of alcohol-based antimicrobial gels by health care workers, may help, because the gels are as effective as soap and water and require only a few seconds to apply. An article published in the *Wall Street Journal* (5 April 2006) described some of the methods used by hospitals to increase hand hygiene among health care workers. The methods included educational programs, monitoring of activities, disciplinary action, and even dismissal for failure to comply. If there was any positive aspect to the 2009–2010 H1N1 influenza pandemic, it was the increased emphasis on hygiene.

And Now for Some Really Good News: You've Got a Bacterial Infection!

Who would have thought that a person could be happy to learn that he or she had a bacterial infection? Yet, this is exactly what has happened to people suffering from ulcers and possibly from some other

chronic infections. The doctrine has been that infectious diseases are acute diseases that develop rapidly and run their courses quickly, whereas chronic diseases—diseases that last for long periods of time without resolving—are caused by an autoimmune response, a genetic disorder, or some environmental factor. Examples of chronic diseases are heart disease, Alzheimer's disease, and cancer. What if these diseases and others like them are caused by microbes? They might then be curable by antimicrobial agents or preventable by vaccines or other measures!

The discovery that many cases of liver cancer are caused by hepatitis B virus made it possible to prevent this type of cancer with an anti-hepatitis B vaccine. Similarly, the discovery that most gastric and duodenal ulcers are caused by bacteria led to a revolution in the way ulcer patients are treated. More recently, cervical cancer was linked to the presence of human papillomavirus, and women are now being encouraged to become vaccinated against human papillomavirus to prevent cervical cancer. Such examples have spawned a revolution that has led virtually every chronic disease, from heart disease to schizophrenia, to be reinvestigated as possibly of bacterial or viral origin.

The *H. pylori* Revolution

Not all is gloom and doom on the infectious disease front. Frustration over the inability of immunologists to find cures for autoimmune diseases or the inability of physiologists to find cures for heart disease has yielded to optimism as scientists begin to suspect that bacteria or other infectious agents may cause many of these diseases. To most people, taking lifelong medication that does not necessarily prevent the disease is not an acceptable "cure." Because microbiologists have by far the greatest track record for cures, the new rallying cry has become "Let's find the microbe that causes this intractable disease so we can cure it!" The landmark discovery that has changed dramatically the way people think about chronic diseases was the discovery that most gastric ulcers are caused by the gram-negative bacterium *H. pylori*. This discovery led to a simple antibiotic combination therapy that cures ulcers. Some people have recurrences, but the rate of recurrences is far lower than that for conventional treatments that addressed the symptoms rather than the cause of the disease. Because having ulcers for a prolonged period increases the risk of developing gastric cancer, a particularly dangerous form of cancer, an effective treatment for ulcers should also help to reduce the incidence of gastric cancer.

This discovery generated great enthusiasm among gastroenterologists, but when this information first came out, there was one sector of the health care profession that did not share in the celebration—the pharmaceutical companies. Ulcer medications, which had to be taken daily for life and which cost patients thousands of dollars a year, were suddenly replaced by a one-time course of antibiotics that cost only about $200. The pharmaceutical companies are now marketing their former ulcer drugs as nonprescription heartburn medications, trying to cut their losses.

The Aftermath

Getting the medical community to accept the idea that bacteria could cause ulcers took years of acrimonious debate, but once the idea was accepted, first by the research community and then by clinicians, it took only a short time for the world at large to understand that the implications of this discovery went far beyond ulcers. Suddenly, they remembered that if a disease is caused by a bacterium, then it could usually be cured with antibiotics, if diagnosed early enough. What followed was a veritable gold rush to find a bacterial cause for more diseases. A partial list of the diseases currently being reexamined as possibly being caused by bacteria is provided in Table 1–1. The scope of this list conveys better than anything else the boundless optimism that surged through the medical community once the implications of *H. pylori* as a cause of ulcers were fully appreciated. Table 1–1 indicates that some of these causal associations are well established, whereas others are still controversial as this text goes to press and may not pan out. Still, if even a few of these diseases become curable because they have bacterial origins, a treatment revolution will have occurred. More details about some of these associations will be given in subsequent chapters. It is striking how rapidly great skepticism (about *H. pylori* as a cause of ulcers) metamorphosed into great optimism about the likelihood of making further discoveries of a similar magnitude.

Microbiota Shift Diseases

A category of bacterial disease that defies conventional wisdom consists of diseases that are not caused by a single bacterial pathogen but rather by a shift in the bacterial population of some part of the human body. Although the natural microbial populations of the human body (microbiota; formerly called microflora) are usually protective, shifts in the composition of these populations can have pathological consequences. Diseases of this type do not yet have a name, so we refer to them as microbiota shift diseases. In chapter 5, "The Normal Human Microbiota," exam-

Table 1–1 Some diseases currently suspected of being caused by bacteria

Disease	Suspected microbe (status of disease association)
Gastric ulcers	*H. pylori* (well-established causal role)
Gastric cancer (some cases)	*H. pylori* (etiologically linked to gastric adenocarcinoma and lymphoma)
Periodontal disease	*Porphyromonas gingivalis, Tannerella forsythia*, and other oral bacteria (established as etiologic agents)
Atherosclerosis	*Chlamydophila pneumoniae* (epidemiological and experimental evidence indicate that it is a possible risk factor)
Low birth weight and preterm babies (some cases)	Bacterial vaginosis and shifts in vaginal bacterial population (associated with increased risk)
Cerebral palsy (some cases)	Placental infection (possible risk factor in multifactorial condition)
Rheumatoid arthritis	Intestinal anaerobic bacteria, *B. burgdorferi* (Lyme disease), mycoplasmas, viruses (possible triggering agents; strong correlation, but causality still controversial)
Crohn's disease	*Mycobacterium avium, Mycobacterium paratuberculosis*, or a shift of the intestinal bacterial population (strong association, but causality unknown)

ples of microbiota shift diseases are described in detail, but for present purposes, one example should suffice: **bacterial vaginosis.**

Bacterial vaginosis is the term used to describe a shift in the vaginal microbiota from a predominantly gram-positive population, dominated by *Lactobacillus* species, to a population of gram-negative anaerobes. For a long time, this condition was not taken seriously by physicians because the only symptoms, if there were symptoms at all, were a sparse discharge, some discomfort, and, in some women, a fishy odor. Two papers in the *New England Journal of Medicine* changed the status of bacterial vaginosis. One of these papers linked bacterial vaginosis with preterm births. This was an epidemiological association, not proof of a cause-and-effect relationship. The second paper described the result of an intervention study in which antibiotics known to target gram-negative anaerobes were administered to pregnant women with bacterial vaginosis, and the effect on the birth weight of the infant was determined. Antibiotic intervention that returned the vaginal microbiota to "normal" was associated with normal full-term births, whereas untreated women were significantly more likely to have preterm infants.

More recently, bacterial vaginosis, like chlamydial disease and gonorrhea, has been linked to a higher risk of contracting human immunodeficiency virus infections and other sexually transmitted diseases. A major challenge for scientists trained in the analysis of diseases caused by a single species of microorganism is to learn how to deal intellectually with polymicrobial diseases caused by shifts in bacterial populations consisting of hundreds of species. Undoubtedly, all of the species present are not equal contributors to the disease state, but the situation is far more complex than single-microbe infections.

Genomics

Breakthroughs in DNA-sequencing technology now allow scientists to sequence entire bacterial genomes in as little as a few days. There are now over 800 complete genome sequences for bacteria, many from disease-causing bacteria, and there are over 4,000 ongoing bacterial sequencing projects. These numbers and the pace of discovery are staggering considering that relatively few complete bacterial genomes were available as recently as the year 2000. At one time, it would have made sense to provide a list of genome sequences, but additions to this list are coming so fast that the best solution is to provide the address of a website that keeps track of genomes that have been or are being sequenced: http://www.genomesonline.org/.

Once a **genome sequence** becomes available, scientists examine the open reading frames (i.e., putative genes) one by one to try to assess their functions. In some cases, this is easy, because the gene has already been characterized. In other cases, the tentative identification is made on the basis of similarities to known genes or proteins in the **DNA** and **protein sequence databases.** Such assignments are useful but should be treated with some degree of caution, because many assignments have been based on relatively poor database matches. The best way to approach DNA sequence data is to realize that DNA sequence similarity to known genes only suggests a hypothesis that needs to be confirmed by more rigorous testing. A sobering fact is that even in the case of sequences of well-studied bacteria, such as *E. coli* and *Salmonella*, a substantial percentage of the genes have no similarity to any known genes. A job for future scientists is to determine the functions of these genes of unknown function.

The way in which DNA sequence information can reveal surprising things about an organism is illustrated by the genome sequence of *Borrelia burgdorferi,* the spirochete that causes Lyme disease. Scientists noted that no genes corresponding to the usual iron-containing proteins normally found in bacteria were present in the genome of this organism. This suggested a radical hypothesis: that *B. burgdorferi* copes with the problem of low iron concentrations in the mammalian host by not using iron at all. Scientists who were trained in an era when every article on iron utilization by bacteria started with words to the effect that all bacteria require iron were startled by this suggestion. Biochemical analyses showed, however, that *B. burgdorferi* apparently does live without iron, thus solving one problem most other pathogens have to confront: how to obtain iron in a host whose iron sequestration proteins keep the supply of available iron very low. As can be seen from this example, genome sequences are excellent hypothesis-generating tools.

The availability of complete genome sequences made possible a new technology called a **microarray.** In a microarray, segments of each gene in the genome (except rRNA genes) are attached to defined spots on a small slide-size "chip." RNA isolated from the bacterium grown under different conditions is converted into cDNA, which is labeled with a fluorescent dye, and the fluorescently labeled cDNA is hybridized to the DNA on the chip. The amount of fluorescent cDNA that hybridizes to the complementary DNA segment on any particular spot of the chip is a measure of the expression of the gene represented by that particular DNA segment. With this technique, thousands of genes can be screened for expression in a single experiment. If the number of bacteria is high enough in a body site of a colonized or infected animal, RNA from bacteria growing under in vivo conditions can be obtained and used to assess the expression of different genes in the animal.

Another form of genomic analysis is being used to detect and identify unknown pathogenic bacteria. This approach takes advantage of the fact that rRNA genes contain highly conserved regions of DNA sequence separated by more variable regions. Primers that detect conserved regions of the rRNA genes, usually 16S rRNA genes, are used to PCR amplify the genes from DNA extracted from tissue suspected to contain the infectious organism. The amplified DNA is called an **amplicon.** Of course, if no bacterium is there or if the level of bacterial DNA is too low, no PCR amplicon will be obtained, but if an amplicon is obtained, its DNA sequence can be determined and compared to the thousands of rRNA gene sequences now in the databases. The variable regions of the rRNA genes are particularly valuable in helping to determine what known microbe is most similar to the one in the tissue. The next step is to establish whether the amplified DNA comes from an organism that is actually infecting the tissue rather than from contaminating DNA from some other source. The first unknown organism to be identified in this way was the bacterium that causes a rare intestinal disease called Whipple's disease *(Tropheryma whipplei)*.

Modeling the Host-Pathogen Interaction in Experimental Animals

Studies of disease-causing bacteria growing under laboratory conditions need to be supplemented by studies in animal models. The most familiar type of infectious disease model is the laboratory rodent. Inbred strains of mice and rats are still widely used as models for the infection process. The availability of animals with known genetic mutations in their defense systems has increased the utility of rodent models. Numerous examples of the use of these models will be seen in later chapters of this book, including discussions of when it is appropriate to use animals in experimental schemes to study infectious diseases and the ethics of using vertebrate animals in experiments.

A new wrinkle on the animal model story is the increasing range of "animals" used, from the nematode *Caenorhabditis elegans* to the fruit fly *Drosophila melanogaster* and the zebra fish, *Danio rerio*. Certainly, one motivation for using such models is the fact that the body of regulations that has grown up around the use of laboratory rodents and other warm-blooded animals is so complicated and so expensive to follow. However, more compelling reasons for the use of these new models are that so much is known about their genetics and the fact that they are much more easily manipulated genetically than mammals.

In the case of *C. elegans,* for example, the developmental origin and fate of every cell in the organism are known. *Drosophila* has a long history of use as a model for insect and human genetics. In fact, a type of receptor on human neutrophils that is important for responding to bacterial infections (Toll-like receptors) was first discovered in *Drosophila* (Toll receptors). The zebra fish is a newer infection model but has some of the same attractive features as the nematode and fruit fly models, e.g., small size, easy maintenance, short generation time, and ease of genetic manipulation. Zebra fish have the added advantage that they have somewhat more advanced host defense systems than the nematode and fruit fly and are thus a better model for the human immune system.

Given the genetic distance between these animals and mammals, a certain degree of care must be used in choosing the experimental question and interpreting the results. For example, although nematodes, fruit flies, and zebra fish have phagocytic cell defenses that have some similarities to those of humans, the systems are not identical. Nonetheless, these simple models can be used to generate hypotheses that can later be tested in laboratory rodents or other animals.

Correlation Studies

Another type of modeling that has been used for a long time in epidemiological studies but is relatively new to pathogenesis studies is the statistical analysis of human and animal populations. At present, this type of modeling is rather primitive and is based on seeking correlations between traits of the organism and outcomes of disease. That is, is the production of a particular protein associated in a statistically significant fashion with various aspects of the disease progression in humans? This approach has the advantage that it is easy to do because one merely needs to apply preexisting statistical methods. There are, however, two rather serious problems with the approach.

First, such "modeling" is not modeling in the sense that the term is used in physics and chemistry, where principles are first expressed mathematically in a way that generates specific predictions about the outcome of an experiment. Correlation studies are usually done without any clear idea of a theoretical connection between the parameters being tested, so one problem is that finding a correlation does not prove cause and effect. There is an urban legend that illustrates this. A gentleman in California happened to pull down a shade in his apartment just before the onset of a particularly severe earthquake and remained convinced for the rest of his life that pulling down the shade had helped to cause the earthquake. A second problem is that the items to be checked for correlation are chosen by the researcher, and there may or may not be some theoretical underpinnings to the choice. These problems do not make the correlation studies inappropriate but need to be appreciated. If the approach is treated as one for potentially generating hypotheses rather than as a method that provides a proof of cause and effect, the objections disappear. As more mathematicians and physicists become interested in infectious diseases, more sophisticated modeling approaches will doubtless emerge.

A Brave New World of Pathogenesis Research

The *H. pylori* revolution captured the public imagination, but an even more important revolution has been the realization by research scientists that new technologies have opened up new opportunities to understand at the molecular level how infectious diseases develop. For several decades after the discovery of antibiotics, during a period in which a number of new vaccines were developed, it seemed sufficient simply to treat or prevent bacterial diseases. As long as antibiotics worked and vaccines were widely available, controlling bacterial infections at the practical level did not require in-depth information about the bacterium-host interaction. As physicians and scientists became concerned about increasing antibiotic resistance, however, there was a growing realization that a better understanding of the detailed interactions between the human body and the bacterial pathogen might suggest new treatment strategies.

Added to this was the recognition that there were some diseases, namely, those whose symptoms are caused by toxins produced by the bacteria, for which antibiotics were not effective. A good example is pulmonary anthrax, a disease caused by *B. anthracis*. The symptoms of this disease are caused by a protein toxin that is produced and secreted as the bacteria multiply in the lungs. If the disease is diagnosed immediately and the right antibiotic is administered, the disease can be controlled. However, antibiotics do not inactivate the toxin, and if antibiotic therapy is delayed for even a few days, enough toxin will have been produced to cause death. Vaccination of the entire population is not practical for such a rare disease, so scientists are seeking chemicals that neutralize the toxin.

In the case of a disease called septic shock that develops when gram-negative bacteria enter the bloodstream, a nonprotein component of the bacterial cell surface, **lipopolysaccharide,** acts as a toxin that leads to organ damage and death. Here, too, antibiotics are effective only if they are administered very early in the infection before the bacteria lyse and release too much of this toxic material. Although anthrax is not that much of a threat (other than as a potential bioweapon), septic shock continues to kill tens of thousands of people each year in the United States alone. As discussed in later chapters, new understanding has recently emerged about how the human body responds to lipopolysaccharide and to nonprotein surface components of gram-positive bacteria, such as **lipoteichoic acid,** and there is hope that this knowledge will make possible new and more effective therapies.

A different type of question is why different people respond differently to the same bacterium. In some cases, the human response can range all the way from a virtual lack of symptoms to death. Clearly, it would be helpful to understand this range of reactions so that people who are most susceptible to an infectious agent could be identified and given preference in treatment.

These and similar practical problems with controlling bacterial infections have driven a new interest in the interaction between bacteria and the human body at the molecular level. Fortunately, a cornucopia of new molecular tools and paradigms have become available that have made it possible to explore the host-pathogen interaction in a detailed way. It has even become more feasible to investigate infections that involve more than one species of bacteria or infections in which the bacterial pathogen acts in an area of the body, such as the mouth or small intestine, where there are many other bacteria that may have an effect on the course of the disease.

As important as the new technologies have been, the most important advance has been a new appreciation for the importance of focusing, not just on the properties of a bacterium in a test tube, but on the myriad ways in which the bacterium interacts with and stimulates responses from the human body. In this book, we place great emphasis on this bacterium-host interaction. It will become clear very quickly that although considerable progress has been made, especially since the 1980s, there is much to be learned and many opportunities for readers of this book to participate in future research in the area of bacterial pathogenesis.

SELECTED READINGS

Ahmed, N., and L. A. Sechi. 2005. *Helicobacter pylori* and gastroduodenal pathology: new threats of the old friend. *Ann. Clin. Microbiol. Antimicrob.* **4:**1–10.

Centers for Disease Control and Prevention. 1998. Preventing emerging infectious diseases: a strategy for the 21st century. *MMWR Recommend. Rep.* **47**(RR-15): 1–14.

Goldenberg, R. L., J. F. Culhane, and D. C. Johnson. 2005. Maternal infection and adverse fetal and neonatal outcomes. *Clin. Perinatol.* **32:**523–559.

Grenfell, B. T., O. G. Pybus, J. R. Gog, J. L. N. Wood, J. M. Daly, J. A. Mumford, and E. C. Holmes. 2004. Unifying the epidemiological and evolutionary dynamics of pathogens. *Science* **303:**327–332.

Gzyl, A., E. Augustynowicz, G. Gniadek, D. Rabczenko, G. Dulny, and J. Slusarczyk. 2004. Sequence variation in pertussis S1 subunit toxin and pertussis genes in *Bordetella pertussis* strains used for the whole-cell pertussis vaccine produced in Poland since 1960. Efficiency of the DTwP vaccine-induced immunity against currently circulating *B. pertussis* isolates. *Vaccine* **22:**2122–2128.

Hauth, J. C., R. L. Goldenberg, W. W. Andrews, M. B. DuBard, and R. L. Copper. 1995. Reduced incidence of preterm delivery with metronidazole and erythromycin in women with bacterial vaginosis. *N. Engl. J. Med.* **333:**1732–1736.

Henderson, D. K. 2006. Managing methicillin-resistant staphylococci: a paradigm for preventing nosocomial transmission of resistant organisms. *Am. J. Med.* **119(6A):**S45–S52.

Hooper, J. 1999. A new germ theory. *Atlantic Monthly* **283(2):**41–53.

Landro, L. 5 April 2006. Hospitals get aggressive about hand washing, p. 93. *Wall Street J.*

Macias, A. E., and S. Ponce-de-Leon. 2005. Infection control: old problems and new challenges. *Arch. Med. Res.* **36:**637–645.

Navon-Venezia, S., R. Ben-Ami, and Y. Carmeli. 2005. Update on *Pseudomonas aeruginosa* and *Acinetobacter baumannii* infections in the healthcare setting. *Curr. Opin. Infect. Dis.* **18:**306–313.

Necly, M. N., J. D. Pfeifer, and M. Caparon. 2002. *Streptococcus*-zebra fish model of bacterial pathogenesis. *Infect. Immun.* **70:**3904–3914.

Pennsylvania Health Care Cost Containment Council. 2005. Hospital-acquired infections. http://www.phc4.org/reports/cabg/04/keyfindings.htm.

Potempa, A. 30 May 2006. States that require infection reporting have varying ways of publicizing data, p. D1. *Anchorage Daily News.*

Relman, D. A. 2002. New technologies, human-microbe interactions, and the search for previously unrecognized pathogens. *J. Infect. Dis.* **186**(Suppl. 2):S254–S258.

Van Beneden, C. A., W. E. Keene, R. A. Strang, D. H. Werker, A. S. King, B. Mahon, K. Hedberg, A. Bell, M. T. Kelly, V. K. Balan, W. R. MacKenzie, and D. Fleming. 1999. Multinational outbreak of *Salmonella enterica* serotype Newport infections due to contaminated alfalfa sprouts. *JAMA* **281:**158–162.

Van Winkelhoff, A. J., and K. Boutaga. 2005. Transmission of periodontal bacteria and models of infection. *J. Clin. Periodontol.* **32**(Suppl. 6):16–27.

Waterfield, N. R., B. W. Wren, and R. H. ffrench-Constant. 2004. Invertebrates as a source of emerging human pathogens. *Nat. Rev. Microbiol.* **2:**833–841.

Woolhouse, M. E. J., J. P. Webster, E. Domingo, B. Charlesworth, and B. R. Levin. 2002. Biological and biomedical implications of the co-evolution of pathogens and their hosts. *Nat. Genet.* **32:**569–577.

QUESTIONS

1. The number of human deaths is often used as a standard for ranking human diseases in terms of importance. What, if anything, is wrong with this classification scheme?

2. Infectious diseases have obvious deleterious effects on the infected individual. Are there other consequences that reach beyond the infected person to his or her family and to society as a whole?

3. The United States and most developed countries have long had a medical community that focuses on therapy rather than prevention. Why is this the case, and under what conditions might this emphasis be appropriate? Why are scientists arguing for a return to a prevention-based health care system?

4. In our classification of emerging or reemerging infectious diseases, we treated antibiotic-resistant bacteria and *E. coli* O157:H7 as new disease sources. Make the case for and against considering a member of an established disease-causing species that acquires a new trait a new disease entity. What is the significance of such changes in bacterial pathogens?

5. The bacteria that cause cholera and tuberculosis are much more infectious than the so-called "opportunists." Why, then, are opportunists currently much more of a health concern in developed countries than cholera and tuberculosis?

6. Under what conditions—assuming no new epidemics—could infectious diseases suddenly move to the second, or even first, most common cause of deaths in the United States?

7. Do you think humans will ever win the battle against disease-causing bacteria? Why or why not? Is the use of warlike language to describe the relationship between humans and bacteria even accurate?

8. Microbiologists are fond of saying that only a tiny minority of bacteria cause disease. Are there reasons for thinking this might not be true?

9. In what sense are bacteria life givers rather than life takers? Is it possible that disease-causing bacteria might have a beneficial role in another context or even in their relationship with the human body?

10. There are many diseases that manifest in a variety of ways in different, apparently healthy individuals (e.g., a bacterium may cause a mild fever and malaise in one person while causing life-threatening disease in another). What are some of the factors that may contribute to this phenomenon?

2

Skin and Mucosa: the First Lines of Defense against Bacterial Infections

The skin and mucosal membranes of the human body are not simply inert barriers that keep good things in and bad things out. They comprise an organ, the body's largest, that has a complex array of activities and functions that are only now beginning to be appreciated.

This chapter and the ones that follow are designed to provide an overview of the defenses of the human body. From the time the first hominids appeared on Earth, they had to survive in a bacteria-dominated world in which bacteria tended to view their hominid hosts as a free lunch. Survival meant developing defenses that kept most of these bacteria at bay. The defenses of the human body against bacterial incursions are so effective that most bacteria are kept out of human tissues and the bloodstream and are instead relegated to the surface of the skin or contained in certain closely guarded areas of the human body, such as the mouth, intestinal tract, and vaginal tract. Fortunately for us, only a tiny minority of bacteria are able to bypass these defenses and cause disease. In order to understand the mechanisms by which disease-causing bacteria cause disease, it is first essential to know what obstacles those few bacteria that can cause disease must overcome.

Barriers: Skin and Mucosal Membranes

Epithelia, the layers of cells that cover all of the external and internal surfaces of the body that are exposed to the external environment, are an important initial defense against pathogenic bacteria. The epithelial cells found in different body sites differ considerably in their properties, but they have some features in common. Skin consists of a lining of living cells called the **dermis.** These cells are continuously replaced as they die and become a tough, dry outer layer called the **epidermis.** The epithelia that line the respiratory, intestinal, and urogenital tracts are the **mucosal epithelia** and consist of tightly packed cells, which are attached to each other by protein structures called **tight junctions** and **des-**

mosomes (Figure 2–1). The tight binding of epithelial cells to each other normally prevents bacteria from transiting an epithelial layer. To get through the epithelial layer, bacteria must either take advantage of wounds or be capable of invading epithelial cells, passing between them or passing through them to get to the underlying tissue.

In contrast, the cells that line the surfaces of the interior of the body **(endothelium),** such as blood vessels and lymphatic vessels, are not tightly bound to each other, so that the cells of the body's defense system can move freely from blood to tissues. Unfortunately, this feature also allows bacteria to move into and out of blood and lymphatic vessels by moving between the cells. Thus, once bacteria gain entry into the body at one site, it is possible for them to gain access to other parts of the body. This is the reason why the epithelia of skin and mucosal surfaces are such important barriers.

A second feature of epithelial cells is that they are attached to a thin sheet of connective tissue, called the basement membrane **(basal lamina),** which consists of **extracellular matrix (ECM) components,** including network-forming collagens and an adhesion protein called laminin. The surface of an epithelial cell that is attached to other cells or to the basal lamina **(basolateral surface)** has a different protein composition from the surface that faces outward **(apical surface).** Cells with this property are said to be **polarized.** Epithelial layers that cover surfaces where absorption or secretion is taking place, e.g., in the intestinal tract,

usually consist of a single layer of epithelial cells **(simple epithelium).** Other surfaces, such as the female cervix or the skin, are composed of many layers of epithelial cells **(stratified epithelium)** (Figure 2–2). Epithelial cells in different sites vary in shape. Some have a flattened shape **(squamous** cells) and form the lining of cavities, such as the mouth, heart, and lungs, and the outer layers of the skin. Some are cube shaped **(cuboidal** cells) and form the lining of kidney tubules and gland ducts and constitute the germinal epithelium that develops into egg cells and sperm cells. Others are tall and thin **(columnar** cells) and form the lining of the stomach and intestine.

Simple epithelia are more vulnerable to bacterial invasion than stratified epithelia because invading bacteria have to pass through only one layer of cells to gain access to the tissue underneath. Most of the surfaces that are exposed directly to the environment (e.g., the skin and mouth) are covered by stratified epithelia, whereas simple epithelia are found in internal areas, such as the intestinal tract or the lungs. We will use the term mucosal layer, mucosal epithelia, or mucosal cells to denote the simple epithelia of these internal areas.

The layer of epithelial cells covers regions of **loose connective tissue** (Figure 2–2), which consists of additional ECM secreted by elongated fibroblasts. The ECM composition varies with the tissue but contains a network of fibrous proteins, such as collagens, gel-like material made up of polysaccharides called glycosaminoglycans, such as chondroitin sulfate, hyaluronan, and heparin sulfate, and adhesion proteins, such as fibronectin and laminins. There are numerous examples of pathogenic bacteria attaching to components of the ECM and manipulating or mimicking it during the course of infection.

Epithelia are protected by an array of innate and adaptive defenses. Some of these defenses are listed in Tables 2–1 and 2–2. Other defenses are more specific to certain areas of the body, such as the eyes, the respiratory tract, and the urinary tract (Table 2–3). For example, tears contain **lysozyme,** an enzyme that degrades the peptidoglycan wall of bacteria. They also provide a washing action that removes bacteria from the eyes. The entry to the respiratory tract is protected by mucus and by specialized ciliated cells that propel bacterium-laden blobs out of the site. The urinary tract epithelium is protected by a sphincter at the end of the urethra, the tube that leads to the bladder. This barrier makes it difficult for bacteria to enter the urethra. Also, the washing action of urine during urination washes out any bacteria that may have gained access to the bladder.

Figure 2–1 Intestinal epithelial cells showing tight junctions and desmosomes.

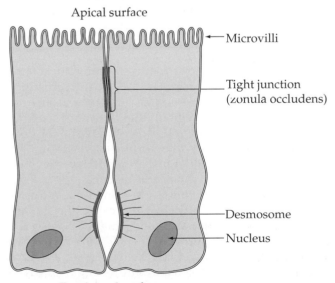

Apical surface

Microvilli

Tight junction (zonula occludens)

Desmosome

Nucleus

Basolateral surface

A Simple squamous epithelium

Basal lamina

B Simple cuboidal epithelium

Basal lamina

C Stratified squamous epithelium

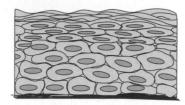

Basal lamina

D Simple columnar epithelium

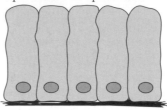

Basal lamina

E Ciliated columnar epithelium

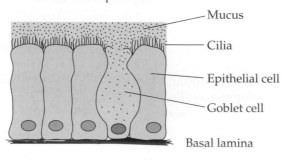

Mucus

Cilia

Epithelial cell

Goblet cell

Basal lamina

F Connective tissue

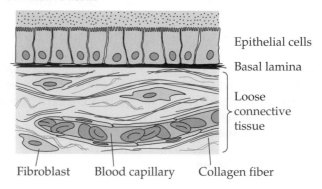

Epithelial cells

Basal lamina

Loose connective tissue

Fibroblast   Blood capillary   Collagen fiber

◯, nucleus

Figure 2–2 Different types of epithelial cells and their relationship to underlying tissue. Shown are simple squamous epithelium; simple cuboidal epithelium; stratified squamous epithelium (the upper layers of cells are dead; typical of skin); simple columnar epithelium; ciliated columnar epithelium, showing goblet cells, which secrete mucus (mucin); and typical structure of tissue under an epithelial cell layer. (Panel F modified from Cooper and Hausman, 2007.)

Defenses of the Skin

Epidermis

Bacteria are unable to penetrate intact skin unaided. That is why skin infections are usually associated with breaches of the skin caused by wounds, burns, or insect bites (Table 2–4). Why is intact skin such an effective barrier to bacterial invasion? A number of characteristics combine to make skin inhospitable to bacterial growth, as well as difficult to penetrate. Skin is composed of two layers, the epidermis (outer layer) and the dermis (inner layer). The epidermis consists of stratified squamous cells, most of which are **kera-**

tinocytes. Keratinocytes produce the protein keratin, which is not readily degraded by most microorganisms. As cells from the dermis are pushed outward into the epidermal region, they produce copious amounts of keratin and then die. This layer of dead keratinized cells forms the surface of the skin. The dead cells of the epidermis are continuously shed **(desquamation).** Thus, bacteria that manage to bind to epidermal cells are constantly being removed from the body (Figure 2–3).

Skin is dry and has an acidic pH (pH 5), two features that inhibit the growth of many pathogenic bacteria, which prefer a wet environment with a neutral pH (pH 7). Also, the temperature of the skin (34 to

Table 2–1   Defenses of the skin

Defense	Function
Dry, acidic environment	Prevents growth of many bacteria
Dead, keratinized cells	Keratin is hard to degrade, and dead cells discourage colonization.
Sloughing of surface cells	Removes bacteria that adhere
Toxic lipids, lysozyme	Protect hair follicles, sweat glands, and sebaceous glands
Normal microbiota	Competes with pathogens for nutrients, colonization sites
Underlying immune cells (Langerhans and other cells)	Combat bacteria that manage to reach the dermis and tissue below it

Table 2–3   Defenses of specific sites

Site	Defense
Eye	Tears (washing action and antibacterial substances; surface of eyeball)
Airway entrance	Mucus traps bacteria; ciliated epithelial cells propel bacterium-laden mucin blobs out of airway.
Stomach	Acidic environment
Small intestine	Rapid flow of contents; sloughing of epithelial cells
Colon	Resident microbiota; sloughing of epithelial cells; flow of contents
Vaginal tract	Resident microbiota; sloughing of epithelial cells; flow of vaginal fluid
Bladder	Sphincter keeps bacteria out of urethra; flushing action of urine washes bacteria out of bladder.

35°C) is lower than that of the body interior (37°C). Accordingly, bacteria that succeed in colonizing the skin must be able to adapt to the very different internal environment of the body if they manage to reach underlying tissue. Interestingly, the causative agent of leprosy, *Mycobacterium leprae,* has an optimal growth temperature of 35°C, which may account for its predilection for the skin.

Hair follicles, sebaceous (fat) glands, and sweat glands are composed of simple epithelial cells and offer sites for potential breaches in the skin that could be used by some bacteria to move past the skin surface. These sites are protected by the peptidoglycan-degrading lysozyme and by lipids that are toxic to many bacteria. Some pathogenic bacteria are capable of infecting hair follicles or sweat glands. That is why skin infections, such as boils (furuncles) and acne (pustules), are commonly centered at hair follicles.

Normal Microbiota

The defenses of the skin do not completely prevent bacterial growth, as is evident from the fact that there are bacteria capable of colonizing the surface of the skin. They consist primarily of gram-positive bacteria, a mixture of cocci and rods. A bacterial population that is found continuously in some body site without causing disease is called the **resident (or commensal) microbiota** of that site. The commensal microbiota of the skin helps to protect against pathogenic bacteria by occupying sites that might be colonized by pathogenic bacteria. It also competes with incoming pathogens for essential nutrients. Some resident bacteria also produce bactericidal compounds, e.g., pore-forming toxins, such as **bacteriocins** or growth inhibitors, which target other bacteria. The commensal mi-

Table 2–2   Defenses of mucosal surfaces

Defense	Characteristic	Protectants	Function
Mucus	Glycoprotein matrix	Lysozyme Lactoferrin	Digests peptidoglycan Sequesters iron, prevents growth of bacteria
Cryptdins and other defensins	Antibacterial peptides produced by the host		Toxic for many bacteria
Antibodies (sIgA)	Protein complexes		Specifically bind certain bacteria
Cells of immune system underlying the mucosal surface or extruded between epithelial cells			Engulf or kill bacteria by bombarding them with toxins

Table 2–4 Some consequences of breaching barrier defenses

Type of breach	Consequences
Venous catheters	Biofilm formation on the catheter; movement of bacteria into bloodstream
Burns	Infection of the burned tissue, sometimes followed by movement of bacteria into underlying tissue and blood
Damage to cornea caused by contact lenses	Infection of cornea
Respirator (tube inserted in the airway)	Allows bacteria to bypass defenses of upper airway
Perforation of the intestinal mucosa during surgery or other abdominal trauma	Release of bacteria into otherwise sterile tissue and blood
Buffering of stomach acid (achlorhydria)	Bacteria that cause intestinal infections are more likely to gain access to small intestine
Indwelling urinary catheter	Constantly drains bladder, eliminating washing action of urine; keeps sphincter open

crobiota does not completely prevent colonization of the skin by potential pathogens but hampers it enough so that the colonization by pathogenic bacteria is usually transient.

Since transient colonization by pathogens can occur and since even normally harmless skin bacteria can cause infections under certain conditions, hand washing and disinfection of the hands adds yet another barrier to infection, as well as to transmission to other people with whom one may come into contact. In the mid-1800s, Ignaz Semmelweis first introduced the concept of hand washing and cleanliness in maternity wards (Box 2–1). The low-tech but very effective protective barrier provided by hand washing and frequent changing of gloves has probably been a key factor contributing to the good laboratory safety record of research scientists. However, despite strong and convincing experimental data and persistent promotion of good hygiene policies by health care officials, compliance of health care workers with the recommended hygiene practices is still unacceptably low, with rates of compliance sometimes less than 50%.

An added source of concern is the correlation between life-threatening hospital-aquired (nosocomial) infections and the wearing of long or artificial fingernails. Several recent infectious outbreaks in intensive-care units (ICU) have been attributed to pathogens (e.g., *Klebsiella pneumoniae, Pseudomonas aeruginosa, Candida albicans,* methicillin-resistant *Staphylococcus aureus,* and *Serratia marcescens*) that were isolated from under the nails of ICU personnel. For example, between 1997 and 1998, the deaths of 16 babies in the neonatal ICU at a hospital in Oklahoma City were linked to a particular strain of *P. aeruginosa* found under the nails of three neonatal ICU nurses.

Defenses of the Dermis

Although the focus of this chapter is the protective barriers provided by intact skin and mucosa, it is important to note that these barriers are backed up by specialized portions of the immune system, which will be described in more detail in subsequent chapters. For example, bacteria that manage to get past the epidermis through cuts or burns encounter a specialized cell type called a **Langerhans cell.** Langerhans cells belong to a class of cells called **dendritic cells** that process the invading bacteria and activate the immune cells of the **skin-associated lymphoid tissue (SALT).**

Members of the skin microbiota normally do not cause human infections unless they are introduced into the body by abrasions, catheters, or surgery. *Staphylococcus epidermidis,* a common skin bacterium, has been implicated in postsurgical and catheter-related infections. (*S. epidermidis* was the bacterial villain in the surgeon-transmitted infections described in Box 2–1.) Relatively nonpathogenic bacteria like *S. epidermidis* would normally be killed rapidly by the defenses of the bloodstream, but if they can reach an area that is somewhat protected from host defenses, such as the plastic surface of a heart valve implant, they can grow and produce quite serious infections. Catheters can provide skin-associated bacteria with a conduit into the bloodstream, thus bypassing the defenses of the epidermis and dermis. Catheter-associated infections have become a serious enough problem in hospitals that catheter companies are developing plastic catheters that are impregnated with antibacterial compounds.

Surgical-wound infections and catheter-associated infections caused by skin bacteria, especially *S. epidermidis,* have become an ever more prevalent problem due to the fact that *S. epidermidis* is now resistant to most available antibiotics. How has this happened? It is clear that at least some antibiotics are exuded in sweat. Also, ointments containing antibiotics are widely used in the treatment of such skin conditions as acne and rosacea (unnaturally red skin). These treatments can last for months or years. Thus, it is not surprising that *S. epidermidis* has become increasingly resistant to a variety of antibiotics.

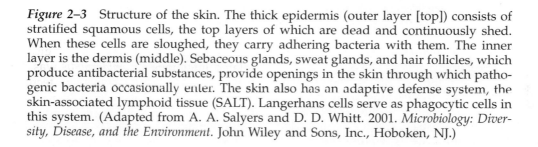

Opening of sweat or sebaceous gland

Bacteria

Sloughing cells and adherent bacteria

Dead cells

Epidermis

Dermis

SALT

SALT

SALT

Subcutaneous area

Blood vessel

Figure 2–3 Structure of the skin. The thick epidermis (outer layer [top]) consists of stratified squamous cells, the top layers of which are dead and continuously shed. When these cells are sloughed, they carry adhering bacteria with them. The inner layer is the dermis (middle). Sebaceous glands, sweat glands, and hair follicles, which produce antibacterial substances, provide openings in the skin through which pathogenic bacteria occasionally enter. The skin also has an adaptive defense system, the skin-associated lymphoid tissue (SALT). Langerhans cells serve as phagocytic cells in this system. (Adapted from A. A. Salyers and D. D. Whitt. 2001. *Microbiology: Diversity, Disease, and the Environment.* John Wiley and Sons, Inc., Hoboken, NJ.)

Defenses of Mucosal Surfaces

The respiratory tract, gastrointestinal tract, and urogenital tract are topologically "inside" the body, but they are exposed constantly to the outer environment and foreign materials. Unlike the many dead layers found in skin, internal surface areas (called mucosal epithelia) are comprised of only one epithelial layer. These mucosal epithelia have a temperature of around 37°C and a pH of 7.0 to 7.4. The role of mucosal cells is absorption or secretion, so they are continuously bathed in fluids. These warm, neutral, and moist conditions are ideal for the growth of bacteria. To protect against bacterial colonization, these vulnerable epithelia have evolved a formidable array of chemical and physical barriers (Table 2–2).

Mucosal cells are regularly replaced, and old cells are ejected into the lumen. In fact, mucosal cells are one of the fastest dividing populations of cells in the body. Thus, bacteria that manage to reach and colonize a mucosal surface are constantly being eliminated from the mucosal surface and can remain in the area only if they can grow rapidly enough to colonize newly produced cells. Chemical and other innate defenses help to reduce the growth rates of bacteria sufficiently to allow ejection of mucus blobs and sloughing of mucosal cells to clear the bacteria from the area.

An important protection of many internal epithelia is **mucus.** Mucus is a mixture of glycoproteins (**mucin**) produced by **goblet cells,** a specialized cell type incorporated into the epithelial layer. (The basic structure of glycoproteins is reviewed in Box 3–2.) Mucus has a viscous, slimy consistency, which allows it to act as a lubricant. It also plays a protective role because it traps bacteria and prevents them from reaching the surfaces of the epithelial cells. Mucus is constantly being produced, and excess mucus is shed in blobs that are expelled. Bacteria trapped in mucus are thus eliminated from the site. In the gastrointestinal and uri-

BOX 2–1 Hand Washing Past and Present: a Lesson in Learning and Forgetting

The idea that physicians and nurses should wash their hands before treating a new patient was a relatively recent innovation. Ignaz Semmelweis, the man credited with making hand washing a standard part of medical practice, lived and practiced medicine in the mid-1800s. Although he was not the first physician to make the connection between contaminated hands and the spread of disease by physicians to their patients, he was the first to prove that proper disinfection of hands could dramatically reduce hospital-acquired infections. Semmelweis had noted that two maternity wards in the Vienna Lying-in Hospital had very different mortality rates. In one, the death rate due to puerperal "childbed" fever (a common cause of death in women of the period) was over 10%, whereas in the second ward, it was less than 3%. This fact was well known to women entering the hospital, who considered assignment to the first ward to be a virtual death sentence. Both wards were equally crowded, with three patients sharing each bed and the sick mixed indiscriminately with the well. Both wards contained women of similar socioeconomic status. The only difference between the two clinics was that the first clinic was used for teaching medical students, who also were dissecting cadavers in between delivering babies, and the second was used for teaching midwives, who were not exposed to potential disease-carrying cadavers. Semmelweis deduced that the medical students were transmitting childbed fever (which we now know is caused most frequently by the bacterium *Streptococcus pyogenes*) to their patients because they failed to cleanse their hands properly. In 1846, he instituted a policy requiring that all midwives and medical students wash their hands with a chlorinated lime solution before examining patients. The mortality rate in both wards promptly dropped to 1%, something the women who came to the ward appreciated but Semmelweis' male detractors did not. When he later added washing of medical instruments to the birthing protocols, the incidence of puerperal fever was nearly eliminated from the hospital ward. Semmelweis' discovery remained controversial for many years, and it was only in the early 1900s that hand washing was universally accepted as an essential medical practice.

Today, proper disinfection of the hands is one of the most basic and firmly entrenched clinical procedures, especially for surgeons. Nonetheless, the advent of antibiotics and the consequent decrease in deaths due to hospital-acquired infections has led some surgeons to neglect this important practice.

A surgeon in a large northeastern U.S. hospital who started bypassing the rigorous surgical scrub procedure because he was troubled by dermatitis on his hands provided a particularly dramatic example of this. He trusted the two pairs of surgical gloves that were commonly worn during operations, but tiny holes in gloves can be made by contact with sharp objects or bone fragments. Also, the surgeon was using mineral oil to ease the irritation to his hands, and mineral oil undermines the integrity of surgical gloves. This physician managed to contaminate heart valve implants in a number of patients with *S. epidermidis* before he was identified as the source of the outbreak. *S. epidermidis* is commonly found as part of the resident microbiota of the skin, where it is not normally pathogenic, but it can cause infections if introduced into the body through wounds. Infections of heart valve implants usually cannot be treated effectively with a simple course of antibiotics, not only because of the high resistance level of *S. epidermidis* strains, but also because of the formation of bacterial biofilms that are more resistant than individual bacteria to antibiotics. Thus, the patients with the infected valves had to endure a second operation to remove and replace the valves, not to mention additional damage to the heart due to the infection.

As is evident from the date on the reference cited at the end of this box, this case occurred in the 1980s. Does this mean that such cases have ceased to occur? Not at all! This case was used because it is a classic example of the hand-washing problem, but there have been many other cases since. The difference between the 1980s and the first decade of the 21st century is that the surgeon in this case would probably have been identified today before he

(continued)

BOX 2–1 Hand Washing Past and Present (*continued*)

infected so many people, because infectious disease surveillance systems in hospitals have improved. However, the attitude and behavior that sparked this sorry episode are still rampant in many hospitals. The silver lining in this particularly black cloud is that the accountants for the insurance agencies have finally figured out how much the lack of hand washing and improper use of gloves is costing them, and they are mounting increasingly vigorous campaigns in favor of hand washing and against health care workers who ignore these simple but effective precautions. In fact, relatives of hospital patients are being urged to question unhygienic practices they witness. The lawyers are circling. Who knows? It might even become safe to enter a hospital in the coming years.

Source: J. M. Boyce, G. Potter-Bynoe, S. M. Opal, L. Dziobek, and A. A. Medeiros. 1990. Tracing the source of a *Staphylococcus epidermidis* "outbreak" in a hospital. *J. Infect. Dis.* **161**:493–499.

nary tracts, peristalsis and the rapid flow of liquids through the area remove the mucus blobs, along with the lumen contents.

In the respiratory tract and in the fallopian tubes, there are specialized cells, **ciliated columnar cells,** whose elongated protrusions **(cilia)** are continuously waving in the same direction. The waving action of the cilia propels mucus blobs out of the area. Hospitals often encounter problems associated with comatose patients, in whom the absence of a normal cough reflex and reduced mucociliary clearance result in increased susceptibility to respiratory infections. In addition, problems can occur with respirators, which introduce air potentially contaminated with a pathogen directly into the lung, bypassing the upper respiratory mucociliary defenses.

Another protective role of mucus is to bind proteins that have antibacterial activity (Table 2–2). **Lysozyme** is one such protein. It targets the peptidoglycan of the bacterial cell wall and hydrolyzes the linkage between *N*-acetylmuramic acid and *N*-acetyl-D-glucosamine (Figure 2–4). Lysozyme is most effective against the cell walls of gram-positive bacteria but can digest the gram-negative cell wall if the bacterial outer membrane is first breached by membrane-disrupting substances, such as the detergent-like bile salts found in the intestine. **Lactoferrin,** an iron-binding protein found in mucus, sequesters iron and deprives bacteria of this essential nutrient.

Lactoperoxidase is another protein with antibacterial activity that is found in secretory fluids, such as milk, tears, saliva, and airway mucus. Lactoperoxidase is a heme-containing peroxidase that uses hydrogen peroxide (H_2O_2) as an oxidant to generate the highly reactive hypothiocyanite ($OSCN^-$) from thiocyanate (SCN^-), which in turn kills bacteria. It is believed that exposed thiol groups ($-SH$) of enzymes and other proteins on the bacterial membrane surface are the primary targets of these reactive oxidants, which convert the thiols into disulfides and thereby disrupt the normal function of the bacterial surface proteins.

Toxic **antimicrobial peptides** called **defensins, cathelicidins,** and **histatins** contain highly cationic (basic) regions that enable them to depolarize or insert into bacterial cell membranes and kill bacteria by forming channels or holes in their membranes, collapsing the proton motive force that is essential for bacterial survival. This type of activity was responsible for the effectiveness of one of the first antibiotics, gramicidin, which is a pore-forming protein that kills bacteria. Defensins and other antimicrobial peptides have been found in the mouth, on the tongue, in the vagina, in the lungs, in skin, and in the crypts of the small and large intestines. In the mouth, defensins may be the reason why infections of the tongue are so uncommon and why animals lick their wounds. In the crypts of the intestinal mucosa, they are presumably protecting the intestinal stem cells, which divide constantly to replenish the cells of the intestinal mucosa. These peptides probably have some antibacterial effects that protect these locations from bacteria. It is worth keeping in mind, however, that the membrane-disrupting activity of these peptides can be inhibited by physiological salt concentrations and by serum. Consequently, their most important antibacterial activities may be exerted largely in another location where they are found: vacuoles inside phagocytic cells that engulf and kill bacteria.

Special Defenses of the Gastrointestinal Tract

Different regions of the gastrointestinal tract have special antibacterial features that serve as barriers to pathogens. The lumen of the stomach is an extremely

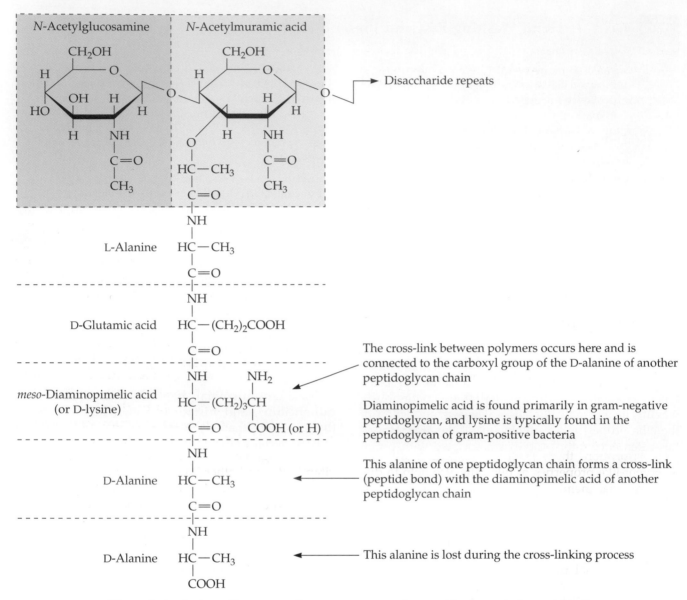

Figure 2–4   Action of lysozyme. Lysozyme targets the peptidoglycan in bacterial cell walls of mainly gram-positive bacteria and hydrolyzes the link between *N*-acetylmuramic acid and *N*-acetyl-D-glucosamine. It can also degrade the peptidoglycan of gram-negative bacteria if the outer membrane is first disrupted by bile salts.

acidic environment (pH ~2). It was previously thought that most bacteria could not survive there, but scientists have recently found DNA from 128 species of bacteria occupying the stomach. The only bacterium inhabiting this harsh environment that has been studied in detail is *Helicobacter pylori,* which causes gastritis, gastric ulcers, and even gastric cancer. *H. pylori* does not actually live in the lumen of the stomach, but rather in the mucin layer that covers and protects the stomach lining. Cells in the stomach lining secrete carbonate, which buffers the mucin layer to near normal pH. *H. pylori* does have the ability to

protect itself long enough to reach the mucin layer and thus is not killed as readily as many bacteria when they are exposed to the highly acidic environment of the stomach.

Other bacteria, including some of the food-borne pathogens (*Escherichia coli, Salmonella,* and *Campylobacter*) have a response to acid **(acid tolerance response)** that makes them better able to survive for short periods of time at pH 4. This is higher than the pH of the stomach interior, so how could it be protective? Speculation is that bacteria ingested in foods are probably protected somewhat from the full impact of stomach acid by the buffering capacity of the food. It is possible that, in the case of bacteria that are able to mount an acid tolerance response, although it does not allow them to live in the stomach for prolonged periods of time, it increases the chance that some of them might survive long enough to reach the small intestine.

The fact that many bacteria do not survive passage through the stomach illustrates the fact that the acidic environment of the stomach lumen does not just contribute to the digestion of food but acts as a protective barrier to prevent bacteria from reaching more vulnerable areas, such as the small intestine and colon, where conditions are more favorable for bacterial growth. An illustration of the protective effect of the acidic environment of the stomach is the fact that people who have achlorhydria (high stomach pH) have increased susceptibility to infections of the lower intestinal tract.

For those bacteria that manage to survive the acid barrier of the stomach, bile salts await them in the small intestine and colon. **Bile salts** are steroids with detergent-like properties that are produced in the liver, stored in the gall bladder, and then released through the bile duct into the intestine. Bile salts help to neutralize the stomach acid and are used by the body to emulsify lipids in food to enable fat digestion and absorption through the intestinal wall. The detergent-like properties of bile salts also help to disrupt bacterial membranes, especially those of gram-negative bacteria.

An equally important protection of the small intestine is the rapid flow of contents through it. This rapid flow, together with the bile salts and the rapid turnover of intestinal mucosal cells, helps to keep high concentrations of bacteria from developing in the small intestine. High concentrations of bacteria would not only increase the chance that bacteria could invade the small-intestinal mucosa, but would also allow bacteria to compete with the human body for the nutrients (simple sugars and amino acids) that the small intestine is designed to absorb.

The importance of the rapid flow of contents through the small intestine as a protection against bacterial colonization is underscored by the fact that bacterial pathogens that cause intestinal infections, such as gastroenteritis (diarrhea and pain), generally are able to swim to the mucosa of the small intestine and attach to the mucosal cells, thus keeping them from being washed out of the colon. Another illustration of the importance of the rapid flow of contents is the fact that people who develop "blind loops," or regions of out-pouching that have rather stagnant contents, have problems due to the buildup of bacteria within those regions. At one time, the intentional surgical introduction of blind loops in the small intestine was tried as a means of weight reduction. Not surprisingly, a side effect of this in some people was the development of sepsis caused by invasion of some of the bacteria that reached high concentrations in the blind loop. In the colon, the flow rate of contents is drastically reduced compared to the flow rate in the small intestine. Some scientists have compared the difference to the passage from a rapidly flowing stream (small intestine) to a stagnant pond (colon).

Most mucosal surfaces are protected by a **normal resident microbiota.** Exceptions are the uterus and upper female genital tract and the urinary tract. The species compositions of the microbiota found at different parts of the body vary appreciably from one site to another. Nonetheless, all have in common the fact that gram-positive bacteria predominate. Shifts in these populations can be pathological, as is seen from diseases such as periodontal disease and bacterial vaginosis. The large intestine (colon) harbors an abundant and rich assortment of normal microbiota, the majority (97%) of which are anaerobes or facultative anaerobes. Many of these bacteria use carbohydrates and fats that are not digested by the stomach or absorbed by the small intestine. In return some resident microbes provide a beneficial function to the host by synthesizing and secreting vitamins (e.g., vitamin K, vitamin B_{12}, and other B vitamins) and other nutrients that the intestine can absorb. Recent experimental evidence indicates that indigenous bacteria play a crucial inductive role in gut and immune development during early postnatal life. They stimulate the development of certain tissues, in particular the cecum and Peyer's patches, which stimulate production of cross-reactive antibodies that prevent infection by related bacteria that may be pathogens (more on this later).

As was the case with the skin, mucosal surfaces have an underlying population of phagocytic cells and immune cells. This mucosal defense system, which is distinct from the system that controls immune cells in blood, lymph nodes, and other organs, is called the **mucosa-associated-lymphoid tissue (MALT).** For instance, the MALT found in the intestinal mucosa is called the **gastrointestinal-associated lymphoid tissue (GALT).** These adaptive defense systems are composed of **M cells (microfold cells** that engulf the lumen contents and present them to underlying antigen-presenting cells), macrophages, T cells, and B cells. Their primary function is to make **secretory immunoglobulin A (sIgA),** an antibody that is secreted into mucus on the apical side of the epithelium. Antibodies are proteins that bind to specific sites on bacteria or other pathogens. sIgA is thought to increase the stickiness of mucin by attaching to mucin sugars at one end, leaving its two other antigen-binding ends free to bind and trap bacteria trying to reach the mucosal layer. The sIgA-trapped bacteria are then expelled along with the mucin. We will discuss the role of phagocytes in chapter 3 and will return to the GALT in chapter 4.

Models for Studying Breaches of Barrier Defenses

Animal models have been widely used to study skin, eye, and mucosal infections. Because some of these animal models involve such breaches as cuts or burns, experimental protocols must include explicit plans to monitor and minimize pain and discomfort to the animals as much as possible and to minimize the number of animals needed to obtain statistically significant results. When available, validated alternative infection models that do not involve animals should be used. In addition, a committee with expertise in animal experimentation must first approve the rationale for experiments on animals and the detailed protocols themselves before the experiments are performed. Some of the ethical and procedural issues that lead to appropriate animal experimentation are discussed in a later chapter. For continuity, some of the models used to study eye, skin, and mucosal infections are mentioned here without this context.

One of the earliest models for studying skin infections was the burned-rodent model. A patch of skin on an animal that is anesthetized is shaved and then burned with an alcohol flame. Just as is the case with human burns, bacteria that could not infect intact skin can infect the burned rodent tissue. The eye is another surface of the body that is remarkably resistant to infection. Eye infections of the sort seen in patients who have been careless with contact lenses or have suffered small cuts in the cornea are mimicked by a rabbit model in which small, shallow cuts are made in the cornea of the animal's eye. Both of these models have been used extensively to study infections caused by *P. aeruginosa,* one of the main causes of burn and eye infections in humans.

In the previous chapter, some unusual lower-animal models were mentioned. *Caenorhabditis elegans* and *Drosophila* are not very useful for studies of skin infections, because the "skin" of these organisms is chitinous rather than epithelial. The zebra fish is a better model, especially for studies of the mucosal defenses. More recently, infection models have been developed based on tissues from, for example, chicken embryos. Rodents have been used widely to investigate pathogens, such as *Salmonella,* that bind to the intestinal mucosa. In rodents, these pathogens can sometimes cause more invasive infections than they cause in humans, but the interaction between the bacteria and the mucosa can nonetheless be followed even in these cases. A rodent model has been developed in which autoclaved feces, containing only the bacterium of interest, is implanted in the intra-abdominal area of the rodent to mimic the effects of surgical penetration of the colonic mucosa.

The impact of toxins, such as diarrheal toxins, on the small intestine can be monitored by using the **rabbit ileal loop model.** The rabbit small intestine is tied off into 5- to 10-cm sections by suture, and the toxin is injected into one of the sections. Many diarrheal toxins cause water to be lost by the intestinal tissues into the lumen of the gut, and this can be observed by a swelling of the section into which the toxin was injected. After 12 to 24 hours, the animal is sacrificed and the loop length and fluid volume (in milliliters per centimeter) are measured as readout. Distension (i.e., swelling) of the ileal loop section indicates release of the fluid into the lumen of the segment as a result of toxin action.

Genetically engineered mice called "transgenic mice" or "knockout mice" are gaining increasing use in experiments to probe the interaction of bacteria and intestinal mucosal cells. In these animals, specific genes have been altered or disrupted. Unexpectedly, some of the mice designed originally for studies of the immune system, mice that were missing genes encoding the cytokines interleukin 1 and 10 (see chapter 3), proved to be good models for a type of intestinal inflammation called inflammatory bowel disease. The presence of the normal bacterial microbiota of the colon seems to be responsible for the inflammatory bowel symptoms seen in some of these mice.

These examples are given to provide an introduction to the types of animal models that are available for studying the protective features of the skin and mucosa and the consequences of breaching these barriers. Additional models used in connection with particular bacterial diseases will be described in later chapters.

SELECTED READINGS

Bik, E. M., P. B. Eckburg, S. R. Gill, K. E. Nelson, E. A. Purdom, F. Francois, G. Perez-Perez, M. J. Blaser, and D. A. Relman. 2006. Molecular analysis of the bacterial microbiota in the human stomach. *Proc. Natl. Acad. Sci. USA* **103:**732–737.

Bullen, J. J., H. J. Rogers, P. B. Spalding, and C. G. Ward. 2006. Natural resistance, iron and infection: a challenge for clinical medicine. *J. Med. Microbiol.* **55:** 251–258.

Cooper, G. M., and R. E. Hausman. 2007. *The Cell—a Molecular Approach,* 4th ed. ASM Press, Washington, DC.

Hooper, L. V. 2004. Bacterial contributions to mammalian gut development. *Trends Microbiol.* **12**:129–134.

Hooper, L. V., M. H. Wong, A. Thelin, L. Hansson, P. G. Falk, and J. I. Gordon. 2001. Molecular analysis of commensal host-microbial relationships in the intestine. *Science* **291**:881–884.

Pronovost, P., D. Needham, S. Berenholtz, D. Sinopoli, H. Chu, S. Cosgrove, B. Sexton, R. Hyzy, R. Welsh, G. Roth, J. Bander, J. Kepros, and C. Goeschel. 2006. An intervention to decrease catheter-related bloodstream infections in the ICU. *N. Engl. J. Med.* **355**:2725–2732.

Salemi, C., M. T. Canola, and E. K. Eck. 2002. Hand washing and physicians: how to get them together. *Infect. Control Hosp. Epidemiol.* **23**:32–35.

Servin, A. L. 2004. Antagonistic activities of lactobacilli and bifidobacteria against microbial pathogens. *FEMS Microbiol. Rev.* **28**:405–440.

Strober, W. 2006. Unraveling gut inflammation. *Science* **313**:1052–1054. (Review of an article in the same issue of the journal.)

Toke, O. 2005. Antimicrobial peptides: new candidates in the fight against bacterial infections. *Biopolymers (Pept. Sci.)* **80**:717–735.

Winslow, E. H., and A. F. Jacobson. 2000. Can a fashion statement harm the patient? *Am. J. Nurs.* **100**(9):63–65.

QUESTIONS

1. In what sense are *S. epidermidis* infections an example of how changing human practices can provide new opportunities for bacterial pathogens? *S. epidermidis* is classified as an opportunist. Why is this the case?

2. Explain why infections of the skin occur more often in folds of the skin or under bandages than in regions of skin exposed to the air.

3. How and why do the defenses of mucosal surfaces differ from those of the skin? How do they resemble each other?

4. Consider a bacterium that is ingested via contaminated water and locally colonizes the small intestine. What host defenses would hamper this type of colonization from occurring initially and from leading to an infection in an unimmunized person?

5. Resident microbiota are essential in preventing the colonization of pathogenic bacteria in certain parts of the body. Name the regions of the body where normal microbiota might be protective. Name some mechanisms by which they accomplish this protection.

6. Why are people with indwelling catheters more susceptible to infection?

7. Explain the role that mucin plays in host defense.

8. Lysozyme is more effective against growing bacteria. Why might that be so?

3

The Innate Immune System: Always on Guard

New bacterial invaders can enter the blood and tissue at any time due to breaches in the skin or mucosal surfaces. The body has to have a way to react quickly, in a relatively nonspecific way, to invaders it may never have encountered before. How would you design a "barcode" for bacteria? Even scientists with advanced degrees might be stumped by that question, but that is just what the human body has done with the cells of its innate immune system. The innate immune system jumps into action if the skin or mucosal membranes are breached, but it does so in a carefully controlled way, which is a good thing, since overreaction by the same defense system can kill the host.

Triggering Innate Immune Defenses

Phagocytes: Powerful Defenders of Blood and Tissue

Skin and mucosal surfaces are highly effective in preventing pathogenic bacteria from entering tissue and blood, but from time to time, bacteria succeed in breaching these surfaces. Bacteria that get this far encounter a formidable defense force, the **phagocytic cells, natural killer (NK) cells,** and the proteins that help organize their activities. **Phagocytes** are cells that ingest and kill bacteria, whereas NK cells attack human cells that are infected with bacteria that grow intracellularly. The phagocytes include **polymorphonuclear leukocytes** (also called **PMNs,** polys, or neutrophils), **monocytes, macrophages,** and **dendritic cells (DCs).** The characteristics of different phagocytes and NK cells and their relationships to cells of the adaptive immune defense system are illustrated in Figure 3–1. A readily available set of illustrations of PMNs and other leukocytes (white blood cells) can be found at http://en.wikipedia.org/wiki/white_blood_cell. These cells, together with a set of blood proteins called **complement** and another set of proteins called **cytokines,** both of which cooperate to organize the activities of the phagocytic cells, are called the **innate immune system** because these cells and proteins are always present and ready to respond

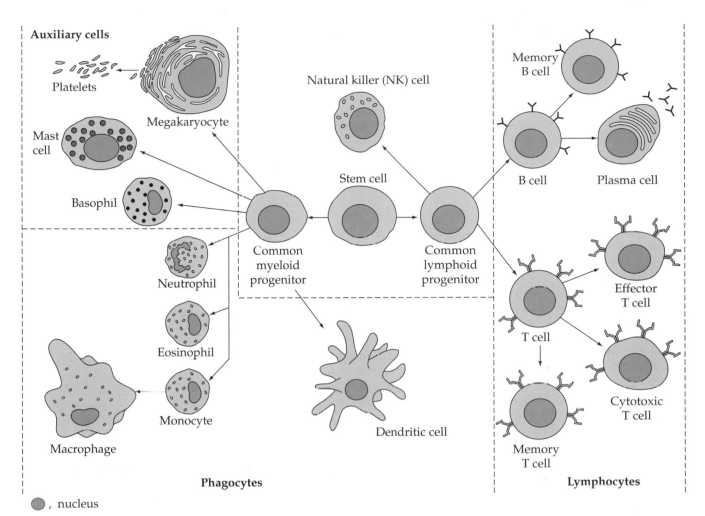

Auxiliary cells

Platelets

Megakaryocyte

Mast cell

Basophil

Neutrophil

Common myeloid progenitor

Eosinophil

Monocyte

Macrophage

Phagocytes

Natural killer (NK) cell

Stem cell

Common lymphoid progenitor

Dendritic cell

Memory B cell

B cell

Plasma cell

T cell

Effector T cell

Cytotoxic T cell

Memory T cell

Lymphocytes

●, nucleus

Figure 3–1 Characteristics and differentiation of the various types of leukocytes of the human body. Leukocytes can be divided into three groups: auxiliary cells (platelets, megakaryocytes, mast cells, and basophils), phagocytes (neutrophils, eosinophils, monocytes, macrophages, and DCs), and lymphocytes (T and B cells and plasma cells).

(Table 3–1). They are the body's first responders to bacterial invasion.

Monocytes are the precursors of macrophages. Macrophages are included here as members of the innate defense system, but they will appear again later as important members of the adaptive defense system (see chapter 4). This is due to the fact that macrophages act to ingest and kill bacteria in a nonspecific way while also presenting antigens to cells of the adaptive immune system. Some examples of macrophages are the **Kupffer cells** of the liver, **alveolar macrophages** of the lung, and **spleen macrophages.**

A cell type closely related to macrophages is the DC. The name dendritic cell comes from the fact that they are covered with spiny projections that look like the dendrites of neurons (Figure 3–1). Some examples

of DCs are the **Langerhans cells** of the skin and **DCs of the lymph nodes** (Table 3–1). DCs are instrumental in initiating and stimulating the second-responder portion of the immune response, called **adaptive immunity.** DCs are found in tissues that are in contact with the external environment or the blood. The forms of DCs that are found in the dermis or in the various organs develop from immature forms that are found in blood. From the bloodstream, they migrate to the locations where they will stand guard to respond to invading bacteria. DCs become activated when they recognize bacteria through shared repeated structural components, such as the bacterial cell wall components **lipopolysaccharide (LPS), lipoteichoic acid (LTA),** or **flagella,** and other molecules specific to the bacteria, such as **CpG-rich DNA** that consists mostly

Table 3–1 Components of the innate immune system and their activities

Component	Characteristics	Function(s)
Cells		
PMNs	Found in blood; short-lived	Ingest and kill bacteria; produce cytokines
Monocytes	Found in blood; migrate to tissue where infection is occurring	Precursor of macrophages
Macrophages		Process bacteria; produce cytokines; activate T cells and B cells
Kupffer cells	Liver	
Alveolar macrophages	Lung	
Spleen macrophages	Spleen	
NK cells	Found in blood and tissue	Kill infected human cells; produce cytokines; organize response to infection
DCs		
Langerhans cells	Dermis	
Lymph node DCs	Lymph nodes	
Proteins		
Complement	Produced in liver; found in blood and tissue	Components opsonize bacteria; other components attract PMNs to infection site
Cytokines	Produced by phagocytic cells and dendritic cells	Organize response of PMNs, cells of adaptive immune system (T cells and B cells)

of alternating cytosine and quanine nucleotide bases (which is not prevalent in mammalian DNA). These "bacterial barcodes" are termed **pathogen-associated molecular patterns (PAMPs)** and play an important part in triggering the adaptive responses. When activated, DCs migrate to the lymphoid tissues, where they present antigens to and stimulate the cells of the adaptive defense system, T cells and B cells. The activation of T cells and B cells by the DCs is described in more detail in chapter 4.

Another type of cell that participates in the defense against bacterial infections, but not as a phagocytic cell, is the **mast cell.** Mast cells congregate around blood vessels. If any foreign material is detected, the mast cells release granules that contain histamine, a compound that makes blood vessels more leaky **(vasodilation).** This helps the PMNs and monocytes, which are normally circulating in blood, to leave the bloodstream and move to the site of infection.

DCs tend to localize in specific parts of the body. By contrast, PMNs and monocytes, which are produced in the bone marrow, migrate constantly through the bloodstream. PMNs are the most abundant and shortest lived of the phagocytic cells. Monocytes are longer lived but less numerous than PMNs. Monocytes also start out in the blood but later leave the bloodstream to migrate into tissues. As they do this, they differentiate into macrophages, a more active phagocytic cell with greater destructive power. There are two types of macrophage: free and fixed. Free macrophages are found in most types of tissue and migrate through them in search of invaders to

ingest and destroy. Fixed macrophages are mostly stationary and are found in organs that filter blood or lymph. When bacteria enter tissue, cytokines stimulate PMNs to become more actively phagocytic as they leave the bloodstream and migrate to an infected area, but unlike monocytes, PMNs do not develop into a different type of cell.

All of these phagocytic cell types are capable of doing a significant amount of damage to tissue. It is not easy to kill bacteria, which are tough little critters, and considerable firepower has to be brought to bear to destroy them. If this firepower is released by the PMNs into the surrounding tissue, as it inevitably is once the battle with invading bacteria is mounted, human cells are vulnerable targets. The body protects itself from these potentially toxic phagocytic cells by keeping them in the bloodstream in a quiescent state unless danger threatens. Only when infection triggers signaling pathways that alert these cells to prepare for battle do the PMNs and NK cells pass through the blood vessel walls and enter tissues by a process called **transmigration** (also referred to as **diapedesis** or **extravasation**).

During an infection, the transmigration process is accelerated. To compensate for this loss of PMNs from the bloodstream, the release of PMNs from bone marrow into the bloodstream is markedly increased. Thus, although passage of these phagocytic cells from the bloodstream into tissue is increased, the net effect is a higher concentration of innate-defense cells in the blood. A high level of PMNs in the blood is a useful diagnostic indicator of infection. During an infection,

PMNs are being produced so rapidly in bone marrow and dumped so quickly into the bloodstream that the immature form of PMNs (called **"bands"** because their nuclei look like bands) is seen in the blood. The presence of bands is another sign of infection.

How do PMNs know when to leave the blood vessels and where to go? Two groups of proteins alert PMNs to leave the bloodstream and guide them to their destination. One is the complement system, which includes the C3a and C5a proteins that are activated by contact with invading bacteria or by interaction with antibodies bound to foreign material. Once the PMN has moved out of the blood vessel, it follows a gradient of complement components to the site where the bacteria have invaded **(chemotaxis).** The second group of proteins is the **cytokine/chemokine** system (Table 3–2). Cytokines are soluble proteins or glycoproteins of 8 to 30 kDa that are produced by PMNs, DCs, and cells of the adaptive immune system (macrophages, T cells, and B cells). The cells of the adaptive immune system and their activities are described in chapter 4. Cytokines mediate the **inflammatory response** (redness, swelling, pain, and fever) to microbial antigens and other types of tissue damage. They also participate in activating cells of the adaptive immune system. Chemokines are small, structurally related glycopeptides of 8 to 10 kDa that are produced by the same cells that produce cytokines. Chemokines, like certain complement components, guide phagocytic cells to the site where an infection is occurring and make them more active. Cytokines and chemokines are produced by a number of human cell types, including monocytes, macrophages, endothelial cells, lymphocytes, and fibroblasts (Table 3–2). How complement and cytokines work together to direct the phagocytes and cytotoxic cells to an infected area and activate them in the process is described in detail below.

Recently, chemokines have been in the news because of their relationship to susceptibility to human immunodeficiency virus (HIV) infection. Chemokines and cytokines can act as messengers because the cells whose activities they direct have receptors that bind them and cause the cell to respond to them. The T cells (specifically, T helper cells) and macrophages that are the targets of HIV have a chemokine receptor, which cooperates with the main HIV receptor, CD4, to allow the virus to bind to these cells in such a way as to gain admission to them so that the virus can replicate. People whose immune cells lack a particular type of chemokine receptor, called CCR5, are somewhat resistant to HIV infection and develop AIDS a lot more slowly than people with the receptor.

Lymph, the fluid that moves through the **lymphatic system,** is also monitored and protected by phagocytes. Lymphatic vessels are tubes composed of overlapping **endothelial cells.** These tubes are organized into a network similar in complexity to the circulatory system that carries blood to all parts of the body. The role of the lymphatic system is to prevent excess buildup of fluid in tissue and to recycle blood proteins. Normally, blood fluids and proteins leak from the blood vessels of capillary beds to feed cells, donate oxygen, and remove carbon dioxide. Blood fluids also provide protective blood proteins, such as complement, cytokines, and chemokines.

The blood vessels do not readily reabsorb fluids that leak from them. Lymphatic endothelial cells are tethered to the muscle bed. Thus, when the level of fluids in an area becomes high enough, the pressure causes muscle cells to separate, pulling the overlapping endothelial cells of the lymphatic vessels apart. This creates openings that allow fluid from the surrounding area to enter the lymphatic vessels. This fluid is then channeled to central holding areas, such as the thoracic duct, where it is returned to the bloodstream.

The inflammatory response to bacterial infection creates a buildup of fluid in tissue, leading to opening of the lymphatic vessels. We experience the accumulation of blood fluids as swelling and redness that appear around infected wounds or sites of infection. In such situations, bacteria can enter the lymphatic vessels. Because the fluid in the lymphatic vessels is returned to the bloodstream, the lymph must be cleansed before it reenters the bloodstream. This task is accomplished by lymph nodes located at strategic points along the lymphatic vessels.

Lymph nodes contain macrophages and the cells of the immune system (DCs, T cells, and B cells). Bacteria that enter a lymph node are usually killed by the macrophages, but there are bacteria that are able to evade this fate. Some of the most dangerous pathogens are the ones that can survive and multiply in the lymph nodes. An example of such a pathogen is *Yersinia pestis,* the cause of bubonic plague. *Y. pestis* growing in lymph nodes creates an inflammatory response so intense that it causes the lymph nodes to become grossly distended (swollen), producing the so-called buboes that give bubonic plague its name. A less pronounced, but still detectable, swelling of lymph nodes **(lymphadenopathy)** occurs during many types of bacterial infections and serves as a diagnostic sign of infection. In chapter 4, we will learn that macrophages and DCs of the lymph nodes also act as antigen-presenting cells and stimulatory cells that potentiate the adaptive defense response against bacteria. Thus, the lymph nodes not only sterilize lymph, but also serve as sites where the adaptive defense system is alerted.

Table 3–2   Selected immune cytokines/chemokines and their activities

Cytokine	Producing cell(s)	Target cell(s)	Function
GM-CSF	Th cells	Progenitor cells	Growth and differentiation of monocytes and DCs
IL-1a	Monocytes	Th cells	Costimulation
IL-1b	Macrophages B cells DCs	B cells NK cells Various	Maturation, proliferation Activation Inflammation, acute-phase response, fever
IL-2	Th1 cells	Activated T and B cells, NK cells	Growth, proliferation, activation
IL-3	Th cells NK cells	Stem cells Mast cells	Growth, differentiation Growth, histamine release
IL-4	Th2 cells	Activated B cells Macrophages T cells	Proliferation, differentiation, IgG1 and IgE synthesis MHC-II Proliferation
IL-5	Th2 cells	Activated B cells	Proliferation and differentiation, IgA synthesis
IL-6	Monocytes Macrophages Th2 cells Stromal cells	Activated B cells Plasma cells Stem cells Various	Differentiation into plasma cells Antibody secretion Differentiation Acute-phase response
IL-7	Marrow stroma, thymus stroma	Stem cells	Differentiation into progenitor B and T cells
IL-8	Macrophages, endothelial cells	Neutrophils	Chemotaxis
IL-10	Th2 cells	Macrophages B cells	Inhibit cytokine production Activation
IL-12	Macrophages B cells	Activated Tc cells NK cells	Differentiation into CTLs (with IL-2) Activation
IL-13	Liver Kupffer cells, lung macrophages, kidney epithelial cells	Macrophages	Inhibit inflammatory cytokine production
IFN-α	Leukocytes	Various	Inhibit viral replication, MHC-I expression
IFN-β	Fibroblasts	Various	Inhibit viral replication, MHC-I expression
IFN-γ	Th1 cells, Tc cells, NK cells	Various Activated B cells Th2 cells Macrophages	Inhibit viral replication Ig class switch to IgG2a Inhibit proliferation Pathogen elimination
MIP-1α	Macrophages	Monocytes, T cells	Chemotaxis
MIP-1β	Lymphocytes	Monocytes, T cells	Chemotaxis
TGF-β	T cells, monocytes	Monocytes, macrophages Activated macrophages Activated B cells Various	Chemotaxis IL-1 synthesis IgA synthesis Inhibit proliferation
TNF-α	Macrophages, mast cells, NK cells	Macrophages Tumor cells	CAM and cytokine expression Cell death
TNF-β	Th1 and Tc cells	Phagocytes Tumor cells	Phagocytosis, NO production Cell death

HOW PMNS RESPOND TO BACTERIA. For many years, the interaction between PMNs and invading bacteria remained mysterious. In recent years, however, this interaction has become a focus of intense research. Much of this research has focused on stimulation of PMNs by components of gram-negative bacteria, such as the cell surface component LPS (Box 3–1) and **flagellin,** a protein involved in bacterial cell motility. However, gram-positive bacteria also trigger the same type of activation of PMNs. In the case of gram-positive bacteria, LTA, the anionic polymer found in the cell walls of these bacteria (Box 3–1), seems to be responsible for this activation. Peptidoglycan fragments from both gram-positive and gram-negative bacteria, as well as certain types of DNA (e.g., CpG-rich DNA), can also act as triggers.

The process by which bacterial surface molecules (PAMPs) trigger cytokine release is best understood in the case of LPS (Box 3–1). LPS is released from the bacterial surface due to lysis of bacteria. LPS binds to **LPS-binding protein (LBP),** an acute-phase protein produced by the liver, and is delivered to **CD14,** a protein receptor on the surfaces of macrophages and other cytokine-producing cells (Figure 3–2). LPS is then transferred to the transmembrane signaling receptor **Toll-like receptor 4 (TLR4)** and its accessory protein **MD-2.** This TLR4–LBP–CD14–MD-2 complex triggers a cellular signal transduction pathway involving the cellular signaling proteins MyD88, TIRAP, TRAM, and Trif, which leads to activation of NF-κB and culminates in cytokine production and release by the stimulated cell.

PAMPs were discovered in a rather unusual way. Scientists working on the fruit fly, *Drosophila melanogaster*, identified a transmembrane receptor protein called **Toll** that was required for resistance to fungal infections. Oddly enough, only one of the Toll receptors in *Drosophila* is linked to a defense against infection. The others are involved in various aspects of fly development. Only later was the human equivalent found, hence the term "Toll-like receptor." To date, 11 mammalian (10 human) paralogs of the original TLR, TLR4, have been described, and there may be more to come. TLRs mediate the activation of the innate immune system through recognition of PAMPs. The TLR gene family and its signaling pathways have been evolutionarily conserved in both invertebrates and vertebrates. This common set of diverse TLRs may reflect the fact that although all gram-negative or gram-positive bacteria have surface "signatures" in common, there are also numerous differences. The different TLRs in various combinations may provide the flexibility to respond to differences between invading bacteria without responding specifically to each particular bacterium, as the antibody-based adaptive immune system does. For example, TLR4 is a receptor for LPS for most enterobacteria, but sometimes a complex of TLR1/TLR2 or TLR2/TLR6 is the receptor for certain pathogens that have modified LPS structures or have surface lipoproteins. TLR9 responds to DNA containing unmethylated CpG sequences. Double-stranded RNA activates TLR3, and bacterial flagellin activates TLR5. Figure 3–3 summarizes the different TLRs and the signaling pathways that they activate.

TLRs are transmembrane protein structures that are exposed on the surface of the phagocytic cell. TLRs have a conserved region of amino acids that is known as the TIR domain because of its similarity to a Toll/interleukin-1 receptor (IL-1R). The TIR domain conveys the signal detected by TLR interaction with LPS or some other bacterial component to the interior of the phagocytic cell. Signaling of the TLR/IL-1R superfamily is mediated through myeloid differentiation primary response gene 88 (MyD88), IL-1R-associated kinases (IRAKs), transforming growth factor beta-activated kinase 1 (TAK1), TAK1-binding protein 1 (TAB1), TAB2, tumor necrosis factor (TNF) receptor-associated factor 6 (TRAF6), and others. This signaling activity triggers a signal transduction cascade that activates PMN functions, such as the oxidative burst that actually kills the bacteria. The signaling cascade may ultimately trigger apoptotic death of the PMN.

The host cell also has a detection system for intracellular bacterial pathogens. The **NOD1** and **NOD2** proteins are leucine-rich-repeat intracellular proteins that function analogously to TLRs to detect bacterial peptidoglycan inside host cells. Human NOD1 detects the *N*-acetylglucosamine-*N*-acetylmuramic acid tripeptide part of the peptidoglycan derived from degradation of the gram-negative cell wall, whereas NOD2 detects the *N*-acetylglucosamine-*N*-acetylmuramic acid dipeptide derived from gram-positive peptidoglycan. Activation of either NOD leads to NF-κB activation and inflammatory responses.

HOW PHAGOCYTES KILL BACTERIA. The steps involved in the killing of a bacterium by a phagocyte are shown in Figure 3–4. The phagocyte first forms pseudopods that engulf the bacterium. **Phagocytosis (engulfment)** requires dynamic rearrangements of actin, a major component of the eukaryotic cytoskeleton. After engulfment, the bacterium is encased in an endocytic vesicle called the **phagosome.** Phagocytes possess two general types of bacterial killing mechanisms: nonoxidative and oxidative. Various lysosomal enzymes, antimicrobial peptides, membrane-permeabilizing proteins, and degrading proteins mediate **nonoxidative killing. Oxidative killing** occurs

BOX 3–1 A Brief Review of the Surfaces of Gram-Negative and Gram-Positive Bacteria

LPS is located at the surfaces of gram-negative cells and is composed of lipid A (made up of fatty acids and a disaccharide), the core oligosaccharide, and the repeating-unit O-antigen polysaccharide. The LPS forms the surface of the outer membrane, whose inner layer is composed of phospholipids. The thin peptidoglycan of gram-negative species is located between the inner and outer membrane. In contrast, the cell wall of gram-positive species lacks LPS and an outer membrane. The thick peptidoglycan is composed of multiple cross-linked layers. A substantial amount of the gram-positive cell wall is made up of another anionic polymer, called TA, and LTA, which contains repeated units of ribitol or glycerol phosphate linked to amino sugars; sometimes to amino acids, such as D-alanine; and to ternary amines, such as phosphorylcholine. TA is covalently bonded to the peptidoglycan, whereas LTA is linked to a lipid anchor in the cellular membrane. Peptidoglycan and LTA activate TLR2, and LPS activates TLR4. (Further details can be found in *The Physiology and Biochemistry of Prokaryotes*, 3rd ed., by D. White, Oxford University Press, Oxford, United Kingdom, 2007.)

A Gram negative

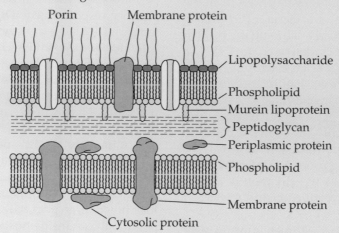

B Gram positive

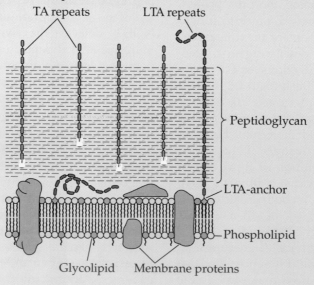

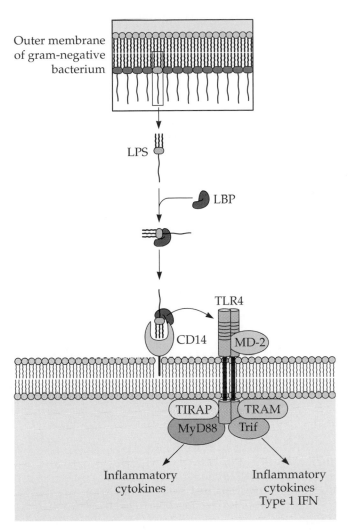

Figure 3–2 Cellular recognition of LPS through TLR signaling.

through the formation of toxic reactive oxygen species.

The phagosomal membrane contains ATPases that pump protons into the phagosome interior, reducing the internal pH to about 5. Phagocytes carry antibacterial proteins that are as toxic to the phagocytes and surrounding tissue cells as they are to their bacterial targets. Accordingly, they are stored in an inactive form in **lysosomal granules.** Fusion of a lysosomal granule with the phagosome to form the **phagolysosome** releases the lysosomal proteins into the phagolysosome interior. The low pH of the phagolysosome generated by the ATPases of the phagosome membrane activates lysosomal proteins. Lysosomal proteins have different types of killing activity. Some lysosomal proteins are **degradative enzymes,** such as proteases, phospholipases, and lysozyme, which destroy surface components of bacteria, as well as nu-

cleases, which degrade the bacterial DNA and RNA released during lysis. Other lysosomal proteins, such as **defensins,** insert themselves into bacterial membranes, creating pores that permeabilize the membrane and allow the bacterial cytoplasmic components to leak into the surrounding environment.

The **granulosomes** of PMNs have another type of lysosomal protein, **myeloperoxidase,** which produces reactive forms of oxygen that are toxic to many bacteria (similar to lactoperoxidase). Myeloperoxidase is activated only when it is brought into contact with an **NADPH oxidase,** which is located in the phagosomal membrane, and when the resulting complex is exposed to the low pH of the phagolysosome interior.

The reaction catalyzed by the myeloperoxidase complex has three steps (Figure 3–5). First, NADPH oxidase generates a **superoxide radical:** $NADPH + 2O_2 \rightarrow 2O_2^-$ (superoxide radical) $+ NADP^+$. The superoxide radical is highly reactive. In fact, it is so reactive that most of it does not react directly with bacterial proteins but converts spontaneously, or in the presence of **superoxide dismutase** rapidly, into hydrogen peroxide (H_2O_2). Myeloperoxidase catalyzes the reaction of H_2O_2 with chloride (Cl^-) or thiocyanate (SCN^-) ions to form **hypochlorite** (OCl^-, the active ingredient in bleach) or **hypothiocyanite** ($OSCN^-$), respectively. The toxicity of these reactive oxygen molecules is due to the fact that they oxidize amino acid side chains on proteins and thus inactivate essential bacterial surface proteins. The two amino acids most prone to oxidative attack are cysteine and methionine, both of which contain susceptible sulfur atoms. Oxidation of cysteine leads to the formation of disulfide bonds; disulfide formation with thiol groups on other molecules, such as glutathione; and thiyl radicals, while the major product with methionine under biological conditions is methionine sulfoxide. Other amino acids that are also susceptible to oxidative attack are histidine, lysine, arginine, proline, threonine, tyrosine, and tryptophan. These oxidative conditions can also damage nucleic acids, because iron (Fe^{3+}) plus H_2O_2 forms hydroxyl radical ($HO\cdot$). This nonenzymatic reaction is called the **Fenton reaction** and can damage DNA. The generation of toxic forms of oxygen by phagocytes is called the **oxidative burst** (or **respiratory burst**). During an infection, cytokines stimulate increased production of lysosomal enzymes, thus increasing the killing potential of the oxidative burst.

JUST SAY NO. Human monocytes and macrophages produce a very simple but powerful antimicrobial compound, **nitric oxide (NO).** NO is toxic in its own right, attacking bacterial metalloenzymes, proteins,

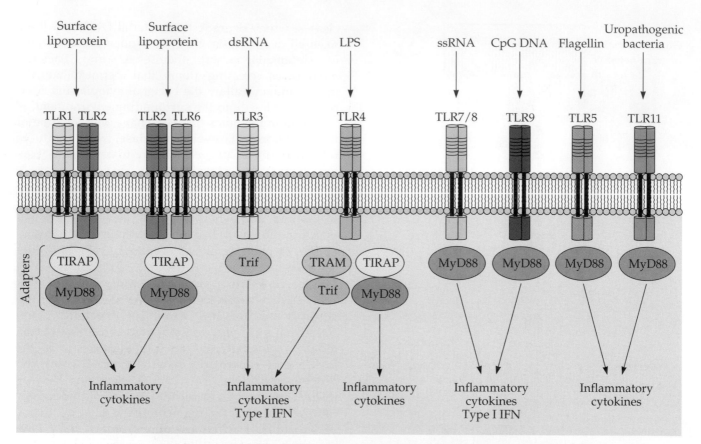

Figure 3–3 The TLR family and its signaling pathways. TLR1, TLR2, TLR6, TLR4, and TLR5 are located on the plasma membrane, whereas TLR3, TLR7, and TLR9 are not located on the cell surface but reside in endosomes. TLRs and IL-1Rs have a conserved region of amino acids, which is known as the TIR domain. dsRNA, double-stranded RNA; ssRNA, single-stranded RNA. (Adapted from Kawai and Akira, 2006, with permission from Macmillan Publishers Ltd.)

and DNA. In addition, it can combine with superoxide to form **peroxynitrite (OONO⁻)**, a very reactive molecule that oxidizes amino acids and is toxic for bacteria and for human cells. Synergistic reactions between NO and superoxides during the oxidative burst may help make the burst more toxic for bacteria. NO may also serve as a signaling molecule to regulate the functions of phagocytic cells, as well as other adaptive immune cells. NO has been implicated in so many areas of human and animal physiology that the journal *Science* chose it as the "molecule of the year" in 1992.

During an infection, cytokines induce NO synthesis by many human cell types. In blood, the stable end products of NO oxidation, nitrite and nitrate, increase, thus indicating that NO is being produced. NO may contribute to some of the symptoms of disease, including vascular collapse and tissue injury. Mice lacking the inducible nitric oxide synthetase responsible for generating NO tolerate bacterial LPS with fewer toxic side effects than normal mice. On the other

hand, these mice are highly susceptible to infections by intracellular bacterial and protozoan pathogens, such as *Mycobacterium tuberculosis* or *Leishmania major,* which cause tuberculosis and leishmaniasis, respectively. The human pathogen *Neisseria meningitidis,* which is responsible for meningococcal disease, has at least two NO detoxification enzymes that enhance survival during nasopharyngeal colonization and during phagocytosis by human macrophages. A decade ago, there was controversy over whether NO was even produced by human cells and whether it had any role in human response to infection. Today, the question has changed to "Is there anything NO is not involved in?" The answer appears to be NO.

Inflammation and Collateral Damage

Inflammation (from Latin *inflammare,* to set on fire) is one important response of vascular tissues to harmful stimuli, such as damaged tissues and the release of irritants, caused by infection. The inflammatory re-

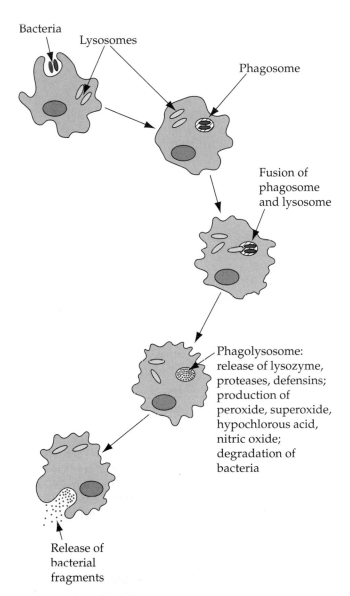

Bacteria

Lysosomes

Phagosome

Fusion of phagosome and lysosome

Phagolysosome: release of lysozyme, proteases, defensins; production of peroxide, superoxide, hypochlorous acid, nitric oxide; degradation of bacteria

Release of bacterial fragments

Figure 3–4 Steps in ingestion and killing of bacteria by phagocytes. Bacteria are first engulfed by endocytosis into a phagosome. The fusion of phagosomes and lysosomes releases toxic enzymes and proteins that kill most bacteria. Debris from dead bacteria is then released by exocytosis.

sponse recruits innate-immune cells, such as macrophages and neutrophils, from the blood vessels to the site of infection. **Proinflammatory cytokines** are induced by the complement cascade (C3a and C5a [see below]) and by mast cells and activated DCs, macrophages, and other phagocytes. Subsequent production of prostaglandins and leukotrienes initiates and regulates this process. Important proinflammatory cytokines include **tumor necrosis factor alpha (TNF-α), interleukin-1 (IL-1),** and **interleukin-6 (IL-6).** The vasoactive peptides **kallidin** and **bradykinin**

increase vascular permeability at the site of infection to allow neutrophil transmigration (extravasation).

Although inflammation serves to protect and control infections, it can also cause further tissue damage, which is manifested as the disease symptoms of redness, swelling, heat, and pain. The increased blood flow due to vasodilation results in redness and increased temperature in the area. The increased vascular permeability causes blood fluids to leak out of the vessels as the phagocytes transmigrate and thereby also causes edema (swelling) of the surrounding tissue. The source of the pain is still not clearly understood, but it is probably due to the combined effects of cytokines, prostaglandins, and coagulation cascade components on nerve endings in the inflamed region. Bradykinin also appears to increase sensitivity to pain. To counter the proinflammatory response, there are a number of **anti-inflammatory cytokines** (including **IL-1 receptor antagonist, interleukin-4 [IL-4],** and **interleukin-10 [IL-10]**) that serve to regulate the immune response by inhibiting the actions of the proinflammatory cytokines.

Although phagocytic cells are effective killers of bacteria and are essential for clearing the invading bacteria from an infected area, the body can pay a high price for this service. During active killing of a bacterium, lysosomal enzymes are released into the surrounding area, as well as into the phagolysosome. Released lysosomal enzymes damage adjacent tissues and can be the main cause of tissue damage that results from a bacterial infection. Also, PMNs kill themselves as a result of their killing activities, and lysosomal granules released by dying PMNs contribute further to tissue destruction. **Pus,** a common sign of infection, is composed mainly of dead PMNs and tissue cells. Because phagocytes cause collateral damage to tissue cells when they combat an invader, it is important for the body to tightly regulate their numbers, locations, and activation states. This is why the body has a complex set of signals (e.g., complement and cytokines) that stimulate phagocytes to leave the bloodstream and enter tissues only in areas of the body where the invading bacteria are located. This same set of signals upregulates the phagocytes' killing ability and occurs only as they approach the site of the infection (see below).

NK Cells

The role of NK cells is to complement the activities of PMNs by killing infected human cells. Once thought to be involved primarily in controlling viral infections, they are now known to be important in controlling infections by bacteria that invade and hide from the immune system inside human cells. Unlike

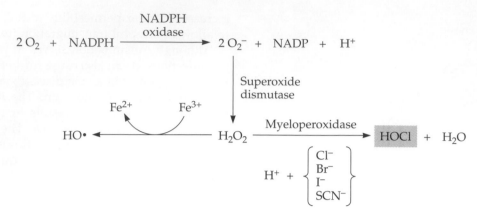

Figure 3–5 Reactions catalyzed by myeloperoxidase that kill bacteria in the phagolysosome. The phagosome contains two unique enzymes: lysosomal myeloperoxidase
and phagosomal membrane-bound NADPH oxidase. Myeloperoxidase produces reactive oxygen molecules when activated by contact with NADPH oxidase. Superoxide dismutase converts the O_2^- generated by NADPH oxidase into H_2O_2. Myeloperoxidase catalyzes the reaction of H_2O_2 with Cl^- or SCN^- ions to form OCl^-, the
active ingredient in bleach, or $OSCN^-$, both of which are extremely toxic to bacteria.
Under these oxidative conditions, iron ions exist predominantly in the oxidized ferric
(Fe^{3+}) state. Inside the body, Fe^{3+} ions are bound to iron-binding proteins, such as
transferrin or lactoferrin, but in phagolysosomes, the low-pH environment causes the
release of Fe^{3+} from the proteins. The freed Fe^{3+} plus H_2O_2 forms hydroxyl radicals
via a nonenzymatic Fenton reaction, which can also damage DNA.

PMNs and other phagocytic cells, NK cells do not ingest their targets, although their mode of killing resembles that of phagocytes in many respects. Like
phagocytic cells, they are produced in bone marrow
and circulate in the bloodstream. Also, they store their
toxic substances in granules. Binding to an infected
human target cell stimulates the release of these granules. Thus, instead of ingesting a bacterium or infected cell, the innate cytotoxic NK cells bombard infected cells. The granule proteins of cytotoxic cells are
not the same as those of the lysosomal granules of
macrophages and PMNs, but they have some similar
functions. For example, cytotoxic-cell granules contain
a protein called **perforin** that inserts into the
membrane of a target cell and causes channels to
form. These channels allow other granule proteins, a
set of proteases called **granzymes,** to enter the target
cell. One effect of this assault appears to be forcing
the target cell to initiate **apoptosis (programmed cell
death),** a process by which the infected human cell
kills itself.

How do NK cells recognize infected human cells?
Like many responses of the immune system, multiple
signals are sensed to direct NK cells to kill infected
cells and to spare normal cells. NK cells use an
opposing-signals mechanism to identify infected cells
(Figure 3–6). When they encounter normal or infected
cells, an activation receptor on the surface of the NK

cell is stimulated by an activating ligand on the surface of the target cell. If left unchecked, this response
will lead to full activation of the NK cell and killing
of the encountered cell. Healthy, normal cells also express proteins on their surfaces called **class I major
histocompatibility complex (MHC-I),** which is discussed in chapter 4. MHC-I binds to a second inhibitory receptor on the NK cell surface and halts the activation of the cytotoxic response. In contrast, infected
cells express much less MHC-I on their surfaces than
normal cells, and the activation response of the NK
cell proceeds, leading to an attack on the infected cell.
Thus, the MHC-I expression level is the signal that
spares normal cells from being killed by NK cells.

The Complement Cascade

Characteristics and Roles of Complement Proteins and the Complement Cascade

The barriers of skin and mucosa, together with the
phagocyte defense system, are a powerful deterrent to
invading bacteria. There is another arm of the innate
defenses, however, that is highly complex but equally
important—the blood proteins that direct the phagocyte defense system and, in the cases of chemokines
and cytokines, the immune response as well. Because
complement, chemokines, and cytokines form links

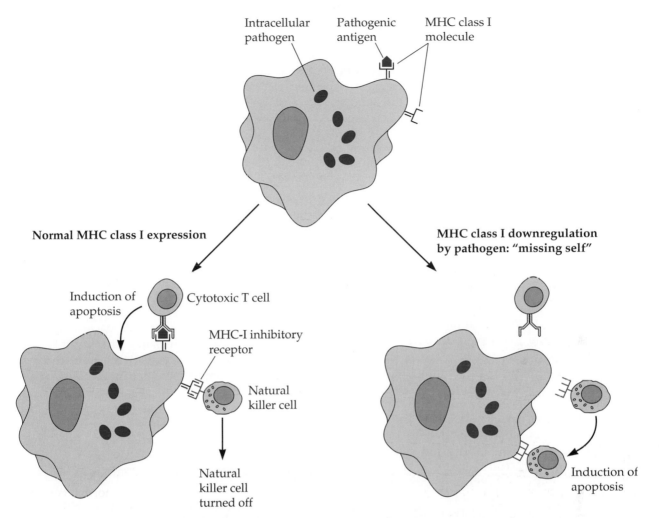

Figure 3–6 Schematic diagram indicating the complementary activities of cytotoxic T cells and NK cells. Host cells infected with intracellular bacteria display antigens on their surfaces through MHC-I complexes (described in detail in chapter 4). Cytotoxic T cells (called CTLs) recognize the infected cells through those complexes and subsequently kill the infected cells. However, since intracellular pathogens can cause infected cells to express many fewer MHC-I molecules on their surfaces than normal cells, NK cells complement this process by recognizing and killing cells that do not have MHC-I complexes on their surfaces.

between the innate phagocyte defense system and the adaptive defenses, it is appropriate to treat them as a separate topic. The inspection of these critically important proteins and their functions also leads naturally to a dark side of the innate defenses—**septic shock.** Septic shock is a major killer of hospital patients who acquire infections while in the hospital.

Complement is a set of proteins produced by the liver. These proteins circulate in blood and enter tissues all over the body. Complement proteins are inactive until proteolytic cleavage converts them to their active forms. The cascade of proteolytic cleavages of complement components that occurs during an infec-

tion is called **complement activation.** Complement components, some of which are multiprotein complexes, are usually designated by a C. There are nine of these components, C1 to C9. Activated proteolytic-cleavage products are indicated by an "a" or "b." In most cases, "a" designates the smaller and "b" the larger of the two proteolytic products, but this rule is not followed uniformly. In the previous edition of this book, we decided to use what had been touted by some leading immunologists as a rational revision of the terminology used to describe the complement cascade. Unfortunately for us, many immunologists preferred to cling to the old nomenclature, so our chapter caused some confusion. In this chapter, we return to the old nomenclature. The confusion concerned complement component C2. Like several of the other components, C2 is cleaved into two different-size components during the activation of the complement cascade. In the case of C3, C4, and C5, the smaller component is designated "a" (e.g., C3a) and the larger segment is designated "b" (e.g., C3b). The exception to this is C2, in which the small C2 cleavage product is designated C2b and the large one is C2a. We will use this designation.

Complement activation can be initiated in three ways. A newly discovered initiation pathway involves protein trimers called **mannose-binding lectins (MBL),** which are members of a family of proteins called **collectins.** Collectins are calcium-binding lectins (proteins that bind very specifically to sugar residues). The mannose-binding lectins bind the mannose residues that are commonly found on the surfaces of bacteria, but not on human cells. Collectins are produced by the liver and are part of what is called the **acute-phase response** (or **acute inflammatory response**) to an infection, the initial onslaught by a variety of proteins, including iron-binding proteins, which make it difficult for bacteria to multiply. Collectins bound to the surface of a bacterium not only sequester the bacteria into clumps that are eliminated from the body by phagocytic cells, but can also activate the complement cascade (Figure 3–7).

Certain molecules found on the surfaces of bacteria can trigger the complement cascade directly without the intervention of collectins. This has been called the **alternative pathway.** The best-characterized complement-triggering bacterial surface molecules are LPS and **teichoic acid (TA),** found on the surfaces of gram-negative and gram-positive bacteria, respectively (Box 3–1). Complement-activating molecules of fungi, protozoa, and metazoa are not as well characterized, but they also appear to be lipid-carbohydrate complexes on the microbial cell surface.

Finally, antibodies of the adaptive defense system can also activate complement similarly to collectins by binding to the surfaces of bacteria and interacting with complement proteins. This antibody-mediated pathway is referred to as the **classical pathway.** Antibodies are blood proteins but differ from collectins in that they are produced by B cells, not liver cells, and they bind to very specific molecules on the bacterial surface, not nonspecifically to a ubiquitous bacterial molecule, such as LPS. Thus, both the innate and adaptive defense systems can trigger the complement cascade. This is yet another example of a link between the innate and adaptive defenses.

Before examining the pathways for complement activation in detail, it is helpful to understand the roles of the activated proteins produced by the series of proteolytic cleavages that comprise the cascade. Regardless of how the complement pathway is activated, the same key activated components, **C3a, C3b, C5a,** and **C5b,** are produced (Figure 3–7).

C3a and C5a are proinflammatory molecules that stimulate mast cells to release their granules, which contain vasoactive substances that increase the permeability of blood vessels and thus facilitate the movement of phagocytes from blood vessels into tissue. C5a also acts together with cytokines to signal phagocytes to leave the bloodstream and to guide them to the infection site. Once PMNs or monocytes have left the bloodstream, they move along a gradient of C5a to find the locus of infection. At the site of infection, C3b binds to the surface of the invading bacterium and makes it easier for phagocytes to ingest the bacterium. This activity is called **opsonization.**

Without opsonization, phagocytes have difficulty ingesting a bacterium unless the bacterium is trapped in a small space. The reason is that most bacteria do not stick to the phagocyte surface, so that the action of pseudopod encirclement can actually propel the bacterium away from the phagocytes, much as a fish can slip from your hands as you try to grab it. Phagocytes have surface receptors that bind C3b (Figure 3–8). Thus, complement component C3b allows the phagocyte to immobilize the bacterium so that it can be engulfed more efficiently. Antibodies, like C3b, can also act as opsonins because a portion of the antibody, the Fc portion, is recognized by phagocyte receptors. However, antibodies bind to specific molecules on the surface of one type of bacterium, whereas C3b binds nonspecifically to surfaces not coated, as human cells are, by sialic acid (Box 3–2). The combined effect of C3b and antibodies is synergistic in stimulating uptake of the bacterium by phagocytes.

Another role of activated complement components is direct killing of the bacterium. Activated component C5b recruits C6, C7, C8, and C9 to form a membrane-damaging complex in the membranes of some types of microorganisms (e.g., enveloped viruses, gram-negative bacteria, and some gram-positive bacteria). This complex is called the **membrane attack complex (MAC).** Formation of the MAC inactivates enveloped viruses and kills bacteria by punching holes in their membranes.

STEPS IN COMPLEMENT ACTIVATION. The steps in complement activation by all three pathways are shown in more detail in Figure 3–9. The so-called classical pathway is initiated when the Fc regions of two immunoglobulin G (IgG) molecules or one IgM molecule bound to the surface of a bacterium are cross-

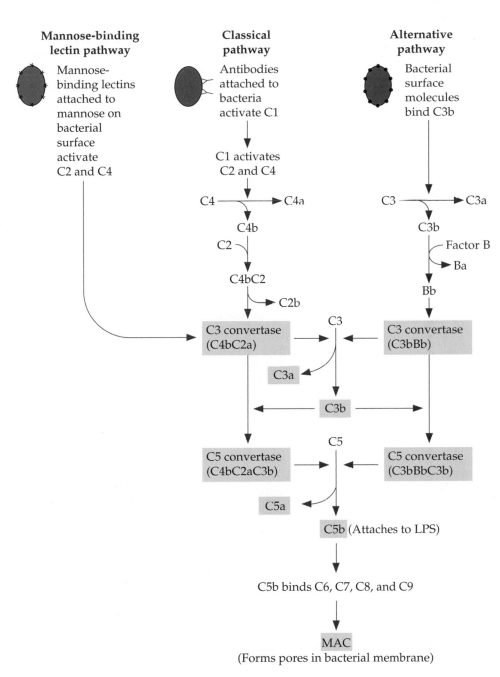

Mannose-binding lectin pathway

Mannose-binding lectins attached to mannose on bacterial surface activate C2 and C4

Classical pathway

Antibodies attached to bacteria activate C1

C1 activates C2 and C4

C4 ⟶ C4a

C4b

C2

C4bC2 ⟶ C2b

C3 convertase (C4bC2a)

C3a

Alternative pathway

Bacterial surface molecules bind C3b

C3 ⟶ C3a

C3b

Factor B ⟶ Ba

Bb

C3 convertase (C3bBb)

C3

C3b

C5 convertase (C4bC2aC3b)

C5a

C5

C5 convertase (C3bBbC3b)

C5b (Attaches to LPS)

C5b binds C6, C7, C8, and C9

MAC
(Forms pores in bacterial membrane)

Figure 3–7 The main steps in activation of complement by the MBL, classical, and alternative pathways. These pathways differ only in the steps that initiate the formation of C3 convertase. Important activated products are C3b (which opsonizes bacteria), C3a (which acts as a vasodilator), C5a (which acts as a vasodilator and attracts phagocytes to the area), and C5b-C9 (the MAC, which inactivates enveloped viruses and kills bacteria).

linked by **C1,** a multiprotein complex. Binding of C1 to the antibodies causes C1 to release one of its components, producing a form of C1 that cleaves **C4** to **C4a** and **C4b.** C4b attaches covalently to the bacterium's surface at a site near C1. C1 also cleaves **C2** to **C2a** and **C2b.** The MBL pathway (also sometimes called the **collectin pathway**) is similar in that it also stimulates cleavage of C2 and C4. C2a binds to C4b to complete the **C3 convertase** complex. C3 convertase cleaves C3 to C3a, which diffuses away from the site and C3b, which binds to the bacterium's surface. Bound C3b has two functions. Some C3b acts as an

opsonin to enhance uptake by phagocytes, and some C3b binds to the C3 convertase complex to form the **C5 convertase** complex, which cleaves C5 to C5a and C5b. C5a diffuses away from the site, whereas C5b binds to the bacterium's surface. C5b recruits C6, C7, C8, and C9 to form the MAC.

Activation via the alternative pathway bypasses C1, C2, and C4 and relies on C3 as the initiating component. As C3 circulates in blood and tissue, it is occasionally activated to produce a form that interacts with water to assume a conformation similar to that of C3b. This activated C3b binds to nearby surfaces.

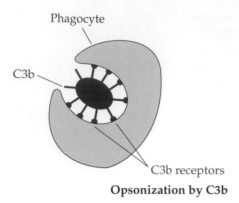

Opsonization by C3b

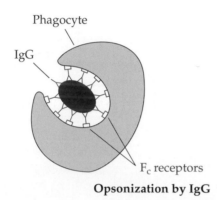

Opsonization by IgG

Figure 3–8 Opsonization of a bacterium by activated complement component C3b and antibodies. Combined opsonization by both C3b and antibodies considerably enhances the uptake of the bacterium by the phagocyte.

Tissues of the body are coated with sialic acid residues, which preferentially bind a serum protein, **factor H.** C3b can interact with H to form a complex that produces C3b, but the C3b remains bound to H. This binding changes the conformation of C3b and targets it for proteolytic cleavage and destruction by serum protein **factor I.**

If C3b instead binds to a bacterium's surface, it is more likely to encounter serum protein **factor B.** B binds better than H to the bacterium's surface. Once C3b binds to B on the bacterium's surface, another serum protein, **factor D,** cleaves B to **Ba** and **Bb.** The resulting C3b/Bb complex produces C3b, which can then bind Bb to form a C3/C5 convertase. The initial C3bBb complex is the C3 convertase, which cleaves C3 to form more C3a and C3b. The newly generated C3b binds to the same bacterial surface, and when this bound C3b comes in contact with the C3bBb complex already on the surface, this new complex becomes the C5 convertase. Some bacteria produce a polysaccharide surface coating, called a **capsule,** which preferentially binds serum protein H, rather than B. The ef-

fect of this is to eliminate C3b as it is deposited on the surface and to prevent effective opsonization of the bacterial surface.

In all three pathways, it is important to keep the accelerated production of C3a, C3b, C5a, and C5b under control. To this end, most C3b molecules on the bacterium's surface are proteolytically cleaved to produce **iC3b.** Although iC3b is still an effective opsonin, it cannot help to form a C3 or C5 convertase complex.

Roles of Cytokines and Chemokines in Directing the Phagocyte Response

Cytokines are glycoproteins (Box 3–2) produced by a variety of cells, including monocytes, macrophages, NK cells, endothelial cells, lymphocytes, and fibroblasts. Chemokines are smaller peptides than cytokines but have many of the same functions, especially attracting phagocytes and activating them, much as complement components C5a and C3a do. Cytokines and chemokines play central roles in regulating the activities of the cells of the innate and adaptive defense systems. Cytokines and chemokines recognize and bind to specific receptors on the surfaces of target immune cells, which set off signal transduction cascades that modify the functions of the immune cells. Just as complement is activated by bacterial surfaces, cytokine release is triggered by interaction between cytokine-producing cells and molecules on the surface of the invading bacterium. Cytokines, chemokines, and the cells that produce them are listed in Table 3–2. In the case of gram-negative bacteria, the outer-membrane LPS that activates complement is also the molecule that stimulates cytokine production. Although the surface molecules of other types of bacteria that activate complement and stimulate cytokine release have not been nearly as well studied, it appears likely that the same surface molecules on these bacteria both activate complement and stimulate cytokine production.

During an infection, macrophages and other cells, such as endothelial cells, release a variety of different cytokines. Some of these appear early in the infection and are responsible for upregulating host defenses. Others appear late in the infection and help to downregulate the defense response. Among the earliest-appearing cytokines are **granulocyte-macrophage colony stimulating factor (GM-CSF)** and **interleukin 3 (IL-3).** These cytokines trigger the release of monocytes and granulocytes (especially PMNs) from the bone marrow into the circulation (Figure 3–10). Other early-appearing cytokines, such as TNF-α, IL-1, **gamma interferon (IFN-γ),** and **interleukin-8 (IL-8),** stimulate the monocytes and granulocytes to leave the

BOX 3–2 Structures of Glycoproteins

Shown are the structures of a typical N-glycosylation structure on human glycoproteins and of the nine-carbon monosaccharide core of sialic acids, which are also called neuraminic acids. The surfaces of human cells are coated with glycoproteins, which, as the name implies, contain proteins covalently linked to oligosaccharides, such as the one shown. These glycoproteins play diverse roles in eukaryotic organisms. Not surprisingly, bacteria produce enzymes on their surfaces that can free the sugars from host glycoproteins, such as NanA, BgaA, and StrH in *Streptococcus pneumoniae,* and provide food to the bacteria. One kind of sugar that is commonly found at the ends of the oligosaccharides in host glycoproteins is sialic acid, whose generic structure is shown. (Figure adapted from S. King, J. Hippe, R. Karen, and J. N. Weiser, *Mol. Microbiol.* **59:**961–974, 2006.)

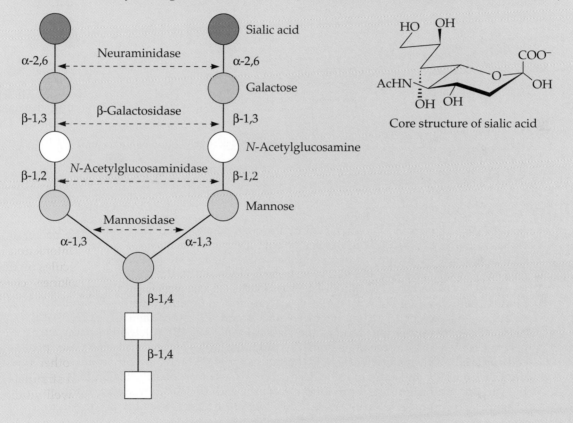

Core structure of sialic acid

bloodstream and migrate to the site of infection. The steps in this process are illustrated in Figure 3–10 for PMNs.

Normally, PMNs move rapidly through the blood vessels, occasionally colliding with one of the vessel walls. TNF-α, IL-1, and IFN-γ stimulate endothelial cells to produce a set of surface proteins called **selectins.** These selectins bind to proteins on the surfaces of PMNs and other blood cells, causing them to bind loosely to the endothelium. This loose binding slows the movement of the blood cells as they assume a rolling motility. Other selectins appear on the endothelial cells, and IL-8 stimulates PMNs to produce proteins called **integrins** on their surfaces. The integrins bind another set of cytokine-stimulated proteins on the endothelial-cell surface, the **intercellular adhesion molecules (ICAMs),** to generate tighter attachment between PMNs and endothelial cells, which stops the movement of the PMNs and causes them to flatten against the blood vessel wall. The slowing and stopping of the PMNs is called **margination.** The PMNs then force themselves between endothelial cells, a process that is assisted by a PMN protein called **platelet-endothelial cell adhesion molecule.**

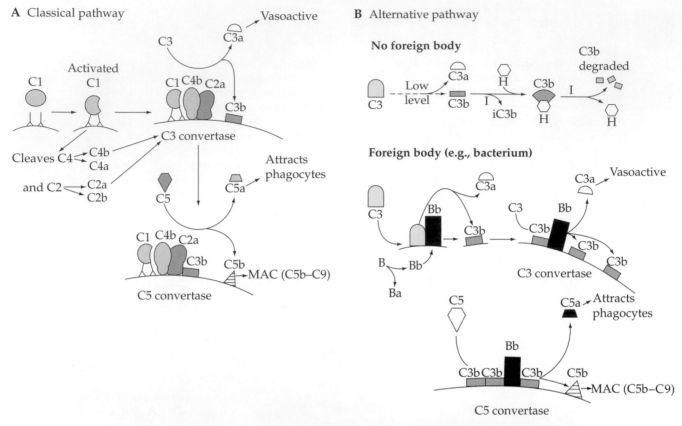

Figure 3–9 Activation of complement by the classical and alternative pathways. **(A)** Classical pathway. Two IgG molecules or one IgM molecule attached to the surface of a bacterium bind complement component C1, causing an autoproteolytic event that activates it. C1, C4b, and C2a bind to each other and to the bacterium's surface to form C3 convertase. Addition of C3b produces C5 convertase. In the mannose-binding pathway (not shown), the MBL activate the complement pathway similarly to antibodies, except that they interact with C4 and C2 rather than C1. After that point, this pathway is the same as the classical pathway. **(B)** Alternative pathway. C3b, an activated form of C3, is normally produced at low levels. If it binds a host cell surface, which preferentially binds serum factor H, H binds to C3b produced by the C3b complex and targets it for destruction by serum protein I. If C3b binds to the surface of a bacterium, it can form a complex with Bb (C3 convertase). Addition of more C3b produces C5 convertase. C5 convertase triggers assembly of the MAC.

The proinflammatory complement components C3a and C5a assist in the process of transmigration (diapedesis, or extravasation) from the bloodstream into tissue by causing mast cells to release vasoactive amines, which dilate blood vessels and make them leakier. Dilation of blood vessels is also assisted by the cytokine **platelet-activating factor (PAF)**. PAF triggers mast cells to produce a number of vasoactive compounds from the membrane lipid arachidonic acid. These compounds include the leukotrienes and prostaglandins. A gradient of complement component C5a leads the PMN to the site of infection (chemotaxis) once it has moved out of the blood vessel and into surrounding tissue. Some bacterial peptides also attract PMNs.

As PMNs move through tissue, the proinflammatory cytokines TNF-α, IL-1, IL-8, and PAF activate the PMNs' oxidative-burst response so that the PMNs arrive at the infection site with their full killing capacity in place. A similar activation occurs in the case of monocytes and the macrophages into which they develop as they move to the infected area. IFN-γ further stimulates the killing ability of macrophages, producing activated macrophages. Note that using cytokines and activated complement components, which are in their highest concentrations near the infected area, as

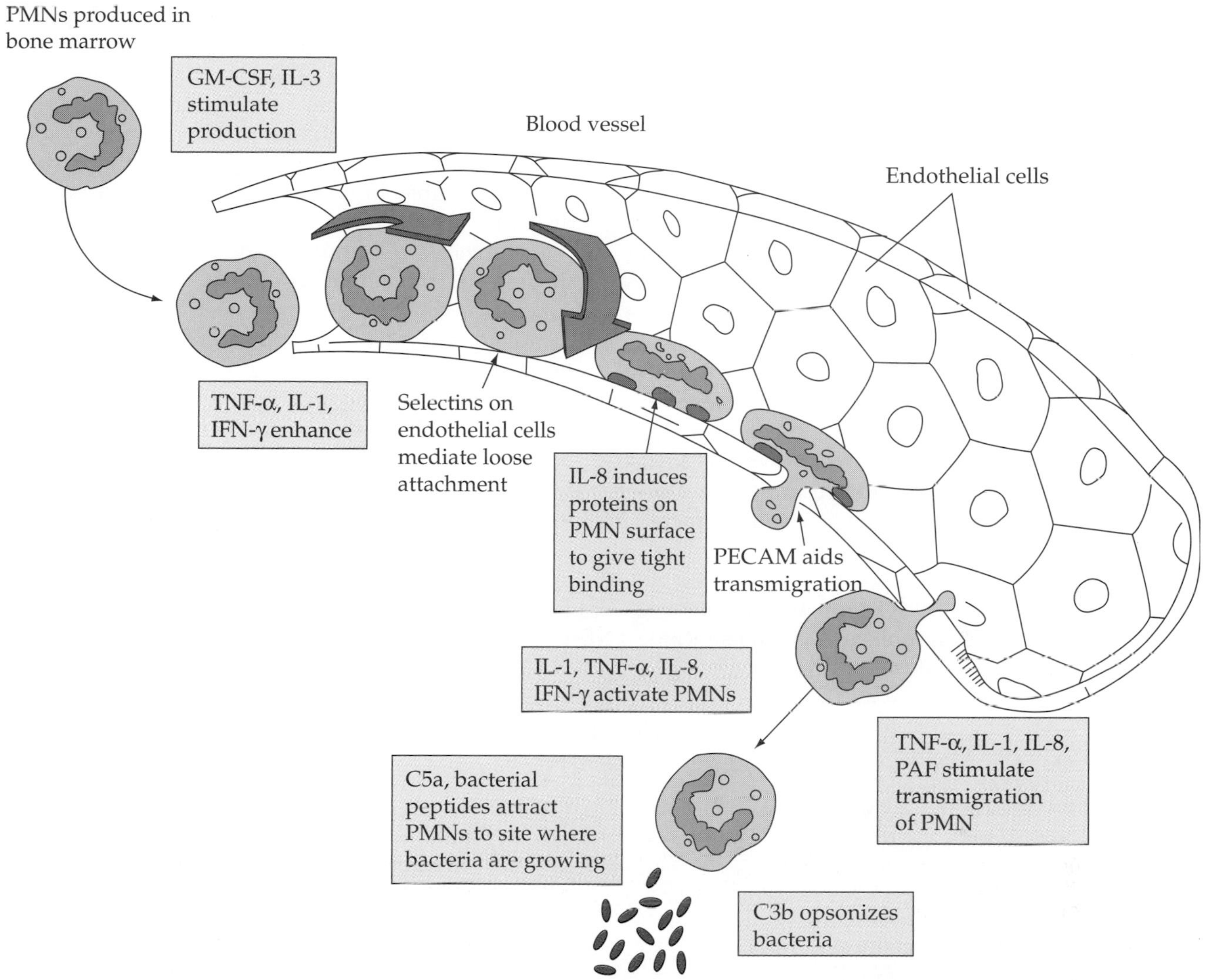

PMNs produced in
bone marrow

GM-CSF, IL-3
stimulate
production

Blood vessel

Endothelial cells

TNF-α, IL-1,
IFN-γ enhance

Selectins on
endothelial cells
mediate loose
attachment

IL-8 induces
proteins on
PMN surface
to give tight
binding

PECAM aids
transmigration

IL-1, TNF-α, IL-8,
IFN-γ activate PMNs

TNF-α, IL-1, IL-8,
PAF stimulate
transmigration
of PMN

C5a, bacterial
peptides attract
PMNs to site where
bacteria are growing

C3b opsonizes
bacteria

Figure 3–10 Roles of various cytokines and chemokines in directing the exit of PMNs from the bloodstream at particular sites. Initially, new proteins are expressed on the surfaces of PMNs and endothelial cells, permitting a loose reversible binding. This gives PMNs a rolling motility as they flow through the blood vessel. Other cytokines cause changes in the cell surfaces, resulting in tighter binding. The PMNs stop moving, flatten against the vessel wall, and force themselves across the endothelial wall. The PMNs then move chemotactically along a C5a gradient. PECAM, platelet-endothelial cell adhesion molecule.

signals to control the activities of PMNs ensures that the PMNs will exit the blood vessel near where the infection is occurring and not in other areas of the body. Also, the fact that activation of the phagocytes occurs only as they are moving into the infected area reduces the chance that they will inadvertently damage tissues outside the infected area.

The result of the signaling pathway just described is that PMNs and other phagocytic cells leave the blood vessels near the infection site in high numbers. Some underlying conditions reduce the effectiveness

of this signaling system and reduce the transmigration of PMNs. These include steroid use, stress, hypoxia, and alcohol abuse. Their inhibitory effect on transmigration may explain why these underlying conditions are frequently associated with increased susceptibility to infection.

If the phagocytes are successful in eliminating the invading bacterium, a second set of cytokines begins to predominate. These anti-inflammatory cytokines (e.g., IL-4, IL-10, and **interleukin-13 [IL-13]**) downregulate the production of TNF-α and reduce the killing activities of phagocytes, thus countering the proinflammatory response and allowing the phagocyte defense system to return to its normal, relatively inactive level.

Other Activities of Cytokines

Some of the other roles of the proinflammatory cytokines are illustrated in Figure 3–11 and explain common symptoms of infectious diseases other than localized infection: fever, somnolence, malaise, anorexia, chills, decrease in blood iron levels, and weight loss. The cytokines IL-1 and TNF-α interact with the hypothalamus and adrenal gland to produce fever and somnolence. The patient interprets somnolence as a feeling of malaise. Indifference to food, which can also characterize this state, explains the anorexia. TNF-α and other cytokines stimulate muscle cells to increase

their metabolic rate and catabolize proteins to provide fuel for the mobilization of host defenses. Increased metabolism of muscle cells may be the cause of the chills seen in some types of systemic infections. If the infection persists, the combination of anorexia and muscle cell breakdown of protein results in weight loss and visible loss of muscle tissue (wasting syndrome).

Cytokine IL-6 stimulates the liver to increase production of **transferrin,** complement components (to regenerate complement components used up during complement activation), LPS-binding proteins (to continue stimulation of cytokine production as long as bacteria are detected in the body), and two general opsonins, MBL, which binds mannose sugars on the bacterial surface, and **C-reactive protein,** which binds to the TA and LTA of some bacterial species. Transferrin is an iron-binding protein that sequesters iron.

Figure 3–11 Overview of innate responses to infection and the effects of cytokines on these responses. Cytokines include TNF-α, IL-1, and IL-6.

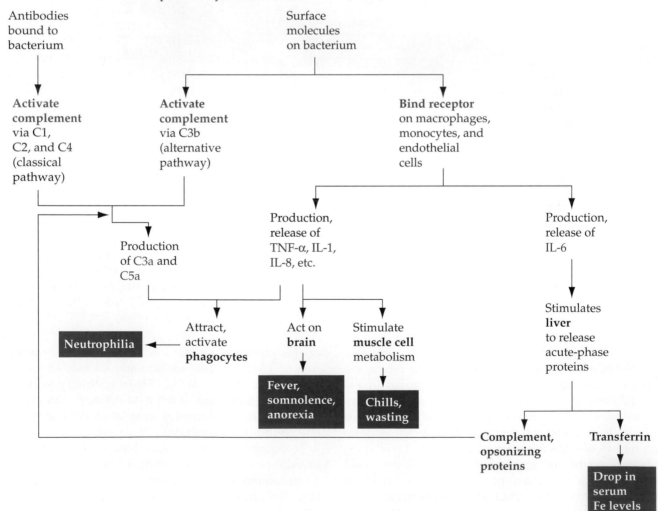

Transferrin-iron complexes are taken up by macrophages, which remove the iron and recycle the transferrin to scavenge more iron. In this way, the levels of iron in the blood drop to an even lower level than normal, which severely limits the growth of most bacteria.

The Dark Side of the Innate Defenses: Septic Shock

Another example of how inappropriate functioning of host defenses can lead to disastrous consequences is provided by septic shock. Septic shock is a form of shock caused by bacterial infection. Other causes of shock include massive crush injuries and burns. In people with septic shock, vascular resistance and blood pressure drop despite normal-to-high cardiac output. The heart rate increases as the body tries to compensate for decreasing blood pressure. Another sign that is frequently associated with septic shock is **disseminated intravascular coagulation (DIC),** which can be seen as blackish or reddish skin lesions **(petechiae).**

Septic shock is a serious condition because severely reduced levels of blood flow deprive essential organs of oxygen and nutrients. The consequent failure of organs, such as the kidneys, heart, brain, and lungs, is the cause of death in septic-shock patients. **Septicemia** (bacteria infecting and growing in the bloodstream) occurs in more than 500,000 patients per year in the United States alone. A quarter of the patients with septicemia die in the hospital. Even those who survive may have long-term aftereffects, such as stroke or permanent damage to the lungs or other organs. Moreover, those who survive an episode of septic shock have a significantly greater risk of dying during the next 5-year period than people with the same underlying conditions but no previous episode of shock. Septic shock is serious business—and seriously expensive business, as hospital administrators and insurance executives are quick to point out.

The all-inclusive term "sepsis" has now been defined more precisely in an effort to aid diagnosis of the various stages of disease. The first stage of shock, called **systemic inflammatory response syndrome (SIRS),** is characterized by a temperature over 38°C or under 36°C, a higher-than-normal heart rate, a higher-than-normal respiratory rate, and an unnaturally high or unnaturally low neutrophil count. The second stage, termed **"sepsis,"** is SIRS with a culture-documented infection, i.e., laboratory results showing the presence of bacteria in the bloodstream. The third stage is **"severe sepsis,"** characterized by organ dysfunction and very low blood pressure. The fourth stage is "septic shock," which is characterized by low blood pressure despite fluid administration.

How does a bacterial infection of the bloodstream produce such a serious condition? The steps involved in septic shock are illustrated in Figure 3–12. From previous sections of this chapter, it is evident that the body goes to great lengths to confine the inflammatory response to certain areas of the body. Septic shock is an example of what happens when an inflammatory response is triggered throughout the body. Shock occurs when bacteria or their products reach high enough levels in the bloodstream to trigger complement activation, cytokine release, and the coagulation cascade in many parts of the body. The effects of this are illustrated in Figure 3–13. High levels of cytokines, especially TNF-α, IL-1, IL-6, IL-8, and IFN-γ, cause increased levels of PMNs in the blood and encourage these PMNs to leave the blood vessels throughout the body. This leads to massive leakage of fluids into surrounding tissue. PMNs and macrophages activated by IFN-γ also damage blood vessels directly, resulting in loss of fluid from blood vessels.

Activation of complement throughout the body further increases the transmigration of phagocytes. The vasodilating action of C3a, C5a, leukotrienes, and prostaglandins contributes to leakage of fluids from blood vessels and further reduces the ability of blood vessels to maintain blood pressure. Some cytokines cause inappropriate constriction and relaxation of blood vessels, an activity that undermines the ability of the circulatory system to maintain normal blood flow and normal blood pressure. Widespread triggering of the coagulation system produces the clots that can plug capillaries (manifested as DIC). More seriously, it depletes the blood of essential clotting factors, so that damage to endothelial cells caused by phagocytes and cytokines leads to hemorrhages in many parts of the body. Hemorrhages not only contribute to hypotension, but also damage vital organs.

Once septic shock enters the phase where organs start to fail, it is extremely difficult to treat successfully, and the death rate exceeds 70%. Treatment is most likely to be effective if it is begun early in the infection. Diagnosis of septic shock in its early stages is not straightforward, however, because the early symptoms of shock (fever, hypotension, and tachycardia) are nonspecific. Also, the transition from the early stages to multiple-organ failure can occur with frightening rapidity. Hundreds of thousands of cases of septic shock occur in the United States each year. Many of these occur in patients hospitalized for some other condition than infectious disease. Accordingly, a massive effort has been made to develop new techniques for treating septic-shock patients more successfully,

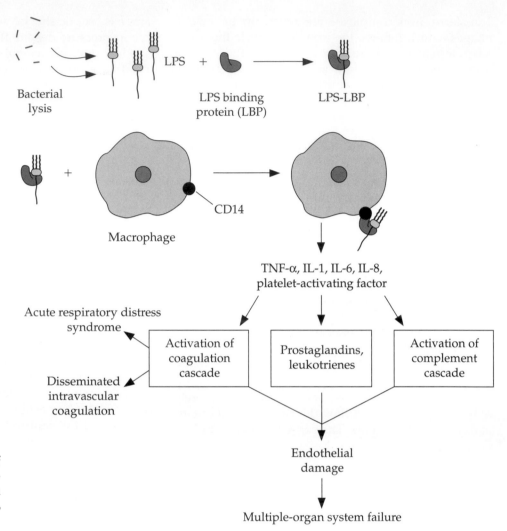

Bacterial lysis

LPS + LPS binding protein (LBP) → LPS-LBP

+ Macrophage → CD14

TNF-α, IL-1, IL-6, IL-8, platelet-activating factor

Acute respiratory distress syndrome

Disseminated intravascular coagulation

Activation of coagulation cascade

Prostaglandins, leukotrienes

Activation of complement cascade

Endothelial damage

Multiple-organ system failure

Figure 3–12 Triggering of cytokine release by gram-negative LPS and its role in septic shock. TNF-α and IL are cytokines.

especially in cases where the disease has reached the point that treatment with antibacterial agents is no longer sufficient to avert disaster. Early efforts to combat septic shock centered on administration of glucocorticoids, which downregulate cytokine production. Physicians have long believed that administration of steroids or ibuprofen would help shock victims because these compounds dampen various aspects of the inflammatory response, but there is now general agreement that glucocorticoid treatment is not effective in treating most types of shock. Clinical trials have now shown no significant effects from either of these therapies.

More recently, attention has focused on the cytokines, such as TNF-α, which seem to play such a central role in the pathology of shock. Antibodies or other compounds that bind and inactivate cytokines have been tested for efficacy in clinical trials. The outcome of early clinical trials has been disappointing, but newer anticytokine agents now being tested appear to

be more promising. Nonetheless, it is clear that this type of therapy will never be as effective as catching septic shock in its very early stages.

Antibiotic therapy administered early enough can stop the shock process, but the right antibiotic must be chosen. This means more laboratory tests, once more bringing physicians into conflict with health management organizations that want to save money by minimizing the use of tests. Microbiological testing tends to be expensive because it requires skilled technicians and is less automated than other types of clinical tests. Although it is true that bacterial identification and antibiotic susceptibility tests cost money, septic shock costs even more. If the causative bacteria are resistant to most antibiotics, the cost can rise by as much as 100-fold. Moreover, patients who have developed shock and survived may have experienced strokes and permanent damage to vital organs. The effect of shock, in terms of risk of untimely death, lasts years beyond the actual shock experience.

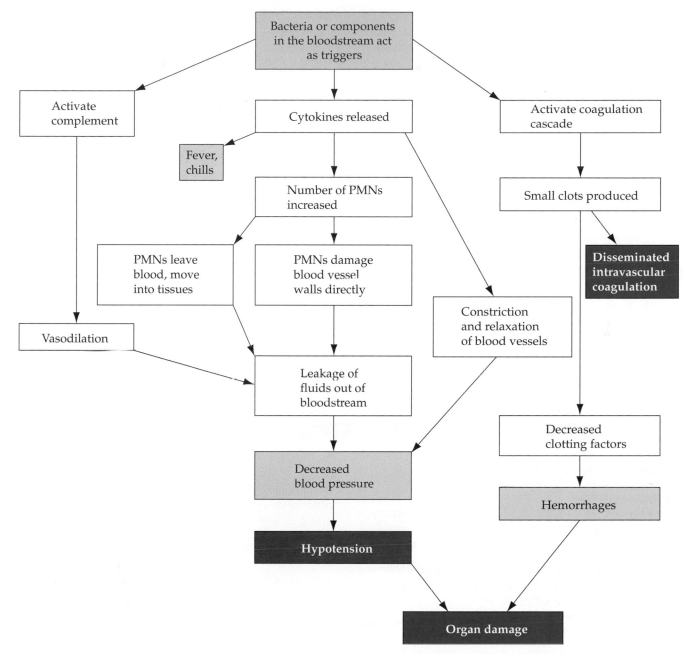

Figure 3–13 Sources of hypotension, DIC, and internal hemorrhages seen in cases of septic shock.

Then there are the lawsuits. Patients who through neglect or misdiagnosis develop strokes or other long-term damage are starting to sue hospitals. Consumer advocates are also suing to obtain information about infection rates in hospitals, especially those leading to serious conditions like septic shock. The rate of post-surgical infections is higher than most people realize, especially in large big-city hospitals. Increasing antibiotic resistance is only making things worse. Not surprisingly, hospitals guard their incidence of hospital-acquired infection figures in strict secrecy. This policy may be changing, as described in chapter 1.

One impediment to early diagnosis of shock has already been mentioned: the nonspecific nature of the signs and symptoms. Another impediment to early diagnosis is the fact that so many different types of bacteria can cause septic shock. A definitive microbial diagnosis cannot be made in about a third of patients with clinical signs of sepsis. Bacteria are the microorganisms most frequently implicated in septic shock (approximately 80% of cases), but many different species of gram-positive and gram-negative bacteria can cause shock. Because no single antibiotic is effective against all of these bacterial pathogens, it is important to determine the species of bacterium causing the infection. Current research efforts are focused on defining a spectrum or profile of molecular biomarkers that are indicative of different sepsis causes. The bad news so far is that rapid detection methods like PCR may not be effective because concentrations of bacteria below the PCR detection level can cause sepsis.

SELECTED READINGS

Abbas, A. K., and A. H. Lichtman. 2009. *Basic Immunology,* 3rd ed. Saunders, Philadelphia, PA.

Beutler, B. 2004. Innate immunity: an overview. *Mol. Immunol.* **40**:845–859.

Creagh, E. M., and L. A. J. O'Neill. 2006. TLRs, NLRs and RLRs: a trinity of pathogen sensors that cooperate in innate immunity. *Trends Immunol.* **27**:352–357.

Dinarello, C. A. 1996. Thermoregulation and the pathogenesis of fever. *Infect. Dis. Clin. N. Am.* **10**:433–449.

Kawai, T., and S. Akira. 2006. TLR signaling. *Cell Death Differ.* **13**:816–825.

Kindt, T. J., B. A. Osborne, and R. A. Goldsby. 2007. *Kuby Immunology,* 6th ed. W. H. Freeman and Company, New York, NY.

Marshall, J. C., and K. Reinhart (for the International Sepsis Forum). 2009. Biomarkers of sepsis. *Crit. Care Med.* **37**:2290–2298.

Medzhitov, R., and C. Janeway. 2000. Innate immunity. *N. Engl. J. Med.* **343**:338–344.

Pancer, Z., and M. D. Cooper. 2006. The evolution of adaptive immunity. *Annu. Rev. Immunol.* **24**:497–518.

Parham, P. 2009. *The Immune System,* 3rd ed. Garland Science Publishing Company, New York, NY.

Rice, T. W., and G. R. Bernard. 2005. Therapeutic intervention and targets for sepsis. *Annu. Rev. Med.* **56**:225–248.

Stevanin, T. M., J. W. B. Moir, and R. C. Read. 2005. Nitric oxide detoxification systems enhance survival of *Neisseria meningitidis* in human macrophages and in nasopharyngeal mucosa. *Infect. Immun.* **73**:3322–3329.

Travers, P., and M. Walport. 2008. *Janeway's Immunobiology,* 7th ed. Garland Science Publishing Inc., New York, NY.

QUESTIONS

1. What are two types of bacterial killing mediated by complement?

2. In the activation of complement by the alternative pathway, the stabilization of one particular component is essential. What is this component? How is it stabilized?

3. Define opsonization and two components of the immune system that can act as opsonins. Are they acquired, or are they innate?

4. How do the functions of various activated complement components resemble the functions of cytokines and chemokines?

5. Why does the body have such a complex system (complement plus cytokines) for directing the activities of the phagocytes? What is this system trying to achieve?

6. Why are there three pathways for activating complement?

7. What is the point of using proteolytic cleavage to activate complement? Why not just have the molecules made in their active form?

8. In septic shock, why does blood pressure fall?

9. Why are antibiotics (chemicals that kill bacteria) ineffective after a certain point in the course of septic shock, even though the causative bacterium is susceptible to them? Make an educated guess as to why so many of the anticytokine therapeutic agents failed.

10. What kinds of evasive action could a bacterium take to prevent itself from being killed by a phagocyte? (Hint: consider the steps in the killing process.)

11. Why do neutrophils circulate in blood? Why not have them migrate permanently into tissue, as the macrophages do?

12. Inflammation near the skin is characterized by redness and swelling. What causes these symptoms?

13. What is the primary function of TLRs and Nod proteins? Where are they located? How do they stimulate immune responses?

Comic Relief

14. How would you attract the attentions of a good-looking phagocyte? (Try complements.)

15. How would you describe a macrophage that is wearing sunglasses and a camera? (Phago-sightseer.)

4

The Second Line of Defense: Antibodies and Cytotoxic T Cells

I f ordinary police are confronted with a situation that they are not able to resolve, they call in the SWAT (special weapons and tactics) team: specialists, such as snipers and hostage negotiators, who are trained to target and deal with a specific crisis. When the innate immune system is faced with a bacterial attack that it is not able to handle, the human body uses a similar strategy. It brings in the specialists, antibodies and cytotoxic T cells (also called **CD8$^+$ T cells** or **cytotoxic T lymphocytes [CTLs]**), which are designed to target and kill a specific invading microbe.

The innate defenses of the body are normally an effective protective force that eliminates invading microbes. Unfortunately, some microbes have developed strategies for evading the innate defense system and surviving in the host. For example, they may be able to resist phagocytosis or neutrophils or even survive and multiply within phagocytic cells if they are ingested. To cope with such microbes, the body has evolved a second defense system, the **adaptive or acquired immune response,** which includes **dendritic cells (DCs), antibodies, CTLs,** and **T helper (Th) cells.** We will use the term "adaptive" for simplicity, but it is worth noting that the reason adjectives like "adaptive" and "acquired" are used to indicate this type of immunity is that specialized cells of the body have to change to target the invader in a highly specific way. Because the adaptive response is designed to target a particular microbe, it takes time to adjust and develop and may take up to a week to appear on the scene. In a sense, the body treats microbes that elicit the adaptive response much like law enforcement treats serial killers, because upon subsequent encounters with the same microbe, the specific defenses appear much more rapidly, within a day or two. Vaccination, which is covered in more detail in chapter 17, is a strategy for eliciting a particular adaptive defense without actually having to endure the first episode of the disease.

Understanding how the components of the adaptive defense system are induced and how they protect the body from infection is important for understanding how vaccines work, how they are designed, and why different vaccines are administered in different ways. Although vaccines have been important in preventing disease, the yield of successful vaccines has been disappointingly small. This is due to the fact that scientists are only now beginning to understand important nuances of the adaptive immune response, which can go wrong and produce an unwanted response in which the immune cells begin to attack the human body itself (autoimmunity). Therefore, just as the innate immune system had its dark side (septic shock), the adaptive immune system also has its dark side (autoimmune reactions). Obviously, a vaccine should not evoke this dark side. New insights into how the adaptive defense response develops, together with new insights into how to better deliver vaccines, may help to break down some of the barriers that have prevented the development of vaccines against diseases such as AIDS, gonorrhea, chlamydial disease, tuberculosis, and malaria, which are serious causes of morbidity and mortality throughout the world.

In this chapter, we start with a description of two of the main results of an adaptive immune response, antibodies and CTLs, which actually attack the invading bacterium. We then delve into the complex series of steps that produce these defenses. Also important is the body's strategy for remembering past infectious experiences so that if the infectious microbe is encountered again, as it almost certainly will be, the specific defenses are prepared to respond rapidly to subsequent attack by the invader.

Antibodies

Characteristics of Antibodies and Their Diverse Roles in Preventing Infection

Antibodies are protein complexes produced by mature **B lymphocytes** (B cells, or **plasma cells**). This rather bland description is accurate as far as it goes, but it fails to convey the amazing variety of tasks performed by these relatively simple molecules. To understand how they perform such diverse tasks, it is first necessary to understand how an antibody is put together. The basic structure of an antibody monomer is shown in Figure 4–1. The monomer consists of two heavy chains and two light chains. (The words "heavy" and "light" refer to the sizes of the proteins, with the heavy chain being the larger of the two.) The heavy and light chains are held together by a combination of disulfide bonds and noncovalent interactions.

Antibodies have two important domains: the **antigen-binding region (Fab),** which contains the end

of the antibody that binds to a substance considered foreign by the body, and the **constant region (Fc),** which is at the other end and interacts with complement component C1 via a glycosylated region or with phagocytic cells via an Fc receptor-binding region. The N terminus of each chain, comprising the Fab domain, contains a constant region and a **variable region (Fv)** that binds to a specific antigen. An **antigen** is defined as any material the body recognizes as foreign that binds to an antibody. For the purposes of this chapter, "antigen" means an infectious microbe or some protein, nucleic acid, or carbohydrate component of it. Examples of other types of antigens are animal proteins, macromolecules on organs from a noncompatible donor, and pollens of some plants.

A detailed description of the mechanism that generates a large number of specific antibodies that can bind to all of these different kinds of antigens is beyond the scope of this book, but briefly, an amazing recombination process shuffles a wide repertory of variable gene segments and fuses them to constant regions during the development of B cells. The transcripts of the resulting mosaic genes are further processed to give expression of the specific antibody produced by each B cell in the population.

An antibody monomer has two **antigen-binding sites,** each of which recognizes and binds to the same specific segment of an antigen. The antigen-binding sites are grooves in the antibody Fv ends that bind tightly only to a molecule having one particular structure, called an **epitope.** An epitope on a protein antigen can vary in size from 4 to 16 amino acids, although most are 5 to 8 amino acids in length. Complex antigens, such as microbes, contain many possible epitopes, each binding to a different antibody. In practice, however, a subset of the epitopes on an antigen dominates the specific response to that antigen. Why some epitopes are highly **immunogenic** (i.e., elicit a robust antibody or T-cell response) whereas others are only weakly immunogenic is still not well understood. Immunogenicity is often based on the size and complexity of the antigen molecule and is reflected in the order of antigenicity of macromolecules, where proteins are better than carbohydrates, which are in turn better than nucleic acids and lipids at generating an immune response. Differences in immunogenicity have important practical consequences. For example, it is now possible to produce epitope-size peptides synthetically. Peptides are not only much cheaper to produce than proteins, which must be purified by time-consuming biochemical procedures, they also make it possible to target an antibody or cytotoxic T-cell response to one or more specific regions of an antigen.

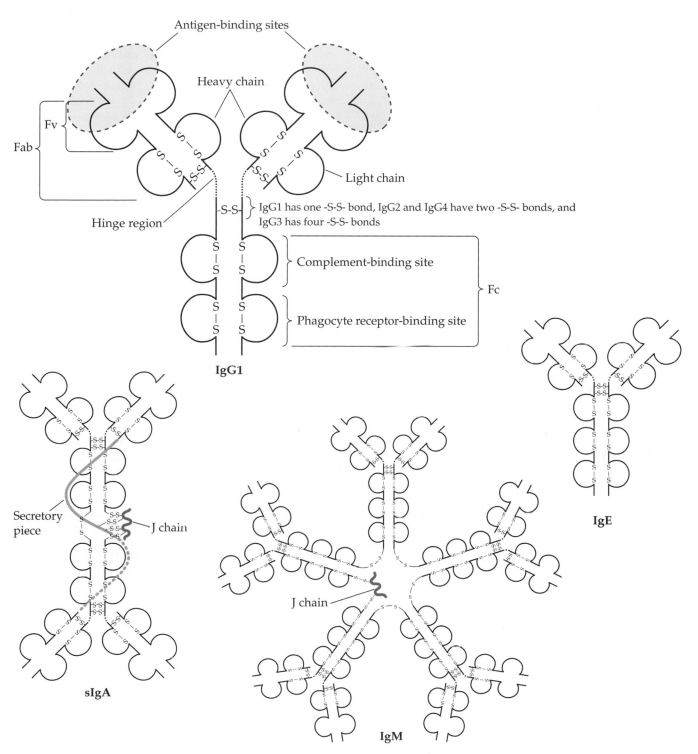

Figure 4–1 Structures of IgG, IgA, IgM, and IgE. The antigen-binding sites are in the variable regions (Fv) of the antibody molecule that recognize the target epitope. The Fc region of the molecule is responsible for complement activation and mediates binding to phagocyte receptors during opsonization.

Directing the adaptive defense response toward particular epitopes is important because not all immunogenic epitopes elicit protective responses. Some immunogenic epitopes, for example, are buried within a folded protein on a microbe's surface or are expressed inside a microbe and are thus not exposed to circulating antibodies. Eliciting an antibody response to such an epitope is useless because the antibody will not be able to bind it. Some microbial proteins have regions that vary considerably from one strain of microbe to another. Antibodies or cytotoxic cells that recognize highly variable regions of microbial proteins are useful against a limited number of strains. A better strategy is to target regions of microbial proteins that are highly conserved, i.e., that are found on all strains of the microbe. Using peptide epitopes as vaccines makes it possible to program a specific defense response directed toward exposed, conserved epitopes.

Serum Antibodies

IgG. Immunoglobulin G (IgG), a monomer, is the most prevalent type of antibody in blood and extravascular fluid spaces (approximately 80% of the circulating antibodies are IgG). There are four different subtypes of IgG antibodies in humans **(IgG1 to IgG4),** all of which have somewhat different functions (Table 4–1). These subtypes differ not only in their functions, but also in their amino acid sequences, in their sugar modifications (posttranslational decoration of sugars on the IgG protein), and in the extent of disulfide cross-linking of their heavy chains (primarily in the Fc portion). IgG1 is the most abundant of the IgG subtypes. IgG1 and IgG3 are called **"opsonizing antibodies"** because these two subtypes are the most effective in opsonizing microbes. IgG2 and IgG4 opsonize poorly, if at all.

Normally, opsonization of an antigen by IgG facilitates the ingestion of the opsonized antigen by phagocytic cells. In the case of intracellular infections of tissues, IgG (and IgA and IgE, described below) mediate a different response called **antibody-dependent cell-mediated cytotoxicity (ADCC).** In ADCC, antigens presented on the surfaces of infected cells bind to the Fab regions of specific IgG molecules. The exposed Fc portions of the IgG molecules are then free to bind to effector cells that contain Fc receptors on their surfaces, including polymorphonuclear leukocytes (PMNs), macrophages, and natural killer (NK) cells. This linkage triggers a cytotoxic bombardment of the infected cell by NK cells, as described in chapter 3. In macrophages and PMNs, a similar killing response is elicited instead of phagocytosis. Thus, ADCC, by linking together players of the innate and adaptive immune responses, serves as an important defense against intracellular pathogens (more on this below).

IgG1 and IgG3, but not IgG2 or IgG4, also activate the classical pathway of the complement cascade. Because an IgG monomer has only one Fc region, two monomers must be bound close together so that C1 can bind to the Fc regions of both IgG monomers. This cross-linking activates C1 and initiates the complement cascade. (Note: the ability of human IgG2 to activate complement has been controversial, but evidence now indicates that it does not.) IgG is the only antibody type that crosses the placenta and is responsible for protecting an infant during the first 6 months of life until the adaptive defenses of the infant are fully developed. (Warning: The nomenclature used to describe human IgG is not the same as that used to describe murine IgG. Thus, IgG1 of mice does not necessarily have the same features as IgG1 of humans. We mention this now because later we will see that this explains why different papers on the develop-

Table 4–1 Protective roles of serum antibodies IgG and IgM

Role	IgG1	IgG2	IgG3	IgG4	IgM
Neutralization of toxins	+[a]	+	+	+	+
Neutralization of microbes (prevents binding of microbe to a target host cell)	+	+	+	+	+
Opsonization; Fc portion binds:					
PMN receptors	+	−	+	−	−
Macrophage receptors	+	−	+	+	+
Complement activation					
Classical pathway	+	−	+	−	+
Alternative pathway	−	−	−	−	+
Cross placenta	+	+	+	+	−

[a] +, present; −, absent.

ment of the immune response seem to contradict each other but actually do not. For a summary of the different types of Ig molecules of humans and mice, see Box 4–1.)

IGM. IgM consists of five or more monomers that are connected to each other and to a peptide called the **J chain** (Figure 4–1). IgM accounts for 5 to 10% of the total serum Ig but is also secreted at mucosal surfaces and in breast milk. IgM predominates in the initial (primary) antibody response against a pathogenic microbe, whereas IgG predominates in the response to sustained or subsequent infections (secondary antibody response) by the same microbe (Figure 4–2). This feature of the immune response is useful for di-

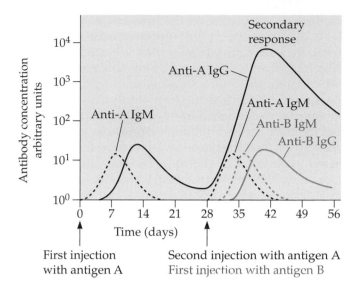

Figure 4–2 Time course of elicitation of antibodies upon initial exposure and subsequent exposure. After exposure to an antigen, such as antigen A, IgM is usually the first antibody detected during an acute infection, but levels decrease again after about 2 weeks. It takes 5 to 7 days for selection and proliferation of B cells producing IgG against antigen A to appear in the blood. After a period of time, the levels of IgG also decrease. Subsequent exposure to antigen A again results in production of IgM at about the same levels as after the first exposure, but the levels of antigen A-specific IgG are now much higher. In contrast, exposure to antigen B, along with the second exposure to antigen A, does not enhance the antigen B-specific IgG levels.

BOX 4–1 Are You a Man or a Mouse?

This archaic challenge, issued to someone showing classical symptoms of soft-spine syndrome in the hope of getting him (or her) to act decisively and courageously, actually applies as well to antibodies. Although humans and mice are much more closely related than many of us would like to admit, there are subtle but real differences between some antibody classes shared by humans and mice. Accordingly, there are also somewhat different nomenclatures used to designate these antibody classes. These different designations can be confusing, especially in the case of the IgG subtypes. A guide to the different antibody designations in humans and mice is provided below:

Human	Mouse
IgG1	IgG2a
IgG2	IgG2b
IgG3	None
IgG4	IgG1
IgM1	IgM
IgM2	None
IgE	IgE

Why such differences occur between closely related species remains a fascinating but unanswered question.

agnostic purposes. Diagnostic tests that detect IgM are used to determine whether a patient is experiencing a first infection with a particular microbe. Because IgG can circulate in serum for long periods, detectable IgG levels can either signal the presence of a current infection that is well under way or simply be the residue of a previous infection, whereas IgM levels are detectable during an initial infection but then go away. IgM is the most effective activator of complement. Because of its polymeric form, a single molecule of IgM is sufficient to activate complement via the classical pathway because it has five Fc regions that can be complexed with C1 or an Fc receptor.

Both IgG and IgM bind to the surfaces of bacteria and viruses and prevent them from attaching to and invading target host cells **(microbe neutralization)** (Table 4–1). Antibodies do this very effectively because they are bulky molecules that block interaction between microbial surface proteins and the receptors they recognize on host cells. Antibodies can also neutralize toxic proteins produced by bacteria **(exotoxins)** by binding to the exotoxins and preventing them from binding to a receptor on the host cell surface before they can exert their toxic effects **(toxin neutralization).** By binding to catalytic sites of bacterial secreted enzymes, such as proteases or glycosidases, antibodies can similarly block those bacterial proteins from degrading host extracellular molecules.

An example of antibody neutralization of a bacterial exotoxin is the antibody response to *Corynebacterium diphtheriae,* the cause of diphtheria. Diphtheria is a serious disease of children. The bacteria often live innocuously in the upper respiratory tract, but when they become infected with a corynebacteriophage encoding the diphtheria toxin gene, they are then able to produce and secrete the toxin, which enters the bloodstream. Diphtheria toxin is one of the most potent bacterial toxins known. It kills many types of human cells. If the toxin makes it through the bloodstream to the heart, it can cause death due to heart failure. The action of the toxin in the throat is evident from a whitish "pseudomembrane" that forms a layer in the upper respiratory tract consisting of dead epithelial cells and mucus. In some cases, this membrane can grow to the point where it causes asphyxiation. The most effective protective response to infection with toxin-producing *C. diphtheriae,* which is also the response elicited by the diphtheria vaccine, is production of antibodies that bind to diphtheria toxin and prevent it from binding to and killing human cells (neutralization of the toxin).

IgE. A serum antibody with a different function from IgG and IgM is **IgE.** Like IgG, IgE is a monomer that is found in extremely low concentrations in serum, usually as a receptor-bound complex on the surfaces of **mast cells** and **basophils.** Many of the symptoms of infections caused by metazoal parasites (i.e., worms, to the uninitiated) are traceable to the effects of increased levels of IgE. IgE levels rise during metazoal infections, but the presence of increased IgE levels associated with infection does not necessarily mean that the response is protective. If two IgE molecules bound to mast cells are complexed by a **polyvalent antigen** (i.e., an antigen with multiple antibody-binding sites), it causes the mast cell to release granules containing potent inflammatory cytokines, such as **histamine** and other **vasoactive compounds.** The release of mast cell granules in the vicinity of the intestinal wall may provoke an allergic response that ejects metazoal parasites from the intestine.

An interesting point to ponder is that the human body evolved over millions of years to cope with worm infestations. IgE seems to be part of that response. In recent times, in developed countries, we have eliminated worms from our intestinal landscape. Thus, in developed countries where metazoal infections are relatively uncommon, IgE is most often associated with noninfectious diseases, such as allergies or asthma. The most serious complication of massive release of mast cell granules is anaphylaxis, which can rapidly kill a person. Is the rise in allergies and asthma seen in all developed countries due to an immunological imbalance caused by elimination of a former enemy, thus leaving the worm-oriented part of the specific and nonspecific defenses with nothing to do except cause trouble? Rest assured that this is not the start of a "bring back the worms" initiative, but it is interesting to think about the potential negative consequences of an abrupt change (in evolutionary terms) in our exposure to invaders that have been with us from antiquity.

Secretory Antibodies—Antibodies That Protect Mucosal Surfaces

IgA. IgA in its monomeric form represents about 10 to 15% of the total serum antibody content. The role of IgA in blood and tissue is to aid in the clearance of antigen-antibody complexes from the blood. By far the most important form of IgA, however, is **secretory IgA (sIgA),** which plays a role in the defense of all mucosal surfaces, including the gastrointestinal (GI) tract, the pulmonary tract, and the urogenital tract. sIgA is the dominant antibody in secretions (e.g., tears, saliva, bile, and milk). sIgA consists of two IgA antibody monomers joined through disulfide bonds to a peptide J chain and to which another peptide, called the **secretory piece,** has been attached (Figure 4–1).

The secretory piece is acquired when IgA is transported through the mucosal epithelial cell into mucosal secretions covering various mucosal surfaces. The secretory piece comes from the poly-Ig receptor that is responsible for the uptake and transcellular transport of the dimeric IgA across epithelial cells. The secretory piece is cleaved from the poly-Ig receptor and stays bound to the sIgA when the sIgA is released into the lumen. sIgA does not activate complement; instead, it is transported out into the mucin layer, where its main role is to attach to incoming microbes or microbial toxins and trap them in the mucus layer, thus preventing them from reaching the mucosal epithelial surface and binding to receptors on the mucosal cells. sIgA can trap microbes in mucus because the Fc portion of sIgA binds to glycoprotein constituents of the mucin. By sequestering the antigen away from the mucosal-cell surface, the sIgA-antigen complex that is bound to mucin is then sloughed off and excreted from the body. sIgA is also secreted into mother's milk. Thus, sIgA, like IgG, serves as an important protection against infection for young infants who have not yet developed their own set of immune responses.

Humans have two subtypes of IgA, IgA1 and IgA2. An interesting evolutionary development is the production of an IgA1-specific protease by a number of pathogenic bacteria (e.g., *Neisseria gonorrhoeae*), which is thought to have provided selective pressure that resulted in humans having another gene encoding IgA2, which lacks the sites recognized by the IgA1 protease. The secretory piece also helps to protect the protease-sensitive sites in the hinge region of the heavy chain from cleavage by bacterial and host proteases. IgA is heavily O glycosylated in the hinge region, which protects it from proteolysis.

Affinity and Avidity

A characteristic of antibody binding to antigens, which is of critical importance in assessing the effectiveness of the antibody, is the **avidity** of the antigen-binding site for the epitope it binds. Avidity is a combination of **affinity** (the strength of the interaction between an antigen-binding site and an epitope) and **valence** (the number of antigen-binding sites available for binding epitopes on an antigen). A single epitope can elicit a mixture of antibodies that vary considerably in affinity. The reason for this variation may be that, since the body cannot "know" in advance what epitopes it will encounter, it produces a variety of antibodies with slightly differing antigen-binding sites, among which will be some binding sites that have a high affinity for a particular antigen, while for others the affinity might be weaker. In fact, as the antibody response to an epitope develops, the B cells producing antibodies with the highest affinity will proliferate the most, and those high-affinity antibodies will eventually predominate.

High affinity is important, but it is not sufficient to ensure that an antibody bound to an epitope retains its hold on the epitope. Since the binding between antigen-binding sites and epitopes is noncovalent, the interaction is reversible. Thus, there is an "off rate," as well as an "on rate," associated with antibody binding to an epitope. The importance of valence is that an antibody with a higher valence will be significantly less likely to detach from the antigen to which it is bound than is an antibody with a lower valence. If two antigen-binding sites of an antibody monomer bind to two adjacent epitopes on an antigen, the probability that both of them will detach at the same time is much lower than the probability that a single antigen-binding site will detach from its epitope. Thus, higher valence can improve the strength of binding of an antibody to an epitope by orders of magnitude. Some types of antibodies, such as IgM and sIgA, have a higher valence number than IgG because they are composed of more than one monomer and can thus bind up to 4 (sIgA) or 10 (IgM) epitopes. These antibodies thus have a greater avidity for antigens than IgG, which is a monomer and can bind only two epitopes.

High avidity for an antigen is more effective for neutralizing microbes and toxins, opsonizing microbes, and activating complement. Moreover, phagocytic cells more rapidly clear from the body antigens that are bound to antibodies with high avidity. This is a desirable feature, because the longer antibody-antigen complexes remain in circulation, the more likely they are to be deposited in the kidneys or other blood-filtering organs, where they can activate complement and cause an inflammatory response that damages the organ.

CTLs

CTLs have a role similar to that of PMNs (in ADCC) and that of NK cells (in innate immunity and ADCC)—to kill infected host cells—but the mechanism is different from that of PMNs. The difference between CTLs and NK cells, which are like T cells in many respects, is that CTLs have receptors that are specific for a particular epitope from a microbial antigen. Thus, whereas NK cells kill host cells infected with a variety of intracellular pathogens (see above and chapter 3), CTLs kill only cells infected with a specific intracellular pathogen (see Figure 3–6). CTLs, ADCC, and NK cells are all an important part of the defense response against intracellular pathogens, because killing infected cells may be the only way to attack these pathogens, which are protected from antibodies and complement by their intracellular location.

Do CTLs also attack and kill intracellular bacteria? Answering this question has long been controversial. In particular, it was unclear whether CTLs simply killed infected human cells or could also kill microorganisms directly. Recent research has shown that CTLs do both. CTLs have two mechanisms for killing infected cells. In the first, the CTL binds to an infected cell using a **T-cell receptor (TCR)** on the surface that recognizes a microbial antigen being displayed on the surface of the infected cell. How the microbial antigen is displayed on the cell surface is described below. The CTL then releases granules that contain two proteins, **perforin** (a pore-forming cytolysin that pokes holes in

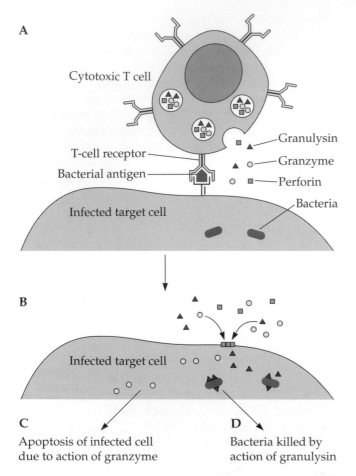

A

Cytotoxic T cell

T-cell receptor
Bacterial antigen

Infected target cell

Granulysin
Granzyme
Perforin
Bacteria

B

Infected target cell

C

Apoptosis of infected cell
due to action of granzyme

D

Bacteria killed by
action of granulysin

Figure 4–3 The roles of perforin and granulysin in the killing of bacteria that have been taken up by macrophages. The CTL recognizes and binds to a target cell (e.g., macrophage) containing bacteria. The CTL releases granules containing perforin and granulysin, and perforin creates pores in the target cell membrane. Granulysin enters the target cell through pores created by perforin, and granulysin binds to and kills bacteria by creating pores. The target cell then processes the degraded bacteria into antigens.

the cell membrane) and proteolytic enzymes (**granzymes,** which were described in chapter 3), that enter the cell through the pore and trigger **programmed cell death (apoptosis)** in the infected cell (Figure 4–3). This type of attack kills the infected cell but not the microbes. The released microbes, however, can then be taken up by activated macrophages that are better able to kill them. In the second mechanism, the CTLs can also release a second pore-forming cytolytic protein, called **granulysin.** Granulysin is rather ineffective in lysing host cells, but it is very effective at killing bacteria. Presumably, granulysin kills bacteria the way perforin kills eukaryotic cells—by creating holes in the bacterial membranes and collapsing the proton motive force that bacteria use to gain energy. Perforin may also help to deliver other antibacterial lysins, which remain to be discovered, into intracellular compartments of host cells.

Production of Antibodies and Activated CTLs

Processing of Protein Antigens by DCs

When microbes or their products first enter the body, professional phagocytes called **antigen-presenting cells (APCs)** engulf, process, and present antigens on their surfaces, which thereby directs other cells in the adaptive immune system to develop into cells with a specific antibacterial function. There are three types of APCs: macrophages, DCs, and B cells. As described in chapter 3, macrophages and DCs are part of the innate immune system, but they also serve as links to the adaptive immune system. Macrophages, as part of the innate immune system, are produced in an immature form (monocytes) and migrate through the bloodstream before moving into tissue where an in-

Figure 4–4 MHC-I, MHC-II, and CD1 pathways of antigen processing by APCs and presentation to T cells lead to activation and increased proliferation of T cells. **(A)** In the MHC-II pathway, extracellular protein antigens (Ag) are endocytosed into vesicles, where they are processed into peptides that displace the invariant chain of the MHC-II molecule. The peptide antigen–MHC-II complex is then transported to the cell surface, where the complex binds to the TCR and CD4 on the surfaces of CD4$^+$ T cells. **(B)** In the MHC-I pathway, protein antigens present in the cytosol are processed by the proteasome, and the resulting peptides are transported to the endoplasmic

reticulum (ER) via the transporter associated with antigen processing (TAP), where they then bind to MHC-I molecules and are transported to the cell surface; there, the complex binds to the TCR and CD8 on the surfaces of CD8$^+$ T cells. **(C)** The CD1 antigen presentation pathway. Glycolipid antigens bind to APCs via pattern recognition molecules, such as CD14 and the mannose receptor. The mannose receptor can traffic such antigens through the endosomal pathway, where at acid pH, lipid portions of the glycolipids bind to CD1. The antigen-CD1 complex then traffics to the cell surface, where it binds to the CD1-specific TCR and CD3.

A Class II MHC pathway

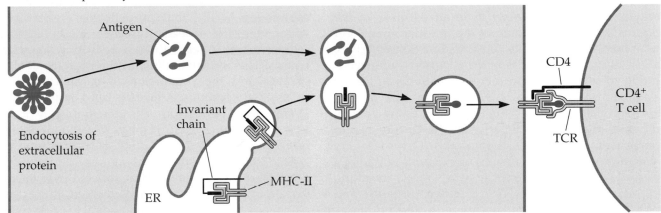

B Class I MHC pathway

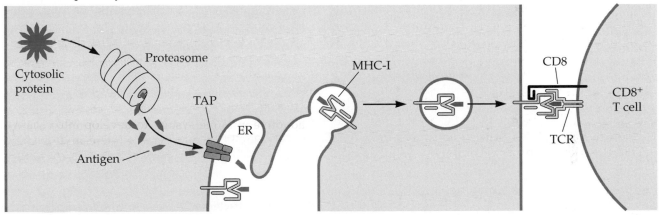

C CD1 pathway

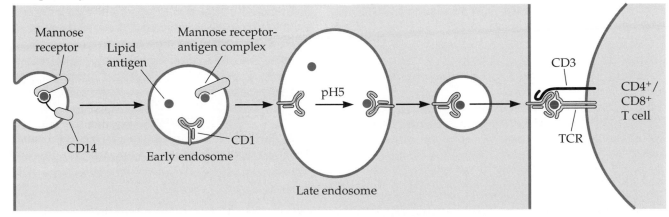

fection is taking place. They help to clear debris from dead human cells that may be circulating in the blood. Macrophages in the lungs help to clear dust particles, but their main role is to initiate and organize the adaptive immune response. DCs, like macrophages, initiate and organize the adaptive immune response, but they are found mainly in localized areas of the dermis, the mucosal lining of the intestinal tract, and lymphoid tissue. B cells can also function as APCs, albeit not as efficiently as macrophages or DCs. In addition to producing and secreting antibodies, B cells also produce **membrane Igs (mIgs)** that they display on their surfaces. When the mIg captures an antigen, the mIg-antigen complex is internalized, and the antigen is processed and presented to the Th cells.

Shown in Figure 4–4 is an overview of how APCs degrade protein antigens and display the resulting peptide antigens on their surfaces as protein complexes called the **major histocompatibility complexes (MHC).** Two types of MHC molecules, **MHC class I (MHC-I)** and **MHC-II,** that form the complexes with the peptide antigens have been known for some time. MHC-I molecules are produced by all nucleated cells in the body, while MHC-II molecules are produced only by professional phagocytic cells, such as macrophages, DCs, and other APCs. Professional APCs produce both classes of MHC molecules.

The type of MHC used to display the peptide epitope determines whether the APC will stimulate activation of CTLs (CD8$^+$ T cells) or stimulate Th cells **(CD4$^+$ T cells)** to produce antibodies and activate macrophages (Figure 4–4). The display of intracellularly derived peptide antigens on MHC-I allows the APCs to activate and stimulate the proliferation of CD8$^+$ CTLs. In addition, as described in chapter 3, NK cells use the amount of MHC-I on cell surfaces as an indicator of cell health, since infected cells generally express less MHC-I than normal cells. Display of an epitope on MHC-II leads to recognition by CD4$^+$ Th cells, which then leads to their activation and proliferation. This in turn stimulates the production of antibodies and the production of gamma interferon (IFN-γ), which activates macrophages and CTLs.

How the APC decides whether to display an epitope on MHC-I or MHC-II has received a lot of attention, because an understanding of the properties that lead to each type of presentation is critical for vaccine design. From cumulative data gathered over the past decade, some basic rules have emerged. Intracellular pathogens, such as viruses and some bacteria, which can enter the cytoplasm or nucleus of an APC, are most likely to elicit an MHC-I display of their antigens (Figure 4–5). Particulate antigens, which do not escape the phagocytic or pinocytic vesicle, however, also seem to elicit primarily an MHC-I-linked display. In contrast, soluble antigens, such as peptides or proteins, and extracellular microbes that evade APC uptake are displayed almost exclusively on MHC-II. Thus, if one wishes to use peptides to elicit a CTL response, it is necessary to present those peptides in particulate form, e.g., bound up in a complex consisting of inert materials that will encourage their pro-

Figure 4–5 Characteristics of an antigen determine whether the antigen is presented on MHC-I (to trigger the cytotoxic T-cell response) or on MHC-II (to trigger the Th response).

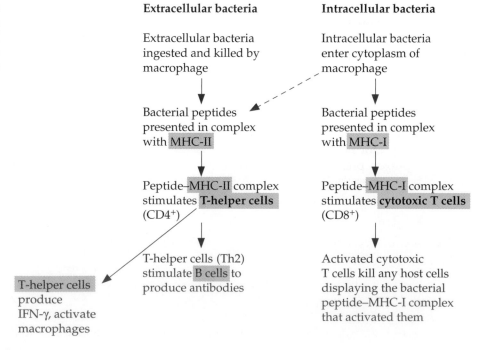

cessing via the MHC-I pathway. Alternatively, those peptides could be expressed intracellularly using mammalian expression vectors (more on this in chapter 17).

Unfortunately, many of the bacterial antigens recognized by the immune system are lipid, carbohydrate, or lipid-carbohydrate antigens. Gram-negative lipopolysaccharide and gram-positive lipoteichoic acid are excellent examples. In the past, immunologists have focused almost exclusively on peptide antigens because they are easier to characterize than carbohydrate- or lipid-containing molecules and because peptides elicit a strong immune response. Peptide antigens consist of different amino acids linked by a single type of bond, the peptide bond. Carbohydrate oligomers, in contrast, can be linked by any of 12 glycosidic linkages. Lipids also contain more than a single linkage. Carbohydrate, lipid, and lipid-carbohydrate molecules present to the scientist enough diversity to make studying them quite challenging, so it is not surprising that immunologists have ignored the processing of carbohydrate and lipid antigens until recently.

Nevertheless, some immunologists have begun to study the processing of lipid-carbohydrate antigens. Their first big find was **CD1 molecules,** which present lipid or glycolipid antigens. *Mycobacterium tuberculosis*, the cause of tuberculosis, provides a cornucopia of lipid-saccharide and lipid-peptide antigens not found in most other bacteria. CD1 was first discovered to present *M. tuberculosis* lipid and lipid-glycan antigens on the immune cell surface. So far, five forms of human CD1 have been found: CD1a to CD1e. The steps in processing and displaying lipid antigens appear to be similar to those shown in Figure 4–4 for peptide antigens, with CD1 taking the place of the MHC molecule for presentation to CD1-specific CTLs (for CD1a, CD1b, and CD1c) or NK cells (for CD1d). CD1 is related at the amino acid sequence level to MHC-I and MHC-II but has obviously diverged from them during evolution. In what follows, we will concentrate our attention on MHC-I and MHC-II, but brace yourself for new literature on alternative pathways to antigen presentation, as intrepid immunologists begin to enter the quagmire of bacterial lipid and lipid-carbohydrate antigens.

Interaction between APCs and T Cells: the T-Cell-Dependent Response

T cells have protein-receptor complexes on their surfaces called TCRs. The repertoire of TCRs is generated during T-cell development by a recombination process of the genes encoding the TCR proteins analogous to the one mentioned above that leads to antibody diversity. The result is a large pool of T cells with surface receptors that recognize a wide variety of different epitopes. When an APC displays a particular MHC-I epitope or MHC-II epitope complex on its surface, only a small number of the vast pool of available T cells will have a TCR capable of recognizing that particular MHC-epitope combination.

Antigen presentation and binding to the TCRs stimulates T cells to become either CTLs or helper T cells. CTLs have a protein on their surfaces, **CD8,** which helps the TCR respond to epitopes displayed on MHC-I. CD8 binds to MHC-I and stabilizes the interaction between MHC-I and the TCR. CTLs are particularly well designed to recognize infected cells because virtually all cells of the body produce MHC-I. If a cell is infected, it displays epitopes from the invading microbe on its surface. Binding of a CTL to the surface of an infected cell causes the CTL to release cytolytic or apoptotic proteins that can kill the infected cell.

If an epitope is presented by the APC in complex with MHC-II rather than MHC-I, a different class of T cells, called Th cells, is stimulated. Th cells have a different surface protein, **CD4,** which helps their TCRs respond to epitopes displayed on MHC-II and stabilizes the interaction between MHC-II and the TCR. Other proteins on the surface of the APC and the Th cell, called costimulatory molecules (e.g., CD54, CD11a/18, CD58, and CD2), must also interact to make the binding between the APC and the Th cell tight enough to stimulate the APC to release cytokines (e.g., interleukin 1 [IL-1] and tumor necrosis factor alpha) that will stimulate the Th cell to proliferate and become activated. The necessity for contacts between different surface proteins of the APC and Th cell helps ensure that only specific binding of a TCR to an MHC-epitope complex will result in T-cell activation.

Once activated, T cells begin to proliferate, with most of the resulting effector Th cells becoming involved in combating the invading microbes. A few of the T cells, however, become quiescent **memory T cells.** Memory T cells persist for long periods in the body. They are generally present in higher numbers than naïve T cells with the same TCR, and they are more easily stimulated to proliferate and produce stronger cytokine responses when they encounter their specific epitopes on APCs during subsequent infections. Memory T cells allow the body to respond to a second encounter with a microbial invader much more swiftly and more strongly than it did after the initial encounter.

Some bacteria and viruses produce proteins called **superantigens** that interfere with the progression of events in an interesting way, described in more detail in chapter 12. Superantigens force a close association between APCs and T cells that would not normally occur unless the MHC-antigen complexes and the TCRs matched. Whereas normally only a fraction of the T cells would interact specifically with a particular APC presenting an antigen on its cell surface, superantigens can cause up to 20% of T cells to participate in such interactions. As the cytokine signaling begins, cytokines are produced at higher levels than normal, and this overreaction can trigger shock. The term "septic shock" is sometimes used to describe this phenomenon; however, as described in chapter 3, lipopolysaccharide or other bacterial surface components

usually initiate what is considered "septic shock." The term "T-cell-mediated shock" or "toxic shock" might be more appropriate to describe the shock initiated by superantigens to distinguish it from septic shock.

The Th1/Th2 View of the Adaptive Immune System

A paradigm for the process by which Th cells influence the development of different types of immunity involving CTLs, activated macrophages, antibodies, and the IgE response has emerged in recent years. The simplest form of this paradigm is the Th1/Th2 model, in which there are two subtypes of Th cells, Th1 and

Figure 4–6 Model for development and roles of Th1, Th2, Th17, and Treg (Th3) cells in humans.

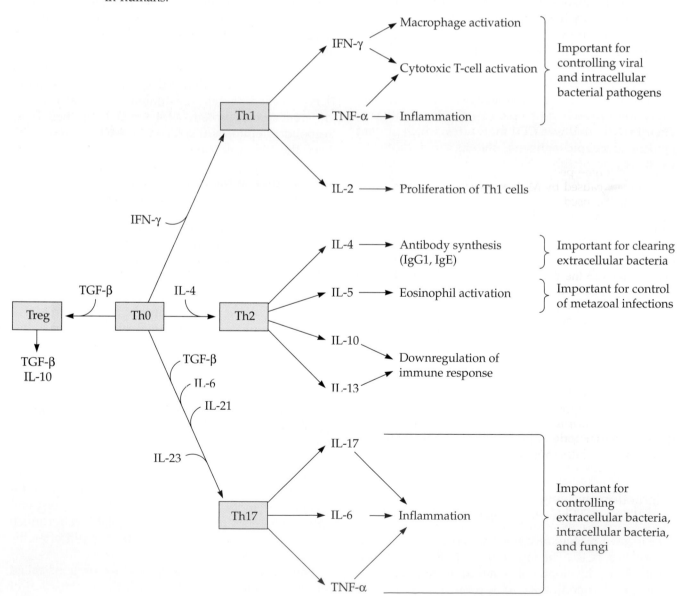

Th2, which control the development of acquired immunity (Figure 4–6). Both Th1 and Th2 cells are descended from the same Th0 cell type. The decision to produce Th1 cells is triggered in part by PMN production of IL-12, which stimulates NK cells to produce IFN-γ, which in turn stimulates Th0 cells to differentiate into the Th1 form. Extracellular antigens stimulate differentiation of Th0 cells to develop into Th2 cells. IL-4 is required for Th2 differentiation and is later produced by mature Th2 cells. Cytokines produced by Th1 cells, such as IFN-γ, or by Th2 cells, such as IL-4, induce a positive feedback loop that leads to further differentiation of Th0 cells into Th1 or Th2 cells, respectively, and inhibits production of the other cell type. Later in the course of combating infection, the balance between Th1 and Th2 differentiation is eventually restored.

A third branch of this pathway that may contribute to this regulation, the so-called Treg pathway (or Th3 pathway), has now been added. The Treg pathway consists of Th cells that produce further regulatory compounds that control the differentiation of Th0. A fourth pathway, more recently discovered, is the Th17 pathway, so called because it produces IL-17 and is stimulated by IL-17. The Th17 pathway is thought to stimulate proinflammatory reactions and could well prove in the future to have a beneficial side.

Th2 cells activate eosinophils and stimulate B cells to produce antibodies of the IgG1 class, which are the most effective opsonizing antibodies, as well as antibodies of the IgE class, which are associated with the allergic or antimetazoal response. Th2 cells also produce IL-10 and IL-13, cytokines that downregulate some cells of the immune system. Thus, the Th1-type response is the most desirable type of response to viral and intracellular bacterial infections. The Th2-type response produces more effective opsonizing antibodies, which are important for clearing extracellular bacterial pathogens and toxins. An example of the type of Th1 or Th2 response that is elicited and its importance in disease outcome is illustrated in Box 4–2.

BOX 4–2   The Th1/Th2 Response and the Outcomes of Leprosy and Tuberculosis

Some people who contract tuberculosis, caused by *M. tuberculosis*, or leprosy, caused by *Mycobacterium leprae*, develop a systemic form of the disease. In the case of tuberculosis, this is called miliary tuberculosis. In the case of leprosy, it is called lepromatous leprosy. Other people develop a more localized form of the disease, and although there is damage in the area of infection, bacteria do not spread so readily to internal organs. What causes some people to develop the localized type of infection and others to develop the more dangerous systemic form of infection? Both *M. tuberculosis* and *M. leprae* have in common the fact that they enter and replicate in macrophages. While a difference in the optimal growth temperature of the mycobacteria (which is lower than the normal internal body temperature of 37°C) can account in part for the limited systemic spread, an explanation based on the Th1/Th2 response has been proposed and seems to be supported by the evidence.

In people who are able to elicit a Th1-type response, which generates a robust activated-macrophage response that kills many of the bacteria, the numbers of bacteria are reduced and the infection tends to be localized. In people who do not mount a strong Th1-type response but instead mount a Th2-type response, which does not contribute to activation of macrophages, the bacteria reach high numbers in macrophages and subsequently break out to spread throughout the body. Thus, the choice the body makes between the Th1- and Th2-initiated paths of response to infection has critical practical implications for the patients. This is one of many reasons immunologists are trying to understand the complex interplay of cytokines that controls this crucial decision. Obviously, understanding this decision is crucial to the development of effective vaccines, but it is also possible, in cases where a vaccine might not be available, that pharmaceutical compounds could be developed that would manipulate the response to infection to favor the more effective one.

Source: R. L. Modlin. 1994. Th1-Th2 paradigm: insights from leprosy. *J. Invest. Dermatol.* **102:**828–832.

Production of Antibodies by B Cells

Two signals are required to activate B cells to differentiate into plasma cells and produce antibodies. This fail-safe mechanism makes sense, considering the dire consequences of not strictly controlling antibody production. One mechanism that activates B cells is the associated Th cells that mature in the thymus gland **(thymus-dependent antigens)**. Resting B cells are not dividing, but they do express a single type of mIg on their surfaces. There is an enormous number ($>10^{10}$) of resting B cells displaying mIg molecules with different antigen specificities on their surfaces! The mIg molecules interact with a membrane-bound transducer protein to form the **B-cell receptor (BCR).** When an antigen binds to the Fab regions of two mIg molecules, linking them together, an intracellular signal is sent to the B cell that can lead to cell proliferation if a second signal is detected (Figure 4–7). The second signal is generated by endocytosis and presentation of the antigen on the MHC-II–antigen complexes on the surface of the same B cell. Only a few cell types, such as B cells, produce MHC-II, so the use of MHC-II helps B cells to find the activated Th cells that are displaying TCRs specific for the displayed antigen. Binding of a Th cell to the B cell involves a set of multiprotein contacts similar to those that occur between other APCs and T cells (Figure 4–7). This binding causes the Th cell to produce additional surface receptors that lead to even tighter binding between the antigen-stimulated B cell and the Th cell. This tight binding stimulates the B cell to produce and display cytokine receptors and the Th cell to produce cytokines, which in turn stimulate the B cell to proliferate and differentiate into the mature form of B cells, called plasma cells, that secrete large quantities of antibodies. A fraction of activated B cells become quiescent memory B cells. In subsequent encounters with

Figure 4–7 A Th cell finds and activates B cells presenting the same epitope on their MHC-II. For simplicity, only some of the other proteins involved in binding and activating the B cell are shown. Stimulation of B cells through interactions with Th cells and production of cytokines results in proliferation of the B cells and conversion into antibody-producing cells, called plasma cells. A fraction of the activated B cells become memory B cells.

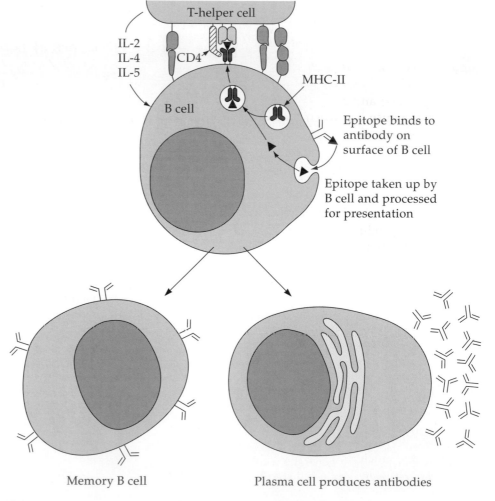

, epitope–MHC-II complex; , T-cell receptors

the same antigen and this time with memory Th cells, the memory B cells respond with a much shorter lag period and produce more antibodies for a longer period of time than in the primary response involving the naïve Th and B cells. In addition, this secondary response produces antibodies with higher affinity, primarily of the IgG isotype.

Links between the Innate and Adaptive Defense Systems

Treating the innate and adaptive defense systems as separate subjects in separate chapters makes pedagogical sense because the human mind has not yet evolved to the point where it can confront the full complexity of the defenses of the human body in one massive dose. However, now that you have come this far and have swallowed most of this complex dose, you may be at the right stage of numbness to look back and face a very obvious fact: for the human body to have the innate defenses acting independently of the adaptive defenses would be counterproductive because the two types of defense need to act synergistically. Some examples of links are **cytokines** and **chemokines,** which are produced by and act on both types of immune responses; PMNs that contribute through cytokine production to the Th1/Th2 decision-making process; and antibodies that are made as an adaptive response but when bound to bacterial cells activate one arm of the complement cascade, i.e., the classical pathway.

We now add yet another link, the interaction between NK cells and B cells. It may be simply a coincidence that NK cells have a number of similarities to T cells in terms of morphology and the types of cytokines they produce. Since T cells interact with B cells, it is somewhat satisfying to learn that NK cells also interact with B cells. Briefly, NK cells can stimulate B cells to proliferate (just as Th cells do), but by some factor that is not one of the known cytokines. B cells in turn induce NK cells to secrete more IFN-γ, which, as we have already seen, is an important factor in macrophage activation.

T-Cell-Independent Antibody Responses

The APC–T-cell pathway processes primarily protein antigens, but this is not the case for nonprotein antigens, such as polysaccharides (e.g., bacterial-capsule components), nucleic acids (e.g., bacterial DNA or RNA), and, to a lesser extent, lipids and glycolipids (e.g., bacterial-membrane components). In addition to the CD1-dependent T-cell pathway discussed above, nonpeptide antigens of these types can also stimulate

T-cell-independent antibody responses. Since this process does not involve T cells, these antigens are often referred to as thymus independent. T-cell-independent antigens provoke an antibody response by directly interacting with B cells (Figure 4–8) instead of through a T-cell-mediated process. Polysaccharides are a well-studied class of T-cell-independent antigens. Those B cells displaying mIg with affinity for the sugar repeat units will bind to the polysaccharide chain. Since these epitopes are repeated many times, numerous mIgs and BCRs are linked together. This acts as a signal to stimulate the B cells to proliferate and differentiate into plasma cells that release antibody. A second signal, such as binding of cytokines produced in the infection, is also required to initiate the T-cell-independent response.

The T-cell-independent response is particularly important for protection against bacterial pathogens that can avoid phagocytosis by covering themselves with a polysaccharide layer (capsule), which is not effectively opsonized by C3b. Such bacterial pathogens are ingested and killed by phagocytes only if antibodies that bind to capsular antigens are elicited and act as opsonins. Although the T-cell-independent response provides protection against capsule-producing bacteria, it has some important drawbacks. First, the anti-

Figure 4–8 T-cell-independent production of antibodies.

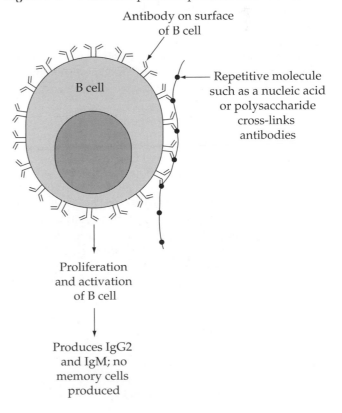

Antibody on surface of B cell

B cell

Repetitive molecule such as a nucleic acid or polysaccharide cross-links antibodies

Proliferation and activation of B cell

Produces IgG2 and IgM; no memory cells produced

body response elicited by T-cell-independent antigens is not as strong as the T-cell-dependent response. It is also not long lasting, because no memory B cells are developed. Second, the main antibodies elicited by T-cell-independent antigens are IgM and IgG2. IgG2 does not opsonize, and IgM does so less effectively than IgG1 and IgG3. Third, young infants do not mount a T-cell-independent response. Polysaccharides and lipids can elicit an immune response in children and adults, but not in infants under the age of 2 years. Thus, the ability to respond to T-cell-independent antigens is acquired after birth. This is an important consideration, because it means that vaccines consisting of T-cell-independent antigens are not effective until an infant has become old enough to respond to these antigens. Unfortunately, infants are one of the highest-risk groups for contracting serious infections due to capsule-producing bacteria (e.g., pneumonia and meningitis).

A strategy for improving the immune response to T-cell-independent antigens and extending this response to infants is to covalently link epitopes of the polysaccharide capsule to a protein. Such a vaccine is called a **conjugate vaccine** (more about this in chapter 17). APCs process conjugate vaccines as if they were proteins and elicit a T-cell-dependent response that culminates in production of antibodies that also recognize the polysaccharide antigens. This immune response is long-lived, since it involves T-cell activation and memory T-cell and B-cell generation and produces opsonizing IgG1 antibodies, as well as IgG3 and IgG4.

The account just given of the T-cell-independent response to polysaccharides is the one currently favored by most immunologists. Questions about the accuracy of this account can be raised, however, when one considers the protection record of the capsular-polysaccharide-based vaccine against pneumonia caused by *Streptococcus pneumoniae*. According to the currently accepted version of the T-cell-independent response, this vaccine should confer only short-term immunity. In fact, in young adults, the vaccine confers immunity for several years. Does this mean that these antibodies persist for several years, or does it indicate some sort of memory response to polysaccharide antigens? Stay tuned for future developments.

Unfortunately, one aspect of the previous account of the T-cell-independent response remains correct. Infants are not protected at all by this polysaccharide vaccine. To overcome this obstacle, an effective conjugate vaccine reached the market in developed countries around 2000. This vaccine was based on the seven most prevalent capsule serotypes of *S. pneumoniae* causing invasive pneumococcal infections in children in developed countries at that time. However, the heptavalent vaccine was not completely effective in all children or in children in developing countries, where other serotypes commonly cause infection. There are about 90 different capsular serotypes of *S. pneumoniae*, and by 2007, pneumococcal infections caused by serotypes not covered by the heptavalent vaccine were already emerging in developed countries. Moreover, it is extremely alarming that these new clinical isolates were already resistant to the antibiotics approved to treat pneumococcal infections in children. A new conjugate vaccine, which includes 13 capsule serotypes, is now being used in a number of countries and is in the final stages of the approval process in the United States. Stay tuned to see how long this new one will last before the bacteria find a way around it.

Mucosal Immunity

An important immune defense against infectious diseases, but one that is much less well understood than the humoral or cell-mediated responses, is the immune system that produces **sIgA**. The first step in many microbial infections is colonization and invasion of a mucosal surface. sIgA can prevent such infections by blocking colonization. Thus, whereas the cell-mediated or humoral antibody response may cause collateral damage to tissues in the area where infection is occurring, the sIgA-mediated defense is completely innocuous to the host, because it occurs in the mucus layer. Skin and mucosal surfaces all have associated lymphoid tissues located just below the epithelial layer in the **lamina propria**. Mucosal immune systems are generally referred to as **mucosa-associated lymphoid tissue (MALT)**. The best studied of these systems is the **GI-associated lymphoid tissue (GALT)** that makes up the collection of **follicles** found in the gut epithelium and lamina propria (called **Peyer's patches**), which are most highly concentrated in the ileum and rectum of the intestine. Similar lymphoid tissues are found in the respiratory and vaginal tracts. Skin also has a similar system, called the **skin-associated lymphoid tissue** or **SALT**. The **Langerhans cells** of the epidermis are the APCs of SALT.

The cells that form the GALT are illustrated in Figure 4–9. The **M (microfold) cell** takes up antigens from the lumen of the intestinal tract and passes them to closely associated GALT macrophages, which act as the APCs of the GALT. M cells have never been successfully cultivated in vitro, so little is known about their activities. The process by which GALT macrophages process antigens and elicit production of

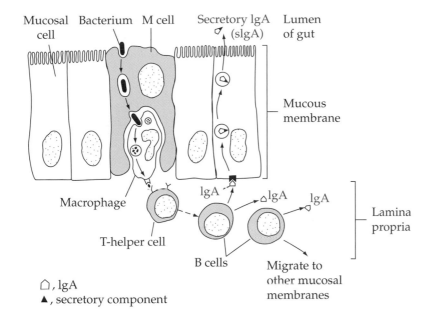

Mucosal cell — Bacterium — M cell — Secretory IgA (sIgA) — Lumen of gut

Mucous membrane

Macrophage

T-helper cell

B cells

IgA — IgA — IgA

Lamina propria

Migrate to other mucosal membranes

☐, IgA
▲, secretory component

Figure 4–9   Cells of the GALT that confer mucosal immunity. M cells and their associated macrophages and lymphoid cells (T and B cells) are sometimes called follicles. Collections of such follicles in the gut are called Peyer's patches. M cells sample the contents of the gut lumen and transfer the antigens to closely associated resident macrophages, which in turn ingest the bacteria and present antigens to the underlying T cells, which then stimulate nearby B cells to produce IgA. The IgA binds to receptors on the basal surfaces of the mucosal epithelial cell and is transcytosed across the cell and secreted into the lumen of the gut as sIgA.

CTLs or antibodies is the same as that described above, except that the bulk of the macrophages, B cells, and T cells of the GALT reside specifically in the lamina propria of mucosal surfaces.

Although they do not appear to be part of the humoral immune system, some T cells and B cells stimulated by antigen processing at the GALT can migrate to other mucosal sites, and vice versa. Stimulation of one of the MALT sites can transfer to other sites and thus results in general mucosal immunity. The first evidence for this came from elegant experiments performed by Husband and Gowans in the late 1970s, in which they excised a segment of the small intestine from a rat, preserving its vascular and lymphatic supplies, and reconnected the ends of the intestinal segment to the skin surface of the animal, forming a so-called Thiry-Vella loop (Figure 4–10). They then introduced an antigen, in this case cholera toxin, into the loop and found that sIgA was secreted not only in the immunized loop segment, but also in a second such loop (when made) and in the main intestine. Their results demonstrated that introduction of an immunogen at one mucosal site could confer mucosal immunity at a remote site. This characteristic of the MALT system is what makes **oral vaccines** feasible. Initially, oral vaccines stimulate the GALT, but sIgA against vaccine antigens is later detectable in other MALT sites. Thus, an oral vaccine can be used to elicit immunity to respiratory and, presumably, to urogenital pathogens.

Currently, efforts are being made to develop vaccines administered by inhalation, so that stimulation of the nasal MALT (**NALT**) would produce an sIgA response at other MALT sites. These vaccines would have the advantage that they do not have to pass through the stomach. Developing vaccines that target the GALT means developing vaccines capable of surviving the low-pH/protease-rich stomach environment, a barrier that has proven problematic in many cases. Administering vaccines by rectal or vaginal suppositories is theoretically possible, but this strategy has not been actively pursued to date. On the other hand, the NALT and SALT are gaining in attraction as a target for vaccine development.

When the GALT is stimulated, one outcome is production of IgA (Figure 4–9). IgA binds to the **poly-Ig receptor** on the basal surfaces of mucosal cells and is then taken up and carried in vesicles to the apical surface, where it is released. Release into the lumen of the gut involves proteolytic cleavage of the receptor, and a portion of the receptor remains attached to the IgA, making it sIgA. Activation of the GALT can also lead to production of CTLs. These cells probably remain on the basal side of the mucosa, although it is possible that during an infection some of them migrate to the apical surface, especially in areas where damage to the mucosa has occurred. GALT CTLs are important for protection against viral infections of the GI tract and some bacterial infections where the bacteria multiply inside mucosal cells.

One of the many mysteries swirling around the intestinal immune system is the role of a particular type of mucosal cell called **gamma-delta T cells.** The majority of gamma-delta T cells are CD8$^+$ T cells and would thus be grouped with CTLs. However, whereas CTLs of the humoral immune system have TCRs composed of alpha and beta protein subunits, the **intestinal epithelial lymphocytes (IELs)** have a TCR com-

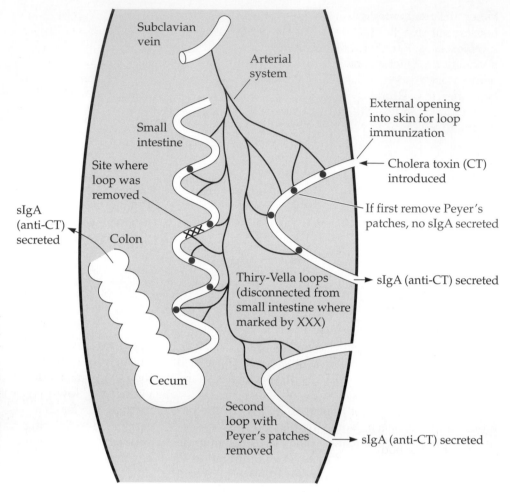

Figure 4–10 Classic experiment by Husband and Gowan demonstrating mucosal immunity at remote sites. (Based on A. J. Husband and J. L. Gowans. 1978. The origin and antigen-dependent distribution of IgA-containing cells in the intestine. *J. Exp. Med.* **148**:1146–1160, with permission of Rockefeller University Press.)

posed of related but somewhat different protein subunits called gamma and delta. Gamma-delta T cells account for less than 4% of circulating CD8$^+$ cells, but they account for as much as 10 to 15% of the mucosal cells in the GI tract, and in some parts, such as the colon, the level may be as high as 40%. Recently, some light has been shed on their role. In fact, gamma-delta T cells have gone very rapidly from being cells in search of a function to cells to which too many functions are now attributed.

A particularly mysterious feature of gamma-delta T cells—at least mysterious until recently—has been that these cells have a very limited repertoire for recognition of antigens displayed on human cells. There are thousands, perhaps millions, of types of cytotoxic and Th cells, each of which recognizes a specific antigen. In contrast, the gamma-delta T cells seem to recognize only a limited number of cell surface antigens. Also, gamma-delta T cells seem to bypass the macrophage antigen presentation step and recognize nonpeptide antigens that have not been processed. This is T-cell heresy at its best. T cells are named to

indicate that these are cells that must first pass through the thymus, an organ that contains APCs, which give the T cells their marching orders. Gamma-delta T cells are not thymus derived and are relatively nonspecific. How, then, could they play any role in the response to infection? Nevertheless, there are certainly a lot of them hanging around the mucosal membranes of the body. As the old saying goes, if there are this many ants, the picnic must be a success.

The mystery surrounding the function of the gamma-delta T cells has been solved to some extent by the discovery that gamma-delta T cells respond primarily to two human protein complexes related to MHC-I, **MICA** and **MICB**. These proteins are displayed on the surfaces of cells that are stressed, e.g., by being infected. Does this relative lack of specificity mean that they should be classified as members of the innate defense system? Not necessarily, but it does illustrate our previously stated caveat that although drawing a line between the so-called adaptive and innate defenses may be useful from a pedagogical point of view, it obscures the many linkages that exist between the two systems. Alternatively, these cells may have specificity that has not yet been determined.

Another role attributed to gamma-delta T cells is production of factors that help downregulate the inflammatory response. Gamma-delta T cells produce cytokines that stimulate alpha-beta CTLs to migrate to the area and eliminate damaged cells. Finally, gamma-delta T cells produce growth factors that may aid in repair of damaged epithelia. Perhaps the best way to describe the role of gamma-delta T cells is as a busboy who doubles as a bouncer. The gamma-delta T cells not only take part in the battle against invaders, they also help with the cleanup and repair process that returns the body to normal and reestablishes the mucosal membrane as an intact defense against invaders. As if that were not enough to keep a T cell busy, gamma-delta T cells also appear to stimulate Th cell activities and to secrete chemokines that help orchestrate the activities of neutrophils and monocytes. Perhaps the busboy-sometimes bouncer is also the manager of the establishment. As you can see, things have been hopping in cellular immunology lately.

The GALT is not an unmixed blessing. Normally, bacteria and other viable antigens that pass through M cells are killed by the GALT macrophages. Some bacterial pathogens, however, have acquired the ability to avoid this fate and exploit the GALT as an entryway into the body. Because the M cell is an antigen-sampling cell, it usually does not take up substantial amounts of an antigen because only a few bacteria or other antigens are sufficient to stimulate a GALT-mediated immune response. The bacteria that use the GALT as an entryway into the body, such as *Salmonella enterica* or *Yersinia pseudotuberculosis,* stimulate M cells to transport them in higher than usual numbers (Figure 4–11).

Development of the Adaptive Response System from Infancy to Adulthood

Human infants do not develop fully effective specific and nonspecific defenses until they are 1 to 2 years of age. For a period up to 1 year of age, maternal antibodies that were transferred through the placenta during gestation or ingested in milk during infancy protect them at least partially. Circulating maternal antibodies can actually interfere with an immune response to some antigens because high levels of antibody to a particular epitope discourage the development of an antibody response to that epitope. This is a consequence of immune regulation that normally protects the body from overreacting to a particular epitope. The dampening effect of circulating maternal IgG is the reason why some vaccines are not given to infants less than 1 year of age. The transition between the protection conferred by maternal antibodies and

Figure 4–11 Scanning electron micrographs of mouse Peyer's patches before **(a)** and after **(b)** incubation with *Y. pseudotuberculosis*. Both images depict a central M cell surrounded by enterocytes (and part of a brush cell in the lower left of panel b). The M cell in panel a lacks adherent bacteria, but in contrast, the association of bacteria with the M cell in panel b is accompanied by disruption of the normal surface morphology of the M cell. Bars, 2 μm. (Reprinted from M. A. Clark, B. H. Hirst, and M. A. Jepson, *Infect. Immun.* **66:**1237–1243, 1998.)

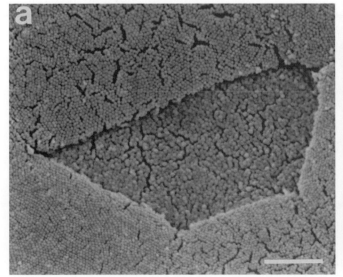
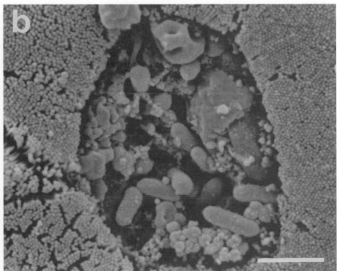

the development of the infant's own immune system provides pathogenic microbes with a "window of opportunity" and is one of the reasons why children under the age of 2 are particularly vulnerable to infectious diseases. The innate defenses of infants are also underdeveloped prior to age 2. Infants have very low levels of NK cells and are deficient in mannose-binding protein, one of the acute-phase proteins produced by the liver that helps to opsonize some types of microbes.

Adaptive Defense Systems in Nonmammals

Scientists have found primitive versions of the mammalian adaptive defense system in many different organisms, ranging from sharks to insects (Table 4–2). Even plants have a primitive defense system in which proteins produced by the plant (R proteins) bind to invading bacteria, a reaction that produces apoptosis in the surrounding plant cells. This area of dead, dry tissue, which appears as black spots on leaves or fruit, traps the bacteria in the area. Bacteria need to move throughout the plant via the xylem and phloem in order to cause serious disease. The plant's defense response is designed to prevent this. Insects, such as *Drosophila melanogaster* (fruit fly), and slime molds, such as *Dictyostelium*, also have primitive defense systems that resemble mammalian defense responses, especially the innate immune response. Zebra fish not only have an innate immune response, but also have a response that parallels the Th1/Th2 adaptive response. Sharks produce antibodies that can bind to human pathogens.

Assertions that systems in nonmammals are "immune systems" should be viewed with some skepticism. In the cases of sharks, plants, and zebra fish, there is evidence that the systems are actually a response that protects the animal from infection. In the case of *Drosophila*, however, the Toll receptors that led to the discovery of mammalian Toll-like receptors, which are a central part of the mammalian innate immune response, appear to be involved primarily in the development of the insect or in feeding behavior, not in defending the insect against infection. Nonetheless, it is intriguing to see hints of what we know as the mammalian innate and adaptive defense systems so widespread in other creatures. Interest in the evolution of the mammalian immune system is a relatively new area of research focus, so there are likely to be many more insights in the future as this research progresses.

The Dark Side of the Adaptive Defenses—Autoimmune Disease

Complex figures, such as Figure 4–7, are not designed just to drive students crazy (although that is certainly a bonus feature). What they illustrate is the manifold interactions that occur between cells of the immune system when they activate each other. All those proteins labeled CD followed by a number have an important role. They help to ascertain that the contacts between the immune cells are occurring in response to true foreign-antigen presentation events and not accidental associations between cells of the immune system and other human cells. They serve as insurance that the immune system will not go out of control and

Table 4–2 Defense systems in higher organisms

Organism	Innate immunity	Adaptive immunity	Invasion-induced signaling	Phagocytic cells	Antimicrobial peptides	Pattern recognition receptors	T and B cells	Antibodies
Plants	+	−	+	−	+	+	−	−
Invertebrates								
Sponges	+	−	?	+	?	?	−	−
Worms	+	−	?	+	?	?	−	−
Insects and crustaceans	+	−	+	+	+	+	−	−
Vertebrates								
Sharks	+	+	+	+	+	+	+	+
Bony fish	+	+	+	+	+	+	+	+
Amphibians	+	+	+	+	+	+	+	+
Reptiles	+	+	+	+	?	+	+	+
Birds	+	+	+	+	+	+	+	+
Mammals	+	+	+	+	+	+	+	+

attack human cells. Unfortunately, this safeguard is sometimes breached, for example, when bacterial antigens mimic human antigens. In such cases, the adaptive defenses can mount an attack that targets human tissues. If the tissue is heart tissue or tissues of other vital organs, the misguided attack can be lethal. In later chapters, we will encounter examples of such misguided attacks by the adaptive immune defenses, a set of conditions that can lead to **autoimmune disease.** Bacteria that elicit an autoimmune response are particularly difficult targets for vaccines because the vaccine can induce the very self-attacking response that can cause the autoimmune disease and can actually make the disease worse rather than preventing it when the vaccinated person is exposed to the bacterium. Thus, one of the concerns of vaccine producers is to make sure that their vaccine does not elicit an autoimmune response.

SELECTED READINGS

Bengten, E., M. Wilson, N. Miller, L. W. Clem, L. Pilstrom, and G. W. Warr. 2000. Immunoglobulin isotypes: structure, function, and genetics. *Curr. Top. Microbiol. Immunol.* **248:**189–219.

Bettelli, E., T. Korn, M. Oukka, and V. K. Kuchroo. 2008. Induction and effector functions of Th17 cells. *Nature* **453:**1051–1057.

Boes, M. 2000. Role of natural and immune IgM antibodies in immune responses. *Mol. Immunol.* **37:**1141–1149.

Clark, M. A., and M. A. Jepson. 2003. Intestinal M cells and their role in bacterial infection. *Int. J. Med. Microbiol.* **293:**17–39.

Cresswell, P., A. L. Ackerman, A. Giodini, D. R. Peaper, and P. A. Wearsch. 2005. Mechanisms of MHC class 1-restricted antigen processing and cross-presentation. *Immunol. Rev.* **207:**145–157.

Dabbagh, K., and D. B. Lewis. 2003. Toll-like receptors and T-helper-1/T-helper-2 responses. *Curr. Opin. Infect. Dis.* **16:**199–204.

Dörner, T., and P. E. Lipsky. 2006. Signaling pathways in B cells: implications for autoimmunity. *Curr. Top. Microbiol. Immunol.* **305:**213–240.

Kindt, T. J., R. A. Goldsby, and B. A. Osborne. 2007. *Kuby Immunology,* 6th ed. W. H. Freeman and Company, New York, NY.

Maizels, R. M. 2005. Infections and allergy—helminths, hygiene and host immune regulation. *Curr. Opin. Immunol.* **17:**656–661.

Lawton, A. P., and M. Kronenberg. 2004. The third way: progress on pathways of antigen processing and presentation by CD1. *Immunol. Cell Biol.* **82:**295–306.

Lieberman, J. 2003. The ABCs of granule-mediated cytotoxicity: new weapons in the arsenal. *Nat. Rev. Immunol.* **3:**361–370.

Parham, P. 2009. *The Immune System,* 3rd ed. Garland Science, New York, NY.

Pichichero, M. E., and J. R. Casey. 2007. Emergence of a multiresistant serotype 19A pneumococcal strain not included in the 7-valent conjugate vaccine as an otopathogen in children. *JAMA* **298:**1772–1778.

Reiner, S. L., F. Sallusto, and A. Lanzavecchia. 2007. Division of labor with a workforce of one: challenges in specifying effector and memory T cell fate. *Science* **317:**622–625.

Suzuki, K., S. Ha, M. Tsuji, and S. Fagarasan. 2007. Intestinal IgA synthesis: a primitive form of adaptive immunity that regulates microbial communities in the gut. *Semin. Immunol.* **19:**127–135.

Szabo, S. J., B. M. Sullivan, S. L. Peng, and L. H. Glimcher. 2003. Molecular mechanisms regulating Th1 immune responses. *Annu. Rev. Immunol.* **21:**713–758.

Tupin, E., Y. Kinjo, and M. Kronenberg. 2007. The unique role of natural killer T cells in the response to microorganisms. *Nat. Rev. Microbiol.* **5:**405–417.

Watanabe, N., F. Bruschi, and M. Korenaga. 2005. IgE: a question of protective immunity in *Trichinella spiralis* infection. *Trends Parasitol.* **21:**175–178.

QUESTIONS

1. In the short statement that began the chapter, the innate immune system was likened to ordinary police and the adaptive immune system was likened to specialized backups, such as snipers. For those of you who follow true crime or mystery stories, you might find it amusing—and helpful in your studies—to try to come up with other criminal justice analogies. For example, you might make an analogy between "antibody production by B cells" and "smart bombs, surveillance, and tracking devices." See what you can do with Th cells, cytokines, APCs, and NK cells.

2. Does the fact that sIgA does not activate complement make it less effective in preventing infection than IgG? Hint: consider its location and function.

3. Explain how the structure of an antibody is designed to facilitate the function of the particular antibody, e.g., why an antibody has at least two active sites, why it has an Fc portion, and why it might be cross-linked as multimers.

4. Why is a Th1 or Th2 cell needed to mediate APC-initiated antibody production? Why do the B cells not bind directly to the APCs?

5. Explain why direct antigen activation of CTLs and B cells is more important than APC-initiated activation in subsequent encounters with a microbe. Why is the APC-initiated pathway needed at all?

6. There are many interactions between the innate and adaptive defense systems. List as many of them as you can and explain how they work. Why are these interactions important? Are all of them beneficial?

7. What are the two ways that polysaccharides, DNA, and lipid antigens can elicit an immune response? How are these responses different than that elicited by a peptide antigen?

8. Why is the level of specific IgM antibodies of diagnostic significance?

SOLVING PROBLEMS IN BACTERIAL PATHOGENESIS

1. A group of researchers have isolated a new pathogenic bacterium Q from lungs and lymph nodes of cystic fibrosis patients that produces an unusual polysaccharide capsule, called QPS. They can see QPS on the surfaces of the bacteria in electron micrographs from fresh isolates (i.e., bacteria just obtained from patients). In addition to the capsule, bacterium Q produces a protease, called QP, which the researchers believe is responsible for degrading host sIgA.

A. The researchers propose that QPS and QP might make good targets for vaccine development against bacterium Q. What led the researchers to propose this?

B. What potential problems could the researchers have in using QPS as a vaccine?

C. How are QP antigens presented to immune cells?

D. The researchers found that the QPS vaccine does not elicit a long-lasting immune response. How does QPS elicit an immune response? What vaccine strategy could be used to generate long-lasting immunity targeting QPS?

E. The researchers believe that they can develop an oral vaccine that will be effective against lung infections with bacterium Q using QP as a vaccine component. What specific experimental evidence supports their rationale for using this proposed approach?

F. If QP were used as a vaccine, what immune response would be primarily responsible for controlling lung infection caused by bacterium Q?

2. *Listeria monocytogenes* is a food-borne, gram-positive bacterium that causes infections (listeriosis) in individuals with compromised immune systems. *Listeria* is predominantly an intracellular pathogen. After phagocytosis by intestinal macrophages, *Listeria* escapes the phagosome and replicates in the cytosol before spreading to other neighboring cells through a special invasion process that allows it to remain intracellular while spreading from cell to cell.

A. What advantage is there for *Listeria* to invade and spread intracellularly?

B. What immune response is primarily responsible for controlling *Listeria* infection in healthy but unimmunized individuals?

C. What immune response is primarily responsible for controlling *Listeria* infection in healthy individuals who have had prior exposure to *Listeria*?

D. How are *Listeria* antigens presented to immune cells?

5 | The Normal Human Microbiota

Although some philosophers assure us that humans are the measure of all things, to the bacteria that live in or on us we are more like the proverbial "free lunch." However, the bacteria that colonize our bodies from shortly after our birth to our death do "pay rent" in various ways, mainly by protecting us from disease-causing bacteria and by contributing to our nutrition and healthy immune status. Our bodies are adapted not only to tolerate them, but also to encourage their presence. Getting in touch with your prokaryotic side is an important part of understanding what it is to be human.

Importance of the Normal Microbial Populations (Microbiota) of the Human Body

A healthy human body harbors more than 10 times as many bacterial cells as human cells. These bacteria, collectively known as the **microbiota,** do not take up as much space as human cells, so their bulk is less than that of the human cells, but they manifest themselves in a variety of ways. For example, the odor of sweat comes from bacterial metabolism of compounds in skin secretions. Flatus is another odiferous indication of the activities of bacteria in the intestine. Bacteria form the scum that develops on unbrushed teeth and the plaque that we pay dental hygienists to remove.

Although these particular manifestations of the presence of bacteria are viewed by most of us as unpleasant, eliminating bacteria from our bodies, even if it were possible, would be a very bad idea. These tiny freeloaders play a number of beneficial roles. Many members of our resident microbiota provide nutrients for us by degrading foodstuffs normally not digested by our own systems and by synthesizing and excreting vitamins, e.g., vitamin K, vitamin B_{12}, and other B vitamins. The normal microbiota also protects us from pathogens by competing for nutrients and attachment sites and thereby preventing colonization by pathogens. Women who have taken antibiotics that affect the balance of the normal microbial population of the vaginal tract can develop vaginitis caused by yeast, whose numbers are normally kept low by the vaginal bacteria. However, if the numbers of these bacteria decrease, the yeasts and other bacteria can overgrow and produce toxic substances that kill members of the microbiota. Antibiotic use can also lead to diarrhea if the normal microbiota of the colon is disrupted. In some cases, a more serious disease is caused by overgrowth of a bacterium, *Clostridium difficile,* that is normally present in very low numbers but can cause damage to the colon lining and death. Recent experimental evidence indicates that indigenous bacteria play a crucial inductive role in gut development during early postnatal life, in particular, the cecum and Peyer's patches, and in the development and maturation of the innate and adaptive immune systems. For example, colonization by the commensal microbiota stimulates the production of cross-reactive antibodies, which prevent infection by related pathogens.

The notion that maintaining "good" bacteria in the intestinal tract is conducive to health has spawned an entire industry, the **probiotics** industry. Probiotics are preparations of live bacteria, usually freeze-dried into pellets or added to foodstuffs, such as yogurt, that are ingested intentionally to bolster the normal population of "good" colonic bacteria or are included in douches introduced into the vaginal tract. Most of the preparations contain *Lactobacillus* or *Bifidobacterium* species, often species isolated from food, not humans. These bacteria do not colonize the colon and must be taken daily to maintain the bacteria in the intestinal environment. Some probiotics for skin may appear on the market soon. Another approach to maintaining a healthy microbiota that is gaining some popularity is **prebiotics,** compounds, such as fructo-oligo-saccharides, that are supposed to foster the growth of "good" bacteria in the gut.

Claims for the probiotics currently on the market tend to be vague, and there is not yet enough credible scientific evidence that these preparations have long-lasting beneficial effects. Nevertheless, the idea that probiotics or prebiotics could help patients whose intestinal bacterial populations have been disrupted by antibiotics or other factors has attracted new interest. New information on the normal microbial populations (microbiota) of our bodies may help in the design of more effective probiotics. Also, accurate diagnosis of changes in the microbiota could help to identify patients who are at higher risk for intestinal disease so that intervention strategies could be initiated.

Figuring out how to maintain bacterial populations that are conducive to health is not a trivial undertaking. The microbial populations found in the mouth, intestinal tract, and vagina are complex, consisting of hundreds of species. The microbial populations are complex, not only in terms of the number of different

species, but also in terms of the metabolic activities of the different species. Until recently, the only way to characterize the microbiota of a site was laboratory cultivation using various growth media. This approach was time-consuming and expensive. It was also unreliable, because the plating efficiencies of the more difficult to cultivate organisms, which were in the majority, were far from 100% and could vary from one batch of medium to another. Some species might not even be cultivatable using available media.

As a result of these difficulties, a number of important questions were not addressed in the past because of the limitations of cultivation. They included the following. How much person-to-person variation is there? Does diet, hormones, or age affect the composition of the microbiota? How does antibiotic therapy affect the microbiota? Are variations from site to site at a particular location within the same person significant? Do animals, such as laboratory rodents that are used as models for human disease, have a microbiota that is similar to that of humans?

An important recent advance, which has changed this picture and reinvigorated interest in the microbiota, is the advent of culture-independent, nucleic acid-based methods for characterizing bacterial populations. Before surveying the microbial populations of different parts of the human body, it is worth reviewing some of the new culture-independent methods.

Nucleic Acid-Based Approaches to Characterization of the Microbiota

There are three main types of questions about the human microbiota that are currently being addressed by nucleic acid-based approaches. These are summarized in Table 5–1. The first questions that arise when scientists are seeking to characterize a complex microbial population are, what microbial species are present and how do they change with conditions? The next question occurs when scientists need to trace outbreaks of disease caused by a specific bacterial species: how is one isolate related to another? Another question concerns the complexity and changes of metabolic capacity of the microbiota.

Taking a Microbial Census by Using Microbial rRNA Sequence Analysis

The most widely used approach to answering the first questions above is to isolate total DNA from the microbial population and then employ the **polymerase chain reaction (PCR)** to amplify the 16S ribosomal RNA (rRNA) genes, which are then cloned and se-

Table 5–1 Questions about the microbiota that are currently being addressed by nucleic acid-based approaches

Question 1. What types of microbes are present? (Microbial census)

Approach

Use primers that bind to conserved regions of the 16S rRNA genes to PCR amplify these genes.

Sequence the 16S rRNA genes and look for matches in the databases to indicate the identities of the microbes.

Advantages

Provides rapid information about what types of microbes are present.

Can serve as a guide to further analysis or cultivation attempts.

Limitations

Some species may be missed if they are not amplified efficiently.

Not quantitative, does not give accurate assessment of microbial abundances.

Seldom gives conclusive information about the physiological features of the organisms.

Question 2. What are the clonal relationships among different isolates of the same bacterial species?

Approach

Use PCR-based or restriction enzyme-digestion profiling methods to compare the DNA banding patterns of microbial communities for their similarities and differences.

Advantages

The methods are cheaper than large-scale sequencing of DNA libraries.

They allow a large number of samples to be analyzed at the same time to distinguish microbial communities that are the same or different.

Limitations

The greatest disadvantage is that they do not provide taxonomic identification of the microbial-community members.

Question 3. What are the physiological potentials of members of the microbial population?

Approach (Metagenomic analysis)

Isolate DNA from the mixed microbial population.

Shear or digest DNA into small fragments, clone, and sequence.

Use database sequence comparisons to identify metabolic genes in the population's genome (microbiome).

Advantages

Potentially gives extensive information about the physiological (genetic) potential of the microbial population.

Limitations

Major sequencing capacity is required (high-throughput sequencing).

Focuses on major members of the population. Many genes may not have identified relatives (genes of unknown function).

May be difficult to associate a particular gene with a particular microbe or to assemble portions of genomes.

Does not give information about what genes are expressed.

quenced. This approach is based on the fact that 16S rRNA genes are a mosaic of regions that are highly conserved among all bacterial species and regions that are less conserved and that contain sequence signatures for different bacterial species. The realization that sequencing the 16S rRNA genes can be used to rapidly and accurately identify bacterial strains has introduced a new era in bacterial detection and identification. The revolution first occurred in environmental microbiology, where nothing equivalent to the detailed identification protocols of clinical microbiology existed and where the vast majority of bacteria could not be cultured using conventional media.

The basic procedure is relatively simple. DNA primers that recognize highly conserved regions at the beginnings and ends of 16S rRNA genes are used to amplify most of the rRNA gene, including the variable rRNA gene region signatures, by PCR using a thermostable DNA polymerase (Figure 5–1). For rRNA genes prepared from isolated cultures, the resulting PCR products, called **amplicons,** are sequenced directly. For rRNA gene amplicons prepared from complex mixtures of bacteria, such as from the human microbiota, the resulting mixture of PCR amplicons is first cloned to form libraries, as described below, and the clones are then sequenced. It should be noted, however, that new DNA-sequencing technologies, discussed below, show promise in streamlining this procedure further by eliminating the need for the cloning step. Computer programs compare the sequenced rRNA gene fragments to the growing rRNA gene sequence databases to find which organisms they most closely match. It is now possible to identify an unknown bacterial isolate within 24 h by this approach. The initial characterization of a complex community can be done in days.

This method uses the same reagents (primers, DNA polymerase, and reaction mixture) for all bacterial species, because the primers recognize conserved regions of the 16S rRNA gene, which are almost universal in bacteria. Even better, the approach can be used to rapidly identify bacteria that are not amenable to cultivation. One of the first successes of this approach in clinical microbiology was the identification of the bacterium that causes a rare form of intestinal disease called Whipple's disease. A bacterium-like form could be seen in tissues of infected people, but attempts at cultivation had been unsuccessful. Finally, using this technique, the gram-positive bacterium associated with Whipple's disease was identified as *Tropheryma whipplei*.

16S rRNA GENE-BASED PROFILING OF COMPLEX MICROBIAL COMMUNITIES. Scientists are often interested in understanding the effects of certain conditions or factors over time on the composition of the

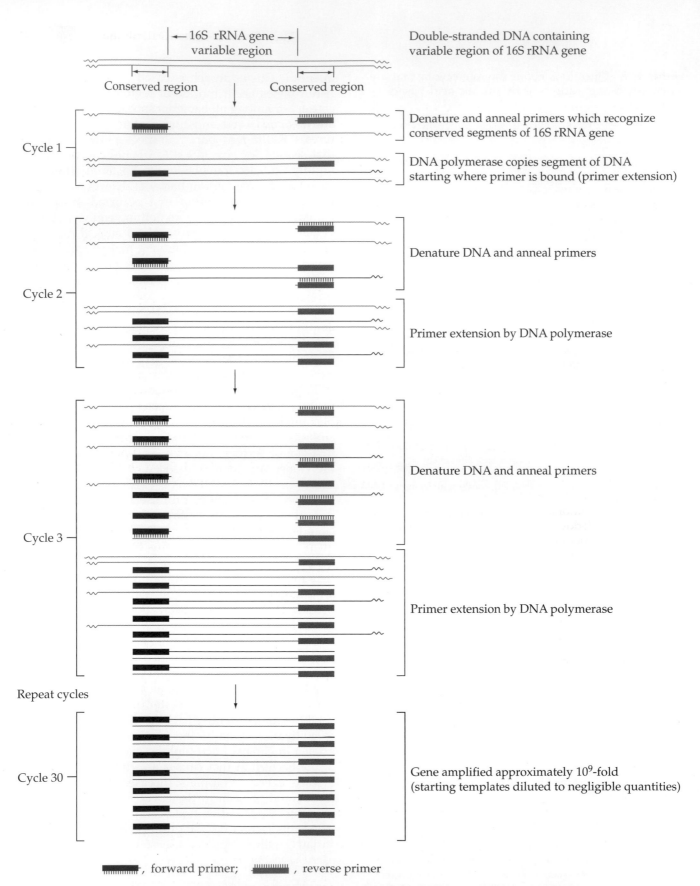

Figure 5–1 Using PCR to detect microbes in a clinical specimen. PCR primers (solid dark bars) recognize segments of DNA on either side of the region to be amplified. The amplified segment can be detected by DNA hybridization or by fluorescently labeled bases incorporated during amplification.

microbial community. To accomplish this, it is necessary to collect at any given time a sample of the microbes that is representative of the whole community (a profile) so that changes in abundance or diversity can be monitored. Although amplification and sequencing are now the preferred approach, cruder methods have also been used. There are a number of popular PCR-based profiling methods that are currently used by researchers to rapidly assess differences in patterns between complex microbial communities. Because cloning and sequencing thousands of 16S rRNA genes can be quite expensive for large numbers of samples, many researchers opt to use profiling methods that do not require sequencing to gain insight into the diversity or dynamics of the microbial

communities without necessarily ascertaining the identities of the component microbes.

One such profiling method is **denaturing/thermal gradient gel electrophoresis (DGGE/TGGE),** in which total genomic DNA is isolated from a complex mixture of bacteria and highly conserved regions within the 16S rRNA genes are amplified to give ~200-bp products that are then separated by electrophoresis on a denaturing or thermal gradient polyacrylamide gel based on the different base compositions (G+C contents). The banding pattern (profile) observed for one mixed sample can be compared to profiles of other samples to evaluate the relative similarity of the microbial communities obtained from different habitats or treatments (Figure 5–2). The dis-

Figure 5–2 PCR-DGGE analysis. The diversity of complex microbial communities observed under different conditions over time can be assessed by PCR amplification of ~200-bp regions of 16S rRNA genes and comparing their banding patterns on polyacrylamide gels containing a gradient of urea and formamide, which denature the DNA duplexes. In the experiment shown here, the researchers were interested in monitoring the effects of antibiotic treatment and a low-residue diet on mouse fecal bacterial populations. PCR-DGGE was performed on microbial DNA isolated from fecal samples from mice on days 1, 2, 7, and 14. Antibiotic administration (25 ppm cefoxitin in drinking water) was begun on day 1. The lanes marked LC contained samples from the control diet, and those marked LR contained samples from a low-residue diet; the minus and plus signs correspond to the absence or presence of antibiotic, respectively. M is the marker lane corresponding to the bacterial-standard ladder. The letters A, B, and C indicate bands differentially expressed in a specific diet or treatment. (Adapted from McCracken et al., 2001, with permission from the American Society for Nutrition.)

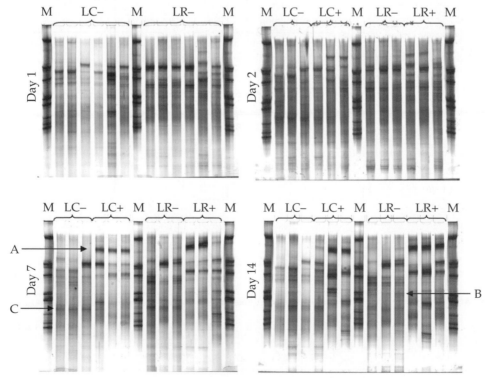

advantage of this method is that it is difficult to obtain taxonomic information and a quantitative assessment of the microbial content.

Terminal restriction fragment length polymorphism (T-RFLP) analysis is another 16S rRNA gene-based method for profiling microbial communities. T-RFLP depends on the PCR amplification of the 16S rRNA gene with a primer set in which the ends of each primer are fluorescently labeled with different colored dyes. The resulting PCR products are cleaved with selected restriction enzymes and separated on a DNA-sequencing column. Due to sequence variations, the terminal restriction sites for amplicons from each bacterial species in the community are different, and a pattern of the output provides information on the size of the product in base pairs (which should cor-

relate with the species) and the intensity of fluorescence (an indication of the relative abundances of the various community members). As illustrated in Figure 5–3, a comparison of DGGE and T-RFLP methods showed that similar qualitative results can be obtained. In both cases, it was possible to distinguish samples that appeared to be similar (samples 26 and 34) versus different (sample 44). T-RFLP is less labor-intensive, faster, and subject to less variation than gel-based methods, such as DGGE. However, a problem sometimes encountered with this method is incomplete digestion of the DNA, which may lead to extra peaks in the profile.

The advantages of the DGGE and T-RFLP methods for microbial-community profiling are that they are both relatively inexpensive compared to sequencing

Figure 5–3 Comparison of the DGGE and T-RFLP methods for microbial-community profiling. The microbial-community profiles for three different human vaginal samples collected were determined using DGGE (left) or T-RFLP (right) analysis. In both cases, similarities or differences in the diversities of the microbial communities among the samples could be distinguished. (Courtesy of N. Nakamura, M. Ho, and B. A. Wilson.)

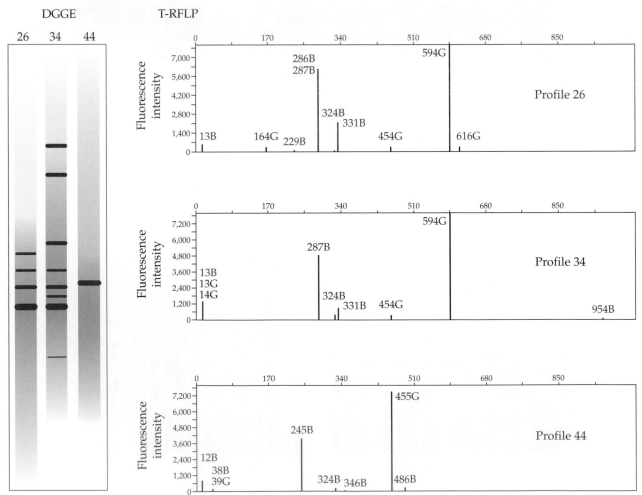

and they allow a large number of samples to be analyzed at one time. Also, they do not require cloning and/or sequencing of amplicons for each sample. However, the disadvantage of these methods is that they do not provide clear-cut taxonomic identification of the microbes due to the lack of sequence information.

Another recently developed profiling method involves **16S rRNA gene microarray chips,** called **phylochips,** comprised of thousands of oligonucleotide-containing spots, each corresponding to a 16S rRNA gene from one of the various microbial species (or **"phylotypes"**) present in samples (Figure 5–4). To design appropriate phylochips, the bacterial species

Figure 5–4 Phylochips for microbial-community profiling. **(A)** General procedure for making a phylochip. Fluor, fluorescence. **(B)** Example of how a phylochip could be used to distinguish microbes at the genus and species levels in a sample. Positive and negative controls were included to ensure that the hybridization steps worked and that there was no background detection, respectively. Some spots are probes targeted toward distinguishing specific microbes at the genus level, while other spots are probes targeted toward distinguishing specific microbes at the species or subspecies level. Custom probes (upper left) can be made for identifying species with particular genes present (e.g., virulence factors).

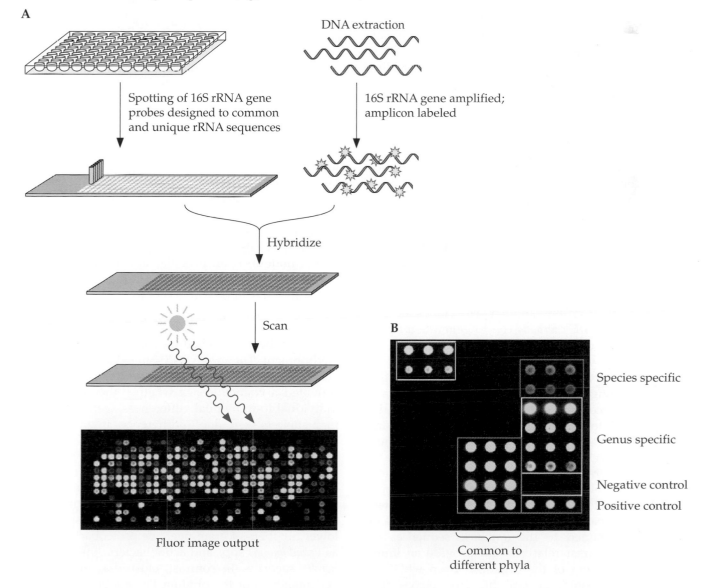

present in environmental niches must first be identified by rRNA gene-sequencing methods from representative samples. Oligonucleotide probes derived from the sequencing data are then attached to the microarray. These phylochips can be used to monitor shifts in microbial-community compositions in environmental and clinical samples. The greatest advantage of this approach is that all microorganisms of interest in an entire community can be detected in a single assay by multiple probes to give reliable taxonomic information.

The current limitation of these phylochips is that they first require the rRNA gene sequence identification of all the microbes expected to be present in the samples. However, the increasing number of powerful high-throughput sequencing facilities available at ever more affordable costs makes the design of phylochips increasingly feasible. As phylochips are developed for various microbial and clinical ecosystems, non-sequence-based profiling methods, such as DGGE and T-RFLP, will undoubtedly be phased out. The breathtaking advances in massively parallel DNA-sequencing methods described below may soon allow cost-effective studies of the changing dynamics of complex microbial communities over time and under different conditions by direct sequencing alone.

SEQUENCE ANALYSIS OF 16S RRNA GENE CLONE LIBRARIES. To complete a census of the species present in a microbial community, sequence analysis of all, or at least the most abundant, 16S rRNA genes present in a sample must be performed. One approach to census taking involves the construction of libraries of 16S rRNA gene clones (Figure 5–5). After PCR has been used to amplify the 16S rRNA genes from the combined genomic DNA of all the microbes present, the amplicons are cloned into plasmid vectors to separate the amplicons from one another. The clones are then transformed into *Escherichia coli* to generate a library of clones with different PCR amplicons. Each *E. coli* cell and colony contains only a single cloned sequence. 16S rRNA genes have the advantage that they are large enough (1,542 nucleotides) to contain adequate sequence information for identification but small enough to be sequenced fairly easily using standard sequencing technology that can now determine at least 800 nucleotides per read. Moreover, there are a large number of sequences from cultivated bacteria that are available in the databases (Box 5–1), although the databases tend to be biased in favor of bacteria that cause disease in humans. Comparing each of the sequences obtained to the database sequences can identify the nearest relatives and provide an immediate identification of the organism from which the sequence originated. Several hundred clones from

each library are sequenced to get a 16S rRNA gene sequence profile of the microbial community present, which can be represented as a phylogenetic tree (Figure 5–6). This approach is currently being used to characterize the contents of complex microbial populations, such as those found in the human colon or vagina.

An example of such a study that may help you to appreciate this process is provided by a study of the microbiota of the baboon vaginal tract. In contrast to the microbiota of the human vaginal tract, no culture-based analysis of this microbiota had ever been done. A recent 16S rRNA gene analysis of the human vaginal tract revealed some differences from the outcomes of cultivation-based studies, but no big surprises, except for the amount of animal-to-animal variation found, which challenges some previous assumptions about what constitutes vaginal health. The cultivation-based studies took many years to complete. The baboon study illustrates that it is now possible to get similar information in a very short time.

Why baboons? Baboons have been widely used as an animal model in studies of the female genital tract. The topics of these studies have ranged from endometriosis to the efficacy of birth control methods. The reason for choosing baboons is that, anatomically and hormonally, the baboon and human uteruses resemble each other. Baboons are also cheaper to house and more accessible to researchers than chimpanzees, another candidate for an animal model.

As part of a larger study of the microbiota of the human vaginal tract, two of the authors of this book, B. A. Wilson and A. A. Salyers, and two University of Illinois graduate students, Angel Rivera and Jeremy Frank, undertook an investigation of the microbiota of the baboon vaginal tract. Through their anthropology colleagues, they had access to vaginal specimens collected from a research colony of baboons housed at the Southwest Regional Primate Research Center in Texas. All the animals had the same diet and the same environment. They were also somewhat inbred due to years of breeding within the colony. This study also provided a chance to ask whether the considerable individual-to-individual differences seen in the human subjects were due to differences in environment and genetics or to other factors.

The 16S rRNA gene analysis of the baboon microbiota yielded a totally unexpected finding: the microbiota of the baboon vaginal tract was quite different from that of the human vaginal tract. The human vaginal tract is dominated by *Lactobacillus* spp., with lower numbers of gram-negative proteobacteria, such as *Pseudomonas* spp., and actinobacteria, such as *Gardnerella vaginalis*. In contrast, clostridia, fusobacteria, and members of the phylum *Bacteroidetes* dominated

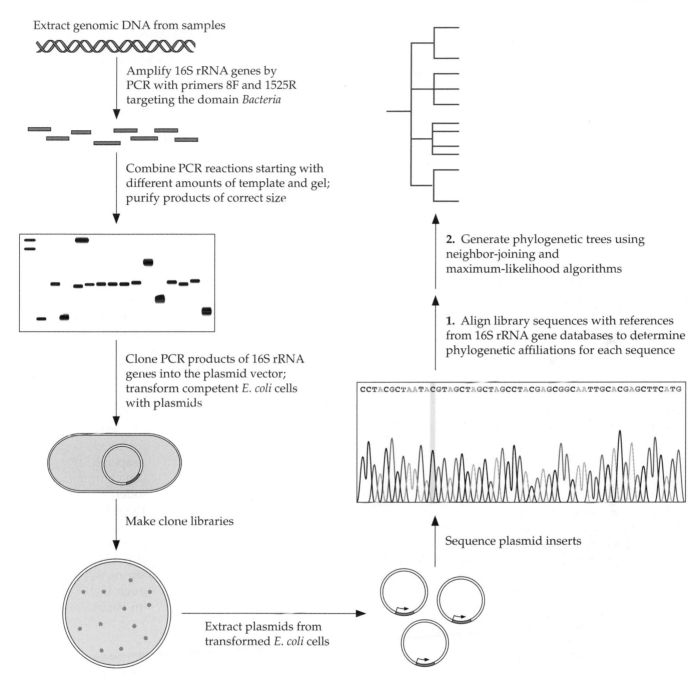

Extract genomic DNA from samples

Amplify 16S rRNA genes by PCR with primers 8F and 1525R targeting the domain *Bacteria*

Combine PCR reactions starting with different amounts of template and gel; purify products of correct size

Clone PCR products of 16S rRNA genes into the plasmid vector; transform competent *E. coli* cells with plasmids

Make clone libraries

Extract plasmids from transformed *E. coli* cells

Sequence plasmid inserts

2. Generate phylogenetic trees using neighbor-joining and maximum-likelihood algorithms

1. Align library sequences with references from 16S rRNA gene databases to determine phylogenetic affiliations for each sequence

CCTACGCTAATACGTAGCTAGCTAGCCTACGAGCGGCAATTGCACGAGCTTCATG

Figure 5–5 Steps for building a 16S rRNA gene clone library to fingerprint a complex microbial community.

the microbiota of the baboon vagina. This difference is illustrated in the clustering analysis shown in Figure 5–7. The difference between the compositions of the human and baboon microbiota is striking, a finding that is surprising in view of the fact that baboons seem to be so closely related to humans at the physiological level. All of the bacterial species found in the baboon vaginal tract have representatives that were isolated from humans, but largely from the human mouth and colon, rather than the vaginal tract. Even

within these groups, however, the human sequences clustered independently from the baboon sequences, indicating that in many cases, the baboon sequences were not closely related to the human sequences and might represent new genera.

This type of analysis has some important limitations. Ideally, enough clones from a particular library of a sample would be sequenced to reach the point where no new clones are found. In practice, given the complexity of the populations found in most parts of

BOX 5–1 Data, Data, Data—What To Do with All That Information?

How does one go about storing and sorting through the massive amounts of sequencing data and information that have been generated over the years? Because of the critical need for researchers to have access to the data and to be able to readily use it, a number of centralized public databases have been formed around the world. These **databases,** most of which are web based and freely available online to the public, consist of libraries of life sciences information, DNA sequencing data, protein structure data, gene expression data, and other computational or scientific data from genomics, proteomics, metabolomics, microarrays, and phylogenetics. An entire new field of **bioinformatics** has emerged that involves the design, development, management, and utilization of these life sciences databases. Databases have become an important tool and resource for scientists studying complex biological systems. Whenever a researcher obtains or publishes a nucleotide sequence or other data in a scientific journal, the researcher is required to deposit that sequence and/or information in one of the databases, and the sequence receives an **accession number,** a tracking number that helps the databases maintain and cross-reference the information.

The largest primary-sequence databases, which form part of the **International Nucleotide Sequence Database** (INSD), are **GenBank** (National Center for Biotechnology Information [**NCBI**]), the U.S. centralized library of various biological data, including nucleotide sequences; **EMBL NDB** (European Molecular Biology Laboratory Nucleotide Database), Europe's library of nucleotide sequence data; **DDBJ** (DNA Data Bank of Japan), Japan's nucleotide database; **UniProtKB** (Universal Protein Resource Knowledgebase), a database that provides protein translations of nucleotide sequences from the nucleotide sequence databases; **Swiss-ProtKB** (Swiss Institute of Bioinformatics), a protein sequence database; and **PDB RCSB** (Protein Data Bank, Research Collaboratory for Structural Bioinformatics), a protein structure model database. There are public genome databases that collect libraries of genome sequences and provide annotation (assigning identification and possible function to the genes), curation (literature citations supporting the annotation), and analysis tools to aid researchers in comparative-genomics studies. For example, the **NMPDR** (National Microbial Pathogen Data Resource) is a curated database of annotated genomic data for a number of bacterial pathogens, and **JGI Genomes** (Department of Energy Joint Genome Institute) is a database for many eukaryotic and microbial genomes. There are also databases that integrate information from multiple databases. For example, **Entrez** is the integrated search and retrieval system used by NCBI for assembling data from major life sciences databases, including literature sources (such as **PubMed**), nucleotide and protein sequences, and protein structure, taxonomy, genome, expression, chemical, and other databases, and making the resulting combined information available to the public through a single platform (http://www.ncbi.nlm.nih. gov/sites/gquery).

One of the greatest challenges with having so much data and information available is that it is difficult for the databases to verify the input data. While some database resources try to maintain oversight, it is often left to the researchers who deposit the data to annotate and curate their data. This is not always a reliable way to ensure that the data are correct, so the end user must also be wary and take care not to use false data. Often what happens is that the researchers deposit large quantities of sequencing data for which no annotation or curation has occurred. This is becoming more prevalent with the large metagenomic sequencing efforts that are currently under way. To deal with this issue, some databases, such as the **Ribosomal Database (RDB)** (http://rdp.cme.msu.edu/index.jsp), which provides online data analysis, alignment, and annotation of bacterial and archaeal small-subunit 16S rRNA gene sequences, have the capability to match sequences from a library of such sequences from known, well-characterized bacteria (so-called type strains, which allows one to link taxonomy with phylogeny) or from the entire collection of sequences regardless of annotation (type and non-type strains). In addition, the RDB provides alignments for sequence comparisons

(continued)

BOX 5–1   Data, Data, Data *(continued)*

and phylogenetic analysis that incorporates information from the conserved secondary structure of 16S rRNAs, which enables improved comparisons of short partial sequences and handles some artifacts that might arise from large-scale sequencing.

The best way to experience the amazing power of bioinformatics is to try it for yourself. Go to the Entrez site or another database and type the name of your favorite protein. You will be amazed at the depth of information that is available about the protein: which species produce it; the phylogenic relationships of the protein in different species; its structure

(often done by different methods and bound to ligands); the possible functions of its domains and how these domains relate to other, related domains in other proteins; how its expression is regulated at the gene and activity levels; where it fits in metabolism or cellular processes; signal transduction pathways that impinge on it; and on and on. The total amount of new biological information may seem daunting, but you can best appreciate the new depth of the current biological revolution by plunging in and looking for yourself. Besides, the structures and relationships are truly beautiful—and it is all free!

---

the body, this ideal state is almost never reached in analyses of complex populations due to the diversity of strains in a population and the cost of sequencing. Thus, a 16S rRNA gene analysis, such as this example, provides at best a representation of the most abundant genera and species in the site. This information, nevertheless, is extremely valuable because it narrows down the number of groups of bacteria in the population and can guide cultivation efforts. For example, the possible presence of anaerobic bacteria means that anaerobic conditions should be included in any attempt to cultivate members of the dominant groups.

A more serious limitation of this approach is the fact that the data from a 16S rRNA gene analysis are only semiquantitative. This is due to the fact that some sequences seem to be amplified and cloned more efficiently than others (often referred to as PCR and cloning bias). Part of the problem is that end point PCR (Figure 5–1) is not strictly quantitative. Nonetheless, the analysis gives an idea of what the leading members of the population are and provides a general assessment of their relative abundances.

Once the members of a microbial community have been identified, the relative representation of different bacterial species can be determined by another method, called **quantitative real-time PCR (qPCR).** In one variation of qPCR (Figure 5–8), genomic DNA from the microbial community is prepared and used directly as the template in PCR amplifications containing primer pairs that anneal specifically to the 16S or 23S rRNA genes of one bacterial species in the population. The course of the PCR is followed by an increase in fluorescence caused by the binding of a dye,

commonly **SYBR green,** to the double-stranded PCR products. The procedure determines the kinetics of the increase in fluorescence intensity after each round of PCR amplification—this is the "real-time" aspect of the method—and relates these kinetics to a parameter called the threshold cycle number (C_T), which is inversely proportional to the starting concentration of template DNA. Computer programs are then used to calculate the relative concentration of each rRNA gene, which is proportional to the relative number of bacteria of each species in the starting microbial community.

MASSIVELY PARALLEL METHODS OF DNA SEQUENCING: THE WAY OF THINGS TO COME. A revolution has occurred within the last few years that will have a lasting impact on profiling bacterial communities and on bacterial genetics in general. Technological advances that allow the determination of millions of base pairs of DNA sequences in single reactions at very reasonable costs have taken place. New bioinformatics methods allow the rapid assembly of these sequences. At this writing, there are several formats and chemistries for this massively parallel sequencing that are marketed by competing companies, such as 454 pyrosequencing (Figure 5–9) and Illumina large-scale sequencing (Figure 5–10). However, each of these examples has certain limitations. Currently, the 454 sequencing method allows longer sequence readings, about 450 contiguous nucleotides, but the chemistry has trouble discerning the lengths of runs of the same base. In contrast, the Illumina chemistry yields robust sequence determinations, even of repeated ba-

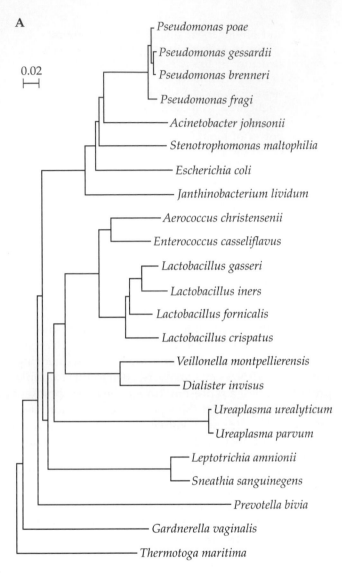

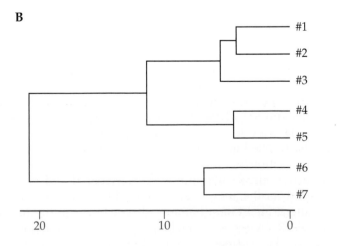

Figure 5–6 Phylogenetic trees showing relationships among microbial communities. The trees shown here were chosen in part to illustrate examples of some of the ways rRNA gene sequence data are displayed. **(A)** Phylogenetic tree displaying the relationships among the microbes found in the vaginal ecosystem of healthy women. The scale bar represents 0.02 nucleotide substitution per site in the 16S rRNA gene sequences. **(B)** Dendrogram showing the phylogenetic relationships of DNA sequence profiles of 16S rRNA gene libraries from seven different samples. The lines denote the phylogenetic distance between each of the samples, which is a measure of the relationship of one sample to the other. For example, samples 4 and 5 are about 10% different from each other (the lines, or branches of the tree, converge at around 5% on the bar index), whereas samples 6 and 7 are about 40% different from all the other samples (1 to 5).

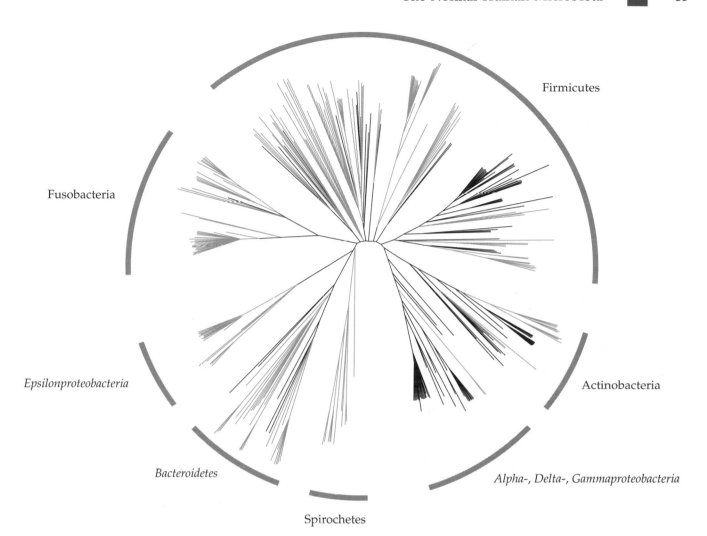

Firmicutes

Fusobacteria

Epsilonproteobacteria

Actinobacteria

Bacteroidetes

Alpha-, Delta-, Gammaproteobacteria

Spirochetes

Figure 5–7 A radial phylogenetic tree showing the affiliation of vaginal microbes to six major taxonomic groups. This radial tree shows relationships based on nearly complete 16S rRNA gene sequences of clones from the vaginal microbial community of the baboons used in the University of Illinois study (blue) and published human sequences (black).

ses, but the sequence lengths (runs) are limited to about 80 contiguous nucleotides. Despite these limitations, these methods are extremely powerful, and their costs are decreasing. Meanwhile, improvements in these and other DNA-sequencing and detection technologies are appearing at an astounding pace.

To understand the power of these methods, let us consider a common, recurring problem in bacterial genetics. Often, interesting mutations arise spontaneously in bacteria whose complete genomes have already been determined. We would really like to know what these mutations are. Classical bacterial genetics provides exceedingly clever ways to map mutations so that they can be located by conventional sequencing of a limited region of the chromosome of the mutant strain. However, these classical methods are often

time-consuming and are far from foolproof, especially in bacterial species lacking powerful genetic systems, such as many bacterial pathogens. In a recent experiment, the Illumina sequencing technology was used to locate point mutations in mutants of *Streptococcus pneumoniae*, whose genome contains about 2.2 million bp. Chromosomal DNA isolated from two mutant strains was sheared into random fragments of about 400 bp. Adaptors required for hybridization during the sequencing method were ligated to the ends of the DNA fragments, and the resulting products were amplified by PCR to give random libraries of genomic fragments from each mutant. However, the adaptors used had slightly different sequences (**"barcodes"**) so that DNA sequences from the two mutants could be distinguished. The two barcoded libraries were mixed

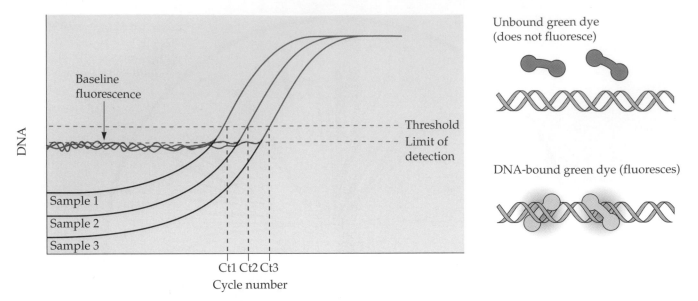

Figure 5–8 qPCR used to quantify specific bacteria in complex samples. C_T, threshold concentration.

and subjected to the Illumina sequencing method (Figure 5–10).

The sequencing run yielded nearly 700 million bp of sequence in 12 million reads, with an average length of 56 bp! The coverage for the two genomes contained in the reaction lane was 174-fold, and comparison with the known genome sequence of *S. pneumoniae* showed that there were no gaps in mutant sequences and located the point mutations that were causing the phenotypes. What was the cost of this analysis? At this writing, the sequence determination and bioinformatics analysis cost about $1,000, which means each mutant sequence was a mere $500. The high level of coverage in this run means that at least two more barcoded fragment libraries could be added to each sequencing lane. The resulting $250 for each mutant genome sequence is comparable to or less than the cost of older sequencing approaches, which yield far less information and require weeks or months compared to less than 1 week for the large-scale sequencing.

Currently, combined 454 and Illumina sequencing methods can be used to determine the complete genome sequence of a bacterial isolate without cloning. The longer reads of the 454 method provide a scaffold and draft sequence that can be rapidly polished by the high-accuracy Illumina sequencing. As these DNA-sequencing methods develop, it will become increasingly quick and cost-effective, not only to take the census of bacterial species in complex microbial communities, but to sequence partial or entire genomes directly from the mixture of DNAs isolated from these populations.

MULTILOCUS SEQUENCE TYPING (MLST). We have considered the problem of identifying different bacterial species in complex mixtures taken from the environment or from sites in the human body, but scientists and epidemiologists are often faced with the problem of distinguishing different isolates of a single bacterial species. Suppose that there is an outbreak of *Staphylococcus aureus* in a hospital and you need to trace how this particular strain got into and around the hospital. Was it carried by a hospital worker or by a family member of a patient? Or suppose there is an outbreak of food-borne disease caused by *Listeria monocytogenes*, which can contaminate food-processing equipment and ready-to-eat meat products. You want to find out how and at what point this strain of *L. monocytogenes* entered the food chain and whether it is similar to strains that caused previous outbreaks. The rRNA genes of all isolates of *S. aureus* have the same sequence, because the rRNA genes change very slowly within a given species. Likewise, the DNA sequences of all *L. monocytogenes* rRNA genes are pretty much identical.

Clearly, rRNA typing will not work to trace infections such as these. However, unlike rRNA genes, the DNA sequences of genes that encode housekeeping enzymes and virulence factors do change with time within a species. That is, isolates of *S. aureus* or *L. monocytogenes* from different locations or sources accumulate slight differences in the sequences in their housekeeping and virulence factor genes over time. This drift arises partly because the genetic code is degenerate. Recall that more than one codon can specify the same amino acid (e.g., there are six specifying leu-

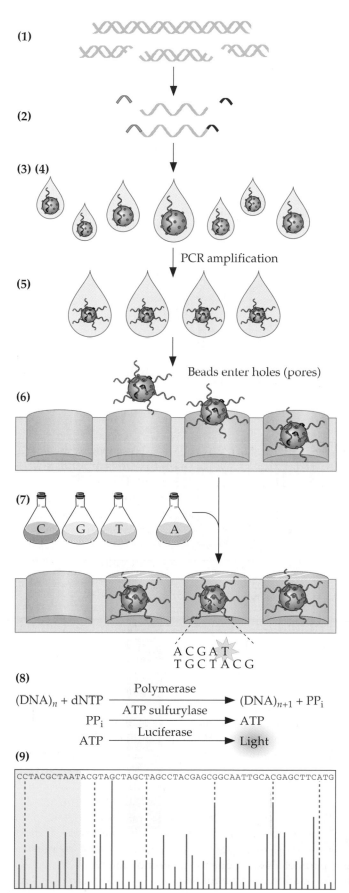

(1)

(2)

(3) (4)

PCR amplification

(5)

Beads enter holes (pores)

(6)

(7)

C   G   T   A

A C G A T
T G C T A C G

(8)

$$(DNA)_n + dNTP \xrightarrow{\text{Polymerase}} (DNA)_{n+1} + PP_i$$

$$PP_i \xrightarrow{\text{ATP sulfurylase}} ATP$$

$$ATP \xrightarrow{\text{Luciferase}} Light$$

(9)

CCTACGCTAATACGTAGCTAGCTAGCCTACGAGCGGCAATTGCACGAGCTTCATG

Figure 5–9 454 Pyrosequencing technology. Shown are the steps taken in sequencing. (1) Genomic DNA is sheared to create fragments. (2) The fragments are denatured, and linkers (one with a biotin tag) are added to the ends of the DNA. (3) One strand of the DNA fragment is bound using the biotin-containing linker to one bead containing an attached streptavidin tag. (4) The beads are emulsified in an oil-water mixture so that one bead with one DNA fragment is contained in one oil droplet. (5) The DNA is amplified by PCR (each oil droplet serves as a microreactor). (6) One bead is placed in each well of a picotiter plate (a fiber optic chip). (7) The DNA is pyrosequenced by reagents flowing across the plate. (8) Each time a nucleotide is added to a complementary nucleotide on the template, a light signal is generated by reaction of the released pyrophosphate with ATP sulfurylase, followed by luciferase. (9) Output readings from the intensity of the light detected are generated on a pyrogram. dNTP, deoxynucleotide triphosphate; PP_i, inorganic pyrophosphate. (Adapted from copyrighted diagrams from Roche with permission.)

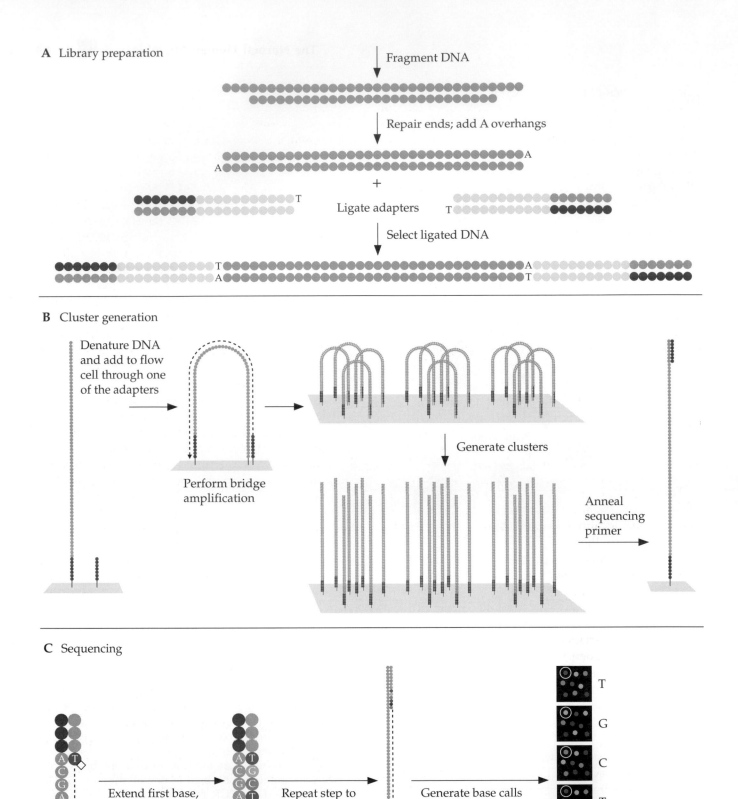

A Library preparation

Fragment DNA

Repair ends; add A overhangs

+

Ligate adapters

Select ligated DNA

B Cluster generation

Denature DNA
and add to flow
cell through one
of the adapters

Perform bridge
amplification

Generate clusters

Anneal
sequencing
primer

C Sequencing

Extend first base,
read, and deblock

Repeat step to
extend strand

Generate base calls
for each spot during
each round of
base extension

T
G
C
T
A
C

Figure 5–10 Illumina method of massively parallel DNA sequence determination. (Adapted from copyrighted publications from Illumina with permission.)

cine), and much of this redundancy occurs at the third positions of codons. Therefore, the DNA sequences of housekeeping genes can show variations in different isolates of a given bacterial species and still specify the same amino acid in the enzyme product. The drift in virulence factor genes often involves changes in amino acids compared to those in housekeeping genes, because the virulence factor genes are subjected to strong selective pressures during infection.

This variation within coding sequences is the basis of MLST, but rather than relying on genetic drift in just one housekeeping or virulence factor gene, the DNA sequences of regions in multiple (usually seven or more) genes are analyzed for each isolate. MLST analysis is easy and has now become cost-effective. Sequencing has been used to identify regions of seven or more housekeeping or virulence factor genes that show variations in a given bacterial species. Pairs of PCR primers are designed to amplify and sequence about 500 bp from each of these variable regions. The sequences of these multiple loci can then be compared among samples or with other isolates in databases to trace the relatedness of the different isolates locally and worldwide. This analysis can also take into account combinations of alleles in bacterial species that exchange genetic material frequently. Going back to our examples, samples collected from staff, visitors, patients, and locations in the hospital would be cultured to identify those that contain *S. aureus*, which is a common commensal bacterium that is easily identified on growth media. The different isolates of *S. aureus* are subjected to MLST. Progeny isolates have the same DNA sequences in most of the multiple loci, whereas strains from a different source may have loci with sequence variations. The resulting profile indicates whether patients are infected with the same strain of *S. aureus* and where this strain may have arisen in the hospital. Similarly, MLST can trace the sources of *L. monocytogenes* from current and previous outbreaks.

Metagenomic analysis to determine the physiological potential of the population. The rRNA gene-sequencing approach gives information about what types of microbes are present in a community. A limitation of this approach is its failure to generate functional genomic information for deciphering the metabolic contributions of microbes to an ecosystem. For example, which microbes use the various substrates found in the site or produce end products that might affect the host? The emerging technology for sequencing large numbers of cloned DNA segments has increased interest in a more ambitious approach to analyzing the potential activities of the

microbial populations of the body (the **microbiomes**). This approach, called **metagenomic analysis,** is designed to go beyond the question of what organisms are present, i.e., the census question, to the question of what is the metabolic potential of the microbiota, i.e., what metabolic genes are also present?

Metagenomic analysis starts with the census information indicating which species are present in bacterial populations. Hundreds of complete bacterial genomes are already in databases, and thousands more are in the works. These resources can provide a huge amount of information about the metabolic potential of the bacterial population, but what about species whose genomes are incomplete or not in the database at all? To answer this question, the next stage of metagenomic analysis involves isolation of DNA from a mixed microbial population, which is sheared or cut with restriction enzymes into small segments that are then cloned and sequenced. Clearly, this means sequencing many more DNA segments than is necessary in the case of the 16S rRNA gene census approach. Indeed, the number of clones needed to represent the entire **metagenome** so as to harvest the remarkable and vast diversity present in a microbial community is staggering. This means it is imperative to have the ability to sequence massive numbers of clones in a reasonable amount of time and for a reasonable cost. This is called **high-throughput or shotgun sequencing** and is being greatly accelerated by robotics, massively parallel sequencing approaches, and bioinformatics assembly methods. In fact, the advances in massively parallel sequencing methods, such as 454 and Illumina sequencing, are leading to attempts to determine complete genomes of microbes directly, without cloning, from the mixed DNA samples isolated from complex microbiota.

Individual genome sequences from many thousands of sequences is challenging for existing bioinformatics programs, but here again, rapid advances are being made in the analysis of the huge volume of new sequence data that is emerging, and these advances may well make what is almost unimaginable today feasible tomorrow. In biotechnology, as in many modern scientific endeavors, "impossible" just means "not possible *yet*." Interpretation of a metagenomic analysis is much more complex than the 16S rRNA analysis, and the details are beyond the scope of this text, but for the adventurous, examples of recent metagenomic analyses of the human colonic microbiota are provided in two of the suggested readings (Gill et al., 2006, and Turnbaugh et al., 2009). What these analyses allow us to do is have a glance at the types of biosynthetic and metabolic pathways that the microbes in any given population might have at their disposal to utilize. This sequence information, combined with census information indicating which species are present and the thousands of reference bacterial genomic sequences already in databases, is making metagenomics possible.

Beyond the Metagenome

A limitation of both 16S rRNA gene and metagenomic-analysis approaches is that they do not provide information about which genes are being expressed. As conditions, such as diet or antibiotic use, change, the genes that are expressed can change. Thus, only a subset of genes is likely to be expressed at any one time. Moreover, even within the same site, changes in conditions, such as changes in diet or hormonal level, may cause increases or decreases in expression of certain sets of genes. Work is now under way to measure gene expression in complex populations using techniques that detect and quantitate mRNA levels, including microarrays and qPCR.

Even this advance may not be enough. Two genes in a microbial population may encode the same type of enzyme or transport protein, but different enzymes and different transport proteins have different affinities for their substrates and different levels of activity. Thus, a gene that is expressed at a higher level than another gene may not be more important metabolically. Moving to the level of physiology in complex environments, such as within hosts, is a challenge that few have dared to think about—yet.

Overview of the Human Microbiota

A human fetus is devoid of microorganisms. Passage through the vaginal tract begins the colonization process, but this process continues as the infant grows, and the final core microbiota is not achieved until the child is about 2 or 3 years old. Once the microbiota of an area assumes its more mature form, different areas of the body harbor very different microbial populations. Even within a single site, such as the mouth, different parts may contain different sets of microbes. This diversity is not surprising in view of differences between conditions that microbes encounter in a particular site.

There are, however, some features that are common to all sites of the human body that are colonized by microbes. First, the numerically predominant microbes are bacteria. Archaea and fungi are often found but are present in much lower numbers. Second, the majority of the bacteria are gram-positive bacteria. In fact, some of the population shifts that are associated with such conditions as periodontal disease and bacterial vaginosis appear to involve a change from a predominantly gram-positive to a predominantly gram-negative population. Why the gram-positive bacteria dominate the human microbiota is not clear. Finally, although the microbiota of different areas of the body are usually protective, some members of the microbiota can cause serious infections if they manage to enter the normally sterile areas of the body, such as blood and tissue.

Skin Microbiota

The surface of the skin is a dry, slightly acidic, aerobic environment. Staphylococci, such as *Staphylococcus epidermidis,* are the predominant bacteria occupying this site, although it is not uncommon to have transient colonization of the skin by soil bacteria or bacteria from other parts of the body. Although the surface of the skin is aerobic, pores and hair follicles can be anoxic enough to support the growth of anaerobic bacteria. A commonly isolated skin anaerobe is *Propionibacterium acnes.* The name of this organism reflects the belief by some scientists that *P. acnes* has a role in acne. This notion is controversial, because it is not clear whether the bacteria play a role in initiating acne or simply colonize the acne lesions after they form. Antibiotic therapy is widely used to treat acne patients. The success of this therapy could be used as an argument to support the hypothesis that bacteria cause or exacerbate acne, but the most commonly used antibiotics, such as tetracycline, also have anti-inflammatory activity, which might also be responsible for their effectiveness.

Skin bacteria like *S. epidermidis* were long assumed to be unable to cause disease. Finding them in a blood specimen, for example, was considered to be proof that a careless health care worker or technician had contaminated the specimen. Today's view is completely different. *S. epidermidis* in particular is now accepted as a serious cause of hospital-acquired infections. Worse still, *S. epidermidis* is increasingly resistant to many different antibiotics. The fact that people taking antibiotics exude the antibiotics in their sweat may explain this. Whether the use of antibacterial soaps contributes to resistance is under investigation. *S. epidermidis* infections are most likely to occur in patients who have indwelling venous catheters or who have surgical wounds that become contaminated either during or after an operation.

Oropharyngeal Microbiota

The human nose is the most common source of *S. aureus,* a major cause of hospital-acquired and community-acquired infections. At any particular time, about one-third of the population harbors *S. aureus*. *S. aureus* is most abundant in the upper part of the nose. The nasopharynx is also the home to *S. pneumoniae*, another human commensal bacterium that can cause a variety of serious invasive diseases elsewhere in the body. About 25% and 40% of healthy adults and children, respectively, carry *S. pneumoniae* at any given time. Several large-scale efforts are currently under way to understand the composition of the microbiota and the dynamics of colonization of the nose.

The microbiota of the mouth and throat is fairly well characterized, largely due to the involvement of the oral microbiota in periodontal disease, a major cause of gum disease and tooth loss in adults. The microbiota of the healthy mouth consists largely of facultative gram-positive bacteria, mainly streptococci, such as *Streptococcus mutans* and *Streptococcus salivarius*. These bacteria ferment sucrose to lactic acid, which in turn contributes to the development of dental caries. Utilization of sucrose also results in production of the polysaccharide dextran, which binds bacteria together and allows plaque to form.

In periodontal disease, this gram-positive microbiota shifts to a gram-negative anaerobic microbiota in the area where the tooth enters the gum. The space between the gums and the lower portion of the tooth surface is called the periodontal pocket. It is a fairly anoxic area and so is able to support the growth of obligate anaerobes, such as *Porphyromonas gingivalis* and *Prevotella* spp. These species produce proteases and other tissue-degrading enzymes, and this may be a major cause of the inflammation that characterizes the disease. More recently, some scientists have suggested that bacteria involved in periodontal disease, such as the gram-negative anaerobe *Fusobacterium nucleatum*, are responsible for preterm birth. The hypothesis is that the bacteria enter the bloodstream through the inflamed gum tissue and lodge in the placenta. The resulting inflammation causes the fetus to be delivered prematurely. Similarly, others have suggested that oral bacteria associated with gingivitis might enter the bloodstream, causing inflammation in the blood vessels and leading to heart disease. This hypothesis is quite controversial, but it illustrates the new thinking about connections between alterations in the normal microbiota and diseases in other areas of the body.

An earlier hypothetical connection between oral bacteria and heart disease posited that dental surgery or other manipulations that might introduce bacteria into the bloodstream would place people with abnormalities of the heart valves, such as those due to rheumatic heart disease or valve implant surgery, at risk for bacterial colonization of and damage to these aberrant valves. Concern about this possibility caused some dentists to administer preventive antibiotic therapy to patients with such risk factors. Whether this prophylactic therapy is actually effective in preventing endocarditis (inflammation of heart tissue) is still uncertain.

Microbiota of the Small Intestine and Colon

The small intestine is characterized by the fast flow of contents. The fast flow helps to wash bacteria out of the site. Throughout most of the small intestine, bacteria have to adhere to the mucosa in order to stay in the site. The fast flow of contents is probably designed to keep down the number of bacteria, which would compete with human intestinal cells for easily digested nutrients. The microbiota of the small intestine is poorly characterized, largely because it is difficult to obtain samples from that area. The samples that are easiest to obtain are those taken from a swallowed tube that works its way into the small intestine. Although such tubes have been used, from a microbiological perspective, they have a significant deficiency: they sample the lumenal, but not the adherent, microbiota. In mice, there is an adherent microbiota that consists of clostridia. Whether such a microbiota exists in humans is controversial.

The colon, in contrast, is characterized by a much slower flow of contents, and so, much higher concentrations of bacteria are found there (Figure 5–11). Bacteria make up a third of the contents of the human colon. These microbes have a complex relationship with us. We provide them with undigested food and fermentable substances, like plant polysaccharides and mucins, whereas they contribute substances to our nutrition. The human small intestine can absorb small molecules, such as mono- and disaccharides, and intestinal enzymes can digest soluble starch, but most of the polysaccharides in the human diet, such as cellulose, xylan, and less soluble starch, pass through to the colon. Colonic bacteria ferment carbohydrates, including these polysaccharides, to produce short-chain fatty acids that are absorbed from the colon and used as carbon and energy sources by colonic cells. Colonic bacteria can also ferment host-produced polysaccharides, such as mucopolysaccharides and mucins.

The colon is an anoxic environment, so it is not surprising that the numerically predominant colonic bacteria are obligate anaerobes. Facultative bacteria, such as *E. coli* and *Enterococcus* spp., are present in much lower numbers. The numerically predominant anaerobes include gram-negative *Bacteroides* spp. and a large number of poorly characterized gram-positive

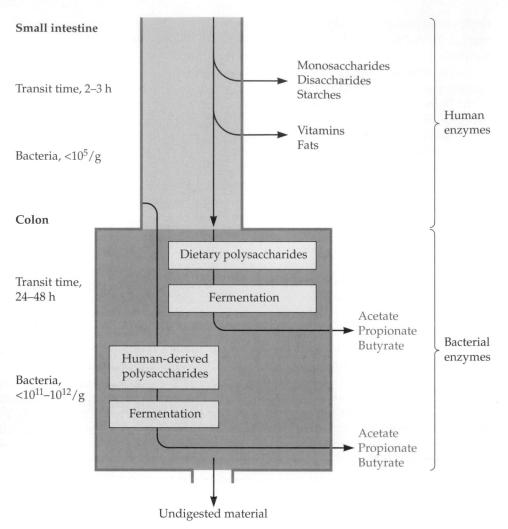

Figure 5–11 Schematic view of activities of the colonic microflora. In the small intestine, concentrations of bacteria are low due to the fast flow of the contents, and human intestinal enzymes mediate most of the digestion. In the colon, the concentrations of bacteria are so high that bacteria account for about 30% of the volume or contents. In this site, polysaccharides from the human diet (plant polysaccharides or dietary fiber) and host-derived polysaccharides (mucins and mucopolysaccharides) are fermented by the bacteria, and the short-chain fatty acids that are produced are absorbed by the human body and used as sources of carbon and energy. (Adapted with permission from A. A. Salyers and D. D. Whitt. 2001. *Microbiology: Diversity, Disease, and the Environment.* John Wiley and Sons, Inc., Hoboken, NJ.)

anaerobes. These anaerobes play a major role in fermenting the dietary polysaccharides that our bodies cannot digest. The products of fermentation include CO_2, H_2, and the short-chain fatty acids (acetate, propionate, and butyrate). The short-chain fatty acids are absorbed by intestinal cells and used by them as a source of carbon and energy.

Acetate, CO_2, and H_2 are also used as carbon and energy sources by minor populations, such as methanogenic archaea. The methane not used by the methanogens is absorbed, though not used, by the human body and is expelled in the breath. About one-fifth of people tested have enough methane produced in their colons to be easily detectable in their breath by gas-liquid chromatography. Methane is also expelled in flatus. Sulfide-reducing bacteria also reside in the colon. Sulfides produced by the sulfate reducers are responsible in part for the odor of feces. Recent 16S rRNA gene analyses of the microbiota of the colons of obese humans or mice compared to those of non-obese individuals, have led to the suggestion that obesity might be caused in part by the composition of the colonic microbiota (Box 5–2). Shifts in the microbiota may also contribute to inflammatory bowel disease.

In judging the energy balance, it is worth realizing that intestinal microbes also take an energy toll from us; they stimulate the immune system and the turnover of intestinal mucosal cells. The constant sloughing of intestinal mucosal cells is a very effective defense that prevents bacteria that have attached to the mucosal cells from staying in the site long enough to invade. Similarly, the intestinal immune system is an important defense. However, these activities require an output of proteins and energy by the human body. On the whole, however, the energy balance seems to go in our favor.

Many colon bacteria, such as *Bacteroides* spp. and numerically minor populations, like *E. coli* and *Enterococcus* spp., are capable of causing serious infections if they escape from the colon as a result of surgery or

BOX 5–2   We Are What We Eat, or Rather, What Our Microbiota Eats

Conventional wisdom has it that obesity is a result of genetics, lack of exercise, or a poor diet. But what if your intestinal bacterial population also makes a contribution? A study that used 16S rRNA analysis found that the microbiota of obese mice and humans differed from that of lean mice and humans. Moreover, when germ-free mice (mice lacking any intestinal bacteria) were colonized with an "obese microbiota," they gained more fat than germ-free mice colonized with a "lean microbiota." The main difference was the ratio of the two numerically predominant groups, the *Bacteroidetes* (*Bacteroides* spp.) and the *Firmicutes* (gram-positive obligate anaerobes). A higher proportion of *Bacteroidetes* was associated with leanness.

The hypothesis that the composition of the colonic microbiota is associated with obesity is, as you might imagine, quite controversial, especially among those committed to theories that give precedence to exercise or diet. Also, the association of obesity with more active colonic fermentation seems to run counter to the belief that high-fiber diets are associated with increased colonic fermentation due to the fact that fiber is primarily composed of polysaccharides that are fermentable by colonic bacteria. The efficiency of the fermentation may be a factor. If so, the prediction from the obesity studies would be that the *Firmicutes* are more efficient fermenters than the *Bacteroidetes*. Since virtually nothing is known about the gram-positive anaerobes and their carbon sources, this is difficult to assess. Another possibility is that some fermenters take a lower energy toll in the form of stimulating mucosal-cell turnover.

A good feature of the hypothesis regarding a connection between obesity and the microbiota composition is that it may prompt more studies of the metabolic activities of gram-positive anaerobes. Moreover, it illustrates the fact that the 16S rRNA gene approach, and even the metagenomics approach, may be a good start for addressing these questions but that work on better understanding bacterial physiology will be critical.

Source: P. J. Turnbaugh, R. E. Ley, M. A. Mahowald, V. Magrini, E. R. Mardis, and J. I. Gordon. 2006. An obesity-associated gut microbiota with increased capacity for energy harvest. *Nature* **444**:1027–1031.

---

some other trauma and get into the bloodstream and tissues. How could an obligate anaerobe like *Bacteroides* cause infection in the human body, which would seem to be a highly aerobic environment? *Bacteroides* prefers to lodge in regions of prior tissue damage. Disruption of the blood supply to such areas causes them to rapidly become anoxic and, as such, fertile ground for an anaerobic infection. Moreover, blood itself is actually a hypoxic environment that is low in free oxygen.

Bacteria in the intestine interact with each other metabolically in the sense that methanogens and sulfate reducers use the end products of the polysaccharide fermenters, but they also interact with each other genetically by exchanging DNA. There is an old idea called the reservoir hypothesis that frames this interaction in terms of antibiotic resistance gene transfer (more on this in chapter 7). Briefly, colonic bacteria exchange DNA with each other. They may also exchange DNA with swallowed bacteria that are present only transiently in the colon as they pass through and are expelled into the environment. In this view, colonic bacteria act as reservoirs of resistance genes in the sense that they are present in high numbers in an area in which other bacteria are transiently present for 24 to 48 h, which is more than enough time for DNA transfers to occur. Only recently has it been possible to test this hypothesis by using molecular methods to follow the movement of resistance genes in the human colon.

The type of exchanges envisioned by the reservoir hypothesis would need to have very broad host ranges. That is, the transfers would have to occur, not only within a species, but also between members of different species or even different genera. The type of gene transfer most likely to mediate broad-host-range transfers of DNA is the direct cell-to-cell transfer of DNA through conjugation. In fact, evidence is mounting that transfers of antibiotic resistance carried on conjugative elements, like plasmids and conjugative transposons, occur frequently in the colon. These transfers occur between different species and genera, including between colonic bacteria and bacteria from different sites, as hypothesized by the reservoir hypothesis.

Why is the reservoir hypothesis suddenly becoming relevant, even though most of the bacteria coexist with us most of the time without causing any problems? The reason is that, as already mentioned, a subset of these bacteria can cause infections. For example, swallowed bacteria include *S. pneumoniae,* the main cause of bacterial pneumonia and a major cause of infectious-disease deaths in developed, as well as developing, countries. Moreover, normal colonic inhabitants, such as *Bacteroides* spp., *E. coli,* and *Enterococcus* spp., are notorious causes of potentially lethal postsurgical infections. Increasingly, these bacteria are becoming resistant to many antibiotics.

Most recently, the reservoir hypothesis loomed large in the debate over possible adverse consequences of the use of antibiotics on the farm. The concern is that antibiotic-resistant bacteria, which arise on the farm due to selection by the use of antibiotics as feed additives or to prevent infections in crowded populations of animals, are moving through the food supply and into the human intestinal tract, where the resistance genes could be transferred to bacteria permanently or temporarily resident in the human colon. How important this process is for the development of resistance in bacteria that are serious causes of human infections has not been conclusively established, but it remains a concern.

Microbiota of the Vaginal Tract

The microbiota of the female vaginal tract has already been introduced in the example given of the application of DNA-based analysis of complex microbial populations, but that description did not explore the special features of the site that presumably explain the composition of the microbiota of the vagina. The vagina is a complex site. Although there are secretions that are constantly bathing the vaginal mucosa, there is not the flow of fluids seen in the intestinal tract, except during menstruation. Thus, most of the time, the bacteria are loosely or strongly associated with the vaginal mucosa. The vaginal tract also experiences hormonal changes associated with the menstrual cycle, not to mention periodic influxes of menstrual blood and fluid exchanges during sexual intercourse.

The traditional view of the vaginal tract is that the microbiota of the healthy vagina consists mainly of *Lactobacillus* spp. These lactobacilli are fermentative bacteria whose main product is lactic acid, which probably contributes to making the normal pH of the vaginal tract less than 5. This low pH was originally

envisioned as a powerful protective barrier against colonization by disease-causing bacteria. Unfortunately for this hypothesis, many pathogenic bacteria can survive and multiply at pH 5. Another possible contribution of lactobacilli to protecting the vaginal tract against disease-causing bacteria is the fact that some of them produce hydrogen peroxide, which is toxic to many microbes, not just bacteria. There is no question that women who take antibiotics that kill or inhibit the growth of lactobacilli often develop yeast infections, but it is not clear what other bacteria are also affected. The availability of new molecular techniques has now allowed a new in-depth analysis of the vaginal microbiota.

The traditional cultivation-based view of the vaginal microbiota is that the healthy vaginal tract is dominated by lactobacilli. A disease called bacterial vaginosis gave rise to a view of the unhealthy vaginal tract as one lacking lactobacilli. **Bacterial vaginosis** was once considered to be a minor disease, which was characterized by a mild inflammation and a fishy odor. More recently, it has been suggested that bacterial vaginosis may be one possible cause of premature birth, although the mechanism of such a connection is still unclear. In contrast to diseases caused by a single pathogen, such as *Neisseria gonorrhoeae* or *Chlamydia trachomatis,* bacterial vaginosis appears to arise from a shift in the microbiota and thus has been used to develop a view of an unhealthy vagina.

Bacterial vaginosis is characterized microbiologically by the switch from a predominantly gram-positive lactobacillus population to a predominantly gram-negative one, principally *Gardnerella vaginalis.* In fact, a woman with a significant concentration of gram-negative bacteria, like *G. vaginalis,* has been assumed by physicians to be exhibiting disease. DNA-based analyses of a number of apparently healthy women have revealed a surprising finding: many of them carried high concentrations of *G. vaginalis* and did not have a predominantly gram-positive population. Also, interestingly, while about 70% of the healthy women had predominantly lactobacilli, about 30% had very few or no lactobacilli. As more detailed analyses are done, not just of person-to-person variation but of variations with hormonal status, different sites in the vaginal tract, and age, it seems likely that the microbiota of the vaginal tract will emerge as a population as complex as that of better-studied microbiota, such as those of the mouth and colon.

The Forgotten Eukaryotes

The content of this chapter has so far focused on bacteria. Therefore, it is appropriate to end the chapter with a brief description of a group of microbes that has been routinely ignored in most studies but which nevertheless has an impact on the bacterial communities and our immune system—the eukaryotic microbes. In developing countries, there is a significant eukaryotic component of the colonic microbiota that consists of fungi, protozoa, and helminths (such as tapeworms). Their numbers are much lower than those of bacteria, but they may affect human health. It now seems likely that even in developed countries there is a eukaryotic component of the colonic microbiota. The vaginal microbiota has long been known to harbor yeasts.

Many people who carry these eukaryotic microbes are not sick. In fact, the eukaryotic component of the microbiota has been a fact of human life for millions of years. Only during the last two centuries, and only in certain parts of the world, has the eukaryotic component of the microbiota been severely reduced due to clean water, better hygiene, and a high-quality food supply.

SELECTED READINGS

Bäckhed, F., R. E. Ley, J. L. Sonnenburg, D. A. Peterson, and J. I. Gordon. 2005. Host-bacterial mutualism in the human intestine. *Science* **307:**1915–1920.

Chen, Y., W. Zhang, and S. J. Knabel. 2005. Multi-virulence-locus sequence typing clarifies epidemiology of recent listeriosis outbreaks in the United States. *J. Clin. Microbiol.* **43:**5291–5294.

Coolen, M. J. L., E. Post, C. C. Davis, and L. J. Forney. 2005. Characterization of microbial communities found in the human vagina by analysis of terminal restriction fragment length polymorphisms of 16S rRNA genes. *Appl. Environ. Microbiol.* **71:**8729–8737.

Gill, S. R., M. Pop, R. T. DeBoy, P. B. Eckburg, P. J. Turnbaugh, B. S. Samuel, J. I. Gordon, D. A. Relman, C. M. Fraser-Liggett, and K. E. Nelson. 2006. Metagenomic analysis of the human distal gut microbiome. *Science* **312:**1355–1359.

Hyman, R. W., M. Fukushima, L. Diamond, J. Kumm, L. C. Giudice, and R. W. Davis. 2005. Microbes on the human vaginal epithelium. *Proc. Natl. Acad. Sci. USA* **102:**7952–7957.

Ley, R. E., D. A. Peterson, and J. I. Gordon. 2006. Ecological and evolutionary forces shaping microbial diversity in the human intestine. *Cell* **124:**837–848.

McCracken, V. J., J. M. Simpson, R. I. Mackie, and H. R. Gaskins. 2001. Molecular ecological analysis of dietary and antibiotic-induced alterations of the mouse intestinal microbiota. *J. Nutr.* **131:**1862–1870.

Palmer, C., E. M. Bik, M. B. Eisen, P. B. Eckburg, T. R. Sana, P. K. Wolber, D. A. Relman, and P. O. Brown. 2006. Rapid quantitative profiling of complex microbial populations. *Nucleic Acids Res.* **34:**e5.

Rawls, J. F., M. A. Mahowald, R. E. Ley, and J. I. Gordon. 2006. Reciprocal gut microbiota transplants from zebrafish and mice to germfree recipients reveal host habitat selection. *Cell* **127:**423–433.

Reid, G., S. O. Kim, and G. A. Kohler. 2006. Selecting, testing and understanding probiotic microorganisms. *FEMS Immunol. Med. Microbiol.* **46:**149–157.

Salyers, A. A., A. Gupta, and Y. Wang. 2004. Human intestinal bacteria as reservoirs for antibiotic resistance genes. *Trends Microbiol.* **12:**412–416.

Samuel, B. S., and J. I. Gordon. 2006. A humanized gnotobiotic mouse model of host-archeal-bacterial mutualism. *Proc. Natl. Acad. Sci. USA* **103:**10011–10016.

Turnbaugh, P. J., V. K. Ridaura, J. J. Faith, F. E. Rey, R. Knight, and J. I. Gordon. 2009. The effect of diet on the human gut microbiome: a metagenomic analysis in humanized gnotobiotic mice. *Sci. Transl. Med.* **1:**6–14.

Urwin, R., and M. C. J. Maiden. 2003. Multi-locus sequence typing: a tool for global epidemiology. *Trends Microbiol.* **11:**479–487.

QUESTIONS

1. How does a DNA-based analysis, such as the 16S rRNA gene analysis, differ from a cultivation-based analysis? In this chapter, emphasis has been given to the DNA-based approaches. What are some advantages of the cultivation-based approach?

2. More and more research groups are seeking to show that changes in the microbiota of a particular site are involved in diseases, such as periodontal disease, inflammatory bowel disease, and premature birth. Critics object that showing an association is not the same as demonstrating cause and effect. In the case of the obesity study, scientists tried to do this by inoculating germ-free mice with different variations of the microbiota. Clearly, in humans, this would not be possible. How might you prove cause and effect in humans?

3. Infants in the first years of life are often more susceptible to certain bacterial infections than older children. How can you explain this? What function of the microbiota does this illustrate?

4. Members of the microbiota cause some quite serious diseases. How could a bacterium that normally lives in a beneficial or neutral association with its human host cause serious disease?

5. Metabolic interactions between members of the microbiota are attracting more attention because two microbes working together can make a reaction catalyzed by one of them more effective. Consider an association between a polysaccharide-fermenting microbe and a methanogen in the colon. Consider also that the overall energy of a reaction depends on the ratio of substrate to end products for a bacterium like a polysaccharide fermenter. Can you explain why a polysaccharide fermenter and a methanogen might team up in the colon?

6. The assertion is made in this chapter that scientists now believe that transfer of DNA by conjugation in the colon is occurring across species and genus lines. Suppose you found the same type of antibiotic resistance gene in members of two different genera. What criteria might you use to show that the gene was transferred horizontally? Why might you suspect that the gene was transferred by conjugation?

7. Conventional wisdom asserts that there are no methanogens in the vaginal tract. If they were present, they would probably be present at low levels. How would you use the 16S rRNA approach to find them? What modification of the approach used to find bacterial sequences would you have to make?

8. PCR combined with sequencing can provide a quick identification of bacteria. What are the limitations of this approach?

9. Resident microbiota provide protection from colonization by some pathogenic bacteria in certain parts of the body. Describe regions of the body where normal microbiota are protective and how they accomplish this protection.

10. Why is the decreasing cost of DNA sequencing making methods like DGGE and T-RFLP less and less popular?

11. How do 454 and Illumina sequencing differ from the earlier cloning and sequencing of amplified rRNA genes? What is the limitation of all of these methods? What type of information does metagenomic analysis give you that sequencing of 16S rRNA genes does not?

12. Why are scientists adding transcriptional analysis to metagenomic analysis?

SOLVING PROBLEMS IN BACTERIAL PATHOGENESIS

1. Dental plaque is a biofilm consisting of a complex community of over 700 different bacterial species. Epidemiological evidence suggests that a population shift toward certain gram-negative anaerobes is responsible for the initiation and progression of periodontal diseases. *Tannerella forsythia,* a gram-negative, filamentous, nonmotile, anaerobic bacterium, is also considered one of the pathogens implicated in contributing to advanced forms of periodontal disease in humans and is strongly associated with cases of severe periodontitis. It is found coaggregated in periodontal pockets with other putative periodontal pathogens, such as *P. gingivalis* and *F. nucleatum.* Infection with *T. forsythia* induces alveolar bone resorption in a mouse infection model, in which the bacterium is inoculated under the gums of mice, followed by measurement for loss of dental bone.

    A. Considering all of the above information, which of Koch's postulates as defined in Table 6–1, if any, have been satisfied so far for *T. forsythia* involvement in periodontal disease? Be sure to state your rationale. Provide at least two additional modern molecular experiments (different from those already described above) that could be performed to help satisfy Koch's postulates for the involvement of each bacterial species in periodontal disease.

    B. Considering that periodontal disease might be a community shift-type disease with multiple microbial participants, how might you use microbial-community-profiling methods to demonstrate the importance of the microbial-community composition in contributing to the onset and maintenance of the diseased state? Set up the experiment first without using DNA-sequencing approaches and then using DNA-sequencing approaches. Be sure to provide your rationale for the choice of method. From your results, how could you distinguish between a model of disease caused by a microbial-community shift involving multiple microbes and one involving a single pathogen, such as *T. forsythia* or *P. gingivalis,* or a combination of both *T. forsythia* and *P. gingivalis?*

2. You are a researcher working for the U.S. Department of Agriculture. After 2 years of effort, you have isolated in pure culture a new, highly virulent bacterium from duck feces that is responsible for several major outbreaks of deaths in mammalian wildlife from contaminated pond water in the South. Based on 16S rRNA sequence comparisons, you have determined that this new bacterium is distantly related to the gram-negative bacterium *Vibrio cholerae,* and you have named the new strain *Vibrio birdsii.* You find that ducks are apparently unaffected by *V. birdsii.* You suspect that *V. birdsii* may be part of the normal microbiota of ducks. To test this hypothesis, you set up an experiment to examine the host response to *V. birdsii* in germ-free ducks. The results are summarized in the graph shown. Provide a detailed explanation and interpretation of the results. Do the results support the hypothesis? Provide a rationale for your answer.

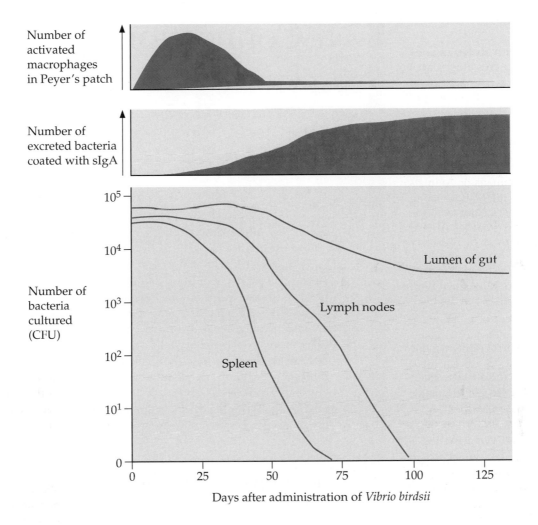

6

Microbes and Disease: Establishing a Connection

One of the most exciting areas of modern infectious disease research was kicked off by ulcer research. For decades, physicians believed that ulcers were a chronic condition caused by excess acid production. Thus, ulcers could not be cured. It was only possible to treat the symptoms. The discovery that most gastric ulcers are caused by bacteria led at long last to a cure for ulcers. Now scientists are enthusiastically exploring the possibility that other chronic conditions, such as inflammatory bowel disease, cardiovascular disease, and atherosclerosis, are caused by bacteria and thus might be curable. The problem is demonstrating that bacteria are causing the disease and are not simply associated with the condition. Scientists have been grappling with this problem for over a century, and even today, finding a clearcut connection between microbe and disease is not always easy.

History and Relevance of Koch's Postulates

The first sightings of microbes through the microscopes invented by Antonie van Leeuwenhoek (1632–1723) occurred in the 1600s, but it took 2 centuries for the connection between microbes and disease to be made. This lag is not surprising, given the relative sizes of humans and microbes. How, asked early skeptics, could something that was too small to be seen by the unaided eye possibly sicken and kill human beings? Today, we take the connection between microbes and infectious diseases for granted, but this connection was not accepted easily at the time. Louis Pasteur (1822–1895), Friedrich Gustav Jacob Henle (1809–1885), and other early advocates of what has been called the **germ theory of infectious disease** realized that the connection between a microbe and a particular set of symptoms needed to be placed on a sound scientific basis. A German microbiologist, Robert Koch (1843–1910), proposed a set of "rules" for establishing a connection between a microbe and a disease; these four criteria for establishing cause and effect came to be known as **Koch's postulates** (Table 6–1).

Koch's first postulate states that the microbe must be associated with the lesions of the disease. That is, the microbe should be pres-

Table 6–1 Koch's postulates

Original postulates
1. The microbe must be associated with symptoms of the disease and must be present at the site of infection.
2. The microbe must be isolated from the lesions of disease and grown as a pure culture.
3. A pure culture of the microbe, when inoculated into a susceptible host, must reproduce the disease in the experimental host.
4. The microbe must be reisolated in pure culture from the experimentally infected host.

Proposed fifth postulate
5. Elimination of the disease-causing microbe from the infected host or prevention of exposure of the host to the microbe should eliminate or prevent disease.

Alternative methods used to satisfy Koch's postulates
- PCR can be used to detect and identify the microbe in diseased tissue.
- Immunohistochemistry can be used to detect and identify the microbe in diseased tissue.
- Antibiotic therapy can by used to eliminate the microbe and thereby cure the disease.
- Vaccination can be used to prevent infection by the microbe and thereby prevent disease.
- Hygiene, disinfection, and health practices can be used to prevent exposure to the microbe and thereby prevent disease.

ent in all cases where symptoms of disease are evident and must be found in diseased tissue but not in healthy tissue. The second postulate directs that the microbe must be isolated from the lesions of the disease and grown as a pure culture. The third postulate states that a pure culture of the isolated microbe should reproduce the symptoms of the disease if it is inoculated into a susceptible host, either humans or experimental animals. The fourth postulate states that the microbe must be reisolated in pure culture from the experimentally infected host that was used to satisfy the third postulate.

Since their first publication in 1882, Koch's postulates have proven invaluable for demonstrating that infectious diseases are caused by microbes. In addition, the experimental methods developed by Koch and his coworkers to satisfy the postulates have had lasting influence on the practice of microbiology, including the isolation of bacterial colonies on agar media contained in the ubiquitous petri dish, which was named after one of Koch's assistants. However, while the postulates sound very reasonable and straightforward and have shown a high success rate for identifying and evaluating a large number of bacterial pathogens, it quickly became clear that they can be difficult to satisfy in some cases and may not apply universally to all infectious diseases. First, they assume that the disease symptoms are dependent en-

tirely on the bacterium, yet host susceptibility due to genetic and other factors, such as age and the proficiency of the immune system, varies in human and animal populations and is now known to play a major role in virulence. Some individuals can be colonized with a potentially disease-causing bacterium yet not develop symptoms of the disease, whereas others are severely affected. Not all individuals respond equally to an infectious microbe, so the disease may not take the same course in all people. In addition, prior infection with one microbe may influence the response of the host to subsequent infection with another microbe.

Second, Koch's postulates assume that a pathogenic bacterium can be readily isolated and cultured. Unfortunately, not all bacteria can be cultured (or at least we have not yet determined how to culture them) under standard laboratory conditions. They may change their properties upon being cultured so that they lose some of the traits that allow them to cause disease. Third, Koch's postulates assume that all members of a bacterial species are equally virulent and that a single species causes each disease. This is clearly not the case for all bacterial pathogens. There can be dramatic and distinct differences in disease-causing abilities, even between closely related bacterial strains.

Finally, Koch's postulates require reinoculation into a susceptible host to reproduce the disease symptoms. For human disease, this requires either brave volunteers in highly structured studies or, more likely, a good animal model. However, an animal model may not be available, and indeed, for many diseases non-human animals are only approximate model systems. This leads to the question of just how closely an animal model should mimic the disease in humans for it to serve as an acceptable model. To illustrate why these issues might make it difficult to satisfy Koch's postulates, it helps to consider some examples of attempts to prove that a particular bacterium causes a particular disease.

The First Postulate: Association of the Microbe with Lesions of the Disease

Koch formulated his postulates based on the elegant and detailed studies that he conducted over many years, beginning with his work, which was started in an improvised home laboratory, on anthrax and the life cycle of the causative bacterium, *Bacillus anthracis*. In addition, his microscopic studies of microorganisms, his advances on improved techniques for staining tissues, and his development of new pure-culture isolation methods all played a role, as did his experimentation with animal models of disease, culminating in his demonstration of the microbial origin of anthrax, and later tuberculosis, using these experimental methods.

It was relatively straightforward for Koch to demonstrate the presence of the large *B. anthracis* bacilli in anthrax lesions and later to demonstrate that the more heat-resistant (and hence more sterile) anthrax spores also caused disease. However, the small, slow-growing bacterium that causes tuberculosis required the development of more sophisticated detection and cultivation methods before Koch was able to satisfy his first postulate. As is now well known to microbiologists, *Mycobacterium tuberculosis*, the causative agent of tuberculosis, is not only difficult to cultivate, it is also difficult to stain. At the time, Koch was convinced that the causative agent was a microbe because he could induce similar disease pathology in guinea pigs by inoculating tuberculous material from various clinical sources, but he could not see the bacteria in stained tissues. This spurred him on to finally find a method of staining the bacilli, a stain called the acid-fast stain. Thus, Koch's success in demonstrating that a bacterium was responsible for tuberculosis depended critically on his discovery of a method for staining and thereby detecting the bacteria in tissue samples. Special cultivation methods were also needed.

Strictly satisfying the first postulate has not always been possible, even when a causal relationship is eventually established between a microbe and a disease. For example, it took over 10 years for the scientific community to accept the idea that ulcers could be caused by *Helicobacter pylori*. Over half the people in developed countries and nearly all of the people in developing countries carry *H. pylori* in their stomachs, but only a small number of colonized people develop ulcers. As an aside, it is worth pointing out that this same pattern is seen in a number of bacterial diseases, where the bacterium colonizes many people, but only those with some predisposing condition develop a symptomatic infection. Koch himself was one of the first to worry about his first postulate, as it became obvious from his studies and those of others that people could be colonized with a pathogenic bacterium without having any symptoms (this is called the carrier state). We now know that the genetic makeup, as well as the current and prior health, of the host plays a critical role in the length and severity of disease and in the recovery of an individual from an infection. We also know that disease outcome is not always predictable and that disease progression is dynamic and depends on many microbial and host factors.

In other cases, a clear-cut causal relationship has not been established, and infection with the microbe may represent a risk factor instead of a cause (Table 1–1). An example of this situation is the suggested link between the bacterium *Chlamydophila pneumoniae* and the heart disease atherosclerosis. Scientists first began to suspect that bacteria might cause atherosclerosis when they isolated *C. pneumoniae* from atherosclerotic plaques but seldom isolated the bacteria from healthy blood vessel tissue. Occasionally, however, *C. pneumoniae* was isolated from healthy tissue. Finding *C. pneumoniae* in healthy tissue is not surprising if the characteristics of the bacteria are considered. *C. pneumoniae*, a common cause of mild respiratory infections, invades and lives inside human cells, thus protecting itself from the body's defenses. Therefore, if bacteria from the throat of a person with a respiratory tract infection leaked into the bloodstream, they might well infect blood vessel cells transiently and thus be found in tissue that looks healthy. The fact that *C. pneumoniae* was occasionally isolated from healthy tissue may be rationalized in this way, but it blurs the clear line implicit in Koch's first postulate. Even worse, although *C. pneumoniae* could be isolated from atherosclerotic plaque samples, it was not isolated from all such samples. Thus, *C. pneumoniae* is often associated with lesions of the disease, but not all the time.

The Second Postulate: Isolating the Bacterium in Pure Culture

Koch was well aware that the presence of a microbe did not, in and of itself, prove that the organism was the actual cause of the disease. He knew he first had to isolate the microbe in pure culture before he could demonstrate that it was the primary cause of disease. However, the second postulate is also not as easy to satisfy as it sounds. As was pointed out in the first chapter, bacteria are extraordinarily diverse with respect to their metabolic traits and growth requirements. Bacteriologists who work with disease-causing bacteria frequently grow these bacteria by streaking them onto a complex agar medium and incubating the agar plates with ordinary air as the atmosphere. Some bacteria are more difficult to cultivate than others. Koch, too, ran into this problem when he began his work on tuberculosis. He had spent over 6 years developing his culture medium formulations and plating techniques, which worked well for *B. anthracis* and other bacteria that caused wound infections. However, these methods did not work for *Mycobacterium*, which did not grow at room temperature and grew only very slowly at body temperature. Because the nutrient gelatin he used for the plates melted at body temperature, he had to develop another method that instead used coagulated blood serum, upon which, after about a week, the bacteria formed tiny colonies that could be seen under a low-power microscope.

Microbiologists have developed a wide variety of formulations for liquid media or agar plates over the years so that we now are able to cultivate a very large number of bacteria. However, although there are some media that can support the growth of many different kinds of bacteria, there is no universal medium on which all bacteria can be cultivated. Indeed, these approaches would fail to cultivate *C. pneumoniae*. Because *C. pneumoniae* grows only inside human cells, it must be cultivated using tissue culture cells and not agar medium. Moreover, an atmosphere that has an elevated CO_2 content is required for optimal growth. The fact that *C. pneumoniae* is more difficult to cultivate than many other disease-causing bacteria raises the question of whether the failure to cultivate *C. pneumoniae* from all atherosclerotic lesions could be due to failure of bacteria actually present in a lesion to grow in all cases. Likewise, although *H. pylori* grows on agar medium, it also requires a special atmosphere, and it took a lot of effort to find conditions that would support its growth.

An extreme example of the difficulty that can arise in attempting to satisfy Koch's second postulate is provided by *Treponema pallidum*, the bacterium that is generally accepted as being the cause of the **sexually transmitted disease** syphilis. The *T. pallidum* bacterium has a distinctive corkscrew shape and can be seen in the lesions associated with the early stages of syphilis. However, *T. pallidum* has never been cultivated as a pure culture in laboratory medium, despite many attempts. The closest thing we currently have to a laboratory medium for *T. pallidum* is the rabbit testicle, which is used as a growth chamber by scientists who work on *T. pallidum*.

The Third Postulate: Showing that the Isolated Bacterium Causes Disease in Humans or Animals

The third postulate demands that the bacterium isolated in pure culture must produce the disease when inoculated into a human or an animal. Around the time that Koch was demonstrating that *B. anthracis* caused anthrax, his colleague, Ferdinand Cohn (1828–1898), had isolated from hay a similar bacterium, *Bacillus subtilis*, that could form spores just like *B. anthracis*. Koch realized early on that just because an organism could be isolated from diseased tissue and grown in pure culture or because it appeared to be the disease-causing bacterium due to the way it looked under a microscope or due to some of its in vitro properties, the proof could come only after the critical demonstration that the isolated bacterium, when inoculated into a susceptible host, caused the disease. Koch subsequently showed that *B. subtilis*, unlike *B. anthracis*, did not cause symptoms of anthrax in experimental animals. Later, Koch was able to satisfy this critical postulate for *M. tuberculosis* using the guinea pig model because, even though guinea pigs do not naturally become infected with or succumb to tuberculosis, they exhibit disease pathology similar to that in humans and die from tuberculosis when inoculated experimentally.

Of all of Koch's postulates, this is the one that has been the most difficult to satisfy for many pathogens. Ironically, one of the first critics of this postulate was Koch himself. Less than 2 years after Koch proposed his postulates, he ran afoul of them when he tried to tackle cholera (Box 6–1). It is impressive that Koch was far more flexible in his approach to proving cause and effect in infectious disease than some modern-day scientists. In recent years, the third postulate has been a stumbling block for scientists working on *H. pylori* and *C. pneumoniae*. Acceptance of the proposal that most ulcers are caused by bacteria was held up for more than a decade because critics of the idea insisted that Koch's third postulate had to be met. There are now good animal models for ulcers, but in the early days of *H. pylori* research, these models were not available. This led one frustrated scientist to use himself as the "guinea pig" in an attempt to satisfy Koch's third postulate (Box 6–2).

Scientists working on the proposed connection between *C. pneumoniae* and heart disease have used a breed of rabbit that is prone to develop atherosclerosis if fed a high-fat diet to show that infection with *C. pneumoniae* increases the development of atherosclerotic plaques. This demonstration has not convinced many critics, who argue with some justification that the rabbit model is not a good replica of the disease in humans. Such objections raise an important issue: how closely does an animal model have to mimic the disease in humans?

One of the early animal models for gastric ulcers was a ferret inoculated with *Helicobacter mustelae*. *H. mustelae* is a close relative of *H. pylori*, but it is not *H. pylori*. Similarly, many researchers today use *Salmonella enterica* serovar Typhimurium infection in mice to mimic systemic disease caused by *S. enterica* serovar Typhi in humans, because the human-specific *S.*

BOX 6–1   Koch Backs Off

People sometimes grumble about criticisms of Koch's postulates. After all, they say, Koch's postulates are all we've got. What they may not realize is that the first person to give the old heave ho to Koch's postulates, at least the third one, was Koch himself. Koch presented his famous postulates in 1882. The ink was barely dry on the paper before, in 1884, Koch was having second thoughts. Koch was passionately interested in the disease cholera and was convinced he had found the causative agent, a curved bacillus ("comma bacillus" to Koch, now known as *V. cholerae*). It turned out that Koch was right, but in 1884, Koch's own postulates were being used against him because there was no animal model for cholera. Here is what Koch had to say to his critics:

> ... no one ever observes animals with cholera. Therefore, I believe that all the animals available for experimentation and those that often come in contact with people are totally immune. True cholera processes cannot be artificially created in them. Therefore, we must dispense with this part of the proof. This certainly does not mean that there is no proof that comma bacilli are pathogenic. I have already mentioned that even without animal experiments, I can imagine nothing other than that the comma bacilli cause cholera. If a cholera process is finally produced in animals, I will be no more convinced than I am now.

Source: A. D. Haffajee and S. S. Socranski. 1994. Microbial etiological agents of destructive periodontal diseases. *Periodontology 2000* **5**:78–111. In this paper, the authors quote from a translation of Koch's essays (K. C. Carter. 1987. *Essays of Robert Koch*, p. xvii–xix, 161. Greenwood Press, Westport, CT) in defense of their proposed alternatives to Koch's postulates.

ent animal species, to satisfy Koch's third postulate? Scientists differ in their answers to this question.

The Fourth Postulate: Reisolating the Bacterium from the Intentionally Infected Animal

Anyone who manages to satisfy the first three postulates will probably be able to satisfy the fourth with little difficulty. Nevertheless, the fourth postulate is important. In satisfying the fourth postulate, a scientist shows that the human or animal used to satisfy the third postulate was actually infected and that the lesions of the inoculated human or animal, like those occurring in the natural disease, contained the microbe. For example, in the case of the rabbit model used to study whether *C. pneumoniae* would cause atherosclerotic plaques to form, one might object that the reason the plaques formed in the rabbit arteries was not because the arteries were infected by the bacteria but because the bacteria caused a mild respiratory illness, which stressed the rabbits enough to increase the rate of plaque formation indirectly. Finding the bacteria in the plaques helps to bolster the contention that they caused the increase in plaque formation.

Modern Alternatives To Satisfy Koch's Postulates

Do the above examples of problems that can be encountered when trying to satisfy Koch's postulates prove that the postulates are useless? Not at all! For many diseases, Koch's postulates can be rigorously met. In other cases, Koch's postulates are very useful for guiding the design of experiments and subsequent discussions about the meaning of experimental results. What these examples also show, and what Koch realized at the time, was that these postulates should not be treated as monolithic requirements. To waive any one of Koch's postulates, scientists should certainly be expected to explain why failure to satisfy that postulate does not invalidate their claim of a cause-and-effect relationship. Not surprisingly, scientists have developed a number of modern molecular approaches that can be used to help prove cause and effect and thereby serve as alternatives to Koch's postulates (Table 6–1).

A modern alternative to detecting and cultivating pathogenic bacteria in diseased tissues is to use molecular biology approaches, such as the PCR-based typing approaches described in chapter 5. People who study *C. pneumoniae* often use this approach, because cultivation of the bacteria is so difficult and time-consuming. Another modern approach is to use **im-**

enterica serovar Typhi does not cause typhoid-like disease in mice. However, *S. enterica* serovar Typhimurium is a well-known cause of intestinal food-borne disease in humans and does not usually cause systemic disease in humans. Is it acceptable to use a different bacterial serovar or species, as well as a differ-

BOX 6-2 Koch's Postulates Are Not Just Words on Paper

How much do you care about proving a theory you believe in but cannot get others to accept? Barry Marshall, a young Australian internist working with J. R. Warren, the discoverer of *H. pylori*, did not exactly put his life on the line to do so, but he placed himself at risk for a very unpleasant condition. Marshall was an early proponent of the hypothesis that many gastric and duodenal ulcers are caused by bacteria. The idea that bacteria and not stress are the cause of ulcers flew in the face of well-established medical dogma. Given this, it is not surprising that Marshall and others who advanced this outrageous notion were not at first taken seriously by others working in the field. They were on firm ground in the beginning because all that had been proven was that *H. pylori* was associated with ulcer lesions and could be isolated in pure culture from these lesions (Koch's postulates 1 and 2). Critics of Marshall's theory insisted that Koch's third postulate must be satisfied as well, i.e., that the bacteria isolated from ulcers could cause ulcers in animals or humans. At the time, Marshall had been trying unsuccessfully to get the bacteria to infect various laboratory animals. Frustrated by this, he turned to the obvious alternative: human subjects—but who would volunteer to be inoculated with a bacterium that might cause ulcers? Keep in mind that this was before a successful antibiotic regimen had been developed. Even if volunteers could be found, approval for the experiment would be denied by the committee overseeing the use of human subjects. Marshall thus turned to his most faithful and reliable supporter: himself. He first had his stomach checked by endoscopic examination to make sure that the stomach mucosa was healthy. Then, he drank a turbid culture of *H. pylori*, which had been recently isolated from the lesions of a patient with peptic ulcers. Within a few hours, his stomach began to growl. A week later, he became nauseated and vomited. During this period, he felt unusually hungry and tired. Near the end of the second week, he underwent a second endoscopic examination of his stomach, and a biopsy specimen was taken. A portion of Marshall's stomach had an inflamed appearance, and *H. pylori* bacteria were found in the mucin layer over the lesion. The bacteria Marshall had ingested were clearly capable of causing disease. Fortunately for Marshall, the infection healed spontaneously, but Marshall's point was proven.

According to an article in the *New Yorker*, which described Marshall's experiment and its aftermath in detail, the initial event of taking the bacterial dose (a potentially momentous occasion in the history of gastroenterology) was a disappointingly undramatic affair. A fellow worker in the laboratory who knew what Marshall was about to do said, "You're crazy." Marshall said, "Here goes," and then drank the culture, remarking that it tasted like swamp water. Apparently, Marshall had spent too much time thinking about the experiment he planned to do to bother about providing himself with a suitably memorable statement to commemorate the occasion. Because of their epochal discoveries showing *H. pylori* as a causative agent of ulcers, Marshall and Warren were awarded the Nobel Prize in Medicine and Physiology in 2005.

Source: T. Monmaney. 1993. Marshall's hunch. *New Yorker*, 20 September 1993, p. 64–72.

munohistochemistry to detect the bacteria in tissue samples using antibodies against the bacteria that are conjugated either to an enzyme that can be used in a colorimetric assay for staining or to a fluorescent dye that could be used for visualization by fluorescence microscopy. In these cases, the bacteria would be detected or visualized in diseased tissue, but not in healthy tissue.

Is a Fifth Postulate Needed?

Technically, scientists working on *C. pneumoniae* have satisfied all four of Koch's postulates, if one accepts the rabbit model as a good model for human disease. Nonetheless, there are still skeptics, as there should be, given some of the problems described above. Similarly, even the use of a human volunteer to show that ingesting *H. pylori* could cause a mild inflammation of the stomach lining did not convince people skeptical of the *H. pylori*-ulcer connection. What finally made true believers out of most scientists and physicians was the development of an antibiotic therapy that eliminated the bacteria and at the same time cured the disease (Box 6–3).

Purists can argue about whether a fifth postulate should be added to Koch's original four, namely, a postulate stating that the information about the microbe should enable scientists to design effective therapeutic or preventive measures for eliminating the disease. In practice, however, this phantom fifth postulate will inevitably be invoked and will be seen by many as the ultimate test of a proposed microbe-disease hypothesis. The use of antibiotics to cure disease as a type of test was not possible in Koch's time, because antibiotics were not discovered until the 1930s. In a sense, however, there was a fifth step taken in Koch's time that helped convince the skeptics that bacteria caused cholera. Knowing the properties of *Vibrio cholerae*, public health officials were able to prevent the spread of cholera by identifying and shutting off contaminated water sources and, later, by treating water to eliminate *V. cholerae* from the water supply. Thus, by eliminating the disease-causing microbe or

BOX 6–3 Old Dogma versus New Model

Although the "discovery" of *H. pylori* is often believed to have occurred in the 1980s, evidence of gram-negative bacteria in the stomachs of patients at autopsy has been reported since the late 19th century. In the 1940s, New York City hospitals treated ulcers with antibiotics, such as tetracycline, but by the 1950s antibiotics ceased to be used because scientists and doctors could not prove all of Koch's postulates for the involvement of bacteria in ulcer formation. For 4 decades, medical texts attributed peptic ulcers to too much stomach acid, stress, smoking, alcohol consumption, and genetic predisposition. The stress-acid-ulcer theory gained a lot of credibility in the 1970s, when safe and effective agents that reduced gastric acid, such as histamine H2 receptor blockers (e.g., Tagamet),

were found to often heal or reduce painful ulcers. However, when people stopped taking these blockers, the ulcers recurred. Therefore, typically, these people had to take the drugs for the rest of their lives. Consequently, the drugs became a very lucrative pharmaceutical business. In 1983, this dogma changed with the first published report by Robin Warren and Barry Marshall connecting *H. pylori* and chronic gastritis. Although it took more than 10 years (in 1994, there was finally an official report confirming the microbial source of peptic ulcers), it is now generally accepted that *H. pylori* is the primary causative agent of gastroduodenal inflammation and peptic ulcer disease, which is readily treatable and even curable with antibiotic treatment.

Comparison of treatment strategies for gastric ulcers

Parameter	Old dogma	New model
Cause	Excess stomach acid damages tissues and causes inflammation.	*H. pylori* secretes toxins that cause inflammation and damage tissues.
Treatment	Bland diet (no citrus or spicy foods, no alcohol or caffeine); histamine H2 receptor blockers; surgery to remove ulcers	Antibiotic regimen (1–4 weeks) of one or two antibiotics plus an antacid
Success	Ulcers recur if H2 receptor blockers are discontinued.	No recurrence after completion of antibiotic therapy
Cost	H2 receptor blockers cost $60–100 per month; surgery costs up to $18,000.	One week of therapy costs less than $200.

by preventing exposure to the microbe, the disease itself was eliminated or prevented, which also supported cause and effect. However, even in Koch's time, there were alternative explanations for the success of improved sanitation in reducing cholera, showing how incorrect explanations can sometimes lead to positive outcomes (Box 6–4).

The proponents of the link between *C. pneumoniae* and heart disease are also taking this route, hoping to find an antibiotic regimen that prevents or cures atherosclerosis. However, finding such a regimen may be difficult. Antibiotics that kill a bacterium in the laboratory may not have the same effect in the human body because of the way the antibiotic is distributed and because of local conditions that may destroy the antibiotic. It took years to find the antibiotic combinations that finally subdued *H. pylori* because of the low pH of the stomach contents. However, once found and demonstrated to be effective in most cases, this effect of antibiotics finally allowed widespread acceptance of the hypothesis that *H. pylori* causes ulcers.

The results to date of clinical trials testing whether antibiotics prevent or cure atherosclerosis have been mixed. Some show an effect, and some do not. The cases in which antibiotic treatment was found not to have an effect have been used to argue against *C. pneumoniae* as the cause of atherosclerosis. This argument was also used in the early days of the search for an antibiotic treatment for ulcers. To complicate things even more, besides inhibiting bacterial growth, many antibiotics have the propitious side effect of reducing inflammation. So, do apparent positive effects of antibiotics in trials to test a causative role of *C. pneumoniae* in atherosclerosis reflect killing of the bacterium or reduction of inflammation that underlies the disease?

Along similar lines, using **vaccination** that induces immunity against a specific pathogen can also be used to prevent or lessen disease and again provide evidence of cause and effect, particularly if the immunized individual is subsequently exposed to the pathogen and does not develop symptoms of disease.

BOX 6–4   Bad Smells and Bad Science: Being Right for the Wrong Reasons

The germ theory appeared in the middle of the 19th century and challenged and eventually replaced the miasma theory, which had been the predominant explanation for the spread of infectious diseases since the Middle Ages. The miasma theory was a belief that infectious diseases were caused and transmitted by foul vaporous emanations that were associated with decaying matter and bad sanitary conditions. We still have a remnant of this notion in the name "malaria," which appeared in 1740 and literally means "bad air" in Italian. In the 21st century, it is somewhat hard to imagine that the germ theory of disease, which was based on scientific method, was so controversial when it was first proposed by Pasteur and Henle and later expanded by Koch. On the other hand, Pasteur was the scientist who disproved the other widely held notion of "spontaneous generation" by his elegantly simple "gooseneck" flask experiments. The germ theory of disease provided a scientific explanation for the horrendous infection rates that occurred during surgery in the 19th century, when hand washing was not even practiced routinely because of the supposed airborne transmission of disease. Joseph Lister (1827–1912) (after whom the mouthwash is named) and others seized on the germ theory to develop antiseptics and sterilization methods in surgery that greatly improved survival rates. However, for the wrong scientific reasons, the miasma theory also led to some notable successes in public health and improved hygiene. Champions of miasma, notably Max J. von Pettenkofer (1818–1901), realized that cleaning up sewage and water significantly reduced diseases, such as cholera and typhoid. Of course, improved cleanliness removed bad-smelling fens and only further reinforced the miasma theory to its proponents. The ill-fated von Pettenkofer is also reported to have tried to infect himself by drinking a culture of *V. cholerae* that he obtained from Koch. When this uncontrolled and dangerous self-experiment failed, it was again interpreted by him as evidence against the germ theory. Tragically, despite his successes in improving public health, albeit for the wrong scientific reasons, von Pettenkofer eventually fell into depression and killed himself.

Vaccination has been used repeatedly to demonstrate bacterial, as well as other microbial, causes of infectious disease in which a microbe was clearly implicated. However, it is now also being used as a way of showing causality for diseases whose etiology has not yet been clearly defined. One recent example of the use of this approach is aimed at preventing cervical cancer. The major risk factor for cervical cancer is infection by human papillomavirus (HPV). There is much excitement about a vaccine against HPV that was recently approved by the FDA and that is now on the market. If it prevents cervical cancer, this will also add to the proof that HPV causes cervical cancer.

As should be evident from the above discussion, it can be very difficult to be absolutely sure, without any doubt, that microbe X causes disease Y. There is enough room for error in each of Koch's postulates to make absolute certainty a virtual impossibility. The fact that there are still people, some with good scientific credentials, who are not convinced that human immunodeficiency virus is the cause of AIDS is an example of this. Moreover, as we have seen in the case of diseases like syphilis, failure to satisfy one or more of Koch's postulates does not necessarily disprove a connection between a microbe and a disease. In the end, the preponderance of evidence (as the lawyers would say) is what convinces people that there is a cause-and-effect relationship.

The Microbiota Shift Disease Problem

If it is difficult to prove a connection between one microbe and a disease, what does one do about polymicrobial diseases, such as periodontal disease (associated with gum erosion and tooth loss) and bacterial vaginosis (associated with higher risk for preterm birth and increased susceptibility to sexually transmitted diseases)? These diseases appear to be caused by shifts in the compositions of bacterial populations of the body (the microbiota) that contain hundreds of species. Scientists working in this area have tried to construct postulates similar to Koch's postulates in which "pure culture" is replaced by "a shift from one specified population to another." The third postulate presents real problems, because it is difficult to produce specific population shifts in laboratory animals. In the cases of periodontal disease and bacterial vaginosis, the phantom fifth postulate may turn out to be of great importance and might become equivalent to Koch's third postulate for pure cultures. That is, if the original composition of the bacterial population can be restored, is the disease prevented? This area will be a challenge for future microbiologists, espe-

cially if more diseases, such as inflammatory bowel disease, are shown to be microbiota shift diseases.

Concepts of Disease

Varieties of Human-Microbe Interactions

The early view of infectious diseases, implicit in Koch's postulates, was that there were microbes capable of causing disease and that exposure to them inevitably caused the disease. Already in Koch's time, however, this simplistic description of infectious diseases was being questioned. Scientists realized that not all people who drank water containing *V. cholerae* developed cholera (Box 6–4). Not all people exposed to someone with tuberculosis developed symptomatic disease, even though it could be shown that they were infected with the bacterium. Clearly, all humans are not equal in their response to an infectious microbe.

As scientists worked with disease-causing microbes, they further discovered that just as there are differences in susceptibility from person to person, there is also variation among different strains of the same bacterial species. Sometimes, a bacterium that had been grown too long in laboratory medium lost the ability to infect, or different isolates of what appeared to be the same bacterium differed in their abilities to cause disease. Through the years, a greater appreciation for the complexity of the interaction between microbes and humans has emerged. Today, scientists view the infection process as a multifaceted interaction between the microorganism and the human body. This interaction may result in clearance of the microbe from the body, in asymptomatic carriage of the microbe, or in the development of symptoms. The outcome of a microbe-human encounter depends on the infected person's defenses against disease and on the traits of the infecting strain. Finally, as described in chapter 1, ecology contributes to infections. Diseases come and go, depending on ecological factors, including human activities.

Views of the Microbe-Human Interaction

Although everyone agrees that the interactions between microbes and the human body are complex, there are differences of opinion about how these interactions should be understood. Perhaps the most widely held view is that disease-causing bacteria evolved specifically to cause human disease. A second view, which has gained more adherents lately, is that disease-causing bacteria are actually trying to achieve an equilibrium with humans that does not result in disease and that disease symptoms result when this equilibrium does not develop. Adherents of this view point to the fact that in many diseases, the number of people who develop serious infections is far smaller than the number of people who carry the bacteria without developing any symptoms. A third view is that humans are more often than not accidental hosts of some bacteria that may be able to cause human disease but have actually evolved to occupy some other niche. In this view, bacteria entering the human body react by activating stress responses, producing disease symptoms in the process.

Probably each of these views is correct for a subset of diseases. For example, a bacterium that causes disease only in humans, has no external reservoir, and causes symptoms in virtually all infected people fits the first view, whereas a bacterium that causes an asymptomatic carrier state in most of the humans it infects fits the second view. A bacterium that spends most of its time outside the human body and only occasionally causes human disease may fit more closely with the third view. The varieties of human-bacterium interactions are so numerous and distinct that there is no one model that fits all diseases.

Molecular Koch's Postulates

Just as microbiologists in the second half of the 19th century struggled with the question of how to prove that a particular bacterium caused a particular disease, microbiologists in the second half of the 20th century struggled with the question of how to prove that a particular gene or genes from a pathogen contributed to virulence. If Koch were alive today, he would no doubt be flattered by the fact that the initial attempts to answer this question used his postulates as a guide, but he would not be surprised to see these postulates argued just as hotly as the postulates he put forward. On the other hand, Pasteur, more than Koch, anticipated and would appreciate the new understanding of the adaptability and evolution of bacteria during the disease process, since Koch viewed bacteria as largely static and unchangeable.

There have been several versions of molecular Koch's postulates, but most of them read as follows. First, the gene (or its product) should be found only in strains of bacteria that cause the disease and not in bacteria that are avirulent. Second, the gene should be "isolated" by cloning. Third, disrupting the gene in a virulent strain should reduce or attenuate its virulence (the concept of attenuation). Alternatively, introducing the cloned gene into an avirulent strain should render the strain virulent. Finally, it should be demonstrated that the gene is expressed by the bacterium when it is in an animal or human volunteer at some point during the infectious process.

Today, this view of virulence genes seems overly simplistic, but at the time, attempts to establish a set of criteria for what would be called a "virulence gene" were instrumental in getting the field off to a running start. This version of molecular Koch's postulates also performed a service because, almost immediately, it generated dissatisfaction with this simple view of virulence factors and fueled discussions that have led to a more sophisticated (but perhaps still not entirely correct) view of virulence.

Virulence as a Complex Phenomenon

In the early days of modern pathogenesis research, things seemed much simpler than they do today. This is probably because the first diseases to be studied in depth at the molecular level were diseases like diphtheria that were caused by bacteria that produced a single toxic protein (diphtheria toxin), which caused the symptoms of the disease. A toxin is clearly a virulence factor. As more complex diseases were probed, however, it became obvious that in some cases, a large number of factors were involved in the ability of a bacterium to cause disease. An experimental definition of a virulence factor has been that loss of the factor by the bacterium results in a decrease in its ability to cause disease. This seems pretty straightforward, but what does one do about cases in which the loss of two factors reduces the pathogenicity of the bacterium but loss of each trait separately does not, or about a trait that is clearly a virulence factor in one bacterium but is not a virulence factor in another bacterium? Then, there are the cases in which loss of the ability to synthesize an amino acid makes the bacterium unable to cause disease. Can the biosynthesis of an amino acid, a trait normally considered to be a routine part of the microbe's basic physiology (generally considered a housekeeping factor), also be considered a virulence factor? Even more confusing are the cases in which the loss of a trait makes a bacterium more virulent.

Throughout this book, you will encounter examples of scientists grappling with this problem of how to define a virulence factor. Either the current definition of a virulence factor is inadequate or the entire concept of the virulence factor, which implies that a small number of discrete traits make the difference between the ability to cause disease and the inability to cause disease, needs to be rethought. Rather than critique the definition of a virulence factor and early versions of molecular Koch's postulates to identify virulence factors, it makes more sense to consider the ways in which modern molecular methods have provided new ways to investigate how bacteria cause disease.

To cope with the multifactorial nature of virulence, many methods, such as the transposon mutagenesis approach, have been designed that do not just focus on one or a few genes but seek mutations that affect some general feature of pathogenesis, such as the ability to survive inside a phagocyte or the expression of a set of genes in an animal or survival in the animal. Two of the best-known examples of this approach are in vitro expression technology, which seeks genes that are expressed only in the animal, and signature-tagged mutagenesis, which seeks transposon insertions that eliminate the ability of a bacterium to survive in the animal. More recently, scientists have been turning to comparative genomics and microarray technologies to identify genes involved in the pathogenesis of bacterial infection. All of these methods generally identify numerous genes as being important in the animal. These approaches, as well as others, are described in more detail in later chapters.

SELECTED READINGS

Brock, T. D. 1999. *Robert Koch: a Life in Medicine and Bacteriology.* ASM Press, Washington, DC.

Casadevall, A. 2005. Host as the variable: model hosts approach the immunological asymptote. *Infect. Immun.* **73:**3829–3832.

Falkow, S. 2006. Is persistent bacterial infection good for your health? *Cell* **124:**699–702.

Grimes, D. J. 2006. Koch's postulates—then and now. *Microbe* **1:**223–228.

Groisman, E. A., and H. Ochman. 1994. How to become a pathogen. *Trends Microbiol.* **2:**289–294.

Higgins, J. P., J. A. Higgins, P. M. Higgins, S. Ahuja, and D. L. Higgins. 2003. *Chlamydia pneumoniae* and coronary artery disease: legitimized linkages? *Exp. Rev. Cardiovasc. Ther.* **1:**367–384.

Koropatnick, T. A., J. T. Engle, M. A. Apicella, E. V. Stabb, W. E. Goldman, and M. J. McFall-Ngai. 2004. Microbial factor-mediated development in a host-bacterial mutualism. *Science* **306:**1186–1188.

Merrell, D. S., and S. Falkow. 2004. Frontal and stealth attack strategies in microbial pathogenesis. *Nature* **430:** 250–256.

Peek, R. M., Jr. 2005. Events at the host-microbial interface of the gastrointestinal tract. IV. The pathogenesis of *Helicobacter pylori* persistence. *Am. J. Physiol. Gastrointest. Liver Physiol.* **289:**G8–G12.

QUESTIONS

1. Three views of bacterial pathogens were presented. In one view, disease-causing bacteria evolved specifically to cause disease in humans. (They're out to get you!) In the second view, bacteria evolved to colonize certain sites in a certain animal and cause disease when they do not establish equilibrium with their host. (They're out to dine at your expense!) In the third view, humans are often accidental hosts for bacteria whose traits were not designed for colonizing or infecting humans. (They're lost and not happy about it!) What sort of traits in a bacterial pathogen would convince you that one of these views was the best explanation for a particular disease?

2. Why are scientists so reluctant to let go of Koch's postulates even though they often chafe—as Koch did—under their restrictions? Why are Koch's postulates, with all their problems, still as relevant today as they were when Koch first proposed them?

3. Could you prove cause and effect using something like Koch's postulates if you could not cultivate the organism you suspect is causing the disease? What type of approach would you use?

4. In the chapter on the microbiota of the human body (chapter 5), we mentioned the possibility that some diseases might be caused by shifts in the microbial population of a site on the human body. For example, some oral microbiologists have pointed to an association between gingivitis (gum disease) and cardiovascular disease. How would you formulate "Koch's postulates" for such a disease? Or could you?

5. Animal models are often used to satisfy Koch's postulates, yet mice and other laboratory animals sometimes react differently than humans to infectious agents. For example, *S. enterica* serovar Typhi, the cause of typhoid fever in humans, a very serious disease, does not cause disease in mice. In other cases, the bacteria localize differently in humans and mice. *S. enterica* serovar Typhimurium causes a fatal bloodstream infection in mice but causes only diarrhea in most humans. What features of an animal model need to be evaluated before the model is used to satisfy Koch's postulates?

6. Are there features of a disease that allow human subjects to be used to satisfy Koch's postulates? What are they?

SOLVING PROBLEMS IN BACTERIAL PATHOGENESIS

1. A group of researchers at the U.S. Department of Agriculture isolate a new bacterium from the lungs and lymph nodes of several young horses that became ill and died at a local stable. Prior to death, their symptoms included disorientation and loss of motor function, so the researchers suspected central nervous system involvement, which was confirmed by the observation of brain lesions in the dead animals. Based on 16S rRNA comparisons, the bacterium was distantly related to *Neisseria meningitidis*, and they subsequently named it *Neisseria equiniae*. The researchers believe that *N. equiniae* may be responsible for the brain lesions observed in a small percentage of older horses that die of apparent dementia. What four criteria must be satisfied in order for the researchers to prove that the brain lesions in these older horses are caused by *N. equiniae*? Provide at least three modern molecular experiments that could be performed to satisfy these criteria.

2. You have just received an urgent call from the Centers for Disease Control and Prevention (CDC) to consult on the following case of an unusual disease outbreak that they have been investigating. A new gram-positive bacterium related to *Listeria monocytogenes* was isolated from an outbreak of food poisoning in Wisconsin due to contaminated cheese and appears to cause painful gastritis and in about half of exposed individuals, along with sudden onset of bleeding ulcers, death from toxic shock within 2 or 3 days. Upon biopsy of infected individuals, it was found that the bacteria were growing on the surfaces of epithelial cells lining the gastric pit of the stomach. Autopsy of individuals who died showed that bacteria were found only in the stomach and not in any of the other body organs. The researchers at the CDC have subsequently determined that, like *L. monocytogenes*, this new species of *Listeria* invades epithelial cells in tissue culture. Which, if any, of Koch's postulates have been satisfied for the involvement of this new *Listeria* strain in gastritis, bleeding ulcers, and toxic shock? What additional measures could the researchers take to help satisfy Koch's postulates?

3. *Porphyromonas gingivalis*, a gram-negative, non-spore-forming, heterodiploid, anaerobic, black-pigmented bacterium, is widely considered to be an important etiological agent of periodontal disease because of its strong correlation with active disease process and abscess formation in animal infection models. Therapy for severe periodontal disease usually involves scraping to remove dental plaque and de-bridement (surgical removal of affected tissue and abscesses) in combination with a 2-week regimen of anaerobe-specific metronidazole and/or a 3- to 5-week regimen of tetracycline. A mouse abscess model of infection has been developed for *P. gingivalis* in which the bacteria are injected into the skin of the animal and allowed to form an abscess over the course of a week. The lesion site where the abscess is forming can be surgically removed, plated on blood agar plates, and incubated in an anaerobic chamber to grow colonies after about 5 days of incubation. There is an emerging paradigm shift in our current understanding of the causes of coronary heart disease from a purely hereditary and nutritional causation to a possible infectious etiology. This shift comes from recent epidemiological studies that have demonstrated a correlation between periodontal disease and coronary heart disease. There is a recent report that *P. gingivalis* can invade oral epithelial tissues during severe disease and exacerbate the inflammatory responses. Inflammatory responses have been linked to atherosclerosis (plaque formation in arteries) that could lead to heart disease. In support of this, another recent report found that *P. gingivalis* could also invade human primary coronary artery endothelial cells in tissue culture. Considering all of the above information, which of Koch's postulates, if any, have been satisfied so far for *P. gingivalis* involvement in heart disease? Be sure to state your rationale. Provide at least four different modern molecular experiments that could be performed to help satisfy Koch's postulates for *P. gingivalis* involvement in heart disease.

4. In 1976, the Bellevue Stratford Hotel in downtown Philadelphia hosted a convention of members of the American Legion. During the convention, 221 Legionnaires developed a new form of pneumonia, and 34 people died. Infectious disease specialists working for the CDC were sent to Philadelphia to identify the cause of this new, mysterious illness, which, not surprisingly, was subsequently named Legionnaire's disease. Using Koch's postulates, list the steps that those specialists had to take to complete their assignment. Briefly explain the significance of each step. The specialists understood that it is not always possible to follow Koch's postulates to the letter. List at least two problems that they might have faced in attempting to fulfill Koch's postulates as they embarked on their

(continued)

quest to identify the causative agent of this new illness.

5. In 1984, Barry Marshall took a unique approach to fulfill Koch's postulates. He drank a culture of the bacterium *H. pylori* and developed a mild case of gastritis, although another volunteer developed a more severe case of gastritis. He and his mentor, Robin Warren, won the Nobel Prize in Physiology or Medicine in 2005 for their discovery of the role of *H. pylori* in the pathogenesis of gastritis and peptic (stomach and duodenal) ulcers. Given that 80% of peptic ulcer disease cases are known to be associated with *H. pylori* infection, briefly describe how each of Koch's postulates is fulfilled for the link between *H. pylori* and peptic ulcer disease. What modern approaches can be taken to help satisfy Koch's postulates for this case?

7

Mechanisms of Genetic Modification and Exchange: Role in Pathogen Evolution

Imagine touching palms with someone for about an hour and acquiring a whole new chromosome, which bestows upon you new powers. Sound like a far-fetched science fiction scenario? It is, for humans, but it is commonplace for bacteria. Bacteria routinely practice a process called conjugation, in which two bacteria make close contact with each other and one transfers segments of DNA to the other. The DNA segments can be quite large, even hundreds of kilobases. This is one-fifth to one-third the size of the average bacterial chromosome. Of course, the sizes of the segments that are more commonly transferred are smaller than this, but most still carry multiple genes. This process is one of a group of processes called horizontal gene transfer. Incredibly enough, such large additions of DNA are well tolerated most of the time and can give the bacteria new, powerful traits, such as resistance to antibiotics, the ability to utilize new carbon sources, and new traits that allow them to cause disease. DNA segments of the last type are called pathogenicity islands. To take our science fiction scenario even further, imagine contacts involving humans and insects. Many of the transfers carried out by bacteria occur across species and genus lines.

Adapt or Perish

Acquiring New Virulence Traits by HGT

The ability of a bacterium to respond to new selective pressures, to survive adverse environmental conditions, or to exploit new environments that it encounters depends on its ability to evolve through modification of gene function (mutation) or acquisition of new genes **(horizontal gene transfer [HGT]).** Until recently, it was thought that changes that alter a microbe's virulence properties would arise by slow processes involving point mutations, gene duplications, gene deletions, or chromosomal rearrangements and that adaptive changes would occur largely through antigenic or phase variation (Table 7–1). The prevailing view was that in order to maintain or amplify these mutations once they arose, it was necessary to have a strong, persistent selective pressure, which could be applied only by specific exposure to the host environment. New evidence, particularly from genome-sequencing analyses of many bacteria, has shown that high-frequency exchange of large DNA sequences between different species and even different genera by HGT contributes significantly to genome variability and accounts for a much more rapid evolution of pathogens than was previously thought. Moreover, DNA transferred by HGT is often stably maintained in the absence of selection.

Likewise, it was commonly believed that disease-causing microbes evolved over long periods of time through complex interactions between the microbe and its host. However, increasing evidence suggests that the evolution of the genes encoding most bacterial toxins and other virulence factors involves HGT mediated by plasmids, transposons, bacteriophages, and other transmissible DNA elements (Table 7–1). HGT allows one-step acquisition, within a matter of hours, of genes encoding antibiotic resistance, toxins, and other virulence factors (see chapters 11 and 12), which increase survival in an animal or human host by helping the bacteria to evade or suppress the host's immune system, providing a means for dissemination, or enhancing survival, and thereby transmission, in the external environment outside the host. (See Box 7–1 for an interesting twist on the implications of HGT and evolution.)

Mechanisms of Genetic Change and Diversification

Point Mutations, Gene Deletions or Duplications, and Chromosomal Rearrangements

Although the focus of this chapter is on HGT, it is worthwhile to review change by mutation. Spontaneous mutations, such as single-nucleotide exchanges **(single-nucleotide polymorphisms, or SNPs)** and insertions and deletions **(indels),** occur at rates ranging from 10^{-6} to 10^{-9} per nucleotide per generation in bacteria growing under normal culture conditions. Recombination can result in gene duplication, deletion, or rearrangement (Figure 7–1). Whether the genetic change occurs in a single base pair or in a large segment of DNA, the resulting mutation is inheritable and can lead to genes or gene products with altered activity, function, regulation, and/or physiological consequences. These changes, in turn, can result in gain or loss of virulence properties. In general, the order or arrangement of genes in an operon is conserved between two closely related species, but modular exchange of large DNA sequence segments

Table 7–1　Mechanisms of genetic change and diversification

Slow processes	Rapid processes
Point mutation	Phase variation
Nucleotide change	Promoter inversion
Nucleotide insertion	Slipped-strand synthesis
Nucleotide deletion	
	Antigenic variation
Gene duplication	Gene shuffling
	Gene conversion
Gene deletion	
	HGT
Chromosomal rearrangement	Intergenic recombination
Inversion	Transformation
Intragenic recombination	Plasmid
	Exogenous DNA
	Conjugation
	Plasmid
	Transposon
	Transduction
	Phage
	Generalized
	Specialized

through chromosomal rearrangements, inversions, tandem gene duplications, deletions, or recombination also occurs at a finite rate and contributes significantly to the genetic diversity of bacterial populations. Finally, recent studies have shown that when bacterial populations are stressed by nutrient deprivation and extended stationary-phase conditions, a small subpopulation of cells becomes hypermutable for spontaneous mutation. This fraction acquires mutations at an increased rate relative to the general population of bacteria. Many of these stress-induced mutations are deleterious, but some help the bacteria to acquire new phenotypes that allow them to cope with the stress and begin to grow again. The pathways that allow stressed bacteria to become hypermutable seem to involve the induction of an error-prone DNA polymerase that has less fidelity than the DNA polymerase that normally replicates the chromosome. Stress-induced mutagenesis seems to occur in many bacterial species and may contribute to the long-term evolution of new traits. However, its contribution to bacterial pathogenesis has not been thoroughly explored.

Phase Variation

Phase variation usually refers to changes in the expression of important virulence proteins that occur at relatively high frequency compared to spontaneous mutations. This process can occur by several different mechanisms. One of the first examples of phase variation reported involved changing the composition of flagella (motility structures) of *Salmonella* species. Pro-

BOX 7–1   War of the Worlds—Revisited

In H. G. Wells' classic science fiction thriller *War of the Worlds*, lethal infection by terrestrial microorganisms saved humankind from annihilation by an extraterrestrial army of killer Tripods. The alien invaders, who, unlike us, had not evolved alongside these microbes over millennia, had no natural immunity against these "ordinary" Earth microbes and thus became mortally ill upon exposure to us and our environment. Could the converse occur? Is it possible that extraterrestrial microorganisms introduced to Earth might be capable of causing human disease?

Astrobiologists are taking this possibility seriously and are concerned about how safe it is to bring back samples from our space exploits. Indeed, NASA has even created a special position entitled "Planetary Protection Officer," whose task is to protect the Earth from extraterrestrial contamination. This person is responsible for determining whether extraterrestrial soil samples collected during space expeditions to the moon or Mars, or other planets, pose a potential biohazard to Earth's inhabitants.

It was previously believed that harmful microbes could not evolve without a human or animal host environment to provide selective pressure for increased virulence. However, the acquisition of large pieces of DNA by HGT provides a means of gaining virulence traits in the absence of strong selection. Also, with the finding that considerable HGT can occur outside the human or animal environment, it is conceivable that genetic exchange of material by bacteria in the external environment might pose unknown risks, particularly if those sources of genetic material evolve in unusual environments, such as those of other planets.

Source: R. L. Mancinelli. 2003. Planetary protection and the search for life beneath the surface of Mars. *Adv. Space Res.* **31:**103–107.

---

teins that make up flagella are highly antigenic because they are present on the surfaces of the bacteria. It is advantageous for a bacterium to be able to periodically change its flagellar proteins to avoid elimination by the host's immune system. In the 1940s, it was observed that some *Salmonella* strains could switch from making flagella with type H1 flagellin protein to flagella with type H2 flagellin protein, and back again, at frequencies of about 10^{-4} per cell, considerably higher than normal mutation rates. This switching between the expression of different versions of the genes encoding flagellin is caused by a DNA invertase (called Hin), which promotes **inversion of sequences** upstream of the H2 flagellin gene. Figure 7–2 depicts the steps involved in the *Salmonella* flagellin H1/H2 phase variation process. The invertible sequence region consists of the invertase gene, its promoter, and an additional promoter that can regulate downstream genes. When it is in one orientation, the second promoter transcribes the H2 gene and a downstream repressor of the H1 gene promoter. When the invertible region is in the other orientation, the H2 gene and the repressor are no longer transcribed and the H1 gene is expressed instead. Phase variation (or phenotypic switching) has also been found to play an integral role in the formation of diverse phenotypes within microbial communities that make up biofilms.

Phase variation in *Bordetella pertussis*, the agent that causes whooping cough, is mediated by a reversible mutation in the *bvgS* regulatory gene, which controls the expression of virulence genes. The mutation adds or deletes a GC base pair in a sequence region containing a string of six Gs, resulting in reversible loss of virulence through a frameshift in the BvgS regulatory protein (Figure 7–3). This type of phase variation results from **slipped-strand misrepair** (also sometimes called **slipped-strand synthesis**) at sites of repeated DNA sequences (e.g., ATG-$[G]_n$-gene or ATG-$[CTCTT]_n$-gene), which can lead to different protein sequences or truncations from frameshifts. Slipped-strand misrepair has emerged as a common mechanism of phase variation in numerous pathogens. Another example of this occurs in *Neisseria* Opa and Opc (Opa-like) proteins. Opa proteins are integral outer membrane proteins expressed in most meningococcal and all gonococcal strains and mediate intimate adherence and invasion during infection. *Neisseria meningitidis* strains have 3 or 4 *opa* genes, and *Neisseria gonorrhoeae* strains have 8 to 13 *opa* genes, any of which may be on or off at different stages during

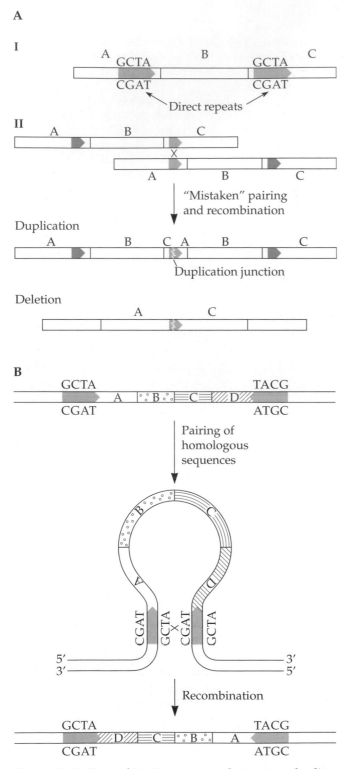

A

I

A GCTA B GCTA C

CGAT CGAT

↖ Direct repeats ↗

II

A B C

X

A B C

↓ "Mistaken" pairing and recombination

Duplication

A B C A B C

Duplication junction

Deletion

A C

B

GCTA TACG

A ∘∘B∘∘ ═C═ ⫽D⫽

CGAT ATGC

↓ Pairing of homologous sequences

↓ Recombination

GCTA TACG

⫽D⫽ ═C═ ∘∘B∘∘ A

CGAT ATGC

Figure 7–1 Recombination can result in gene duplication or deletion **(A)** or gene rearrangement **(B)**. (Adapted from L. Snyder and W. Champness, *Molecular Genetics of Bacteria*, 3rd ed., ASM Press, Washington, DC, 2007.)

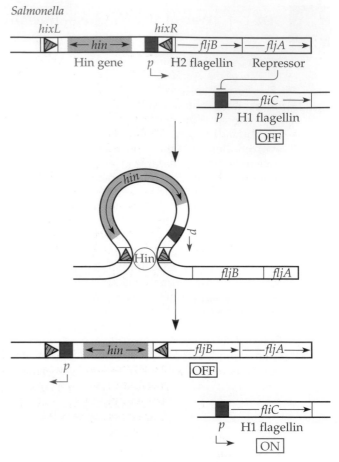

Salmonella

hixL hin hixR fljB fljA

Hin gene p H2 flagellin Repressor

p fliC

H1 flagellin OFF

hin

d

Hin fljB fljA

hin fljB fljA

p OFF

fliC

p H1 flagellin ON

Figure 7–2 Phase variation in *Salmonella* by Hin invertase-mediated inversion. The Hin invertase inverts a sequence upstream of the gene for H2 flagellin *(fljB)* and a repressor of the H1 flagellin gene *(fljC)*. In one orientation, the H2 flagellin gene and the *fljA* repressor gene are transcribed from the promoter *(p)*. In the other orientation, these two genes are not transcribed, so the repressor no longer represses transcription of the H1 flagellin gene, and H1 is expressed. The Hin invertase is constitutively expressed under its own promoter. (Adapted from L. Snyder and W. Champness, *Molecular Genetics of Bacteria*, 3rd ed., ASM Press, Washington, DC, 2007.)

infection. All *opa* genes contain 5′ tandem repeats of $(CTCTT)_n$ in the signal peptide coding region of the gene that cause a high frequency of slipped-strand frameshifts and loss or gain of expression (Figure 7–4).

In a third example, *Neisseria* species also produce a distinct type of membrane surface sugar-lipid molecule called lipooligosaccharide (LOS), which differs from lipopolysaccharide (LPS) in having short, highly branched sugar units attached to the lipid core and in having no repeating O-antigen units. LOS is a very

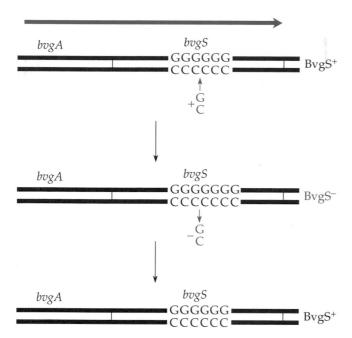

Figure 7–3 Phase variation in *Bordetella* by slipped-strand misrepair. The toxin and other virulence factor genes of *B. pertussis* are controlled by a two-component regulatory system consisting of the BvgS and BvgA proteins. Phase variation results from a frameshift mutation in a region of the *bvgS* gene containing a string of Gs. (Adapted from J. F. Miller, S. A. Johnson, W. J. Black, D. T. Beattie, J. J. Mekalanos, and S. Falkow. 1992. Constitutive sensory transduction mutations in the *Bordetella pertussis bvgS* gene. *J. Bacteriol.* **174**:970–979.)

toxic molecule and accounts for most of the local inflammation and tissue damage during infection. The surface structure of LOS is highly variable. Many of the genes encoding enzymes (glycosyltransferases) that add sugars to LOS have a tract of poly(G)s in the 5' region of the gene's sequence that undergoes slipped-strand misrepair at high frequency. In addi-

tion, LOS can be sialylated during different stages of infection, which also contributes to serum resistance. Sialic acid groups are covalently attached to galactose residues on the LOS by sialyltransferases, which undergo slipped-strand misrepair phase variation. Depending on which glycosyltransferases and sialyltransferases are being expressed at any given time, the LOS will have different surface structures.

Antigenic Variation

Pathogenic microbes can avoid the host immune system by changing their surface antigens through gene-shuffling events **(antigenic variation)**. The pili of *N. gonorrhoeae*, which are responsible for attaching to host epithelial cells, are highly antigenic and undergo both phase and antigenic variation by recombination between different pilin genes. In addition to the major pilin gene *pilE*, which has a promoter, each genome contains multiple (10 to 20) silent copies of the pilin gene *pilS* that lack a promoter. The *pilE* and *pilS* genes share regions that are highly conserved but also have regions that differ significantly, the so-called hypervariable regions. Recombination events between the conserved regions of any of the *pilS* genes and *pilE* result in a new version of the *pilE* gene (Figure 7–5) and, hence, a new pilin protein on the surface, which is not recognized by existing antibodies in the host. Pilin recombination in *Neisseria* requires a functional RecA protein, as well as RecO and RecQ (RecF-like recombination), and is a type of **gene conversion,** since it is nonreciprocal (occurs in only one direction).

In *Neisseria*, phase and antigenic variation are so prevalent and important for survival in the host that they are considered the major mechanisms of neisserial pathogenesis. Because of this type of variation, the host's immune system cannot keep up. Repeated infections with *Neisseria* are common, and indeed, var-

Figure 7–4 All *opa* genes contain 5' tandem repeats in the signal peptide coding region of the gene. Slipped-strand misrepair occurs at the $(CTCTT)_n$ repeats in the N terminus of the *opa* gene, resulting in alteration of the reading frame and loss of expression of the protein. In addition, each *opa* gene contains semivariable (SV) and hypervariable (HV) regions. (Adapted from G. L. Murray, T. D. Connell, D. S. Barritt, M. Koomey, and J. G. Cannon. 1989. Phase variation of gonococcal protein II: regulation of gene expression by slipped-strand mispairing of a repetitive DNA sequence. *Cell* **56**:539–547, with permission from Elsevier.)

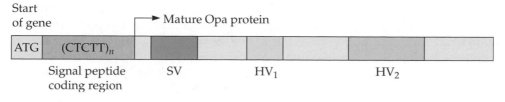

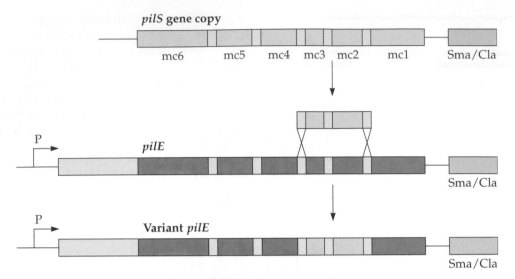

Figure 7–5 Antigenic variation in *Neisseria* by gene conversion. The pilin genes of *N. gonorrhoeae* can undergo nonreciprocal homologous recombination events between the expressed *pilE* locus and different silent copies of the *pilS* gene at other loci. The major conserved regions of *pilE* and *pilS* are represented by the shaded gray boxes, and the variable sequences are represented by either dark-blue boxes for *pilE* or light-blue boxes for *pilS*. At the 3′ ends of all pilin loci is a conserved DNA sequence, called the Sma/Cla repeat, which has sequence similarity to several recombinase-binding sites and is involved in efficient pilin recombination. (Adapted from S. A. Hill and J. K. Davies. 2009. Pilin gene variation in *Neisseria gonorrhoeae*: reassessing the paradigm. *FEMS Microbiol. Rev.* **33**:521–530, with permission from John Wiley and Sons.)

iation often occurs within a single infected individual. There is usually a vigorous antibody response, yet genetic variability in pili and other surface antigens always occurs in a significant fraction of the bacterial population at any given time. Although pressure from the immune response may allow a particular pilin variant to become the dominant form in a population, once antibodies are made against that variant, it is rendered obsolete, and one of the minor variants then takes over and becomes the next dominant form. Since millions of possible pilin variants can arise from phase and antigenic variation, no effective antibody response can be mounted against *N. gonorrhoeae*, and repeated infections occur.

HGT: Mobile Genetic Elements

Natural Transformation

It turns out that while *Neisseria* pilin recombination can occur with significant frequency within a single bacterium **(intragenic recombination),** most of the variation results from recombination between the genes from different bacteria. DNA containing *pilS* is released when the donor cell lyses. The recipient cell

picks up this DNA, and the *pilS* of the donor recombines with the *pilE* of the recipient cell **(intergenic recombination).** This is possible because *Neisseria* cells frequently undergo autolysis to release their genomic DNA and other *Neisseria* cells are then able to take up the released DNA and incorporate it into their genomes through homologous recombination. This mechanism of exchanging DNA through intergenic recombination allows even greater antigenic variation than what might occur through intragenic recombination alone.

Bacteria can exchange DNA in three ways: transformation, conjugation, and transduction. The first mechanism of gene exchange in bacteria to be discovered was **natural transformation:** DNA from a **donor cell** is released into the environment and is then taken up by a **recipient cell.** The resulting progeny are called **transformants.** Most bacteria will not take up DNA efficiently unless they have a little help from an external force (such as a scientist) by exposure to certain chemical or electrical treatments that perturb their membranes and make them more permeable (bacteria that are capable of taking up DNA are said to be **competent**). The field of molecular biology depends on the scientist's ability to introduce recombinant DNA into chemically competent or electrocompetent cells.

There are, however, a number of gram-negative and gram-positive bacteria, including some notable pathogens (*Haemophilus influenzae, Haemophilus parainfluenzae, N. gonorrhoeae, N. meningitidis, Campylobacter jejuni, Helicobacter pylori, Streptococcus pneumoniae, Legionella pneumophila,* and *Aggregatibacter* [formerly *Actinobacillus*] *actinomycetemcomitans*), which can become **naturally transformable (naturally competent)**. Natural competence allows the uptake of DNA from the environment and incorporation of the homologous DNA into the chromosome.

Frederick Griffith (1879–1941) in 1928 performed the first experiment demonstrating natural transformation in the pathogen *S. pneumoniae*. Griffith observed that *S. pneumoniae* came in two varieties, a pathogenic one (S strain) that formed smooth colonies on agar plates (now known to be due to secretion of a polysaccharide capsule) and a nonpathogenic one (R strain) that formed rough colonies. In his classic experiment (Figure 7–6), Griffith found that if he injected mice with the S strain, they became sick and died, but if he first heat killed the S strain before injection or if he injected the R strain, then the mice did not die. Surprisingly, if he mixed the live R strain with the heat-killed S strain before injection, then the mice died. He also found that the bacteria recovered from the dead mice formed smooth colonies on agar plates. That is, the R strain had converted into an S strain. He concluded that there must be a "transforming principle" in the sample of dead S-strain bacteria that transferred the pathogenic trait to the live R-strain bacteria. Sixteen years later, Oswald Avery (1877–1955), Colin MacLeod (1909–1972), and Maclyn McCarty (1911–2005) showed that the "transforming principle" in Griffith's experiment was actually DNA.

Many naturally competent bacteria have short (10- to 12-bp) **DNA uptake sequences (DUS)** scattered throughout their genomes that serve as recognition sites for binding and uptake of the DNA into the cell (Figure 7–7). Some bacteria, e.g., *N. gonorrhoeae* and *H. influenzae*, have specific receptors that recognize their own DUS and allow them to take up DNA only from their own species. By contrast, other bacteria, e.g., *Bacillus subtilis* and *S. pneumoniae*, seem to be able to take up any DNA. In many bacteria, competence is highly regulated. This regulation often involves the bacteria sensing a signal during growth, as well as other factors, such as pH. For example, in *S. pneumoniae*, the signal is a small peptide (competence-stimulatory peptide). Bacteria that sense competence signals initiate a complex cascade of gene expression that produces the proteins needed to take up DNA. In addition, some bacterial species can also undergo a form of fratricide, in which competent cells kill noncompetent cells in the population, releasing even more DNA.

The mechanism of DNA uptake across the inner membranes of most naturally competent gram-negative bacteria is similar to that of gram-positive bacteria, and many of the proteins involved in competence are conserved in both systems. Recently, a second type of DNA uptake mechanism has been reported for *H. pylori* and *C. jejuni*, which is related to type IV secretion and *Agrobacterium tumefaciens* Ti plasmid transfer systems. However, as expected, the initial interaction of DNA with the bacterial cell surface (due to the presence of an outer membrane in gram-negative bacteria) is different. The general pathway of natural transformation for each type of bacteria is depicted in Figure 7–8.

In gram-positive bacteria, the first step is nonspecific binding of the double-stranded DNA to the cell surface, followed by nicking and/or cleavage of the DNA and uptake of one of the DNA strands across the cell membrane into the cytosol, while the other strand is degraded by nucleases. Homologous single-stranded donor DNA integrates into the chromosomal DNA of the recipient to form a heteroduplex that replaces the corresponding recipient strand. Efficient DNA uptake in gram-negative bacteria requires binding of specific DUS to some protein on the surface of the cell and transport of the DNA into the periplasm. As for gram-positive bacteria, transport of DNA across the inner membrane involves nicking and/or cleavage, followed by uptake and integration of one strand into the chromosome of one strand and degradation of the other strand.

Conjugation: Plasmids and Transposons

A number of bacteria are able to transfer DNA (plasmids and other DNA elements) directly from one cell to another in a process called **conjugation.** Joshua Lederberg (1925–2008) and Edward Tatum (1909–1975) first suspected in 1947 that bacteria could exchange their DNA when they observed that mixing two different strains of *E. coli* resulted in new progeny that were unlike either of the parental strains, which they proposed was some sort of mating process. During conjugation, the two DNA strands separate and one strand moves from the donor bacterium into the recipient bacterium through a pore in the mating bridge that forms between the two bacteria. The two single strands in each cell then serve as replication templates to generate double-stranded DNA again (Figure 7–9). The progeny that result from conjugation are called **transconjugants.**

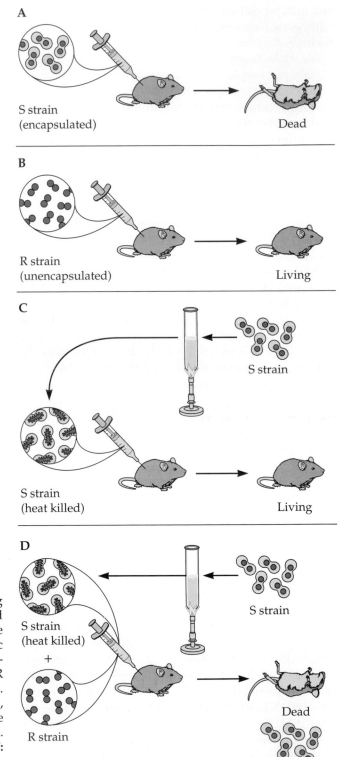

Figure 7–6 Griffith's classic experiment demonstrating natural transformation in *S. pneumoniae*. **(A)** Mice injected with a pathogenic encapsulated strain (S strain) became sick and died. **(B)** Mice injected with a nonpathogenic strain (R strain) did not die. **(C)** Mice injected with a heat-killed S strain did not die. **(D)** Mice injected with an R strain mixed with a heat-killed S strain got sick and died. Living S-strain bacteria were isolated from the dead mice, indicating the conversion of the avirulent bacteria into the virulent, encapsulated form. (Adapted from F. Griffith. 1928. The significance of pneumococcal types. *J. Hyg.* **27:** 113–159, with permission from Cambridge University Press.)

| *Haemophilus influenzae* | 5'- AAGTGCGGTCA - 3' |
| *Neisseria gonorrhoeae* | 5'- GCCGTCTCAA - 3' |

Figure 7–7 Bacterial DNA uptake sequences. Some bacteria will take up only DNA from the same species with specific DUS.

Plasmids are extrachromosomal DNA that can replicate and partition independently of the chromosome. Many genes required for virulence, including antibiotic resistance genes, some toxin genes, and genes that promote adherence to host cells, reside on plasmids and are frequently acquired by and transferred among pathogenic bacteria, as well as some commensal organisms that serve as "reservoirs" for new genetic material. Naturally occurring plasmids are either self-transmissible or mobilizable. **Self-transmissible** plasmids possess all the genetic information needed for them to be transferred via conjugation from one bacterium to another (Figure 7–10). They contain an *oriT* site (origin of transfer sequence), where transfer initiates, and they carry all the genes, called *tra* genes or *mob* genes, needed to assemble the machinery to transfer the DNA from the donor bacterium to the recipient. **Mobilizable** plasmids have an *oriT* site but lack the *tra* genes and so need help to move their DNA (Figure 7–10). These plasmids must use the *tra* machinery from another transfer system to transfer their DNA. Mobilizable plasmids are very useful in biotechnology for transferring cloned genes into recipient bacteria containing promiscuous transfer machinery.

Transposons were first discovered in bacteria in the 1970s. Transposons are DNA elements (segments) that can move (transpose) from one place in the bacterial DNA to another. The smallest bacterial transposons are called **insertion sequence (IS)** elements and consist of a **transposase** gene, encoding an enzyme that promotes their transposition, and inverted-repeat sequences at their ends, which are used to target IS sites (short direct repeats) in the target DNA (Figure 7–11). These end sequences can also contain promoters that point outward from the IS element and drive the expression of adjacent genes on the chromosome. Larger transposons can encode other genes for selectable markers (e.g., antibiotic resistance genes) or virulence factors (e.g., biosynthetic genes for polysaccharide capsule or toxin genes). Some transposons, called **conjugative transposons,** also carry *tra* genes that can promote transfer of their own DNA, as well as other small mobilizable DNA elements that are present in

the same strain. Conjugative transposons also have genes for proteins that promote integration into or excision from the chromosome. Many conjugative transposons and plasmids have promiscuous transfer systems that enable them to move DNA between unrelated species, as well as between like species. Interspecies transfer of plasmids and transposons is most likely responsible for the emerging prevalence of antibiotic resistance (see chapter 16).

Phage Transduction

Another mechanism by which bacteria can exchange DNA is through transfer by bacterial viruses (**bacteriophages,** or "phage" for short) in a process called **transduction.** In 1927, Martin Frobisher and J. Howard Brown (1884–1956) showed that a filterable agent from scarlet fever isolates could convert non-scarlet-fever-inducing strains of *Streptococcus pyogenes* into strains that could induce scarlet fever. However, it was not until 1951 that Victor Freeman (1919–2002) first demonstrated that the diphtheria toxin gene was carried on a β-bacteriophage from *Corynebacterium diphtheriae*. We now know that many toxins and other virulence factors are encoded on integrated bacteriophages found in a wide range of gram-negative and gram-positive bacteria.

Transduction allows a bacterium to acquire new DNA segments, usually in the range of 20 to 100 kb in length. Phages can have two phases as part of their life cycles. During the **lytic phase,** the phages replicate and lyse their bacterial hosts, but during the **lysogenic phase,** the phages integrate into the bacterial chromosome and can express their encoded genes, thereby conferring new biochemical properties on their hosts. **Generalized transducing phages** can transfer their DNA from one bacterium to another when they go into lytic phase. However, while packaging their own DNA, they may also accidentally package some of the bacterial cell's chromosomal DNA instead (Figure 7–12A). Those mispackaged phages can still infect bacteria but cannot replicate because they do not have a phage genome. Instead, they deliver DNA segments from the donor bacterium that can subsequently recombine through homologous regions into the chromosome of the recipient bacterium. **Specialized transducing phages,** on the other hand, move their own phage genes but sometimes can also package segments of bacterial DNA that flank the phage attachment site (Figure 7–12B). Specialized transducing phages undergo both lytic and lysogenic phases in their life cycles (Figure 7–12C). A **lysogenic phage** can revert to the lytic phase. When the phage

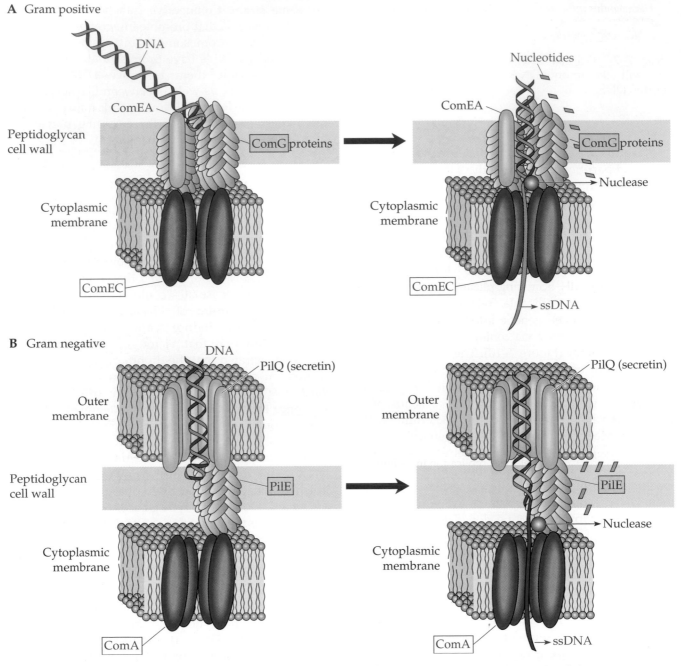

A Gram positive

DNA

ComEA

Peptidoglycan cell wall

ComG proteins

Cytoplasmic membrane

ComEC

Nucleotides

ComEA

ComG proteins

Nuclease

Cytoplasmic membrane

ComEC

ssDNA

B Gram negative

DNA

PilQ (secretin)

Outer membrane

Peptidoglycan cell wall

PilE

Cytoplasmic membrane

ComA

PilQ (secretin)

Outer membrane

PilE

Nuclease

Cytoplasmic membrane

ComA

ssDNA

Figure 7–8 Comparison of general pathways of transformation in gram-positive **(A)** and gram-negative **(B)** bacteria. (Adapted from L. Snyder and W. Champness, *Molecular Genetics of Bacteria*, 3rd ed., ASM Press, Washington, DC, 2007.)

genome excises once again from the bacterial chromosome, the potential for specialized transduction occurs.

Lysogenic phages that have integrated their genomes into the bacterial chromosome are called **prophages.** When prophages stay in the chromosome and replicate with the bacteria, they may mutate and lose the ability to undergo a lytic phase. Such defective

prophages that become dormant (i.e., inactive and no longer able to enter a lytic phase) once integrated into the bacterial chromosome, however, can often still express many of their virulence genes. Over time, these virulence gene-carrying prophages, whether defective in reentering the lytic phase or not, have been recognized as a form of **pathogenicity island (PAI)** (see below).

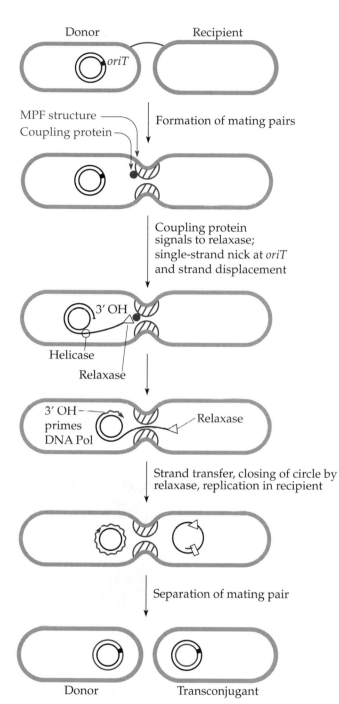

Figure 7–9 Mechanism of DNA transfer during conjugation (self-transmissible plasmids). (Adapted from L. Snyder and W. Champness, *Molecular Genetics of Bacteria,* 3rd ed., ASM Press, Washington, DC, 2007.)

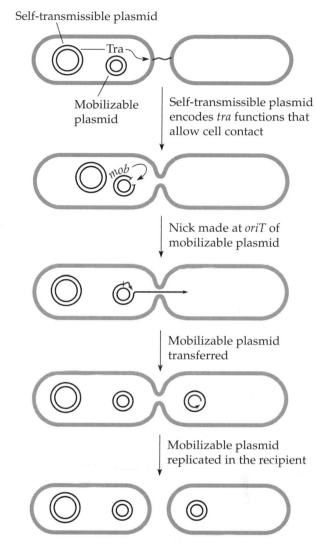

Figure 7–10 Mobilizable plasmids. Shown is the mechanism of DNA transfer of a mobilizable plasmid.

PAIs and Pathogen Evolution

Properties of PAIs

In the early 1980s, studies of *E. coli* strains involved in urinary and intestinal disease revealed that the *hly* genes, encoding the pore-forming toxin α-hemolysin, were located on large chromosomal DNA regions with a percent G+C content (G+C%) and codon usage that were different from those of the rest of the bacterial chromosome. The G+C% of these DNA regions was found to be 41%, whereas the overall G+C% of the *E. coli* chromosome was 51%. Further studies showed that additional virulence genes, encoding P fimbriae (adherence proteins), a toxin called cytotoxic necrotizing factor 1, and other virulence-associated proteins, were also located within these regions. The regions were flanked by direct repeats of 16 to 18 bp, indicating that these DNA sequences had been acquired through HGT (i.e., recombination, phage insertion, or transposition). In the early 1990s, these DNA segments of the genome containing one or

A

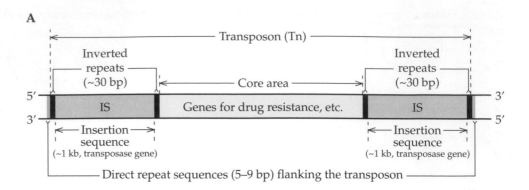

B

1   IS element

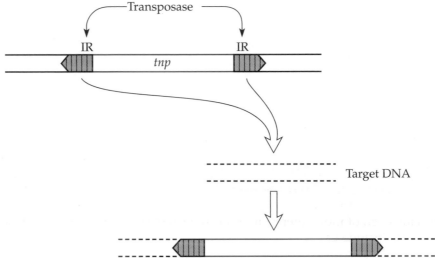

2   Composite: 2 IS + gene *A*

Figure 7–11 Mechanism of DNA transposition. **(A)** General structure of a transposon. A common feature of transposons is the presence of flanking short direct repeats of 3 to 9 bp. **(B)** During transposition, a transposon moves from one location to another nonhomologous location. The target DNA contains only one copy of the IS, but during transposition, the sequence is duplicated. Most transposons have little or no target sequence specificity. (Panel B is adapted from L. Snyder and W. Champness, *Molecular Genetics of Bacteria*, 3rd ed., ASM Press, Washington, DC, 2007.)

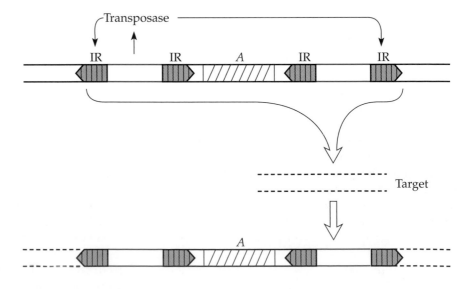

IR, indirect repeats;
DR, direct repeats

more virulence genes acquired through HGT were named PAIs.

Comparison of a large number of genome sequences, including multiple sequences from multiple strains of the same bacterial species, has revealed the importance of PAIs in the diversification of strains within a bacterial species. We now know that the genomes of most bacterial pathogens contain multiple PAIs, whereas their nonpathogenic counterparts do not. Many PAIs appear by sequence analysis to be prophages or remnants thereof, but some appear to be integrated conjugative elements. PAIs can constitute as much as 10 to 20% of the bacterial genome. PAIs have been found in both gram-positive and gram-negative bacteria and are usually inserted into defined locations within highly conserved DNA regions on the chromosome, such as phage attachment *(att)* sites, bacterial tRNA genes *(leuX* and *selC)*, or ISs (direct repeats) (Figure 7–13). Nevertheless, many PAIs are still parts of active mobile DNA elements (plasmids, transposons, or phages) or of mobilizable DNA elements and have a tendency to be deleted from the chromosome at frequencies as high as 10^{-4} to 10^{-5} or to undergo duplications and amplifications.

PAIs that are still mobilizable (plasmids, transposons, or prophages) carry genes needed for transmission, although many have lost them or have defective ones. However, of more interest in terms of virulence, many PAIs include genes that confer a variety of new functions. These genes may provide the host with a selective advantage under certain environmental conditions. For example, they may provide new types of pili for altered adhesion, allowing the bacteria to bind to different tissues and enhance colonization. They may provide means of acquiring iron and other nutrients. They may allow increased survival in the host by carrying genes for novel surface structures, such as LPS, and increased serum resistance; for capsular biosynthesis to prevent phagocytosis; or for delivery of proteins that enhance bacterial invasion, modulate intracellular signaling processes, or dampen immune responses (e.g., type III secretion systems and their secreted proteins—more on these later).

Genes encoding most bacterial protein toxins are located on PAIs. Indeed, phages or lysogenic prophages carrying toxin genes are thought to serve as a major natural reservoir for toxin genes. HGT of toxin genes in natural environments may account for the prevalence of related toxins among diverse pathogens (Table 7–2). For example, the cholera toxin gene *(ctx)*, which is carried on a phage in *Vibrio cholerae*, is closely related to the heat-labile enterotoxin genes *(elt* and *etx)* found in different strains of *E. coli* that cause di-

arrhea. These enterotoxin genes can be found either on a plasmid or on the chromosome, depending on the strain of *E. coli*. Other examples (shown in Table 7–2) are the widespread occurrence of Shiga toxin-related *(stx)* genes in *Shigella* and *E. coli* strains and the closely related botulinum or tetanus neurotoxin *(bot* or *tet)* genes found in different strains and species of *Clostridium*. Considerable evidence suggests that phages containing *stx* from *Shigella* and *E. coli* can be transmitted, not only between different bacteria in the intestines of humans and other animals, but also in external aquatic environments, such as sewage or water and soil contaminated with feces. The diverse locations of the PAIs carrying *Clostridium botulinum* and *Clostridium tetani* neurotoxin genes, which in different strains are found on plasmids and lysogenic prophages or as gene clusters on the chromosome (as putative transposons or prophages), illustrate the degree of HGT that has contributed to their evolution.

Pathogen Evolution in Quantum Leaps

Point mutations, genomic rearrangements, and antigenic variation lead to slow adaptive evolutionary changes, but acquisition of a single PAI can convert a nonpathogenic bacterium into a pathogen in a single step. One of the first demonstrations of this was that of the phage-encoded diphtheria toxin, which is the primary virulence factor responsible for disease symptoms. Nontoxigenic strains of *C. diphtheriae* are avirulent and, indeed, are often found colonizing the upper respiratory tract. Conversion into virulent toxigenic strains can occur by acquisition of the toxin gene via phage transduction. Plasmids, phages, and transposons are thus means of rapid evolutionary change. These mobile elements, sometimes resulting in PAIs, can contribute in a single step to the ability of the newly formed strains to enhance colonization, survival, and dissemination in the host. Sometimes more than one type of trait is acquired in a single step. Thus, mobile elements have high potential for generating new pathogens in relatively short periods of time. Since gene clusters in PAIs are acquired as a unit in a single HGT integration event, virulence genes carried on PAIs found in different bacterial genera have remarkable sequence similarity and are often found arranged in the same order, i.e., they share a common genetic organization and evolutionary origin.

Stepwise acquisition of PAIs can lead to progressive increases in virulence and the rapid emergence of new pathogens (Figure 7–13). For example, two large PAIs have been identified in *Salmonella*, each contributing to a specific step in the course of infection. The *Salmonella* pathogenicity island 1 (SPI-1) con-

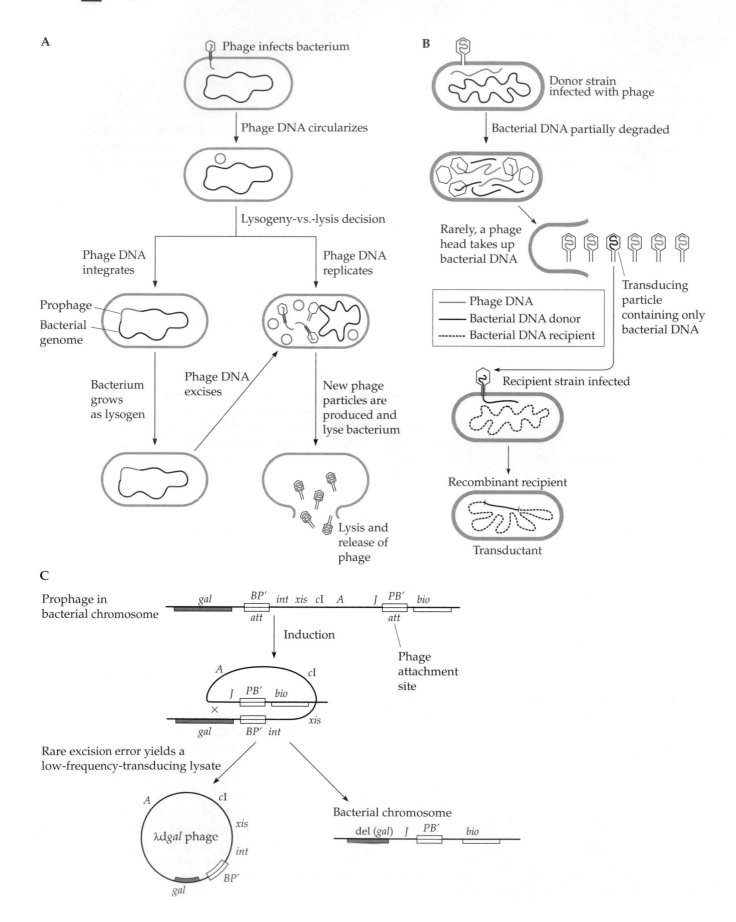

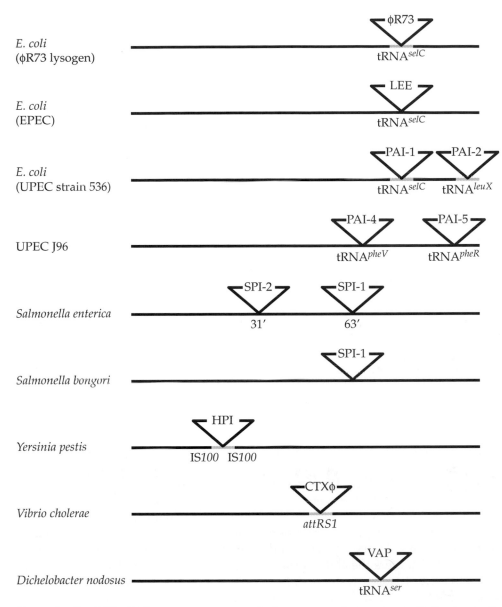

Figure 7–13 PAIs. PAIs can carry many virulence genes in compact, distinct genetic units that are inserted into defined locations on the chromosome, flanked by tRNA genes *(leuX* and *selC)*, phage attachment sites *(att)*, or plasmid/transposon ISs. Stepwise acquisition of PAIs can lead to progressive increases in virulence. EPEC, enteropathogenic *E. coli*; UPEC, uropathogenic *E. coli.* (Adapted from E. A. Groisman and H. Ochman. 1996. Pathogenicity islands: bacterial evolution in quantum leaps. *Cell* **87:**791–794, with permission from Elsevier.)

Figure 7–12 Mechanisms of phage transduction. **(A)** Phage cycle, including lytic phase and lysogenic phase. **(B and C)** Bacteriophages can transfer DNA from one bacterial cell to another through two different mechanisms: generalized transduction, where the phage randomly packages some of the bacterial DNA along with its own DNA **(B),** and specialized transduction, where the phage genes, and sometimes genes adjacent to the phage attachment site in the bacterial chromosome, are transferred **(C).** (Adapted from L. Snyder and W. Champness, *Molecular Genetics of Bacteria,* 3rd ed., ASM Press, Washington, DC, 2007.)

Table 7–2 Diverse locations of PAI-encoded toxin proteins[a]

Protein toxin	Gene	PAI location	Bacterial host
Diphtheria toxin (DT)	*tox*	Phages	*C. diphtheriae, C. ulcerans, C. pseudotuberculosis*
Clostridial neurotoxins			
BoNT/A1	*botA1*	Chromosome	*C. botulinum*
BoNT/A2	*botA2*	Chromosome	*C. botulinum*
BoNT/B	*botB*	Chromosome	*C. botulinum*
BoNT/C1	*botC1*	Prophage	*C. botulinum*
BoNT/D	*botD*	Prophage	*C. botulinum*
BoNT/Dsa	*botC/D*	Phage?	*C. botulinum*
BoNT/E	*botE*	Plasmid, phage	*C. botulinum, C. butyricum*
BoNT/F	*botF*	Chromosome	*C. botulinum, C. baratii*
BoNT/G	*botG*	Plasmid	*C. botulinum*
TeNT	*tet*	Plasmid	*C. tetani*
Shiga toxin (ST)	*stx*	Phages	*Shigella dysenteriae, Escherichia coli*
Shiga-like toxins (SLT)	*stx*	Phage	*Shigella sonnei*
	stx1	Phage	*E. coli*
	stx1c	Phage	*E. coli*
	stx2	Phage	*E. coli*
	stx2c	Phage	*E. coli*
Cholera toxin (CT)	*ctxAB*	Phage	*Vibrio cholerae, V. mimicus*
Heat-labile enterotoxin (HLT)	*elt, etx*	Plasmid, chromosome	*E. coli*

[a] Adapted from B. A. Wilson and M. Ho. 2006. Evolutionary aspects of toxin-producing bacteria, p. 25–43. *In* J. E. Alouf and M. R. Popoff (ed.), *The Comprehensive Sourcebook of Bacterial Protein Toxins.* Elsevier, Amsterdam, The Netherlands.

sists of about 25 genes, including a type III protein secretion system and various effector proteins that are delivered to the host cells via the type III secretion system (see chapter 13). SPI-1 confers the ability to invade epithelial cells. SPI-2 consists of about 15 genes, including a second type III protein secretion system, a two-component regulatory system, and other effector proteins. SPI-2 confers the ability to survive within macrophages and cause systemic infections. Acquisition of these two PAIs was critical in the development of *Salmonella* as an intracellular pathogen because the PAIs enabled *Salmonella* to invade host cells, evade host defenses, and cause systemic infections. SPI-1 is present in strains from all subgroups of *Salmonella enterica*, but SPI-2 is not found in *Salmonella bongori* strains, which are of intermediate virulence and are thought to represent an intermediate step in *Salmonella* evolution.

A major take-home lesson of HGT and the multiple mechanisms of genetic exchange in bacteria is that there is tremendous diversity within bacterial species, as well as the more traditional view of differences between bacterial species. Two different clinical isolates that are the same bacterial species based on rRNA sequence analysis can show considerable differences in the numbers and types of PAIs, lysogenic bacteriophages, and extrachromosomal elements they contain.

These differences underlie the fact that single bacterial species cause a multitude of infections. For example, there are multiple diseases caused by different kinds of *E. coli*, *Staphylococcus aureus*, *S. pneumoniae*, and *S. pyogenes*, to name a few. This diversity within bacterial species and mechanisms of exchange of large chromosomal segments, such as occurs with natural transformation, bacteriophage-mediated transduction, and conjugation, means that bacterial pathogens actually have a much larger gene pool, called a **supragenome** or **pangenome,** than is in the chromosome of any given bacterial cell. The supragenome within a bacterial species and the potential to exchange genes between different bacterial species (often called the **metagenome**) discussed in this chapter present a formidable challenge. Bacterial diseases will continue to emerge, reemerge, and change due to the tremendous genetic plasticity of supragenomes and the vast number of genes available in the metagenome.

SELECTED READINGS

Bender, J. B., C. W. Hedberg, J. M. Besser, D. J. Boxrud, K. L. MacDonald, and M. T. Osterholm. 1997. Surveillance for *Escherichia coli* O157:H7 infections in Minnesota by molecular subtyping. *N. Engl. J. Med.* **337:**388–394.

Brüssow, H., C. Canchaya, and W.-D. Hardt. 2004. Phages and the evolution of bacterial pathogens: from genomic rearrangements to lysogenic conversion. *Microbiol. Mol. Biol. Rev.* **68:**560–602.

Casjens, S. 2003. Prophages and bacterial genomics: what have we learned so far? *Mol. Microbiol.* **49:**277–300.

Chen, I., and D. Dubnau. 2004. DNA uptake during bacterial transformation. *Nat. Rev. Microbiol.* **2:**241–249.

Claverys, J. P., B. Martin, and L. S. Havarstein. 2007. Competence-induced fratricide in streptococci. *Mol. Microbiol.* **64:**1423–1433.

Deitsch, K. W., E. R. Moxon, and T. E. Wellems. 1997. Shared themes of antigenic variation and virulence in bacterial, protozoal, and fungal infections. *Microbiol. Mol. Biol. Rev.* **61:**281–293.

Hinnebusch, B. J., M.-L. Rosso, T. G. Schwan, and E. Carniel. 2002. High-frequency conjugative transfer of antibiotic resistance genes to *Yersinia pestis* in the flea midgut. *Mol. Microbiol.* **46:**349–354.

Medini, D., D. Serruto, J. Parkhill, D. A. Relman, C. Donati, R. Moxon, S. Falkow, and R. Rappuoli. 2008.

Microbiology in the post-genomic era. *Nat. Rev. Microbiol.* **6:**419–430.

Mehr, I. J., and H. S. Seifert. 1998. Differential roles of homologous recombination pathways in *Neisseria gonorrhoeae* pilin antigenic variation, DNA transformation and DNA repair. *Mol. Microbiol.* **30:**697–710.

Moxon, R., C. Bayliss, and D. Hood. 2006. Bacterial contingency loci: the role of simple sequence DNA repeats in bacterial adaptation. *Annu. Rev. Genet.* **40:**307–333.

Schmidt, H., and M. Hensel. 2004. Pathogenicity islands in bacterial pathogenesis. *Clin. Microbiol. Rev.* **17:**14–56.

Shoemaker, N. B., H. Vlamakis, K. Hayes, and A. A. Salyers. 2001. Evidence for extensive resistance gene transfer among *Bacteroides* spp. and among *Bacteroides* and other genera in the human colon. *Appl. Environ. Microbiol.* **67:**562–568.

Wilson, B. A., and A. A. Salyers. 2003. Is the evolution of bacterial pathogens an out-of-body experience? *Trends Microbiol.* **11:**347–350.

Winstanley, C., and C. A. Hart. 2001. Type III secretion systems and pathogenicity islands. *J. Med. Microbiol.* **50:**116–126.

QUESTIONS

1. Should microbiologists be alarmed about the unknown risks posed by bacteria gaining potential virulence traits from external environmental sources?

2. Is acquisition of a PAI sufficient to transform an organism into a pathogen?

3. What is the advantage of incorporating virulence genes into the chromosome over retaining them on plasmids?

4. You have identified an operon on a new strain of *E. coli* that carries three genes involved in capsule biosynthesis, as well as a regulatory gene. After careful analysis of the operon and its flanking DNA sequences, you suspect that the operon is located on a PAI that is a lysogenic prophage. What possible evidence might have led you to propose this?

5. *Neisseria* species are known to undergo both phase and antigenic variation. Define phase versus antigenic variation and describe how each might occur. What advantage do they provide for a pathogen?

6. What role have bacteriophages played in pathogen evolution?

7. Some researchers have tried to identify the PAIs in *Salmonella* by comparing the *E. coli* K-12 genome with that of *S. enterica* serovar Typhimurium. What are the advantages and disadvantages of this approach?

8. Which types of variation in surface antigens would still be seen in a strain of *N. gonorrhoeae* that was deficient in the ability to carry out homologous recombination?

9. Bacterial cells have the ability to change the set of genes that they express in response to changes in the environment. Explain why the ability to "turn on" only those genes that are needed at a given time is an advantage to the bacterium.

10. Joe is a sexually promiscuous 22-year-old who has had gonorrhea five times in the past 2 years. Each time he went to the doctor, he was given antibiotics, and the infection cleared up. List at least two reasons why *N. gonorrhoeae* was able to cause recurrent infections in Joe. Could Joe be vaccinated to avoid these recurring infections? Why or why not?

SOLVING PROBLEMS IN BACTERIAL PATHOGENESIS

1. Autolysis and natural competence contribute to the ability of *Neisseria* to undergo antigenic variation, an important virulence factor that allows it to avoid an effective immune response by the host and that also contributes to cell and tissue tropism. To determine the importance of RecA-dependent recombination in antigenic variation in *Neisseria,* you make a mutant of *Neisseria* lacking the *recA* gene and you develop a PCR method for detecting whether homologous recombination has occurred. The method you develop involves using a primer to the N-terminal region of the *pilE* gene and a set of primers to the C-terminal regions of several of the *pilS* genes. You find that the PCR using these primers yields a product for the wild-type *Neisseria* strain, but not for the Δ*recA* mutant strain. You conclude that RecA-dependent recombination is important for antigenic variation to occur. Explain what is the basis of the experimental design that allowed you to come to this conclusion (i.e., how could you tell whether antigenic variation occurred using this PCR method?).

2. In the early days of the study of genetic variation in *Neisseria,* there was a debate about whether most of it was due to intragenic or intergenic recombination. Design a simple experiment to distinguish between these two possibilities. How would the outcome look if both were making comparable contributions?

3. You do a transposon mutagenesis experiment on bacterium X. You are interested in mutants that have decreased ability to produce a toxin. In several cases, you find that toxin production has disappeared. In one case, however, the toxin is produced, but at a lower level than by the wild-type strain. Also, whereas toxin production in the wild-type strain was triggered by the exposure of the bacteria to low-iron conditions, the mutant makes as much toxin in high-iron medium as in low-iron medium. How could you explain these two types of mutants? Why might the second type of mutation (constitutive expression of the toxin gene) reduce the ability of the bacterium to cause disease?

4. The production of hemagglutinins, which are bacterial adhesins for mammalian cells, is a well-established virulence factor for a number of bacterial pathogens. Thus, it was not surprising that *Porphyromonas gingivalis,* a gram-negative anaerobic bacterium associated with dental plaque and periodontal disease, also expresses several hemagglutinins on its cell surface. It was found that one of the hemagglutinin genes *(hagA)* of *P. gingivalis* contains four large, contiguous direct repeats varying from 1,318 to 1,368 bp in length, which together encode a 2,628-amino-acid protein of 283.3 kDa (a very large protein!). The repeat unit (denoted *HArep*), which contains the hemagglutinin adhesin domain, was also found to be present in several other protease and hemagglutinin genes in *P. gingivalis*. The beginning amino acid sequence encoded by the first repeat *(HArep1)* is PNPNPGTTT, while that of the other three repeats *(HArep2 to -4)* is GTPNPNPNPGTTT. The amino acid sequence at the C terminus of the fourth repeat *(HArep4)* is GTPNPNPNP. Provide a possible mechanism that could account for the presence of this repeat unit four times in *hagA* and also in the other protease and hemagglutinin genes (from a molecular evolutionary point of view).

5. Brucellosis, caused by gram-negative *Brucella* species, is a zoonotic disease with serious impact on the livestock industry. In animals, brucellosis can lead to abortions in females and sterility in males. In humans, brucellosis is rarely fatal, but it does cause systemic febrile (fever-causing) disease that can be debilitating. *Brucella* bacteria enter macrophages but are able to evade phagolysosomal fusion and so are able to survive intracellularly. *Brucella,* unlike other gram-negative bacteria, such as *E. coli,* has an unusual LPS that does not activate the innate immune responses. Comparison of the sequenced genomes of several strains that are nonpathogenic to humans (*B. neotomae* and *B. ovis*) to those that are pathogenic to humans (*B. melitensis, B. abortus,* and *B. suis*) has revealed that *Brucella* species are highly conserved overall (>90% of the genes share 98 to 100% sequence identity at the nucleotide level, with a G+C% of 57.3), except for certain distinct regions, called genomic islands (GIs). A group of researchers hypothesized that host specificity and virulence differences must stem from the limited genome diversity found in five of these regions that are present in all pathogenic strains: GI-1 (8.1 kb; 9 open reading frames [ORFs]; G+C% = 53.2), GI-2 (15.1 kb; 20 ORFs; G+C% = 51.3), GI-3 (21 kb; 30 ORFs; G+C% = 52.3), GI-5 (44.1 kb; 42 ORFs; G+C% = 57.2), and GI-6 (7.5 kb; 10 ORFs; G+C% = 54.2). Comparing these GIs among the species to that of the human pathogen *B. melitensis,* the researchers found that GI-6 is absent in *B. neotomae* and GI-1, GI-2, and GI-5 are absent in *B. ovis*. Below is a schematic diagram of the genetic organization of the ORFs within GI-2. Based on sequence homology to other genes in the DNA database, the researchers were able to annotate some of the ORFs present. The arrows indicate the direction of transcription. OMP, outer membrane protein; Tnpase, transposase.

A. Considering all of the above information, name at least 5 features of GI-2 that led the researchers to propose that GI-2 might be a PAI. What is the most likely mode of HGT for GI-2 if it is a PAI? Be sure to provide your rationale.

B. The researchers suspect that GI-2 may contain genes that contribute to the unusual LPS found in *Brucella* species. What led them to suspect this? (Hint: how does *E. coli* LPS stimulate innate immunity?)

C. Describe an experiment that the researchers could perform to demonstrate that GI-2 indeed contains genes responsible for making the unusual LPS and that this unusual LPS is important for dampening the innate immune responses and enhancing survival in macrophages.

D. Describe an experiment that the researchers could perform to demonstrate that only GI-2 and not any of the other GIs (GI-1, GI-3, GI-5, or GI-6) is involved in pathogenesis.

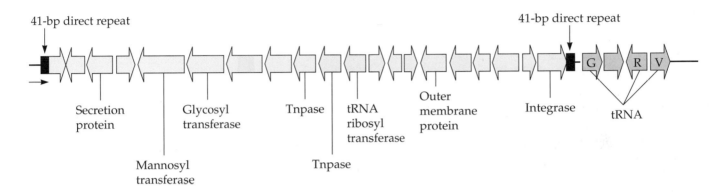

8

Identification of Virulence Factors: Measuring Infectivity and Virulence

A re you a man or a mouse? So goes the traditional challenge to stand up and fight. The answer is supposed to be "a man, who will be strong." But today, that question has a different answer. The new answer is, "Well yes, sometimes I am a mouse. But also, sometimes I am a fish or a fruit fly or even a worm." This answer is not meant to impugn the integrity of the person asked but rather to address a different issue. In standing up for humanity, it is necessary to understand an old and persistent threat, worse than any human terrorist—infectious disease. Identifying the cause of an infectious disease requires satisfying Koch's third postulate, that inoculation of a new host with the isolated organism causes the disease. Moreover, this is only the beginning of the journey. Once a disease agent is identified, it is important to determine how the agent causes disease. This usually means some sort of animal model that can be used to identify virulence factors and steps in disease process and progression, and just as the variety of methods for probing mechanisms of virulence has expanded drastically, so the variety of animal models has expanded beyond the traditional laboratory rodent.

Because virulence is a microbial attribute that is exhibited only in the context of a susceptible host, it is critical that when designing experiments a suitable infection model system be used. First, the animal host must be susceptible to the pathogen. Ideally, infection using this animal model should mirror all of the disease symptoms that occur during natural infection. Alas, this is not always possible, and alternative infection models must then be found. Moreover, to identify factors that contribute to virulence (**virulence factors**) and to understand their roles in the infection process that leads to disease, one must also have some way to accurately and reproducibly measure virulence using the infection model. In this chapter, we describe various animal models that have been developed and that are being used for studying infectious diseases caused by bacteria. We also describe various methods that can be used to measure virulence and infectivity (i.e., the degree to which a host has been infected).

Animal Models of Infection

Human Volunteers

We start with the best possible model for studying human disease: humans. Human volunteers can be, and have been, used in many infectious disease studies. Ethical considerations make this approach problematic unless the disease is not life threatening, is easily treatable, or can be prevented with a vaccine. Human volunteers have played an important role in testing of vaccines against cholera, chancroid (genital ulcers), and gonorrhea, all diseases for which there is no suitable animal model. Also, antibiotic intervention trials, such as those to assess the efficacy of antibiotic treatments for ulcers or atherosclerosis, are also ethically defensible, because the treatment presumably does no harm to the participants and may even benefit them.

Some human trials, on the other hand, have raised ethical issues. A recent example is provided by the large trials in Thailand and Africa in the 1990s of a short course of zidovudine (AZT) therapy to prevent mother-to-infant transmission of human immunodeficiency virus (HIV). Developing countries cannot afford to use the long-term AZT regimen used in developed countries to prevent maternal transmission of HIV, a regimen that starts early in pregnancy and continues for months after childbirth. Physicians had begun to suspect that a much shorter course of AZT would protect newborns nearly as well and a lot more cheaply.

The controversy arose because the control group received a placebo, whereas the trial group received the short-term course of AZT. Although this seems at first glance to be a rational design, it would not be approved in any developed country. The ethically acceptable test design in developed countries would be to compare women treated with the long course of AZT, a treatment known to be effective, with those treated with the shorter course. Scientists in the United States and Europe questioned the ethics of the short-course trial on the basis that the trial applied an ethical standard to people in developing countries different from the one applied to people in developed countries.

Why were scientists in developed countries, especially the United States, so concerned over the ethical issues raised by this test? One reason is a human trial called the Tuskegee syphilis experiment, which was conducted in the United States under the auspices of the precursor of today's National Institutes of Health. In this trial, men with syphilis were not treated but instead were merely observed as the disease took its course (Box 8–1). This "experiment" was initiated in the 1930s but continued until the first accounts of it appeared in the news in 1972. Two other unethical experiments took place in the 1960s. One was the Jewish Chronic Disease Hospital study, in which chronically ill patients were injected with live cancer cells to see if the cancer cells would be rejected. In the second study, mentally handicapped children newly admitted to the Willowbrook State School in New York were injected with viable hepatitis virus to study the course of infection.

These are only a few examples of historical events that eventually led to the current ethical guidelines and federal regulations concerning human research subjects, which are based on the Belmont Report of 1979. This report, written by a National Commission for the Department of Health and Human Services, identified three ethical principles that must be maintained in the conduct of research involving humans: respect for persons (informed consent), beneficence (maximize benefit and minimize harm), and justice (fair treatment and risk distribution). Although most scientists practicing today were not around when these infamous examples of unethical science were perpetrated, the chilling implications of them are still very real to older scientists, who are determined that no such thing or anything remotely like them will ever be repeated. Thus, today, there are very strict guidelines for conducting research with human subjects, and all experiments must be approved in advance by institutional committees made up of experts in performing and interpreting these types of experiments, including physicians, statisticians, and bioethicists.

A clinical-trial study involving the inoculation of humans with a disease agent or the treatment of infected humans is called a **"prospective"** study because it starts in the present and moves into the future. Another type of human clinical study is called a **"retrospective"** study. This type of study is done on infectious disease outbreaks that have already occurred accidentally or naturally and that are studied in retrospect (looking back at past events) to obtain information about disease transmission or progression in humans. One example is the case of a school bus driver with tuberculosis who managed to infect a number of school children before his disease was diagnosed. This study, in retrospect, provided valuable information about the factors affecting transmission of tuberculosis. A retrospective study of various aspects of this case revealed, for example, that the likelihood of infection was directly linked to the amount of time a child spent on the bus. The fact that 40 minutes a day versus 10 minutes a day made a discernible dif-

BOX 8–1 The Tuskegee "Experiment"—a Shameful Chapter in Infectious Disease History

Syphilis, a sexually transmitted bacterial disease, was widespread in the southeastern United States in the 1930s. The first treatment for syphilis, arsphenamine (an arsenic compound) and bismuth, was introduced in the early 1930s. In 1932, a decision was made by the U.S. Public Health Service in Alabama to withhold treatment from 399 black men with syphilis for the purpose of learning more about the development and pathology of the disease. The men, most of whom were poor sharecroppers who could not afford medical care, were offered "free health care," something unknown in their experience. This care consisted of regular visits to a clinic, where they were treated for any ailments (other than syphilis) and where detailed records could be kept of the progression of the disease. The first decision to withhold therapy has been defended on the basis that arsphenamine treatment had some serious side effects and was not a foolproof cure, although it cured many treated individuals. However, when penicillin became available to the public in 1947 and proved to be an effective and nontoxic cure for syphilis, the decision to continue to withhold therapy from the men was simply criminal. This experiment continued until 1972. By that time, 28 of the men had died of syphilis and 100 others had died of syphilis-related complications. The study was dubbed the "Tuskegee experiment" because some of the laboratory facilities at Tuskegee University were used, but the study was designed and carried out by people working for public health agencies. When this experiment was discovered and made public in 1973, it provoked extensive congressional hearings, which formed the basis for the current ethical rules governing the use of human volunteers. The U.S. government, after a long delay, offered free medical care to the men involved, but not to their infected wives or children, who were infected because of being born to syphilitic mothers. Grudgingly, health care was later extended to family members, but it took a lawsuit on behalf of the men to prod the government to offer a cash settlement, small by today's standards. To the end, the Public Health Service officials refused to acknowledge any wrongdoing and continued to defend the study on the basis that the men would not have had treatment anyway because of the lack of knowledge of medical services and their poverty, an argument that infuriated rather than convinced critics of the study. This incident is considered a tragic example of research abuse, racial oppression, and profound social injustice. In May 1997, President Clinton officially apologized for the gross injustice, but the Tuskegee study has become a prime symbol of unethical research and governmental lack of concern about the fate of some of its most vulnerable citizens. This painful chapter in U.S. medicine had almost been forgotten—a fact that raised some serious questions about pretensions to train students in bioethics— when it resurfaced in connection with the controversial AZT trial design. Critics of the AZT study invoked the Tuskegee precedent in their outrage over the fact that different standards of medicine were being applied to people in the developing world than to people in the developed world. Defenders of the study pointed out that the women involved in the trials would not, under normal conditions, have received any therapy, but that defense was unacceptable to critics of the study design.

Source: J. H. Jones. 1993. Bad Blood. The Tuskegee Syphilis Experiment. Simon and Schuster, New York, NY.

ference gave scientists a new appreciation of how infectious tuberculosis actually is. Similarly, retrospective studies of food-borne disease outbreaks have helped to determine what types of people are most likely to develop a life-threatening disease in such cases.

Nonhuman Animal Models

For most studies of infectious diseases, however, nonhuman animals are the models of choice, when possible. Since the time of Koch and Pasteur, laboratory rodents have been the most widely used models for infectious disease research because they are small and are thus more easily (and cheaply) housed and cared for than larger animals, such as pigs and baboons. Although rodents are closely related to humans on the evolutionary scale, there are a number of important differences. Anatomically, they are similar to humans, except that they have fur and tails. They also have a more prominent cecum, whereas humans have a vestigial appendix, and a very different microbiota. Rats do not have a gall bladder. A fact that is frequently overlooked but that could be a factor in the use of rodents to study intestinal disease is that rodents practice coprophagy (routine ingestion of one's own feces), whereas humans do not (except unintentionally via unwashed hands). There are undoubtedly many other differences, as is evident from the very different course some human diseases have in mice. For example, *Salmonella enterica* serovar Typhimurium, which causes diarrhea in humans, causes a systemic disease in mice that closely resembles the human disease typhoid fever that is caused by *S. enterica* serovar Typhi. However, *S. enterica* serovar Typhi, which can be deadly in humans, does not infect mice. There are a number of other examples of this type of host specificity.

Because of these differences, for certain diseases, scientists have developed more exotic animal models that may not mimic the human anatomy in many ways and may not mimic the disease in every respect but still provide unique insight into certain aspects of the disease. Examples are ferrets as a model for gastric ulcers caused by *Helicobacter,* guinea pigs as models for tuberculosis, armadillos for leprosy (skin lesions caused by *Mycobacterium leprae*), chinchillas for otitis media (ear infections caused by *Haemophilus influenzae* or *Streptococcus pneumoniae*), and zebra fish for necrotizing fasciitis ("flesh-eating" disease caused by gram-positive streptococci) or infections by fish-specific *Mycobacterium* species related to *Mycobacterium tuberculosis*, which causes tuberculosis in humans. Recently, the nematode *Caenorhabditis elegans* has been used as an animal model for certain pathogenic bacteria, such as *Pseudomonas aeruginosa,* that infect many different hosts. In fact, some scientists have been looking into plants as a model for infectious disease studies. An interesting, or at least entertaining, exotic-animal model is discussed in Box 8–2.

Ideally, an **animal model** for a human infection should contract a disease whose symptoms and distribution of bacteria in the body mimic the human form of the infection. Similarly, animals should acquire the disease by the same route as humans. How far to trust an animal model that does not satisfy these criteria is a matter of judgment. Rather than establishing rigid criteria for whether an animal model is "good" or "bad," a variety of factors need to be considered when choosing such a model for human disease. Is it easy and inexpensive to maintain the animal? For some diseases, nonhuman primates are the only suitable animal models, yet these animals are difficult to obtain, particularly in large numbers, and are expensive to house and maintain. Another consideration is whether studies using the animal model have produced insights that are consistent with observations of the disease in humans. Has research on the animal model led to effective interventions in human disease? There are also ethical considerations. Does the model involve causing significant pain to the animal, which might be considered unwarranted or unethical? This important topic is discussed further below.

More recently, another criterion has surfaced for choosing an animal model: the ease with which the animal model can be manipulated genetically. One of the primary reasons for putting forth the nematode *C. elegans* as an animal model, despite the fact that primitive worms are a long way evolutionarily from humans, is that there are many mutant strains of *C. elegans,* making it possible to investigate the effects of host traits in the development of a disease. There are few scientists who would fail to agree that an imperfect animal model is better than no animal model at all. On the other hand, the course of disease can be very different in distant models, such as *C. elegans,* that have innate immunity but lack adaptive immunity (see chapter 4).

Recently, scientists have focused on modifying conventional animal models, such as mice, to gain unique insights into particular infectious processes. Infant mice have immature immune systems compared to adult mice, so they are often more susceptible to infection. Likewise, irradiated mice are immunocompromised because X rays have destroyed their immune cells, nude mice are genetically defective in their ability to produce T cells, neutropenic mice are defective in their ability to produce neutrophils, and SCID (severe-combined-immunodeficient) mice are genetically defective in their ability to produce B cells and T cells, and hence all of these types of mice are more susceptible to infection.

A good reason for remaining attached to the laboratory-rodent-as-model issue is the increasing number of available mouse strains with specific genetic alterations (i.e., **transgenic mice**). These include not only the so-called **"knockout mice,"** which have disruptions in specific genes, but what we might call **"knock-in mice,"** mice that have had human genes introduced into their genomes. Both knockout and

BOX 8–2 A Blood-Curdling Animal Model

Despite repeated assurances from scientists that vampires do not exist, a segment of the public seems to place more credence in novels and movies than in the opinion of scientists and persists in holding onto a belief in vampires. Accordingly, a group of Scandinavian scientists decided to do the superstitious public a service (or perhaps pull its leg) by testing the old hypothesis that vampires can be repelled with garlic. Unfortunately for the scientists who wanted to conduct the studies, vampires are not easy to come by, so they had to resort, as scientists often do, to an animal model. They chose as a vampire substitute the blood-sucking medicinal leech, *Hirudo medicinalis*. Leeches are readily available because the medicinal leech has recently become respectable once again in medical circles due to its ability to prevent the formation of clots and internal accumulations of blood (hematomas). Hematomas and clots interfere with the healing of surgically reattached fingertips or ear segments because they block the circulation of blood to the reattached area. A person with a reattached fingertip or other limb has only to place a

medicinal leech on that appendage for a short time every few days to eliminate clotting and hematoma formation. End result? Happy patient, happy leech! The Scandinavian scientists borrowed a few leeches and set up an experiment in which leeches were given the opportunity to choose a garlic-smeared hand or a garlic-free hand for their next meal. The results of this experiment demonstrated that leeches actually preferred the garlic-smeared hand to a clean one. The difference was statistically significant. The scientists conducting this study modestly declined to extrapolate their results too recklessly, but their findings definitely raise questions about the efficacy of garlic as a vampire repellant. Does anyone think these scientists have a little too much time (not just garlic) on their hands? At least this study shows that, contrary to their usual portrayal in the media (especially movies), scientists do know how to have a little fun.

Source: H. Sandvik and A. Baerheim. 1994. Does garlic protect against vampires? An experimental study. *Tidsskr. Nor. Laegeforen* **114**:3583-3586.

knock-in mice have been used to study the attachment of *Helicobacter pylori* to the gastric epithelium. Knockout mice that lack **decay-accelerating factor (DAF)** have been used to demonstrate the binding of *H. pylori* to DAF, thus leading to inflammation of the gastric mucosa. Alternatively, a knock-in mouse, the Le^b mouse, which carries the human Lewis b antigen (one of the blood group carbohydrate antigens on the surfaces of human cells), has been used to study *H. pylori* infections because the Lewis b antigen appears to be another receptor for attachment of *H. pylori* to gastric cells and gastric mucin. Such mutant animals offer vast new possibilities for research on infectious diseases, possibilities that are only beginning to be explored.

The availability of these animals is particularly important given the slowly accumulating examples of specific genetic defects that predispose people to certain types of infectious diseases. For example, heterozygous carriage of the sickle-cell trait confers partial immunity to malaria. Homozygous carriage of defective chemokine receptor genes confers resistance to

HIV infection. Heterozygous carriage of the cystic fibrosis gene confers reduced severity of cholera diarrhea. Humans, particularly males, with a certain ε4 allele of the apolipoprotein E (ApoE) gene are more prone to atherosclerosis associated with *Chlamydophila pneumoniae* than those who do not possess this allele. More recently, variant alleles of several complement proteins, such as properdin (also called factor P) and mannose-binding lectin (MBL), have been shown to be linked to an increased childhood susceptibility to *Neisseria meningitidis* and other bacterial infections. As studies of human genetic disorders expand in scope, we will undoubtedly gain a better understanding of how human genetic variation affects patterns of susceptibility to bacterial infections.

Two other useful animal models have recently emerged as important tools for studying infection processes and the host immune response. **Gnotobiotic (germ-free) animals** are raised in sterile environments and have no bacteria in or on them. As a result, they have severely underdeveloped mucosa-associated-lymphoid tissue and no partial immunity due to prior exposure. Unfortunately, they are very expensive to buy and to maintain. **Specific-pathogen-free animals** are animals that are raised in an environment free of a particular pathogen but are otherwise exposed naturally to other microbes. These models eliminate pre-existing immunity due to prior exposure to a microbe that might complicate the interpretation of host response to the pathogen of interest. Gnotobiotic animals have been particularly useful for studying the development of immunity to a pathogen in a "truly naïve" (unimmunized) animal. Such animals have provided invaluable insight into the nature of commensalism and those characteristics that determine whether a microbe will become a commensal or a pathogen.

Measuring Bacterial Infection in Animal Models

Ethical Considerations

Experiments involving animals also involve numerous ethical issues. Disease models invariably require infection of animals, most often rodents, with bacteria that cause disease symptoms, distress, and sometimes death. Therefore, there must be truly compelling reasons for carrying out these experiments. The genetic and physiological properties of the bacterial strains to be tested should be fully characterized before animal experiments are considered. The animal experiments should be designed to test critical hypotheses that can provide useful information for understanding the bacterial disease being studied, even if negative results are obtained. Most important, all procedures used in animal experiments must be approved in advance by a duly appointed institutional committee whose members have experience performing animal experiments, including veterinarians. Unlike many other types of research, even if you have a great idea involving the use of an animal, you need to get the procedure approved in advance before performing the experiment.

Approved animal protocols contain a variety of important details about the experiments and conform to stringent standards of experimental design. They need to argue persuasively that no alternative model can provide information equivalent to that provided by the proposed animal model. The protocol must contain extensive information about the choice of the animal model to be used, ways to minimize the number of animals required that will still give statistically significant results, precedents from the scientific literature, and documentation of appropriate training of laboratory personnel. Besides detailed experimental methods, these documents must contain extensive information about care of the animals before and during the experiments, anesthesia, minimization of pain and discomfort, and euthanasia. In general, death of the animals should not be used as an end point, and complete monitoring rubrics to judge pain and discomfort are compiled, including how to judge whether the animals are sufficiently sick or moribund to require euthanasia. Finally, these protocols must be reviewed and renewed annually.

ANIMAL MODEL BASICS. Because virulence factors are defined as factors that allow a bacterium to infect, to cause symptoms, and sometimes to cause death, measuring properties such as infectivity and lethality is an important part of virulence studies. There are several guiding principles in the design and execution of animal models of infectious disease. First, an animal species that reflects the disease process in humans or that will answer the questions being asked should be chosen. There must be a truly strong rationale to use species higher than mice, rats, or zebra fish. Next, a route of infection must be chosen. Bacterial broth or plate cultures are diluted to defined starting doses based on numbers of bacteria, usually reported as **colony-forming units (CFU).** In the simplest models of systemic infection, the bacterial dose is simply injected with a syringe into the peritoneal cavity of a mouse. In simple models of invasive respiratory infection, a small drop containing the bacteria is placed on the nose of an anesthetized mouse whose mouth is gently held closed for a moment. The mouse will then inhale the drop into its nasopharynx and lungs. In other simple models of infection, the dose of bacteria is delivered orally or directly to the lung or stomach of anesthetized animals through fine tubes. Other, more complicated modes of infection are used for specific models. Once the species and route of infection are chosen, methods for monitoring the infection must be chosen and planned. In some cases, very tiny amounts of blood are removed from tail veins at different intervals, and the numbers of CFU in the blood are determined by spreading dilutions onto plates. Bacteremia can lead to the accumulation of well over 10^8 to 10^9 CFU per ml of blood. However, in many cases, the number of CFU per organ needs to be determined, and this paradigm involves euthanizing the animal, removing the organ, grinding it in physiological saline solution, and spreading a dilution onto plates to determine the number of CFU. To determine colonization in the nasopharynx, saline is injected from the dissected trachea of the mouse and collected at the nose. In all of these models, several animals need to be sacrificed for each dose and time point following infection to obtain statistically significant results, and the number of animals can become quite high. New methods to image infections in live mice **(biophotonic imaging),** which substantially reduce the number of animals needed for some of these experiments, are described below.

SURVIVAL-CURVE ANALYSIS AND BIOPHOTONIC IMAGING.

Survival-curve analysis is a commonly used approach that determines the median survival time of animals following infection with a wild-type bacterial pathogen and mutants that are being tested for their virulence properties. Advantages of this approach are that it can provide statistically robust results using a relatively small number of animals and that it can be combined with biophotonic imaging. In the example shown in Figure 8–1, a wild-type strain of *S. pneumoniae* and a Δrel_{Spn} mutant containing a deletion of the gene whose protein product catalyzes the synthesis of the signaling molecule, guanosine-pentaphosphate and -tetraphosphate [(p)ppGpp], were inoculated intranasally at a dose of about 10^7 CFU into a specific strain of outbred mouse. A control was included in which the Δrel_{Spn} mutation was complemented by a copy of the wild-type rel^+_{Spn} gene located elsewhere in the bacterial chromosome. Ten mice were inoculated per strain tested. The infected

mice were examined for moribundity every few hours following inoculation. As noted above, death was not used as an end point, but rather, mice that were moribund (i.e., showing obvious signs of acute illness) were euthanized. The survival curve was generated by counting the moribund and nonmoribund mice at each time point. A special kind of statistics, Kaplan-Meir analysis with log rank tests, was used to generate the median survival times and the *P* values, which are an indication of statistical difference. The survival curve showed that complemented and wild-type bacteria caused statistically similar median survival times of about 60 hours at this dosage in this mouse strain. In contrast, the median survival time of mice inoculated with the Δrel_{Spn} mutant was about 140 hours, which means that the virulence of the $\Delta relA_{Spn}$ mutant was significantly attenuated. Another important lesson from this study is that the median survival time depends on the bacterial dose and the type of mouse used.

What of the infection, though? We could have drawn a tiny drop of blood or removed the lungs to determine the number of CFU at each time point of the experiment. However, the load of bacteria in the blood would probably not be that informative, because the blood CFU tracks with the severity of infection, and moribund animals are bacteremic. To deter-

Figure 8–1 Disease progression and survival of mice infected with *luxABCDE rel^+_{Spn}*, *luxABCDE Δrelₛₚₙ*, and complemented *luxABCDE Δrelₛₚₙ bgaA::relp-rel^+_{Spn}* strains of serotype 2 *S. pneumoniae*. The mice were inoculated intranasally with 6×10^6 CFU, and disease progression was followed in real time by survival-curve analysis **(A)** and biophotonic imaging **(B).** (Adapted from K. M. Kazmierczak, K. J. Wayne, A. Rechtsteiner, and M. E. Winkler. 2009. Roles of *rel^{Spn}* in the stringent response, global regulation, and virulence of serotype 2 *Streptococcus pneumoniae. Mol. Microbiol.* **72:** 590–611, with permission from John Wiley and Sons.)

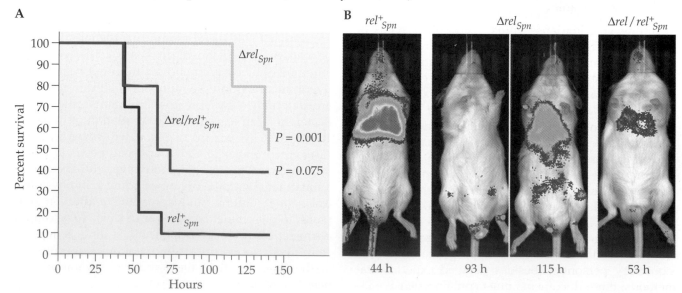

mine the number of CFU per lung per strain, we would need at least five animals per time point, which would increase the number of animals needed from 10 mice per bacterial strain tested to about 100 mice per strain. Although both of these approaches are legitimate and would yield meaningful results, new technologies have made other approaches possible.

One new alternative approach is biophotonic imaging. In this approach, a bacterial luciferase operon *(luxABCDE)* is transplanted into bacterial strains to be tested at a chromosomal location that does not affect virulence. This luciferase operon from a *Vibrio* species contains all of the genes for bacterial luciferase and its substrate needed to make the bacteria glow in the dark. The bacteria carrying the luciferase operon are used to infect the mice as described above. However, now, at intervals of about 8 hours, the mice not only are checked for disease progression, they are lightly anesthetized in a light-tight chamber connected to a supersensitive digital camera that can detect and quantitate light produced by the bacterial infection in the mouse. The camera literally looks right through the mouse to follow the course of infection (Figure 8–1B). In the example shown, wild-type and complemented strains of *S. pneumoniae* caused localized pneumonia in the lungs of the mice. However, deletion of the rel_{Spn} gene changed the course of infection completely in most of the mice. The bacteria initially localized to the lower abdomen, probably in lymph nodes, instead of in the lungs. Later, the bacteria moved to the peritoneal cavity before moving to the lungs and bloodstream. Thus, the rel_{Spn} gene is not only required for full virulence, it dictates the normal course of infection of *S. pneumoniae*. This change in the course of infection would not have been detected by simply dissecting lungs, and the experiment required far fewer animals.

LD$_{50}$ and ID$_{50}$ Values

Two other common measures widely used to define virulence in animal models are the 50% lethal dose (LD$_{50}$) and 50% infectious dose (ID$_{50}$). The LD$_{50}$ value is the dose at which 50% of the animals are moribund, i.e., the number of bacteria needed to cause terminal acute infections in 50% of the animals. LD$_{50}$ values measure a much later event in the disease process, namely, moribundity of the host that will lead to eventual death. However, not all pathogens kill the host, even during a natural infection. In such cases, it is necessary to have some other means of measuring the extent of infection caused by the bacterium. The ID$_{50}$ value is the infectious dose at which 50% of the animals are infected, i.e., the number of bacteria nec-

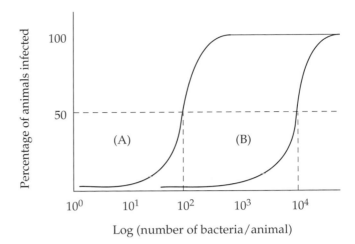

Figure 8–2 Typical curves obtained when determining ID$_{50}$ values for bacteria with different levels of infectivity. For bacterium A, only 10^2 organisms per animal are required to cause disease in 50% of the animals (ID$_{50}$ = 10^2). In comparison, for bacterium B, 10^4 microbes per animal are required to cause disease in 50% of the animals (ID$_{50}$ = 10^4). Therefore, bacterium A is more infectious than bacterium B because it requires fewer cells of bacterium A to cause disease.

essary to infect 50% of the animals exposed to the bacterium. ID$_{50}$ values measure the ability of the bacterium to colonize the host, establish infection, and manifest disease symptoms that are measurable. The most common method for determining an ID$_{50}$ value is to determine the number of CFU present at a certain site in the animal (e.g., blood, liver, spleen, lung, lymph node, or brain) after a certain time of infection, but there are other disease indicators that can also be used, such as the host's temperature (fever), the size of the necrotic lesion, increased swelling from edema, or loss of motor function. Biophotonic image analysis, instead of CFU, can also be used to follow the progress of infections for ID$_{50}$ value determinations.

Plots of the number of animals infected (or that become moribund) versus the number of bacteria in the inoculum (dose) are sigmoidal, not linear (Figure 8–2). To locate the 50% point precisely, statistical methods, such as the procedure of Reed and Muench or probit analysis, are used. These methods not only determine the 50% point mathematically, but also provide a measure of experimental error that is important for deciding whether two ID$_{50}$ or LD$_{50}$ values differ from one another by a statistically significant amount. The reason a 50% value is determined rather than a 100% value is that it is much easier to determine a 50% value accurately because it is in a region of the curve where maximal change occurs. Students who have trouble with the concept that the lower the LD$_{50}$

or ID_{50} value, the more lethal or infectious, respectively, the bacterium is, might make use of the following reminder: when it comes to LD_{50} or ID_{50}, less is worse.

LD_{50} and ID_{50} values have proven to be useful measures of lethality and infectivity, but they have some important limitations. These parameters reflect the cumulative effect of the many steps involved in colonization and production of symptoms. Therefore, they lack a certain level of sensitivity, and the failure of a mutation in a bacterial gene to increase the LD_{50} or ID_{50} value does not always mean that the mutation did not affect some virulence determinant. Moreover, LD_{50} and ID_{50} determinations require the use of a reasonably large number of animals at each bacterial dose.

To determine the role of a particular virulence factor in a pathogen that has multiple virulence factors, it is sometimes necessary to eliminate one virulence factor genetically in order to be able to determine the role of a second virulence factor. Once the first virulence factor is eliminated, one can compare the effect on virulence of deleting the second virulence factor to the virulence of the attenuated strain lacking the first virulence factor. However, difficulties can be encountered in attempting to study highly attenuated pathogens (including opportunistic pathogens). In these cases, large numbers of bacteria (up to 10^9) are injected into the animal model. Because the LD_{50} or ID_{50} value is already so high and such large numbers of bacteria are already needed to observe virulence, it is difficult to see the effects of mutations that might affect virulence by looking for increases in the LD_{50} or ID_{50} value.

When generating the data for LD_{50} or ID_{50} plots, it is important to define clearly what the measurement is that is being determined. For example, infectivity might be defined by CFU detected in the spleen of a mouse after 3 days of infection. The ID_{50} value determined using this definition of infectivity might differ significantly from one in which the bacteria are counted after 5 days of infection. Likewise, the LD_{50} value that is determined as the number of moribund animals after 3 days most likely will differ from the value determined after 5 days. In addition, it is extremely important to remember that these infectious parameters will depend on the animal model used. In some cases, outbred mice with robust immune systems are used. In other cases, animal models are chosen to give the bacteria as much of an edge as possible, so that infection can be obtained reproducibly in most or all of the animals. A trait that might be critical to establishing an infection in an animal model where the animal was chosen for its high susceptibility

might not be as critical in the more challenging environment of the body of an immunocompetent human.

Another limitation is that LD_{50} and ID_{50} values at best provide relative measures of virulence when different strains of a bacterial species or different mutants of the same strain are compared, and indeed, they provide the most useful information when clearly defined isogenic strains are used. They can be misleading when misused to compare two different diseases. For example, the bacterium that causes cholera has an ID_{50} value of about 10,000 bacteria when ingested by humans, whereas the bacterium that causes bacterial dysentery has an ID_{50} value of 10 to 20. At first glance, this might appear to indicate that the bacterium that causes dysentery leads to a more serious disease than the bacterium that causes cholera, but this is not the case. In fact, cholera is often a fatal disease whereas bacterial dysentery seldom causes death. The difference in ID_{50} values is due to the relative abilities of the two species of bacteria to survive passage through the acid environment of the stomach, the first step in infection by an ingested pathogen. Thus, comparisons of LD_{50} or ID_{50} values must be made with care and are best applied to assessing the relative infectivity or lethality of closely related strains of bacteria.

Competition Assays

A way to make infection experiments more sensitive is to use a competition assay and to determine the **competitive index (CI)**. The CI is defined as follows: **CI = [output ratio ($CFU_{mutant}/CFU_{wild\ type}$)]/[input ratio ($CFU_{mutant}/CFU_{wild\ type}$)].** In this assay, the animal is infected with a mixture of mutant and wild-type bacteria (input ratio = $CFU_{mutant}/CFU_{wild\ type}$). After the bacteria are given time to establish an infection, samples are taken from various parts of the animal, and the ratio of the number of mutant to the number of wild-type bacteria (output ratio = $CFU_{mutant}/CFU_{wild\ type}$) is determined. If the ratio of mutant to wild-type bacteria is the same as in the infecting dose (CI = 1.0), the mutation had no detectable effect, but if the wild-type outcompetes the mutant (CI < 1.0), the mutation clearly had a negative effect on virulence (i.e., the mutant was less virulent than the wild-type bacterium), or if the mutant outcompetes the wild type (CI > 1.0), the mutation caused an increase in virulence. One reason this type of assay is more sensitive than using LD_{50} or ID_{50} values is that the mutant and wild-type bacteria are competing for the same turf, so we are comparing the fitness of the mutant to compete with the wild-type bacterium for survival in the host environment. A second reason is that it is often easier to quantify the ratio of the wild type to the mutant accurately than it is to determine the LD_{50} or ID_{50} value accurately. Also, it allows the investigator to use fewer animals, since it is not necessary to use multiple animals for each dose to generate a dose curve.

However, there are some serious issues to consider when interpreting competition experiments. Mutants that grow more slowly in broth cultures will likely lose in competition assays in an animal. But this does not mean that the process affected in the mutant is necessarily contributing directly to the virulence process. It can also be difficult to interpret what subtle differences obtained in competition experiments actually mean in terms of the mechanism of virulence. The competition experiment paradigm precludes the use of some detection approaches, such as biophotonic imaging, because the light levels produced from the parent and the mutant are indistinguishable. A very serious concern is *trans* effects that may allow mutants defective in virulence to grow in the presence of the wild-type parent strain. For example, suppose a mutant cannot produce a secreted toxin that is required for disease symptoms. Using single-strain approaches, such as survival-curve analysis, the parent strain would appear virulent, whereas the toxin mutant would be strongly attenuated. However, in the competition experiment, the parent strain may secrete sufficient toxin for the mutant to grow fully. In this case, approximately equal numbers of CFU of the parent and mutant would be recovered from the infection, and the conclusion from the competition experiment would be that the mutant was fully virulent. This is clearly the wrong conclusion. Usually a combination of single-strain and competition infection experiments is needed to begin to understand processes as complicated as bacterial virulence in animal models.

Tissue Culture and Organ Culture Models

Tissue Culture Models

Although animal models are the gold standard of research on bacterial virulence, animals present a complex system in which many variables cannot be controlled. Cultured mammalian cells are commonly used to provide a more easily controlled system for investigating host-bacterium interaction. Tissue culture cells can be grown in defined medium under reproducible conditions with only one or a limited number of cell types represented, which makes measurements and interpretations easier and more re-producible. It is also easier to perform experiments involving radioactive compounds and to introduce foreign DNA into tissue culture cells. Cultured cells can be readily visualized by microscopic techniques, and cells expressing certain fluorescently labeled marker proteins can be sorted using high-speed devices. They also cost less per day to house than laboratory rodents, do not fight with each other, and have seldom been known to escape from their cages. Because of the very important role tissue culture cells have played in molecular investigations of bacterium-host cell interactions, it is important to understand their limitations, which must be kept in mind when interpreting the results of experiments.

Primary cultures of mammalian cells that are not derived from tumors can be obtained from animal tissue, biopsy material, or the blood of human volunteers (e.g., macrophages), but these primary cell types usually undergo a limited number of divisions in culture and can only be maintained for a couple of weeks. Consequently, "immortalized" cells that continue to divide in culture are often used in routine experiments. These immortalized cells are derived directly from a tumor, by fusing a primary cell type with a tumor cell, or by continuous selection of primary cell lines for unregulated growth. Propagation of mammalian cells that are unregulated for growth in culture leads to the accumulation of numerous mutations, gene rearrangements, and gene duplications. These mutations are not only uncharacterized because they are so numerous and complex, they are also not easily reproducible. Two separate sets of primary cells from the same source that are treated in exactly the same way to produce immortalized cell lines will have different combinations of mutations and rearrangements. Another way of immortalizing the cells is to introduce certain viral proteins into them. Thus, not only are cultured cells far from being genetically identical to cells in the organ from which they were derived, but different lines developed from the same type of cell are not necessarily identical. Cell lines that are passed repeatedly in culture will continue to develop new mutations. Therefore, the fact that two different investigators are using a cell line with the same name does not necessarily mean that they are using genetically identical cell lines.

Cells in culture are no longer in the same environment as in the organ of origin, and many genes that were expressed by cells in an intact organ may not be expressed in cultured cells. Culture medium does not represent the natural conditions and biochemical composition found in the body, where fluids and cellular secretions may significantly influence cellular processes in ways that are not reproducible in a culture system. Consequently, tissue culture cell lines lose many traits of the original tissue from which they were derived and may not display the same surface markers in culture. To make the transition from a differentiated, nondividing cell to a rapidly dividing cell, cultured cells must be stripped of many of the properties that made them the sort of cell they were in the first place. A consequence of this is that bacterial and viral pathogens that are highly specific for a particular tissue when causing an infection in an intact animal are frequently able to invade cultured cells derived from tissue they do not normally infect. This may be due to the expression of different receptors on the cell surface. These differences in gene expression can be corrected to some extent by the addition of growth stimulants or hormones to the culture medium or by providing an artificial **matrix** (also called a **substrate**) for the cells to grow on, but no one has yet managed

to make an immortalized cell line into a completely faithful replica of the same cells in the body.

A related problem is that most cultured cells lose their normal shape and distribution of surface antigens. Cells in an intact animal are usually **polarized.** That is, different regions of the cell surface membrane are exposed to different environments, e.g., lumen, adjacent cells, underlying blood, and other tissue types. Membranes on different sides of polarized cells contain different sets of proteins, a feature that is presumably important for their function. This is illustrated in Figure 8–3 for a layer of mucosal cells, but the same considerations apply to cells in other parts of the body. In a layer of normal mucosal cells, the **apical surface** is exposed to the external environment (e.g., the lumen contents in the gastrointestinal tract), whereas the **basal** and **lateral surfaces** are in contact with the **extracellular matrix** (a combination of pro-

Figure 8–3 Differences between cells of an actual mucosal membrane and tissue culture cells. **(A)** Actual membrane in vivo. **(B)** Nonconfluent, nonpolarized tissue culture cells. **(C)** Polarized monolayer of tissue culture cells attached to a semipermeable membrane.

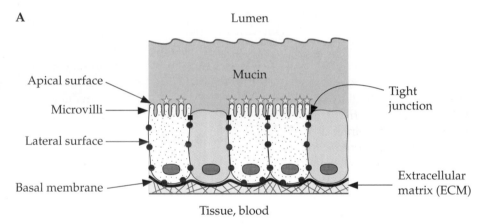

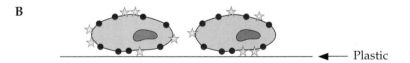

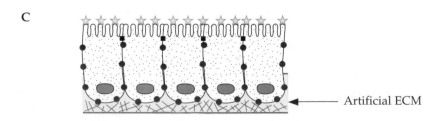

☆, apical antigens;   ●, basolateral antigens

teins and polysaccharides that "glues" the cells together). Mucosal cells in the gastrointestinal tract and some other tissues are also connected to each other by **tight junctions,** which are made of specialized tightly binding protein complexes that make an impermeable connection between adjacent cells.

In contrast, tissue culture cells that are grown as nonconfluent monolayers do not have differentiated surfaces, and proteins that are found only on the apical or on the basal-lateral surface of cells in vivo may be distributed over the entire surface of such cells, assuming they are produced at all. Allowing the cells to grow to confluence does not necessarily solve this problem. It is usually necessary to provide an extracellular-matrix substitute and to provide hormones to obtain a polarized monolayer in culture. Production of a polarized monolayer in culture has been achieved for some types of cell lines, but the expression and distribution of relevant surface molecules should still be checked before concluding that a polarized cell monolayer in culture is the same as tissue in the intact animal. In practice, this is seldom done.

Still another problem with using cultured cells as, for example, representatives of human mucosal surfaces is that real mucosal surfaces are covered with mucus and bathed in solutions that are difficult to mimic in an in vitro system. For example, the fluid bathing the small-intestinal and colonic mucosal cells is anaerobic and contains bile salts. The fluids bathing the vaginal mucosal cells and the bladder mucosal cells also have a low oxygen content and high concentrations of compounds, such as urea or lactic acid, that could have an effect on mucosal-cell physiology. Finally, real tissues consist of multiple cell types, not of a single cell type. Today coculturing more than one cell type can be done for a number of systems to mimic the natural interactions that might occur. For example, adding activated T cells or B cells, which serve as cellular sources of cytokines and chemokines, to a culture of dendritic cells can induce them to undergo maturation.

The fact that there are problems with existing tissue culture cell lines does not mean that such cell lines are not extremely useful. If their limitations are kept in mind, cultured cell lines can be marvelous tools for discovery. Once a new phenomenon has been discovered in cultured cells (e.g., attachment of bacteria to a mammalian cell receptor or reorganization of the host cell cytoskeleton), experiments can be designed that use organ cultures or animals to test the importance of the phenomenon in vivo. It is not uncommon for a mutation that affects the ability of a bacterium to infect a tissue culture cell to have no effect when tested

in animals, so taking the study from tissue culture cells to the animal is an important step. In other words, tissue culture cells can be important for generating hypotheses, which can then be tested in the intact animal. It is a mistake, however, to take the results obtained from studies of tissue culture cells and extrapolate directly from them to the disease in humans.

Gentamicin Protection Assay

Tissue culture cells have been widely used to study the adherence properties of bacteria and are particularly useful for identifying virulence factors involved in binding and invasion by intracellular pathogens. One such assay that researchers frequently use to distinguish between mutants that are defective in attachment and those that are defective in invasion is the **gentamicin protection assay** (Figure 8–4). In this assay, a monolayer of mammalian cells is incubated with bacterial cells at a certain **multiplicity of infection (MOI),** which is defined as the number of input bacteria per mammalian cell, in the well of a culture plate. Following incubation for a period of time to allow binding and invasion to occur, which may vary depending on the particular organism being examined, the samples are divided into three sets. For the first set, which will provide the total number of bacteria (CFU) that bound to the mammalian cells, the medium containing unattached bacterial cells is removed and placed in a separate tube. The mammalian cells are then lysed using mild detergent or gentle scraping (disruption that lyses the mammalian cells but does not lyse the bacteria), and the mixture is added back to the tube containing the medium with unattached bacteria. This suspension is plated on agar plates in serial dilutions, and the bacteria are allowed to grow to form colonies, which are counted. This represents the total number of CFU in each well at the end of the experiment. For the second set, which will provide the number of adherent bacteria (CFU), the infected mammalian cells are first washed several times with a buffered solution to remove any nonadherent bacteria, the mammalian cells with adherent bacteria are lysed, and serial dilutions of the suspension are plated to determine the number of cell-associated bacteria (in CFU). The ratio of cell-associated CFU to total CFU at the end of the experiment is defined as the **adhesion frequency.** For invasion assays, a third set of wells is washed as described above to remove nonadherent bacteria, and fresh culture medium containing the antibiotic **gentamicin** is added to each well. Gentamicin cannot enter the mammalian cells, so it kills only extracellular bacteria and does not kill bacteria that have already entered the mammalian cells (i.e., internalized bacteria are protected from the antibiotic and therefore survive and can be counted). The cells are incubated with the gentamicin-containing medium for anywhere from 30 minutes to 2 hours, at which time the medium is removed, and the cells are washed and lysed as described above. Serial dilutions are then plated to determine the number of intracellular bacteria (as gentamicin-resistant CFU). The ratio of gentamicin-resistant CFU to cell-associated CFU at the end of the experiment is defined as the **invasion frequency.** This allows invasion to be measured as an event separate from adherence. It is also possible to report the ratio of the number of gentamicin-resistant CFU to the total number of CFU in the well, which may vary greatly depending on the frequency of adherence of a given

bacterial strain for the given mammalian cells used. There may be some variation in the details of the experiment. For example, if the mammalian cells do not form adherent monolayers and are instead cultured in suspension, then centrifugation steps must be added during the wash steps (as shown in Figure 8–4).

If a bacterial mutant is defective in an adhesion factor that prevents it from attaching to the mammalian cell, then no colonies will form on the agar plates for the second and third sets of wells. Mutants that can still adhere but are defective in invasion factors, so that they cannot enter cells or survive once inside, will produce colonies on plates from the second set of wells but will have no colonies on plates from the third set of wells. By comparing the adherence frequencies and the invasion frequencies for wild-type bacteria versus mutants, it is possible to determine the

Figure 8–4   Gentamicin protection assay. Shown are the steps in a gentamicin protection assay for mammalian cells that are grown in suspension.

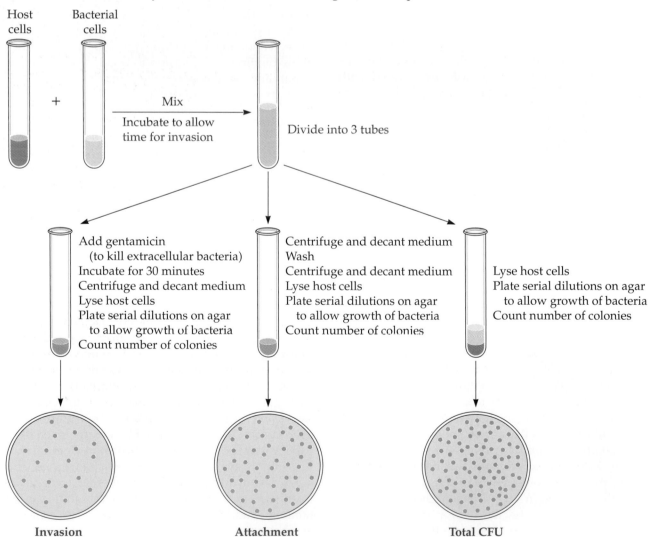

nature of the virulence factor that was defective in the mutant bacteria.

Studying pathogenic mechanisms of *Campylobacter jejuni*, a food-borne pathogen that causes diarrhea, is hampered by the lack of simple animal models that mimic human disease, so cell culture assays have provided useful alternative ways to investigate *C. jejuni* interaction with host epithelial cells. Use of the gentamicin assay has shown that invasiveness varies considerably depending on the *C. jejuni* strain, as well as the human cell lines, used; the number of bacteria in the inoculum (MOI); and other assay conditions. Researchers in the field have consequently sought to standardize assay conditions and to set up tests for comparisons between strains. One such test is to generate an **invasion success curve** (Figure 8–5), which plots the log(MOI/cell) versus the log(invaded bacteria/cell) to give the minimum MOI (minMOI) required to obtain the maximum number of internalized bacteria per cell (BI_{max}). In comparing strains, those with lower minMOIs are more invasive.

Figure 8–5 Invasion success curve. Shown is an invasion success curve for an invasive *Campylobacter* strain. The average number of invaded bacteria per mammalian cell was calculated for each well in a plate over a wide range of MOIs, starting with a low MOI and relatively small numbers of internalized bacteria, until a maximal invasion plateau (BI_{max}) was reached, where increasing the MOI no longer increased the number of internalized bacteria. Each data point represents the results obtained for one plate well. The BI_{max} and minMOI (lowest MOI required to reach the BI_{max}) values are indicated by arrows. (Adapted from Friis et al., 2005, with permission from Elsevier.)

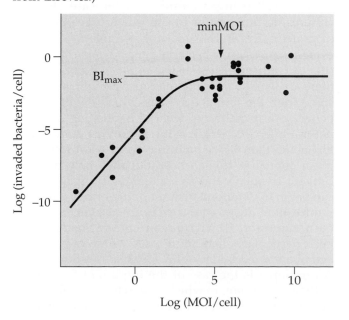

Plaque Assay for Cell-to-Cell Spread

Some intracellular pathogens, such as *Legionella* and *Chlamydia*, are able to invade eukaryotic cells, multiply within the eukaryotic cell until they burst from the host cell, and then spread from one host cell to the adjacent host cell. During this process (called **cell-to-cell spread**), the bacteria replicate within each invaded host cell, killing the host cell in the process and creating a cleared zone of killed cells (called a **plaque**) around the initial cell that was invaded. Some pathogens, such as *Listeria* and *Shigella*, are also able to spread laterally in a monolayer by propelling themselves into the adjoining cell without being released into the medium. Researchers have developed a modified gentamicin tissue culture assay that allows the assessment of an intracellular pathogen's ability to spread from cell to cell. In this **plaque assay** (Figure 8–6), tissue culture cells are grown on plates to form an even confluent monolayer, and bacteria are added to the medium and incubated with the mammalian cells for a short time to allow invasion to occur. The medium with unattached bacteria is removed, and the monolayer is incubated with gentamicin to kill any remaining extracellular bacteria. The monolayer is then gently covered with another layer of agar containing gentamicin (to prevent diffusion of the bacteria through the medium), and the cells are further incubated. After a while, the living cells are stained, and plaques (cleared areas) in the monolayer can be observed where bacteria have invaded and spread from cell to cell, killing the mammalian cells in the process. Mutants that are defective in efficient cell-to-cell motility form small plaques, while mutants that are defective in factors necessary for intracellular survival or replication do not form any plaques.

Fluorescence Microscopy Techniques for Assessing Effects of Pathogens on Host Cells

The development of new reagents for probing the interior of a mammalian cell or changing its chemical environment has generated opportunities for sophisticated new approaches to studying the bacterium-host cell interaction. For example, fluorescent dyes attached to specific reagents (chemicals or proteins) or monoclonal antibodies that bind specifically to host cell cytoskeletal components, such as actin, have been used to follow by fluorescence microscopy the cytoskeletal rearrangements caused by bacteria attaching to and invading host cells. These methods have also been used to observe the changes in cellular morphology caused by toxins or effector proteins produced by the bacteria.

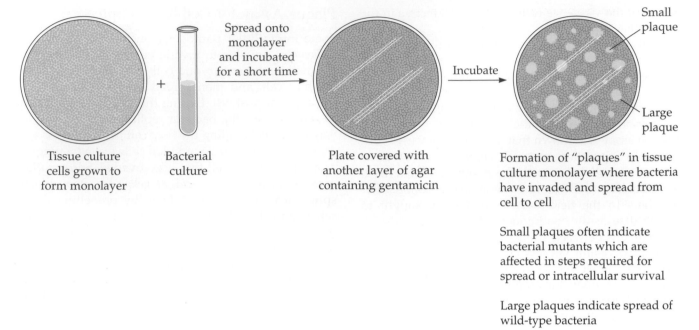

Figure 8–6 Plaque assay for assessing cell-to-cell spread by an intracellular pathogen.

One particularly useful reagent for visualization of cytoskeletal rearrangements caused by the formation of actin filaments in cells is the marine metabolite **phalloidin,** which is a compound that binds very tightly to polymerized F actin (which forms the actin filaments) but not to monomeric, free G actin. When phalloidin is linked to a fluorescent dye, such as **rhodamine** (red) or **fluorescein isothiocyanate** (green), fluorescence microscopy can be used to monitor actin cytoskeletal changes and actin stress fiber formation. Calcium release from intracellular stores (called **calcium mobilization**) is another intracellular response often triggered by interactions of a cell with bacteria or bacterial toxins. A number of fluorescent dye reagents (such as **Fura-2**) that can detect calcium levels inside tissue culture cells are available.

A number of host cellular markers have been developed that allow researchers to visualize where bacteria or bacterial proteins are located inside host cells. These cellular markers are mammalian proteins that are known to localize to particular compartments within the host cell. They are introduced into the mammalian cells by transfection with mammalian expression vectors (usually plasmids or retroviruses) carrying genes for these cellular marker proteins that have been modified with tags, such as green or red fluorescent proteins or epitope tags that are recognized by antibodies conjugated to fluorescent dyes. The tags can be used to visualize the proteins inside the host cells by fluorescence microscopy. These are only a few examples of reagents and approaches that are currently being generated by cell biologists to study intracellular processes but which are proving equally applicable to studies of bacterium-host cell interactions.

Organ Culture Models

Organs or portions of them can now be kept viable in vitro for longer and longer periods of time. Such tissues are called **organ cultures.** In the case of organ cultures, techniques for in situ quantitative detection of proteins and mRNA make it possible to detect elevated expression of bacterial genes in a particular tissue. An advantage of using organ cultures is that there are usually multiple cell types present, including some from the immune system, which might allow a better approximation of what is happening during a natural infection. Organ cultures provide a much better model than tissue culture cells of what transpires in an animal or human host, but they may be more difficult to obtain and to maintain. An organ culture can begin to deteriorate within hours or days, making it difficult to do long-term experiments, yet there have been successes. Ex vivo organ culture models have been particularly helpful in cases where research has been hampered by lack of a suitable infection model. An example is the case of the important food-borne pathogen *C. jejuni*, in which researchers developed an organ culture model using human gastrointestinal tissue, obtained from endoscopic biopsies of patients, to highlight the surprising propensity of *C. jejuni* to adhere to mucosal tissue via its flagellum (not pili).

Scientists who study the adherence of bacteria to skin cells have benefited enormously from the popularity of cosmetic surgery, which generates large amounts of skin tissue. Similarly, hysterectomies make fallopian tube and uterine tissues available, although the availability of these tissues has dropped considerably in recent years due to a trend away from performing complete hysterectomies. However, the number of people lining up to donate portions of their liver or heart is rather limited.

Because of the limited supply of organ cultures from donors, scientists have begun to develop artificial organ cultures. One example of this is artificial skin equivalents, which are made by culturing a suspension of human foreskin fibroblasts in a matrix of native acid-soluble collagen- and serum-supplemented medium at 37°C, where the collagen polymerizes and traps the cells, which then elongate and spread for several days. Then, a freshly isolated suspension of human skin-derived keratinocytes is seeded onto the surface of the collagen-fibroblast matrix, and the keratinocytes are allowed to grow to cover the surface, depositing basement membrane beneath them, differentiating into epidermal cells, and leading to formation of skin layers. Other connective-tissue and bone, heart, liver, and neuronal-tissue equivalents are also being developed, and one day, hopefully, we will see these systems utilized for host-bacterium interaction studies.

The Continuing Need for Reliable and Plentiful Information about Disease Pathology

A topic that is seldom discussed but that is very important for making progress in pathogenesis research is the quality and quantity of information available about how a disease progresses and—perhaps even more important—the extent to which basic scientists in the field are aware of this information. There are some fields, like tuberculosis research, in which scientists are fortunate to have a vast store of information about how the disease progresses in the human body. Also, scientists working in the tuberculosis field are almost all conversant with this body of knowledge and appeal to it on a regular basis to assess whether their experimental findings make sense in the real-world context.

Unfortunately, this is not the case in all areas of pathogenesis research. To completely understand the complex nature of host-pathogen interaction and its disease outcome, it is extremely important for researchers to acquire key knowledge of human physiology or the physiology of whatever animal they are using as a model and to know as much as possible about the normal distribution of bacteria in the body during infection and about how, and what type of, damage occurs as the disease progresses. Against this background, when physiological relevance is taken into consideration, the results of experiments on tissue culture cells and other ex vivo model systems are more likely to be interpreted correctly and are more able to provide new insights that could lead to improved treatments.

SELECTED READINGS

Auerbuch, V., L. L. Lenz, and D. A. Portnoy. 2001. Development of a competitive index assay to evaluate the virulence of *Listeria monocytogenes actA* mutants during primary and secondary infection of mice. *Infect. Immun.* **69:**5953–5957.

Chiavolini, D., G. Pozzi, and S. Ricci. 2008. Animal models of *Streptococcus pneumoniae* disease. *Clin. Microbiol. Rev.* **21:**666–685.

Friis, L. M., C. Pin, B. M. Pearson, and J. M. Wells. 2005. In vitro cell culture methods for investigating *Campylobacter* invasion mechanisms. *J. Microbiol. Methods* **61:**145–160.

Grant, A. J., J. Woodward, and D. J. Maskell. 2006. Development of an *ex vivo* organ culture model using human gastro-intestinal tissue and *Campylobacter jejuni*. *FEMS Microbiol. Lett.* **263:**240–243.

Kharat, A. S., and A. Tomasz. 2003. Inactivation of the *srtA* gene affects localization of surface proteins and decreases adhesion of *Streptococcus pneumoniae* to human pharyngeal cells in vitro. *Infect. Immun.* **71:**2758–2765.

Machin, D., Y. B. Cheung, and M. K. B. Parmar. 2006. *Survival Analysis—a Practical Approach,* 2nd ed. John Wiley and Sons, Ltd., West Sussex, England.

O'Brien, D. P., D. A. Israel, U. Krishna, J. Romero-Gallo, J. Nerud, M. D. Medof, F. Lin, R. Redline, D. M. Lublin, B. J. Nowicki, A. T. Frane, S. Ogden, A. D. Williams, D. B. Polk, and R. M. Peek, Jr. 2006. The role of decay-accelerating factor as a receptor for *Helicobacter pylori* and a mediator of gastric inflammation. *J. Biol. Chem.* **281:**13317–13323.

Orihuela, C. J., G. Gao, K. P. Francis, J. Yu, and E. I. Tuomanen. 2004. Tissue-specific contributions of pneumococcal virulence factors to pathogenesis. *J. Infect. Dis.* **190:**1661–1669.

Timmerman, M. M., J. Q. Shao, and M. A. Apicella. 2005. Ultrastructural analysis of the pathogenesis of *Neisseria gonorrhoeae* endometrial infection. *Cell. Microbiol.* **7:**627–636.

Zak, O., and M. A. Sande. 1999. *Handbook of Animal Models of Infection.* Academic Press, New York, NY.

QUESTIONS

1. Discuss at least three ideal attributes of an animal model.

2. Using leeches as models for vampires is a humorous example of an animal model, yet it illustrates some of the problems encountered when choosing an animal model and then deciding how far to believe the results of animal model experiments. What problems and what advantages of animal models does this rather silly example illustrate?

3. What should the criteria be for a nonhuman animal model to be acceptable as a stand-in for humans? Under what conditions would you favor allowing human volunteers to be used as guinea pigs?

4. Describe the type of disease that would have a low ID_{50} but a high LD_{50}. Is it possible to have a disease for which the LD_{50} is an appropriate measure but not the ID_{50}? Is there a combination of high/low ID_{50} and high/low LD_{50} that should not be possible?

5. When two bacterial strains are orally ingested, bacterium A has an ID_{50} of 10,000 and bacterium B has an ID_{50} of 10. However, when the bacteria are injected into a mouse model of infection, bacterium A (LD_{50} = 10) is more virulent than bacterium B (LD_{50} = 10,000). Explain how these results might be possible.

6. What limitations would organ cultures have as model systems?

7. Explain how it is possible to differentiate bacteria able to invade cultured epithelial cells from those only able to attach to the surface, and to determine the number of internalized bacteria (without using a microscope!).

8. Fill in the blank: the relative ability of a pathogen to cause disease of greater or lesser severity is that pathogen's _____.

9. Which of the following statements is true for LD_{50} and ID_{50} measurements regarding non-toxin-mediated bacterial diseases?

A. The LD_{50} is a more accurate measure of virulence than the ID_{50}.

B. The ID_{50} is a more accurate measure of virulence than the LD_{50}.

C. If the LD_{50} is low, then the ID_{50} is also low.

D. If the LD_{50} is high, then the ID_{50} is always low.

E. If the LD_{50} is high, then the ID_{50} is always high.

SOLVING PROBLEMS IN BACTERIAL PATHOGENESIS

1. A group of researchers at the USDA isolated a new bacterium from lungs and lymph nodes of several young horses that became ill and died at a local stable. Prior to death, their symptoms included disorientation and loss of motor function, so the researchers suspected central nervous system involvement, which was confirmed by the observation of brain lesions in the dead animals. Based on 16S rRNA comparison, the bacterium was distantly related to *N. meningitidis*, and they subsequently named it *Neisseria equiniae*.

> **A.** The researchers developed a horse model of infection. During their studies, they isolated two avirulent mutants (NeMut1 and NeMut2) that were still able to grow as well as wild-type *N. equiniae* in vitro, but not in vivo. They also found that these mutants had deletions in genes encoding putative virulence factors. To confirm that the disrupted genes in the two mutants indeed encoded proteins important for virulence, the researchers compared the mutants to wild-type (wt) bacteria in the horse model of infection. The results are summarized in the table that follows:

Strain	Dose (CFU)	Mortality on day 3 (no. dead/total no.)	No. (CFU) of bacteria recovered from lymph nodes on day 3
Wt	10	2/10	10^3
Wt	10^2	3/10	10^5
Wt	10^3	6/10	10^8
Wt	10^4	9/10	10^{11}
Wt	10^5	10/10	10^{12}
NeMut1	10^5	0/10	10
NeMut1	10^7	0/10	10^3
NeMut1	10^9	0/10	10^5
NeMut2	10^5	0/10	10^3
NeMut2	10^7	2/10	10^5
NeMut2	10^9	3/10	10^8

Determine the LD_{50} and ID_{50} values for each of the wild-type and mutant strains. Interpret your results.

B. The researchers decided to determine the CI of each of the mutants, again using the horse infection model. The results are summarized in the table below:

Strains used	No. of bacteria in inoculum (CFU mutant:CFU wt)	No. of bacteria recovered from lymph nodes on day 3 (CFU mutant:CFU wt)
NeMut1:wt	10^5:10^5	10:10^{12}
NeMut2:wt	10^5:10^5	10^2:10^9
NeMut1:NeMut2	10^5:10^5	10^3:10^6

Determine the CI for NeMut1:wt, NeMut2:wt, and NeMut1:NeMut2. Interpret your results. Compare the results obtained from the LD_{50} and ID_{50} values with those obtained from the CI studies.

2. A group of researchers have isolated a new bacterium, bacterium W, from the lymph nodes of several patients who returned from a camping trip and presented in the emergency room with high fever, rash, and swollen lymph nodes. Based on a 16S rRNA sequence comparison, they determined that this new bacterium is distantly related to the gram-negative bacterium *Francisella tularensis*. The researchers developed a tissue culture model of invasion using a phagocytic cell line and a mouse model of infection (via injection). They found that bacterium W binds to and invades phagocytic cells. Using a mouse model of infection, the researchers isolated two avirulent mutants of W that they call Wmut1 and Wmut2, both of which grew as well as the wild type in vitro. In the tissue culture model of invasion, Wmut2 could still bind to and invade the phagocytic cells just as well as wild-type W, while Wmut1 could not.

A. To confirm that the genes disrupted in the two mutants indeed encoded proteins important for virulence, the researchers compared the mutants to wild-type (wt) bacterium W in the mouse model of infection. The results are summarized in the table below:

W strain	Dose (CFU)	Mortality on day 3 (no. dead/total no.)	Mortality on day 7 (no. dead/total no.)	No. (CFU) of bacteria recovered from lymph nodes on day 3
Wt	10	2/10	10/10	10^3
Wt	10^2	3/10	10/10	10^5
Wt	10^3	6/10	10/10	10^8
Wt	10^4	9/10	10/10	10^{11}
Wt	10^5	10/10	10/10	10^{12}
Wmut1	10^5	0/10	0/10	10
Wmut1	10^7	0/10	0/10	10^3
Wmut1	10^9	0/10	0/10	10^5
Wmut2	10^5	0/10	0/10	10^3
Wmut2	10^7	0/10	1/10	10^5
Wmut2	10^9	1/10	2/10	10^8

(continued)

Determine the LD_{50} and ID_{50} values for each of the wild-type and mutant strains. Interpret your results.

B. The researchers decided to determine the CI of each of the mutants, again using the mouse model. The results are summarized in the table below:

W strains used	No. of bacteria in inoculum (CFU mutant:CFU wt)	No. of bacteria recovered from lymph nodes on day 3 (CFU mutant:CFU wt)
Wmut1:wt	10^5:10^5	10:10^{12}
Wmut2:wt	10^5:10^5	10^3:10^9
Wmut1:Wmut2	10^5:10^5	10^3:10^8

Determine the CI for Wmut1:wt, Wmut2:wt, and Wmut1:Wmut2. Interpret your results. Compare the results obtained from the LD_{50} and ID_{50} values with those obtained from the CI studies.

3. Your research laboratory has been working for several years on a new bacterium that was isolated from the lungs of two emergency room patients who became ill several days after a vacation trip to a popular resort hotel called Paradise. According to a 16S rRNA comparison, this bacterium is related to the gram-negative *Klebsiella pneumoniae*, and you named this new bacterium *Klebsiella paradisiae*. Your laboratory developed a tissue culture model of invasion using a bovine lung cell line and a rabbit lung model of infection, and during their studies, your students isolated five avirulent mutants (Mut1 through Mut5) in the rabbit lung model of infection, which had genes encoding putative virulence factors deleted. Your students conducted a series of experiments to determine the role of each of these virulence factors in pathogenesis. Based on the results shown in the table below, predict the possible function(s) of each of the putative virulence factors whose genes were deleted in Mut1 through Mut5 and provide at least one experiment that you could perform for each that could be used to confirm your prediction.

Bacteria	Gentamicin assay		Plaque assay phenotype observed
	No. of colonies without gentamicin	No. of colonies with gentamicin	
Mut1	10^9	10^6	Large plaques
Mut2	10^9	10^5	No plaques
Mut3	10^9	10^2	Large plaques
Mut4	10^9	10^5	Small plaques
Mut5	10^2	1	No plaques
Wild type	10^9	10^6	Large plaques

9

Identification of Virulence Factors: Molecular Approaches for Bacterial Factors

You have a patient who has a complex set of symptoms associated with the patient's antisocial behavior. First, you might ask a lot of questions. Based on the answers, you might suspect the patient's problem is physical or genetic in origin. Then, you might order psychological or blood tests, perhaps followed by an MRI (magnetic resonance imaging). Eventually you might turn to genetic screening. From all this information, you formulate a diagnosis that suggests an explanation for the patient's symptoms. Now, consider, in contrast, a "patient" that is a bacterium, and that bacterium makes a living by causing infection. This is actually a rare trait among bacteria, most of which are benign or neutral with respect to humans. How would you examine this bacterium? You can't ask questions or take a blood sample. You can order an MRI, but the technicians will tell you, with a scarcely disguised smirk, that your patient is not only too small but does not have a brain. How do you diagnose the source of your tiny patient's aberration? In this chapter, we explore some of the molecular methods, both biochemical and genetic, that are being developed to aid you in your diagnosis. It is important to know the root cause of your micropatient's problem, just as it is in the case of a human with aberrant behavior, because based on that knowledge, you might be able to design interventions or treatments, and maybe even a cure.

Twenty-five years ago, recombinant DNA technology started a line of research that has enabled the discovery and characterization of a wide array of genes associated with virulence in pathogenic bacteria. It was no longer necessary to painstakingly isolate and purify a bacterial component to demonstrate that it was toxic or caused certain disease symptoms in order to verify that it was a virulence factor. Advances in molecular technologies allowed more rapid discovery of virulence genes. These technologies also made possible the discovery of virulence traits previously not suspected. Table 9–1 lists a number of popular technologies that have been developed and used in recent years to discover new virulence factors.

The field of bacterial pathogenesis is once again being transformed, this time through whole-genome sequencing and microar-

Table 9–1 Approaches used to identify new virulence factors

Traditional biochemical and genetic approaches
  Biochemical approaches
    Purify virulence factors, such as toxins, and use them to reproduce symptoms associated with infection.
  Molecular genetic approaches
    Clone genes from pathogens into avirulent *E. coli*, and show that the resulting strain has become virulent.
  Transposon mutagenesis
    Introduce transposon with selectable marker into pathogen to generate collection of mutants.
    Screen for loss of virulence.
    Selectable marker permits identification of mutated gene.
  Reporter fusions
    Used to identify potential virulence genes by their regulatory properties.
    Reporter protein synthesized only under conditions in which virulence gene would normally be expressed.

Finding genes that are expressed in vivo
  STM
    Mixture of bacterial isolates containing transposons with unique oligonucleotide tag used to inoculate animal.
    After infection develops, organ or tissue of interest is removed.
    Tags are amplified, digested, and labeled.
    Mixture of labeled tags is used to probe original mutants.
    Any mutant not hybridizing with probe is examined to determine why it is less able to survive in the host.
  IVET or RIVET
    Used to detect genes that are turned on when the bacterium is in the host.
    Many genes detected so far are housekeeping genes.

Genomic methods for identifying virulence genes
  GSH
    Used to identify genes that are present in virulent but not avirulent strains.
    Genomic DNA from both strains is digested, and primers are ligated to the ends for later cloning into plasmids.
    DNA from nonpathogen labeled with biotin, mixed with DNA from pathogen, and treated so complementary fragments from the two strains can hybridize.
    Streptavidin beads bind biotin; only DNA common to both strains will bind.
    Unbound DNA is amplified with PCR, cloned, and sequenced.
  SCOTS
    mRNA from bacteria recovered from host is reverse transcribed into cDNA; adaptors are attached and labeled for later cloning into plasmids.
    Control mRNA from bacteria grown in vitro is isolated and labeled with biotin.
    Control mRNA is hybridized to cDNA; common genes are removed with streptavidin.
    Remaining cDNA represents mRNA from genes turned on only in vivo.
    cDNA is cloned into plasmids and sequenced.
  IVIAT
    Pooled serum expected to have antibodies against a specific pathogen is absorbed with cells (or extracts) from bacteria grown in vitro.
    Remaining serum, with antibodies against antigen that is induced in vivo, is used to probe genomic expression library to identify putative antigens expressed in vivo—potential vaccine candidates.
  Microarray technology
    Uses chips that contain DNA oligonucleotides (or expressed protein) corresponding to all genes in the genome.
    mRNA purified from bacteria growing under specific conditions or from mutants is reverse transcribed and hybridized to chip.
    Used to identify relative transcript amounts of genes expressed under certain conditions or in mutants.
    Protein arrays are used in IVIAT.
  Comparative genomics for vaccines and therapeutics
    Microbial genomes are sequenced.
    Sequences of different strains are compared to look for serovar-specific virulence genes.
    The unique sequences are tested as potential vaccine candidates.

ray technologies, combined with automation and powerful new bioinformatics tools. These technologies have revolutionized the way we approach the identification of virulence factors. The new data provided by complete knowledge of a bacterium's genetic makeup and the differential patterns of gene and protein expression revealed through microarrays dwarf previous methods of virulence gene discovery. Indeed, the vast and exhaustive list of new candidates revealed by comparative genomic analysis of virulent and avirulent strains of the same bacterium or closely related species is overwhelming. The challenges now for researchers are to characterize these candidates, to annotate the gene functions, and to determine their roles in pathogenesis.

However, just because we have this new genomic technology, does this make all of the previous approaches obsolete? Recent experience suggests not. It is becoming increasingly clear that a combination of approaches, including new ones, like genomics and microarray analyses (see below), as well as older genetic and classical biochemical techniques, will be most successful. Those who become trapped in a single new technology generally fall by the wayside scientifically, so it is important to understand a variety of old and new techniques. Not only are many of the older ones still useful, but they are also constantly being updated and improved to keep them relevant and useful. Most interesting of all, a recent trend has been a return to the consideration of bacterial physiology. The reason is that genetic and biochemical analyses alone do not take the final step in understanding how all this information plays out in the living cell.

Traditional Biochemical and Genetic Approaches

Traditional Biochemical Approaches

The first virulence factors to be characterized at the molecular level were bacterial toxins, proteins that damage host cells. The activities of these proteins on eukaryotic cells clearly mark them as factors that allow bacteria to cause disease. Bacterial toxins, and later other secreted or released proteins (proteases, nucleases, glycosidases, and lipases), carbohydrates (lipopolysaccharides), and lipids (mycolactones), were the first virulence factors to be isolated from cultures of pathogenic bacteria and purified using biochemical approaches, such as filtration, centrifugation, selective precipitation, chromatography, and other separation methods. The purified factor was then shown to reproduce some or all of the disease symptoms associated with infection upon reintroduction into a susceptible host. Once a protein toxin has been purified, its amino acid sequence can be determined by using new mass spectroscopy methods, and the genes encoding the toxin can rapidly be identified in the genome sequence of the bacterial pathogen or on a plasmid harbored by the pathogen. Likewise, the compositions and structures of carbohydrates and lipids can give hints about the genes that mediate their synthesis in the bacterial genome.

The protein toxins produced by *Corynebacterium diphtheriae* (diphtheria toxin) and *Vibrio cholerae* (cholera toxin) are examples of such virulence factors that could be purified from culture filtrates of the bacteria and shown to cause toxic effects in animals in vivo or in host cells in vitro. Soon after methods for isolating and growing bacteria in pure culture were developed in the late 19th century, the discovery of diphtheria toxin in culture medium from *C. diphtheriae* heralded a new era in medical microbiology. Not only could culture filtrates be shown to produce characteristic death of internal organs upon injection into experimental animals and to kill sensitive animal cells in tissue culture, but vaccination with sublethal doses or with chemically or heat-inactivated toxin afforded complete protection from subsequent disease upon exposure to the bacteria. However, it was not until the 1930s that sufficient quantities of purified diphtheria toxin could be obtained to conclusively show that the toxic substance present in the culture filtrates from *C. diphtheriae* was a protein, and it was not until the 1960s that its mechanism of action could finally begin to be understood at the molecular level.

Interestingly, purified cholera toxin at first did not appear to have any obvious effects on cells, and even intramuscular injection with active toxin did not seem to cause harm to cells or to provide protection from cholera diarrhea, which misled many researchers to believe that it was not the virulence factor present in *V. cholerae* culture filtrates that was responsible for the massive diarrhea. It was not until much later, when genetic approaches were used and a better diarrhea model system using a rabbit ileal loop assay (described in chapter 2) was developed, that it could finally be demonstrated that cholera toxin was indeed the etiological agent of cholera diarrhea. The problem stemmed in part from the fact that cholera toxin changes the metabolism of the host cells yet does not damage or kill them; that the diarrheal effect of cholera toxin is very specific for the type of cells present in the intestine; and that injection with cholera toxin leads to development of humoral immunity, which is less important against a gut pathogen, such as *V. cholerae*, than a mucosal immune response.

Other potential virulence factors, such as cell surface proteins or carbohydrates, that could confer on the bacteria the ability to adhere to human tissues and/or the ability to trigger uptake of a bacterium by a normally nonphagocytic cell, could also be shown by biochemical approaches to contribute to disease. When genetic approaches proved intractable, biochemical approaches were used to identify a particularly elusive and novel virulence factor from *Mycobacterium ulcerans*, an emerging human pathogen harbored by aquatic insects and the causative agent of Buruli ulcers, a devastating disease that has become a serious health threat in western Africa. Infection with *M. ulcerans* causes progressive necrotic lesions that, if untreated, can extend over 15% of a victim's body and lead to lifelong disability and occasionally death. Surprisingly, there is little inflammation or pain associated with the large skin lesions even in advanced stages of the disease, and it was believed that the extensive necrosis was due to a toxin that diffused beyond the infection site that the extracellular bacteria had colonized.

Early attempts to define the biochemical properties of this toxin suggested that it was a heat-stable substance, but the nature of the toxin remained unknown until 1998, when a group of investigators at the Rocky Mountain Laboratories used biochemical approaches to extract and eventually purify the lipid-soluble toxins from the bacterial filtrates and then performed structural analyses using nuclear magnetic resonance and mass spectroscopy to determine that the biologically active substances were a mixture of lipid-like polyketide-derived macrolides, which they called mycolactones (Figure 9–1).

Molecular Genetic Approaches

One approach to identifying the traits that contribute to virulence is to clone the genes from the pathogen of interest into a strain of *Escherichia coli* that is avirulent and then to screen for mutants of *E. coli* that have become virulent. An example might be to clone the toxin gene from a pathogenic strain into a non-toxigenic, avirulent strain and to demonstrate that the resulting strain produces the toxin and is now virulent. In another example (Figure 9–2), ordinary laboratory strains of *E. coli* do not adhere to or invade tissue culture monolayers. Selecting for clones that contain DNA segments that enable *E. coli* to adhere to or invade the tissue culture cells can identify potential adhesins and invasins.

Cloning strategies have worked in some cases to identify candidate virulence genes, but they have some important limitations. Because standard cloning procedures isolate only relatively small portions of the total bacterial genome, fewer than 30 kbp, this approach works best if only one or a few contiguous genes from the pathogen are sufficient to give the desired phenotype. Also, this approach requires that the gene(s) of interest be expressed in *E. coli* (or whatever avirulent bacterium is being used). Not surprisingly, this approach has been most successful when applied to bacterial species that are closely related (to *E. coli* or to the avirulent bacterium).

Transposon Mutagenesis

Genetic approaches can be used to generate strains attenuated in virulence compared to the wild-type parent strain. A classical genetic approach is to create a library of mutants using chemical mutagens or UV irradiation and then to screen the mutants for those that still grow normally in vitro but are no longer virulent in an infection model. While it is relatively easy to generate the mutants and identify which ones are avirulent, until recently (see chapter 5), it has often been nontrivial and labor-intensive to find the actual genes that have been altered in these mutant bacteria, because these mutagenesis methods do not provide an easily selectable marker or other indication of where the mutation has occurred.

To circumvent this problem, **transposon mutagenesis** was developed, and it has become a widely used strategy for identifying virulence genes (Figure 9–3). In this approach, insertion mutations are generated by nearly random insertion of a transposon (chapter 7) into the genome of a pathogen, and the resulting transposon insertion mutants are screened for loss of virulence. The transposons commonly used in these approaches carry a selectable marker, for example, an antibiotic resistance gene or a reporter gene, such as *lacZ*, encoding β-galactosidase, or *lux*, encoding luciferase. By introducing a transposon into the bacterial genome and selecting for colonies expressing the selectable marker, the investigator generates a collection of mutants, each of which contains a single transposon insertion. This approach has the advantage that every selected colony carries some type of mutation, and many of these mutations disrupt a gene. Another advantage of transposon mutagenesis is that the transposon serves as a marker to locate the gene of interest and can be used to clone the gene later. This has been an important trait in bacterial species lacking sophisticated genetic mapping tools needed to locate and clone point mutations. Still another advantage of transposon mutagenesis is that it can be used to identify virulence genes that are not expressed in *E. coli* or are not closely linked to other virulence genes on the chromosome. These advantages explain why transposon mutagenesis is so widely used in bacterial pathogenesis research today. Moreover, before whole-genome sequencing recently became widely available (chapter 5; see below), linking transposon insertions to point mutations was one of the only ways to locate the point mutations.

Two limitations of the transposon mutagenesis approach should be kept in mind. First, transposons can carry transcriptional terminators. If a transposon lands in the first gene in an operon, it will eliminate expression, not only of that gene, but also of downstream genes, i.e., transposon insertions are often **polar.** The avirulent phenotype of the mutant could thus be due either to loss of expression of the gene interrupted by the transposon or to loss of expression of

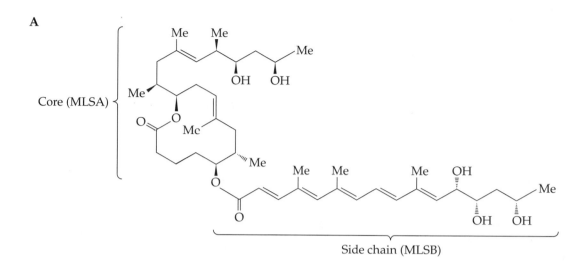

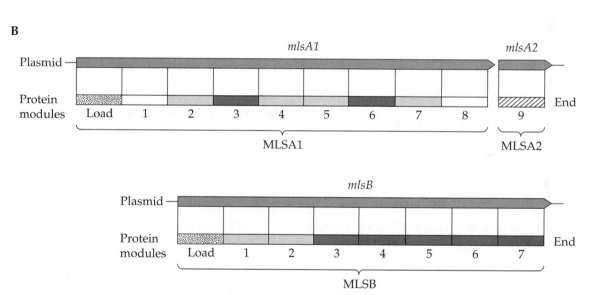

Figure 9–1 Mycolactone toxin responsible for Buruli ulcer formation. Mycolactones are polyketide-derived macrolide toxins that play a major role in the tissue destruction and immune suppression that occur in cases of Buruli ulcer caused by various strains of *M. ulcerans*. The three giant polyketide synthase genes responsible for the biosynthesis of the mycolactones are carried on a large, 174-kb plasmid. The 12-member core structure of the mycolactone is produced by two giant modular polyketide synthases, MLSA1 (1.8 MDa) and MLSA2 (260 kDa), whereas the side chain is synthesized by MLSB2 (21.2 MDa). The hallmark of these large multienzyme complexes is that they are arranged in a linear "assembly line" of catalytic-domain modules that polymerize and modify the polyketide units of the macrolide core (MLSA1 and -2) and side chain (MLSB). Shown are the domain and module organizations of the polyketide synthase genes. Within each of the genes, different domains are represented by colored blocks. **(A)** Mycolactone structure. **(B)** Giant plasmid encoding the polyketide synthases responsible for biosynthesis of the mycolactone core (MLSA1 and -2) and the side chain (MLSB). (Adapted from T. P. Stinear, A. Mve-Obiang, P. L. C. Small, W. Frigui, M. J. Pryor, R. Brosch, G. A. Jenkin, P. D. R. Johnson, J. K. Davies, R. E. Lee, S. Adusumilli, T. Garnier, S. F. Haydock, P. F. Leadlay, and S. T. Cole. 2004. Giant plasmid-encoded polyketide synthases produce the macrolide toxin of *Mycobacterium ulcerans*. *Proc. Natl. Acad. Sci. USA* **101:**1345–1349, with permission from the National Academy of Sciences.)

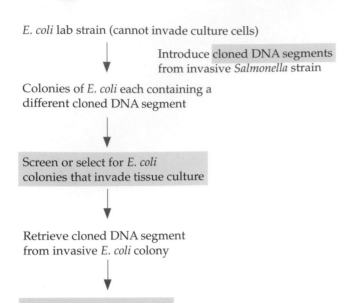

E. coli lab strain (cannot invade culture cells)

Introduce cloned DNA segments from invasive *Salmonella* strain

Colonies of *E. coli* each containing a different cloned DNA segment

Screen or select for *E. coli* colonies that invade tissue culture

Retrieve cloned DNA segment from invasive *E. coli* colony

Sequence and characterize cloned DNA segment

Figure 9–2 Identifying virulence genes by cloning and expressing them in *E. coli*. The example shows cloning of a gene(s) from a *Salmonella* strain that allows *E. coli* to invade tissue culture cells.

downstream genes in the operon. Standard genetic **complementation** tests, described below, can be used to distinguish polar effects. A second limitation is that transposon insertion mutations can be obtained only in genes that are not essential for growth on the selective medium. A transposon insertion in a gene essential for growth will not be isolated, because the bacteria will not survive to form a colony. This could be considered an advantage if one assumes that the most interesting virulence genes are the ones that are not expressed by bacteria growing in laboratory medium but that are induced specifically under conditions that mimic those found in an animal host.

Measuring Virulence Gene Regulation: Gene Fusions

You will learn more about the regulation of virulence genes later in this book, but since identification of virulence genes often requires the exploitation of their differential regulation under selective conditions (e.g., off outside the host and on when in the host), we will introduce here a few basic concepts to assist with understanding how reporter systems work in identifying virulence genes. **Coordinated regulation** is the regulation of multiple genes in response to a particular signal. In an **operon,** the genes are all transcribed as part of a single transcript controlled by a single pro-

moter upstream of the genes. Genes are also frequently organized in **regulons,** which are controlled by the same regulatory proteins (Figure 9–4). Virulence genes can often be identified based on their control by a common regulatory protein, which may repress transcription by binding to an operator region in a promoter. Many regulatory proteins activate transcription, often by binding immediately upstream of a promoter. Inactivation of genes encoding common regulatory proteins alters the virulence properties of the mutant bacteria.

Experimentally, identification of potential virulence genes by their regulatory properties is done using **reporter fusions** (also called **operon fusions**). When the virulence gene has already been cloned, a **transcriptional fusion** can be constructed. In a transcriptional fusion, a hybrid gene is created that contains the promoter and regulatory regions of a virulence gene fused to a structural gene encoding some reporter enzyme **(reporter gene)** that can be assayed using easily measured colorimetric, spectrophotometric, fluorescence, luminescence, or chromatographic methods (Table 9–2). The most popular reporter gene is *lacZ*, the gene that encodes β-**galactosidase.** Other commonly used reporter genes are *uidA*, the gene encoding *E. coli* β-**glucuronidase;** *cat*, the gene encoding **chloramphenicol acetyltransferase (CAT);** *gfp*, the gene encoding **green fluorescent protein (GFP);** *phoA*, the gene encoding **alkaline phosphatase (PhoA);** *luc*, the gene encoding firefly **luciferase (Luc);** and *lux*, the gene encoding **bacterial luciferase (Lux)** and enzymes needed to synthesize its substrates.

The fusion approach hooks the regulatory circuit of a virulence gene that we want to study to a reporter gene whose output is easy to measure or detect. In the hybrid construct (Figure 9–5), the ribosome-binding site and ATG start site are provided by *lacZ*. Although RNA polymerase starts transcription from the regulated promoter region of the virulence gene, the β-galactosidase protein is translated from the segment of the mRNA transcript beginning at the AUG site that corresponds to the start of the *lacZ* gene. Thus, β-galactosidase expression responds to conditions to which the virulence gene would normally respond. For example, if the original virulence gene was expressed at higher levels at 37°C than at 25°C, the β-galactosidase activity of the fusion strain will be produced at higher levels at 37°C than at 25°C. β-Galactosidase activity is readily detected on plates or in liquid assay systems by using **chromogenic substrates.** A popular chromogenic substrate is **5-bromo-4-chloro-3-indolyl-β-D-galactopyranoside (X-Gal),** which turns blue when the galactosyl bond is cleaved by the enzyme. In the example mentioned above, col-

Transposon ($\overset{Q^r}{\blacksquare\!\!-\!\!\blacksquare}$) introduced into chromosome of invasive bacterial cells

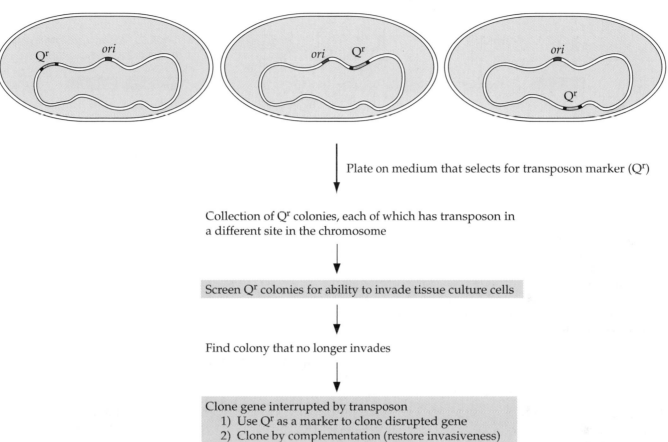

Plate on medium that selects for transposon marker (Q^r)

Collection of Q^r colonies, each of which has transposon in a different site in the chromosome

Screen Q^r colonies for ability to invade tissue culture cells

Find colony that no longer invades

Clone gene interrupted by transposon
  1) Use Q^r as a marker to clone disrupted gene
  2) Clone by complementation (restore invasiveness)

Figure 9–3 Identifying virulence genes by transposon mutagenesis. The example shows identification of a gene(s) needed for invasion of tissue culture cells by *Salmonella* by using a transposon with a quinolone resistance (Q^r) gene as a selection marker.

onies containing a fusion of *lacZ* to the temperature-regulated virulence gene would be blue at 37°C and white or light blue at 25°C. Not only is it easier to assay β-galactosidase activity than it is to assay most virulence proteins, but β-galactosidase poses none of the hazards associated with, for example, assaying a toxin.

The type of medium used to determine whether the gene is expressed during growth on laboratory medium can cause some problems. Many scientists use lactose-MacConkey agar to differentiate Lac$^+$ from Lac$^-$ colonies. This medium contains bile salts, which may be an important signal in vivo for gastrointestinal pathogens. Thus, some insertions identifying new virulence genes may be discarded as positive in vitro when in fact the decision is based on an artifact of medium composition. Also, body temperature is the

inducer of some important virulence genes. Mutations in such genes could be missed because the plates are incubated at 37°C and thus induce gene expression in vitro. Careful attention must be given to the means used to determine which mutants to study further and which to discard.

Often investigators use transposons carrying a promoterless *lacZ* gene in addition to a selectable antibiotic resistance marker to locate genes that respond to specific conditions or signals (Figure 9–6). Selection for antibiotic resistance generates a set of random insertions of the transposon into the bacterial chromosome. Colonies carrying transposon insertions are then screened for regulated expression of *lacZ*. Only a fraction of the transposon insertions will fuse *lacZ* to promoters, and only a fraction of these fusions will be regulated in the desired manner. Suppose the genes

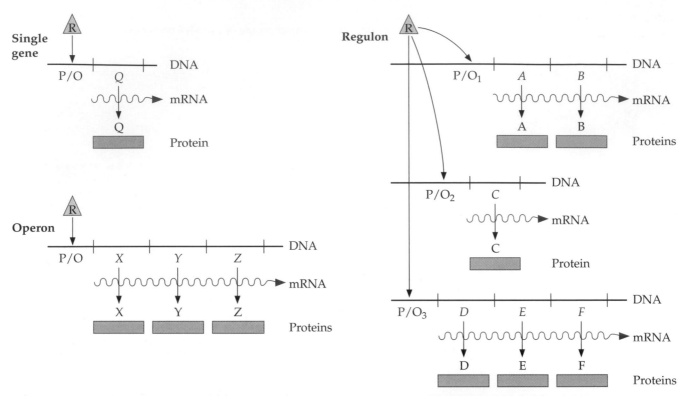

Figure 9–4 Description of a gene, an operon, and a regulon. Regulatory proteins (R) regulate the transcription of genes that are part of a single operon, which can be made up of a single gene (Q) or multiple genes (X, Y, and Z) under the control of an upstream promoter/operator (P/O) or genes (A to F) that are part of multiple operons (regulon), all under the control of upstream promoters that bind the same regulatory protein.

being sought are expressed at high levels only under low-iron conditions. In this case, colonies carrying transposon insertions would be replica plated onto high-iron medium and low-iron medium, both of which contain X-Gal. Colonies containing the desired fusions will be blue on the low-iron plates but white on the high-iron plates. The resulting strain containing the *lacZ* reporter can then be used to find regulatory genes that control the expression of that virulence gene. This is illustrated in Figure 9–7. The strain is mutagenized by using another transposon, and the resulting colonies are screened either for those exhibiting aberrant regulation, i.e., colonies that do not turn blue under inducing conditions (possible loss of an activator) or colonies that turn blue under both inducing and noninducing conditions (possible loss of a repressor).

When a gene has been tagged by a *lacZ* fusion or by a transposon insertion, it can readily be identified. Several commonly used PCR methods (inverse PCR and touchdown PCR) can be used to amplify part of the transposon DNA and the junction region of the chromosomal DNA. Sequencing these amplicons

gives the exact point of insertion and identifies the virulence gene by comparison to sequenced genomes. Identification of the gene opens up a variety of additional experimental approaches. One of the first things to determine is whether changes in phenotypes are caused by disruption of the gene containing the transposon or by polarity on the expression of downstream genes. This issue is addressed by putting a second intact copy of the gene being tested into the mutant and determining whether the wild-type gene complements the mutant phenotype. This second copy of the wild-type gene can be introduced into the mutant on a plasmid or at another (ectopic) site in the bacterial chromosome.

Finding Genes That Are Expressed In Vivo

Signature-Tagged Mutagenesis

A major limitation of standard transposon mutagenesis is that the selections or screens involve growth on bacteriological medium (so-called "in vitro" ex-

Table 9–2 Commonly used reporter gene systems

Reporter gene	Reporter enzyme	Visualization assay
lacZ	β-Galactosidase (β-Gal)	This is a colorimetric assay in which cells that express β-Gal appear blue when grown on medium containing a substrate analog, such as X-Gal. X-Gal is cleaved by β-Gal into galactose and 5-bromo-4-chloro-3-hydroxyindole, which is then oxidized to 5,5'-dibromo-4,4'-dichloro-indigo, an insoluble blue product. For spectrophotometric analysis, the substrate *o*-nitrophenyl-β-D-galactoside (ONPG) is often used. There are also fluorescent dye-labeled substrates for fluorescence microscopic analysis.
uidA	*E. coli* β-glucuronidase (GUS)	The most common substrate for GUS histochemical staining is 5-bromo-4-chloro-3-indolyl glucuronide (X-Gluc); the product of the reaction is blue. Other common substrates are *p*-nitrophenyl-β-D-glucuronide for the spectrophotometric assay and 4-methylumbelliferyl-β-D-glucuronide (MUG) for the fluorometric assay.
cat	CAT	CAT covalently attaches an acetyl group from acetyl-coenzyme A to the antibiotic chloramphenicol. The conversion of a less hydrophobic substrate to a more hydrophobic product can be detected chromatographically.
phoA	PhoA	PhoA is a hydrolase enzyme responsible for removing phosphate groups from various substrates, such as the chromogenic substrate 5-bromo-4-chloro-3-indolyl phosphate (XP, or BCIP), the product of which turns dark blue in the presence of an oxidant, such as nitroblue tetrazolium chloride (NBT). Bacterial PhoA is a periplasmic protein, so it is often used as a reporter for exported or secreted proteins.
gfp	GFP	GFP is a protein from the jellyfish *Aequorea victoria* that fluoresces green when exposed to blue light. Cells that express GFP glow green under UV fluorescent light.
luc	Luc	The enzyme luciferase converts luciferin substrate to produce oxyluciferin and energy in the form of light (chemiluminescence).
luxCDABE	*Photorhabdus luminescens* luciferase (Lux)	A bacterial luciferase reporter that is also used extensively in bacterial systems. The operon mediates synthesis of the bacterial luciferase substrate, so the substrate does not need to be added.

Figure 9–5 Transcriptional fusion of a virulence gene promoter/operator (P/O *vir*) to the promoterless reporter gene encoding β-galactosidase *(lacZ)*. Although the DNA segment encoding β-galactosidase lacks the *lacZ* promoter and instead the mRNA transcript is made using the promoter for the virulence gene *(vir)*, it still has its ribosome-binding site (rbs) and start codon (ATG/AUG). Thus, β-galactosidase will be regulated in the same way as the virulence gene.

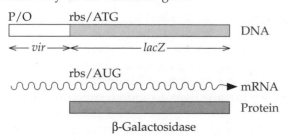

periments in pathogenesis parlance). Because no laboratory medium is a perfect mimic of the environment inside the human body, the best way to select or screen for interesting mutations would be to do it in an animal (i.e., an "in vivo" experiment). Of course, for this to work, a good animal infection model must first exist or be developed (see chapter 8). For a description of an interesting animal model that can be used for screening transposon-generated mutants, see Box 9–1.

An approach to discovering in vivo-expressed virulence genes is **signature-tagged mutagenesis (STM)** (Figure 9–8), which combines in vitro transposon mutagenesis with in vivo selection using an animal model of infection to screen for mutants that do not grow in the host. Instead of a single transposon being used to generate a library of mutants, STM uses a mix-

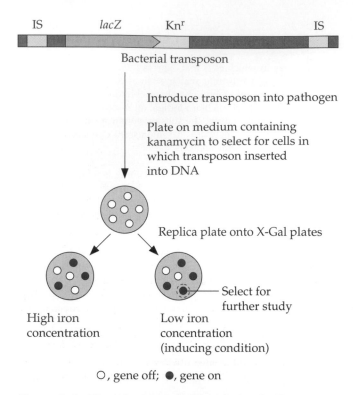

Figure 9–6 Use of a transposon carrying *lacZ* to generate *lacZ* fusions. Transposase allows the DNA carrying the promoterless *lacZ* to integrate into the genome. The selectable marker for the kanamycin resistance (Kn^r) gene allows the selection of colonies that have the transposon. IS, insertion sequence.

Figure 9–7 Screening mutants of a *lacZ* fusion strain for mutants with aberrant regulatory properties.

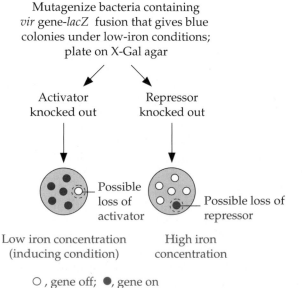

BOX 9–1 Worms As Animal Models Revisited

Researchers interested in identifying virulence factors of *Pseudomonas aeruginosa*, a common gram-negative pathogen that causes infection in burn patients and people with cystic fibrosis, have turned to the worm *Caenorhabditis elegans* as a stand-in for mice and humans. Screening random transposon-generated mutants in mice would be enormously expensive and time-consuming, but *C. elegans* is much smaller and easier to handle. Interestingly, *P. aeruginosa* kills *C. elegans*, and inactivation of some of the same virulence genes important for infection of mice is also important for infection in *C. elegans*. Accordingly, *C. elegans*, which has heretofore served mainly as a model for the development of complex animals, has now been proposed as a model for *P. aeruginosa* infections in humans and is being used as an initial screening tool for identifying mutants that are less virulent, as indicated by their inability to kill *C. elegans*.

Source: M. W. Tan, L. G. Rahme, J. A. Sternberg, R. G. Tompkins, and F. M. Ausubel. 1999. *Pseudomonas aeruginosa* killing of *Caenorhabditis elegans* used to identify *P. aeruginosa* virulence factors. *Proc. Natl. Acad. Sci. USA* **96:** 2408–2413.

ture of transposon variants generated from a single transposon to generate a library of mutants, each with a different variant inserted. The transposon mixture is obtained by cloning a mixture of random oligomers (usually ~40 bp in length) into a transposon so that each transposon has its own individual "tag" or "barcode," the oligomer it carries. A library containing a mixture of transposons, each with its individual tag, is available from laboratories that have used this method. The transposon library is transformed (or electroporated) into the target bacterial strain, and transformants (electroporants) that received a transposon insertion somewhere in their chromosome are detected by selecting for the antibiotic resistance gene carried on the transposon. Individual colonies, each representing a different random mutation, are saved on master plates or in wells of a plate. STM is a way to screen many knockout mutants at a time, depending on how many unique "signature" tags are available in the pool. The DNA of the bacteria from each well (each well has a different mutant) is transferred onto two replicate membranes, one for the input pool and the other for the output pool.

The mutants containing transposons are pooled and grown in vitro. The resulting culture is then split, and one part is used to identify the input pool of mutants that grew in vitro in culture and the other part is used to inoculate an animal and perform the in vivo output experiment. After sufficient time has passed for the infection to develop, an infected organ or blood is removed and the bacteria that replicated in the host are recovered to give the output pool. DNA is extracted from the input (in vitro growth) and output (animal growth) pools of bacteria, and PCR primers that recognize the DNA flanking the 40-bp oligomer tag (P3 and P5 in Figure 9–8) are used to amplify the mixture of tags from the pooled DNA. Restriction enzyme digestion (e.g., with HindIII, as in Figure 9–8) then removes the primer regions, leaving only a mixture of unique signature tags. The PCR is performed with radiolabeled nucleotides or with nonradioactively labeled nucleotides (e.g., biotinylated nucleotides), so that the mixture of tags can be used to probe membranes containing replicates of DNA from the original collection of transposon-generated mutants.

Those mutants from the output pool that do not hybridize with the mixture of probe tags represent mutants that were lost during the infection process, presumably because they had a mutation that made them less able to survive in the host environment. By comparing the input and output probes, it is possible to identify those mutants that can replicate in vitro but not in the host. These "lost" mutants are recovered from the original master plate and examined further to determine why they are less able to survive in the host, keeping in mind that the transposon insertion could have introduced polar effects. Since the transposon insertion marks the site of mutation, the gene disrupted by the transposon can be identified as described above using PCR and DNA sequencing of amplicons containing chromosomal flanking regions. Some of the mutants identified will have resulted from loss of a gene required for growth in the host, such as a gene needed to acquire, biosynthesize, or metabolize a nutrient. Other mutants will have resulted from loss of a virulence gene required for colonization, invasion, or dissemination in the host, and still others will have resulted from loss of a toxin or other effector protein. To verify that the lost genes indeed encode virulence factors, it is necessary to construct nonpolar deletion mutants and to individually assess their virulence.

An advantage of STM is that it is a negative-selection method for identifying virulence genes based on mutations in genes that do not allow survival in the host animal. The technique is applicable to a wide range of bacteria (Table 9–3). Another advantage is that once the transposon library is made, the technique is fast and requires only small numbers of host animals. Finally, it generates and identifies attenuated strains that can be used for further analysis. A potential disadvantage is that STM relies on there being a method of insertional mutagenesis that can be used to randomly knock out the genes. Also, STM can be used only for studying haploid pathogens (i.e., bacteria that have only one chromosome and thus only one copy of each gene). Transposons are not available for all bacteria, and many transposons are not completely random in their insertions (i.e., most have preferred insertion sites). Another complication is that the STM approach is a competition experiment in which multiple mutants are vying for growth in the same animal host. Strong *trans* effects caused by secreted factors, such as toxins, can potentially be masked in competition experiments (see chapter 8).

IVET

Another in vivo method for identifying genes that are expressed only when the bacteria are infecting an animal is **in vivo expression technology (IVET).** This method uses a positive-selection, gene expression strategy to select for promoters of genes or operons that are active only during infection in an animal and are not turned on when bacteria are grown on laboratory medium. The key to IVET is the construction of a reporter system that contains several important features (Figure 9–9). First, it requires a promoterless in vivo selection gene, which allows survival in the animal only when it is expressed. This selection gene, which is usually contained on a plasmid that can integrate into the bacterial chromosome, can be a critical biosynthetic gene, such as *purA* (described below). A deletion mutant of this biosynthetic gene will grow in vitro on medium containing the end product of the biosynthetic pathway (e.g., purines), but it will not grow in an animal infection model. In an IVET experiment, DNA fragments containing bacterial promoters expressed in the animal that drive the selection gene and allow survival in the host are selected. In addition to biosynthetic genes, a promoterless antibiotic resistance gene can be used, in which case the animal is fed the antibiotic and the bacterium can survive in vivo only if the antibiotic resistance gene is turned on from a cloned bacterial promoter expressed during infection.

DNA sequence tag

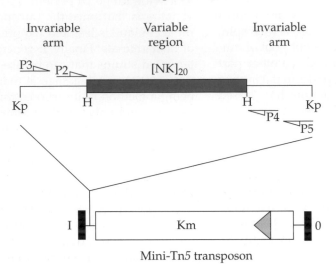

Invariable arm | Variable region | Invariable arm

P3   P2   [NK]$_{20}$

Kp   H   H   Kp
                P4
                P5

Mini-Tn5 transposon

I   Km   0

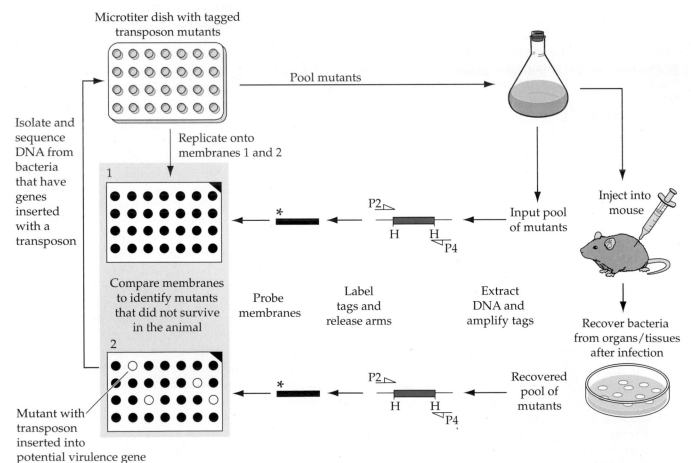

Microtiter dish with tagged transposon mutants

Pool mutants

Isolate and sequence DNA from bacteria that have genes inserted with a transposon

Replicate onto membranes 1 and 2

1

Compare membranes to identify mutants that did not survive in the animal

2

Mutant with transposon inserted into potential virulence gene

Probe membranes

Label tags and release arms

Extract DNA and amplify tags

Input pool of mutants

P2
H   H
        P4

Recovered pool of mutants

P2
H   H
        P4

Inject into mouse

Recover bacteria from organs/tissues after infection

Second, there needs to be a restriction site in front of the promoterless in vivo selection gene to allow insertion of random DNA fragments generated by digestion of the genomic DNA of the pathogen with the same restriction enzyme. This generates a library of bacteria with different DNA fragments, some of which have promoters within the DNA insert. If those promoters are turned on in vivo, then the in vivo selection gene will be expressed, and the mutant bacterium will survive in the host (positive selection). Third, to select for bacteria that integrate the promoter selection plasmid into their chromosomes, another antibiotic resistance gene with its own promoter is used. Finally, downstream of the in vivo selection gene is a second promoterless reporter gene (different from the in vivo selection gene), such as *lacZ*, which allows color screening in vitro of bacteria that survived in the animal. The rationale for this step is that promoters that are expressed constitutively (i.e., are always on) will be Lac$^+$ on indicator plates (i.e., colonies that are blue on X-Gal plates have promoters that are turned on both in vitro and in vivo). Since the goal is to identify genes that are not expressed during growth on laboratory medium but are specifically expressed in the animal, the desired colonies are those that are Lac$^-$ (i.e., white) on X-Gal indicator plates.

IVET was first used to search for genes that permit *Salmonella enterica* serovar Typhimurium to cause a typhoid-like disease in mice. The approach was based on the observation that purine auxotrophs, such as *purA* deletion mutants, of *S. enterica* serovar Typhimurium are unable to infect mice. To generate the pool of fusion clones, *S. enterica* serovar Typhimurium chromosomal DNA was digested with restriction enzymes into small fragments, which were cloned upstream of an artificial operon consisting of a promoterless *purA* gene fused to a promoterless *lacZY* reporter gene (*lacZ* is the indicator gene, and *lacY* encodes a permease that allows bacterial uptake of the colorimetric substrate X-Gal) incorporated into a plasmid (pIVET1 [Figure 9–9]). Cloning was done in *E. coli* using β-lactamase (encoded by the *bla* gene, which imparts ampicillin resistance) as a selection marker. A pool of these plasmids was introduced into a strain of *S. enterica* serovar Typhimurium that contained a deletion in its *purA* gene by using conjugation (the plasmid also included *mob* mobilization genes). The plasmid used for cloning had a replication origin (*oriR6K*) that allowed it to exist as an autonomous plasmid in *pir*$^+$ strains of *E. coli*, but it could not replicate in *Salmonella*. Thus, in *S. enterica* serovar Typhimurium, the plasmids integrated into the chromosome in response to selection for ampicillin resistance. The integration occurred by homologous recombination between the chromosomal fragments cloned into the plasmids and the corresponding region in the *S. enterica* serovar Typhimurium chromosome. This recombination resulted in a collection of bacteria in which the plasmid was integrated into different places in the *S. enterica* serovar Typhimurium chromosome. Some of these integrated plasmids placed the *purA* gene downstream of a promoter that is turned on in vivo, and these were the clones that survived when the mixture was inoculated into a mouse.

The homologous recombination via cloned fragments of *S. enterica* serovar Typhimurium DNA was used instead of the transposon method described above to generate fusions for a very good reason. Transposon insertion disrupts the gene it enters, whereas homologous recombination leaves an intact copy of the wild-type gene (i.e., a merodiploid is formed). This was important because the investigators were seeking genes that are essential for survival in the animal and did not want to disrupt the genes to which fusions were made. A pool of integrated *purA-lacZY* fusions was injected into the mouse. The bacteria that survived in the mouse were then plated on medium on which the expression of *lacZ* could be

Figure 9–8 STM used for simultaneous identification of *S. enterica* serovar Typhimurium virulence genes based on transposon mutagenesis and negative selection. Shown on top is the general design of the transposon containing unique signature tags. A library of transposon insertion mutants is generated, and a master plate is used to store the mutants and to make an array of the mutants on replicate membranes. The pooled mutants are then inoculated into a mouse infection model, and the input and output pools of mutants are compared to determine which mutants were not able to survive in the mouse. Identification is based on hybridization of the labeled tags with their complementary tags on the replicate membranes. NK, nucleotides whose tag regions vary; P, primers; Kp, KpnI; H, HindIII; I and O, ends of mini-Tn*5*; Km, kanamycin resistance gene; *, labeled tag. (Adapted from M. Handfield and R. C. Levesque. 1999. Strategies for isolation of in vivo expressed genes from bacteria. *FEMS Microbiol. Rev.* **23**:69–91, with permission from John Wiley and Sons.)

Table 9–3 Examples of pathogens studied using non-microarray-based expression technologies to identify virulence genes (through 2007)

Technology	Bacterium studied	Infection model used	Yr reported
STM	*Actinobacillus pleuropneumoniae*	Pig endotracheal infection model	2003
	Actinobacillus suis	Pig upper respiratory tract infection model	2005
	Brucella abortus	Mouse acute and chronic infection models	1998, 2000
	Brucella melitensis	Mouse and goat infection models	2000, 2003, 2006
	Burkholderia cenocepacia	Chicken and mouse infection models, rat lung infection model	2003, 2004
	Burkholderia pseudomallei	Mouse melioidosis infection model	2002
	Campylobacter jejuni	Chicken gastrointestinal tract infection/commensalism model	2004, 2005
	Desulfovibrio desulfuricans	Anaerobic sediment survival in the environment	2005
	Escherichia coli	Chicken systemic infection model	2005
	Haemophilus influenzae	Rat infant systemic infection model	2002
	Helicobacter pylori	Mongolian gerbil gastric colonization model	2003
	Klebsiella pneumoniae	Continuous-flow culture model for biofilm formation	2006
	Klebsiella pneumoniae	Mouse gastrointestinal tract colonization/infection model, intranasal infection model	2003, 2005
	Listeria monocytogenes	Mouse brain infection model	2001
	Mycobacterium marinum	Goldfish model of tuberculosis	2004
	Pasteurella multocida	Mouse septicemia model and chicken fowl cholera model	2003
	Proteus mirabilis	Mouse urinary tract infection model	2004
	Pseudomonas aeruginosa	Nematode *Caenorhabitis elegans* model of infection	2007
	Pseudomonas aeruginosa	Rat chronic respiratory infection model	2003
	Salmonella enterica serovar Cholerasuis	Pig oral and systemic infection model	2003, 2005
	Salmonella enterica serovar Gallinarum	Chicken fowl typhoid infection model	2005
	Salmonella enterica serovar Typhimurium	Mouse and pig infection models	1996, 2006, 2007
	Shewanella oneidensis	Anaerobic sediment survival in the environment	2005
	Streptococcus agalactiae	Rat neonatal sepsis model	2000
	Streptococcus pneumoniae	Mouse pneumonia model	2002
	Streptococcus sanguis	Rabbit endocarditis model	2005
	Vibrio cholerae	Mouse infant small intestine infection model	2002
	Yersinia pestis	Mouse infection model	2004
IVET	*Bacillus cereus*	Insect larva model of oral infection	2006
	Salmonella enterica serovar Typhimurium	Mouse infection model	1993, 1995
	Actinobacillus pleuropneumonia	Pig pneumonia infection model	1999
	Erwinia chrysanthemi	African violet plant leaf infection model	2004
	Klebsiella pneumoniae	Mouse infection model	2001
	Lactobacillus plantarum, L. reuteri	Mouse gastrointestinal tract	2003, 2004
	Listeria monocytogenes	Mouse infection model	2000
	Pasteurella multocida	Mouse infection model	2001, 2004, 2005
	Porphyromonas gingivalis	Mouse abscess model	2002
	Pseudomonas aeruginosa	Mouse burn infection model	2004
	Pseudomonas aeruginosa	Respiratory mucus derived from cystic fibrosis patients	1996, 2004
	Pseudomonas fluorescens	Soil environment	2004
	Pseudomonas putida	Maize root rhizosphere model	2005
	Pseudomonas syringae	Tomato plant infection model	2002
	Pseudomonas viridiflava	Plant soft rot infection model	2006
	Ralstonia solanacearum	Tomato plant infection model	2004
	Shigella flexneri	Intracellular survival in human epithelial cells and mouse macrophage-like cells, mouse lung infection model	2002, 2003, 2004
	Streptococcus pneumoniae	Lung and intraperitoneal infection models	2006, 2007

Table 9–3 (continued)

Technology	Bacterium studied	Infection model used	Yr reported
	Streptococcus gordonii	Rabbit endocarditis model	1999
	Streptococcus suis	Pig infection model	2001
	Yersinia pestis, Y. enterocolitica, Y. pseudotuberculosis	Mouse infection model	2005
IVIAT	*Actinobacillus actinomycetemcomitans*	Human periodontal infection, human epithelial cell model of infection	2002, 2005
	Escherichia coli O157:H7	Human gastrointestinal infection	2005
	Mycobacterium tuberculosis	Human tuberculosis infection	2002
	Porphyromonas gingivalis	Human periodontal infection	2002
	Salmonella enterica serovar Typhi	Human typhoid infection	2006
	Vibrio cholerae	Human cholera infection	2003
	Vibrio vulnificus	Human septicemic infection	2003
GSH	*Actinobacillus pleuropneumoniae*	Pig chronic lung infection	2007
	Borrelia burgdorferi	Human Lyme disease	2003
	Burkholderia cenocepacia	Alfalfa infection model	2005
	Bukholderia mallei	Glanders infection model	2001
	Escherichia coli	Human and mouse infection	2000, 2001, 2002, 2006
	Helicobacter pylori	Human peptic ulcer disease	2006
	Klebsiella pneumoniae	Mouse lung infection model	2007
	Mycobacterium avium	Human macrophage model, bird infection model	2002, 2003
	Mycobacterium bovis	European wild-boar tonsil and mandibular lymph node infection, macrophage infection model	1996, 2001, 2006
	Mycobacterium gallisepticum	House finch (songbird) infection	2006
	Mycobacterium tuberculosis	Guinea pig and mouse models of infection	2007
	Mycobacterium ulcerans	Human Buruli ulcer infection	2003
	Neisseria meningitidis, N. gonorrhoeae	Human meningitis, gonorrhea, toxic septicemia	1995, 2000
	Phytophthora infestans	Potato blight infection	2007
	Phytophthora sojae	Soybean infection	2007
	Pseudomonas aeruginosa	Human mast cell inflammation model, lung infection model	2004, 2005
	Salmonella cholerasuis	Pig infection model	2006, 2007
	Salmonella enterica serovar Enteriditis	Chicken infection model	2001
	Salmonella enterica serovar Typhimurium, *S. enterica* serovar Typhi	Mouse and pig infection models	1999, 2007
	Sinorhizobium meliloti	Plant infection model	2007
	Staphylococcus aureus	Amphioxus infection model	2007
	Streptococcus agalactiae	Human group B streptococcal infection	2003
	Streptococcus sanguis	Platelet aggregation (thrombosis) model	2005
	Vibrio parahaemolyticus	Amphioxus infection model	2007
	Vibrio penaeicida	Shrimp infection	2005
	Xylella fastidiosa	Pierce's grape disease	2007
	Yersinia pestis	Macrophage infection model	2003

detected. For the resulting white colonies, the genomic DNA fragments upstream of the *purA* gene were sequenced to identify the promoters that were driving the expression of the *purA* gene in vivo. Interestingly, many of the genes identified by IVET for *S. enterica* serovar Typhimurium have proved to be metabolic "housekeeping genes."

IVET has been successfully used for the study of many pathogens (Table 9–3). One limitation of the IVET approach is that it is a positive-selection approach that identifies only promoters. More work is needed to examine the putative virulence genes downstream of the promoter to determine which of those genes are responsible for virulence. In addition,

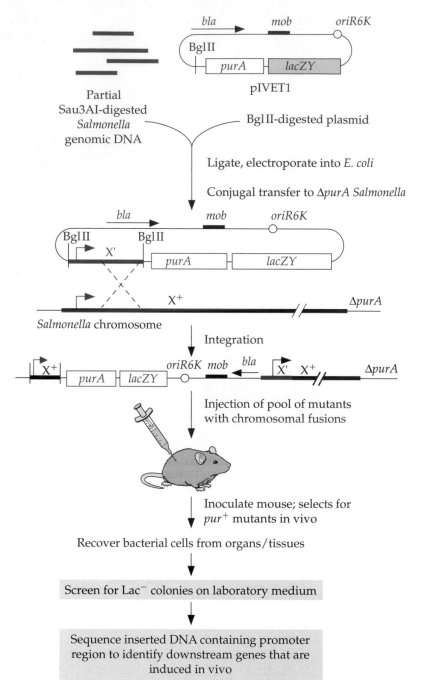

Figure 9–9 IVET for detecting promoters of genes that must be expressed for the bacteria to survive in vivo. Shown is a schematic of the steps involved in using IVET to identify *S. enterica* serovar Typhimurium promoters that are turned on only in the mouse. *bla*, gene for ampicillin resistance during cloning of the genomic library in *E. coli*; *purA*, promoterless in vivo selection gene necessary for purine utilization; *lacZY*, β-galactosidase indicator gene plus β-galactoside permease gene that transports lactose into cells; *X*, gene into which the plasmid construct was integrated; *X'*, *Salmonella* DNA fragment containing a partial copy of *X*. (Adapted from M. J. Mahan, J. M. Slauch, and J. J. Mekalanos. 1993. Selection of bacterial virulence genes that are specifically induced in host tissue. *Science* **259:** 686–688, with permission from AAAS.)

IVET detects only promoters of genes that are transcribed at elevated levels in the host. Gene products that may be activated posttranscriptionally in response to host signals will be missed.

Several recent modifications to IVET have broadened its applicability. Since it is not always easy to make auxotrophic strains of different pathogens (such as the *purA* mutant of *Salmonella* in the example above) for in vivo studies, a promoterless antibiotic resistance gene different from the one used to maintain the plasmid or merodiploid (*bla* in the above example) can be used for the in vivo selection. The antibiotic is administered to the host during infection. Only mutants that have promoters that turn on in the host will express the antibiotic resistance selection gene and thereby survive in the host. However, auxotrophy-based and antibiotic resistance-based selection methods suffer from the inability to isolate the genes that are only transiently or weakly expressed in vivo during infection. Use of **recombinase-based IVET (RIVET)** has overcome this problem.

In RIVET, the promoterless selection gene encodes a **site-specific DNA resolvase,** such as the TnpR protein from the transposon Tnγδ. When expressed, the TnpR protein causes the excision of DNA segments between two specific recombinase recognition sequences, the so-called res1 sites. The res1 sites are constructed to flank a promoter and gene imparting antibiotic resistance different from that used to maintain the plasmid or merodiploid. When an active cloned promoter drives expression of TnpR, the antibiotic resistance gene between the two res1 sites is excised, and the resulting bacterium becomes sensitive to the antibiotic. By first being plated in vitro on plates containing the antibiotic, mutants that have promoters that are active in vitro will be eliminated prior to infection. After infection, bacteria are reisolated on plates lacking antibiotic, and the colonies are checked for antibiotic resistance. Those colonies that grow on plates lacking antibiotic but not on plates with antibiotic are the ones that were expressed in vivo. RIVET is applicable even to those bacteria that are difficult to manipulate genetically and does not expose the bacteria to auxotrophic selection during infection. On the other hand, the extreme sensitivity of the RIVET approach (even low levels of constitutive promoter expression lead to excision of the resistance marker) sometimes limits the identification of in vivo-induced promoters.

Genomic Methods for Identifying Virulence Genes

GSH

Genomic subtractive hybridization (GSH) is a PCR-supported method for isolating genomic DNA sequences that are unique to particular strains of closely related bacteria (Figure 9–10). GSH can be used to identify genes that are present only in pathogenic or only in nonpathogenic strains. Genomic DNA from the pathogenic and nonpathogenic strains is isolated and digested into manageable-size fragments with a specific restriction enzyme. Linker or adaptor primers are ligated to the ends of the DNA fragments for later cloning into plasmids. The DNA fragments from the nonpathogenic strain are labeled with biotin. The two pools of DNA fragments are mixed, denatured, and annealed, thereby allowing complementary fragments from the two strains to hybridize to each other. Streptavidin beads, which bind tightly to the biotin label,

are added to the hybridized DNA and used to separate the DNA fragments that the two bacterial strains have in common from the DNA that is unique to the pathogen (i.e., DNA fragments from the pathogenic strain that are not present in the nonpathogenic strain will not hybridize to biotin-labeled DNA and will not bind to the beads). After several rounds of this selection process, only the unique DNA remains, which is then amplified using PCR, cloned into a plasmid, and sequenced to determine which putative virulence genes are carried by the unique DNA segments.

GSH can also be used to identify genes that are associated with host specificity in closely related but host-specific strains of a particular bacterium (Figure 9–11). For example, GSH has been used to identify *Salmonella* serovar-specific genes. Chromosomal DNA from various *Salmonella* strains was digested with the restriction enzyme EcoRI, and sequences that were shared with biotin-labeled DNA fragments from *S. enterica* serovar Typhi were removed using streptavidin beads. The remaining DNA from each of the samples was separated on a polyacrylamide gel and then transferred to a membrane. The subtracted genomic DNA fragments from *S. enterica* serovar Typhimurium, which has broad host specificity, were also radiolabeled and used as a probe to hybridize with the DNA on the membrane. This allowed the identification of DNA bands that were not present in human-specific *S. enterica* serovar Typhi but were present in other host-specific or host-adapted *Salmonella* strains.

In another example, to search for the genetic basis of virulence in *M. ulcerans*, researchers took advantage of the close genetic relationship between *M. ulcerans* and *Mycobacterium marinum* to perform GSH. They found several DNA fragments specific to *M. ulcerans*, but in particular, they found one locus containing a cluster of polyketide synthase genes with a highly repetitive modular arrangement (Figure 9–1). Knowing the polyketide-like structure of the mycolactone toxin in *M. ulcerans*, they deduced that this polyketide synthase gene cluster was responsible for the synthesis of the mycolactone toxin. Unexpectedly, the researchers also found that these polyketide synthase genes were carried on a large, 174-kb plasmid. GSH has been a useful method, but the availability of massively parallel sequencing methods (see chapter 5) means that the genome sequences of different bacterial strains can be determined rapidly and compared. This high-resolution approach not only indicates differences, but provides the actual DNA changes.

Selective capture of transcribed sequences (SCOTS) is a reverse transcriptase PCR-based method often combined with GSH (as such, it is sometimes

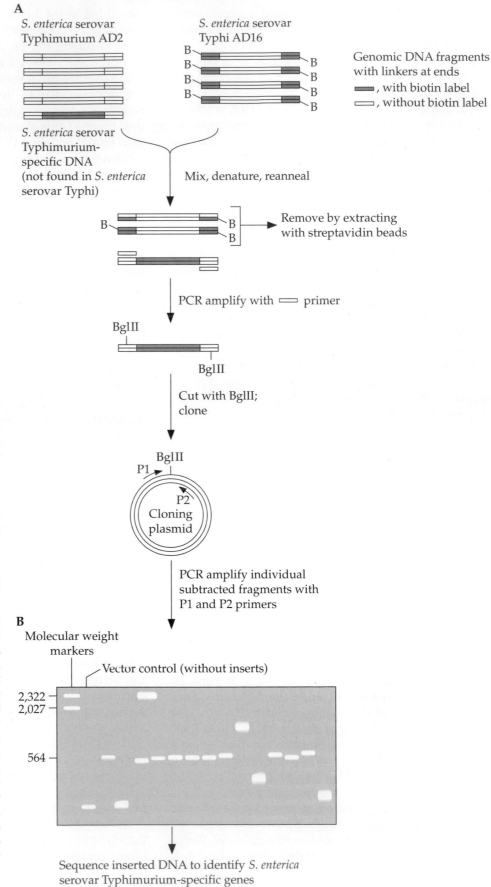

Figure 9–10 Steps involved in PCR-supported GSH, the genomic-subtraction procedure used to isolate *S. enterica* serovar Typhimurium-specific DNA fragments. **(A)** Schematic representation of the subtraction procedure. A fragment of *S. enterica* serovar Typhimurium DNA not present in *S. enterica* serovar Typhi is shown in gray. Biotinylated adaptor sequences are indicated in blue. **(B)** Agarose gel showing individually amplified *S. enterica* serovar Typhimurium DNA fragments after subtraction and cloning. The PCR product obtained by using the cloning vector without an insertion as the template is in the lane adjacent to the size marker lane. (Adapted from M. Emmerth, W. Goebel, S. I. Miller, and C. J. Hueck. 1999. Genomic subtraction identifies *S. enterica* serovar Typhimurium prophages, F-related plasmid sequences, and a novel fimbrial operon, *stf*, which are absent in *S. enterica* serovar Typhi. *J. Bacteriol.* **181:**5652–5661.)

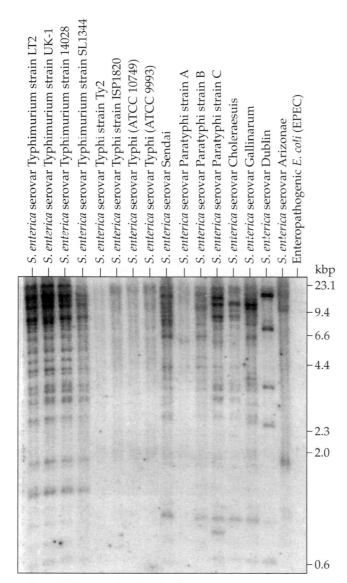

Figure 9–11 Host-specific DNA banding patterns identified by GSH. In this experiment, GSH was used to detect the presence of *S. enterica* serovar Typhimurium-specific genomic sequences in host-adapted *Salmonella* serovars and *E. coli*. Two micrograms of EcoRI-digested chromosomal DNA from each tested strain was hybridized with radiolabeled *S. enterica* serovar Typhimurium genomic sequences that had the DNA sequences in common with *S. enterica* serovar Typhi subtracted. (Reprinted with permission from B. J. Morrow, J. E. Graham, and R. Curtiss III. 1999. Genomic subtractive hybridization and SCOTS identify a novel *S. enterica* serovar Typhimurium fimbrial operon and putative transcriptional regulator that are absent from the *S. enterica* serovar Typhimurium genome. *Infect. Immun.* **67:**5106–5116.)

called **suppressive subtractive hybridization),** which can be used to identify genes that are transcribed only under in vivo or in culturo conditions (Figure 9–12). In this method, total mRNA from bacteria recovered from the host is reverse transcribed into cDNA, which is then labeled with adaptors at the ends for later cloning into plasmids. Control mRNA is similarly isolated from bacteria that are grown in vitro under laboratory conditions and then labeled with biotin and hybridized to the cDNA isolated from the bacteria exposed to host conditions. Streptavidin beads are used to remove the common genes (i.e., those transcribed in the bacteria both in vivo and in vitro). The remaining cDNA corresponds to the mRNA from genes that are turned on only under in vivo conditions. This cDNA is then cloned into plasmids, and sequencing and comparison of the DNA with those in DNA databases can be used to identify the genes. In this way, SCOTS, which involves GSH of cDNA fragments, can be used to identify bacterial genes that are expressed only in the host.

IVIAT

An antibody-based genomic method that can be used to identify genes induced during human infections yet avoids the use of animal infection models is **in vivo-induced antigen technology (IVIAT).** The method (illustrated in Figure 9–13) takes advantage of pooled sera from one or more patients who have been exposed to a particular pathogen and have developed a protective immune response. In the version of this method shown, the pooled sera are absorbed with whole cells or cellular extracts from bacteria grown in vitro. The remaining serum, which contains a subpopulation of antibodies reactive against in vivo-induced antigens, is then used to probe a genomic expression library of *E. coli* clones that express genes from the pathogen. Often, this step is done by using **phage display,** which incorporates regions from the pathogen proteins into the coats of bacteriophage. This display method lends itself to high-throughput screening methods. Those clones expressing proteins that are cross-reactive with the remaining antibodies in the sera are identified as putative in vivo-expressed antigens that might generate protective immune responses and, hence, are potential vaccine candidates. The potential candidate proteins are then purified and inoculated into an animal model, if available, to determine whether they are antigenic. If antibodies are produced, the vaccinated animals can then be challenged with the pathogen to learn whether the antibodies are protective against the infection. In addition, IVIAT has been useful for identifying virulence factors

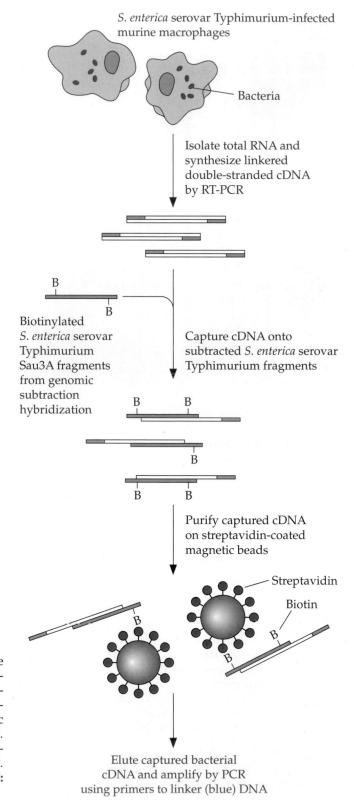

Figure 9–12 SCOTS. Shown is a schematic diagram of the SCOTS technique to identify subtracted genomic sequences, which are transcribed by bacteria following macrophage phagocytosis. B, biotin. (Adapted from B. J. Morrow, J. E. Graham, and R. Curtiss III. 1999. Genomic subtractive hybridization and SCOTS identify a novel *S. enterica* serovar Typhimurium fimbrial operon and putative transcriptional regulator that are absent from the *S. enterica* serovar Typhimurium genome. *Infect. Immun.* **67:** 5106–5116.)

S. enterica serovar Typhimurium-infected murine macrophages

Bacteria

Isolate total RNA and synthesize linkered double-stranded cDNA by RT-PCR

Biotinylated *S. enterica* serovar Typhimurium Sau3A fragments from genomic subtraction hybridization

Capture cDNA onto subtracted *S. enterica* serovar Typhimurium fragments

Purify captured cDNA on streptavidin-coated magnetic beads

Streptavidin

Biotin

Elute captured bacterial cDNA and amplify by PCR using primers to linker (blue) DNA

involved in human infection and for identifying virulence factors of human-specific pathogens for which there is no suitable animal model (Table 9–3). Finally, in some approaches related to IVIAT, such as antigenic fingerprinting, the initial sera are not preabsorbed by cells or extracts grown in vitro. This approach identifies antigenic surface proteins that are expressed both in vitro and in vivo, including essential surface proteins involved in cell division and signaling.

Microarrays

A **DNA (or protein) microarray** consists of an orderly arrangement of DNA oligonucleotides (or expressed proteins) corresponding to all the genes in a genome of an organism (Figure 9–14). On the **DNA chip (or gene chip),** each DNA oligonucleotide representing a gene is assigned a specific, discrete location on the array (usually on a glass slide or a membrane) and then microscopically spotted onto that location. Fluorescently labeled cDNAs (generated by reverse transcription of bacterial mRNA samples) or DNAs (from different bacterial strains) isolated from experimentally infected cells are hybridized to the DNA probe spots on the array, which contain complementary DNA. These arrays can be used to identify genes that are expressed under different conditions in vitro or in vivo.

Microarray technology can provide a profile of the expression patterns of thousands of genes in parallel in a single experiment and can be used to ask questions, such as which genes are turned on or off under one condition compared to another condition. For bacteria, the total RNA obtained from cells grown under the two conditions is first converted into cDNA (Figure 9–14). Random mixtures of primers that will hybridize to most of the mRNAs in the samples are used. The cDNAs corresponding to the mRNAs are synthesized by reverse transcription using reaction mixtures containing special nucleotides labeled with different colored fluorescent dyes, such as Cy3 or Cy5. A different colored dye is used for each cDNA reaction, and the different colored cDNAs corresponding to two different RNA preparations are mixed in equal amounts. The mixture is hybridized to the DNA oligonucleotide spots on the microarray chip. The reference sample is from a strain incubated under untreated conditions, and the experimental sample is from a strain treated in some way, such as limited for iron. Alternatively, the global transcriptome expression patterns of a mutant bacterial strain and the wild-type parent strain can be compared. After the competitive hybridization with the mixture of two cDNAs, followed by washing, the relative amounts of fluorescent dyes bound to each DNA oligonucleotide spot on the microarray chip are quantified. The results are compared using sophisticated software that normalizes the large data sets and performs statistical analyses of the changes based on independently repeated experiments. Most spots show the same relative ratio of both dyes, indicating that the relative transcript amounts of that mRNA did not change under the experimental condition or in the mutant compared to the untreated or wild-type parent strain. However, some spots show an excess of one dye compared to the other, indicating increases or decreases in relative transcript amounts. Often, these large data sets are displayed graphically to indicate the subset of genes whose relative transcript amounts change under the two conditions.

Microarrays are extremely versatile and have been extended to many applications. Analyses of the relative transcript amounts that change during iron limitation of pathogenic bacteria have not only revealed genes that mediate iron uptake, but have indicated possible virulence factor genes that use low iron as a signal for expression in hosts. Microarray analyses have been performed on bacteria recovered from animal tissues, such as blood, and compared to the same bacteria grown in culture. This type of comparison has indicated differences in relative transcript amounts of bacterial genes that are turned on in the animals compared to those that are turned on in vitro. Finally, microarrays have even been applied to track where proteins bind to DNA inside bacterial and other cells. In this sophisticated ChIP (for chromatin immunoprecipitation)-on-chip approach, the microarray is used to identify which segments of DNA in the genome correspond to DNA segments that were chemically cross-linked to the protein inside cells.

Microarray technology has recently been extended to proteins, as well. For example, **protein microarrays (proteoarrays)** have been combined with IVIAT technology for determining complete antigen-specific humoral immune response profiles from vaccinated or infected humans and animals. Once the genome sequence of a pathogen has been determined, high-throughput PCR can be used to clone every protein reading frame of the pathogen into an expression vector. The corresponding proteins are then expressed using an *E. coli*-based cell-free in vitro transcription-translation system (which avoids the need to purify the expressed proteins), and the in vitro-expressed proteins are printed (spotted) onto nitrocellulose membrane microarrays. Alternatively, the phage display approach mentioned above can be used to express the proteins from the pathogen. The proteoarrays can then be used to determine the anti-

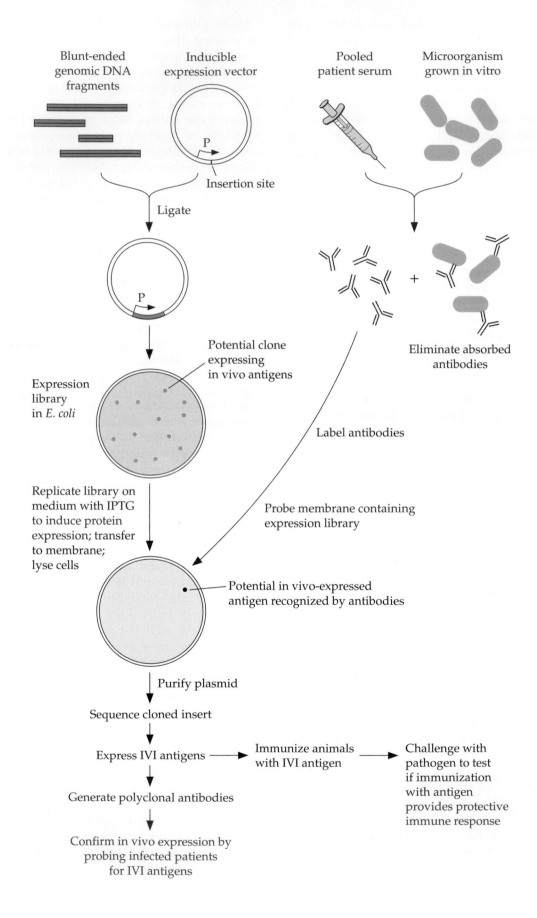

Blunt-ended genomic DNA fragments

Inducible expression vector

Insertion site

Pooled patient serum

Microorganism grown in vitro

Ligate

Eliminate absorbed antibodies

Potential clone expressing in vivo antigens

Expression library in *E. coli*

Label antibodies

Replicate library on medium with IPTG to induce protein expression; transfer to membrane; lyse cells

Probe membrane containing expression library

Potential in vivo-expressed antigen recognized by antibodies

Purify plasmid

Sequence cloned insert

Express IVI antigens → Immunize animals with IVI antigen → Challenge with pathogen to test if immunization with antigen provides protective immune response

Generate polyclonal antibodies

Confirm in vivo expression by probing infected patients for IVI antigens

body-binding profiles of sera (serological screening) from vaccinated humans or animals, which can then be used to identify cross-reactive antigens that might serve as potential vaccine candidates. It should be noted that because the in vitro expression system contains *E. coli* lysate, it is important to first remove the anti-*E. coli* antibodies that are normally present in human serum before performing the serological screening. This is not necessary for mouse serum, which lacks significant *E. coli* cross-reactivity.

Comparative Genomics for Vaccines and Therapeutics

Major genome-sequencing projects (Figure 9–15) (see chapter 5) are currently under way with the goal of conducting comparisons between different strains to identify unique sequences that might be serovar-specific virulence genes. Sequencing of microbial genomes can be monitored at the following websites: http://www.ncbi.nlm.nih.gov/sites/genome and http://www.ncbi.nlm.nih.gov/genomes/lproks.cgi. With the increasing speed and efficiency, yet decreasing cost, of sequencing microbial genomes, more and more genome sequences of nonpathogenic and pathogenic bacterial strains are becoming available for genome comparisons. From these comparative-genomics approaches, it is clear that the bacterial genome is dynamic, with multiple factors having contributed to its evolution, including functional diversification and adaptation through gene mutation, gene duplication, genome rearrangement, gene loss (genome reduction), and/or gene gain (acquisition of new functions through horizontal gene transfer).

Results from these comparative genomics approaches are rapidly advancing the field of microbial pathogenesis, leading to the development of novel di-agnostics, improved vaccines, and the identification of new drug targets and other antimicrobial therapeutics. Microbial genome sequence information and the ability to analyze the expression activity of every gene in a cell are powerful tools that are accelerating vaccine development (Figure 9–16). Once a genome has been completely sequenced, all potential antigens can be identified using software analysis programs that predict proteins that might be secreted or expressed on the cell surface. These putative vaccine targets are then expressed as recombinant proteins in *E. coli*, and the purified proteins are used to immunize mice. The immune sera are then screened for antibody binding to the bacteria and bactericidal activity. Those vaccine candidates that test positive in these assays are then further tested for the ability to provide protective immunity. In a recent example of the power of this technology, from a total of 570 putative secreted or surface proteins identified from the genome sequence of *Neisseria meningitidis*, about 350 recombinant proteins were expressed in *E. coli* and used to immunize mice. Seven proteins were selected from the immune serum screen, and of those, two proteins were chosen as vaccine candidates for further clinical trials.

The Importance of Bacterial Physiology

In retrospect, a critical experiment done on *S. enterica* serovar Typhimurium presaged what is now being learned from genomics and other ultramodern approaches. The investigators introduced transposons nearly randomly into a virulent strain of *S. enterica* serovar Typhimurium. For the purposes of this study, the investigators defined "virulence" as the ability of *S. enterica* serovar Typhimurium to grow inside macrophages. They screened thousands of transposon-generated mutants for the ability to survive and grow

Figure 9–13 IVIAT. Sera from patients who have experienced an infection caused by the pathogen under study are pooled and exhaustively absorbed with cells of the pathogen grown in vitro, leaving antibodies against antigens that are expressed only in vivo (step 1). An expression library of the pathogen's DNA is generated in a suitable host (step 2), and clones are probed with the absorbed serum (step 3). Reactive clones, which are producing antigens that are expressed during a natural infection but not during in vitro cultivation, are purified, and their cloned DNA is sequenced (step 4). Genes are identified in this fashion as encoding in vivo-induced (IVI) antigens. These antigens are purified and used to verify that the IVI antigen is expressed by the pathogen during an infectious process (step 5). This can be done using fluorescently labeled antibodies raised against the purified IVI antigen to probe biological samples taken from infected patients. (Adapted from M. Handfield, L. J. Brady, A. Progulske-Fox, and J. D. Hillman. 2000. IVIAT: a novel method to identify microbial genes expressed specifically during human infections. *Trends Microbiol.* **8:** 336–339, with permission from Elsevier.)

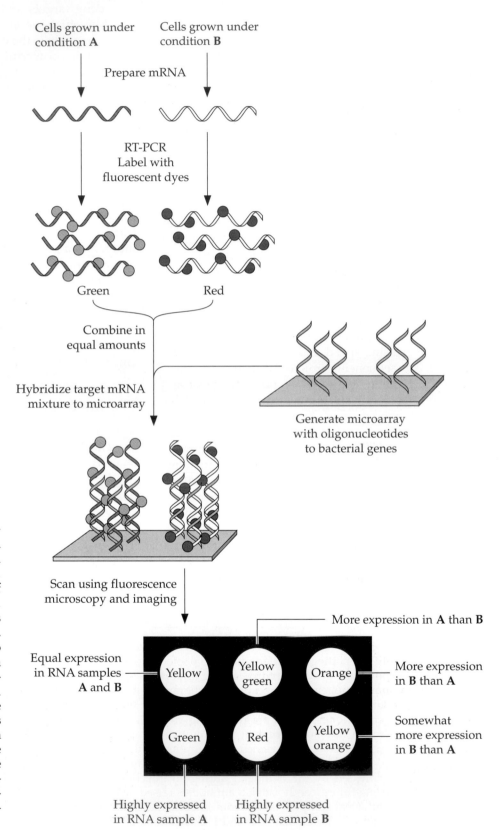

Figure 9–14 Microarrays. Microarrays are chips that contain DNA segments representing all or selected genes in a genome. Each spot contains a segment of DNA from a particular gene. mRNA is prepared from cells exposed to different conditions, and the mRNA is converted into cDNA by reverse transcription (RT) and amplified by PCR, incorporating different colored dyes into the PCR products. The different colored PCR products are mixed and hybridized with the DNA chip. The spots are then visualized by fluorescence microscopy, and images are analyzed by computer to determine differences in gene expression.

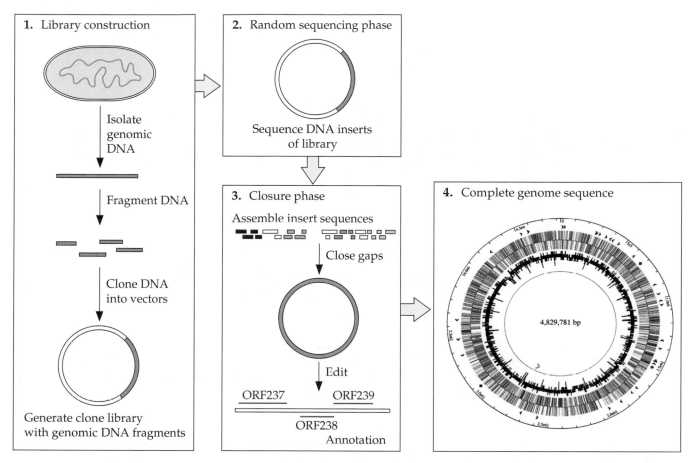

Figure 9–15 Steps in whole-genome shotgun sequencing. (Adapted from C. M. Fraser, J. A. Eisen, and S. L. Salzberg. 2000. Microbial genome sequencing. *Nature* **406:** 799–803, with permission from Macmillan Publishers Ltd.)

inside macrophages. Their results led them to conclude that at least 200 genes (over 5% of the chromosome) were involved in the ability of *S. enterica* serovar Typhimurium to survive in macrophages, and these genes were scattered all over the bacterial chromosome. Today, comparisons of the genome sequences of, say, *S. enterica* serovar Typhimurium and *E. coli* K-12 (an avirulent *E. coli* strain) or of *S. enterica* serovar Typhimurium and *S. enterica* serovar Typhi (the cause of typhoid fever in humans) show that the differences are many and are scattered throughout the chromosome. Clearly, virulence is multifactorial. In addition, it has become clear that genes important for virulence in one strain may also be found in a nonvirulent strain of the same species, and hence, the assumption that only virulent organisms contain virulence genes is flawed. More often than not, the genes identified by the approaches we have described in this chapter prove to be genes that would be considered housekeeping genes or stress response genes. Perhaps what this is telling us is that it is the entire physiology

of the bacterium, not just a few genes, that is important.

Another striking finding that has come out of the genome-sequencing data is that about one-third of all the genes are not recognizable, i.e., they code for proteins of unknown function. And, significantly, the majority of the genes identified as putative virulence genes have no significant matches in the known databases. (How "significant" is defined varies somewhat from one scientist to another, but usually it means that the percentage of amino acid identities should be at least 20%.) In addition to the sequences that are not recognized at all are the ones for which the tentative identification is questionable. For instance, a deduced amino acid sequence that contains a consensus ATP-binding site may be identified as an ATPase when it actually has a completely different function. Again, about one-third of the identified genes fall into this category. Ultimately, it is going to take a new initiative to advance our knowledge of bacterial physiology to the point that the sequence in-

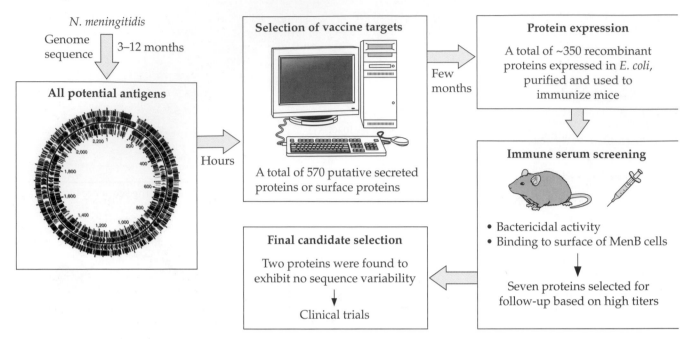

N. meningitidis

Genome sequence → 3–12 months

All potential antigens

Hours →

Selection of vaccine targets

A total of 570 putative secreted proteins or surface proteins

Few months →

Protein expression

A total of ~350 recombinant proteins expressed in *E. coli*, purified and used to immunize mice

Immune serum screening

- Bactericidal activity
- Binding to surface of MenB cells

Seven proteins selected for follow-up based on high titers

Final candidate selection

Two proteins were found to exhibit no sequence variability

Clinical trials

Figure 9–16 Whole-microbial-genome sequencing to accelerate vaccine development. (Adapted from C. M. Fraser, J. A. Eisen, and S. L. Salzberg. 2000. Microbial genome sequencing. *Nature* **406:**799–803, with permission from Macmillan Publishers Ltd.)

formation becomes truly decipherable. More than that, the interactions between different pathways remain to be elucidated. As subsequent chapters discussing individual microbes show, the expression of different virulence genes and the relationships of the proteins they encode are highly interactive. How are we to understand this second layer of complexity, beyond the function of gene products to the interaction of these products with those of other genes?

Bacterial physiology is the field most likely to get at the essence of what it means to be a living organism. For instance, the genome sequence of a tiny bacterium called *Mycoplasma genitalium*, which contains about 300 genes, currently stands as the simplest genome of a free-living organism. However, over 100 of these genes are not recognizable as having a known function. What is exciting about this small genome is that it raises the hope that if scientists could understand the function and interactions of this small number of gene products, they might finally have an insight into what defines "life." Again, these future insights must come from more sophisticated studies of bacterial physiology, coupled with high-throughput methods. For example, collections of knockout mutations have recently been constructed for every nonessential gene of *E. coli*. In fact, two insertion mutations were constructed in every gene, each imparting resistance to two different antibiotics. Therefore, it is the-

oretically possible to construct double mutants to test the genetic interactions between every pair of nonessential genes in *E. coli*, but the numbers would be very large (e.g., just 20 knockouts would lead to 190 unique pairwise combinations). However, high-throughput methods coupled with robotic handling are making this approach possible, and we are beginning to learn the possible functions of many of the unknown genes mentioned above. These approaches are beginning to be applied to pathogens, such as *Streptococcus pneumoniae*, as well.

SELECTED READINGS

Angelichio, M. J., and A. Camilli. 2002. In vivo expression technology. *Infect. Immun.* **70:**6518–6523.

Darwin, A. J. 2005. Genome-wide screens to identify genes of human pathogenic *Yersinia* species that are expressed during host infection. *Curr. Issues Mol. Biol.* **7:**135–149.

Davies, D. H., X. Liang, J. E. Hernandez, A. Randall, S. Hirst, Y. Mu, K. M. Romero, T. T. Nguyen, M. Kalantari-Dehaghi, S. Crotty, P. Baldi, L. P. Villarreal, and P. L. Felgner. 2005. Profiling the humoral immune response to infection by using proteome microarrays: high-throughput vaccine and diagnostic antigen discovery. *Proc. Natl. Acad. Sci. USA* **102:**547–552.

Faucher, S. P., R. Curtiss III, and F. Daigle. 2005. Selective capture of *Salmonella enterica* serovar Typhi genes that are expressed in macrophages but are absent from the *Salmonella enterica* serovar Typhimurium genome. *Infect. Immun.* **73:**5217–5221.

Fraser-Liggett, C. M. 2005. Insights on biology and evolution from microbial genome sequencing. *Genome Res.* **15:**1603–1610.

George, K. M., L. P. Barker, D. M. Welty, and P. L. C. Small. 1998. Partial purification and characterization of biological effects of a lipid toxin produced by *Mycobacterium ulcerans. Infect. Immun.* **66:**587–593.

Giefing, C., A. L. Meinke, M. Hanner, T. Henics, M. D. Bui, D. Gelbmann, U. Lundberg, B. M. Senn, M. Schunn, A. Habel, B. Henriques-Normark, A. Ortqvist, M. Kalin, A. von Gabain, and E. Nagy. 2008. Discovery of a novel class of highly conserved vaccine antigens using genomic scale antigenic fingerprinting of pneumococcus with human antibodies. *J. Exp. Med.* **205:**117–131.

Handfield, M., and R. C. Levesque. 1999. Strategies for isolation of in vivo expressed genes from bacteria. *FEMS Microbiol. Rev.* **23:**69–91.

Pizza, M., V. Scarlato, V. Masignani, M. M. Giuliani, B. Arico, M. Comanducci, G. T. Jennings, L. Baldi, E. Bartolini, B. Capecchi, C. L. Galeotti, E. Luzzi, R. Manetti, E. Marchetti, M. Mora, S. Nuti, G. Ratti, L. Santini, S. Savino, M. Scarselli, E. Storni, P. Zuo, M. Broeker, E. Hundt, B. Knapp, E. Blair, T. Mason, H.

Tettelin, D. W. Hood, A. C. Jeffries, N. J. Saunders, D. M. Granoff, J. C. Venter, E. R. Moxon, G. Grandi, and R. Rappuoli. 2000. Identification of vaccine candidates against serogroup B meningococcus by whole-genome sequencing. *Science* **287:**1816–1820.

Rollins, S. M., A. Peppercorn, L. Hang, J. D. Hillman, S. B. Calderwood, M. Handfield, and E. T. Ryan. 2005. In vivo induced antigen technology (IVIAT). *Cell. Microbiol.* **7:**1–9.

Schoolnik, G. K. 2002. Functional and comparative genomics of pathogenic bacteria. *Curr. Opin. Microbiol.* **5:**20–26.

Stabler, R. A., G. L. Marsden, A. A. Witney, Y. Li, S. D. Bentley, C. M. Tang, and J. Hinds. 2005. Identification of pathogen-specific genes through microarray analysis of pathogenic and commensal *Neisseria* species. *Microbiology* **151:**2907–2922.

Stoughton, R. B. 2005. Applications of DNA microarrays in biology. *Annu. Rev. Biochem.* **74:**53–82.

Townsend, C. A. 2004. Buruli toxin genes decoded. *Proc. Natl. Acad. Sci. USA* **101:**1116–1117.

Typas, A., R. J. Nichols, D. A. Siegele, M. Shales, S. R. Collins, B. Lim, H. Braberg, N. Yamamoto, R. Takeuchi, B. L. Wanner, H. Mori, J. S. Weissman, N. J. Krogan, and C. A. Gross. 2008. High-throughput, quantitative analyses of genetic interactions in *E. coli. Nat. Methods* **5:**781–787.

QUESTIONS

1. Scientists have proposed a molecular version of Koch's postulates for associating a bacterium with a disease. What would such postulates look like, and what would be their limitations?

2. The genomes of a number of bacterial pathogens have now been completely sequenced. How might the availability of this information affect the definition of virulence factors?

3. Critique the following statement: if you compare the genomes of a pathogen and a closely related nonpathogen (e.g., *S. enterica* serovar Typhimurium and *E. coli* K-12), the differences will be the virulence genes.

4. Why are techniques like IVET and STM continuing to turn up housekeeping genes rather than the "virulence genes" the inventors of these methods envisioned? (If you have a good explanation for this, publish it right away. A lot of us are waiting for the answer.)

5. If you were working with a newly isolated bacterium, what would you have to be able to do before you could use gene fusions to identify virulence genes? Do these same restrictions apply to microarrays? What would you need to have to use microarray technology?

6. If you have the genome sequence of a bacterial pathogen, would you still need to clone genes, or does cloning become obsolete?

7. What basic assumption is made about the regulation of potential virulence factors when using IVET?

8. How does IVET differ from STM? Would the two methods come up with the same set of potential virulence genes? Explain your answer.

9. What are some of the advantages and disadvantages of the IVET, RIVET, and STM approaches? Compare these approaches with genomic-sequencing and microarray approaches. Would you say that IVET, RIVET, and STM are or are becoming obsolete? Why?

10. Suppose you perform a microarray experiment comparing RNA samples prepared from a bacterium isolated from the blood of an infected animal and grown in rich culture medium in vitro. You detect changes in the relative transcript amounts of several genes for the bacterium grown in blood versus culture medium. How do you interpret these results? What does it tell you about virulence of the bacterium?

SOLVING PROBLEMS IN BACTERIAL PATHOGENESIS

1. Comparison of the genomes of a wild-type bovine-tuberculosis strain of *Mycobacterium bovis* and contemporary *M. bovis*-derived bacillus Calmette-Guérin (BCG) vaccine strains with that of a virulent reference strain of *M. tuberculosis* indicated that almost 100 *M. tuberculosis* genes were found to be "missing" from all of the *M. bovis* and BCG isolates that were used for making the vaccines. Describe how you might isolate and identify those genes that are unique to *M. tuberculosis*.

2. You are a researcher working for the USDA. After 2 years of effort, you have isolated in pure culture a new, highly virulent bacterium from duck feces that is responsible for several major outbreaks of deaths in mammalian wildlife from contaminated pond water in the South. Based on 16S rRNA sequence comparison, you have determined that this new bacterium is distantly related to the gram-negative bacterium *V. cholerae*, and you have named the new strain *Vibrio birdsii*. You have subsequently determined that *V. birdsii* is sensitive to tetracycline (i.e., it cannot grow in its presence) but is resistant to ampicillin. You obtained a *Vibrio* plasmid that you then used to construct a suitable *E. coli* shuttle vector for *V. birdsii*, which you could use for genetic manipulation in *E. coli* and then transfer into *V. birdsii* by transformation. You have also determined that guinea pigs, which have good innate immune systems, are an excellent animal model for the disease, which results primarily from uptake through drinking contaminated water, followed by colonization of the gut and bloody diarrhea, and then by invasion of intestinal cells with spreading to lymph nodes and spleen, and finally death by dehydration and/or organ failure. You are interested in identifying virulence factors associated with disease in guinea pigs. You have decided to use IVET as a strategy to identify these virulence factors. After infection and harvesting of the intestine, spleen, and lymph nodes of the guinea pigs, you have isolated a number of colonies on MacConkey agar plates from your initial screen: 10 white colonies, 15 pink colonies, and 75 red colonies.

A. Draw a diagram of the shuttle vector that you constructed. Be sure to label all the key features (genes, promoters, etc.) that are necessary for it to work with IVET for *V. birdsii*.

B. How do you interpret the results from the initial screen? (Be sure to explain why there are three different colony colors.) What would you do with the pink colonies?

C. Assuming that you have successfully identified several potential genes from the IVET screen that

may encode virulence factors, describe how you would verify that the putative virulence factors identified by your method are indeed involved in pathogenesis. Be sure to state what specific criteria must be satisfied.

3. You are a researcher working for the National Wildlife Health division of the U. S. Geological Survey. After 2 years of effort, you have isolated in pure culture a new, highly virulent bacterium from fish that is responsible for several major outbreaks of fish deaths along the coasts and in the Great Lakes. Based on 16S rRNA sequence comparison, you have determined that this new bacterium is distantly related to the spirochete *Cristispira clone*, which is thus far noncultivatable, and you have named it *Cristispira fisherii*. You have subsequently determined that *C. fisherii* has an unusual polysaccharide capsule and that it is sensitive to chloramphenicol, tetracycline, and erythromycin but is resistant to ampicillin, penicillin, and gentamicin. You have also determined that it contains two similar circular chromosomes and three endogenous plasmids, one of which has some homology with plasmids from gram-negative bacteria. You have used this plasmid to construct an *E. coli* shuttle vector for *C. fisherii*, which you can use for genetic manipulation in *E. coli* and transfer into *C. fisherii* by electroporation. You have also determined that zebra fish, which have good innate immune systems, are an excellent animal model for the disease, which results primarily from uptake through the gills into the fish lungs, followed by invasion and dissemination through the body, and then death.

A. You are interested in identifying virulence factors associated with infection in fish. Considering all of the above information, describe *in detail* a strategy that you might use to identify these virulence factors. Be sure to provide a rationale for your choice of strategy, the appropriate reagents that you will need to use, the overall experimental design, and how you will determine the identities of the virulence factors.

B. Describe how you will verify that the putative virulence factors identified by your method are indeed involved in pathogenesis.

C. From your screening, you identified two genes encoding putative virulence factors, which you have named *cff1* and *cff2* for *C. fisherii* factors 1 and 2. You have created mutant strains with deletions in these two genes. The wild-type bacterium has 50% infectious dose (ID_{50}) and 50% lethal dose (LD_{50}) values of 10. When administered to the zebra fish via the water, both mutant strains, the $\Delta cff1$ and $\Delta cff2$ strains, have ID_{50} and LD_{50} values of 10^7. However, when injected into the dorsal muscle of the fish, the $\Delta cff1$ strain has an ID_{50} value of 10^2 and an LD_{50} value of 10^4, while the $\Delta cff2$ strain has an ID_{50} value of 10^2 and an LD_{50} value of 10. Provide an explanation for these findings.

4. From infected rabbits, you have isolated a new, highly virulent gram-positive bacterium related to *Listeria monocytogenes*, which you have named *Listeria leporine*. Pathologic findings are most prominent in the intestine and spleen. Clinical signs are generally mild or absent in healthy adult animals, but you find that young and old animals have symptoms of increased thickness of the lining of the gut due to overproliferation of epithelial cells, and swelling of lymph nodes and spleen, and those animals often succumb to systemic infection, including brain lesions and death in about 70% of cases. You wish to better understand the pathogenesis of the disease and to identify potential virulence factors. However, because the need for these results is rather urgent, instead of developing a new approach, you decide to use an existing IVET approach based on reagents that have already been developed by other researchers for *L. monocytogenes*, which includes a temperature-sensitive plasmid (i.e., it integrates into the chromosome when the temperature is shifted from 37°C to 42°C for a brief time). The plasmid, which can be electroporated into *L. lep-*

orine, contains an erythromycin resistance gene *(erm^r)* with a constitutively "on" promoter and a promoterless listeriolysin O gene, *hly*, downstream of a BglII site, but no other genes. From your IVET screening using a young-rabbit infection model, you identified seven genes encoding putative virulence factors, which you have named *llp1* through *llp7*. Two of the genes *(llp1* and *llp2)* were found adjacent to each other on a two-gene operon, the genes *llp3* and *llp4* were part of an operon consisting of four genes, the gene *llp5* was part of an operon consisting of five genes, and the other two genes *(llp6* and *llp7)* were found on separate single-gene operons.

A. Describe *in detail* how you identified those six genes as genes encoding putative virulence factors using this IVET approach. Be sure to include a description of all the reagents, conditions, and experimental procedures used, as well as your rationale. (Hint: it might be helpful to include one or more diagrams showing the plasmid with all of its features, indicating how the plasmid integrates into the chromosome, showing how you would distinguish between in vivo and in vitro virulence gene expression, and describing how you determined in which operons the genes were located.)

B. You would like to know at which point during the in vivo infection process each of these putative virulence genes gets turned on. Describe an experiment that you could perform to determine this.

10

Identification of Virulence Factors: Molecular Approaches for Host Factors

A nyone who has ever read a mystery novel or watched a mystery program on TV knows that criminals have an MO (modus operandi) that needs to be understood before the evildoer can be identified and arrested. An important part of that MO is how the perpetrator attacks the victim. In the area of bacterial pathogenesis, the identification of bacterial virulence factors, together with an understanding of how these factors are designed to combat human defenses against infection, is tantamount to defining the MO of a bacterial pathogen. As with human criminals, there is no single MO for bacteria. Also, the detection and interrogation practices that establish this MO occur increasingly at the molecular level. In a sense the scientists who first applied molecular methods to the investigation of bacterial pathogens prefigured what is now standard in the mystery genre: CSI. Welcome to CSI microbiology!

Approaches to Identifying Host Factors Required for Infection

Transgenic Animal Models

The ability to manipulate animal hosts has greatly expanded the potential insights that can be gained from in vivo models of infection (the stand-in for the "victim" in our crime analogy). In particular, the ever-expanding library of mouse mutants with defined immunodeficiencies provides a powerful set of tools for studying bacterial disease processes (Table 10–1). Of particular use is the combined strategy of comparing the host responses in immunodeficient transgenic animals to infection with wild-type bacteria or with mutant bacterial strains that have specific deletions in bacterial virulence factors. This approach allows researchers to better understand the role of the bacterial virulence factor in pathogenesis. For example, Nramp1 (natural-resistance-associated macrophage protein 1) is a host factor that provides resistance against several intracellular pathogens, including *Salmonella enterica* serovar Typhimurium. Researchers found that mice with a mutant *Nramp1* gene, which results in a defect in macrophage metal ion transport and function, need to have neutrophils in order to defend against wild-type *S. enterica* serovar Typhimurium infection.

In another example, immunodeficient mice were used to determine the in vivo role of the *Bordetella* adenylyl cyclase toxin (CyaA) in pathogenesis. In immunocompetent mice, both wild-type bacteria and bacteria with a deletion in the *cyaA* gene (Δ*cyaA*) readily colonized the respiratory tracts of the animals but produced no overt symptoms of disease. However, in mice deficient in T cells and B cells (e.g., SCID [severe combined immunodeficient] or RAG-1 knockout mice), infection with wild-type bacteria led to lethal infection. The *cyaA* deletion mutant was still able to colonize the mice, although it did not kill them. These results suggested that the adaptive immune response of the host is important in providing protection against the CyaA toxin. In vitro studies had suggested that CyaA targets neutrophils and the innate immune response early in infection. Neutropenic mutant mice or mice chemically treated to deplete them of neutrophils were killed by both the wild type and the Δ*cyaA* mutant. This result confirmed that neutrophils and innate immunity are important for the early stages of infection, whether toxin is produced or not, but before the adaptive response kicks in.

Table 10–1 Examples of animal infection models

Infant mice have immature immune systems, so they are more susceptible to infection.

Irradiated mice are immunocompromised because their immune cells have been destroyed, so they are more susceptible to infection.

Nude mice are genetically defective in the ability to produce T cells, so they are more susceptible to infection.

Neutropenic mice are defective in the ability to produce neutrophils, so they are more susceptible to infection.

SCID mice are genetically defective in the ability to produce functional B cells and T cells, so they are more susceptible to infection but are good for examining innate immunity.

Transgenic mice are genetically defective in specific immune cells or immunity genes, so they are more susceptible to infection. Today, there are large resources of transgenic mice available for these types of studies.

Gnotobiotic (germ-free) animals are raised in a germ-free environment. Because of the absence of resident microbiota, they have severely underdeveloped mucosa-associated lymphoid tissue (MALT). They also have no partial immunity due to prior exposure, so they are "truly" naive (unimmunized).

Specific-pathogen-free (SPF) animals are raised in an environment free of particular pathogens but are exposed to other microbes, including other potential pathogens.

SCID mice, which lack a competent immune system, have also been used as recipients for xenografts of human tissues or human stem cells. These SCID-human chimeric animals end up with an amazing complement of human immune cells, including T cells, B cells, lymphocytes, monocytes, macrophages, and dendritic cells, as well as organs and tissues with human cells in them, including liver, lung, and gastrointestinal (GI) tract cells. These chimeric animals can serve as in vivo models for infection studies and preclinical drug screening and toxicity studies.

The construction of transgenic mice that harbor human tissues or express specific human genes can provide a better model of human infection than normal mice. For example, "humanized" transgenic mice expressing the human enterocyte-associated protein E-cadherin receptor were used to study the intestinal response to the food-borne intracellular pathogen *Listeria monocytogenes*, which has a surface protein (InlA) that binds to a host cell receptor protein called E-cadherin and thereby acts as an adhesin. It was proposed that this interaction might allow *L. monocytogenes* to translocate across the gut epithelial layer directly, rather than having to invade through Peyer's patches. When *L. monocytogenes* is internalized inside a cell within an endosome, it normally escapes from the endosome using a membrane-disrupting toxin called listeriolysin O (LLO) (see chapter 11). To determine the importance of translocation mediated by InlA–E-cadherin-mediated translocation in invasive infections, transgenic mice were colonized with either wild-type *L. monocytogenes* or mutants that do not produce the internalins InlA and InlB. Invasion by the wild-type *L. monocytogenes* was compared with invasion by a nonpathogenic *Listeria* species, *Listeria innocua*, which lacks LLO. In these humanized E-cadherin transgenic mice, wild-type bacteria and InlA or InlB mutants were still virulent, but the bacteria lacking LLO were avirulent. Contrary to what was anticipated, these studies suggested that escape from the endosome via LLO, rather than bacterial invasion mediated by InlA or InlB, was the more critical factor in determining the host response in this model of infection.

Transgenic animals harboring inducible reporter gene fusions in their genomes can be used to identify cells in which specific host responses have been activated during infection. For example, transgenic mice were constructed containing a transcription fusion between the bacterial *lacZ* (β-galactosidase) reporter gene (see chapter 9) and the control region of a gene regulated by NF-κB, which is a mammalian transcription factor that regulates several genes involved in host immunity. This NF-κB-responsive *lacZ* reporter was then used to monitor NF-κB activation during infection with *L. monocytogenes*. Activation of NF-κB was visualized by cleavage of chromogenic substrates by the β-galactosidase that was synthesized when the genes were turned on. During infection with wild-type bacteria, NF-κB activation was visualized strongly in endothelial cells of the spleen, liver, and brain. However, this activation was not observed for a mutant lacking LLO, suggesting that this virulence factor is important for dissemination of the bacteria to these organs.

Comparative Genomics of the Host Response

Just as the availability of genomic sequences of bacterial pathogens has greatly increased our understanding of host-pathogen interactions, the sequences from the host species that they infect have also led to the identification of host factors that contribute to bacterial virulence. This has been particularly useful in gaining insight into underlying differences in host susceptibility to bacterial infections and the severity of disease. For example, inbred mouse strains exhibit striking differences in the susceptibilities of their macrophages to the effects of lethal toxin produced by *Bacillus anthracis*, the causative agent of anthrax. By comparing the two mouse strains, it was possible to show that this difference in susceptibility lies downstream of anthrax lethal-toxin entry into macrophages. Subsequently, a locus controlling this phenotype, called *Ltxs1*, was mapped to chromosome 11, and the responsible gene was identified as a gene encoding Kif1C, a kinesin-like cellular-motor protein. Multiple alleles of this gene were found to determine the susceptibility or resistance of mice to anthrax lethal toxin. Using a similar strategy, a novel genetic locus in BALB/cJ mice that confers resistance to *Yersinia pestis*, the causative agent of plague, was recently mapped to the major histocompatibility complex on chromosome 17.

Changes in a single nucleotide (single-nucleotide polymorphisms [SNPs]; see chapter 7) account for the majority of interindividual genetic variations found in responses to pathogens. The increased ability to perform whole-genome scans and SNP analyses has enabled large-scale genotyping studies to map candidate genes with infectious-disease outcomes. For example, an SNP that introduces a stop codon into the gene for Toll-like receptor 5 (TLR5), a protein in the membranes of host defense cells that controls the cellular response to bacterial flagella, is associated with increased susceptibility to Legionnaire's disease caused by the flagellated bacterium *Legionella pneumophila*.

Two SNPs in another TLR, TLR4, which recognizes bacterial endotoxin (lipopolysaccharide [LPS]) increase predisposition to gram-negative bacteremia. Not all SNPs result in increased susceptibility to infection. For example, a common SNP found frequently in individuals of European descent that is located within TLR1, which mediates host responses to a variety of bacteria, including mycobacteria, causes aberrant trafficking of TLR1 to the cell surface and thereby loss of TLR1 function. This SNP, however, is associated with decreased incidence of leprosy, caused by *Mycobacterium leprae*.

Transcriptional and Proteomic Profiling To Identify Host Factors Required for Infection

Transcriptional and proteomic profiling of host responses to infection can be used to identify host proteins involved in pathogenicity and to monitor host response to colonization and infection by bacteria on a global scale. **Microarray technology** provides a profile of the expression patterns of thousands of genes in parallel in a single experiment and can be used to ask questions such as which genes are turned on or off under one condition compared to another condition. For example, numerous innate immune factors are consistently up- or downregulated in host cells infected with pathogenic bacteria compared to cells without exposure to any microbe. Even in germ-free mice, reconstitution with commensal bacteria has been shown to upregulate expression of colonic epithelial cell genes associated with growth and innate immune responses.

Indeed, there is evidence that the **transcriptional response (transcriptome;** see chapter 9) of the host can be specific and distinct for the particular microbial challenge, whether commensal, pathogen, or a complex microbial community. There is also growing evidence that the host transcriptome, in response to the presence of microbial communities, may influence the phenotypic outcome in a way that is different from the influence of individual microbes alone and could explain certain discrepancies observed between some clinical manifestations and those observed in laboratory infection models.

The utility of using microarrays to identify and monitor genes differentially regulated by certain immune cytokines was first demonstrated in 1998 for the regulation of genes in human cells that respond to treatment with alpha interferon (IFN-α), as well as IFN-β and IFN-γ. Within a short time thereafter, a burst of reports appeared exploiting this strategy for observing inflammatory cytokine and chemokine regulation, as well as apoptotic and other signaling pathways. By 2000, microarrays were gaining considerable popularity for profiling the host response to bacterial pathogens, such as *S. enterica* serovar Typhimurium, where it was shown that not only the live bacteria, but also certain bacterial factors, such as LPS, could significantly impact the host transcriptome.

The contributions of specific virulence factors to the host response and vice versa can be characterized by comparing changes in host gene expression profiles following adherence of bacteria to host cells (for an example, see Figure 10–1) or following infection with wild-type or mutant bacteria. For instance, individual or combined effects of host gene expression were examined in response to infection with *Pseudomonas aeruginosa* producing one or more effector toxin proteins, ExoS, ExoT, and/or ExoU, which are injected directly into host cells by a type III secretion system (see chapter 13). The host responses to these proteins revealed transcriptome patterns consistent with the suspected intracellular functions of these bacterial effector proteins.

Figure 10–1 Identification of differentially regulated host genes by microarray expression profiling of human epithelial cells upon adherence of *P. aeruginosa*. A microarray was used to analyze mRNA transcript levels in cultured lung cells after incubation with *P. aeruginosa*. Total RNA was extracted from the cells at time zero and after 3 hours of incubation with *P. aeruginosa*. Each RNA sample was used as a template for synthesis of cDNA probes, which were incorporated with either Cy3- or Cy5-dCTP. The sample at time zero was labeled with Cy5-dCTP (a red dye), and the sample after 3 hours of incubation was labeled with Cy3-dCTP (a green dye). The probes were mixed and hybridized to a microarray slide, and the slide was scanned in a dual-laser confocal microscope. Shown is an enlargement of a region of the microarray showing adherence-dependent upregulation of two genes (blue spots indicated by arrowheads 1 and 2). The cross-hatched spots (indicated by arrowhead 3) have a mixture of the two dyes, indicating no difference in gene expression between the two time points. (Adapted from Ichikawa et al., 2000; copyright [2000] National Academy of Sciences of the United States of America.)

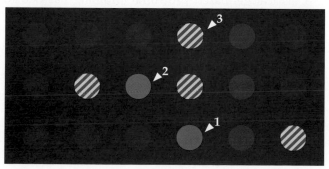

Similarly, the virulence roles of various toxin proteins (called Yop effector proteins) secreted by the type III secretion system of *Yersinia enterocolitica* were delineated by monitoring the concerted actions of the different Yop effectors on the expression of host genes. Using mutants of *Y. enterocolitica* with deletions in genes that encoded two different Yop proteins, YopP and YopM, it was possible to determine from the type of gene regulated that the primary function of the YopP effector was to counteract the host cell proinflammatory responses. At the time, it was not known what the function of YopM was, but based on the microarray analysis, it was found that YopM induced regulation of genes involved in control of the cell cycle and cell growth, hinting for the first time at a possible role for YopM in gene regulation. This was later confirmed when YopM was found to have a nuclear localization signal sequence and subcellular fractionation studies showed that it indeed localized to the nuclei of host cells.

Many challenges remain today for microarray technology, including standardization of replicates within and between experiments, universal agreement on what level of gene expression changes signifies "real" differences, and how to organize, interpret, and extrapolate the enormous amounts of data obtained from even a single experiment into useful information about biological processes that can be translated into new targets for disease diagnosis, intervention, and treatment. One recommendation from those who have used microarrays extensively is to use quantitative reverse transcription-PCR to verify interesting differences rather than rely on microarray analysis alone. Another problem with microarray analysis is the large number of "genes of unknown function." The bottom line is that there is nothing magical about microarray analysis. The yield from this analysis is still only as good as the ability of the user to think deeply in biological terms.

Nonetheless, despite these issues, global transcriptional profiling offers an amazing avenue into further understanding host-microbe interactions, which researchers have just begun to explore. As bioinformatics and data-mining tools develop to meet the growing demands, this technology will undoubtedly play an even greater role in the future.

Most host cell transcriptional- and proteomic-profiling studies have been performed in in vitro infection models using cell cultures and have provided unique insights into the complexity of host-microbe interactions during infection. More recent efforts have turned to **in vivo infection models** using two different types of techniques, **whole-body biophotonic imaging** and **laser capture microscopic dissection**

(LCM). These in vivo studies of gene expression have shed new light on the dynamics of infection and disease progression.

The ultimate goal of infection biology is to study a pathogen in real time during infection and in its natural infectious environment, the living host. Whole-body biophotonic imaging, introduced in chapter 8, provides a noninvasive technique that allows researchers to do just that—study bacterial infections in vivo, including monitoring of disease profiles and conducting real-time drug efficacy studies in live animals over the entire course of the disease. For this technique, the bacterial pathogen or the host animal (e.g., a transgenic mouse) is engineered with reporter genes to express bioluminescence (visible light), either constitutively or upon induction during infection. The animal is anesthetized, and the bioluminescence is monitored using ultrasensitive photon-counting video cameras that record images at extremely low light levels (Figure 10–2). The bacterial bioluminescent reporter is often a constitutively expressed bacterial luciferase operon that encodes all of the enzymes required to make the bacteria glow in the presence of ATP and oxygen (see chapter 8).

Sometimes other reporters are used, such as firefly (*Photinus pyralis*) luciferase, which uses D-luciferin, oxygen, and ATP as substrates to produce oxyluciferin, AMP, inorganic pyrophosphate, water, and light (at 547 to 617 nm). In this case, the D-luciferin substrate needs to be injected into the animal before each reading. Fusions between bacterial proteins and fluorescent proteins have also been used as bacterial reporters in biophotonic imaging (Figure 10–2). Red fluorescent protein is often preferable to green fluorescent protein (GFP) for this application, because animal bodies readily absorb more green light than red light. Conversely, host genes can be fused to luciferase or fluorescent proteins to follow the responses of these genes in different organs and tissues in response to bacterial infections. Other methods of following infections by biophotonic imaging are being developed for pathogenic bacterial species or animal hosts that are not amenable to genetic manipulation. One approach is to tag molecules that bind specifically to bacterial surfaces, such as antibodies and lectins, with synthetic fluorescent labels. These tagged molecules can then be injected into the animals to locate and image the bacteria as the infection progresses.

Another problem encountered when using in vivo models of infection is the inability to measure the host responses at the site of infection because the response of the cells directly in contact with the microorganism may be drowned out by the responses of other, surrounding cells. This interference can sometimes be circumvented by the powerful method of LCM. LCM is a technique that allows isolation of selected single cells or small groups of cells from tissue sections by means of microdissection (Figure 10–3). A tissue section is placed in contact with a thermoplastic polymer coating (ethylene vinyl acetate) attached to a transparent film. Using a microscope, the desired cells or cell clusters are identified, and the targeted area is exposed to a low-energy near-infrared laser pulse, which attaches the targeted cells to the polymer coating. The captured cells can then be extracted, and their DNA, RNA, or proteins can be analyzed by PCR or real-time PCR for analysis of DNA levels, reverse transcription-PCR for RNA analysis, or sodium dodecyl sulfate (SDS)-polyacrylamide gel electrophoresis and/or mass spectroscopy for protein analysis.

Proteomic profiling has emerged as a technology complementary to transcriptome studies. In classical proteomic analysis, the proteins in a specimen are separated on a two-dimensional (2D) gel to form a pattern of protein "spots" on the stained gel. Ideally, each spot represents a single protein. Typically, the proteins are first separated on the basis of their isoelectric point (pI) and then separated in the second dimension on the basis of size (on a denaturing SDS gel). Besides being used to determine patterns of gene and protein expression in the bacteria or host, classical proteomic profiling can be used as a diagnostic tool. For example, changes in the normal protein pattern in amniotic fluid can indicate the presence of possible infection, due to the presence of bacteria, and thus can be used as a predictor of increased risk of preterm delivery, early-onset neonatal sepsis, or other pregnancy complications due to infection. Proteomic profiling can also be used to identify factors that are recognized by different hosts during infection (Figure 10–4) and thus can be used to distinguish between different host responses to the same or similar pathogens. The degree of inflammation is determined by measuring the levels of certain indicator proteins (called **biomarkers**), such as proinflammatory cytokines.

Classical gel-based proteomic profiling has had to overcome several disadvantages. Host cell signaling patterns are often complex, and there may be numerous splice variants of some proteins. Therefore, the spot patterns, even for bacterial cells and especially for host cells, which produce many more proteins, are often complicated. Some researchers have overcome this complexity problem by purifying cellular compartments from host cells, such as phagosomes. Usually, only a portion of the total proteins can be visualized in one range of isoelectric points or molecular

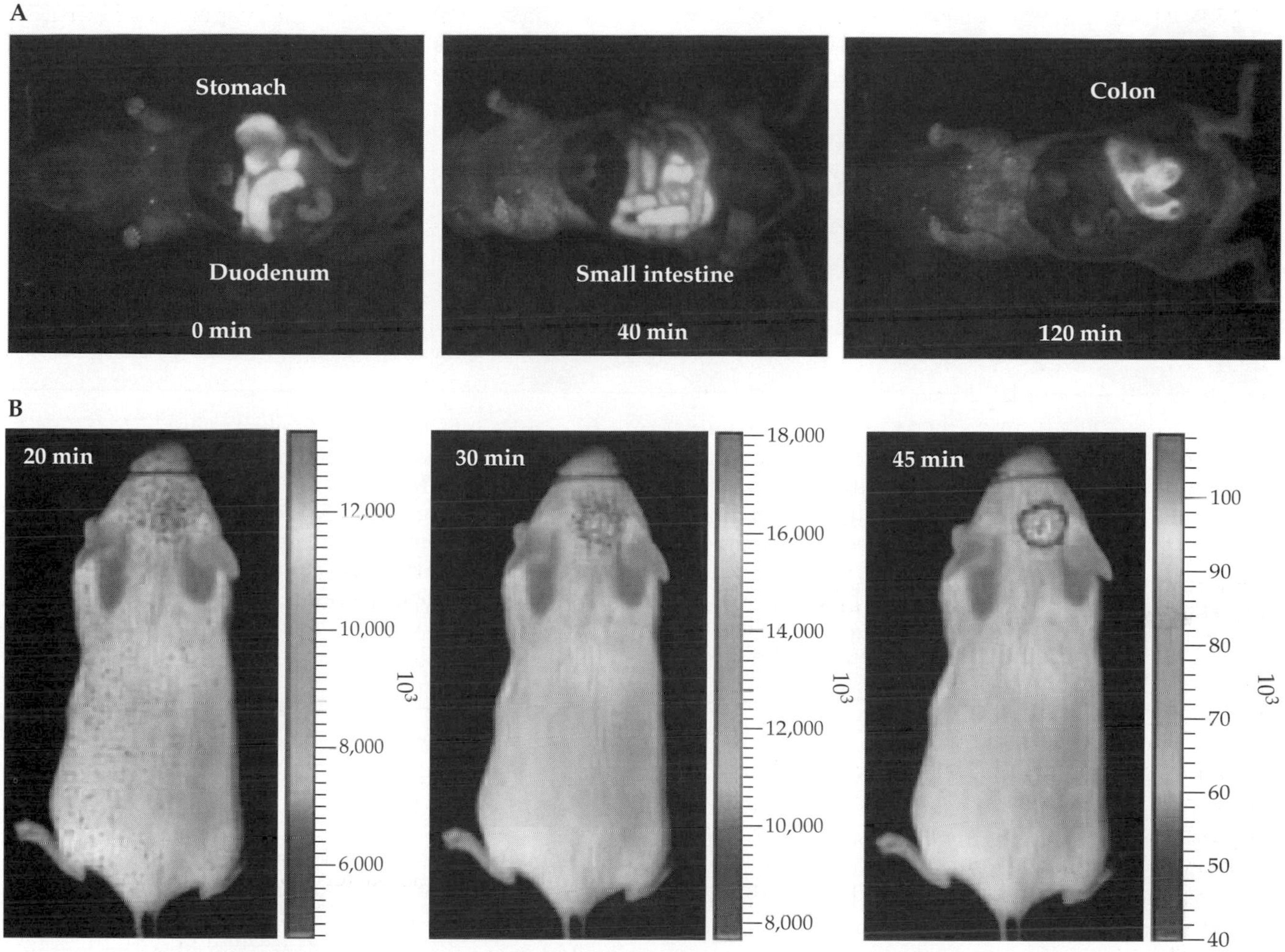

A

B

Figure 10–2 Noninvasive whole-body monitoring of bacterial infections using biophotonic imaging. **(A)** Imaging of GFP-expressing *E. coli* from outside intact infected animals to monitor the spatial-temporal flow of the bacteria through the GI tract. The green bacteria were introduced into the stomach via gavage, and images were taken over time to reveal the progression of the bacteria through the GI tract. (Adapted from R. Hoffman and M. Zhao. 2006. Whole-body imaging of bacterial infection and antibiotic response. *Nat. Protoc.* **1:**2988–2994, with permission from Macmillan Publishers Ltd.) **(B)** Bioluminescence detected in mice inoculated in the skin over the head/neck area with *B. anthracis* spores containing a plasmid with the *lux* reporter downstream of germination-specific promoters. The mice were monitored with an in vivo biophotonic imaging system to observe the bioluminescence emitted upon germination of the spores. Shown are the images 20, 30, and 45 minutes postinjection. (Adapted from P. Sanz, L. D. Teel, F. Alem, H. M. Carvalho, S. C. Darnell, and A. D. O'Brien. 2008. Detection of *Bacillus anthracis* spore germination in vivo by bioluminescence imaging. *Infect. Immun.* **76:**1036–1047.)

sizes, and minor proteins may not be detected. On the other hand, there have been numerous technological advances in gel-based proteomics. Similar to microarray analysis, proteomic analysis is done as a comparison of protein amounts from a mutant or stressed bacterial or host cell to those of the parent or unstressed cell. Previously, this comparison was made by trying to compare and quantify the protein spots from two separate 2D gels. Alignment of the separate gels was difficult, and quantification was linear over a limited range. New differential 2D gel approaches have overcome these serious problems. Proteins from the two conditions are labeled separately with different fluorescent dyes, mixed, and resolved on the same

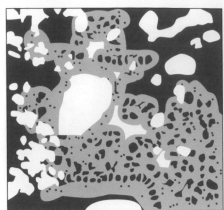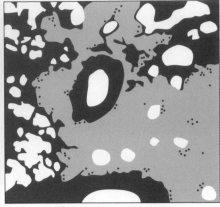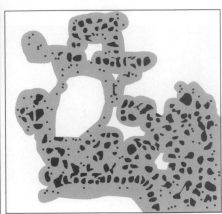

Figure 10–3 LCM of granulomatous tissues from mice infected with *Mycobacterium tuberculosis*. Stained sections of paraffin-embedded lung tissue cells from mice at 42 days postinfection with *M. tuberculosis* were subjected to LCM to retrieve the characteristic granulomas that form during infection. **(Left)** Before LCM. **(Middle)** After LCM. **(Right)** Captured granuloma tissue cells. The results from subsequent real-time PCR analysis of the dissected tissue cells in the right panel provided evidence that conventional use of whole infected lungs to study granuloma-specific gene expression can yield data that may not reflect the actual physiological events that are occurring at the site of the granuloma lesions.

2D gel. As with microarray analysis of transcript amounts (see chapter 9 and above), quantifying changes in relative amounts of spots becomes much easier by this approach. The intensity of fluorescence of each spot for the two dyes can be reliably determined, and those spots whose relative amounts increased or decreased can be rapidly identified compared to most spots whose relative amounts did not change.

In addition, routine identification of proteins in spots detectable on 2D gels is now rapid and cost-effective. The spots are cut out (sometimes by spot-picking robots!), digested with trypsin, and subjected to highly sensitive mass spectrometry. The translated genome sequence is used to predict the tryptic peptide profiles of all the proteins encoded by the organism (bacterium or host) under study. Computer programs use these profiles to rapidly identify the protein in the spot from the 2D gel. Finally, advances continue to be made in gel-less proteomics. These approaches are based on high-resolution mass spectroscopy of highly complex protein or tryptic peptide mixtures. The protein mixture initially extracted is usually first separated into different fractions using a series of **high-performance liquid chromatography (HPLC)** columns that run directly into a series of mass spectrometers. In many cases, these cutting-edge, highly sensitive gel-less approaches have provided insights not easily obtained by gel-based approaches.

Using Genome-Wide RNAi Screening To Identify Host Factors Required for Infection

In 1998, Andrew Fire and Craig Mello discovered that exposing cells of the nematode *Caenorhabditis elegans* to gene-specific double-stranded RNA (dsRNA) resulted in specific and efficient gene silencing that was more potent than using sense or antisense RNA. They called this silencing process **RNA interference (RNAi).** Since then, this new RNAi technology, for which they received the Nobel Prize in 2006, has enabled the screening of large libraries of host genes from any eukaryotic organism to identify host genes required for bacterial pathogenesis, particularly those for intracellular pathogens.

In RNAi analysis, a synthetic 21- to 23-nucleotide-long dsRNA molecule, called **short interfering RNA (siRNA),** is introduced through **transfection** into mammalian cells. Its presence inside the cell induces selective degradation of any mRNA that has homology to the siRNA. The result is posttranscriptional gene silencing (Figure 10–5). To overcome problems with poor transfection efficiencies of synthetic or plasmid vector-derived siRNA, several virus-based vectors have been developed to increase delivery of the siRNA into mammalian cells.

One of the first applications of RNAi technology was for the identification of host factors required for infection by *L. monocytogenes*. Norbert Perrimon's lab-

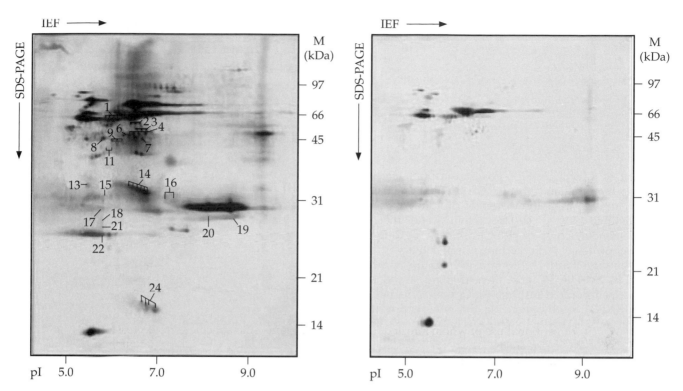

Figure 10–4 Proteomic profiling to identify protein antigens of *H. pylori* showing greater recognition by antisera from gastric cancer patients than by those from duodenal ulcer patients. Extracts of cell surface proteins from *H. pylori* were separated by 2D electrophoresis using isoelectrofocusing with a linear pH 3-to-10 gradient in the first dimension to separate proteins based on charge, followed by 12.5% SDS-polyacrylamide gel electrophoresis in the second dimension to separate proteins based on size. The separated proteins were then transferred to immunoblot membranes and probed with serum from a patient with gastric cancer (left immunoblot) or from a patient with duodenal ulcer (right immunoblot). Comparison of immunoblots from 15 patients with gastric cancer with those from 15 patients with duodenal ulcers revealed a number of protein antigens that were recognized by the sera from gastric cancer patients but not by the sera from duodenal ulcer patients. (Reprinted from Y. F. Lin, M.-S. Wu, C.-C. Chang, S.-W. Lin, J.-T. Lin, Y.-J. Sun, D.-S. Chen, and L.-P. Chow. 2006. Comparative immunoproteomics of identification and characterization of virulence factors from *Helicobacter pylori* related to gastric cancer. *Mol. Cell. Proteomics* **5**:1484–1496. Copyright [2006] American Society for Biochemistry and Molecular Biology.)

oratory generated a large dsRNA library, in which siRNA targeted nearly all of the open reading frames of the *Drosophila* genome for silencing. Darren Higgins' laboratory then used this valuable resource to develop a fluorescence microscopy-based high-throughput RNAi screen, in which particular phenotypes associated with specific defects in the infection process were identified (Figure 10–6). In this screen, *Drosophila* cells were treated with siRNA for about 4 days to downregulate (silence) the target genes. This was followed by infection for 24 hours with *L. monocytogenes* that expressed GFP as a fluorescent marker for visualization of the bacteria. The mammalian cells

were then fixed, and the host nuclei were stained with blue Hoechst dye. An automated fluorescence microscopy system was set up to record the images of green bacteria with blue host nuclei.

Normally, *L. monocytogenes* invades a host cell and escapes the vacuole in which it is internalized. Then, the bacteria replicate inside the invaded host cell and subsequently move to adjacent cells. Accordingly, images were discriminated based on five different phenotypes observed: (i) the normal phenotype, control cells (no siRNA) having bacteria mostly inside the cells; (ii) the absent phenotype, with no bacteria inside cells, presumably due to a defect in a host factor

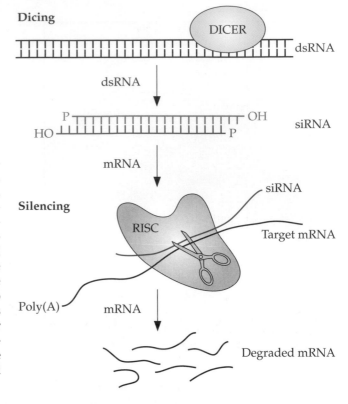

Dicing

DICER

dsRNA

dsRNA

P ————————————— OH

HO ————————————— P

siRNA

mRNA

Silencing

siRNA

RISC

Target mRNA

Poly(A)   mRNA

Degraded mRNA

Figure 10–5 Mechanism of posttranscriptional gene silencing in mammalian cells by RNAi. In mammalian cells, RNAi occurs through the action of the enzyme DICER, which exhibits RNase III-like activity, and the RNA-induced silencing complex (RISC), which is a group of cytoplasmic proteins with endonuclease activity that degrades the target mRNA. Long dsRNA is first processed into siRNA by DICER, and then a single strand of the siRNA is incorporated into the RISC, which degrades the target mRNA, thereby silencing the gene expression so that no protein is made. In RNAi technology, this process can be mimicked by introduction of synthetic siRNA or transfection with plasmid or viral vectors that are designed to produce siRNA through use of RNA polymerase III-dependent promoters. (Adapted from Vanhecke and Janitz, 2005; copyright [2005] Elsevier Ltd.)

needed for adhesion or invasion; (iii) the decreased-internalization phenotype, with only a few bacteria inside cells, presumably due to a defect in a host factor that allows bacteria to escape from the vacuole and/or to replicate in the vacuole or host cell; (iv) the "spots" phenotype, in which bacteria were clustered into spots within the cell, presumably due to a defect in a host factor needed for escape from the vacuole but still allowing the bacteria to replicate in the vacuole; and (v) the "up" phenotype, with an increased number of bacteria inside each cell, presumably due to a defect in a host factor that allows cell-to-cell spread so that bacteria accumulate inside the cell. The putative host factors required for *L. monocytogenes* infection identified through this screen spanned a wide range of cellular functions, including proteins involved in endocytic and vesicular trafficking, signal transduction, and cytoskeletal organization.

Initially, gene-silencing studies were performed in lower eukaryotes, such as worms *(C. elegans)* or fruit flies *(Drosophila melanogaster)*, but now this technology can be applied to most mammalian systems. Indeed, RNAi technology has now been adapted for high-throughput genome-wide screening of genes from any organism for which an annotated genome sequence is available and a bacterial infection model has been developed. As more host factors involved in bacterial pathogenesis are identified for different bacteria and

different hosts, comparative analyses will undoubtedly reveal common pathogenic mechanisms used by different bacteria, as well as mechanisms unique to a single pathogen or disease, and they could provide insights into new therapeutic strategies.

Because bacteria do not have the mammalian machinery needed for the RNAi effect, siRNA specifically inhibits only the host gene products that are expressed during infection. This makes RNAi technology ideal for identifying potential host targets for therapy. For example, one study showed that RNAi could be applied in vivo in adult mice to delay the onset of LPS-induced sepsis by administration of liposomes containing siRNA that inhibited expression of tumor necrosis factor alpha, which is normally turned on during sepsis.

Using the Host's Immune Response To Find Bacterial Origins of Disease

We mentioned in an earlier chapter that infection with *Helicobacter pylori* is an important risk factor for developing gastritis, duodenal ulcers, and gastric cancer, yet duodenal ulcers and gastric cancer are considered to be clinically quite distinct diseases. How, then, can one go about reconciling these distinctions and proving Koch's postulates for *H. pylori* as the origin of these clinically different diseases? In a variation of in

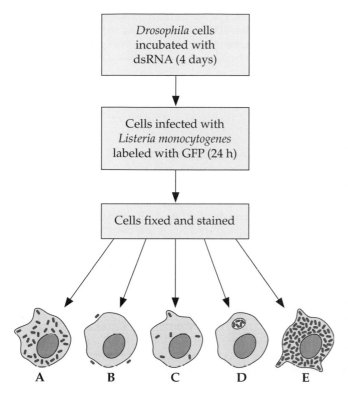

```
┌─────────────────┐
│ Drosophila cells │
│ incubated with   │
│ dsRNA (4 days)   │
└─────────────────┘
         │
         ▼
┌──────────────────────┐
│ Cells infected with   │
│ Listeria monocytogenes│
│ labeled with GFP (24 h)│
└──────────────────────┘
         │
         ▼
┌─────────────────────┐
│ Cells fixed and stained│
└─────────────────────┘
```

A B C D E

Figure 10–6 Genome-wide RNAi screening procedure for identifying host factors involved in *L. monocytogenes* pathogenesis and phenotypes observed. *Drosophila* S2 cells were incubated with siRNAs for 4 days to allow inhibition of host protein expression. The cells were then infected with GFP-labeled bacteria. The infected cells were fixed and stained and then followed by fluorescence microscopy and image analysis of the infection phenotypes. Shown are candidates representative of each of the observed phenotypes. **(A)** Control (no dsRNA). **(B)** Absent phenotype with defect in adhesion or invasion. **(C)** Decreased-internalization phenotype with few cells inside. **(D)** Spots phenotype with clustering of bacteria inside cellular vacuoles. **(E)** Up phenotype with increased numbers of bacteria accumulating inside cells. (Adapted from Agaisse et al., 2005, with permission from the American Association for the Advancement of Science.)

vivo-induced antigen technology described in chapter 9, one approach is to use comparative **immunoproteomics** to identify bacterial antigens that give common versus distinct host responses associated with the three diseases.

As an example of how this might work, serum samples from a group of patients with duodenal ulcers and a group of patients with gastric cancer were used by a group of researchers to probe 2D gels of *H. pylori* cell extracts that had been probed with serum samples taken from patients with ulcers or with gastric cancer (Figure 10–4). While many of the protein spots were recognized by more than one serum sample, a number of the spots observed were more strongly recognized or were only recognized by the sera from gastric cancer patients. The researchers found one particular spot that was especially pronounced in gastric cancer patients, a finding which suggested that this virulence factor from *H. pylori* might specifically contribute to gastric carcinogenesis in a subpopulation of presumably predisposed individuals.

The Promise and the Caution

Despite the appeal of all of these new genomic tools and high-tech approaches to identification of virulence factors from both the bacterium and the host, it is important to realize that there are significant limitations and caveats that must be considered. A clear hypothesis and critical biologically oriented thinking are still necessary elements to translate the data obtained through these new approaches into coherent models of disease and virulence mechanisms. Ultimately, understanding the complex interactions between pathogens and their hosts will require a comprehensive knowledge of the role of bacterial virulence factors in disease progression, the host factors that provide resistance to disease, the prior exposure history of the host to other or similar bacteria, and the ensuing interplay that occurs between them during infection. We are clearly at a critical juncture, where the excitement of discovery, tempered a bit with caution, could lead to very promising advances in our understanding of host-microbe interactions.

SELECTED READINGS

Agaisse, H., L. S. Burrack, J. A. Philips, E. J. Rubin, N. Perrimon, and D. E. Higgins. 2005. Genome-wide RNAi screen for host factors required for intracellular bacterial infection. *Science* **309:**1248–1251.

Banus, S., R. J. Vandebriel, J. L. A. Pennings, E. R. Gremmer, P. W. Wester, H. J. van Kranen, T. M. Breit, P. Demant, F. R. Mooi, B. Hoebee, and T. G. Kimman. 2007. Comparative gene expression profiling in two congenic mouse strains following *Bordetella pertussis* infection. *BMC Microbiol.* **7:**88–100.

El-Omar, E. M., M. T. Ng, and G. L. Hold. 2008. Polymorphisms in Toll-like receptor genes and risk of cancer. *Oncogene* **27**:244–252.

Hasegawa, Y., J. J. Mans, S. Mao, M. C. Lopez, H. V. Baker, M. Handfield, and R. J. Lamont. 2007. Gingival epithelial cell transcriptional responses to commensal and opportunistic oral microbial species. *Infect. Immun.* **75**:2540–2547.

Hooper, L. V., M. H. Wong, A. Thelin, L. Hansson, P. G. Falk, and J. I. Gordon. 2001. Molecular analysis of commensal host-microbial relationships in the intestine. *Science* **291**:881–884.

Ichikawa, J. K., A. Norris, M. G. Bangera, G. K. Geiss, A. B. van't Wout, R. E. Bumgarner, and S. Lory. 2000. Interaction of *Pseudomonas aeruginosa* with epithelial cells: identification of differentially regulated genes by expression microarray analysis of human cDNAs. *Proc. Natl. Acad. Sci. USA* **97**:9659–9664.

Jenner, R. G., and R. A. Young. 2005. Insights into host responses against pathogens from transcriptional profiling. *Nat. Rev. Microbiol.* **3**:281–294.

Mans, J. J., R. J. Lamont, and M. Handfield. 2006. Microarray analysis of human epithelial cell responses to bacterial interaction. *Infect. Disord. Drug Targets* **6**:299–309.

Sauvonnet, N., B. Pradet-Balade, J. A. Garcia-Sanz, and G. R. Cornelis. 2002. Regulation of mRNA expression in macrophages after *Yersinia enterocolitica* infection: role of different Yop effectors. *J. Biol. Chem.* **277**:25133–25142.

Tannu, N. S., and S. E. Hemby. 2006. Two-dimensional fluorescence difference gel electrophoresis for comparative proteomics profiling. *Nat. Protoc.* **1**:1732–1742.

Tapping, R. I., K. O. Omueti, and C. M. Johnson. 2007. Genetic polymorphisms within the human Toll-like receptor 2 subfamily. *Biochem. Soc. Trans.* **35**:1445–1448.

Vanhecke, D., and M. Janitz. 2005. Functional genomics using high-throughput RNA interference. *Drug Discov. Today* **10**:205–212.

Xia, M., R. E. Bumgarner, M. F. Lampe, and W. E. Stamm. 2003. *Chlamydia trachomatis* infection alters host cell transcription in diverse cellular pathways. *J. Infect. Dis.* **187**:424–434.

Zhu, G., H. Xiao, V. P. Mohan, K. Tanaka, S. Tyagi, F. Tsen, P. Salgame, and J. Chan. 2003. Gene expression in the tuberculous granuloma: analysis by laser capture microdissection and real-time PCR. *Cell. Microbiol.* **5**:445–453.

QUESTIONS

1. What sort of host factors might be expected to influence the evolution of a pathogen? Explain briefly how each of the techniques described in this chapter could be used to probe the role of a newly discovered host factor (factor X) in the infection process.

2. Following up on your answer to question 1, explain how the results of each technique covered in this chapter could *disprove* the involvement of host factor X in the response to infection by a particular bacterial pathogen.

3. Are any of the experimental approaches described in this chapter useful only for bacterial infections? In that case, where might the approach have to be modified if you were interested in studying the host response to a viral or fungal infection?

4. What obstacles do researchers face in the search for host factors that contribute to virulence?

5. Critique this statement: the virulence factors of a pathogen are established by determining how strong or weak a patient's innate and acquired defenses are.

6. What host factors might be involved in determining whether a microbe becomes a pathogen in one individual and a commensal in another? How might these host factors be identified through some of the screening assays discussed in this chapter?

SOLVING PROBLEMS IN BACTERIAL PATHOGENESIS

1. Research animals serve as essential components with which advances in biomedical research are made. It is now clear that the validity and value of research findings derived from the use of animal models are directly dependent upon their uniform care and physiological health. Thus, it is important for researchers to work with animal care professionals to know and understand what potential infections and/ or diseases their animals currently have or have been exposed to in the past. A researcher at your university is interested in developing a new animal model for rheumatoid arthritis that would confirm research findings they have already made in rodents. This researcher has just ordered two different colonies of animals (ferrets and rabbits) to test for suitability in their studies. Upon their arrival at the animal care facility, however, the university veterinarian discovers that most of the animals in each of the animal colonies have contracted bacterial infections, and the veterinarian and researcher have called you, as a microbiologist with expertise in bacterial pathogenesis, to consult and assist with evaluating the impact of these infections on the use of the animals for their research.

In the case of the ferrets, symptoms of disease appear to be anorexia, weight loss, and a massive inflammation of the lining of the gut, with overproliferation of epithelial cells and swelling of local mesenchymal lymph nodes. While several of the animals died due to profound weight loss and blockage of the intestine, most of the animals recovered after 1 to 2 months. You have identified a new gram-negative bacterium that was detected by PCR in biopsy specimens of the intestines and lymph nodes of the infected animals. Based on 16S rRNA sequence comparison, you have determined that this new bacterium is related to the bacterium *Lawsonia intracellularis*,

which is thus far noncultivatable, and you have named it *Lawsonia ferretii*. So far, you cannot grow *L. ferretii* in laboratory medium, but you can grow it in mouse tissue culture cells, where it appears to be an obligate intracellular pathogen.

A. What immune response was primarily responsible for controlling *L. ferretii* infection in the animals that survived? Be specific.

B. How are *L. ferretii* antigens presented to immune cells? Provide a detailed schematic diagram of the pathway with clear labeling.

C. To further understand the pathogenesis of the intestinal disease in ferrets, you set up an experiment to examine the host response during infection after feeding *L. ferretii* to unexposed ferrets. The results are summarized in the graphs shown. Interpret the results in terms of the host response to infection (i.e., provide a detailed model that explains each of the results shown). Be sure to include your response to question A in your model.

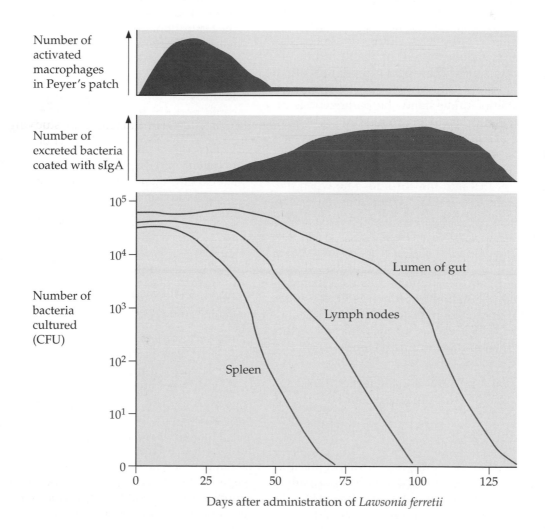

(continued)

D. You find that reinfection of animals that have had prior exposure to *L. ferretii* and have cleared the infection results in massive inflammation of the gut, followed by blockage of the intestine and rapid progression to death. Provide a plausible explanation with your rationale for this result.

E. When you take serum from animals that survived infection with *L. ferretii* and use the antibodies in Western blots, you find that there are several cross-reactive host proteins in uninfected animals. What possible implications do these results have in terms of using these animals for arthritis research? Be sure to provide your rationale.

F. Based on the results from all of the above-mentioned experiments, do you recommend that the arthritis researchers use ferrets for their animal studies? Do you recommend that they use this particular colony? Provide your rationale.

In the case of the rabbits, you have isolated a new, highly virulent gram-positive bacterium related to *L. monocytogenes*, which you have named *Listeria leporine*. Pathologic findings are most prominent in the liver and spleen, where you see swelling and necrosis. Clinical signs are generally absent in healthy adult animals, but you find that young and old animals often succumb to systemic infection, including brain lesions and death. To better understand the pathogenesis of the disease, you have developed an in vivo expression technology (IVET) approach based on existing reagents that have already been developed for *L. monocytogenes*, which include a temperature-sensitive plasmid (i.e., it integrates into the chromosome when the temperature is shifted to 42°C) containing an erythromycin gene and a promoterless LLO gene, *hly*. From your screening, you identified six genes encoding putative virulence factors, which you have named *llp1* through *llp6*. You then created mutant strains with in-frame deletions in each of these six genes. When administered to the rabbits through their feed, the wild-type bacterium has 50% infectious dose (ID_{50}) and 50% lethal dose (LD_{50}) values of 10, while all of the $\Delta llp1$ through $\Delta llp6$ mutant strains have ID_{50} and LD_{50} values of $>10^9$.

G. Describe *in detail* how you identified those six genes using this IVET approach. Be sure to include a description of all the reagents, conditions, and experimental procedures used, as well as your rationale.

H. To verify that the putative virulence factors identified by IVET are indeed involved in pathogenesis, what four criteria must be satisfied? (State them in terms of the rabbit infection model.) Provide three *different* modern molecular experiments that could be performed to satisfy these criteria. Be specific!

I. You decide to determine the competitive index of each of the mutants using a young-rabbit model. The results for two of the mutants are summarized in the table below (wt, wild type). Determine the competitive index for $\Delta llp1$:wt, $\Delta llp2$:wt, and $\Delta llp1$:$\Delta llp2$. Be sure to show how you derived your answer. Interpret your results.

Strain used	No. of bacteria in inoculum (CFU mutant/ CFU wt[a])	No. of bacteria recovered from brain on day 10 (CFU mutant/ CFU wt)
$\Delta llp1$:wt	$10^5/10^5$	$10/10^{12}$
$\Delta llp2$:wt	$10^5/10^5$	$10^2/10^9$
$\Delta llp1$:$\Delta llp2$	$10^5/10^5$	$10^3/10^6$

[a] wt, wild type.

J. To gain some insight into the roles of these virulence factors in pathogenesis, you decide to conduct additional experiments. The results are summarized in the table below. Based on these results, predict the possible virulence factors deleted in $\Delta llp1$ through $\Delta llp6$. Provide your rationale.

K. Provide one additional experiment that you could perform to confirm your prediction for each gene.

L. Describe a strategy you could use to identify a potential transcriptional regulator(s) of the expression of the gene *llp1*. Be sure to provide the reagents and conditions, and explain how you will know if the regulator(s) is a transcriptional activator or repressor.

M. Based on the results from all of the above-mentioned experiments, do you recommend that the arthritis researchers use rabbits for their animal studies? Do you recommend that they use this particular colony? Provide your rationale.

2. *Burkholderia pseudomallei* is a gram-negative bacterium that causes the disease melioidosis. Clinical manifestations of melioidosis vary greatly from acute or chronic pneumonia to overwhelming lethal septicemia. The most common clinical presentation is chronic pneumonia, which is found in about half of the patients showing symptoms of melioidosis. An important risk factor for melioidosis was identified in a number of near-drowning survivors following the tsunami of 26 December 2004 in southern Thailand. A group of researchers hypothesized that direct aerosol delivery of *B. pseudomallei* into the lungs may result in the enhanced ability of the pathogen to cause dis-

Mutant	Gentamicin assay			Plaque assay (plaques observed)	Brain lesions (no. of colonies on day 10)	Swelling of spleen (increased thickness of spleen section [cm] on day 10)
	No. of colonies without gentamicin	No. of colonies with gentamicin	No. of colonies with galactose and gentamicin			
Wild type	10^9	10^8	10^2	Large	10^4	1.5
$\Delta llp1$	10^9	10^8	10^2	Large	10^4	0
$\Delta llp2$	10^9	10^2	10^2	Large	10	1.6
$\Delta llp3$	10^9	10^8	10^2	Small	0	1.5
$\Delta llp4$	10^2	1	0	None	0	1.7
$\Delta llp5$	10^9	10^8	10^2	Large	0	1.6
$\Delta llp6$	10^9	10^8	10^2	Large	0	0

ease. The researchers wanted to develop two animal models of disease, one that mimicked the acute form of disease, which usually results in lethal septicemia if untreated, and one that mimicked the chronic form of disease. To do this, they tested three different clinical isolates of *B. pseudomallei* (Bp1, Bp2, and Bp3) in a rat lung model of infection, where the bacterium was directly introduced into the left lung using laparoscopy and disease progression from the left lung to the right lung, as well as systemic infection in the blood, was monitored via isolation of bacteria from the lungs and blood of the survivors at each time point, followed by plating them on agar and counting the bacterial colonies (CFU per left or right lung or per milliliter of blood). The results are shown in the table below.

A. Determine the LD_{50} and ID_{50} values for each of the isolates (Bp1, Bp2, and Bp3) at 2 days and 28 days postinoculation (use the left-lung data for the ID_{50} values). Be sure to show how you derived your answers. Interpret your results.

B. Considering all of the above information, which of the *B. pseudomallei* isolates would be most suitable for use in an infection model of chronic melioidosis? Which would be best for an infection model of acute melioidosis? Be sure to state your rationale for each.

C. When isolates of each of the bacteria were injected directly into the rats either intraperitoneally or intravenously, Bp1 and Bp2 did not result in death of any of the rats, even at doses as high as 10^8 CFU, and the animals showed no signs of dis-

Isolate	Dose	No. of survivors at:		No. of CFU					
		2 days (20/group)	28 days (20/group)	Left lung at:		Right lung at:		Blood at:	
				2 days	28 days	2 days	28 days	2 days	28 days
Bp1	10^1	20	0	10^8		10^8		10^8	
	10^2	18	0	10^8		10^8		10^9	
	10^4	8	0	10^{10}		10^8		10^{10}	
	10^6	4	0	10^{10}		10^{10}		10^{10}	
	10^8	2	0	10^{10}		10^{10}		10^{10}	
Bp2	10^1	20	18	10^2	10^3	0	10^2	0	0
	10^2	20	16	10^3	10^4	0	10^2	0	0
	10^4	20	4	10^7	10^8	0	10^3	0	0
	10^6	20	2	10^8	10^7	10^2	10^6	10	0
	10^8	20	1	10^8	10^9	10^2	10^6	10	0
Bp3	10^1	20	20	10^2	10^3	0	0	0	0
	10^2	20	20	10^3	10^3	0	0	0	0
	10^4	20	20	10^8	10^3	0	0	0	0
	10^6	20	9	10^{10}	10^3	0	0	0	0
	10^8	20	18	10^{10}	10^3	0	0	0	0

(continued)

tress or infection after 3 days. However, Bp3 killed all of the rats by 2 days after injection. Are these results consistent with those found for direct lung inoculation? Provide your rationale and suggest an explanation for this observation. Do these results support the hypothesis put forth by the researchers regarding the role of aerosolization in disease outcome? Be sure to state your rationale.

D. From initial inoculation until 28 days postinoculation, describe *in detail* the processes that you would expect to be involved in the host immune response to isolate Bp3 in an unimmunized tsunami near-drowning victim. Be sure to include both the innate and adaptive immune responses. (Hints: Use the results from the animal infection model as a guide. How are the immune pathways activated upon initial exposure? How are the antigens presented? How are the bacteria cleared or controlled? You may use a schematic diagram so long as everything is clearly labeled.)

E. Using one of the host-based molecular approaches described in this chapter, how could you identify host factors that contribute to the differential responses to the three *B. pseudomallei* isolates? Be sure to provide the reagents, conditions, and experimental procedures used, as well as your rationale.

11

Bacterial Strategies for Evading or Surviving Host Defense Systems

I n serious crimes, police sometimes call in profilers, specialists who analyze what is known about the crime, especially the MO of the criminal, and attempt to deduce the characteristics of the perpetrator. Microbiologists who investigate infectious diseases have recourse to similar types of specialists, namely, the basic scientists who attempt to deduce the detailed traits that explain what characteristics of a particular "microperp" explain the type of disease (MO) caused by the bacterium. The traits that go into such a microbial profile go beyond superficial traits, such as Gram stain status, to more detailed characteristics called virulence factors. These factors explain why in many cases members of the same species do not all cause disease: only a subset of the strains possess the necessary virulence factors. Also, as expected from the fact that different microperps have different MOs, the set of virulence factors employed by one microperp differs from the set of virulence factors employed by another.

Overview of Bacterial Defense Strategies

Just as the body has evolved multilayered strategies for defending itself against bacterial invasion, bacteria have evolved their own strategies for countering the host defenses of the body. The ability of pathogenic bacteria to cause infection and disease is the sum of structural, biochemical, and genetic characteristics that contribute to a microbe's pathogenicity. As noted in chapter 9, **virulence factors** are molecules produced or strategies used by disease-causing bacteria that enable the bacteria to achieve attachment to and colonization of the host, evasion or survival of the host's immune system, invasion and dissemination in the host, acquisition of nutrients and growth within the host, and release and spread to other hosts. In this chapter, we explore some common themes seen in bacterial strategies for penetrating, evading, and surviving the defenses of the human body. We first cover some general features of bacterial pathogens that enable them to colonize and infect the human body (Table 11–1). We have relegated one group of bacterial virulence factors—toxins and other factors that are produced, released into the medium, or delivered to the host and that then damage eukar-

Table 11–1 Mechanisms used by bacteria to facilitate colonization and survival in a host

Virulence factor/strategy	Function
Biofilm formation	Ability to bind to surfaces and establish multilayered bacterial communities; reduced susceptibility to antibiotics; bacteria continue to dislodge and disseminate through body
Motility and chemotaxis	Reaching mucosal surfaces (especially areas with fast flow or with nutrients)
Siderophores, surface proteins that bind transferrin, lactoferrin, and other iron-binding proteins	Iron acquisition
Iron abstinence	A few bacteria (e.g., *Borrelia burgdorferi*) have replaced iron-requiring enzymes with similar ones that use manganese instead.
Capsules (usually polysaccharide)	Prevent phagocytic uptake; reduce complement activation
Lengthened or shortened LPS O-antigen	MAC not formed; serum resistance
sIgA proteases	Prevent trapping of bacteria in mucin
C5a peptidase	Interferes with signaling function of complement component C5a
Pili and fimbriae	Adherence to mucosal surfaces
Nonfimbrial adhesins	Tight binding to host cells
Binding to M cells	M cells used as natural port of entry into body
Variation in surface antigens	Evade antibody response
Invasins	Force nonphagocytic cells to engulf bacteria
Actin rearrangements in host cells	Induce phagocytosis, movement of bacteria within host cells, cell-to-cell spread
Catalase, superoxide dismutase	Reduce strength of oxidative burst to allow survival in phagosomes
Elastase and other proteases	Degrade extracellular matrix proteins to allow dissemination through body
Nucleases	Degrade DNA released from host cell to reduce viscosity and to allow dissemination through body
Toxic proteins (toxins)	Kill phagocytes; reduce strength of oxidative burst; poke holes in host cells to release cell contents for iron and nutrients; modify host cell signaling to dampen immune response; cause cell morphology changes that increase dissemination; modify host cell processes to allow intracellular survival and replication

yotic cells and tissues—to a chapter of their own (see chapter 12). Many bacteria have also developed or acquired resistance to antibiotics or disinfectants. These virulence properties are discussed in chapters 15 and 16.

In order to survive in a host, a pathogenic microbe must be able to do the following.

1. Attach to host cells for colonization.

2. Evade the host's innate and adaptive immune defenses and persist in the host.

3. Obtain iron and other nutrients, especially those that are essential for growth but may be limiting within the host, in order to multiply.

4. Disseminate or spread within a host and to other hosts (this is critical for the survival of the species).

5. Produce symptoms of disease in the host in order to be considered pathogenic (although production of symptoms is not necessarily a requirement in and of itself, disease is often a result of the presence of the microbe and/or its products or the host response to the presence of the microbe).

A useful paradigm for thinking about the three stages of bacterial infection (colonization, persistence, and spread) is a comparison to an in vitro growth curve (Figure 11–1). Typically, bacterial growth in culture can be divided into a lag phase; a logarithmic, or exponential, phase; transition to stationary phase; and a stationary phase, which is sometimes followed, after a period of time, by a death phase. For some bacterial species, genes needed for colonization, such as those encoding adhesins (attachment proteins), are turned on during exponential growth. Genes required for persistence, such as those encoding the adhesins needed for close association with host cells or the proteins required to withstand the initial immune response, are turned on in the exponential-stationary transition phase. Genes encoding toxins and degradative enzymes required for spread are often turned on in stationary phase. In addition, some bacteria enter a long-term-survival state upon entering stationary phase, others form endospores that can germinate later, and still others lyse themselves (autolysis), presumably to exchange genetic material and to trigger host inflammatory responses. This model provides the perspective that bacterial virulence progresses in distinct stages, similar to those of a standard growth curve.

Many strategies used by bacterial pathogens to establish infection in a host probably began evolving long before animals and humans were even blips on the evolutionary radar screen. The majority of bacterial defense strategies would also have been useful for survival in natural environments (e.g., adherence to surfaces to stay close to a promising food source) or as defenses against voracious protozoa (e.g., the ability to avoid or survive phagocytosis by neutrophils). Perhaps the reason that intact skin and the antibody response are such effective defenses is that they are very recent evolutionary developments that have no counterpart in protozoa. Still, even these defenses can be compromised. Microbes that use insect vectors or take advantage of wounds have ways to bypass the barrier of human skin, and microbes that can rapidly change their surface proteins or produce antibody-degrading enzymes have developed methods for bypassing the specific defenses of the immune system.

Figure 11–1 Model showing the correlation between virulence gene expression and the stages of a typical growth curve. The model emphasizes the ordered progression of bacterial gene expression in *S. pyogenes* (group A streptococcus, which causes several serious human diseases, including "strep throat") that enables colonization, persistence, and spread. Mga, Nra/RofA, and Rgg/RopB are "standalone" transcriptional regulators that mediate gene expression at each stage of growth or infection. The up and down arrows indicate increased and decreased expression, respectively, of the indicated genes. MSCRAMM stands for microbial surface components recognizing adhesive matrix molecules of the host. (Adapted from Kreikemeyer et al., 2003; copyright 2003, with permission from Elsevier.)

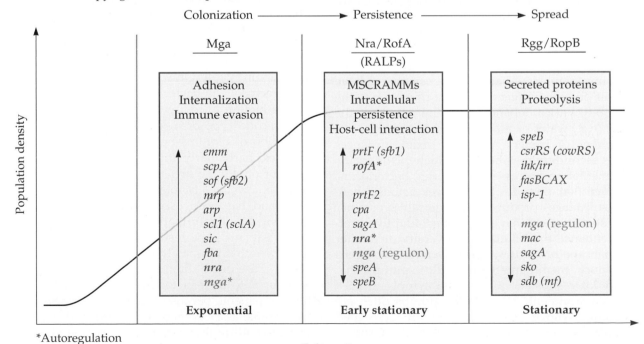

Preinfection

Survival in the External Environment

Some bacterial pathogens move directly from human to human or between humans and animals, but a number of them spend long periods of time in the external environment. They must not only be able to survive in the external environment, but must also be able to adapt to an abrupt transition to the very different environment of the human body. In natural environments outside the host body, bacteria encounter four major problems: availability of essential nutrients, lack of adherence sites similar to those found in the host, exposure to noxious chemicals and predatory protozoa, and exposure to sunlight and weather extremes.

To deal with some of these stressors, some gram-positive bacteria, such as *Clostridium* and *Bacillus* species, form **endospores** that protect them against environmental extremes, including temperature extremes, desiccation, UV light, high or low pH, and many potentially harmful chemicals. Even bacteria that do not produce spores can be resistant to desiccation and other environmental stressors. Examples of these are staphylococci and enterococci, which are inherently able to limit protein oxidation and DNA damage during dehydration and are thus more tolerant of desiccation.

Bacteria that can attach to abiotic surfaces or to each other through formation of colonies or biofilms (see below) are better able to survive in the environment and form a significant reservoir. These reservoirs can serve as the sources for many pathogens that we encounter. Bacteria can also protect themselves by converting to a metabolically reduced or dormant state. Finally, some bacteria simply parasitize another host, such as an insect or an amoeba, which protects and sustains them. How pathogenic bacteria that have a reservoir outside humans make the transition to the human body varies depending on the nature of the reservoir (e.g., water or an invertebrate).

A number of bacteria produce toxic substances known as **secondary metabolites,** which include acids from fermentation processes, antibiotics, bacteriocins, or antimicrobial peptides. To protect against these noxious substances produced by themselves or other bacteria, some bacteria have developed adaptive strategies to enhance their metabolic diversity or to alter their membrane properties. Many bacterial pathogens also produce membrane-bound efflux pumps that transport toxic substances from the cytoplasm back into the environment. These efflux pumps often are recruited to remove antibiotics and are associated with antibiotic resistance (see chapter 16). Genome plasticity, an evolutionary feature of some bacterial genomes, is thought to contribute to the development of increased metabolic diversity and altered membrane properties. Since nutrients are limiting in most external environments, most bacteria that spend at least some of their time outside the host body have the ability to metabolize a variety of substrates and have evolved uptake mechanisms to enhance nutrient acquisition where nutrient concentrations are very low.

Biofilms

Many bacteria are able to form **biofilms.** As the name suggests, biofilms are dense, multiorganismal layers of bacterial communities attached to surfaces; they are refractory to treatment with antibiotics or disinfectants, and they protect against phagocytic attack by protozoa in the external environment and by phagocytes of the host immune system. In biofilms bacterial attachment to a surface and to each other is mediated by an extracellular polysaccharide slime that acts as a type of "glue" to hold the community together. Sometimes, this slime is also composed of secreted proteins or released DNA. A biofilm is more than just a bunch of glued-together bacteria, however. At one time, scientists thought of biofilms as relatively simple, uniform, multilayered communities of bacteria, but it is now clear that biofilms exhibit a developmental sequence and have a complex architectural structure that looks more like mushroom-shaped islands of bacteria separated by channels that allow nutrients to reach most of the biofilm community (Figure 11–2).

Biofilms contribute to different types of medical problems. For example, in the case of Legionnaires' disease, where air-conditioning units created aerosols containing *Legionella pneumophila*, which caused a new type of lung disease, biofilms on the walls of the air-conditioning cooling towers were probably the source of the aerosolized bacteria. It is very difficult to eliminate a biofilm once it is established.

A second type of medical biofilm problem is the formation of biofilms on external and internal body surfaces, particularly mucosal surfaces, such as the lung, or on wound areas, such as severe burns. *Pseudomonas aeruginosa* is notable for its ability to form biofilms and has been used as a model organism for studying biofilm formation and structure. Cystic fibrosis patients have recurrent lung infections with *P. aeruginosa* that ultimately result in enough lung damage to cause death. Biofilm formation by *P. aeruginosa* is thought to be a factor in lung infections of cystic fibrosis patients because the bacteria appear to form biofilm-like assemblages in the lung that are difficult to treat with antibiotics. *P. aeruginosa* is also known for its ability to cause death in burn patients, who

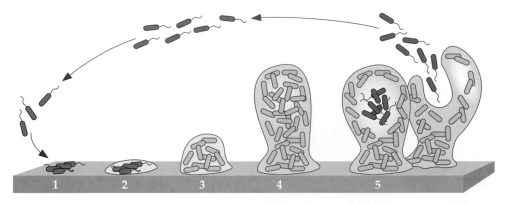

Figure 11–2 Developmental life cycle of biofilms. Biofilms have a developmental sequence with a complex architectural structure that includes attachment to a surface and to each other mediated by an extracellular polysaccharide slime that holds the community together, as well as a free-swimming planktonic stage and, for some bacteria, sometimes a spore stage. (Courtesy of David G. Davies.)

survive the initial burn trauma but succumb to later infection. In addition to lung infections caused by lung mucosal damage due to inhaling fire-heated air, the badly burned skin surface area can also be colonized by *P. aeruginosa* to form a biofilm, and since the normal defenses of the skin are breached at these sites, the bacteria can leave the biofilm and enter the body.

The **plaque** that causes dental caries or periodontal disease is actually a biofilm containing mainly aerobic bacteria when it is above the gums and anaerobic or facultative anaerobic bacteria when it is below the gum line. Plaque that is not removed becomes mineralized over time to form calculus (tartar). Dislodging plaque is not a trivial task, as anyone who has visited a dentist's office for tooth cleaning or gum scraping knows very well. If not removed regularly, the plaque can lead to dental caries, irritated and inflamed gums (gingivitis), or even periodontal disease. Those bacteria closest to the tooth surface switch to anaerobic respiration, which leads to release of acids that in turn can lead to demineralization of the tooth surface and dental caries. Irritation of the gums around the teeth where the plaque is located could also lead to gingivitis, periodontal disease, and eventually tooth loss.

Biofilms can also cause problems on medical devices, such as catheters, that breach the skin. Biofilms also form on plastic tubing (e.g., venous catheters and urinary catheters) and on contaminated plastic implants (e.g., heart valve or hip joint replacements). Because catheters breach important defensive barriers, such as skin or the urethral sphincter, they can transmit bacteria into the body and can serve as a constant source of infection. Although plastic is a modern human invention, bacteria rapidly adapted to this novelty and have become quite adept at forming biofilms on it. Biofilms on plastic implants, whether caused by

contamination at the time of insertion or seeded by transient bacteremia (bacteria in the bloodstream), create a serious medical problem. The bacterial communities within these biofilms are not only refractory to antibiotic treatment, but formation of the biofilm gives the bacteria a measure of protection against phagocytic cells. Bacteria in the interior of the biofilms are also protected from the forces of fluid flow in many body sites and are thus resistant to expulsion from the body and can persist for long periods. However, because cells can break off from the biofilm and enter the bloodstream, biofilm-encrusted implants are ticking time bombs that may "explode" into septic shock at any moment.

Plastic implants that acquire a bacterial biofilm usually must be removed and replaced surgically because antibiotics are not effective at sterilizing them. It is still not well understood why bacteria that are susceptible to an antibiotic if they are free living become more resistant to the same antibiotic when growing in a biofilm. One possible explanation is that bacteria in a biofilm are in a relatively inert metabolic state that makes them less susceptible to antibiotics, most of which work best on rapidly growing cells. Also, those antibiotics that simply stop bacterial growth rather than killing the bacteria outright rely on the accessibility of the stalled bacteria to phagocytic cleanup for effectiveness. The extended size of the biofilm structure also makes engulfment by phagocytes almost impossible.

Motility and Chemotaxis

Movement through fluids, such as urine and intestinal contents, is mediated by bacterial structures called **flagella,** long (up to 20-μm) helical structures that extend outward from the surface of the cell (Figure 11–3). A

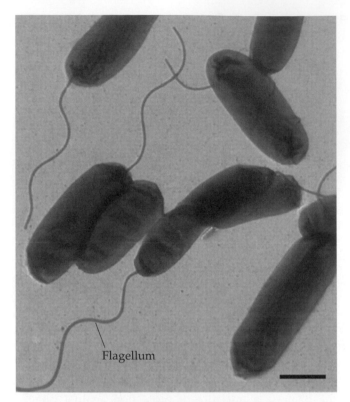

Flagellum

Figure 11–3 *V. cholerae* has a single polar flagellum for motility. (Reprinted with permission from D. E. Cameron, J. M. Urbach, and J. J. Mekalanos. 2008. A defined transposon mutant library and its use in identifying motility genes in *Vibrio cholerae*. *Proc. Natl. Acad. Sci. USA* **105:**8736–8741; copyright [2008] National Academy of Sciences of the United States of America.)

flagellum consists primarily of polymerized subunits of a protein called flagellin. Each flagellum is attached to the cell surface by a hook-shaped protein complex. This hook structure not only connects the flagellum to the cell, but since the structure is embedded in the cytoplasmic membrane, it has access to the ATP in the cytoplasm that provides energy for flagellar rotation. Since the flagellum has a loose helical structure, rotation makes it act like an oar that propels the bacterium through the environment. The helical structure of the flagellum is critical for this movement to occur, and hence, flagellin proteins are very similar among different flagellated bacteria. Vertebrate innate immune systems have taken advantage of this fact by evolving Toll-like receptor 5 (TLR-5), which specifically binds to this conserved protein (see chapter 3). Flagellins are also highly antigenic (constituting the H antigens of motile bacteria) and are often regulated in invasive pathogens so that they are not expressed once inside the host. In addition, flagellin biosynthesis is regulated by phase variation mechanisms, which

change the type of flagellin protein expressed on the bacterial surface to outfox the adaptive immune system (see chapter 7).

The number and arrangement of flagella on the bacterial surface vary among species, ranging from a single polar flagellum to multiple flagella arranged in tufts (lophotrichous) to even distribution over the entire cell surface (peritrichous). Bacteria have developed different flagellar systems to move in liquid or over surfaces. *P. aeruginosa*, *Vibrio cholerae*, and *Helicobacter pylori* express single or multiple flagella located on one end of the bacterium (polar flagella). This provides motility for propelling the bacteria through liquids (swimming). Polar flagella work best in dilute solutions. Other bacteria, such as *Escherichia coli*, *Salmonella enterica*, and *Proteus mirabilis*, express numerous lateral, constitutive flagella for motility in viscous medium or over surfaces (swarming). A limited number of bacteria, including *Aeromonas*, *Azospirillum*, *Rhodospirillum*, *Vibrio parahaemolyticus*, *Vibrio alginolyticus*, *Helicobacter mustelae*, and *Plesiomonas shigelloides*, have dual flagellar systems that are used for movement under different environmental conditions. Swimming in dilute liquid is promoted by a single, constitutively expressed polar flagellum. Swarming is enabled by numerous inducible lateral flagella, which are produced only under conditions that hamper the function of the polar flagellum. In the water-borne pathogen *V. parahaemolyticus*, a single polar flagellum is important for propulsion through the water, but after attachment to a surface, lateral flagella are induced, which contribute to microcolony and biofilm formation on the chitinous shells of crustaceans in the external environment. In the host environment, *Vibrio* uses the polar flagellum to move through the mucous lining of the gut to the epithelial layer, where it then attaches and secretes toxins.

Another type of flagellar motility that is adapted to movement through viscous media, such as mucin and even cellular barriers, is the corkscrew motility of the spiral-shaped spirochetes, where the flagellum is actually located within the periplasmic space of the bacterium. The spirochete *Treponema pallidum*, which is the causative agent of syphilis, seems to be able to move readily from blood to tissue through the endothelial cells lining the blood vessels. In contrast, when placed in a drop of saline solution, *T. pallidum* merely vibrates in place.

Movement by bacteria often involves the ability to sense chemical, light, oxygen, or magnetic-field gradients (**chemotaxis, phototaxis, aerotaxis,** or **magnetotaxis,** respectively), followed by directional swimming toward or away from the signal being sensed. *E. coli*, the major cause of urinary tract infections, provides an example of the importance of having flagella and chemotaxis. Urine is a rich source of nutrients for bacteria, so it is to the advantage of a bacterium to stay in the bladder. However, since the bladder is periodically cleansed by urination, which washes bacteria out of the site, the only way a bacterium can stay in the bladder is either to divide fast enough to counter dilution by new urine or to attach to the bladder mucosal cells. *E. coli* has multiple lateral flagella that allow it to swim through the mucin to the mucosa, where they can then adhere. This process involves a combination of motility and chemotaxis. Although the signals being sensed by these sensing systems are not yet fully known, we will learn more about how the signals are transmitted in the chapter on regulatory systems (see chapter 14).

Colonization of Host Surfaces

Penetrating Intact Skin

Except for a few parasitic helminths (worms), there are no microbes known that can penetrate human skin unaided by surgery, catheters, or other events that breach the skin's normal integrity. Thus, the true first line of host defense a microbe must overcome is the skin. It could be argued, however, that some microbes have created their own skin trauma opportunities by colonizing biting arthropods. The bacterium that causes Lyme disease is one such pathogen. Unable to penetrate skin on its own, *Borrelia burgdorferi* enters the human or animal body through the damaged area created by the tick that carries it. Similarly, the pathogen responsible for plague, *Yersinia pestis*, penetrates the skin and enters the host bloodstream through a flea bite. Other bacteria, such as staphylococci, that can survive on the skin are in a good position to take advantage of catheters, burns, wounds, or surgical cuts as a means of bypassing the intact skin defense.

Penetrating the Mucin Layer

In many parts of the body, mucosal cells are protected by **mucin,** a complex meshwork of protein and polysaccharide. One role of mucin is to act as a lubricant, but another vital function is to trap bacteria and prevent them from gaining access to mucosal cells. One of the least understood of all virulence factors is the ability of some bacteria to transit the mucin layer. Why do scientists interested in how bacteria cause intestinal or vaginal disease almost never consider the mucin layer or include it in their diagrams? One reason is that until recently, the procedures to fix biological samples for microscopy collapsed the mucin layers. Photographs of intestinal sections that show a layer of bacteria lying directly on the epithelial cells illustrate this artifact, yet such images have a powerful impact and have led many people to imagine intestinal bacteria as all adhering to the mucosal cells. This is almost certainly an incorrect view. Most bacteria passing through the intestine probably never come close to the mucosa because they are carried along by lumen contents or get trapped in mucin, which is then replaced, so that the trapped bacteria are carried out of the site. In the respiratory tract, the ciliated cells expel bacteria caught in mucin. A second reason for ignoring the mucin layer has been that it is very difficult to study due to its complex structure.

Bacteria that lack surface proteins or carbohydrates that bind mucin could penetrate the mucin without being trapped. The fact that most bacteria have such surface components is what makes mucin so effective in trapping them. Mucin is also highly viscous, and virtually all of the motile bacteria that cause disease in the intestinal tract are motile by means of flagella, so they have difficulty swimming through this viscous layer. Some bacteria partially digest mucin, but this is a slow and difficult process due to the complexity of the mucin structure. Remember, the role of mucin is to keep bacteria away from the mucosa, so there would have been strong selection for a structure that is resistant to enzymatic digestion.

The mucin layer is not a uniform mat. Mucin is expelled in thick streams from **goblet cells.** Thus, the mucin layer is probably more like a field of closely spaced mucin strands than a solid mat. If so, bacteria could transit the mucin layer by moving in the spaces between the mucin strands, a trick that would also help to guide bacteria to the mucosal-cell surface and keep them from wandering in the mucin layer parallel to the mucosal surface.

Another possibility is suggested by the observation that the **M cells** of the small intestine are the targets of most of the bacteria that cause intestinal disease. The normal role of the M cells is to sample the material passing through the intestine and to deliver it to the immune system associated with the gastrointestinal mucosa (see chapter 4), so they are not covered with a thick mucin layer. Many pathogens use the M cells as a portal through which they can transit the mucosal layer and enter underlying tissue and blood.

There is one area of infectious-disease research in which scientists have been forced to take the mucin layer into account: *H. pylori* infection. In the stomach, the mucin layer creates a buffer zone of nearly neutral pH that protects the mucosal cells from stomach acid. This zone also serves as a safe haven for *H. pylori*. If steps are taken to maintain the integrity of the mucin layer during sample preparation for microscopy, bacteria can be seen within the mucin layer, as well as adhering to the mucosa. *H. pylori* has four to six flagella located at one end. Although *H. pylori* moves by means of flagella, the flagella, combined with the corkscrew shape of the bacterium, apparently enable *H. pylori* to penetrate the mucin layer. The importance of the flagella for colonization of the stomach was demonstrated in studies in which a nonmotile strain was used to challenge gnotobiotic (germ-free) piglets. Only 2 of 8 piglets were colonized by the mutant strain, while 9 of 10 piglets were colonized by the wild-type strain. Moreover, further studies showed that on those occasions when infection by nonmotile strains does occur, it does not last long, and the mutant strains are cleared much sooner than wild-type strains.

Secretory IgA Proteases

Bacteria attempting to colonize mucosal surfaces have to solve the problem of how to avoid being trapped in the mucin layer. The stickiness of mucin is due in part to the presence of secretory immunoglobulin A (sIgA) molecules that simultaneously bind bacterial antigens via their antigen-binding sites and interact with mucin via their Fc portions. A bacterial strategy that may be designed to avoid sIgA-mediated trapping in mucin is production of an extracellular enzyme that cleaves human sIgA in the hinge region. This cleavage would separate the part of the sIgA that binds bacteria from the part that interacts with mucin. Such enzymes are remarkably specific for sIgA and have thus been called **sIgA proteases.** Most sIgA proteases are specific for a particular human isotype, sIgA1. Another human isotype, sIgA2, which does not have the same hinge region, is not cleaved. The fact that in many mucosal sites sIgA1 is the predominant isotype may explain this specificity. The actual role of sIgA proteases in virulence is still not well understood, and there is some controversy about their importance, but the unusual specificity of these enzymes suggests that they probably play some role in colonization.

Resistance to Antibacterial Peptides

Antibacterial peptides can insert into bacterial membranes and create pores that allow essential cytoplasmic molecules to escape, thus killing the bacterium. Such peptides are produced in a number of body sites, ranging from the base of the tongue to the crypts of the intestinal tract. Antibacterial peptides were first discovered in frogs by scientists who were curious about why frogs seldom get wound infections despite the lack of a highly developed immune system. The answer was that they produce antibacterial peptides, called magainins, or natural antibiotics. Similar antibacterial peptides have now been found in almost every animal in which scientists have looked for them, and it is now clear that they are an important innate defense of the human body (see chapters 2 to 4).

Human antibacterial peptides, called **defensins,** are characterized by their cationic nature and the presence of multiple cysteine residues that, through disulfide cross-linking, could give the peptides a circular shape. This model of their structure accords well with the hypothesis that the antibacterial peptides interact with the negatively charged bacterial cell membrane to either depolarize it or form pores in it. The difference in lipid composition between bacterial and mammalian membranes is likely a major factor in the strong activity preference that defensins show for bacterial targets instead of the host. What is hard to understand is that the defensins first must transit the outer membrane (in the case of gram-negative bacteria) and peptidoglycan to reach the vulnerable cytoplasmic membrane. Presumably, defensins are small enough to diffuse through porins and through the porous net formed by the peptidoglycan. They must be able to get through somehow, because they are toxic to many types of bacteria, gram positive and gram negative alike. Nonetheless, some bacteria are resistant to defensins. One mechanism of resistance is a capsular polysaccharide layer, which protects bacteria from antimicrobial peptides by limiting their diffusion to the bacterial cell surface.

A second mechanism is the lipopolysaccharide (LPS) layer of gram-negative bacteria. The negatively charged LPS molecules bind the cationic defensins, preventing them from reaching their membrane target. This hypothesis is supported by experiments showing that changes in LPS that make it less able or more able to bind the cationic defensins confer less or more, respectively, resistance to defensins. For example, various gram-negative bacteria that are resistant to the antibiotic polymyxin B have extensive cationic modifications on their LPS that reduce the negative charge of the LPS and thereby reduce the affinity of the bacterial surface for polymyxin B. Acylation of the lipid A component of LPS and lipooligosaccharide (LOS) has also been shown to increase resistance to antimicrobial peptides. Peptidases that degrade the defensins are produced by some defensin-resistant bacteria, providing yet another possible resistance mechanism. In other cases, the bacteria produce cytoplasmic proteins that counter the permeabilizing effects of the defensin channels.

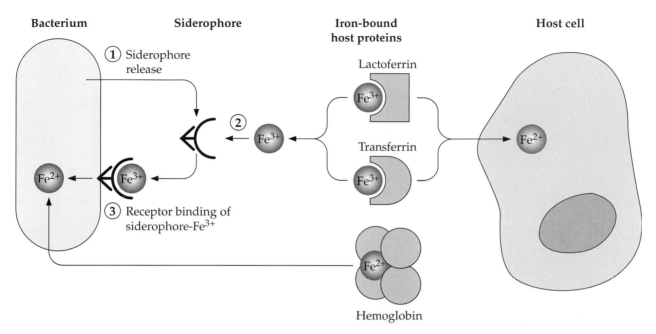

Figure 11–4   Iron acquisition mechanisms.

The fact that bacterial resistance to defensins can occur is discouraging, because some scientists are currently developing similar peptides for use as antibiotics. They had assumed that it would be difficult, if not impossible, to develop resistance to peptides that form channels in phospholipid bilayer membranes. Clearly, however, bacteria have evolved more than one strategy for dealing with the toxic-peptide problem. There are other significant challenges to developing antimicrobial-peptide antibiotics, including poor availability in mammals, potential immunogenicity and toxicity, and high production costs. These problems have led some researchers to try to develop small-molecule drugs that act as mimics of antimicrobial peptides.

Iron Acquisition Mechanisms

Most articles on iron acquisition by bacteria start with the statement that iron is essential for the growth of all bacteria. It should come as no shock by now that bacteria—no respecters of human paradigms—defy this one, too. It is now becoming apparent that there are some bacteria that solve their iron acquisition problem by not using iron at all. For example, the causative agent of Lyme disease, *B. burgdorferi,* uses this iron abstinence strategy. It lacks all of the iron-dependent cofactors and enzymes and uses manganese instead of iron for the chemistry it needs for survival and replication.

Most pathogenic bacteria, however, do require iron and have to cope with the fact that iron concentrations in nature are generally quite low. Estimates of the free-iron concentrations at sites in the human body range from 10^{-18} to 10^{-9} M, depending on the body site. The concentration of free iron is particularly low in the host body due to the actions of host proteins, such as **lactoferrin** (a secreted protein used for chelating iron), **transferrin** (made in the liver; the serum protein responsible for transport of iron), **ferritin** (the major protein used for intracellular iron storage), and **heme** (in humans, ~70% of total body iron is found in **hemoglobin**), that bind most of the available iron. To survive in the body, bacteria must have some mechanism for acquiring this sequestered iron, and consequently, bacteria have developed a number of possible ways to gain access to this vital mineral (Figure 11–4).

The best-studied type of bacterial iron acquisition mechanism is the production of **siderophores.** Siderophores are low-molecular-weight compounds that chelate iron with high affinity, which allows them to bind iron when it is present only in extremely low concentrations, such as that found in the host environment. The structure of one type of siderophore and its uptake mechanism is shown in Figure 11–5. There are two main classes of siderophores, **catechols** and **hydroxamates,** which differ in structure. Both have the same property, namely, that they form tight chelated complexes with iron.

Siderophores are excreted into the medium by bacteria. The siderophores bind to free iron in the medium, and then the iron-siderophore complex is taken back up into the bacteria by special siderophore re-

A

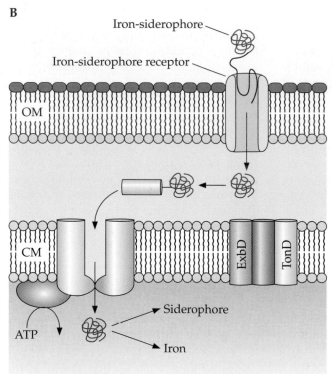

Enterobactin

Anguibactin

B

Figure 11–5 Iron acquisition by bacterial siderophores. **(A)** Siderophores differ considerably in structure (compare the catechol enterobactin from *E. coli* to the hydroxamate anguibactin from *Vibrio anguillarum*) but are similar in that they are basically just iron chelators with a very high affinity for iron. **(B)** The iron-siderophore complex is taken up through a receptor in the outer membrane (OM) of gram-negative bacteria and transported into the periplasm and then through an uptake pump in the cytoplasmic membrane (CM) into the cytoplasm.

ceptors on the bacterial surface. These siderophore-iron transport systems utilize the cytoplasmic-membrane proton motive force as an energy source. Periplasmic binding proteins and ABC-type transporters facilitate the transport of the siderophore-iron complex across the periplasm and inner membrane. The internalized iron-siderophore complex is cleaved to release the iron molecule inside the bacteria. Some bacteria not only produce their own siderophores, but also produce receptors capable of binding siderophores produced by other organisms, the bacterial equivalent of freeloading. This property is thought to contribute to the synergism that sometimes occurs during coinfections, where one bacterium that does not make siderophores but does make siderophore receptors benefits from the presence of another bacterium that makes siderophores.

Although siderophore-based iron acquisition has been shown to contribute to the virulence of many bacterial pathogens, mutations that eliminate siderophore production or uptake by a bacterial pathogen do not always significantly decrease virulence. This finding could be explained by the fact that bacteria often have more than one iron-sequestering system, and a mutant deficient in one system may still be able to survive by relying on the remaining system. Nonetheless, it is also possible that siderophore-based iron acquisition systems are adapted mainly for survival of the bacteria outside the body, in soil and water, whereas other strategies of iron acquisition are more important in the human body.

In soil and water, bacteria have access to free iron, whereas in the human body, virtually all of the iron is already bound to proteins, such as hemoglobin, transferrin, or lactoferrin. A number of pathogenic bacteria have now been shown to be able to use these proteins as a source of iron. In gram-negative bacteria with these capabilities, specific outer membrane receptors bind the iron-containing host molecule as part of the iron acquisition process. How they remove the iron from the host proteins is not clear, but the most common mechanism appears to be direct binding and processing of the iron-containing protein and release of the iron into the periplasm, followed by transport of the iron into the cell. As more and more pathogens are studied, it is becoming apparent that acquisition of iron from transferrin or lactoferrin, once thought to be an unusual mechanism of iron acquisition, is much more common than previously suspected. Bacteria can also produce proteases that degrade transferrin or lactoferrin to release iron. Acquisition of iron directly from hemoglobin and heme has also been detected in gram-positive bacteria, such as *Staphylococcus aureus*, which possesses systems that take up heme into cells

and heme oxygenases that strip off iron in the cytoplasm.

Another possible iron acquisition strategy of bacteria is production of toxic factors, such as hemolysins and other pore-forming cytolysins, that kill host cells. As will be seen in the next chapter, some bacterial toxins are produced only when iron levels are low. Because these toxic proteins kill host cells, they might be part of an iron acquisition strategy in which host cells are killed by the toxin to release their iron stores (primarily ferritin- or heme-bound iron), which can then be acquired by the bacteria via ferritin- or heme-binding proteins. Of course, it probably should be noted that low levels of iron as a signal for regulation of virulence gene expression are not necessarily directly linked to iron acquisition; it could merely be a signal that the bacterium uses to sense that it is in a host environment.

Adherence

ROLE OF ADHERENCE. One of the dogmas of bacterial-pathogenesis research is that adherence is an essential first step in the disease process. As we have just seen, however, bacteria such as *H. pylori* can stay in a particular site by being trapped in the mucin layer. These bacteria need not have special structures for adherence. Moreover, there is some evidence that certain strains of *E. coli*—the most common cause of urinary tract infections—could divide rapidly enough in urine to stay in the bladder without adhering. Nevertheless, although adherence may not be critical for all pathogens, it is an essential first step for most bacterial colonization. In the mouth, small intestine, and bladder, mucosal surfaces are constantly washed by fluids that keep down the number of bacteria present in the site. In such locations, bacteria capable of adhering to mucosal cells have a notable advantage. Even in other, slower-moving areas, such as the colon and vaginal tract, Brownian motion can move a bacterium that has made contact with a mucosal cell away from the surface of the cell. Virtually all known bacterial pathogens—and a lot of nonpathogenic bacteria, for that matter—have ways of attaching themselves firmly to host cells. The main strategies bacteria use to attach themselves to host cells are illustrated in Figure 11–6.

PILI AND FIMBRIAE. The best-understood mechanism of adherence is attachment mediated by rod-shaped, filamentous protein structures called pili or fimbriae. Proteinaceous pili (or fimbriae) are surface **adhesins** most often found on gram-negative bacteria and differ in thickness and length. The term **"fimbriae"** has been

A Pili

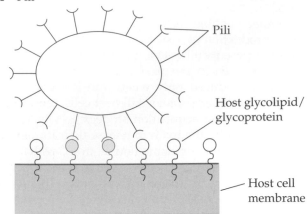

Pili

Host glycolipid/ glycoprotein

Host cell membrane

B Afimbrial adhesins

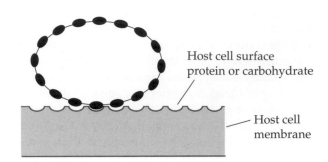

Host cell surface protein or carbohydrate

Host cell membrane

C

D

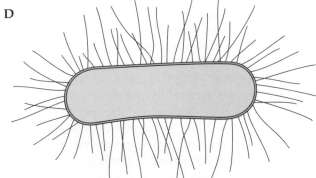

E

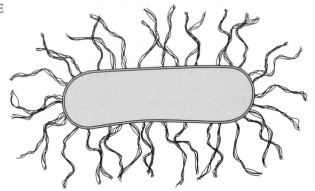

F

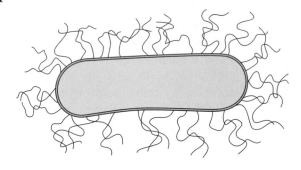

G

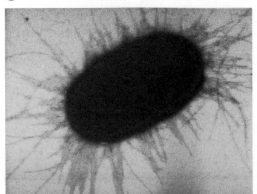

used to describe shorter, thinner structures, whereas **"pili"** is the term used for longer, thicker surface structures. However, this convention is not always observed, and the two terms are often used interchangeably. Still other researchers reserve the term "pilus" for the sexual appendage required for bacterial conjugation. In this book, "pili" will be used as the default term unless people in a particular field have made a point of using the term "fimbriae." There are many different types of pili that have been characterized in both gram-negative and gram-positive bacteria, and various forms are involved in attachment to various matrices and surfaces of host cells, as well as to each other in bacterial cell aggregation and microcolony formation or to each other and to other types of microbial cells in biofilms.

A pilus is a rod-shaped chain of polymerized subunits of the protein called **pilin.** Pilin protein subunits are usually about 20 kDa in size and are packed in an ordered helical array to form a flexible cylindrical structure of varying length (ranging from less than 0.5 μm up to 10 μm or more, depending on the bacterial species and type of pilus) that extends outward from the bacterial surface. For instance, over 30 different types of pili have been identified for different *E. coli* strains, many of which contribute to host cell- or tissue-specific interactions, called **tropism.** Pili act to establish contact between the bacterial surface and the surface of the host cell. In Figure 11–6A, pili are shown evenly distributed over the surfaces of the bacteria, but in some cases, they may be located preferentially on one part of the bacterial surface.

The tip of the pilus is the actual adhesin that attaches to a molecule on the host cell surface. Host cell receptors for pili are commonly carbohydrate residues of glycoproteins or glycolipids. Such host molecules are often involved in targeting of the host cell to its ultimate site, mediating cell-cell contact, or serving as part of the host cell's signal transduction mechanisms. Bacteria have thus subverted these host cell molecules, which because of their importance for host function are usually strongly conserved, for their own use. Binding of a pilus to its host cell target is quite specific. This specificity is important because the availability of suitable receptors often determines what body site is infected by the bacterium. The carbohydrate residues bound by the pilus can often be identified by simple competition assays. Addition of the free carbohydrate in excess saturates the binding sites on the pilus and prevents binding to cultured host cells.

In many cases, the specific binding between the pilus tip and the host cell carbohydrate is mediated by a specialized tip structure that consists of several pilin proteins distinct from the shaft pilin. In other cases, binding of the tip appears to be mediated by the pilin protein itself. It is not always easy to determine whether a pilus has a specialized tip structure, because the main structural pilin subunit accounts for over 99% of the protein in the pilus. Thus, minor pilin proteins that form a specialized tip structure may be missed in the initial biochemical analysis of the pilus composition. Now that DNA sequencing has become so prevalent, genes coding for minor proteins are often detected as unexplained genes located in operons containing the main pilin subunit gene and genes that encode proteins that help to assemble the pilus.

PILI (FIMBRIAE) OF GRAM-NEGATIVE BACTERIA. The major types of pili found in gram-negative bacteria are based on their structures, how they are made, and the nature of the adhesins they use. Type 1 pili, often called fimbriae, are rigid rod-like structures that are assembled through a chaperone-usher system (see

Figure 11–6 Two types of bacterial adherence mechanisms. **(A)** Mechanism by which pili, the rod-shaped protein structures that extend from the bacterial surface, bind to host cell surface molecules, usually carbohydrates. The tip structure is magnified to emphasize its presence. In reality, the tip structure is much smaller. **(B)** Mechanism by which afimbrial adhesins, bacterial surface proteins that are not organized in a rod-like structure but instead are shorter and closer to the surface, mediate tight binding between bacteria and host cells. **(C)** Afimbrial adhesins are embedded in the surface in bacteria that contain no pili. **(D)** P pili are thin filaments that protrude from the surface of the bacterium. **(E)** Type IV bundle-forming pili form rope-like structures made of many individual "threads" intertwined into bundles and tangled with each other. **(F)** Curli pili are curved or curled thin, aggregative, intertwined coiled structures. (Panels C through F are adapted from B. B. Finlay and M. Caparon. 2005. Bacterial adherence to cell surfaces and extracellular matrix, p. 105–120. *In* P. Cossart, P. Boquet, S. Normark, and R. Rappuoli [ed.], *Cellular Microbiology,* 2nd ed. ASM Press, Washington, DC.) **(G)** Electron micrograph of pili of *E. coli* (from *PLoS Biol.* **4:**e314. doi:10.1371/journal.pbio.0040314.g001; image: Manu Forero).

below) and can undergo hemagglutination in a mannose-sensitive manner (i.e., addition of mannose blocks the binding of red blood cells and prevents aggregation, or clumping, of the cells). The best-characterized pili in gram-negative bacteria are the type 1 and P pili of *E. coli*, encoded by the *fim* and *pap* operons, respectively (sometimes called *pil* genes in other bacteria). Type 2 pili are similar to type 1 pili but do not induce hemagglutination. Type 3 pili are more flexible than type 1 or type 2 fimbriae and are resistant to mannose. They are commonly found in *Enterobacteriaceae*, including *E. coli*, *Klebsiella pneumoniae*, and *Salmonella* species. Type 4 pili (fimbriae) have an *N*-methylphenylalanine at the N terminus of the major pilin subunit and are assembled through a type II secretion system. They are flexible and are often found as bundles at the poles of a wide variety of gram-negative pathogens, including *Neisseria* species, *Pasteurella multocida*, *Dichelobacter nodosus*, *Moraxella bovis*, *Actinobacillus actinomycetemcomitans*, and *P. aeruginosa*, where they mediate adhesion and twitching motility. Type 5 pili are similar to mannose-sensitive type 1 fimbriae but are much thinner. Curli pili form long, amyloid-like coiled structures and are formed extracellularly through nucleation-dependent polymerization that occurs on the cell surface. They are found in some strains of *E. coli*, where they adhere to extracellular matrix proteins and contribute to aggregation and microcolony and biofilm formation.

The assembly of type 1, 2, 3, and 5 pili by gram-negative bacteria occurs via the **chaperone-usher system** and is a complex process that requires the participation of a number of auxiliary proteins. A general model for how this process occurs is shown in Figure 11–7. Much has been learned about pilus assembly in recent years, leading to more and more complex models. Despite the diversity in pilus structure and biogenesis, the shaft of the pilus is formed by noncovalent polymerization of the major pilin subunit protein, with other, minor pilins (tip proteins), which often serve as host cell adhesins, added. Here, we provide a simplified version of the process for the common

type 1 pili of *E. coli*. The first step is secretion of pilin (FimA) and specialized tip proteins (FimH, FimG, and FimF) across the inner membrane and into the periplasmic space. In this case, secretion of pilin and the tip proteins appears to be mediated by the normal SecAB protein secretion system of the bacteria (see chapter 13), but for some other kinds of pili, a special pilus-specific secretion system is used. Thus, whereas proteins that are secreted through the normal secretion machinery have an amino-terminal signal sequence that is proteolytically removed during processing, pilin subunits of some pili still have a methionine residue at the amino terminus, indicating that normal proteolytic processing did not occur. Still other types of pilin have a methyl group attached to the amino-terminal amino acid (often a modified phenylalanine residue) after secretion and processing.

Once in the periplasm, special proteins called **chaperones** (FimC) prevent the pilin subunits from folding into their final configuration and aggregating. In support of this role, pilin subunits aggregate in the periplasm of mutants that do not produce the chaperone, setting off a drastic stress response. The chaperones convey the bound pilin subunits to an outer membrane **usher** protein dimer (FimD), where assembly and export of the pilin structure begin. The adhesive tip structure is assembled and extruded first; then the shaft of the pilus is made by sequential addition of pilin subunits to the base of the pilus, pushing the already assembled tip portions outward from the bacterial surface. This handoff of subunits during pilus biogenesis occurs by a mechanism of complementary donor strand exchange. Finally, a termination protein (called PapH of type P pili) signals the end of the extrusion process and presumably stabilizes the resulting pilus on the cell surface. How the pilus is actually stabilized in the outer membrane and how the large pore of the usher protein functions are areas of active research. The requirement for pili in urinary tract infections by *E. coli* and structural knowledge of pilus biogenesis has led to the design and testing of new classes of potential antibiotics called **pilicides.**

Figure 11–7 Model for assembly of type 1 and P pili with adhesive tip structures in gram-negative bacteria. The size of the tip structure is exaggerated to emphasize its location and different structure from the pilus shaft itself. The assembly of a type 1 pilus is shown, along with a completed P pilus, in which the subunits have slightly different designations. The adhesive tip components of the type 1 pilus (FimH, FimG, and FimF) are assembled first, and then the main pilus subunit (FimA) is assembled to form the pilus shaft. Chaperones (FimC) prevent aggregation of the pilin subunits in the periplasm and deliver the subunits to the dimeric FimD usher protein embedded in the outer membrane. The assembly process of type P pili is terminated by the PapH protein. (Adapted from Li and Thanassi, 2009, with permission from Elsevier.)

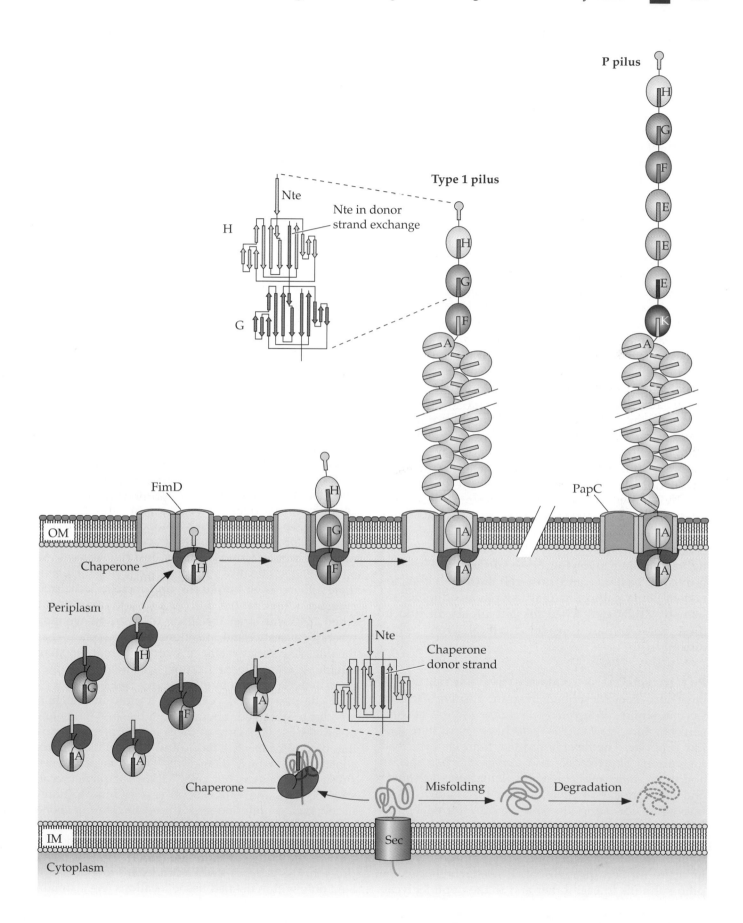

Bacteria growing in the body are constantly losing and re-forming pili. Continuous production of new pili is necessitated in part by the fragility of the pili, which are easily broken off through mechanical shearing. An equally important function of pilus replacement, at least for some bacteria, is that it provides them a way to evade the host's immune response. Host antibodies that bind the tips of pili physically block the pili from binding to their host cell targets. Once the host begins to produce antibody to a particular type of pilus, that pilus type is no longer useful to the bacteria. Replacing one type of pilus with another renders the host's antibody response obsolete. Some bacteria, such as those that cause urinary tract infections (*E. coli*) and those that cause gonorrhea (*Neisseria gonorrhoeae*), can change pilin types often enough by antigenic variation mechanisms (see chapter 7) to make it virtually impossible for the host to mount an effective antibody response that prevents colonization.

Why do bacteria use such long, fragile structures for adherence? One plausible explanation is that, since the surfaces of bacteria and host cells are both negatively charged, pili allow the bacteria to bind host cells without getting close enough for electrostatic repulsion to prevent attachment. Another possibility is that pili allow the bacteria to make an initial loose contact with a host cell surface, which then triggers production of surface proteins needed to mediate tighter binding. Indeed, many pathogenic bacteria appear to use a two-step process for attachment, in which a loose association is first made using pili. This is then followed by depolymerization of the pili and tighter binding involving other surface proteins that bring the bacterial and host cell surfaces closer together. Some pathogens that use pili as adhesins are severely attenuated when the pili are absent, and in those cases, pili are considered critical virulence factors.

PILI (FIMBRIAE) OF GRAM-POSITIVE BACTERIA. Pili have also been described in gram-positive bacteria. Certain strains of *Streptococcus salivarius, Streptococcus gordonii,* and *Streptococcus oralis* have short, thin, rod-like fimbriae. The flexible rod-like fimbriae described for various *Streptococcus* and *Corynebacterium* species and the primary oral colonizers *Actinomyces naeslundii* and *Streptococcus parasanguis* have been shown to be important for adhesion and biofilm formation. In pathogenic strains, they are known to play key roles in the adhesion-and-invasion process. Unlike the pili of gram-negative bacteria, those of gram-positive bacteria are formed through covalent attachment of the pilin subunits to each other and to the peptidoglycan

cell wall. For the streptococcal species, this has been shown to occur through recognition of an LPXTG amino acid sequence motif (where X denotes any amino acid) by an enzyme called **sortase,** which is a transpeptidase that catalyzes covalent peptide bond formation between the pilin subunits themselves and ultimately a cross bridge in the peptidoglycan. A model for pilus assembly in gram-positive bacteria is shown in Figure 11–8. Specific sortases required for pilus assembly in gram-positive bacteria are often encoded by pathogenicity islands that encode the pilin gene. Finally, it should be noted that gram-positive bacteria produce at least one "housekeeping" sortase that covalently links many surface enzymes, adhesins, and virulence factors to the cross bridges of the peptidoglycan (Figure 11–8). This covalent linkage anchors these proteins to the surfaces of the gram-positive bacteria and prevents their secretion into the medium.

NONFIMBRIAL ADHESINS OF GRAM-POSITIVE BACTERIA. Most studies of pili have focused on gram-negative bacteria. Some gram-positive bacteria are also covered by hair-like protrusions that resemble pili, and it was natural to assume that these structures might play the same adhesive role as the pili of gram-negative bacteria. In at least one case, however, the assumption that fibrillar structures on a gram-positive bacterium are adhesins now appears to be incorrect. The gram-positive pathogen *Streptococcus pyogenes* has an adhesin that does not have a pilus-like structure. This adhesin consists of many monomers of a protein called **M protein** that mediates attachment to **fibronectin,** a protein found on many host cell surfaces. At one time, the pilus-like structures were thought to mediate adherence to fibronectin and thereby to help *S. pyogenes* adhere to the mucosa of the human throat when it causes strep throat or to tissue in wounds when it causes wound infections. We now know that the pilus-like structures have a very different function: inhibiting complement fixation and helping *S. pyogenes* avoid ingestion by phagocytes (see below).

In the past several years, much more information has been obtained about the adhesins of gram-positive bacteria. Many of them are fibrillar adhesins that bind to connective tissue and extracellular matrix constituents, which themselves serve as adherence substrates for eukaryotic cells. Various staphylococcal, streptococcal, and enterococcal species bind to fibronectin, collagen, fibrinogen, vitronectin, laminin, bone sialoprotein, elastin, or thrombospondin. Many of these extracellular matrix proteins are broadly distributed in the body (e.g., fibronectin is found in many different tissue types), while others are found associated only within defined tissue locations (e.g., laminin

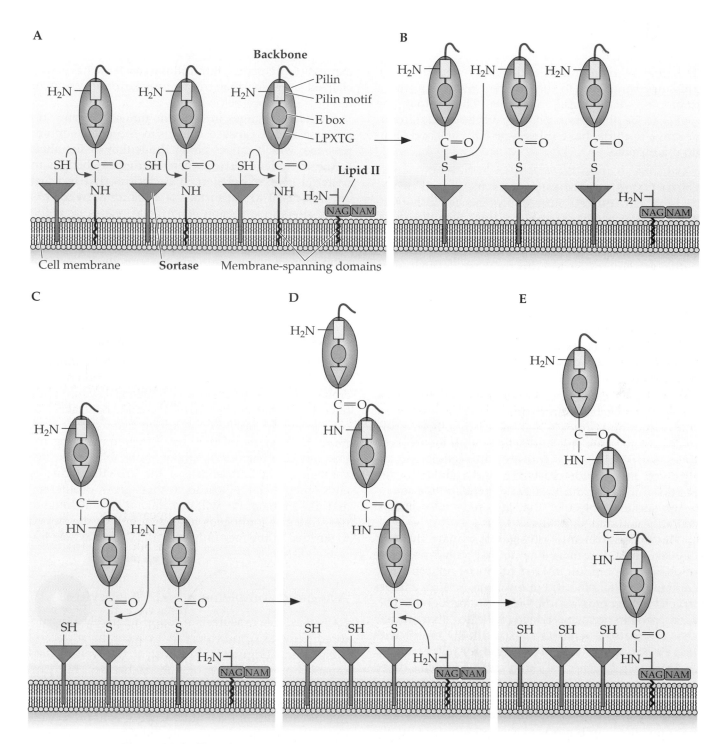

Figure 11–8 Functions of sortase transpeptidase in pilus assembly in gram-positive bacteria. **(A)** Pilin subunits that contain the LPXTG motif are targeted to the cell membrane by the general Sec-dependent secretion system. Sortase cleaves in the motif between the T (threonine) and G (glycine) residues. **(B)** A covalent thiol ester bond is transiently formed between the active-site cysteine residue of the sortase and the carboxyl group of the threonine residue. **(C)** Oligomerization occurs through attack of the thiol ester by the epsilon amino group of the K (lysine) residue in the pilin subunit motif. **(D)** The thiol ester is then attacked by an amino group side chain of the stem peptide in a peptidoglycan precursor. **(E)** The final precursor is a membrane-bound, covalently linked, covalently polymerized pilus. NAG, *N*-acetylglucosamine; NAM, *N*-acetylmuramic acid. This precursor is then incorporated into the peptidoglycan. (Adapted from Telford et al., 2006, with permission from Macmillan Publishers, Inc.)

in basement membranes and sialoprotein in bone). This may contribute to the different tissue tropism of gram-positive bacterial infections. The bottom line seems to be that gram-positive bacteria, like gram-negative bacteria, have pilus-type adhesins and afimbrial adhesins.

OTHER TYPES OF ADHESINS. As mentioned above, bacteria can use cell surface proteins for adherence that do not assemble themselves into pilus-like structures. These adhesins have been called **afimbrial adhesins** and are probably the proteins that mediate the tighter binding of bacteria to the host cell that often follows initial binding via pili, rather than that mediated by pili (Figure 11–6B). Bacterial surface proteins are important components of the systems that allow bacteria to attach to and invade host cells (see below). The structures and functions of only a small number of the afimbrial adhesins have been solved. At least some of them may recognize proteins such as integrins and cadherins, rather than carbohydrates, on host cell surfaces. Integrins are αβ heterodimeric transmembrane proteins that maintain adhesive cell-to-cell interactions and interactions with the extracellular matrix, as well as transmit intracellular signals involved in receptor clustering and cellular invagination. Cadherins are calcium-dependent transmembrane surface molecules that also play roles in host cell adhesion and signaling.

The majority of integrin ligands contain the Arg-Gly-Asp (RGD) motif as the minimal protein sequence necessary for binding. Many afimbrial adhesins also contain this RGD motif. For example, *Yersinia* adheres to cells by recognizing α5β1-integrin through the bacterial protein invasin, which indicates that invasin shares structural features (i.e., the RGD motif) with fibronectin, the natural host ligand for α5β1-integrin. *Listeria monocytogenes* uses two surface proteins called internalins A and B (InlA and InlB) to bind to host cells. InlA binds to E-cadherin, a transmembrane glycoprotein involved in the formation of intercellular tight junctions. InlB binds to the receptor for C1q, the first component of the complement cascade (see below).

ATTACHMENT TRIGGERS HOST AND BACTERIAL SIGNAL TRANSDUCTION. The original view of pili was that their only role was to form an attachment to human or animal cell surface molecules. It is now well established that attachment of adhesins causes conformational changes in eukaryotic cell surface molecules that can trigger signaling cascades inside the eukaryotic cell, which can lead to cytoskeletal responses, including uptake of the bacteria, as well as altered expression of genes. Intracellular bacteria have specialized adhesins that trigger uptake or invasion of the host cells (see below).

Similarly, changes in conformation in the pilin tip proteins might cause a change in the shaft subunit proteins, which moves along the length of the shaft to the bacterial surface, where it triggers changes in bacterial gene expression. A bacterium that has just attached itself to the surface of a mammalian cell exhibits altered expression of virulence genes, as if the act of binding triggered the activation or repression of the virulence genes. We will explore this phenomenon more when we describe systems for delivery of virulence factors to eukaryotic cells (see chapter 13) and regulation of virulence factors (see chapter 14).

Evading the Host Immune Response

The host immune response is an important parameter with which all incoming microbes must contend. Microbes are able to become pathogens because they have traits that enable them to avoid, neutralize, or counteract the innate or adaptive defenses, or both immune defenses, of the host. By circumventing the host immune response, pathogens are able to take advantage of the nutrients and the growth-conducive environment of the host to survive, grow, proliferate, and disseminate. We will now take a look at some of the strategies pathogens have acquired or developed to survive in the face of hostile attack by the host immune system.

Avoiding Complement and Phagocytosis

CAPSULES. A **capsule** is a loose, relatively unstructured network of polymers that covers the surface of a bacterium (Figure 11–9). Most of the well-studied capsules are composed of polysaccharides, but capsules composed of proteins or protein-carbohydrate mixtures have also been described (Table 11–2). The role of capsules in bacterial virulence is to protect the bacteria from the host's inflammatory response (complement activation and phagocyte-mediated killing). Recall that an essential first step in the alternative pathway is assembly of C3bBb (C3 convertase) on the bacterial surface. Some capsules prevent the formation of C3 convertase by failing to bind serum protein B. Other capsules have a higher affinity for serum protein H than for B. If C3b complexes with H rather than with B, then C3b is degraded by serum protein I. Capsules that are rich in sialic acid have a high affinity for protein H.

By preventing C3bBb formation on their surfaces, encapsulated bacteria gain some important advan-

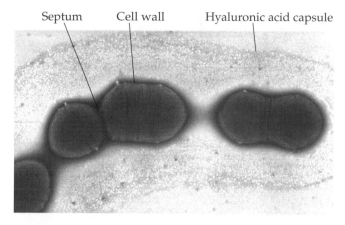

Septum   Cell wall   Hyaluronic acid capsule

Figure 11–9   Capsule. Shown is a transmission electron micrograph of *S. pyogenes* (magnification, ×28,000). The halo around the bacteria is the hyaluronic acid capsule. (Courtesy of Vincent Fischetti.)

tages. The capsule itself is less likely to be opsonized, i.e., phagocytes will not engulf the bacteria as efficiently. Some C3b may diffuse through the loose capsule network and bind to the bacterial surface under the capsule. C3 convertase may even form at this site, but the C3b molecules that attach to the bacterial surface under the capsular layer are prevented from making contact with phagocyte receptors by the thick intervening capsular network.

Less C3bBb formation also means that less C5b will be produced, and hence, the membrane attack complex (MAC) is less likely to form on the bacterial surface, an important consideration for gram-negative bacteria. However, this is not always the case, and the MAC can still form on the surfaces of some encap-

sulated bacteria because proteins can diffuse through the loose network of the capsules. Thus, encapsulated bacteria do not automatically become serum resistant.

MIMICKING THE HOST. An effective host response against an encapsulated bacterium is to produce antibodies that bind the capsule. Antibodies bound to the capsular surface not only provide sites for phagocyte binding so that bacteria can be ingested, but also support activation of complement by the classical pathway, thus increasing the amount of MAC formed. Vaccines consisting of capsular material have proven highly effective in preventing infections caused by encapsulated bacteria, such as *Haemophilus influenzae*, but some bacteria subvert this type of protective host response by having capsules that resemble host polysaccharides. Examples are capsules consisting of **hyaluronic acid** (an extracellular matrix polysaccharide), such as that of *S. pyogenes* (Figure 11–9), or **sialic acid** (a common component of host cell glycoproteins), found in some strains of *Neisseria meningitidis*. Because this type of modified capsule is not immunogenic, the host does not produce antibodies that opsonize the capsular surface.

One of the principal targets of complement on gram-negative bacteria is LPS. LPS not only serves as a site for attachment of C3b, the signal for activation of the alternative pathway, it also binds C5b and serves as a nucleation site for formation of the MAC. Two types of LPS modification affect this interaction between LPS and complement components. First, attachment of sialic acid to **LPS O antigen** prevents the formation of C3 convertase, similar to what happens with the sialic acid-modified LOS of *N. gonorrhoeae*. Second, changes in the length of the LPS O-antigen side chains can prevent effective MAC formation. It is not clear how O-antigen side chain length prevents MAC killing, since C5b and some MAC components still attach. Possibly, the MAC forms too far from the bacterial outer membrane to exert a bactericidal effect. Bacteria that are not killed by the MAC are called **serum resistant.** An indication of the importance of this trait is that many of the gram-negative bacteria that cause systemic infections are serum resistant (Box 11–1).

Table 11–2   Compositions of selected bacterial capsules

Bacterium	Capsule composition
Gram-positive bacteria	
Bacillus anthracis	Poly-D-glutamic acid
Bacillus megaterium	Poly-D-glutamic acid, amino sugars, sugars, polypeptide
Streptococcus mutans	(Dextran) glucose
Streptococcus pneumoniae	Sugars, amino sugars, uronic acids
Streptococcus pyogenes	(Hyaluronic acid) N-acetyl-glucosamine and glucuronic acid
Gram-negative bacteria	
Acetobacter xylinum	(Cellulose) glucose
Escherichia coli	(Colonic acid) glucose, galactose, fucose, glucuronic acid
Pseudomonas aeruginosa	Mannuronic acid
Azotobacter vinelandii	Glucuronic acid
Agrobacterium tumefaciens	(Glucan) glucose

OTHER STRATEGIES TO AVOID COMPLEMENT AND PHAGOCYTOSIS. Bacteria have evolved strategies designed to prevent migration of phagocytes to the site of bacterial colonization. A bacterial enzyme that specifically degrades C5a, the chemoattractant for phagocytes, has recently been found in gram-positive bacteria and may be more widespread as a strategy for interfering with the signaling function of complement.

BOX 11–1   Bacterial Meningitis: an Example of the Power of Capsules

Bacterial meningitis is not a common disease, but it is a much-feared one, because it can kill an infected person within a few days. People who manage to survive a case of meningitis frequently have irreversible neurological damage, resulting in blindness, deafness, and learning deficiencies. A striking feature of meningitis is how rapidly it develops once the infection begins. This is due to the ability of the causative bacteria to divide rapidly in blood, a trait that is due to production of antiphagocytic capsules and to the ability to avoid being killed by complement. The most common causes of meningitis are *N. meningitidis* (a gram-negative bacterium that causes epidemic meningitis), *S. pneumoniae* (a gram-positive bacterium that is also a common cause of bacterial pneumonia), and *H. influenzae* type b (a gram-negative bacterium that was the most common cause of meningitis in children until a vaccine was developed a few years ago).

What do these diverse types of bacteria have in common? All produce polysaccharide capsules, and the most dangerous strains are also serum resistant. Once the bacteria gain access to the bloodstream, the capsule protects them from being killed by phagocytes. As mentioned in the text, possession of a capsule does not automatically make a bacterium serum resistant, but some strains of *N. meningitidis* become relatively serum resistant by covalently attaching sialic acid residues to their LPS molecules. *H. influenzae* type b strains can also become serum resistant by modification of LPS O-antigen side chains. *S. pneumoniae*, being gram positive, is naturally serum resistant, since the MAC does not form on most gram-positive bacteria due to the lack of an LPS-type molecule that can precipitate the MAC. A strain of *E. coli* that causes meningitis in newborns (*E. coli* K1) also produces a capsule. This capsule is composed of sialic acid residues and thus does not bind C3b. In this case, the capsule has the same ultimate effect as serum resistance because it prevents MAC components from being made in the first place. The combination of a capsule and the ability to avoid being killed by complement renders ineffective all of the innate defense mechanisms of serum and blood except transferrin. Unfortunately, *N. meningitidis* and *H. influenzae* can obtain iron from transferrin. The only defense that can check the growth of these bacteria in blood and cerebrospinal fluid is the antibody response. Antibodies to capsular polysaccharides opsonize the bacteria so that phagocytes can ingest them. Polymorphonuclear leukocytes and macrophages readily kill all the bacteria mentioned above. People who make antibodies against capsular antigens are fully protected from the disease. In fact, the reason *H. influenzae* type b is mainly a disease of children (aged 5 months to 5 years) is that most people who survive past age 5 have developed an antibody response against *H. influenzae*, probably as a result of transient nasopharyngeal colonization during childhood. Newborn infants are protected by maternal antibodies. There is now an effective vaccine against *H. influenzae* type b, which consists of capsular polysaccharide attached to a protein (a conjugate vaccine) to make it a better antigen (see chapter 17). This vaccine is now being used to protect children in the vulnerable younger age groups. There is also an older capsular polysaccharide vaccine against *S. pneumoniae* that is fairly effective. However, a newer conjugate vaccine against seven capsular antigens is highly effective in children.

Developing a vaccine against *N. meningitidis* proved a bit more challenging because the most common type of capsule in some parts of the world is type B, a capsule which (like the capsule of *E. coli* K1 strains) consists largely of sialic residues and thus does not evoke an antibody response. Vaccines against other types of *N. meningitidis* are available because these capsules are more immunogenic. There are now two different meningococcal vaccines licensed for use in the United States. MPSV4 is a polysaccharide-based vaccine used for children 2 to 10 years of age who are in high-risk groups, such as those who travel to countries where meningitis is endemic, those with terminal complement deficiencies, or those with a damaged spleen or whose spleen has been removed. MCV4 is a conjugate-based vaccine that was licensed in 2005 and is recommended for all children 11 years old or

(continued)

BOX 11-1 Bacterial Meningitis (continued)

older and for adults who have never been vaccinated but fall into the high-risk groups listed above, who live in college dormitories or military barracks, or who might be exposed to meningococci at work (research or clinical). As a conjugate vaccine, immunity through MCV4 lasts longer than the MPSV4 vaccine, which needs 3- to 4-year boosters.

During outbreaks of meningitis caused by *N. meningitidis*, as many as 80% of people in an exposed population may have their noses and throats colonized by *N. meningitidis*, but only a small percentage of the population colonized by the bacteria actually develop the disease. One possible explanation is that most adult members of a population may have some level of acquired immunity to *N. meningitidis* infection. A factor that clearly affects host susceptibility is the integrity of the mucous membranes of the nose and throat, through which the bacteria usually gain access to the bloodstream. Outbreaks of *N. meningitidis* meningitis in Africa are almost always associated with the dry season, when dry conditions undermine the protective barriers of the mucous membranes. Similarly, outbreaks in other countries appear to coincide with dry periods of the year. Acute viral respiratory infections, such as influenza, may also be a predisposing factor, since such infections kill ciliated cells and undermine the integrity of the mucosal barrier. Interestingly, people with deficiencies in late complement components, who are unable to form the MAC, are also more prone to meningitis. Thus, although meningitis-causing strains are often somewhat serum resistant, the MAC may still exert some protective effect, at least in the case of gram-negative pathogens.

Why do people with meningitis die so quickly? The answer is that release of toxic cell wall components, such as LOS and peptidoglycan fragments, triggers a cascade of events leading to shock and death. This type of bacterial toxicity is discussed in detail in chapter 3. Since the compounds that cause the symptoms of meningitis are released when bacteria lyse, administration of antibiotics, such as penicillins and cephalosporins, that lyse bacteria can actually temporarily worsen the condition of a patient who has developed high numbers of bacteria in the blood or cerebrospinal fluid. To counter this type of effect, which is most likely to occur in children because of the unusually high levels reached by bacteria infecting this age group, clinicians routinely administer corticosteroids along with antibiotics in an attempt to counteract the inflammatory effect of toxic compounds released from lysing bacteria.

Many bacteria produce toxic proteins (see chapter 12) that kill phagocytes, prevent their activation, inhibit their migration, or reduce the strength of the oxidative burst. Some bacteria can directly deliver these inhibitory toxins into host immune cells and block signaling pathways that elicit cytokine and chemokine production (see chapter 12). Such toxins may protect the bacteria from phagocytes in the body.

Another strategy for evading complement activation is to have an LPS-type molecule that does not elicit the sort of strong host responses normally elicited by classical *E. coli* LPS. *H. pylori* is an example of a bacterial pathogen that has such an LPS. The O antigen of *H. pylori* LPS from certain strains contains carbohydrate moieties that are identical to human Lewis antigens x and y, both of which are found on the surfaces of gastric epithelial cells. This immune mimicry could help prevent the host from mounting an effec-tive immune response. Despite this, *H. pylori* elicits an inflammatory response that in some individuals can result in gastritis and even gastric ulcers (Figure 11-10). If the host did respond by recognizing the Lewis antigen portions of the LPS molecule as foreign, the antibodies would cross-react with gastric mucosal cells. Such cross-reacting antibodies could contribute to inflammation by activating complement or stimulating the phagocytic cells to attack gastric epithelial cells. This may explain why patients with symptomatic disease often have antibodies that cross-react with antigens on gastric mucosal tissue.

Invasion and Uptake by Host Cells

One successful strategy used by a number of pathogens to avoid the host immune system is to hide from it by invading and living inside the host cell. Patho-

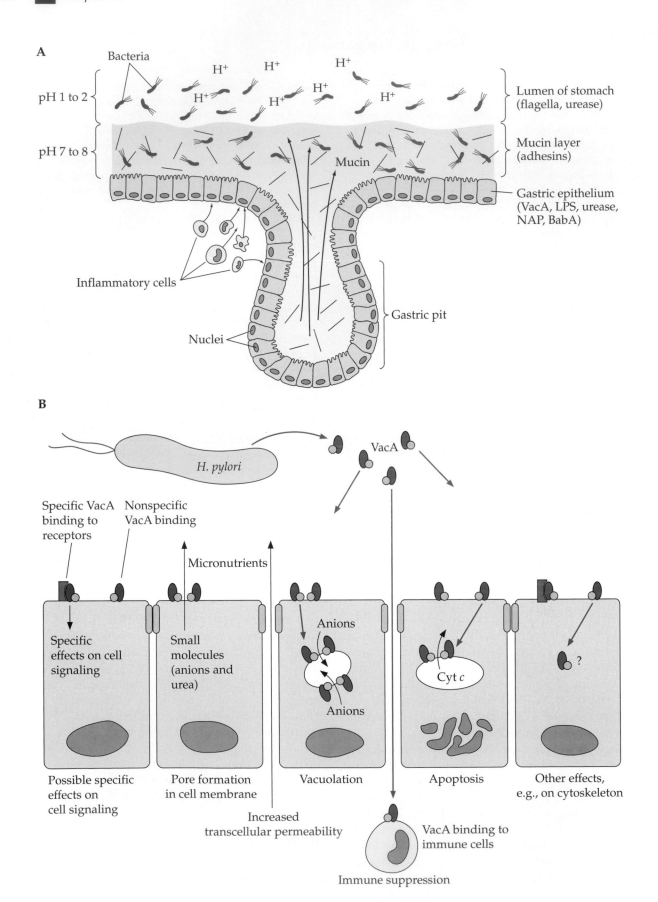

A

Bacteria

H+   H+   H+

pH 1 to 2

H+   H+   H+   H+

Lumen of stomach
(flagella, urease)

pH 7 to 8

Mucin

Mucin layer
(adhesins)

Gastric epithelium
(VacA, LPS, urease,
NAP, BabA)

Inflammatory cells

Gastric pit

Nuclei

B

H. pylori

VacA

Specific VacA
binding to
receptors

Nonspecific
VacA binding

Micronutrients

Specific
effects on cell
signaling

Small
molecules
(anions and
urea)

Anions

Anions

Cyt *c*

?

Possible specific
effects on
cell signaling

Pore formation
in cell membrane

Vacuolation

Apoptosis

Other effects,
e.g., on cytoskeleton

Increased
transcellular permeability

VacA binding to
immune cells

Immune suppression

gens that reside within host cells are referred to as intracellular pathogens. Many of these bacteria have evolved mechanisms for entering host cells, such as epithelial cells, that are not naturally phagocytic. They do this by attaching to the host cell surface and causing changes in the host cell cytoskeleton that result in their engulfment by the host cell. In actively phagocytic cells, cytoskeletal rearrangements involving polymerization and depolymerization of actin occur as an integral part of pseudopod formation. By causing similar actin rearrangements to occur in normally nonphagocytic cells, the bacteria are in effect forcing phagocytosis by eliciting the formation of pseudopod-like structures that mediate engulfment. Bacterial surface proteins that provoke phagocytic ingestion of the bacterium by host cells are called **invasins.**

Some proteins that are essential for invasion of cultured cells by bacteria have now been identified. One of the first invasins discovered was that used by *Yersinia pseudotuberculosis* to gain entry into M cells of Peyer's patches and thereby transit into the underlying tissues. The *Yersinia* invasin was found by randomly cloning *Yersinia* genes in *E. coli* and isolating clones that allowed the normally noninvasive *E. coli* strains to enter cultured mammalian cells. The mammalian receptors that are partners for invasin are members of the integrin family of cell adhesion molecules. A number of cytoskeletal proteins are known to interact with the integrin receptors, including focal adhesion kinase (FAK), talin, α-actinin, other kinases, and GTP-binding proteins, but it is not yet entirely clear how these proteins work to promote actin rearrangements that result in bacterial uptake.

Another example of a bacterium that triggers its own uptake by host cells is *L. monocytogenes,* a foodborne gram-positive pathogen that in susceptible individuals can invade intestinal cells; infect mesenteric lymph nodes, bloodstream, spleen, and liver; and ultimately cross the blood-placental and blood-brain barriers. The *Listeria* invasion proteins are called internalins (InlA and InlB), and their host receptors include E-cadherin, an adhesion molecule found in tight junctions; Met, a hepatocyte growth factor tyrosine ki-

nase receptor; gC1qR, a receptor for complement protein C1q; and glycosaminoglycans found on mammalian cell surfaces (Figure 11–11).

Eukaryotic cell biologists, especially those interested in the body's defense systems, view the process of invasion somewhat differently. Whereas bacteriologists see the invasion process as mediated by bacteria, with the purpose of entering and infecting a human cell, gastrointestinal cell biologists view the mucosal cells as sentinels monitoring bacterial invasion. Bacterial invasion of mucosal cells triggers these cells to release chemokines and cytokines that are known to be important for mobilizing the phagocyte and immune defenses. Although there are examples of attachment of pili or other bacterial surface proteins (such as type III secretion systems [discussed in chapter 13]) triggering signaling in the host cell, bacterial attachment alone is not always sufficient to start the signaling process that causes the body to mount an inflammatory response. However, any more intimate connection, especially disruption of normal actin polymerization processes and the presence of bacteria within mucosal cells, sets off the cytokine/chemokine alarm system that initiates the inflammatory response. In particular, as we learned in chapter 3, bacterial products, such as LPS, can trigger these inflammatory responses through activation of Toll-like receptors without the involvement of uptake by phagocytes or bacterial invasion of host cells. In fact, LPS that has been shed from bacteria can also trigger inflammatory responses.

Surviving Phagocytosis

Bacteria that have been ingested by host cells are enclosed in a membrane vesicle called a **phagosome,** which then fuses with lysosomes to start an intensive attack on the bacterium called the **oxidative burst.** The vast majority of bacteria are killed by this very successful offensive strategy. There are bacteria, however, that manage to survive and even multiply inside phagocytes. These are some of the most dangerous pathogens. The only effective host responses left

---

Figure 11–10 Proposed mechanism of ulcer formation by *H. pylori.* (**A**) Overview of the mechanism of ulcer formation. *H. pylori* uses urease to protect it from stomach acid during transit through the stomach to the mucin layer. It uses its flagellum to move to and colonize the mucin layer and may adhere to the gastric mucosa. Products of the bacteria, including ammonia (from conversion of urea by urease), proteases, catalases, phospholipases, and toxins, provoke an inflammatory response that ultimately damages the mucosa. Virulence factors thought to be involved at each stage are indicated in parentheses. (**B**) View of the mechanism of ulcer formation at the epithelial layer. Cyt *c*, cytochrome *c*.

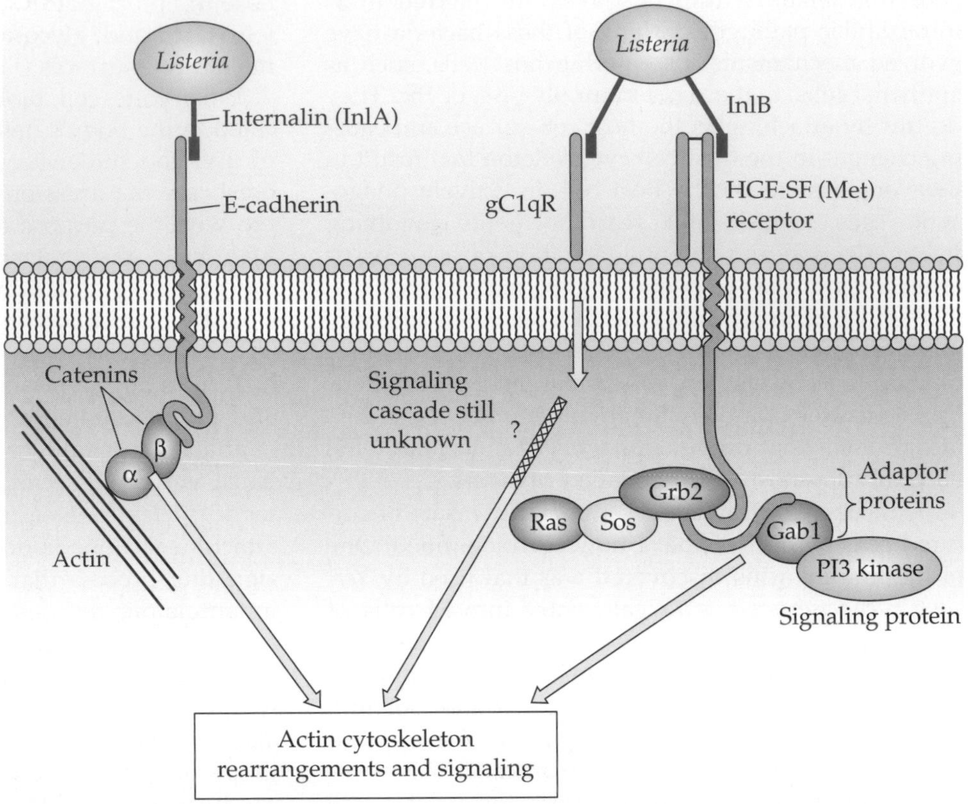

Figure 11–11 *L. monocytogenes* internalins and their host cell receptors. Binding of the internalins to the host receptors triggers signal transduction pathways that lead to cytoskeletal rearrangements and uptake of the bacteria. HGF-SF, hepatocyte growth factor and scatter factor. (Adapted from P. Cossart. 2001. Met, the HGF-SF receptor: another receptor for *Listeria monocytogenes*. *Trends Microbiol.* **9:**105–107, with permission from Elsevier.)

against bacteria that can survive inside normal phagocytes is the activated macrophage response or the cytotoxic-T-cell response (see chapters 3 and 4).

NEUTRALIZING THE PHAGOLYSOSOMAL COMPONENTS. One strategy for surviving phagocytosis is to acquire traits that reduce the effectiveness of the toxic compounds released into the phagolysosome after fusion occurs. Examples include bacterial-cell membranes with altered structures, modified LPS, or capsules that provide resistance to killing by defensins or that are refractory to destruction by lysosomal proteases and lysozyme.

RESISTANCE TO REACTIVE OXYGEN SPECIES AND NITRIC OXIDE. Many bacteria produce enzymes, such as catalase and superoxide dismutase (SOD), that neutralize reactive forms of oxygen; cell surface polysaccharides that interact with and detoxify oxygen radicals; or cell surface proteins that reduce the strength of the oxidative burst. For example, *S. enterica* serovar Typhimurium produces four SODs, two of which are cytoplasmic and two periplasmic. Recent studies suggest that at least one of the periplasmic SODs, SodC1, which is encoded by a prophage, is important for virulence, presumably by enhancing survival within the macrophage during infection.

Nitric oxide appears to play an important antibacterial role in the human body. As described in chapter 3, nitric oxide is a reactive form of nitrogen that is produced by a number of cells of the human antibacterial defense systems. Recently, scientists have begun to learn how some bacteria resist being killed by nitric oxide. In *E. coli*, resistance to nitric oxide is mediated by a **flavohemoglobin** that is normally part of the respiratory system. This flavohemoglobin, also called nitric oxide dioxygenase, uses NADPH, flavin adenine dinucleotide, and oxygen to convert nitric oxide into NO^{3-} by a reaction that is still not well understood. Because the importance of reactive nitrogen intermediates has only recently been recognized, little is known about mechanisms that might protect bacteria from these compounds. Although many bacteria produce proteins that make them somewhat resistant to reactive forms of oxygen and nitric oxide, this alone is usually insufficient to protect them from the full oxidative burst.

PREVENTION OF PHAGOLYSOSOMAL FUSION. Many intracellular bacteria produce factors that allow them to prevent phagosome-lysosome fusion and promote sequestration of the bacteria in specialized vacuoles within the cytosol of the host. The mechanism by which phagosome-lysosome fusion is prevented is poorly understood. The best-studied examples are *Legionella, Salmonella,* and *Mycobacterium* (Figure 11–12).

Legionella is remarkably sophisticated in its ability to manipulate mammalian phagocytic cells. This is evident from a study of attachment and invasion and from what happens to the phagosomes after invasion has occurred. *Legionella* can enter macrophages even in the absence of opsonization by C3b or antibody. The macrophage invasin Mip appears to facilitate the uptake process. Cells are induced to undergo a coiling type of phagocytosis, resulting in a vacuole that is covered with ribosome-studded endoplasmic reticulum. Once the bacteria are internalized inside a vacuole, a complex developmental process begins (Figure 11–12A). Not only do the vacuoles not acidify to the same extent as normal phagosomes, they also leave the normal pathway that leads to phagolysosomal fusion. This seems to be due to a combination of pH homeostasis and elimination of lysosomal membrane proteins, such as LAMP-1 and LAMP-2, which are required for fusion of lysosomes with phagosomes. The resulting vacuole becomes an incubator for bacteria, and eventually they escape from the vacuole into the cytoplasm, where they replicate until the cytoplasm is depleted of nutrients, and then they lyse the host cell and spread to other cells.

S. enterica serovar Typhimurium (Figure 11–12B) triggers actin rearrangements that lead to the formation of ruffle-like structures in the host cell membrane. The ruffles grow into pseudopods that encircle and engulf the bacteria, taking them up into spacious vacuoles. In these special *Salmonella*-containing vacuoles, the bacteria replicate. *Salmonella* actively remodels this *Salmonella*-containing vacuole compartment, which is distinct from a phagosome, using an entourage of bacterial proteins, called effectors, that it delivers into the host cell (more on this in chapter 13) to establish a protected, growth-conducive environment where the bacteria can replicate while evading the host immune response. The vacuoles contain some lysosome-associated membrane protein markers, such as LAMP-1 and LAMP-2, but phagolysosomal maturation is stalled and the vesicles do not acquire other markers, such as the mannose 6-phosphate receptor and lysosomal hydrolytic enzymes.

Mycobacterium tuberculosis (Figure 11–12C) binds directly to macrophage surface protein CR3, the normal receptor for iC3b, or CR4 receptor. Another component that is important for mycobacterial entry is cholesterol, which is relatively abundant in mammalian plasma membranes. Evidence supporting the importance of cholesterol for entry comes from experiments in which membranes depleted in cholesterol had reduced ability to take up mycobacteria. Once again,

binding is followed by internalization of the bacteria into vesicles, which do not fuse with lysosomes. Fusion is prevented by bacterial recruitment of a host protein, termed TACO (for tryptophan-aspartate-containing coat protein), to the surfaces of phagosomes. The bacteria also prevent endocytic acidification and have a reduced oxidative burst and reduced production of interleukin-12, a cytokine that stimulates Th1 responses (see chapter 4).

ESCAPE FROM THE PHAGOSOME.   Another strategy for avoiding killing once inside a phagocyte is for the invading bacterium to escape from the phagosome before it merges with the lysosome. There are many advantages to be gained by escaping the vesicle and growing in the cytoplasm of host cells. They include an abundance of nutrients, protection from antibodies and complement, and partial protection from some antibiotics, as well as transport to other body sites if the infected cell is an immune cell or a blood cell that travels. The only effective host defense against bacteria that do this appears to be the cytotoxic-T-cell (CD8$^+$) response or the natural killer (NK) cell response, which kills the host cells infected by the bacteria (via major histocompatibility complex class I antigen recognition) and thus exposes the bacteria to extracellular defenses, like complement. Note that in this case activated macrophages would not necessarily be effective because the bacteria are in the cytoplasm.

Escape from the phagosome is mediated by a bacterial protein toxin that disrupts membranes either by degrading membrane lipids or by forming pores in the membrane. An example of a bacterium that induces its own uptake and then quickly escapes the phagosome is *L. monocytogenes* (Figure 11–13). One of the proteins responsible for the escape from the phagosome is a pore-forming hemolysin called listeriolysin O (LLO). LLO is responsible for the zone of β-hemolysis seen around bacterial colonies when *Listeria* is grown on blood agar plates. LLO has considerable amino acid sequence similarity to cholesterol-binding cytotoxins produced by other gram-positive pathogens, such as *S. pyogenes* (streptolysin O [SLO]), *Streptococcus pneumoniae* (pneumolysin [PLO]), and *Clostridium perfringens* (perfringolysin O [PFO]).

The importance of LLO for escape from the phagocytic vesicle was demonstrated by an experiment in which the LLO gene, *hly*, was expressed in the common soil bacterium *Bacillus subtilis*. When the resulting *B. subtilis* strain was incubated with a macrophage-like cell line, the bacteria were able to escape the vesicle and grow rapidly in the cytoplasm. The optimal pH for the membrane-lytic activity of

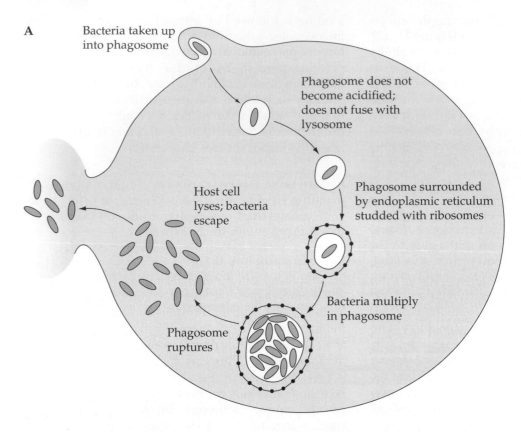

A

Bacteria taken up into phagosome

Phagosome does not become acidified; does not fuse with lysosome

Phagosome surrounded by endoplasmic reticulum studded with ribosomes

Bacteria multiply in phagosome

Host cell lyses; bacteria escape

Phagosome ruptures

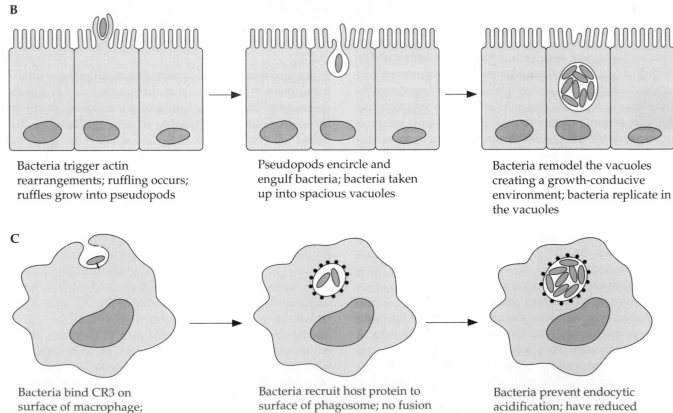

B

Bacteria trigger actin rearrangements; ruffling occurs; ruffles grow into pseudopods

Pseudopods encircle and engulf bacteria; bacteria taken up into spacious vacuoles

Bacteria remodel the vacuoles creating a growth-conducive environment; bacteria replicate in the vacuoles

C

Bacteria bind CR3 on surface of macrophage; bacteria taken up in vesicle

Bacteria recruit host protein to surface of phagosome; no fusion of lysosome with phagosome

Bacteria prevent endocytic acidification; have reduced oxidative burst; bacteria replicate

LLO is 5.5, similar to the acidic pH of the phagolysosome. LLO has very little cytolytic activity at pH 7. This is in contrast to other cytolysins, such as PFO, which do not have a low pH optimum. An advantage of LLO having this difference in pH optimum is that once *Listeria* has escaped from the phagosome, the lytic activity of LLO is reduced, so that the host cell's membrane is not also lysed and the bacteria can replicate in the protective environment of the cytoplasm.

Evading the Host's Antibody Response

One way to evade the host's antibody response, antigenic variation of surface structures, has already been mentioned. In this case, the bacterium makes the host's antibody response obsolete by providing a new antigenic type or variant, which is not recognized by the antibodies. Some bacteria also alter other surface proteins that can serve as targets for antibodies. Another way to avoid the host's antibody response is for the bacterium to be mistaken for part of the host itself. We have also already mentioned that some bacterial capsules that are composed of polysaccharides do not trigger an antibody response because they resemble carbohydrates that are ubiquitous host tissue polysaccharides (sialic acid and hyaluronic acid). Similarly, it is conceivable that bacterial binding of lactoferrin, transferrin, and other host iron-binding proteins serves a dual function: to use them as a protective coat and to acquire iron from them.

Bacteria can also coat themselves with host proteins, such as fibronectin. An interesting example of this type of misdirection that might help the bacteria to evade the immune system is a set of bacterial proteins, such as protein A of *S. aureus* and protein G of *S. pyogenes*, which bind the Fc portion of antibodies, thus coating the bacteria with antibodies, but in a way that does not lead to opsonization of the bacteria because the antigen-binding portions are facing outward while the Fc portion is near the bacterial surface. This antibody coat may prevent recognition of the bacteria

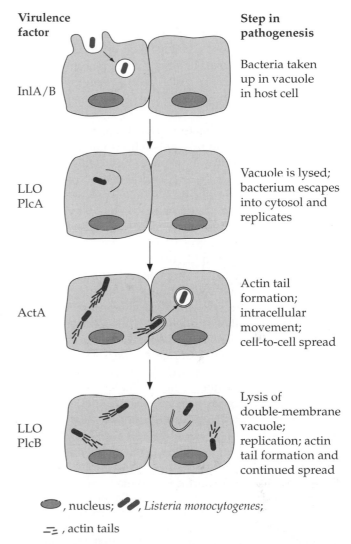

Virulence factor		Step in pathogenesis
InlA/B		Bacteria taken up in vacuole in host cell
LLO PlcA		Vacuole is lysed; bacterium escapes into cytosol and replicates
ActA		Actin tail formation; intracellular movement; cell-to-cell spread
LLO PlcB		Lysis of double-membrane vacuole; replication; actin tail formation and continued spread

⬭ , nucleus; ▰▰ , *Listeria monocytogenes*;

⁻⁼ , actin tails

Figure 11–13 Steps in *L. monocytogenes* entry, escape from the phagosome, actin tail formation, and cell-to-cell spread. The virulence factors thought to be involved at each step are indicated.

Figure 11–12 Schematic showing various mechanisms employed by intracellular pathogens to prevent phagolysosomal fusion and formation of specialized bacterium-containing vacuoles. **(A)** *L. pneumophila* cells are taken up in a vacuole (phagosome) that does not become acidified or fuse with a lysosome. The vacuole becomes surrounded by endoplasmic reticulum. The bacteria replicate in the vacuole, escape, and multiply in the cytoplasm. **(B)** *S. enterica* serovar Typhimurium is taken up into a phagosome in the macrophage. Phagosome-lysosome fusion is prevented, and the bacteria replicate in the phagosome. **(C)** *M. tuberculosis* cells are taken up into a phagosome in the macrophage, where they block calmodulin. Phagosome-lysosome fusion occurs, but the bacteria prevent recruitment of lysosomal hydrolases, so no acidification occurs. The bacteria replicate in the phagolysosome.

by the immune system. For example, *S. aureus* has a number of different types of surface proteins that serve to bind to host cells, as well as to evade host immune cells (Figure 11–14).

Cell-to-Cell Spread

A couple of unique bacteria that force their own ingestion by controlling host cell actin organization continue to interact with actin once they enter the cytoplasm of the host cell. Here, condensation of actin on one end of the bacteria propels them through the host cell cytoplasm and into adjacent cells. The study of the various interactions between invasive bacteria and the host cell cytoskeleton is one of the most exciting areas of current research on virulence mechanisms. One of the most fascinating aspects of this type of motility within host cells is that to date there are only three bacterial species, all pathogens and all from different genera, that are able to use this **actin-based motility** to promote **cell-to-cell spread.** One is the gram-positive food-borne pathogenic *Listeria* species *L. monocytogenes;* the second is the gram-negative causative agent of dysentery, *Shigella flexneri;* and the third is the causative agent of Rocky Mountain spotted fever, *Rickettsia rickettsii* (although not all *Rickettsia* species form actin tails). The ability of these intracellular

Figure 11–14 Cell surface proteins of *S. aureus.* Fibronectin-binding protein and collagen-binding protein mediate attachment to extracellular matrix proteins. Protein A provides protection from the immune system by binding the Fc portion of IgG so that the antibody cannot bind receptors on phagocytic cells.

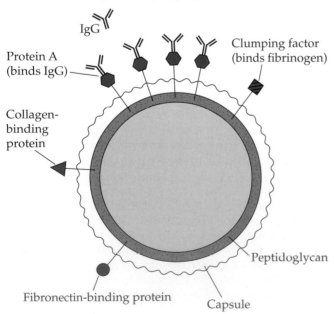

pathogens to spread within host cells and tissues very effectively enables them to evade the humoral immune response of the host.

The process of actin-based motility is illustrated here for *L. monocytogenes* (Figure 11–13). *L. monocytogenes* induces its own uptake into nonphagocytic cells and then escapes from the phagosome into the cytoplasm via the pore-forming toxin LLO, as described above. Once in the cytoplasm, the bacteria are free to exploit the nutrient-rich environment to grow and replicate. The bacteria interact with host structural components to form actin tails.

The process of actin nucleation by the bacterium requires only one bacterial protein, a surface protein called **ActA.** ActA assembles actin filaments by interacting with the host cytoskeletal proteins profilin and Arp2/3. The bacterium polymerizes actin only at one end of the cell (where ActA is anchored) to form an actin tail (sometimes called an actin comet tail), so that only unidirectional movement occurs. Growth of the tail by actin polymerization at the end of the bacterium propels it through the cytoplasm, while the actin tail distal to the bacterium is subsequently depolymerized. In this way, *L. monocytogenes* can move rapidly inside cells at rates of about 1.5 μm/s and then use this actin-based motility to propel itself through the cytoplasm to penetrate into other, adjacent cells (cell-to-cell spread). Once in the adjacent cell, the bacteria again use membrane-damaging toxins to escape from the double-membrane vacuole into the cytoplasm, where they once again continue to grow and spread from cell to cell.

Tissue Penetration and Dissemination

Many pathogenic bacteria produce factors that help them disseminate from the site of entry into the body to other body sites. A phagocytic attack on an invading bacterium generates a region of dead cells. The material in the region is viscous **pus** consisting of DNA and proteins from dead phagocytes and other cells. This viscous pus material traps some bacteria. Many bacteria, however, secrete **DNases** that degrade DNA, thereby thinning the pus. Some bacteria produce enzymes similar to collagenases, elastases, and other proteases that act as "meat tenderizers" to degrade connective tissue matrix proteins, making it easier to spread outward from the area and into areas that contain healthy tissues. Other bacteria produce **hyaluronidases** that degrade the charged polysaccharide hyaluronic acid in connective tissue, thereby also degrading the extracellular matrix. Still other bacteria produce **proteases** that have a similar function. The term "spreading factors" is sometimes used to refer to these enzymes collectively.

BOX 11–2   Disease without Virulence Factors: Subacute Bacterial Endocarditis

The title of this box seems self-contradictory. After all, virulence factors are associated with microbes that cause disease, yet this anomaly has arisen in the case of infections caused by opportunistic pathogens. Subacute bacterial endocarditis is a good example of such opportunistic infections and illustrates the difficulty in defining what is a "virulence factor" for such bacteria. Endocarditis is an infection of the heart valves that can be fatal due to destruction of the valves and surrounding heart tissue. One type of bacterial endocarditis, called subacute because it develops more slowly than endocarditis caused by more virulent bacteria, is caused by a group of streptococci that are normally found in the human mouth (α-hemolytic streptococci). The α-hemolytic streptococci are part of the resident microbiota of the mouth and are not normally able to cause disease. However, in people with prior heart valve damage due to rheumatic fever or congenital valve defects, the α-hemolytic streptococci can be deadly. During dental surgery, large numbers of oral bacteria can enter the bloodstream. Phagocytes in the blood rapidly destroy these bacteria, including the α-hemolytic streptococci. In people with heart valve abnormalities, the turbulent flow of blood near the abnormal valve causes loose clots consisting of fibrin and platelets to form on the valve surface. These clots are called "vegetation." Oral streptococci that manage to reach the heart valve and enter these vegetations are protected from phagocytes, which cannot penetrate the protein network of the vegetation. Accordingly, the bacteria can survive and grow in these sites. It is not clear what causes the damage to the heart valve.

Complement activation triggered by the bacteria may cause some damage, and phagocytes attracted to the area but unable to get to the bacteria may also cause tissue destruction. Proteases produced by the bacteria themselves may also make a contribution. Thus, a completely "avirulent" bacterium can cause a fatal infection in the right kind of host, and a person who dies of subacute bacterial endocarditis is just as dead as someone who dies of typhoid fever or other infections caused by the so-called "virulent" bacteria, which have classical virulence factors. When the connection between dental surgery, heart valve damage, and subacute bacterial endocarditis was first made, dentists were advised to give patients with damaged heart valves penicillin prophylactically before and after dental surgery as a way of preventing the disease. The α-hemolytic bacteria in the mouth are very susceptible to killing by penicillin. Nonetheless, a recently completed clinical trial came to the startling conclusion that penicillin prophylaxis was ineffective in preventing subacute bacterial endocarditis. A possible explanation is that the heart surface is not very highly vascularized. Since antibiotics are delivered to a site by blood vessels, it may be difficult to achieve high concentrations of antibiotics in some regions of the heart despite the fact that large quantities of blood flow through the chambers of the heart itself. Also, the bacteria may form a biofilm, which has reduced susceptibility to antibiotics. Whatever the reason, it appears that α-hemolytic streptococci not only provide an example of disease without special virulence factors, but also failure of antibiotic therapy without special antibiotic resistance mechanisms.

The streptococcal streptokinase protein illustrates another kind of bacterial protease action. This protein acts as a plasminogen activator. A plasminogen activator is a serine protease that cleaves plasminogen into plasmin, which then degrades fibrin clots and allows bacteria to escape from blood clots. Normally, the human body produces plasminogen activators for the same purpose, but with the aim of furthering wound healing.

Beyond Virulence Factors

Bacteria that appear to possess few or none of the virulence factors described above can nonetheless cause serious infections. Such bacteria are generally unable to cause infections in healthy people and preferentially infect people whose defenses are compromised in some way. Such bacteria are called **opportunists.** Despite their initial apparently harmless coexistence with us, opportunists can cause life-threatening infections. In fact, the most common causes of serious infections in hospitalized patients or cancer patients are the opportunistic pathogens. They include members of the body's normal microbiota, such as *Staphylococcus epidermidis, Enterococcus faecalis,* oral streptococci, clostridia, and *Bacteroides fragilis,* as well as common soil bacteria, such as *Burkholderia cepacia* and *P. aeruginosa.*

As some of these opportunistic pathogens are studied in more detail, familiar virulence factors are often uncovered. For example, *B. fragilis,* a gram-negative anaerobe found in the colon, produces a capsule, and *S. epidermidis* has cell surface adhesins. In some cases, such virulence factors play a somewhat different role than they do in the classical pathogens. For example, the capsule of *B. fragilis* seems to function less as an antiphagocytic mechanism than as a factor that elicits an inflammatory response. The surface adhesins of *S. epidermidis* allow it to bind tightly to plastic rather than to mammalian cells.

Most of the bacteria that normally reside in the human body or in soil and water are not capable of acting as opportunistic pathogens, so the bacteria that do act as opportunists must have some special features that enable them to play this role. One is that they are constantly present in high numbers in the body or in the environment. That is, they are on the spot to take advantage of any breach that occurs in the defenses of the human body. Another feature is that many of them are able to take advantage of locations in the body that are somewhat protected from the immune system. For example, *B. fragilis* and other opportunistic *Bacteroides* species tend to gravitate to damaged tissue. Damaged tissue is quite anoxic (lack of oxygen) because it is cut off from the blood supply, so strict anaerobes, such as *Bacteroides* species, are quite happy to grow in these areas. *S. epidermidis* colonizes plastic implants, which are protected from phagocytes because phagocytes do not migrate as efficiently across the plastic surface as they do in tissue. An example of a very serious type of opportunistic infection, which illustrates these principles, is described in Box 11–2.

Still another feature that many opportunists have in common is their resistance to multiple antibiotics (see chapter 16). This not only gives them an advantage in the antibiotic-laden hospital environment, it also makes the infections they cause difficult to treat. This is the main reason why a bacterium, like *S. epidermidis* or *E. faecalis,* that is not very virulent can nonetheless kill the infected person. The combination of impaired host defenses and a multidrug-resistant bacterium is a dangerous one and one that is being seen more and more commonly in very sick hospital patients.

SELECTED READINGS

Barbour, A. G., Q. Dai, B. I. Restrepo, H. G. Stoenner, and S. A. Frank. 2006. Pathogen escape from host immunity by a genome program for antigenic variation. *Proc. Natl. Acad. Sci. USA* **103:**18290–18295.

Blaser, M. J., and J. C. Atherton. 2004. *Helicobacter pylori* persistence: biology and disease. *J. Clin. Invest.* **113:** 321–333.

Brown, M. R. W., and J. Barker. 1999. Unexplored reservoirs of pathogenic bacteria: protozoa and biofilms. *Trends Microbiol.* **7:**46–50.

Chen, L., J. Yang, J. Yu, Z. Yao, L. Sun, Y. Shen, and Q. Jin. 2005. VFDB: a reference database for bacterial virulence factors. *Nucleic Acids Res.* **33:**D325–D328.

Eaton, K. A., S. Suerbaum, C. Josenhans, and S. Krakowka. 1996. Colonization of gnotobiotic piglets by *Helicobacter pylori* deficient in two flagellin genes. *Infect. Immun.* **64:**2445–2448.

Frederickson, J. K., S. W. Li, E. K. Gaidamakova, V. Y. Matrosova, M. Zhai, H. M. Sulloway, J. C. Scholten, M. G. Brown, D. L. Balkwill, and M. J. Daly. 2008. Protein oxidation: key to bacterial desiccation resistance? *ISME J.* **2:**393–403.

Kraus, D., and A. Peschel. 2006. Molecular mechanisms of bacterial resistance to antimicrobial peptides. *Curr. Top. Microbiol. Immunol.* **306:**231–250.

Kreikemeyer, B., K. S. McIver, and A. Podbielski. 2003. Virulence factor regulation and regulatory networks in *Streptococcus pyogenes* and their impact on pathogen-host interactions. *Trends Microbiol.* **11:**224–232.

Li, H., and D. G. Thanassi. 2009. Use of a combined cryo-EM and X-ray crystallography approach to reveal molecular details of bacterial pilus assembly by the chaperone/usher pathway. *Curr. Opin. Microbiol.* **12:**326–332.

Lyczak, J. B., C. L. Cannon, and G. B. Pier. 2000. Establishment of *Pseudomonas aeruginosa* infections: lessons from a versatile opportunist. *Microbes Infect.* **2:**1051–1060.

McClelland, E. E., P. Bernhardt, and A. Casadevall. 2006. Estimating the relative contributions of virulence factors for pathogenic microbes. *Infect. Immun.* **74:**1500–1504.

Merino, S., J. G. Shaw, and J. M. Tomas. 2006. Bacterial lateral flagella: an inducible flagella system. *FEMS Microbiol. Lett.* **263:**127–135.

Molmeret, M., M. Horn, M. Wagner, M. Santic, and Y. Abu Kwaik. 2005. Amoebae as training grounds for intracellular bacterial pathogens. *Appl. Environ. Microbiol.* **71:**20–28.

Nobbs, A. H., R. J. Lamont, and H. F. Jenkinson. 2009. *Streptococcus* adherence and colonization. *Microbiol. Mol. Biol. Rev.* **73:**407–450.

Pizarro-Cerda, J., and P. Cossart. 2006. Bacterial adhesion and entry into host cells. *Cell* **124:**715–727.

Portnoy, D. A., R. K. Tweten, M. Kehoe, and J. Bielecki. 1992. Capacity of listeriolysin O, streptolysin O, and perfringolysin O to mediate growth of *Bacillus subtilis* within mammalian cells. *Infect. Immun.* **60:**2710–2717.

Reniere, M., and E. P. Skaar. 2008. *Staphylococcus aureus* haem oxygenases are differentially regulated by iron and haem. *Mol. Microbiol.* **69:**1304–1315.

Rotem, S., and A. Mor. 2009. Antimicrobial peptide mimics for improved therapeutic properties. *Biochim. Biophys. Acta-Biomembranes* **1788:**1582–1592.

Scott, J. R., and D. Zähner. 2006. Pili with strong attachments: gram-positive bacteria do it differently. *Mol. Microbiol.* **62:**320–330.

Telford, J. L., M. A. Barocchi, I. Margarit, R. Rappouli, and G. Grandi. 2006. Pili in gram-positive pathogens. *Nat. Rev. Microbiol.* **4:**509–519.

QUESTIONS

1. Different parts of the host defense system work together to eliminate invading bacteria. Give some examples of how virulence factors could work together to make a pathogen better able to cause infection.

2. It was stated in the text that virulence factors might have arisen long before animals appeared on Earth. Of course, this is just speculation, but what arguments could be made to support this statement?

3. If the statement made in question 2 proves to be correct, what implications could it have for the number of bacteria able to cause human disease?

4. A bacterium has pili that allow it to attach to intestinal cells, after which it can invade the body. What type of vaccine could help to prevent infections by such a pathogen?

5. In the case of an opportunist, what traits substitute for virulence factors? Could these traits be called virulence factors?

6. How could an actin structure, such as the actin tail produced by *L. monocytogenes,* cause movement of the bacteria?

7. A transposon insertion in ActA stops tail formation by *L. monocytogenes.* Does this fact alone prove that ActA is an actin nucleator? What were some other possibilities scientists had to consider before concluding that ActA is the only bacterial protein that participates in the actin polymerization process? How could you determine whether the actin polymerization factor was produced by the bacteria as opposed to being a host cell factor or protein?

8. How did scientists decide whether LLO was important for escape from the phagosomal vacuole?

9. Infection with *L. monocytogenes* produces an antibody response, but this response is not very protective. Explain why this is so. Also, make an educated guess as to why adults exhibit only mild symptoms whereas intrauterine infections or infections of immunocompromised people can be serious.

10. One perplexing question that needs to be answered is why many people who are colonized with *H. pylori* never develop disease. What types of characteristics might explain why only a small percentage of people who are colonized by *H. pylori* ever develop ulcers?

11. How does the LPS of *H. pylori* differ in function from the LPS of most other pathogens?

12. Why has *P. aeruginosa* continued to be a major problem in hospitals?

13. Speculate on why eliminating adhesin genes genetically might have a much less drastic effect on 50% lethal dose (LD_{50}) values than eliminating genes for invasion or escape from the phagosome.

14. In the introduction to the chapter, we state that bacteria with different disease MOs have different sets of virulence traits. From the examples given in the chapter, identify a few examples illustrating this statement.

15. In 1990, a northeastern U.S. hospital noted an unusually high incidence of sepsis from *S. epidermidis* infections in their cardiac patients. The source of this miniepidemic proved to be a surgeon. How might this episode have happened? (Hint: the surgeon had dermatitis. Also, in cardiac surgery, it is necessary to saw through the rib cage, creating many sharp bone fragments.)

16. Like *Pseudomonas, S. aureus* can form biofilms. What are the clinical implications if *S. aureus* growing in a biofilm is less susceptible, not just to penicillin, but also to vancomycin?

17. Bacteria able to grow intracellularly are often able to survive being phagocytosed by macrophages. What general strategies or properties might such a bacterium use to survive inside a host cell? What advantages might be provided by growing within a phagocytic cell?

18. Consider a bacterium that is ingested via contaminated water and locally infects the small intestine. What types of virulence factors would be most useful to this bacterium?

19. Describe at least four mechanisms that would allow a bacterium Y to compete with its host for iron.

1. A research group is studying a bacterium X that binds to mucosal cells in the lung and invades. Wild-type X has an LD_{50} value of 10 bacteria when administered to mice by inhalation. Using transposon mutagenesis, the researchers have isolated two mutants of X that they call Xmut1 and Xmut2, both of which have LD_{50} values of 10^5 when inhaled by mice. However, in tissue culture cells, Xmut1 can invade the cells just as well as wild-type X, while Xmut2 cannot. Provide a possible explanation for these results.

2. A group of researchers at the USDA have isolated a new gram-positive bacterium related to *Listeria ivanovii* from a recent outbreak of food poisoning due to contaminated cheese. They have determined that the wild-type bacterium can invade epithelial cells and spread from cell to cell. The researchers find that for intestinal epithelial cells, binding and uptake of *L. ivanovii* can be blocked by coincubation with galactose and lactose, as well as a mixture of crude gangliosides. Describe how they might go about experimentally identifying potential host cell surface receptors (glycolipids) involved in mediating the cellular uptake of the bacteria.

3. *Porphyromonas gingivalis* is associated with periodontal disease, including abscess formation and dental caries. Using conventional biochemical and genetic methods, a number of research groups have identified a variety of putative virulence factors thought to be involved in *P. gingivalis* pathogenicity. Among these are several fimbriae (e.g., FimA), a number of secreted proteases (e.g., PrtA, which cleaves IgA; PrtB, which cleaves IgG; PrtH, which cleaves C3; and PrtP, which

cleaves fibrinogen), a couple of hemin-binding outer membrane proteins (e.g., HemR), and a unique capsule that binds factor H (CPS, encoded by a large biosynthetic gene cluster).

A. What is the possible role of each of these seven bacterial factors (FimA, PrtA, PrtB, PrtH, PrtP, CPS, and HemR) in *P. gingivalis* pathogenesis?

B. To verify that these factors are important for pathogenesis, a group of researchers made in-frame deletions of the genes encoding four of the putative virulence factors (FimA, PrtB, HemR, and CPS) and tested the resulting mutants (Mu1 to Mu4, respectively) in a mouse abscess infection model. Four groups of 30 mice each were injected subcutaneously with a 1:1 ratio of *P. gingivalis* wild type (WT) and one of the mutants. The surviving bacteria were then recovered daily for 5 days from the lesion sites of six mice in each group. The percentage of wild-type bacterial cells in the total number of bacterial cells recovered was then determined. The results are shown in the figure below. Interpret these results for each of the virulence factors tested in terms of their putative roles in pathogenesis using this infection model.

C. Another group of researchers found that sera and inflamed gingival tissues of periodontal patients exhibited a positive antibody response to the cytoplasmic and surface-localized heat shock protein (Hsp60) of *P. gingivalis*, which has significant sequence homology to human HSP60. Name possible implications of this finding with regard to immunity, pathogenesis, and vaccine development.

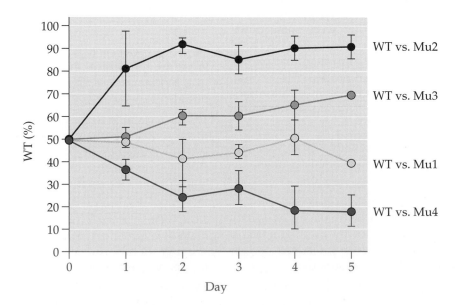

12

Toxins and Other Toxic Virulence Factors

Returning to the criminal investigation metaphor, the parallels continue to emerge. The two foundations of a criminal's strategy are evading discovery and use of the appropriate weapon. In chapter 11, we considered the evasion aspect of a criminal's MO, a strategy that profiling is designed to counter. In this chapter, we consider the choice of weapon. Different pathogens face different challenges and thus choose different approaches to protect themselves or to create a comfortable environment for themselves at our expense.

Bacterial Toxins

Transparent Mechanisms, Mysterious Purposes

On the surface, diseases caused by bacterial toxins seem to be the simplest and easiest to explain. The bacterium produces a substance that is toxic to human cells (the toxin), the toxin causes the symptoms of disease, and the benefit of toxin action to the bacterium is clear. Oddly enough, this type of straightforward explanation has been surprisingly difficult to achieve. Toxins were the first virulence factors to be identified and studied in detail. In fact, scientists who studied toxins were the first to propose the concept of the virulence factor—a single molecule that could cause disease symptoms. For a long time, toxins were the only virulence factors that could be clearly isolated and defined as having a role in pathogenicity. Since those early days, a large number of bacterial toxins have been identified and studied, and our understanding of their biogenesis, regulation, structures, and functions has advanced considerably.

Because many toxins are released or excreted into the extracellular fluid, they have been relatively easy to isolate, purify, and characterize. Modern biochemical and analytical technology has made it increasingly easier to determine the cellular targets of toxins and to reveal the mechanism by which toxins bind to cellular receptors, penetrate or traverse membranes, and recognize and/or modify their intracellular targets to exert their toxic effects. The crystal structures of many toxins, as well as toxins in complex with their cellular targets, have been solved. It is now clear that there is staggering diversity in toxin properties and modes of action.

Toxins often have very specific modes of action and clearly defined targets in mammalian cells, so not surprisingly, toxins have proven to be useful reagents for scientists studying cell biology and are increasingly being used as tools for probing signaling pathways and metabolic processes.

A number of toxins have been made into effective vaccines through physical inactivation or protein engineering. For example, the common childhood DTaP vaccine is a combination of the corresponding toxoids (inactivated, nontoxic versions) of *d*iphtheria, *t*etanus, and *p*ertussis (whooping cough) toxins. Toxins have even been used for beneficial purposes to treat conditions like crossed eyes and to help children with cerebral palsy walk.

Many different bacteria produce toxins. Not all of these toxins are directed at human cells. For example, a toxin produced by *Bacillus thuringiensis,* the so-called Bt toxin, is being used widely as an insecticide in agriculture because it attacks only a small subset of harmful insects, leaving beneficial insects alone. Given the widespread production of toxins and the number of toxins that are associated with human and animal disease, it would seem to be a foregone conclusion that toxins are important for the survival and propagation of the bacteria that produce them. There are an increasing number of cases in which this has been shown to be true. For example, toxins that kill neutrophils and macrophages can protect bacteria from the phagocytic cells that might clear them. Toxins that kill human cells can also release iron stores or carbon sources that the bacteria need to survive and multiply. Toxins can act in more subtle ways as well by dampening the immune response through modulation of the signaling pathways that regulate cytokine production, adhesion, proliferation, or migration.

However, there are also cases in which it has been difficult to discern benefits for the bacteria that produce toxins. Toxins often have a clear function in causing disease symptoms in humans, but sometimes the role of a toxin in the propagation and evolution of the bacteria is not so obvious. A classic example of this is **botulinum neurotoxin** (BoNT), the protein toxin produced by the gram-positive bacterium *Clostridium botulinum*, which is responsible for the paralytic symptoms of the disease botulism. It is a toxin that attacks neurons, causing paralysis and death due to collapse of the respiratory system. BoNT has impacted humans because it is produced when the bacteria are growing in food. Most cases of botulism result from food-borne infection caused by improper canning practices or

from ingestion of clostridial spores present in raw foods, such as bee honey, or on vegetables that have not been washed adequately. Thus, humans who ingest food containing the toxin can die from the effects of toxin that enters their tissues and damages nerves. However, the bacteria do not colonize the intestinal tract of adults and so are eliminated rapidly from the human body. By the time the toxin begins to exert its effects, the bacteria are on the way out. In the case of food-borne disease from canned foods, the bacteria are long gone by the time of ingestion.

Why, then, does *C. botulinum* produce BoNT? *C. botulinum* normally lives in the soil and in lower anaerobic layers in ponds or noncirculating water. In this environment, *C. botulinum* encounters fish and small mammals that may serve as nutrient sources when killed by a strong toxin. Other animals that eat these dead animals killed by BoNT may die themselves as collateral damage, one step removed from the environmental niche of the bacterium. Likewise, when accidentally introduced during food preparation, the bacteria multiply in the food. When subsequently ingested, the clostridia do not colonize the human body and pass right on through. However, the toxin produced by the bacteria in food enters the bloodstream and acts on neurons, producing a flaccid paralysis that can lead to death. Identifying the environmental niches where toxins foster the fitness of the bacteria that produce them remains an intriguing, and sometimes unsolved, mystery of modern pathogenesis research.

The plot thickens when one considers that in many bacteria, the toxin genes are not normal components of the bacterial genome. Indeed, the abundance of toxin genes associated with pathogenicity islands on extrachromosomal plasmids or in the genomes of the sequenced toxin-producing bacteria (Table 12–1), compared to their nontoxigenic counterparts, suggests that horizontal gene transfer is a major driving force in the evolution of toxins and toxin-producing pathogens. For example, when the toxin genes are carried on lysogenic bacteriophages, which integrate into the bacterial genome, only a bacterium that has been infected with one of these phages produces the toxin (Figure 12–1). The gene encoding botulinum toxin is often carried on a bacteriophage, as is the one encoding **diphtheria toxin** (DT), the toxin produced by *Corynebacterium diphtheriae*, the cause of diphtheria. Other toxin genes are found on plasmids that may have come into the bacterium from other bacterial species. As noted above, the solution to these mysteries undoubtedly lies in realizing that we are almost certainly viewing toxins from the wrong perspective when we assume that they are produced with the sole

Table 12–1   Properties of selected protein or peptide toxins

Toxin name	Biological activity	Gene name	Gene location	Bacteria that produce the toxins
Single-peptide-chain AB-type toxins				
Diphtheria toxin (DT)	ADP-ribosyltransferase	*tox*	Corynephages	*Corynebacterium diphtheriae, Corynebacterium ulcerans, Corynebacterium pseudotuberculosis*
Clostridial neurotoxins	Zinc-dependent metalloproteases			
BoNT / A1, A2		*botA1, botA2*	Chromosome	*Clostridium botulinum*
BoNT / A3, A4		*botA3, botA4*	Plasmid	*C. botulinum*
BoNT / B		*botB*	Plasmid	*C. botulinum*
BoNT / C1		*botC1*	Prophage	*C. botulinum*
BoNT / D		*botD*	Prophage	*C. botulinum*
BoNT / Dsa		*botC/D*	Phage?	*C. botulinum*
BoNT / E		*botE*	Chromosome	*C. botulinum, Clostridium butyricum*
BoNT / F		*botF*	Chromosome	*Clostridium botulinum, Clostridium baratii*
BoNT / G		*botG*	Plasmid	*C. botulinum*
TeNT		*tet*	Plasmid	*Clostridium tetani*
Dermonecrotic toxins				
PMT	G-protein deamidase	*toxA*	Phage	*Pasteurella multocida*
DNT	G-protein transglutaminase	*dnt*	Chromosome	*Bordetella bronchiseptica, Bordetella parapertussis, Bordetella pertussis*
Cytotoxic necrotizing factors	G-protein deamidase			
CNF1		*cnf1*	Chromosome	*Escherichia coli*
CNF2		*cnf2*	Plasmid	*E. coli*
CNF3		*cnf3*	Chromosome	*E. coli*
CNFY		*cnfY*	Chromosome	*Yersinia enterocolitica, Yersinia pseudotuberculosis*
Shiga toxin (ST)	RNA *N*-glycosidase	*stx*	Phages	*Shigella dysenteriae, Shigella sonnei, E. coli*
Shiga-like toxins (SLT)	RNA *N*-glycosidase	*stx1*	Phages	*E. coli*
		stx2		*E. coli*
		stx2c		*E. coli*
		stx1c		*E. coli*
Clostridial toxins	UDP-glucosyltransferases		Chromosome	
TcdA		*tcdA*		*Clostridium difficile*
TcdB		*tcdB*		*C. difficile*
TcsH		*tcsH*		*Clostridium sordellii*
TcsL		*tcsL*		*C. sordellii*
Tcnα		*tcnA*		*Clostridium novyi*
Adenylate cyclase (Cya)	Calmodulin-activated adenylate cyclase and pore-forming cytolysin	*cya*	Chromosome	*B. pertussis, B. parapertussis, B. bronchiseptica*
Multisubunit AB-type toxins				
Cholera toxin (CT)	ADP-ribosyltransferase	*ctxAB*	Phage	*Vibrio cholerae, Vibrio mimicus*
Heat-labile enterotoxin (HLT)	ADP-ribosyltransferase	*elt, etx*	Plasmid, chromosome	*E. coli*
Pertussis toxin (PT)	ADP-ribosyltransferase	*ptxA-E*	Chromosome	*B. pertussis*
Pseudomonas exotoxin A (ExoA)	ADP-ribosyltransferase	*exoA*	Chromosome	*Pseudomonas aeruginosa*
Cytolethal distending toxins				
CdtA	Binding subunit	*cdtA*	Prophages, plasmids	*E. coli, Campylobacter jejuni, Haemophilus ducreyi, Shigella dysenteriae, Helicobacter hepaticus, Actinobacillus actinomycetemcomitans*
CdtB	DNA nuclease	*cdtB*		
CdtC	Binding subunit	*cdtC*		
Anthrax toxins			Plasmid	*Bacillus anthracis*
Protective antigen (PA)	Binding subunit	*pag*		

(continued)

Table 12–1 Properties of selected protein or peptide toxins (*continued*)

Toxin name	Biological activity	Gene name	Gene location	Bacteria that produce the toxins
Lethal factor (LF)	Zinc-dependent metalloprotease	*lef*		
Edema factor (EF)	Adenylate cyclase	*cya*		
Pore-forming toxins				
Staphylococcal				
Leukotoxins / leukocidins	Hexameric membrane channels	*pvl, lukD, lukE, lukF, lukM, lukS*	Chromosome, phage	*Staphylococcus aureus*
Hemolysins	*hla, hld, hlg, hly*		Chromosome, plasmid	*Staphylococcus aureus*
Streptolysin O (SLO)	Heptameric membrane channels	*slo*	Chromosome	*Streptococcus pyogenes*
Streptolysin S (SLS)	Heptameric	*sls*	Chromosome	*Streptococcus pyogenes*
Listeriolysin O (LLO)	Large multimeric membrane channels	*hly*	Chromosome	*Liseria monocytogenes*
Perfringolysin O (PFO)	Large multimeric membrane channels	*pfo*	Chromosome	*Clostridium perfringens*
Extracellular proteins, exoenzymes, and peptides				
Staphylococcal enterotoxins	Superantigen (SAg)	*sea*	Phage	*S. aureus*
		sed, sej	Plasmid	
		seg, sen, sei, sem, seo	Chromosome	
		sel, sek, sec1 to *sec3*	Prophages	
Toxic shock syndrome toxin (TSST)	Superantigen (SAg)	*tst*	Prophage	
Streptococcal superantigens exotoxins	Superantigen (SAg)	*sea*	Phage	*S. pyogenes*
		speA, speB	Chromosome	
		speC, speG, speH, speI, speK, speL, speM	Prophages	
Cytolytic phospholipase C (PLC)				
Clostridial				
α-toxin	Zinc-dependent PLC	*plc*	Chromosome	*Clostridium perfringens*
β-toxin	PLC	*plc*	Chromosome	*C. novyi, Clostridium haemolyticus*
γ-toxin	PLC	*plc*	Chromosome	*C. sordellii, C. novyi*
Hemolysins	PLC	*hly*	Chromosome	*C. haemolyticus, P. aeruginosa, P. aureofaciens, Bacillus cereus, Aeromonas hydrophila, Acinetobacter calcoaceticus*
Cytolytic phospholipase D (PLD)				
PLD	PLD	*pld*	Chromosome	*Vibrio damsela*
Lethal toxin	PLD	*pld*	Chromosome	*Corynebacterium ovis*
Murine toxin	PLD	*pld*	Chromosome	*Yersinia pestis*
Hemolytic phospholipase A (PLA)	PLA	*pla2*	Chromosome	*Rickettsia prowazekii*
Sphingomyelinase (SMase)	SMase	*hib*	Chromosome	*S. aureus*
Staphylococcal				
β-toxin	Serine protease	*eta, etb, etc, etd*	Chromosome	*S. aureus*
exfoliatin	Serine protease	*sheta, shetb, exhA, exhB, exhC, exhD*	Chromosome	*Staphylococcus hyicus*
Toxic effector proteins delivered into host cells by bacterial secretion systems				
Pseudomonas			Chromosome, prophage	*P. aeruginosa*
ExoS, ExoT	ADP-ribosyltransferase plus Rho GTPase-activating protein (GAP) domain	*exoS, exoT*		
ExoU	Phospholipase A2 (PLA2)	*exoU*		
ExoY	Adenylate cyclase	*exoY*		

Table 12–1   *(continued)*

Toxin name	Biological activity	Gene name	Gene location	Bacteria that produce the toxins
Salmonella			Chromosome prophage, phage	*Salmonella enterica*
SopA	E3 ubiquitin ligase	*sopA*		
SopB	Phosphatidylinositol phosphatase	*sopB*		
SopE	RhoGAP	*sopE*		
SopE2	RhoGAP	*sopE2*		
SipA	Binds actin	*sipA*		
AvrA	Cysteine protease	*avrA*		
SptP	RhoGAP plus tyrosine phosphatase	*sptP*		
SifA	Filament formation along microtubules	*sifA*		
Yersinia			Plasmid	*Y. enterocolitica, Y. pestis, Y. pseudotuberculosis*
YopE	RhoGAP	*yopE*		
YopH	Tyrosine phosphatase	*yopH*		
YopJ	Cysteine protease	*yopJ*		
YopM	Anti-inflammatory leucine-rich repeat protein	*yopM*		
YopT	Cysteine protease	*yopT*		
YkpA / YopO	Serine / threonine protein kinase	*ykpA/yopO*		
Shigella			Plasmid	*Shigella flexneri*
IpaA	Actin depolymerization	*ipaA*		
IpaB	Membrane pore-forming	*ipaB*		
IpaC	Actin polymerization	*ipaC*		
IpgD	Phosphatidylinositol phosphatase	*ipgD*		
Lipopolysaccharides / bacterial membrane components				
Endotoxin (LPS)	Bacterial cell wall component that binds Toll-like receptors	*los, lps* biosynthetic gene clusters	Chromosome	Gram-negative bacteria
Lipoteichoic acid (LTA)	Bacterial cell wall component that binds Toll-like receptors	*lta* biosynthetic gene clusters	Chromosome	Gram-positive bacteria
Peptidoglycan (PG)	Bacterial membrane surface coating component that binds Toll-like receptors	Peptidoglycan (PG) biosynthetic gene cluster	Chromosome	All bacteria
Tracheal cytotoxin (TCT)	Binds Toll-like receptors	LPS biosynthetic gene cluster	Bacterial PG component	*Bordetella* species
Mycolactone	Immuno- and neurosuppressant; secreted polyketide-derived lipid; intracellular target unknown	*mlsa1, mlsa2* biosynthetic genes	Plasmid	*Mycobacterium ulcerans*

aim of damaging the human body. They may play roles in the physiology of the bacteria and their viruses, possibly regulation of cellular or phage functions or cell-cell signaling, with their impact on the human body merely an accidental side effect of their action. In some cases, animals other than humans or lower eukaryotes, such as insects or even protozoa, may very well be the targets of these toxins in the environment. This notion would explain why most toxins are either excreted from the bacterial cells that make them or delivered directly from bacteria into host cells through special secretion systems and why toxins are often so specific for certain types of animal cells.

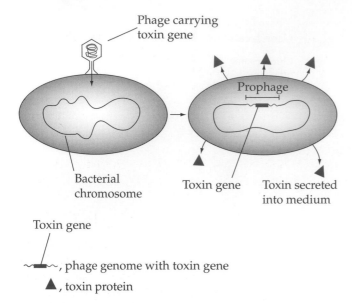

Toxin gene

~~■~~, phage genome with toxin gene

▲, toxin protein

Figure 12–1 Lysogenic bacteriophages carry toxin genes that integrate into the bacterial chromosome, thus making the bacterium capable of producing the toxin.

Characteristics and Nomenclature

As evidenced in Table 12–1, bacterial toxins come in a large variety of forms, ranging from relatively small lipid-like compounds, such as **endotoxin** and the **macrolactones,** and peptides, such as the so-called **superantigens,** to large proteins, such as the single-chain **DT** and the multisubunit **anthrax toxins.** A term that has been used to designate the protein toxins of bacteria is **exotoxin.** The word "exotoxin" was chosen to emphasize the fact that the toxins are excreted from the cell into the medium, in contrast to an endotoxin, which is embedded in the bacterial surface and is usually released only upon cell lysis. The term "exotoxin" has been falling out of use because some protein toxins are not excreted but rather accumulate inside the cell and are released by cell lysis. Others are injected directly into human or animal cells, thereby bypassing the extracellular fluid entirely. These toxins are often referred to as **"effector proteins"** or **"exoenzymes"** and are delivered directly into the host cell via specialized secretion systems, which we will cover in more detail in chapter 13, when we talk about secretion systems and delivery of virulence factors. For convenience, we will use the word "toxin" to mean all types of virulence factors of bacteria, whether excreted or not, that are toxic to the human or animal host.

Bacterial toxins vary considerably, not only in their structures and activities, but also in the host cell types they attack. The names given to different toxins reflect this diversity. Some toxins are simply given letter des-

ignations, e.g., **exotoxin A** of *Pseudomonas aeruginosa,* while other toxins are named for the numerical or Greek or Roman alphabetical order in which they are isolated or discovered from a bacterial species, such as the **α-, β-, γ-, δ-,** and **ε-toxins** produced by *Clostridium perfringens.* Some toxins are named to indicate what type of host cells they attack. Toxins that attack a variety of different cell types are often called **cytotoxins,** whereas toxins that attack specific cell types may be designated according to the cell type or organ affected, e.g., **neurotoxin** (attacks nerve cells), **leukotoxin** (attacks leukocytes), **hepatotoxin** (attacks liver cells), **cardiotoxin** (attacks heart cells), **enterotoxin** (attacks intestinal mucosal or enteric cells), and **verotoxin** (attacks Vero cells, which are a green monkey kidney cell line frequently used in toxicity assays). Toxins are also sometimes named for the bacterial species that produces them or for the disease with which they are associated. Examples are **cholera toxin (CT),** produced by *Vibrio cholerae,* the cause of cholera (watery diarrhea); **Shiga toxin (ST),** produced by *Shigella* species, a cause of bacterial dysentery (bloody diarrhea); **diphtheria toxin (DT),** produced by *C. diphtheriae,* the cause of diphtheria (respiratory and heart failure); and **tetanus toxin (TENT),** produced by *Clostridium tetani,* the cause of tetanus (lockjaw).

Toxins can be named on the basis of their enzymatic activities, e.g., **adenylate cyclase** (a toxin produced by *Bordetella pertussis,* the cause of whooping cough), and **lecithinase** (a toxin with phospholipase activity produced by *C. perfringens,* a cause of gangrene, that cleaves lecithin phospholipid molecules). Still other toxins are named for their biochemical properties, such as the **heat-labile toxin (HLT)** and **heat-stable toxins (HST)** from *E. coli.* The BoNTs are further designated by letters, based on their immunological cross-reactivity, into seven major serotypes (A through G). In this case, antibodies generated against one serotype, e.g., BoNT/A, do not bind to or neutralize the other neurotoxin serotypes, BoNT/B through BoNT/G.

Some toxins have more than one name. For example, a toxin produced by *E. coli* O157:H7 (the "killer *E. coli*") has been called **Shiga-like toxin (SLT),** because it is closely related to the Shiga toxin produced by *Shigella* species, or verotoxin, because it is toxic to Vero cells. An exotoxin produced by *C. perfringens* is called both α-toxin and lecithinase. A potent toxin from *Pasteurella multocida* has been called osteolytic toxin because it causes bone degradation, dermonecrotic toxin because it causes necrotic skin lesions, and mitogenic toxin because of its proliferative effects on certain tissues or cells, as well as simply ***Pasteurella multocida* toxin (PMT).** A notorious source of confusion for students first encountering toxin names is the term "enterotoxin," which should not be mistaken for "exotoxin." Enterotoxin is a specific term that denotes a toxin that causes diarrhea or vomiting, i.e., enteric symptoms.

Recently, another layer has been added to the nomenclature nightmare of toxin designation, although this one has the virtue of separating toxins based on the mechanism of action, type I to III toxins. Normally, the first step in toxin action is binding to the target cell. This binding step may be followed by direct action on a cell surface component or by internalization of the toxin and action on an intracellular target. **Type I toxins** bind to the target cell surface, but they are not translocated into the target cell; instead, they act extracellularly. An example of this type of toxin is the **superantigens (SAg),** which bind to surface molecules on macrophages and T cells, forcing them into an unnatural interaction in which they produce copious amounts of toxic cytokines. **Type II toxins** are the ones that act on eukaryotic cell membranes **(phospholipases, or pore-forming cytotoxins)** and exert their effect by destroying the integrity of the mammalian cell membrane. **Type III toxins** are the classical **A-B-type toxins,** which have two functional components, an enzymatic component or domain (denoted as A) that activates or inactivates some intracellular target or signaling pathway to cause its toxic effect on the cell and a binding component or domain (denoted as B) that recognizes a specific receptor on the host cell and mediates introduction of the A portion into the cell cytoplasm. A-B-type toxins come in two main flavors, single-chain peptides with multiple domains, such as DT and the BoNTs, and multisubunit complexes, such as cholera toxin and anthrax toxins. Below, we consider examples of each of the different types of bacterial toxins.

Nonprotein Toxins

ENDOTOXINS. In chapter 3, the steps in the development of shock in response to **lipopolysaccharide (LPS)** or **lipoteichoic acid (LTA)** were described. Excess cytokine production played a key role in this process. Endotoxins are integral bacterial membrane components (LPS, LTA, and **phosphatidylglycerol [PG]**) that are released into the medium when bacteria lyse. Endotoxin produced by gram-negative bacteria, such as *E. coli*, consists of a polysaccharide (sugar) chain and a lipid moiety, known as lipid A (Figure 12–2). The polysaccharide chain is highly variable among different bacteria. Toxicity is associated with the lipid A component, and immunogenicity is associated with the polysaccharide components. LPS acts as the pro-

totypical endotoxin because it binds to LPS-binding protein and interacts with Toll-like receptors (TLRs) on the surfaces of host macrophages, dendritic cells, and neutrophils (see chapter 3), thereby promoting the secretion of proinflammatory cytokines (such as interleukin-1 [IL-1], IL-6, IL-8, tumor necrosis factor alpha [TNF-α], and platelet-activating factor) and nitric oxide, resulting in **endotoxic shock** (also called **septic shock**).

LPS bound to LPS-binding protein interacts with surface signaling complexes of CD14 receptors, signal-transducing TLRs (mainly TLR4 for most enteric bacteria, but sometimes TLR2/TLR6 for certain pathogens that have modified lipid A cores), and the recently discovered MD-2 protein that associates with TLR4 and is required for optimal TLR-mediated signaling induced by LPS. LPS-induced cytokine release causes activation of the alternative complement cascade, increased vascular permeability, and activation of the coagulation cascade. This in turn causes activation of factor XII and formation of blood clots in small vessels, known as **disseminated intravascular coagulation,** resulting in an overall decrease in blood pressure and blood supply to critical organs. This ultimately results in multiple organ system failure and accumulation of fluid in the lungs, called **acute respiratory distress syndrome.**

Unfortunately, treatment of a systemic bacterial infection with antibiotics, which cause the lysis of bacterial cells and release of cell wall components, such as LPS, can sometimes make the disease worse. For example, treatment of syphilis, caused by the spirochete *Treponema pallidum,* with certain antibiotics can result in the Jarisch-Herxheimer reaction (also known as lesions of syphilis) due to the acute increase in the release of inflammatory cytokines (IL-6, IL-8, and TNF-α).

Certain pathogens actively release parts of their cell walls as membrane blebs during the logarithmic growth phase. For example, *B. pertussis*, the causative agent of **pertussis (whooping cough),** releases a low-molecular-weight glycopeptide derived from its peptidoglycan. This toxin, called *Bordetella* **tracheal cytotoxin (TCT),** is responsible for the respiratory cytopathology observed during whooping cough, including ciliostasis (arrest of ciliary movement), specific extrusion of ciliated cells from the respiratory epithelium, and release of the inflammatory cytokine IL-1. Without ciliary activity, coughing becomes the only way to clear the airways of accumulating mucus, bacteria, and inflammatory debris. TCT thereby triggers the violent coughing episodes symptomatic of pertussis. In addition, the absence of ciliary clearance predisposes the patient to secondary pulmonary infections, which are the primary cause of pertussis

Salmonella lipopolysaccharide (LPS)

O-specific chain	Core	Lipid A

```
                                                          EtN
                                                           |
                                                EtN        P
                                                 |         |
                                                 P        KDO        P
                                                 |         |         |
       Gal              GlcNAc GlcNAc   Gal  Hep  P        KDO       GlcN
        |                  |      |                |         |         |
       Gal — GalNAc — GalNAc — Glc — Gal — Glc — Hep — Hep — KDO — GlcN
                             n                    |                   |
                                                  P                   P
```

> 5–7 fatty acids attached to GlcN

Repeat units
$n \geq 10$

Lipid A moiety from LPS of
Salmonella enterica serovar Typhimurium and *E. coli*

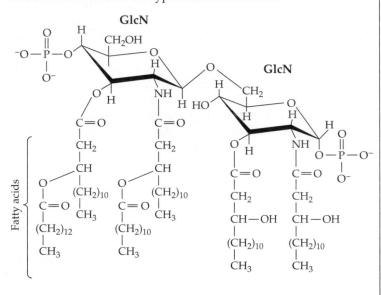

Peptidoglycan (PG) monomer

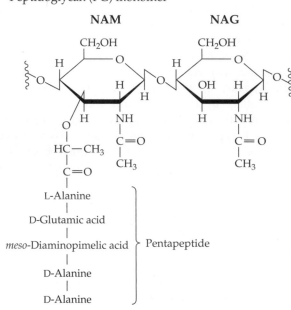

L-Alanine
|
D-Glutamic acid
|
meso-Diaminopimelic acid } Pentapeptide
|
D-Alanine
|
D-Alanine

Bordetella tracheal cytotoxin (TCT)

TCT (GlcNAc-1,6-anhydro-MurNAc-L-Ala-γ-D-Glu-*meso*-A$_2$pm-D-Ala)

mortality. How the cough persists for months after clearance of the bacteria and toxins is still a mystery.

MYCOLACTONE TOXINS. In contrast to endotoxin, other lipid-like toxins, such as the polyketide-derived **mycolactones,** have both cytotoxic and immunosuppressive properties. *Mycobacterium ulcerans* is an emerging human pathogen harbored by aquatic insects. It is the causative agent of Buruli ulcer, a devastating skin disease rife throughout tropical and subtropical regions of Central and West Africa, Central and South America, Asia, and Australia. *M. ulcerans* strains make a variety of mycolactones, which are polyketide-derived macrolides that are made as secondary metabolites. Infection with *M. ulcerans* causes progressive necrotic lesions that, if left untreated, can extend to cover up to 15% of a victim's body surface and can lead to lifelong disability and occasionally death. The most remarkable feature of the disease is that even advanced disease is characterized by very little inflammation and no physical pain. The emergence of *M. ulcerans* as a pathogen most likely reflects the acquisition of a large 174-kilobase plasmid (pMUM001) that is not found in the closely related *Mycobacterium marinum.* The plasmid carries a cluster of biosynthetic genes encoding three giant **polyketide synthases** *(mlsA1, mlsA2,* and *mlsB)* and three polyketide-modifying enzymes (Figure 12–3). The domains appear to be functionally identical, since there is an extremely high level of sequence identity within the different domains of the *mls* (macrolactone synthase) gene cluster (>97% amino acid identity), which undoubtedly caused considerable difficulty in their sequencing and which suggests that acquisition of this virulence factor took place quite recently.

Protein Exotoxins

SUPERANTIGENS (TYPE I TOXINS). In the 1970s, a new and frightening disease appeared in young women, a disease called **toxic shock syndrome** because the afflicted woman developed shock-like symptoms. Some of the women died. As more was learned about toxic shock syndrome, it became clear that this was shock caused not by gram-negative or gram-positive bacteria in the bloodstream but by a toxin that proved to be a superantigen. The toxin was produced by the gram-positive bacterium *Staphylococcus aureus.* The

bacteria colonized the vagina and produced the toxin there. The toxin entered the bloodstream and spread throughout the body. The reason the disease was seen only in women was that it was associated with a type of superabsorbent tampon that could be left in place for an extended time. The bacteria were growing in the tampon, where conditions were conducive to toxin production. Not all strains of *S. aureus* produce this toxin, and this is one of the reasons there were not more cases of the disease. The tampons were taken off the market, and toxic shock syndrome decreased significantly.

Toxic shock syndrome is still of interest, however, because it shows that protein toxins can cause a type of shock that is very like the shock induced by circulating LPS or LTA. Although the toxic shock toxin, **toxic shock syndrome toxin (TSST),** has properties very different from those of LPS or LTA, its ability to force unnatural associations between macrophages and T cells, the hallmark of a superantigen, causes an outpouring of cytokines that trigger the shock process. In addition, superantigens appear to act synergistically with LPS to increase the ability of LPS to elicit cytokine release.

This unusual type of bacterial toxin exerts its effect by binding to the major histocompatibility complex class II (MHC-II) of macrophages and the receptors on T cells that interact with the MHC (Figure 12–4). As was discussed in chapter 4, antigen-presenting cells (APCs), such as macrophages, normally process antigens by cleaving them into peptides and displaying one of the resulting peptides in a complex with MHC-II on the macrophage surface. Only a few helper T cells have receptors that recognize this particular MHC-peptide complex, so only a few T cells are stimulated. Superantigens are not processed by proteolytic digestion inside macrophages but rather bind directly to MHC-II on the macrophage surface. Since they do this rather indiscriminately, many different macrophages with or without peptide antigen bound will have superantigen molecules bound to their MHC molecules. The superantigen also binds T-cell receptors, again rather indiscriminately, and thus forms many more macrophage-T helper cell pairs than would normally form. Thus, instead of a macrophage stimulating 1 in 10,000 T cells (the normal response to an antigen), as many as 1 in 5 T cells can be stimulated by the bridging action of superantigens. (This is why they are called superantigens.) When

---

Figure 12–2 Structures of cell wall components and the corresponding endotoxins. P, phosphate group; EtN, ethanolamine; Hep, L-glycero-D-mannoheptose; KDO, 2-keto-3-deoxy-D-mannooctulosonic acid; A_{2pm}, diaminopimelic acid.

A

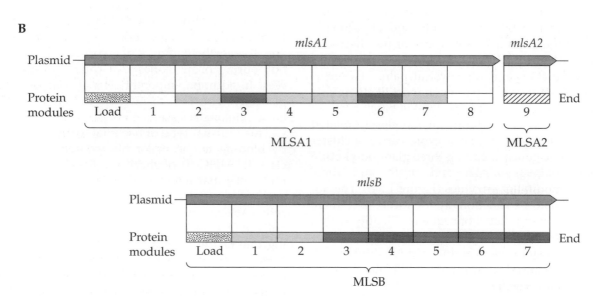

Figure 12–3   The polyketide-derived mycolactone from *Mycobacterium ulcerans* is encoded on a large plasmid. The 12-member core structure of the mycolactone is produced by two giant modular polyketide synthases, MLSA1 (1.8 MDa) and MLSA2 (260 kDa), whereas the side chain is synthesized by MLSB (21.2 MDa). Shown are the domain and module organizations of the polyketide synthase genes. Within each of the genes, different domains are represented by colored blocks. (Adapted from T. P. Stinear, A. Mve-Obiang, P. L. C. Small, W. Frigui, M. J. Pryor, R. Brosch, G. A. Jenkin, P. D. R. Johnson, J. K. Davies, R. E. Lee, S. Adusumilli, T. Garnier, S. F. Haydock, P. F. Leadlay, and S. T. Cole. 2004. Giant plasmid-encoded polyketide synthases produce the macrolide toxin of *Mycobacterium ulcerans*. *Proc. Natl. Acad. Sci. USA* **101:** 1345–1349, with permission from the National Academy of Sciences of the United States of America.)

helper T cells are stimulated by macrophages, one result is that the T cells release cytokines, especially IL-2. Superantigen action thus causes excessively high levels of IL-2 to circulate in the bloodstream, giving rise to a variety of symptoms, including nausea, vomiting, malaise, and fever. The role of superantigens in disease will be discussed more in a later chapter on staphylococcal and streptococcal infections. Superan-

tigens are classified as type I toxins because they do not enter the cell.

MEMBRANE-DISRUPTING TOXINS (TYPE II TOXINS).   A second class of exotoxins lyses host cells by disrupting the integrity of their plasma membranes. Membrane-disrupting toxins have two different roles in virulence. In some cases, their primary role appears to be

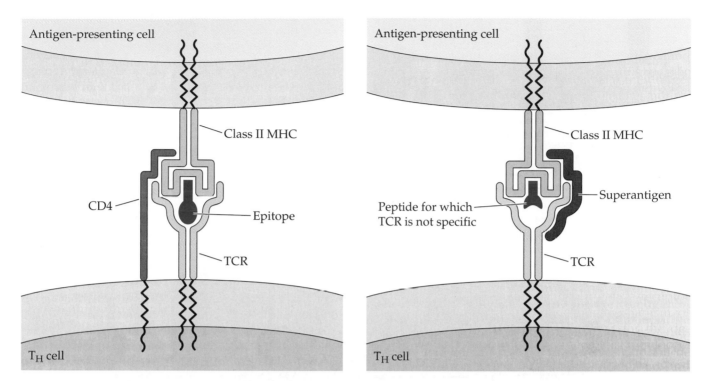

Figure 12-4 Normal interaction of an APC and a T helper cell (left) compared with the interaction of an APC and a T helper cell mediated by a superantigen (right). Superantigens form a bridge complex between the MHC molecules on APCs and T-cell receptors with or without the peptide antigen present, which in turn overstimulates the T cells to release cytokines (especially IL-2), resulting in T-cell and B-cell proliferation and toxic shock.

killing of host cells, especially phagocytes. In others, invasive bacteria use them to escape from the phagosome and enter the host cell's cytoplasm before phagolysosomal fusion.

There are two different types of membrane-disrupting toxins (Figure 12–5). One type is a protein that forms channels in the membrane. Since the osmotic strength of the host cell cytoplasm is much higher than that of the surrounding environment, holes in the membrane allow a sudden inrush of water into the cell. The cell swells and because the membrane is not strong enough to contain the sudden inrush of fluid, the cell lyses. Pore-forming toxins are classified by the nature of the structures that form the pores. The α-pore-forming toxins tend to be highly α-helical in their water-soluble state and form pores in membranes using helices. The β-pore-forming toxins are rich in β-sheets in their water-soluble state and form a β-barrel in membranes. The basic steps involved in forming the pores have been known for quite some time and are depicted in Figure 12–5A. Interaction with a cell surface receptor is generally required to target the toxin to the host cell. Membrane insertion is often facilitated by acidic pH, which leads to a large conformational change in the protein's ter-

tiary structure to an insertion-competent state, followed by membrane penetration. The α-pore-forming toxins do not appear to always require oligomerization to form the membrane pore, but oligomerization appears to be an essential step in pore formation by the β-pore-forming toxins. Most of the A-B (type III) toxins (discussed below) have domains or subunits made up of structural components resembling α- or β-pore-forming toxins, which enable them to penetrate and translocate across host membranes to deliver their toxic components.

The second type of membrane-disrupting toxin is an enzyme that compromises the integrity of membrane phospholipids. Such enzymes go by a variety of names, such as phospholipase, hemolysin, or cytolysin, but they all act in similar ways by destroying the integrity of host cell membrane lipids. Some phospholipases remove the charged head group from the lipid portion of the molecule. Because the charged head group stabilizes the lipid bilayer structure of the host cell plasma membrane, removal of this group destabilizes the membrane, and cell lysis results. Other phospholipases cleave at other sites (Figure 12–5B), but their effects are the same, to destabilize the host cell membrane. Membrane-active toxins are fre-

A

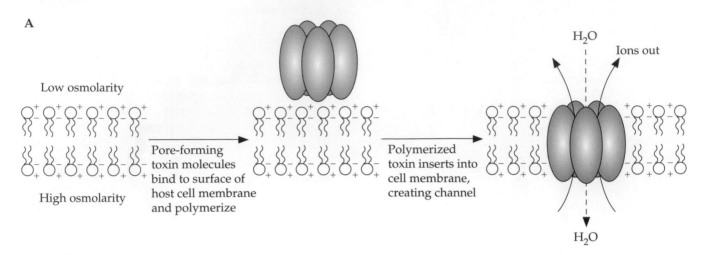

B

quently called "hemolysins" because red blood cells are a convenient cell type to use as an assay system. However, such toxins are toxic for many types of cells, because their target, the cell membrane, is the same, so the term "cytolysin" is also used frequently.

An extreme form of wound infection is **gangrene,** caused by *C. perfringens*. These bacteria produce an expanding zone of dead tissue. The skin blackens and may appear swollen and cracked, as if the area had been cooked. The damage can be so severe that the limb has to be amputated. Tissue damage is caused in part by a toxin called **α-toxin,** which hydrolyzes the lipid lecithin in mammalian cell membranes. If the toxin enters the bloodstream, it can cause damage to organs, such as the kidney.

A-B TOXINS (TYPE III TOXINS). A-B toxins were the first toxins to be studied in detail at the molecular level, so they have come to be the paradigm toxins. A-B toxins usually contain two functional domains or subunits, one (the A part) with toxic (enzymatic) activity and one (the B part) that binds to receptors on host cells and gets the A part across the plasma membrane and into the host cell cytoplasm (a process referred to as **translocation**), where the A part gains access to its intracellular target. Structures of common types of A-B toxins are illustrated in Figure 12–6A. The simplest type of A-B toxin is synthesized as a single polypeptide, which has one binding (B) and one enzymatic (A) domain. Frequently, the A and B domains of such toxins are separated during processing of the toxin by a proteolytic cleavage event, although the two domain-containing fragments remain connected by a disulfide bond. The disulfide bond is then broken under the reducing conditions inside the host cytosol. Detachment of the A portion from the B portion is often necessary for the A portion to become enzymatically active.

A more complex type of A-B toxin, a multisubunit A-B toxin, has an enzymatic (A) portion that is a separate polypeptide from the binding portion (B), which is composed of multiple subunits that are identical in some cases (e.g., cholera toxin), but not in others (e.g.,

pertussis toxin) (Figure 12–6B). In these toxins, the A subunit is attached to the rest of the toxin through noncovalent interactions, which are disrupted when the A subunit is internalized by the host cell. Proteolytic cleavage of the catalytic domain of the A subunit from the portion of the A subunit that interacts with the B subunit complex occurs for many of these toxins (e.g., cholera toxin and pertussis toxin).

Both types of A-B toxins bind to and enter host cells, as illustrated in Figure 12–7. The B portion binds to a specific host cell surface molecule. Often, the molecule recognized by the B portion is the carbohydrate portion of a host cell surface glycoprotein or glycolipid, but some B portions bind to proteins. The B portion determines the host cell specificity of the toxin. For example, a toxin whose B portion binds to a glycoprotein that is found only on the surfaces of neurons will function in the body as a neuron-specific toxin, even though the A portion has the sort of activity that would enable it to kill other types of host cells if it could gain entry into their cytoplasm.

After the B portion attaches the exotoxin to the host cell, the A portion is translocated through the host cell membrane into the host cell's cytoplasm. In some cases, the bound toxin is taken up by endocytosis prior to internalization of the A portion into the cytoplasm. Acidification of the endocytic vacuole may play a role in translocation for toxins that use this route of entry by stimulating the separation of A and B portions, insertion of the toxin into the membrane, and internalization **(translocation)** of the A portion. Translocation is a complex process that is only beginning to be understood. One model posits that the B portion not only binds the host cell surface, but also forms a pore through which the A portion enters the host cell cytoplasm, but this model is somewhat controversial. Recent evidence suggests that there are host cell proteins that act as chaperones to facilitate the process.

Once the A portion has entered the host cell cytoplasm, it becomes enzymatically active and exerts its toxic effect. Incredibly enough, host cell proteins are now known to aid in activation of the A portions of

Figure 12–5 Two kinds of membrane-disrupting cytotoxins (also called hemolysins or cytolysins). **(A)** A channel-forming (pore-forming) type of protein (e.g., α-toxin of *S. aureus*) inserts itself into the host cell membrane and makes an open channel (pore). Formation of multiple pores causes leakage of cell interior components and an inrush of water, leading to lysis. **(B)** A phospholipid-hydrolyzing (phospholipase) type of membrane-disrupting toxin removes the polar head group, as shown here, or otherwise compromises the phospholipid structure, which destabilizes the membrane and causes the host cell to lyse. Different phospholipases act at different sites. Phospholipase C (PLC) is used as an example.

A Example—diphtheria toxin

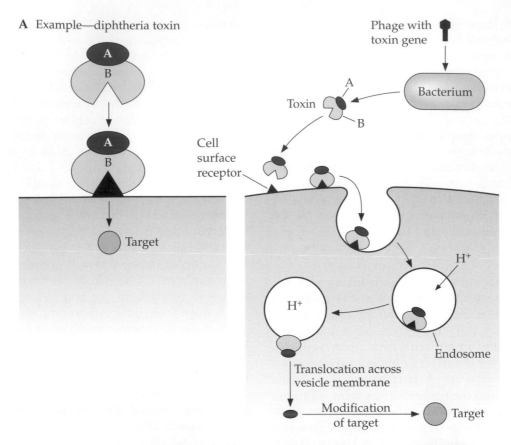

B Examples—cholera toxin (left), pertussis toxin (right), and anthrax toxin (bottom)

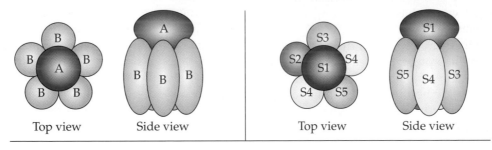

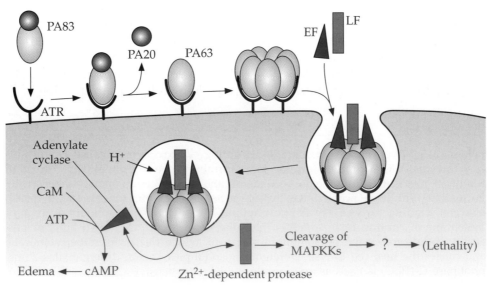

some toxins. The A portions of A-B-type exotoxins may enter very different cell types, but most of them catalyze the same reaction. For example, the A parts of a relatively large number of A-B toxins have **ADP-ribosyltransferase** activity, i.e., they covalently transfer the ADP-ribosyl group from NAD to some host cell protein (Figure 12–8A). **ADP-ribosylation** of the host cell protein either inactivates it or causes it to behave abnormally. The effect of this ADP-ribosylation step on the host cell depends on the role of the protein that is ADP-ribosylated. For example, the A portion of DT ADP-ribosylates an unusual post-translationally modified amino acid, called diphthamide, on **elongation factor-2 (EF-2),** a protein that plays an essential role in host cell protein synthesis (Figure 12–8B). Thus, the effect of the A portion of DT is to kill the host cell by stopping protein synthesis. In contrast, the cholera toxin A portion ADP-ribosylates a regulatory GTP-binding protein that controls cyclic AMP levels in the host cell. This causes the host cell to lose control of ion flow and results in a massive loss of host cell water, which is seen macroscopically as diarrhea.

Not all A-B-type toxins have A proteins that catalyze ADP-ribosylation of host cell proteins. The Shiga toxin A subunit cleaves a host cell rRNA molecule. This results in a shutdown of protein synthesis, presumably because ribosomes containing the nicked molecule no longer carry out translation. BoNTs are zinc-dependent metalloproteases that cleave neuron-specific proteins. Examples of A-B toxins and their catalytic toxic activities are listed in Table 12–1.

Protein Exoenzymes/Effectors

Effector proteins (also called exoenzymes) are bacterial protein toxins that are directly injected into the host cell through specialized bacterial secretion systems that directly contact the host cell. These toxic proteins consist only of the catalytic components of toxins without the receptor-binding and translocation components. As indicated by the examples listed in Table 12–1, many of the toxic activities of these effector proteins are very similar to those of A-B toxins. The primary difference is that these protein toxins are nontoxic by themselves and require the secretion system for entry into host cells in order to exhibit their toxicity. We will learn more about the delivery of these toxins in chapter 13.

Examples of Diseases Mediated by Toxins

The best way to understand how toxins participate in the disease process is to examine a few examples of diseases in which toxins are the primary virulence factor. These examples also illustrate the continuing mystery of why bacteria produce toxins in the first place.

DT

Diphtheria. Diphtheria is caused by *C. diphtheriae*, a gram-positive, non-spore-forming, nonmotile, aerobic rod with a distinctive club-shaped appearance that occurs in V, L, or Y shapes and forms what is often

Figure 12–6 Structures of A-B toxins. **(A)** Single-chain A-B toxins have one A and one B domain. An example is DT, which is encoded by a corynebacteriophage and is secreted into the medium. The toxin then binds to a protein receptor on host cells and is internalized via receptor-mediated endocytosis. At some point, the toxin is proteolytically cleaved between the A and B domains, but the two fragments are held together by a disulfide bond. Upon lowering of the pH in the endocytic vesicles, the toxin undergoes a dramatic conformational change, inserts into the vesicle membrane, and translocates the catalytic A fragment into the host cell cytosol, where it ADP-ribosylates its intracellular target, EF-2. **(B)** Multisubunit complex A-B toxins have one or more A subunits and multiple B subunits. Examples are cholera toxin, which has one catalytic A subunit, which ADP-ribosylates the α subunit of the Gs type of heterotrimeric GTP-binding proteins, and five identical B subunits in its receptor-binding complex (left); pertussis toxin, which has one catalytic A subunit (named S1), which ADP-ribosylates the α subunit of the Gi and Go types of heterotrimeric GTP-binding proteins, and five B subunits (S2 to S5), only two of which are the same, in its receptor-binding complex (right); and anthrax toxin, which has two catalytic A subunits—edema factor (EF), which is an adenylyl cyclase, and lethal factor (LF), which is a zinc-dependent protease—and seven B subunits called protective antigen (PA83) that must be cleaved before forming a heptameric complex that binds up to three subunits of EF or LF. ATR, anthrax toxin receptor; CaM, calcium calmodulin; cAMP, cyclic AMP; MAPKKs, mitogen-activated protein kinase kinases.

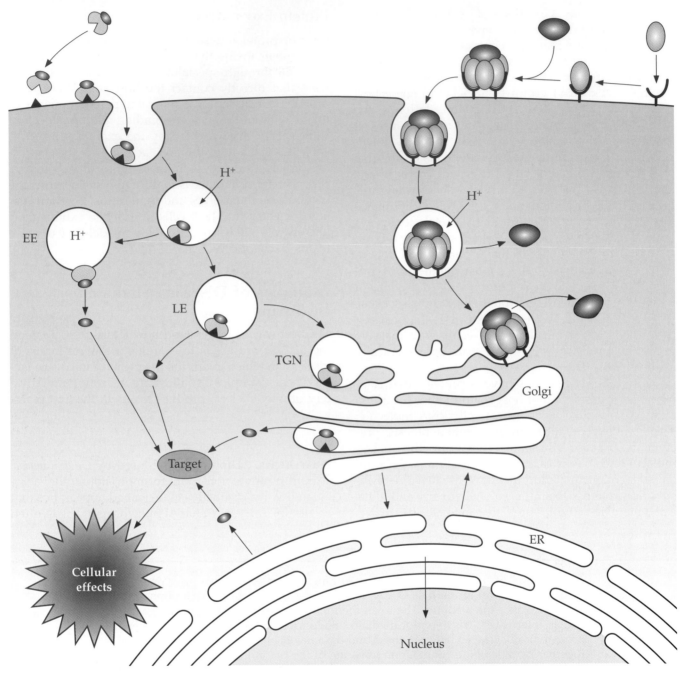

Figure 12–7 Binding and entry of A-B toxins into host cells. Toxin A subunits are in dark blue, and toxin B subunits are in light blue. There are two main mechanisms for cell entry by A-B toxins. Single-chain A-B toxins (e.g., DT and BoNTs) enter by binding to a cell surface receptor (usually a protein or glycoprotein), and their A domains are usually translocated through a pH-dependent process after endocytosis. Multisubunit A-B toxins (e.g., cholera toxin, pertussis toxin, and anthrax toxins) enter by binding to multiple cell surface receptors (usually gangliosides or glycolipids), and their A subunits are translocated either through a pH-dependent process after endocytosis or through retrograde transport to the endoplasmic reticulum (ER), where they are translocated through a secretory transport system into the cytoplasm. Some toxins even have nuclear localization signal sequences that target them to the nucleus. EE, early endosome; LE, late endosome; TGN, *trans*-Golgi network. (Adapted from B. A. Wilson and A. A. Salyers. 2002. Ecology and physiology of infectious bacteria—implications for biotechnology. *Curr. Opin. Biotechnol.* **13:**267–274, with permission from Elsevier.)

A

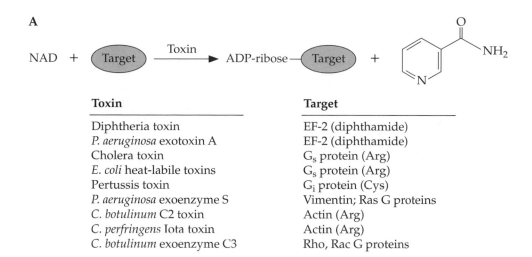

Toxin	Target
Diphtheria toxin	EF-2 (diphthamide)
P. aeruginosa exotoxin A	EF-2 (diphthamide)
Cholera toxin	G_s protein (Arg)
E. coli heat-labile toxins	G_s protein (Arg)
Pertussis toxin	G_i protein (Cys)
P. aeruginosa exoenzyme S	Vimentin; Ras G proteins
C. botulinum C2 toxin	Actin (Arg)
C. perfringens Iota toxin	Actin (Arg)
C. botulinum exoenzyme C3	Rho, Rac G proteins

B Diphtheria toxin-catalyzed ADP-ribosylation

Figure 12–8 ADP-ribosylation of a target host protein. **(A)** The ADP-ribosyl group is transferred from NAD to a side chain group of a host cell target protein. **(B)** DT-mediated ADP-ribosylation of the diphthamide residue on eukaryotic EF-2 is catalyzed by a glutamic acid at the toxin active site, which serves as a base to increase the nucleophilicity of the diphthamide.

described as "Chinese characters" when viewed under a microscope. Diphtheria is normally a disease of children that can be fatal if not treated. Diphtheria is no longer a serious public health problem in developed countries because infants are routinely vaccinated against it, but it remains a major cause of childhood death worldwide. The diphtheria vaccine is a toxoid version of DT, the "D" in the trivalent DTaP vaccine. This vaccine has few side effects and confers long-term protection against diphtheria. With periodic booster shots, the vaccine gives lifelong immunity. Unfortunately, there are still many countries where routine vaccination against diphtheria is not feasible for economic reasons. In these countries, diphtheria remains a significant cause of disease and death in infants and children. Recently, we have been re-

minded that diphtheria can also kill adults (see Box 12–1).

DT is one of the best-studied bacterial toxins and thus serves as a model for understanding how toxins are transmitted, regulated, and produced, as well as how they enter the cytoplasm of eukaryotic cells to elicit their toxic effects and even be exploited for good purposes. In fact, diphtheria provided the first paradigm for a disease in which the symptoms could be explained by the action of a single molecule, DT. This helped to give rise to the concept of the virulence factor and to usher in the new era of molecular analysis of bacterial disease. From that time to the present day, investigations of DT have continued to produce important new insights into toxin structure, function, and interaction with host cells. Also, there has long

BOX 12–1 Diphtheria as a Disease of Adults

In textbooks, diphtheria is generally described as a disease of children. This age distribution reflects the fact that in countries where there is no vaccination program, most people who survive infancy and childhood have acquired immunity to the disease. Under some circumstances, however, diphtheria can be a disease of adults. For example, during the diphtheria epidemics that occurred in the early days of European settlement of the Americas, both the colonists and the natives they encountered had a high adult mortality rate. George Washington is thought to have died of diphtheria at the age of 67. If most of the members of the population have had no prior exposure to diphtheria, adults do not have the protection of immunity acquired by experiencing the disease in childhood. Diphtheria appears to have been a completely new disease in the Americas. Surprisingly, at least some of the European settlers also did not have protective immunity, despite the fact that diphtheria was widespread in Europe.

An unpleasant reminder that diphtheria can kill adults occurred much more recently when diphtheria made an unexpected appearance during the early 1990s in the former Soviet Union, causing an epidemic in which many of the victims were adults. This epidemic illustrates another set of circumstances that can shift the age range of the disease. As the former Soviet Union began to break up after the end of the cold war and civil strife between various ethnic and religious groups became all too common, the public health programs that had been in place in previous years were disrupted, including vaccination programs. Also, defective lots of the vaccine were probably being used even in the days before the breakup of the Soviet Union.

The result was that many adults were not immune and were not protected by the herd immunity conferred by a population of vaccinated children. In some areas, the fatality rate was high for adults, as well as for children. It would be nice to be able to report that this common enemy caused the various warring factions within the former Soviet Union to forget their national and ethnic rivalries as they struggled against a nonhuman foe, but unfortunately, this did not occur. Disease-causing bacteria thrive on human stupidity.

Source: C. R. Vitek and M. Wharton. 1998. Diphtheria in the former Soviet Union: reemergence of a pandemic disease. *Emerg. Infect. Dis.* **4:**539–550.

---

been an interest in using DT as a prototype for targeted killing of tumor cells or, more recently, human immunodeficiency virus (HIV)-infected cells.

Diphtheria starts with colonization of the throat by *C. diphtheriae.* The bacteria are acquired by inhalation of aerosols from an infected person or an asymptomatic carrier. Humans are the only known reservoir of *C. diphtheriae.* The first symptoms are relatively nonspecific: malaise, low-grade fever, sore throat, and loss of appetite. Colonization of the throat causes considerable damage to the mucosal cells due to release of DT, which kills the exposed cells. The bacteria remain primarily in the throat; bacteria are not usually found in the bloodstream. As the bacterial colonization progresses, a grayish membrane begins to form in the throat and may extend into the lung. The membrane, called a **pseudomembrane,** consists of fibrin, bacteria, and inflammatory cells. The pseudomembrane adheres to underlying tissue, which bleeds when attempts are made to remove the pseudomembrane. This is a useful sign for diagnosing diphtheria, because pseudomembranes caused by other infectious agents are nonadherent. Also, the presence of the pseudomembrane and the bleeding of underlying tissue attest to the fact that bacterial colonization of the throat causes considerable damage to the mucosal cells. Sudden death can sometimes occur from suffocation when large sections of the pseudomembrane separate and block air passages.

DT is also released into the bloodstream and causes damage to internal organs, eventually resulting in organ failure. The most serious form of the disease is characterized by irregular heartbeat, difficulty in swallowing, stupor, coma, and finally death, usually from heart failure or respiratory paralysis. Although the throat colonization form of diphtheria is the most common, *C. diphtheriae* can also cause skin infections, and such infections are seen in countries where diph-

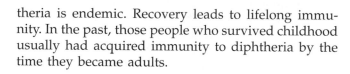

BOX 12-2 A Dog Named Balto, the Iditarod, and Diphtheria Antitoxin...What Is the Connection?

On 21 January 1925, several Inuit children in Nome, Alaska, were diagnosed with diphtheria. The only doctor in Nome, Curtis Welch, had only enough diphtheria antitoxin to treat up to five patients, but he knew that that would not be enough to prevent the rapid spread of the deadly disease throughout the isolated, ice-bound Nome community. Welch used a wireless telegraph to plead for more antitoxin. The only antitoxin available turned out to be in a hospital in Anchorage, almost 1,000 miles away. Ice prevented transport via ship, and blizzards prevented transport via airplane. On 27 January, a train was able to bring the antitoxin part of the distance, to Nenana, but after that, snow and ice prevented further travel. The only choice for delivering the antitoxin was to make use of an old dog sled mail and supply delivery route, called the Iditarod Trail, using relay teams of up to 20 dogs to travel the remaining 674 miles between Nenana and Nome, which was estimated to take about 13 days. More than 20 dog teams participated in this desperate race through blinding arctic blizzards and subfreezing (below –40°F) conditions. Ending the trek in a record 6 days, the last team that delivered the antitoxin in the wee hours of 2 February was led by a musher named Gunnar Kassen, with Balto as his lead dog.

News agencies outside of Alaska learned of Nome's plight from the telegraph message sent by Welch, and the story of the brave dog sled teams became front-page news all over America. The mushers received special hero medals, and a statue of Balto was erected in Central Park in New York City, where it still stands today to commemorate the heroic race that saved the epidemic-stricken children of Nome. The story of Balto even inspired a popular animated children's movie called *Balto*, produced by Steven Spielberg in 1995. The Iditarod Dog Sled Race, which was revitalized in 1973 and since then runs the entire 1,150 miles along the Iditarod Trail between Anchorage and Nome, has become one of Alaska's renowned sporting events, with the best mushers receiving thousands of dollars a year from corporate sponsors.

Source: http://www.iditarod.com.

theria is endemic. Recovery leads to lifelong immunity. In the past, those people who survived childhood usually had acquired immunity to diphtheria by the time they became adults.

DIPHTHERIA ANTITOXIN. In many diseases where exotoxins either are solely responsible for or make a major contribution to the symptoms of the disease, antibodies to the exotoxin provide effective protection against the disease. Binding of antibodies to the B portion of the toxin physically interferes with binding of the toxin to its target cell and thus prevents the toxin from exerting its toxic activity. Prior to the era of antibiotics, the only means to treat a toxin-mediated disease such as diphtheria was to neutralize the circulating toxin through a single massive injection of **antitoxin,** i.e., antibodies against DT obtained from horse serum, and hope that the host immune system could handle the rest. Even today, injection with antitoxin **(passive immunization)** is an effective means for treating toxin-mediated diseases, including diph-

theria and botulism (see below). For an interesting account of how a desperate race to deliver diphtheria antitoxin immortalized a dog and inspired a sport, see Box 12–2.

DIPHTHERIA VACCINE. Before any detailed information was available about toxin structures and mechanisms of action, vaccines were created by a hit-or-miss approach involving genetic mutations, chemical modification, or heat treatment of a toxin to render it nontoxic but still capable of eliciting antibodies that would bind and neutralize the toxin. Such preparations are called **toxoids.** The highly effective vaccines against diphtheria and tetanus are products of this type of approach. Today, it is possible to design **toxoid vaccines** in a more rational way. For example, the fact that the action of A-B-type toxins is dependent on binding of the B region to a host cell receptor molecule suggests that the B portions, which are not toxic by themselves, would make good vaccine candidates.

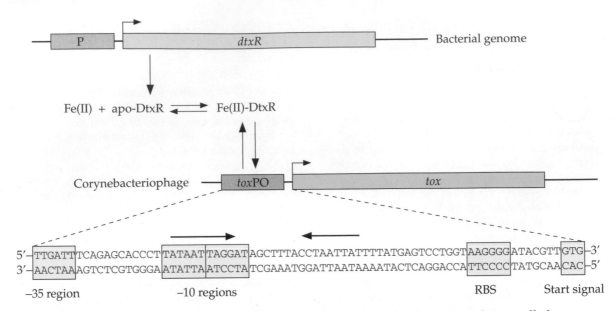

Figure 12–9 Regulation of DT expression. The product of a chromosomal gene called *dtxR* regulates the expression of DT. DtxR is a transcriptional repressor that binds to the operator region of the DT *tox* gene only in the presence of ferrous ions (Fe^{2+}). When the level of iron is high, the Fe(II)-DtxR complex binds the *tox*PO (*tox* promoter/operator) sequence and prevents (represses) transcription. When iron becomes the growth rate-limiting nutrient, which is the prevailing condition within the host, the complex dissociates, allowing transcription to proceed. Fe(II), Fe^{2+}. (Reprinted, with permission, from A. White, X. Ding, J. C. vanderSpek, J. R. Murphy, and D. Ringe. 1998. Structure of the metal-ion-activated diphtheria toxin repressor/*tox* operator complex. *Nature* **394:**502–506. Copyright [2008] Macmillan Publishers Ltd.)

The current vaccine against anthrax toxins is based on this approach.

C. diphtheriae can still colonize the throats of people who have been immunized against DT or who have become immune due to natural exposure to the bacteria. However, no pseudomembrane develops in the throats of immune people who are colonized by the bacteria. For this reason, formation of the pseudomembrane is presumed to result from killing of mucosal cells by DT. It is interesting that in people immunized against DT, the colonizing strain of *C. diphtheriae* usually does not produce toxin.

DT production. The gene encoding DT is carried on a group of related **lysogenic corynebacteriophages (β-phage and ω-phage).** Only strains infected with these phages produce the toxin (Figure 12–1). The fact that β- and ω-phages can be induced to become lytic facilitated cloning of the toxin gene. Isolation of phage DNA narrowed down the DNA segment carrying the gene and thereby enabled its identification. Also, another bacteriophage was found that was closely related to β- and ω-phages but did not encode the toxin **(γ-phage).** Comparison of the three phages helped to locate the toxin gene.

Production of toxin by lysogenized *C. diphtheriae* is enhanced considerably when the bacteria are grown in low-iron medium. A repressor protein called **DtxR (DT regulation),** which is related to an iron-regulatory protein of *E. coli* called **Fur (Fe utilization regulator),** mediates regulation of the DT gene (Figure 12–9). The fact that synthesis of the toxin is enhanced when iron levels are low has led to the suggestion that the purpose of DT may be to kill host cells and thus release iron for use by the bacteria. The only problem with this explanation is that strains that do not produce toxin colonize the human throat just as well as toxin-producing strains. For bacteria that reside only in human hosts, spread from person to person is essential. Perhaps, the symptoms caused by toxin production help to increase spread between people, especially in crowded situations.

Mechanism of DT action. Processing of DT occurs in two steps. The translated form of DT contains a leader region that is removed by proteolytic nicking during secretion of the toxin through the cytoplasmic membrane and into the extracellular fluid, producing a 58.3-kDa mature polypeptide. After secretion, this polypeptide is further cleaved by proteolytic nicking into an A fragment and a B fragment, which remain joined by a disulfide bond until translocation into the reducing atmosphere of the host cell cytosol occurs. The steps involved in binding, endocytic uptake, and translocation of the toxin are shown schematically in Figure 12–10.

The B fragment of DT is actually made up of two domains. One domain, called the **R domain,** binds a protein receptor on the host cell surface. This receptor has been identified as the **heparin-binding epidermal growth factor precursor (HB-EGF).** Epidermal growth factors of various types are important signals for cell growth and differentiation. Thus, it is not

Figure 12–10 Production, secretion, binding, endocytic uptake, and translocation of DT by eukaryotic cells. RSH, mercaptoethanol; ADPR, ADP-ribose; Nic, nicotinamic acid.

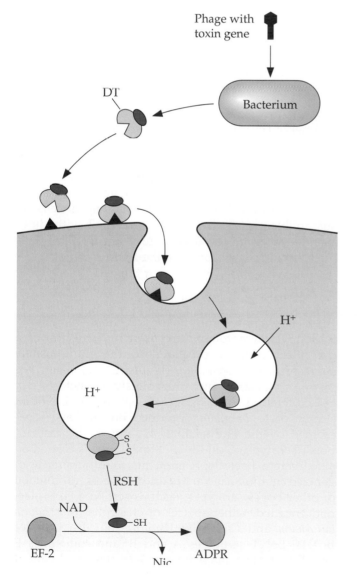

surprising that the receptor for the B fragment of DT is found on so many different cell types and also that its abundance varies so much from one cell type to another. A model for the structure of HB-EGF in the host cell membrane, which is based on the deduced amino acid sequence of the cloned gene, is shown in Figure 12–11. Now that the identity of the toxin receptor is known, it will be possible to investigate further the complex interaction that occurs between the B fragment and its receptor and how this affects the interaction between the toxin and the cell membrane during the translocation step.

After the toxin has bound to the host cell surface, the host cell takes up the toxin into an endocytic vesicle. Endocytosis is an important step in toxin action, because the decrease in pH that occurs in the endocytic vesicle after it is formed makes possible the translocation process. In the endocytic vesicle, as the pH drops to around 5, acidic amino acids are protonated, making them less hydrophilic. The change in charge distribution is associated with a change in the conformation of the toxin and also allows partial unfolding of both the A and B fragments, exposing hydrophobic regions that are normally found in the interior of the toxin. The exposed hydrophobic regions, together with stretches of protonated acidic residues on the T domain, allow regions of the T domain to insert into the vesicle membrane. The current model posits that partial unfolding of the A and B fragments and insertion of the helix-loop-helix regions of the T domain trigger the insertion of toxin through the membrane and expose the A fragment on the cytoplasmic side. Reduction of the disulfide bond that joins the A and B fragments then frees the A fragment, releasing it into the cytoplasm. An interesting feature of the T domain is that it is structurally related to the pore-forming toxins, and indeed, after translocation

Figure 12–11 Proposed model for insertion of HB-EGF, the receptor recognized by the toxin B fragment, in the host cell membrane.

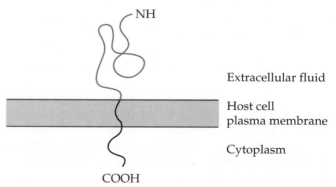

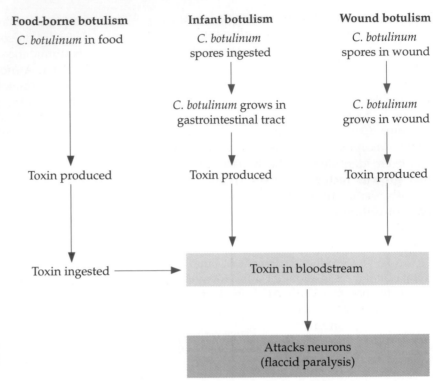

Food-borne botulism
C. botulinum in food

↓

Toxin produced

↓

Toxin ingested ⟶

Infant botulism
C. botulinum
spores ingested

↓

C. botulinum grows in
gastrointestinal tract

↓

Toxin produced

↓

Wound botulism
C. botulinum
spores in wound

↓

C. botulinum
grows in wound

↓

Toxin produced

↓

Toxin in bloodstream

↓

Attacks neurons
(flaccid paralysis)

Figure 12–12 Comparison of food-borne botulism, infant botulism, and wound botulism.

and release of the A fragment, the B fragment remains behind in the vesicle membrane as a pore.

In the cytoplasm, the A fragment catalyzes the ADP-ribosylation of EF-2, an essential factor in the protein synthesis machinery of eukaryotic cells (Figure 12–8B). EF-2 participates in the elongation step of protein translation. Attachment of an ADP-ribosyl group to EF-2 renders it inactive. A single molecule of the catalytic A fragment can ADP-ribosylate enough copies of EF-2 to halt protein synthesis completely and irreversibly. Ultimately, this causes the death of the cell. Attachment of the ADP-ribosyl group occurs at an unusual derivative of histidine called **diphthamide** [3-carboxyamido-3(trimethyl-amino)propyl histidine]. The modification of this particular histidine residue to form diphthamide occurs after EF-2 is translated and occurs in all types of eukaryotic cells (even *Archaea* have a deamidated version of this unusual residue, called **diphthine**). The role of diphthamide is not known, but the existence of this unusual form of histidine in EF-2 and not in other cellular proteins explains why DT specifically inactivates EF-2.

Despite this universality of action, different mammalian cell lines differ considerably in their susceptibility to killing by exogenously added DT. Differences in the susceptibilities of different cell lines can be explained by differences in the numbers of toxin receptors on the cell surface. These differences also suggest

why the disease has its most drastic effects on heart and nerve cells, whose surfaces contain high concentrations of the molecule bound by the B fragment. Preferential attack of these tissues by the toxin produces heart failure and neurological symptoms (e.g., difficulty swallowing), which are symptoms of the severe form of the disease. Not all mammalian cells are susceptible to DT. Despite having EF-2 with diphthamide, rats are remarkably resistant to DT. This is due to the fact that rats lack the toxin receptor.

Botulism and Tetanus

BOTULISM. Botulism is not an infection but an intoxication. Its symptoms are caused by ingestion of a neurotoxic exotoxin produced by the gram-positive spore former *C. botulinum* (Figure 12–12). There are seven serologically and antigenically distinct types of BoNT, referred to as serotypes A through G. Not all strains of *C. botulinum* produce BoNT. This is explained by the finding that, at least in the cases of BoNT serotypes C and D, the toxin gene is carried on a temperate bacteriophage. Thus, only strains lysogenized by the phage are toxin producers. Production of other BoNT serotypes, such as serotype G, is plasmid encoded. Other species of clostridia (*Clostridium butyricum* and *Clostridium baratii*) can also produce BoNTs. BoNT serotypes A and B, and sometimes E and F, are most often involved in natural forms of human botulism, while serotype C is a source of avian botulism, serotype E is commonly associated with fish and aquatic sources, and serotypes B and D are primary sources of equine botulism. BoNT serotype A is considered the most potent protein toxin known for humans, with an intravenous lethal dose estimated to be 1 to 5 ng/kg of body weight.

C. botulinum normally grows in soil or in lake sediments. Spores of *C. botulinum* have been found in soil samples taken from all parts of the United States and are widely distributed in all parts of the world. Spores are also found on plants growing in heavily contaminated soil. Since bees moving from plant to plant accumulate spores along with pollen, honey frequently contains *C. botulinum* spores. The concentration of spores in honey is not high (usually fewer than 7 spores per 25 g) but can be a source of problems for infants (see below). Contamination of human foods by *C. botulinum* spores with consequent ingestion of the spores is a common occurrence, but simple ingestion of spores is generally not sufficient to cause botulism. Germination of ingested spores may occur, especially in the colon, but *C. botulinum* is unable to compete with the resident microbiota of the colon and thus does not grow to high enough concentrations in the colon to cause disease.

Human botulism usually occurs when spores have the chance to germinate in foods, leading to bacterial growth and production of BoNT. Fortunately, *C. botulinum* does not grow readily in most foods. It is a semiobligate anaerobe, and most foods contain enough dissolved oxygen to discourage germination of the spores and growth of the bacteria. Growth is most likely to occur in foods, such as home-canned foods, that have been heated and then cooled and stored for long periods at room temperature. Heating reduces the solubility of oxygen, and most of the dissolved oxygen is lost. When the food is cooled, some oxygen redissolves, but lower regions away from the surface can remain anaerobic enough to support the growth of *C. botulinum*.

Many cases of botulism have been associated with consumption of home-canned foods. Temperatures achievable by boiling are not always sufficiently high to kill *C. botulinum* spores. This is why pressure cookers are used to sterilize canned foods. In home canning, however, the temperature in the pressure cooker sometimes does not reach the level required to kill spores, and surviving spores can germinate in the cooled canned food. Since canned foods are prepared in jars that are filled to the top and sealed before they are cooled, the environment in the sealed jars can be quite anaerobic. Even if BoNT is produced in a jar of canned food, botulism need not result, because boil-

ing the canned food for 10 to 15 minutes before eating it inactivates the toxin. Thus, most cases of botulism are associated with consumption of canned food without prior heating.

Ingested BoNT is absorbed from the stomach and enters the bloodstream. It is specific for neurons and attacks peripheral nerve endings at the neuromuscular junction and autonomic synapses. Inside the neuron, the toxin blocks neurotransmitter release. Thus, nerve impulses cannot be transmitted and muscles connected to the nerves are not stimulated. The result is a generalized flaccid paralysis, where the body goes limp. Symptoms begin to appear in 4 to 36 h after ingestion of contaminated food. The rapidity with which symptoms appear and the severity of the symptoms are directly proportional to the amount of toxin ingested. Initial symptoms include nausea and vomiting, as well as headache, double vision, slurred speech, and other neurological symptoms. Death occurs if the general flaccid paralysis is severe enough to interfere with breathing function, but nowadays, death seldom occurs due to medical intervention, such as the use of mechanical ventilators.

Currently, there is no effective antidote available for preventing or reversing paralysis once exposure to BoNT has occurred and symptoms have initiated. Once BoNT has bound to the surface of a neuron and entered it, external intervention with antitoxin becomes ineffective. Prompt administration of neutralizing antitoxin can only reduce the severity of disease and slow its progression. The good news is that the BoNT does not kill the nerve cell, so nerve regeneration will occur if the affected person survives the initial onslaught of toxin. Unfortunately, recovery from botulism is quite prolonged, with the length of time required to restore neuromuscular function depending on the type of nerve terminal affected and the serotype of BoNT involved. Recovery of the mammalian neuromuscular junction after BoNT serotype A intoxication typically requires 4 to 6 months, but it can take much longer (>1 year) for the human autonomic nervous system.

Although most cases of botulism are due to ingestion of preformed toxin, two rare forms of botulism actually involve transient colonization of the body by the bacteria, followed by toxin production: **infant botulism** and **wound botulism** (Figure 12–12). Infants less than 1 year old have not yet developed a complete colonic microbiota. Since the colon provides an anaerobic environment, *C. botulinum* can sometimes colonize the infant colon and cause infant botulism. Almost all cases of infant botulism are caused by serotypes A and B. The symptoms of infant botulism are essentially the same as those of food-borne botulism, except that nausea and vomiting are not seen and constipation, which is not usually a symptom of botulism, is common. The fecal microbiota of infants with botulism contain *C. botulinum* in relatively high numbers, exceeding 3% of the total microbial content, presumably resulting from in vivo spore germination and bacterial outgrowth due to the slow progression of the fecal contents. Infant botulism has a slower onset than food-borne botulism, with an incubation period of 3 to 30 days from the time of first exposure, but death can still occur. Because treatment with antitoxin derived from horse serum can cause hypersensitivity, special human-derived antitoxin, called BabyBIG (botulism immunoglobulin) is available for infant botulism through the California Department of Health. The fee for obtaining this antitoxin is quite high ($45,000), but with supportive care, treatment with BabyBIG can cut a hospital stay in half.

C. botulinum can also colonize a deep wound, because the environment of such a wound is anaerobic enough to allow *C. botulinum* to grow (Figure 12–12). A wound becomes anaerobic because tissue destruction cuts off the blood supply to the area and residual oxygen is rapidly depleted by body cells. BoNT leaks into the blood from the wound area and causes a disease called wound botulism, which has symptoms similar to those of food-borne botulism. Wound botulism is normally quite rare and is most often seen in wartime. The few civilian cases have occurred in people with severe, deep wounds that were heavily contaminated with soil. Wound botulism should be differentiated from **tetanus,** a neurotoxin-associated disease caused by another *Clostridium* species, *C. tetani* (see below).

Recently, a new form of botulism has emerged. With the increased prevalence of BoNT/A (commercially sold as Botox) and BoNT/B (commercially sold as Myobloc) for therapeutic use to treat neuromuscular disorders or for cosmetic applications (see Box 12–3), inadvertent injection-related botulism has shown a marked increase in occurrence (see Box 12–4 for a twisted tale of the perils of searching for Botox beauty at the hands of a greedy boyfriend).

BoNT. The specificity of BoNT for peripheral neurons arises both from the specificity of toxin binding (due to ganglioside and protein receptors found only on neuronal cells) and from the neuron-specific action of the toxin (Figure 12–13). The bound toxin is internalized and inhibits the release of **acetylcholine (ACh),** a neurotransmitter. BoNT is a large protein (150 kDa) that is part of an even larger complex containing other proteins besides the toxin. The complex is called **progenitor toxin,** and the toxin itself is called **derivative toxin.** Derivative toxin is less effective than progenitor toxin when given orally but is as active as progenitor toxin when injected. This has led to the suggestion that the nontoxic components of progenitor toxin help to protect the derivative toxin from stomach acid and proteases. They could also help the toxin bind to and transit the stomach mucosal surface.

Derivative toxin is originally synthesized and secreted from the bacteria as a single 150-kDa protein, which is cleaved into two protein fragments, the heavy chain (100 kDa) and the light chain (50 kDa), which are connected by a disulfide bridge (Figure 12–13). The binding site of the toxin for its receptors is located near the carboxy-terminal region of the heavy chain. The toxin is taken up by endocytosis into neuronal cells. Following acidification of endosomes containing the toxin, the amino-terminal region of the heavy chain is thought to form a channel in the endosome membrane, allowing the catalytic light chain to enter the cytoplasm of the cell. BoNT catalytic light chains are endoproteases that possess a highly conserved zinc-binding motif that is critical for catalytic activity. The intracellular targets of BoNTs are neuron-specific **SNARE proteins** involved in synaptic vesicle fusion, which releases the ACh neurotransmitter. BoNT serotypes B, D, F, and G, as well as the related tetanus neurotoxin (TeNT) (see below), cleave **synaptobrevin** (also called **VAMP**), a membrane protein found in small synaptic vesicles. BoNT serotypes A and E cleave **SNAP25,** and BoNT serotype C specifically cleaves **syntaxin.** SNAP25 and syntaxin are proteins associated with the plasma membrane that form a helix bundle complex with synaptobrevin and thereby facilitate vesicle-plasma membrane fusion. The net effect of SNARE protein cleavage is that vesicle fusion does not occur, the ACh neurotransmitter is not released to synapses, muscles stop contracting, and flaccid paralysis sets in.

TETANUS. Tetanus is a toxin-mediated disease caused by another clostridial species, *C. tetani*. Tetanus is also known as "lockjaw," because people with classical tetanus suffer muscle spasms that can lock the jaws together, cause stiffening of many muscles in the body, and ultimately cause death. In one sense, tetanus is the opposite of botulism, because whereas in botulism the patient suffers a flaccid paralysis, in tetanus, the patient develops a spastic paralysis, in which the muscles contract and do not relax. Tetanus starts when the bacteria colonize a deep puncture wound that becomes anoxic due to tissue damage in the area. *C. tetani,* an obligate anaerobe, can grow under these conditions and produce TeNT, which diffuses away from the wound site and enters the blood-

BOX 12–3   Bacterial Toxins—Double-Edged Swords—Turning the Bad Guy into a Good Guy

The term "toxin" has very negative connotations, yet scientists have managed to find beneficial uses for bacterial toxins. Since toxins often manipulate eukaryotic-cell pathways that are important for reacting to external stimuli or for intracellular communication, eukaryotic-cell biologists have found some toxins very useful for elucidating regulatory pathways. By poisoning certain pathways, toxins allow biologists to interrupt these pathways biochemically and thereby study the importance of particular steps in the pathway for cellular functions. Studies of this type have provided much information about the functioning of eukaryotic cells, information that should prove useful in a variety of medical areas.

Several years ago, the white knight of toxin-based therapies was DT, and many articles appeared touting numerous innovative applications of the toxin to cure everything from HIV disease to cancer. These applications are still in the developmental phase, but another toxin has rushed in to fill this niche: BoNT. For a number of years now, BoNT (Botox) has been used to treat painful, disabling muscle spasms of various types (dystonias). Examples of such dystonias are strabismus (crossed eyes) and painful spasms of the face and neck. The FDA has approved Botox therapy as a general therapy for dystonia. Dystonias were once thought to be psychosomatic in origin, but attempts to treat them using various forms of psychotherapy were unsuccessful. Their cause is still not known. Until the advent of Botox therapy, the only therapy that worked was surgical destruction of the nerve endings in the area affected by the spasms. Surgical treatment was expensive, dangerous, and not invariably successful.

Botox therapy is generally acknowledged to be far superior to surgical destruction of nerve endings and thus represents the first successful therapy for dystonias. Since BoNT prevents transmission of nerve impulses, small injections of toxin in the affected area counter the spasms and give temporary relief. The injections must be repeated at intervals, because some regeneration of the nerve endings occurs.

Botox has also found favor with cosmetic surgeons, who are using it to reduce the depth of wrinkles. More recently, Botox has been administered to cerebral palsy patients in an effort to help them control the movement of their limbs. The next time you read an article about bioterrorists threatening to dump BoNT into your water supply, remember that BoNT can also be your friend.

Sources: R. M. Kostrzewa and J. Segura-Aguilar. 2007. Botulinum neurotoxin: evolution from poison, to research tool—onto medicinal therapeutic and future pharmaceutical panaceae. *Neurotox. Res.* **12:** 275–290; J. Carruthers and A. Carruthers. 2007. The evolution of botulinum neurotoxin type A for cosmetic applications. *J. Cosmet. Laser Ther.* **9:**186–192.

---

stream. TeNT acts on neurons that control the neural feedback that tells flexed muscles to relax after having performed a task. In effect, it prevents these neurons from signaling the relaxation after a muscle contraction, leading to the spastic paralysis that characterizes tetanus.

Tetanus was never a very common disease in the developed world, but it was much feared because of its high fatality rate and gruesome symptoms. Victims develop painfully bowed spines, clenched arm and leg muscles, and locked jaws. Tetanus is almost invariably fatal. In developed countries today, tetanus is virtually unknown because of the tetanus vaccine (the T in the common DTaP shot). Worldwide, tetanus remains a major cause of infant death. In any listing of the top killers of infants in the developing world, infant tetanus is in the top 15, perhaps not so common as HIV or diarrheal disease, but still pervasive. The custom in many parts of the world of packing the umbilical stump of a newborn with cattle dung is doubtless the major cause of this grim statistic. Possibly, animal dung prevents some other condition, but it is also a prime source of *C. tetani*.

DIFFERENT TOXINS, SAME MECHANISM. For years, botulism and tetanus were treated as completely different diseases, even though the toxins responsible for the symptoms were both neurotoxins. Thus, it came

BOX 12–4   A Lesson in Why Not To Trust a Greedy Beau with Your Beauty

The Botox craze has intoxicated (pun intended) many a middle-aged glamour seeker desirous of a youthful glow. No longer restricted to the Hollywood stars, Botox injections are a new and booming market. However, the demand for this rejuvenating cosmetic has spurred some dubious activities, as well. Botox is manufactured by Allergan, Inc., which is the only firm licensed by the FDA to do so. Botox is normally sold in 100-unit (five-dose) vials at a cost of $1,000 to $2,000 per vial, and one treatment lasts for 4 to 6 months. The problem is that some individuals do not want to pay that much for a cosmetic, which has led to sales of non-FDA-approved Botox knockoffs. Consequently, there has been a rise in the number of cases of misuse of the toxin.

One such case occurred in 2004, when an osteopathic physician working at a health salon in Florida injected his coworker and girlfriend, Alma Jane Hall, and another couple with a non-FDA-approved Botox knockoff. The injections were several thousand times over the dose normally administered for antiwrinkle treatment. Shortly after the treatments, Hall and her boyfriend traveled to New Jersey to visit his family for the Thanksgiving holiday. Hall had started to show symptoms of botulism on the plane trip, but her boyfriend declined to get her help and instead attempted to treat her with ice packs and then an ice bath at his mother's house before a family member finally called 911. Once it was determined that she indeed had botulism, Hall received antitoxin and was transferred to the University of Medicine and Dentistry of New Jersey under the care of Steven Marcus, Professor of Preventive Medicine and Community Health and executive director of the New Jersey Poison Information and Education System (i.e., the New Jersey Poison Control Center). Hall's ordeal and recovery period lasted more than 2 years, all of which was thoroughly documented (with Hall's consent) and has provided enormously valuable information for medical practitioners and biomedical researchers. Back in Florida, the other two victims fell seriously ill that same weekend and were hospitalized, where they received antitoxin and supportive care. The boyfriend, who claimed to have also injected himself and was also hospitalized in New Jersey, received 3 years in prison for "causing reckless harm to others." Oh, and he is no longer Hall's boyfriend.

Source: A. Kuczynski. 2004. Is it Botox, or is it bogus? *The New York Times.* http://www.nytimes.com/2004/12/05/fashion/05BOTU.html.

---

as a surprise when examination of the deduced amino acid sequences of the two toxin genes revealed that the proteins shared a considerable amount of sequence similarity with zinc-requiring endopeptidases. Ultimately, this insight from sequence gazing led to the discovery that TeNT, like some of the BoNT serotypes, cleaves synaptobrevin—but if the two toxins have exactly the same mode of action, how could they cause such different types of effects (flaccid paralysis versus spastic paralysis)? The answer appears to lie in the different cell specificities of the binding regions of the two toxins. BoNTs target peripheral neurons. If TeNT is administered so that it acts only locally on peripheral nerves, it too causes flaccid paralysis. In large enough quantities, and especially if it enters the circulation, however, TeNT acts on the central nervous system and causes spastic paralysis. TeNT also targets different neuronal receptors than the BoNTs. Although TeNT and BoNTs bind to the neuromuscular junction, their intracellular actions take place at different levels of the nervous system. Unlike BoNTs, TeNT undergoes retrograde transport to the cell bodies of spinal cord motor neurons, where it is translocated and cleaves synaptobrevin in inhibitory synapses. This differential trafficking within the neuron has been interpreted to be a consequence of binding to different surface receptors, which trigger sorting to different locations in the cell.

The discovery of the targets of these two toxins illustrates a common theme in toxin research: bacterial toxins can become powerful tools for the study of mammalian cell function. The finding that TeNT and BoNTs cleave different neuron-specific SNARE proteins excited researchers interested in neurobiology, because the functions of the SNARE proteins had previously been unknown. Thus, TeNT and BoNTs pro-

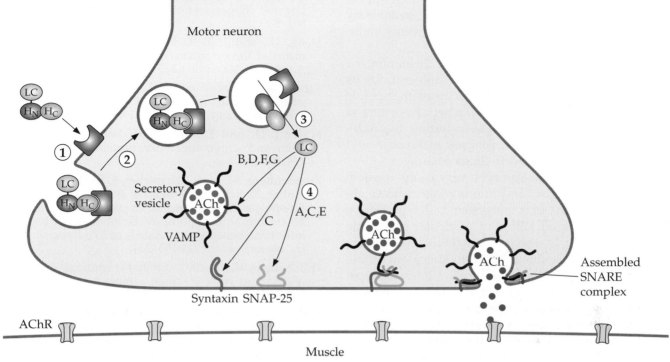

Figure 12–13 Action of BoNTs at the motor neuron synapse. The action of BoNTs occurs in four steps: (1) binding of the C-terminal portion of the heavy chain (H_C) to presynaptic membrane receptors, (2) uptake of the toxin into an endocytic vesicle, (3) translocation of the catalytic light chain (LC) into the cytosol via the N-terminal portion of the heavy chain (H_N), and (4) BoNT-LC-mediated proteolytic cleavage of neuron-specific SNARE proteins involved in synaptic vesicle fusion with the plasma membrane, which prevents membrane fusion and release of the neurotransmitter, acetylcholine (ACh). LCs from different BoNT serotypes target different SNARE proteins. (Adapted from K. Turton, J. A. Chaddock, and K. R. Acharya. 2002. Botulinum and tetanus neurotoxins: structure, function, and therapeutic utility. *Trends Biochem. Sci.* **27**:552–558, with permission from Elsevier.)

vided a new reagent that could help us to understand how these proteins function.

Immunotoxins: Toxin-Based Therapeutics and Research Tools

Because of DT's potent cell-killing ability, it was used as the basis for new therapeutic approaches against cancer cells or virus-infected cells. The prototype for these approaches used hybrid fusion proteins between the catalytic domain (the A fragment) of DT and receptor-binding proteins, such as antibodies or receptor ligands that selectively target the fusion protein to particular cancer cells or virus-infected cells. These hybrid proteins are referred to as **immunotoxins** because they contain a cell-killing part (the A fragment of DT) and a cell-targeting part (an antibody or receptor ligand). For example, the A fragment of DT fused

with IL-2, which could recognize, enter, and kill IL-2 receptor-expressing lymphoma and leukemia cells, showed good response in phase I and II clinical trials. In other trials, a conjugate of human transferrin with the A fragment of DT showed good clinical efficacy when it was administered intratumorally or when it was infused into the brain tissue surrounding a malignant glioma (brain tumor).

From clinical trials, researchers have learned that leukemia cells in the blood are exposed to high immunotoxin concentrations. In addition, the immune system is impaired in many hematologic malignancies. Therefore, since many cycles of immunotoxin therapy can be given without antibody formation, immunotoxins are more effective against leukemia-like diseases. For solid tumors, immunotoxins can reach the cells within the tumors only by slow diffusion and in low concentrations, except when directly injected into the center of the tumor. Also, the immune system is still intact in patients with solid tumors, so antibody formation against the immunotoxins prevents multiple treatments.

Unfortunately, because most people are immunized against DT and have antibodies that can neutralize its activity, use of DT is not ideal for most therapeutic applications. Consequently, researchers have turned to other toxins as alternatives, including the ADP-ribosylating domain of *Pseudomonas* exotoxin A and ricin (a plant toxin related to Shiga toxin).

Immunotoxins have also been very useful research tools for cell biologists and immunologists, since ablation (elimination) of a particular cell type can be achieved by targeted killing of particular cells using the immunotoxin. In developmental biology, researchers have also used protein toxins to knock out certain cell types by expression of the toxin under cell-type-specific mammalian promoters that turn on only during certain stages of development. Elimination of that cell type can provide information about the role and importance of the cell type in development.

SELECTED READINGS

Alfano, M., C. Rizzi, D. Corti, L. Adduce, and G. Poli. 2005. Bacterial toxins: potential weapons against HIV infection. *Curr. Pharm. Des.* **11:**2909–2926.

Alouf, J. E., and M. R. Popoff (ed.). 2006. *The Comprehensive Sourcebook of Bacterial Protein Toxins*, 3rd ed. Academic Press, London, United Kingdom.

Burns, D. L., J. T. Barbieri, B. H. Iglewski, and R. Rappouli. 2003. *Bacterial Protein Toxins*. ASM Press, Washington, DC.

Debinski, W. 2002. Local treatment of brain tumors with targeted chimera cytotoxic proteins. *Cancer Invest.* **20:**801–809.

Deng, Q., and J. T. Barbieri. 2008. Molecular mechanisms of the cytotoxicity of ADP-ribosylating toxins. *Annu. Rev. Microbiol.* **62:**271–288.

Fabbri, A., S. Tarvaglione, L. Falzano, and C. Fiorentini. 2008. Bacterial protein toxins: current and potential clinical use. *Curr. Med. Chem.* **15:**1116–1125.

Fraser, J. D., and T. Proft. 2008. The bacterial superantigen and superantigen-like proteins. *Immunol. Rev.* **225:**226–243.

Freudenberg, M. A., S. Tchaptchet, S. Keck, G. Fejer, M. Huber, N. Schütze, B. Beutler, and C. Galanos. 2008. Lipopolysaccharide sensing an important factor in the innate immune response to Gram-negative bacterial infections: benefits and hazards of LPS hypersensitivity. *Immunobiology* **213:**193–203.

Holzheimer, R. G. 2001. Antibiotic induced endotoxin release and clinical sepsis: a review. *J. Chemother.* **1:**159–172.

Hong, H., C. Demangel, S. J. Pidot, P. F. Leadlay, and T. Stinear. 2008. Mycolactones: immunosuppressive and cytotoxic polyketides produced by aquatic mycobacteria. *Nat. Prod. Rep.* **25:**447–454.

Iacovache, I., F. G. van der Goot, and L. Pernot. 2008. Pore formation: an ancient yet complex form of attack. *Biochim. Biophys. Acta* **1778:**1611–1623.

Kreitman, R. J., and I. Pastan. 2006. Immunotoxins in the treatment of hematologic malignancies. *Curr. Drug Targets* **7:**1301–1311.

Rainov, N. G., and A. Söling. 2006. Clinical studies with targeted toxins in malignant glioma. *Rev. Recent Clin. Trials* **1:**119–131.

Tweten, R. K. 2005. Cholesterol-dependent cytolysins, a family of versatile pore-forming toxins. *Infect. Immun.* **73:**6199–6209.

QUESTIONS

1. The B subunits of A-B toxins are used as vaccine components without any chemical treatment. Why is it safe to use B subunits as they are?

2. What would happen if you injected the A chain of DT into your bloodstream? Could the A chain be used as a vaccine component?

3. How could you use the DT A subunit to kill cancer cells, as some scientists are trying to do? What would be the advantages of such a treatment? What would be the disadvantages?

4. Why is vaccination usually the best way to prevent diseases caused by A-B toxins?

5. Could vaccination help to prevent botulism, gangrene, or streptococcal cellulitis? Is vaccination a feasible solution to these problems?

6. Can you think of any way that BoNT or TeNT could aid the bacterium that produces it to live a longer and fuller life?

7. How can a toxin that has a single target inside a mammalian cell produce different symptoms? Use two of the toxins listed in Table 12–1 as examples.

8. How do A-B toxins resemble or differ from cytotoxins that affect bacterial membrane structure or superantigens?

9. What is the significance of the fact that subcytotoxic doses of membrane-active toxins might contribute to septic shock?

10. How does endotoxin differ in its mode of action from an A-B-type toxin?

11. If a science writer or media interviewer asked you to explain briefly what a toxin is, how would you answer that question? That is, are there any unifying features of toxins that can be easily understood?

12. For the following statements, indicate whether the statement is true or false:

Bacterial endotoxin is the lipid A moiety of LPS and acts by inhibiting the host inflammatory response.

A few microorganisms are pathogenic solely because of the toxins they produce. (In other words, invasion of the host tissue is not required for pathogenicity.)

The pertussis toxin is the cause of the characteristic paroxysmal cough seen in patients with whooping cough.

Cholera toxin causes diarrhea by destroying G proteins.

13. The staphylococcal enterotoxin and toxic shock syndrome toxin are known as _____ that can nonspecifically stimulate large populations of T cells to produce _____.

14. Botulinum toxin is a _____ that binds to the synapses of motor neurons, cleaves _____, and prevents the release of the neurotransmitter, _____.

15. Toxins that promote nonspecific interaction of MHC-II cells and T cells and cause excessive activation of T cells are known as _____.

16. Will heating food contaminated with BoNT protect you from food-borne disease? Why or why not?

SOLVING PROBLEMS IN BACTERIAL PATHOGENESIS

1. Suppose you have two A-B-type exotoxins, A-B and A'-B'. A-B binds specifically to neurons and prevents a neuron-specific function, which is essential for nerve pulse transmission. A'-B' binds specifically to kidney cells and stops protein synthesis.

A. What would a hybrid toxin, A'-B, do to kidney cells? Provide your rationale.

B. What would a hybrid toxin, A-B', do to kidney cells? Provide your rationale.

C. What would a hybrid toxin, A'-B, do to neurons? Provide your rationale.

D. What would a hybrid toxin, A-B', do to neurons? Provide your rationale.

2. Suppose you have two A-B-type exotoxins, A-B and A'-B'. A-B binds specifically to all MHC-II$^+$ cells and blocks vesicle trafficking. A'-B' binds specifically to all MHC-I$^+$ cells and stops protein synthesis.

A. What would a hybrid toxin, A'-B, do to fibroblast (skin) cells? Provide your rationale.

B. What would a hybrid toxin, A-B', do to fibroblasts? Provide your rationale.

C. What would a hybrid toxin, A'-B, do to macrophages? Provide your rationale.

D. What would a hybrid toxin, A-B', do to macrophages? Provide your rationale.

3. Give a possible set of bacterial virulence factors and what properties they have that would explain each of the following observations:

A. A 50% infectious dose value of 10^5 but a 50% lethal dose (LD$_{50}$) value of 10.

B. An LD$_{50}$ value of 10^5 in mice but an LD$_{50}$ value of 10 in humans under similar exposure conditions.

C. Development of an autoimmune response after the infection.

4. In this chapter, we have studied a number of toxins that are responsible for many of the disease symptoms associated with potent pathogens. For each of the following toxins, describe the mechanism of action on host cells in terms of the structure of the toxin, how it gains access to its target, what it does to its target, and the outcome.

A. DT

B. Shiga toxin

C. *E. coli* heat-stable toxin

D. Cholera toxin

E. Superantigen

F. YopE

G. Listeriolysin O

5. You are a researcher characterizing two newly discovered toxins. You have been able to generate research data from some preliminary studies that will allow you to further determine how these toxins function.

(continued)

Toxin 1: In cell culture experiments, you find that when you add the purified toxin to host cell monolayers, the host cells die. When you fluorescently label this toxin and add it to cultured cells, you find it as multimeric complexes within host cell membranes by fluorescence microscopy. Based on these results, answer the following questions.

A. What type of toxin have you identified?

B. How does your toxin function to kill host cells?

Toxin 2: In cell culture experiments, when you add the purified toxin to host cell monolayers, the host cells die. The toxin-infected host cells do not lyse, but the cells have collapsed and show a significant loss of water and ions. The toxin is translated and produced by the bacterium as a single polypeptide. However, cellular fractionation of the host cells into cytosolic and membrane fractions reveals that a portion of the toxin is present in the membrane fraction and another portion is in the cytosolic fraction. Based on these results, answer the following questions:

C. What type of toxin have you identified?

D. How does this toxin function to kill host cells?

6. Both DT and anthrax toxin have been used as prototypes for the development of novel therapeutics.

A. DT was the first toxin to be exploited as an immunotoxin, and many other toxins have since been used as immunotoxins for specific purposes. Name at least two problems with the use of DT as an immunotoxin. What strategies have been employed to overcome these limitations?

B. Describe how you would design an immunotoxin based on DT that might work to cure a tumor that overexpresses on its surface the ATR (anthrax toxin receptor) protein, which has an extracellular von Willebrandt factor A domain that binds to anthrax protective antigen.

7. You have just isolated a new bacterium from a recent hospital outbreak that, based on 16S rRNA comparison, is closely related to *Acinetobacter baumannii*, which you name *Acinetobacter newbii*. *A. baumannii* is a gram-negative, opportunistic, nosocomial (hospital-acquired) pathogen that is able to colonize patients in intensive-care units, causing pneumonia, urinary tract infections, septicemia, or meningitis, depending on the route of infection. Epidemic strains of *A. baumannii* are often resistant to multiple drugs. Using in vivo expression technology in a swine lung model of infection, you have identified two operons that may encode potential virulence factors.

A. One operon consists of two genes in tandem, which you have named *antBA*. Sequence compari-

son with protein databases reveals that *antA* encodes a protein of 25 kDa that has significant sequence similarity to the catalytic N-terminal 200 amino acid residues of DT, while *antB* encodes a protein of 10 kDa that has significant sequence similarity to the B subunit of cholera toxin. You suspect that the Ant protein is an AB-type toxin and decide to test this using a swine lung tissue cell model, and, indeed, the Ant protein is cytotoxic to the cells. Further studies reveal that Ant action on lung cells requires a low-pH-dependent step, since agents that prevent acidification, such as basic buffers, block the toxin effects. You make antibodies against the culture filtrate from the *A. newbii* cells and use the antisera in Western blots of the cell lysates from the bacterial cells, the media, and the intoxicated lung cells after separation of the proteins by sodium dodecyl sulfate-polyacrylamide gel electrophoresis. The results are shown in the figure below. Provide a detailed explanation (with rationale) and interpretation for the results shown in the Western blot in the figure. Was your conclusion that Ant is an AB-type toxin correct? Explain.

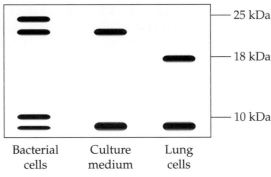

B. Draw a diagram depicting a possible model that accounts for all of the experimental observations, i.e., predict the structure of the toxin and show the mode of action of the toxin. Be sure to justify your model; it must be consistent with your answer to part A.

C. The second operon consists of four genes in tandem, which you have named *anrCDEF*. The genes *anrCDE* have no sequence homology to any genes in the known databases and also have no obvious sequence motifs that could be used to determine their functions. However, analysis of the protein sequence encoded by *anrF* revealed an interesting pattern. Shown in the figure below is a hydrophobicity plot of the predicted AnrF protein. Based on this information, you predict that AnrF might be a pore-forming toxin. What was your rationale for this prediction? Design an experiment that would allow you to determine if your prediction is correct.

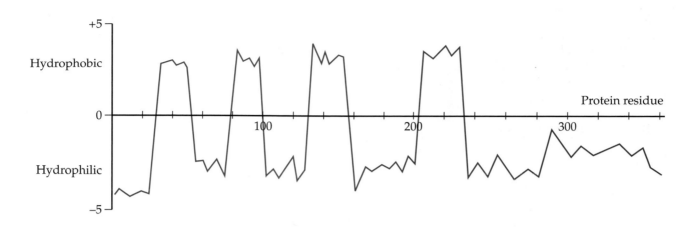

13

Delivery of Virulence Factors

P athogenic bacteria are passive-aggressive. While it is true that in many cases they are passive in the sense that they simply respond to environmental signals, they are also aggressive in that they act to control their external environment in ways that benefit them. An important mode of aggression is the secretion of proteins that allow the bacteria to stick to surfaces or to kill mammalian cells. Of course, all bacteria use protein secretion to some extent to control their surfaces and their environments, but bacterial pathogens seem to be particularly adept at aggressive behavior, even having specialized injection systems to pump their toxic proteins directly into eukaryotic cells.

Bacterial Secretion Systems and Virulence

To deliver their virulence proteins, bacteria must have ways to transport the proteins from the bacterial cytoplasm to the extracellular environment. Although transport of proteins to the cell surface and secretion of proteins into the medium are normal processes that all bacteria have, secretion of virulence factors is itself a virulence determinant of pathogenic bacteria. Without a means to selectively deliver their virulence factors, many bacteria are not virulent. Thus, most bacterial pathogens have evolved specialized, and often dedicated, protein secretion systems for the transport of adhesins, toxins, exoenzymes, proteases, and other virulence factors out of the cytoplasm onto the cell surface; for release into the medium; or for direct delivery into host cells.

Because gram-negative bacteria have two lipid bilayers (see Box 3–1) across which proteins must be transported in order to be secreted into the extracellular medium, the protein secretion systems of gram-negative bacteria are more numerous, more complicated, and more diverse than those of gram-positive bacteria. Protein export from the cytoplasm across the cell membrane in gram-positive bacteria or the inner membrane of gram-negative bacteria is mediated by some secretion systems that are common to both, so we will begin with those. In the case of gram-negative bacteria, the secreted proteins end up in the periplasm, and additional secretion systems must be utilized to further transport the proteins across the outer membrane. Gram-negative bacteria also have other specialized secretion systems that do not utilize the general secretion systems common to both types of bacteria. These additional secretion systems have different sets of secretion proteins that enable transport across both the inner membrane and the outer membrane. Finally, we will consider some usual features of the protein export systems used by several gram-positive bacteria, including secretion systems that allow protein effectors to be transported through hydrophobic surface layers.

Common Secretory Systems

The **general secretory system (Sec system)** is common to gram-positive and gram-negative bacteria and has been well characterized in *Bacillus subtilis*, *Escherichia coli*, and several other eubacterial species. The secreted proteins are synthesized as precursors with a 15- to 26-amino-acid **signal sequence** (also called a **leader sequence**) at the N terminus, which contains 10 to 12 hydrophobic residues preceded by 1 or 2 charged residues and followed by a protease cleavage site. The signal sequence is cleaved by a leader peptidase during the process of transport from the cytoplasm across the membrane. The hydrophobic residues of the leader sequence trap the cleaved peptide in the membrane. Mutations in the protease cleavage site do not prevent transport across the membrane but result in the protein being tethered to the membrane instead of being released into the medium or periplasm. In some cases, lipids are attached to proteins at a cysteine residue located downstream of the cleavage site. Following export and cleavage by a different leader peptidase than that used for exported proteins, these lipoproteins remain anchored to the cell membrane.

The signal sequence is required for the proteins to be recognized and transported by the Sec system (Figure 13–1). The hallmark of the Sec system is that it exports unfolded proteins posttranslationally. Two cytoplasmic proteins are responsible for binding and directing the protein to be secreted. **SecB** is a protein chaperone that binds to proteins containing signal peptides as they exit the ribosome during translation. The SecB tetramer contains a chamber that is thought to bind segments of the proteins and keep them unfolded until they can be delivered to **SecA**. The SecA dimer is a molecular motor whose ATPase activity provides part of the energy for the translocation process (the rest of the energy seems to be provided by the proton motive force across the membrane). SecA binds to the heterotrimeric **translocase complex** composed of **SecYEG**, which forms a **protein-conducting channel** through the cellular membrane. SecB delivers the unfolded protein to the ADP-bound SecA-SecYEG complex. Additional proteins (SecD, SecF, and YajC) help to stabilize the entire complex. The amino terminus of the protein containing the signal peptide is then threaded through a pore in SecA and remains bound to SecA at its interface with SecYEG. The bound ADP is exchanged for ATP, which induces a conformational change that releases SecB from the complex. ATP hydrolysis is then coupled to driving the polypeptide chain through the channel in SecYEG. The signal peptide remains tethered in the complex and is cleaved by the activity of a leader peptidase, after which the peptide is released and refolds. ADP is then exchanged again for ATP, and the binding and hydrolysis cycle is repeated until the entire protein is secreted across the membrane. In gram-positive bacteria, the mature protein folds as it emerges through SecYEG and is released to the surfaces of the bacteria or directly into the medium. In gram-negative bacteria, the transported protein folds in the periplasmic space, where it may remain or be transported across the outer membrane by one of the transport mechanisms described below.

In some gram-positive bacteria, including some species of *Streptococcus*, *Listeria*, and *Mycobacterium*, there is another specialized export system, called the **accessory Sec system** (Figure 13–1B), which appears to contribute to virulence in those bacteria. In these cases, there are two SecA proteins: **SecA1** is closely related to the canonical (i.e., usual) SecA of other systems, and **SecA2** is part of the accessory Sec system. The accessory Sec system also has a homolog of the SecY translocase (SecY2) and five other accessory Sec proteins (Asp1 through Asp5).

Many proteins in gram-positive and gram-negative bacteria are folded into the cell membrane, where they play various roles in transport, metabolism, and signaling. Integral membrane proteins consist of several hydrophobic transmembrane domains joined together by hydrophilic linker regions that face into the cytoplasm or outside of the cell membrane. In addition, some membrane proteins are anchored by one or two transmembrane domains and may face the cytoplasm or are on the cell surface. The SecYEG complex plays a critical role in transporting proteins destined to reside in the membrane. In this case, protein transport is coupled to cotranslation of the peptide by the ribosome (Figure 13–1A). As integral membrane proteins exit from the ribosome, they are sorted so that their signal peptides bind to the **signal recognition particle (SRP)** complex, which consists of 4.5S RNA and a protein called Ffh. The SRP in this complex then docks to the membrane-bound SRP receptor protein,

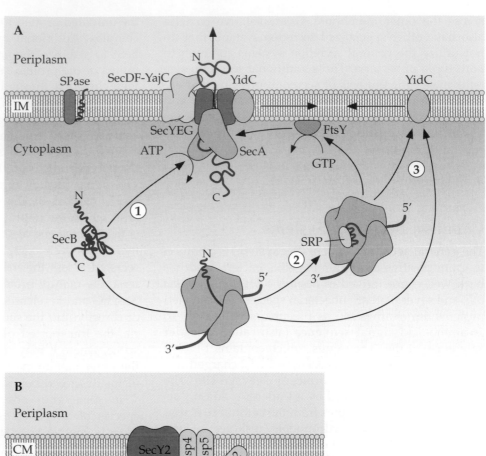

Figure 13–1 The general secretory system. **(A)** The Sec system common to both gram-negative and gram-positive bacteria. SecA and SecB bind to the signal sequence-containing precursor peptide and deliver the protein to the SecYEG translocase complex in the inner membrane of a gram-negative bacterium. SecA undergoes an ATP-dependent conformational change that drives translocation of the precursor protein through the SecYEG pore. The SecDF-YajC complex stabilizes the process. A leader peptidase cleaves the signal sequence, and the mature protein is released and refolds. CM, cell membrane. **(B)** The specialized accessory Sec system found in some pathogens.

called FtsY, which hydrolyzes GTP and delivers the signal peptide to the SecYEG complex. Further translation on the ribosome drives the nascent membrane protein from the SecYEG complex into the cellular membrane. Cotranslational protein export involving the SRP is generally used in a similar mechanism by eukaryotic cells. Finally, prokaryotic cells have another protein called YidC, which is also involved in exporting a number of important proteins to the cell membrane. In some cases, YidC may also utilize the SecYEG complex for this export (Fig. 13–1A).

The **twin-arginine transport (TAT) system** is a protein export pathway found in both gram-positive and gram-negative bacteria that was discovered about 10 years ago. The unique feature of the TAT system is that it is dedicated to the transmembrane translocation of proteins that are fully folded in the cytoplasm

(Figure 13–2). Some of the proteins that are exported by the TAT system contain cofactors, and cytoplasmic folding may optimize insertion of the cofactors in these proteins. However, not all TAT substrates contain cofactors, nor is the TAT system found in all bacterial species. Proteins are targeted to the membrane-embedded TAT translocase by N-terminal twin-arginine signal sequences. The TAT signal sequence includes a polar N-terminal region of variable length, followed by a region bearing the motif SRRXFLK and then a hydrophobic region of 12 to 20 amino acid residues, sometimes with a few basic residues. The TAT translocase consists of three proteins, TatABC. The TAT signal sequence is recognized by TatBC, which induces assembly of the proteins into a large pore complex. The proton motive force across the membrane is thought to energize the transport

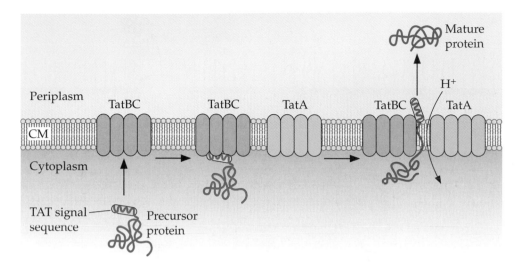

Figure 13–2 The TAT system. Shown is a model for the membrane transport of proteins through the TAT system. Precursor proteins bind to TatBC through their TAT signal sequences. TatA is recruited to the resulting complex, and a major conformational change, possibly induced by a proton electrochemical gradient, induces transport of the protein across the membrane. The TAT signal sequence is then cleaved, and the mature protein is released.

step. Following transport, the TatABC complex dissociates again. Fusion of a TAT signal sequence to the amino terminus of recombinant proteins can be used as a method to get folded proteins exported out of the cytoplasm of gram-positive or gram-negative cells.

Secretion Systems Specific to Gram-Negative Bacteria

Because of their outer membrane, gram-negative bacteria have at least six different mechanisms for secretion. These secretion systems, termed type 1 through type 6 (also denoted type I through type VI), come in two flavors: those that depend on the general Sec system for transport from the cytoplasm into the periplasm (Sec-dependent systems: type 2 and type 5) and those that do not (Sec-independent systems: type 1, type 3, type 4, and type 6).

Sec-Dependent Secretion Systems

T2SS. Proteins secreted through the type 2 secretion system (T2SS) use the general Sec system to reach the periplasm and then traverse the outer membrane through a channel made by special pore-forming proteins (Figure 13–3). The T2SS involves a total of 12 to 14 proteins that are encoded by a cluster of genes. The T2SS apparatus resembles systems that catalyze the biogenesis of type IV pili, filamentous-phage assembly, and the competence apparatus for DNA uptake. The first T2SS characterized was the *pul* system from *Klebsiella oxytoca*. Other examples are the *out* systems of *Erwinia* species, the *xcp* system of *Pseudomonas aeruginosa*, the *exe* system from *Aeromonas hydrophila*, the *xsp* system from *Xanthomonas campestris*, and the *eps* system from *Vibrio cholerae*. The T2SS is used to secrete

the A and B portions of some A-B-type toxins. For example, the ADP-ribosylating toxin from *P. aeruginosa* (ExoA) is exported through the *xcp* system, and cholera toxin is exported through the *eps* system of *V. cholerae*.

T5SS. In the type 5 secretion system (T5SS), the protein to be exported, called an **autotransporter,** is delivered to the periplasm via the general Sec system, where it then transports itself across the outer membrane. The autotransporter is made as a single precursor protein with multiple domains (α, β, γ, and protease) and an N-terminal Sec signal sequence. Once the protein is funneled through the inner-membrane pore of the Sec apparatus, the signal sequence is cleaved, and the β domain inserts into the outer membrane and forms a pore structure that then translocates the remaining domains through the pore to the bacterial cell surface, where the protein may or may not undergo further processing. For some autotransporters, autoproteolysis (catalyzed by the protease domain) releases the other secreted domains into the medium. For others, autoproteolysis does not occur, and the protein remains associated with the bacterial cell surface. The mechanism of secretion by autotransporters (Figure 13–4) was first discovered in 1984, when Tom Meyer's laboratory found that the immunoglobulin A (IgA) protease from *Neisseria* was the only gene product, besides the proteins making up the general Sec apparatus, that was necessary for secretion of active IgA protease from *E. coli* transformed with a plasmid carrying a gene encoding the IgA protease. Other examples of autotransporters are the IgA protease from *Haemophilus influenzae*, the serine protease from *Serratia marcescens*, the vacuolating cytotoxin (VacA) from *Helicobacter pylori*, and pertac-

A

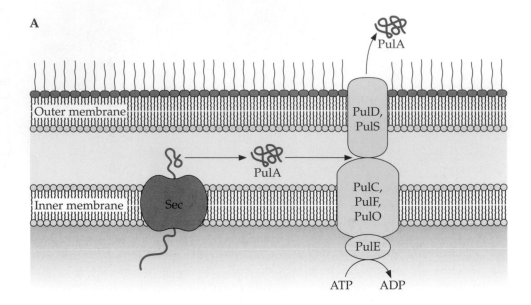

B

Figure 13–3 The T2SS. **(A)** Model for a Sec-dependent T2SS based on the Pul system from *K. oxytoca*, which directs the secretion of pullulanase (PulA), a lipoprotein that catalyzes starch debranching. Transport of the folded protein from the periplasm out of the cell involves a secretion apparatus consisting of 14 Pul proteins, some of which have sequence homology to proteins involved in type IV pilin assembly. **(B)** T2SS of cholera toxin secretion through the *eps* system of *V. cholerae*.

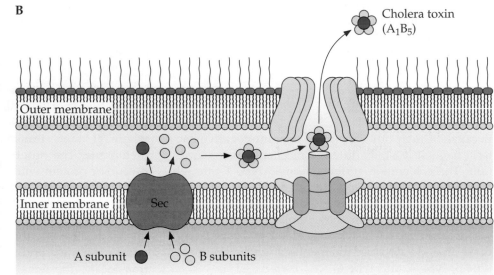

tin, the adhesin from *Bordetella* used in current DTaP vaccines (see chapter 18).

Sec-Independent Secretion Systems

T1SS. Proteins secreted by the type 1 secretion system (T1SS) cross directly from the cytoplasm to the cell surface, bypassing the general Sec system and the periplasm. The T1SS is capable of transporting proteins up to 800 kDa in size out of the cell in only a few seconds. Secretion by the T1SS involves a complex of three proteins that span the inner membrane, the periplasm, and the outer membrane. The complex secretes the proteins directly into the medium in a single step. The inner-membrane-spanning component is

an **ATP-binding cassette (ABC) transporter** whose ATPase activity supplies the energy for traversing the membrane. The secreted protein binds to the ABC transporter protein via a C-terminal signal sequence of ~50 amino acid residues, which contains an amphipathic helix with a glycine-rich GGXGXD sequence motif that is repeated up to 36 times. Unlike the Sec signal sequence, the signal sequence for the T1SS is not cleaved during export. The outer-membrane-spanning component is also a pore-forming protein that is made up of three monomers. A periplasmic connecting protein, called an **accessory factor,** holds the two pore-forming components together, so that the secreted protein is funneled directly from the cytoplasm into the medium. The α-

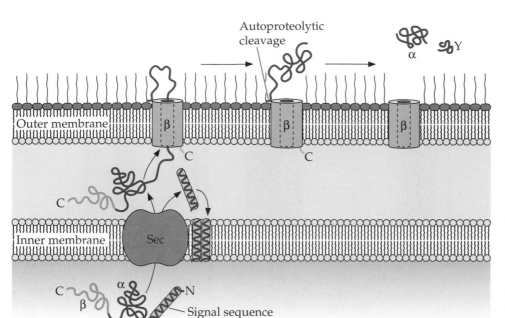

Figure 13–4 The T5SS. The autotransporter protein is delivered into the periplasm via the general Sec system. Once in the periplasm, the β domain of the autotransporter inserts into the outer membrane and forms a pore through which the other domains of the protein traverse. In some autotransporters, autoproteolytic cleavage releases the other domains from the β domain. In other autotransporters, cleavage does not occur, and the protease and other domains remain attached to the β domain.

hemolysin (HlyA) from *E. coli* is secreted by a T1SS made up of the ABC transporter HlyB, the accessory factor HlyD, and the outer-membrane pore protein TolC (Figure 13–5). TolC is also utilized by the **multidrug efflux system** in *E. coli,* which consists of TolC plus an ABC transporter, AcrB, and an accessory factor, AcrA.

T3SS. One of the most exciting developments in the toxin delivery area has been the discovery that there

are bacteria that do not simply excrete their toxins into the extracellular medium but rather inject their toxins directly into mammalian cells through a pore that opens between the bacterial cytoplasm and the host cell cytosol. One such type 3 secretion system (T3SS), depicted in Figure 13–6, was first discovered in *Yersinia pestis,* the causative agent of bubonic plague, and this mechanism of toxin delivery has proved to be surprisingly ubiquitous. In addition to *Yersinia,* a large number of pathogens (e.g., pathogenic

Figure 13–5 The T1SS. Shown is a model of the T1SS from *E. coli* that secretes α-hemolysin (HlyA), which consists of three proteins, an inner-membrane (IM)-spanning ABC transporter (HlyB), a periplasmic connecting protein (accessory factor; HlyD), and an outer-membrane (OM)-spanning pore protein (TolC).

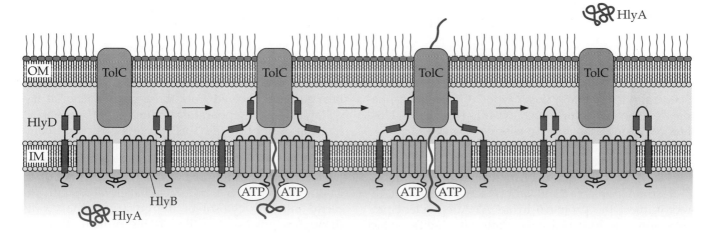

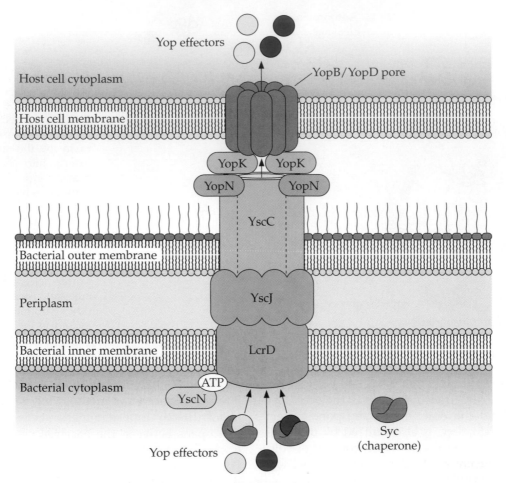

Figure 13–6 The T3SS. One of the best-characterized T3SS injectosomes is the *Yersinia* Ysc system. *Yersinia* species secrete a set of virulence-associated effector proteins, which for historical reasons are called Yops (for *Yersinia* outer proteins). Over 20 Ysc proteins (for Yop secretion) make up the secretion apparatus and mediate Yop secretion into the host cell. In the presence of Ca^{2+}, Yop proteins are not secreted because YopN blocks the secretion pore on the bacterial cell surface and LcrG blocks the pore on the cytoplasmic side. In the absence of Ca^{2+} or upon contact with the host cell, the pore is opened and Yops are secreted into the medium or translocated into the host cell, respectively. YopB and YopD form the pore through which the Yop effectors enter the host cell. LcrV and YopK regulate the translocation process. The Yop effector proteins have specific cytoplasmic chaperones, usually called Syc (for specific Yop chaperone) that bring the Yops to the secretion apparatus. Secretion through the inner- and outer-membrane pores is energized by the YscN ATPase activity.

E. coli, Salmonella, Shigella, Bordetella, Pseudomonas, Xanthomonas, Ralstonia, Erwinia, and *Chlamydia*) have been found to contain these types of secretion apparatuses, the genes for which are often located in clusters on large pathogenicity islands (PAIs).

The T3SS has a contact-dependent mechanism involving an elaborate complex of over 20 proteins that form a delivery apparatus (often referred to as an **injectosome**), which acts like a molecular syringe to directly inject the proteins, in an ATP-dependent process, into the host cell upon intimate (tight) contact

with the host cell surface. The overall morphology and organization of the virulence-associated T3SS appear to have evolved from those of the bacterial flagellar assembly apparatus (Figure 13–7). Just as in flagellar assembly, the T3SS exports structural components responsible for constructing the extracellular structures on the bacterial surface (needle, needle extension, and translocation pore subunits). Then, upon assembly, the T3SS translocates bacterial proteins directly into the cytosol of the eukaryotic host cells through a process that threads the proteins through

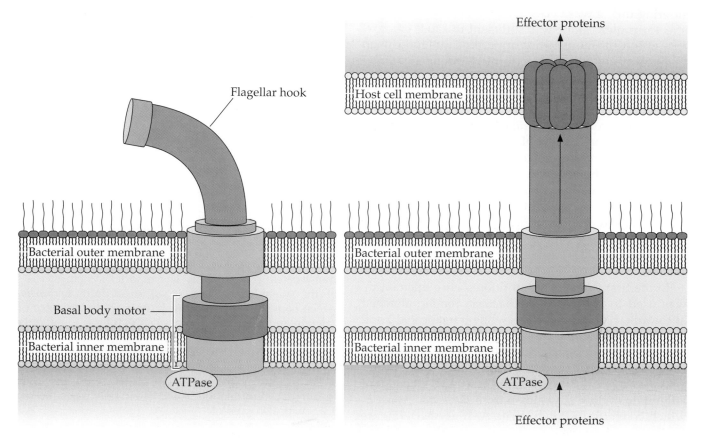

Figure 13–7 Comparison of the flagellar assembly apparatus with the T3SS. The overall structures and organizations of the bacterial flagellar apparatus and the T3SS are highly similar: each consists of inner- and outer-membrane-spanning ring structures, a membrane-associated ATPase that drives translocation, and helical extracellular structures.

the needle extension and another translocation pore that is formed on the host cell membrane. The exported proteins have specific bacterial cytoplasmic chaperones that bind them and guide them to the injectosome for secretion and translocation.

The importance of this secretion strategy becomes clear when one thinks of the usual protective immune response to toxins: production of antitoxin antibodies that prevent the toxin from binding to its target cell type. Clearly, if the toxin is injected directly into the host cell, such a protective response becomes ineffective because the toxin is never truly extracellular. The discovery that toxins could be directly injected into the cytosol of host cells also explained a long-existing conundrum about the function of the putative virulence factors called exoenzymes, which were often found in the extracellular medium and were thought to be the catalytic A subunit of A-B toxins but for which there appeared to be no B subunit. It was assumed that the B subunits of these so-called "orphan"

toxin A subunits were there but were not yet identified. With the discovery of the T3SS, the need for a B subunit was eliminated. Because these toxins do not have B subunits/domains yet clearly have profound effects on the host cell once delivered, they are often referred to as **"effector proteins."**

One of the first questions that you might ask is, how was it demonstrated that these effector proteins were indeed directly injected from the bacterial cytoplasm into the cytosol of host cells? The next question that comes to mind is, how was it determined which of the many exoenzymes secreted from the bacterial cell were also effector proteins that were delivered into the host cell? Remember, at the time (ca. 1994), the prevailing model was that the effector proteins were the A subunits of A-B toxins and that the "missing" B subunits had just not been identified yet. It was known that simply adding the effector proteins, even at relatively high concentrations, to the medium was not enough to cause cytotoxicity in the host cells. Considering this, it was necessary to demonstrate that the effector protein alone was sufficient to cause cytotoxicity as long as the secretion apparatus was functional.

Two key experiments, performed independently by two different research groups working on *Yersinia* pathogenesis, gave the first clear evidence for direct translocation of the Yop effector protein YopE from the bacterial cell into the eukaryotic cell. YopE, along with YopH, YopT, and YpkA (also known as YopO), acts to prevent phagocytosis by host macrophages and polymorphonuclear leukocytes by targeting host proteins involved in actin cytoskeletal rearrangement. YopE disrupts the cell actin microfilament structure and causes rounding of host cells by acting as a GTPase-activating protein (GAP) to inhibit the RhoA family of small GTPases, which regulate actin cytoskeletal assembly.

In the first study, confocal fluorescence microscopy was used to show that the bacteria remain attached at the eukaryotic cell surface, while translocated YopE is found only within the host cell cytosol. In the second study, the researchers constructed a recombinant *Yersinia* strain that produced YopE fused to a reporter enzyme, the **calcium-calmodulin (CaM)-dependent adenylate cyclase (Cya)** toxin from *Bordetella pertussis*, which converts ATP into cyclic AMP (cAMP) (Figure 13–8). Since bacteria do not have calmodulin and calmodulin is located only in the cytosol of eukaryotic cells, cAMP production could occur only if the Cya protein were delivered into the host cell. Mutants lacking components of the T3SS did not allow Cya delivery to the host cell and so showed no cAMP production. The researchers further demonstrated that bacterial adhesion to the host cells was necessary, since mutants lacking the adhesin did not cause cAMP production. However, uptake of the bacteria into the cells was not necessary, since inhibitors of actin polymerization, such as cytochalasin D, had no effect on cAMP production. In more recent studies, other researchers have combined and modified these two approaches by using confocal fluorescence microscopy and other reporters, such as green fluorescent protein, to show the delivery of effector proteins into the host cell.

Another approach is to use subcellular-fractionation and immunoblotting techniques to determine the location of the translocated effector protein. Using this method, it was found that the *Yersinia* effector protein YopM was targeted to the host cell nucleus, while YopE was localized to the cytosol (Figure 13–9). Digitonin, a nonionic detergent that permeabilizes the eukaryotic plasma membrane, but not the nuclear membrane, was used to lyse the host cells, followed by subcellular fractionation using centrifu-

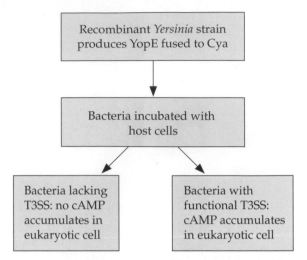

Figure 13–8 Yop-Cya reporter strategy for identifying translocated Yop proteins. The reporter system consists of the adenylate cyclase (Cya) toxin from *Bordetella*, which converts ATP into cAMP only in the presence of calcium and calmodulin. Since calmodulin is found only in the cytosol of the host cell, cAMP production is increased only when the Yop-Cya fusion protein is delivered into the cytosol. Incubation of a *Yersinia enterocolitica* strain producing the YopE-Cya protein with host cells resulted in accumulation of cAMP, but only when the Ysc T3SS was also present.

gation. Immunoblotting with anti-YopM or anti-YopE revealed the YopM protein in the nuclear fraction, whereas YopE was in the cytosolic fraction.

Yet another approach, developed by Greg Plano's laboratory to measure the translocation of Yops into cultured eukaryotic cells, is to detect the delivery of effector proteins into host cells using phosphorylatable peptide tag-based reporters. The first reporter system used a peptide tag containing a nuclear localization signal (NLS) sequence, which directed the effector protein into the nucleus of the host cell, and a portion of the eukaryotic transcription factor Elk-1 (amino acids 375 to 392) that is recognized and phosphorylated at serine-383 by eukaryotic protein kinases, such as Erk2. Translocation of the Elk/NLS-tagged effector protein into the host cell results in transport of the effector protein into the host nucleus, where a host kinase phosphorylates it. Phosphorylation can be detected with antibodies that specifically recognize the phosphorylated peptide sequence on the tagged effector protein (Figure 13–10). More recently, the laboratory has developed another reporter tag that utilizes a peptide derived from residues 1 to 13 of glycogen synthase kinase 3β (GSK). Translocation of a GSK-tagged effector protein into a eukaryotic cell results in host cell protein kinase-dependent phos-

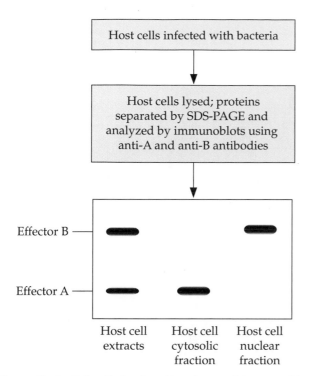

Figure 13–9 Subcellular fractionation and immunoblotting to localize delivered effector proteins. Shown are expected Western blot results from subcellular fractionation of cellular extracts of host cells exposed to a bacterium possessing a T3SS and two effector proteins, one that is delivered into the host cytosol (Effector A) and another that has an NLS, which targets it for the host cell nucleus (Effector B). Antibodies against the two T3SS effector proteins were used for detection of the proteins in the immunoblot. SDS-PAGE, sodium dodecyl sulfate-polyacrylamide gel electrophoresis.

Figure 13–10 Detection of translocated Elk/NLS-tagged effector proteins. Shown are the expected results from Western blotting of cellular extracts from host cells exposed to a bacterium possessing a T3SS and carrying a plasmid encoding an Elk-tagged effector protein either without or with an NLS sequence. For the left blot, antibodies recognizing the Elk peptide tag were used, whereas for the right blot, antibodies specifically recognizing the phosphorylated Elk peptide tag were used. Note that the bands for the Elk/NLS-tagged proteins ran higher on the gels than those without the NLS due to their increased size.

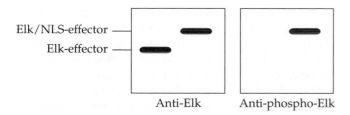

phorylation of the GSK tag at serine-9, which can be detected with antibodies. The advantage of this tag is that it does not require an NLS sequence tag, since it is phosphorylated by cytosolic protein kinases. They further showed that this tag works for monitoring translocation of effector proteins from both the T3SS and T4SS (discussed below).

T4SS. The type 4 secretion system (T4SS) involves an elaborate complex of 12 or more proteins that form a tunnel-like apparatus that directly transfers proteins and/or DNA into host cells in an ATP-dependent manner. The T4SS is related to conjugal-transfer *(tra)* systems used for transfer of DNA (usually plasmids) and proteins from one bacterium to another. The conjugal-transfer systems, which include the F pilin, IncN, and IncP conjugative plasmids that carry the *tra* genes, are well characterized. One of the best-studied T4SSs is the Ti complex (VirB/VirD system) from *Agrobacterium tumefaciens* (a plant pathogen that causes crown gall tumors), which transfers oncogenic Ti plasmid DNA and proteins from the bacterium directly into the plant cells, resulting in tumor formation in the plants (Figure 13–11).

Other T4SSs have been identified on PAIs in a number of pathogenic bacteria, including the *cag* system in *Helicobacter pylori* (which causes gastritis, ulcers, and stomach cancer); the Dot/Icm systems in the intracellular pathogens *Legionella pneumophila* and *Coxiella burnetii;* and the VirB/VirD4/Trw systems in the intracellular bacteria *Brucella* (which causes brucellosis), *Bartonella* (which causes cat scratch fever), *Rickettsia* (which causes typhus and spotted fever), *Mesorhizobium loti* (a symbiont of plant cells), and *Wolbachia* (a symbiont of nematodes that causes river blindness). Although the T4SSs in these bacteria are important for virulence and intracellular survival, not all of the effector proteins have been identified. Some T4SS components of the Ptl system in *B. pertussis,* the causative agent of whooping cough, have evolved to secrete pertussis toxin into the medium rather than to directly inject it into the host cell. Even though pertussis toxin subunits have Sec signal sequences and use the general Sec system to be transported to the periplasm, the Ptl system is used to transport the assembled pertussis toxin complex out of the periplasm and into the medium in a contact-independent manner.

T6SS. First discovered in *V. cholerae* and *P. aeruginosa* in 2006 by the laboratory of John Mekalanos, the type 6 secretion system (T6SS) is the "new kid on the block." Sequence analysis indicates that the T6SS is conserved in many gram-negative proteobacteria, including *Salmonella enterica,* pathogenic *E. coli, Franci-*

A Translocation pore assembly

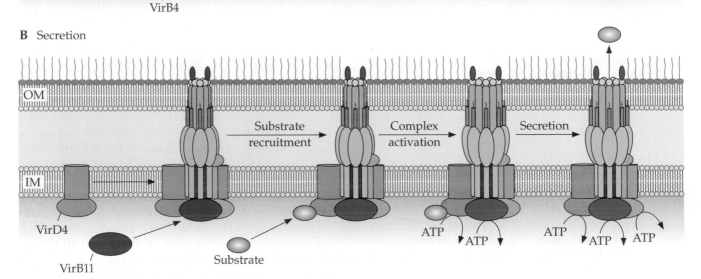

B Secretion

Figure 13–11 The T4SS. T4SSs are related to conjugal-transfer *(tra)* systems used for transfer of DNA from one bacterium to another. Secretion involves an elaborate complex of 12 or more proteins that form a tunnel-like structure through which DNA and proteins can be directly transferred into host cells in an ATP-dependent process. Shown is the T complex, made up of Vir proteins, used by *A. tumefaciens* to transfer proteins and oncogenic T-plasmid DNA into plant cells. IM, inner membrane; OM, outer membrane. (Adapted from R. Fronzes, P. J. Christie, and G. Waksman. 2009. The structural biology of type IV secretion systems. *Nat. Rev. Microbiol.* **7:**703–714, with permission from Macmillan Publishers Ltd.)

sella tularensis, A. tumefaciens, Rhizobium leguminosarum, A. hydrophila, Edwardsiella tarda, Burkholderia mallei, and *Y. pestis.* Some of the inner-membrane-associated proteins appear to have some sequence similarity to T4SS proteins of the Dot/Icm family (Figure 13–12). Indeed, the PAIs containing the T6SS were identified through genome-wide screens for IcmF homologs. There is no obvious N-terminal hydrophobic signal sequence needed for translocation, so the system appears to be Sec independent. The T6SS appears to be involved in ATP-dependent secretion of virulence proteins. In *P. aeruginosa,* the ClpV1 ATPase is presumed to supply the energy needed for translocation of at least one exported protein, Hcp1, a pore-

forming toxin that is found in the sputum of cystic fibrosis patients with chronic *P. aeruginosa* infections. Although the process is not yet clearly defined, it appears that the T6SS can also directly inject effector proteins into host cells in a contact-dependent manner. In *V. cholerae,* the T6SS translocates an Hcp homolog and three related proteins, VgrG-1 through VgrG-3, into host cells, where VgrG-1 catalyzes actin cross-linking. Although the details are still unclear, these proteins can form trimers or hexamers and may play a role in puncturing the host target cell membrane. Support for this notion comes from the finding that the proteins VgrG, Hcp, and a T4 gp25-like protein are structurally similar to the membrane-

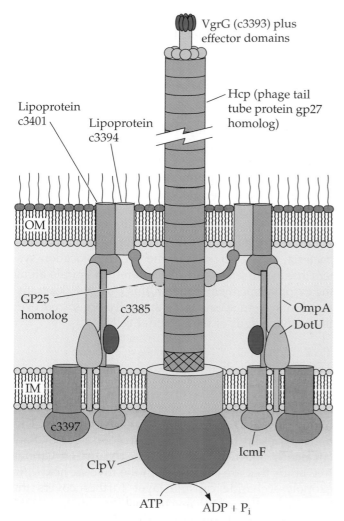

Lipoprotein c3401

Lipoprotein c3394

VgrG (c3393) plus effector domains

Hcp (phage tail tube protein gp27 homolog)

OM

GP25 homolog

c3385

OmpA
DotU

IM

c3397

ClpV

IcmF

ATP   ADP + P$_i$

Figure 13–12   The T6SS. The T6SS transports virulence-associated proteins in a Sec-independent but ATP-dependent manner. The inner-membrane (IM)-spanning proteins of the apparatus have some similarity to the Dot/Icm proteins of the T4SS. Effector proteins, VgrG and Hcp, appear to be secreted into the medium and/or injected directly into the host cell via a contact-dependent process. OM, outer membrane.

penetrating bacteriophage tail proteins. The gp25 protein is a highly conserved protein in all bacteriophages that contain tails. It is important for forming a baseplate structure. The T4 phage baseplate structure is necessary for initiating polymerization of the tube, and T4 gp25 serves as a connector protein with the VgrG needle protein and the Hcp tube during assembly of the T6SS apparatus.

Interestingly, there is recent evidence that suggests at least some T6SS can also deliver the toxin proteins of bacterial **toxin-antitoxin systems** into other bacterial cells. Toxin-antitoxin systems are often found encoded on large, low-copy plasmids, where they are

thought to contribute to plasmid stability and persistence. The antitoxin component, when expressed in the same cell, binds to the toxin with high affinity and blocks its activity. One possibility is that this may be a form of bacterial warfare, where the T6SS-delivered toxin from one bacterium kills other bacteria that do not have the cognate antitoxin component. However, they are also found on bacterial chromosomes, where they are speculated to have a number of different functions related to the activities of the toxin component, including acting as DNA gyrase inhibitors, phosphotransferases, site-specific RNases, ribosome-dependent RNases, and other bacteriostatic activities. Clearly, we are still at an early stage in understanding T6SS, toxin-antitoxin systems, and the roles that they play in bacterial physiology, in interactions with each other and the host, and in pathogenesis.

Secretion Systems Specific to Gram-Positive Bacteria

Gram-positive bacteria lack an outer membrane, and proteins exported by the Sec and Tat translocation systems appear directly on the cell surface. This arrangement causes several problems not encountered by gram-negative bacteria. Foremost among these, the secreted proteins must interact strongly with cell surface components to prevent secretion into the surrounding medium. Some proteins are covalently linked to the peptidoglycan by the sortase system (see chapter 11). Other proteins contain transmembrane anchors or are attached to membrane lipids to become lipoproteins attached to the cell surface. Still other proteins bind to the peptidoglycan or teichoic acids on the surface through binding domains, such as LysM or CBP (choline-binding protein), respectively. Finally, some proteins form ionic or hydrophobic contacts to specific proteins embedded in the cell membrane or to the membranes themselves. Nevertheless, there are indications that some gram-positive species may directly transport effectors into host cells. In addition, some gram-positive bacteria, such as *Mycobacterium*, have unusual waxy cell walls that require special export systems.

CMT and ExPortal

About 10 years ago, Michael Caparon's laboratory discovered a potential new function for pore-forming toxins in gram-positive bacteria, such as *Streptococcus pyogenes*, namely, **cytolysin-mediated translocation (CMT).** In this capacity, CMT may be a mechanism by which effector proteins (toxins without B domains/subunits) are secreted by the general Sec system and directly delivered into the eukaryotic cell through a pore in the eukaryotic membrane that is created by the pore-forming toxin, in this case, **streptolysin O (SLO)** (a cholesterol-dependent cytolysin). A model for how this might occur is shown in Figure 13–13. Their studies indicated that both SLO and an effector protein, SPN (for *S. pyogenes* NAD-glycohydrolase, which converts NAD into cyclic ADP-ribose, an intracellular signaling molecule), contain N-terminal Sec signal sequences that are removed during transport out of the bacterial cell. They found that the mature SLO and SPN proteins were associated with the host cell membrane and cytosol, respectively. They also found that SLO and SPN acted synergistically to trigger cytotoxicity in host cells, but only if the bacteria expressed a protein adhesin on their surfaces which would allow the bacteria to bind tightly to the host cells. Delivery of SPN into the host cells occurred only when all three virulence factors (SLO, SPN, and the adhesin) were expressed together in the bacteria. Coinfections with separate bacterial mutants expressing only one or two of the virulence factors did not result in delivery of SPN into the host cells and did not cause toxicity. Whether CMT is a delivery system common to other gram-positive bacteria remains to be determined.

Figure 13–13 Cytolysin-mediated translocation in gram-positive bacteria. In this model, direct delivery of effector molecules from the gram-positive bacterium into the cytoplasm of host target cells is dependent on contact between the cells through a receptor on the host cell surface and an adhesin on the bacterial cell surface. The effector molecules are secreted from the bacteria through the general Sec system and then translocated into the target host cell through the pore formed by the cholesterol-dependent cytolysin (CDC), where the effector can act on its substrate target within the host cytosol.

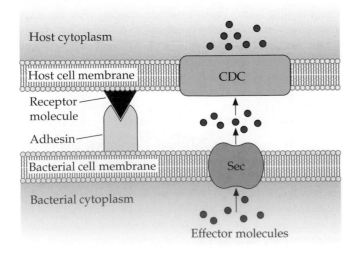

Another unusual twist observed in coccoid *S. pyogenes* is the apparent localization of the Sec translocation system to a membrane microdomain, termed the **ExPortal**. The ExPortal has been proposed to function as an organelle that coordinates interactions between some proteins secreted by the Sec system and membrane-associated chaperones. This type of coordinated secretion is important in gram-positive bacteria that lack a periplasmic space that allows folding of secreted proteins. ExPortal seems to be localized near the division septum and could also play a role in targeting some secreted proteins to subcellular locations, such as the poles or equators. It is unknown whether other ovoid gram-positive species contain ExPortal structures, but in rod-shaped gram-positive bacteria, such as *B. subtilis*, the Sec translocation system has been reported to be located in spirals along the cytoplasmic membrane. In this case, the Sec translocation system seems to be specially organized over the whole length of the bacterial cells.

T7SS

High-GC gram-positive mycobacteria, such as *M. tuberculosis*, the causative agent of tuberculosis, have highly unusual surfaces called mycomembranes. This barrier consists of hydrophobic mycolic acids that are covalently attached to the peptidoglycan and various free lipids that are embedded in the mycolic acids. An exopolysaccharide capsule layer further covers the mycomembrane. The mycomembrane makes DNA uptake and cell staining difficult, yet mycobacteria secret proteins into their growth media. These proteins, such as CFP-10 and ESAT-6, are important T-cell-antigenic targets that are required for virulence. Examination of the genomic regions around the genes encoding these two effector proteins revealed a cluster of genes that likely encoded a putative export system, dubbed the type 7 secretion system (T7SS) (the next number after the gram-negative T6SS described above). Initial experiments confirmed that the T7SS is involved in delivering mycobacterial effector proteins to the medium. The mechanism of the T7SS is not fully understood, but some features have been worked out (Figure 13–14). CFP-10 and ESAT-6 form a complex in the cytoplasm that contains a T7SS signal in the C terminus of CRP-10. The transport signal is found in other secreted effectors, such as Rv3615c, which piggybacks on another protein (EspA) for transport. The signal binds to a recognition protein (Rv3871) that assembles with a membrane-anchored protein (Rv3870). The Rv3870-Rv3871 complex is a member of the FtsK/SpoIIIE family of ATPases that are involved in many bacterial transport systems.

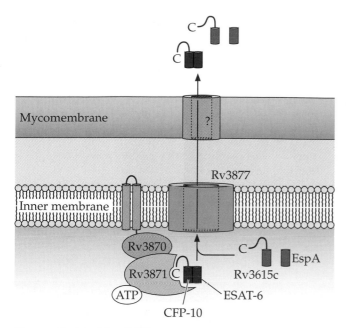

Figure 13–14 The T7SS.

ent and are presumed to have evolved independently of each other, some of them have overlapping functional properties and share some homologous protein components. Many researchers are thus hopeful that targeting the more conserved components of the secretion apparatus rather than the more diverse effector proteins that are transported by these systems would increase the likelihood of successful therapeutic intervention, as well as broaden the spectrum of therapeutic applications.

The effector complex is then transported through a membrane channel protein (Rv3877), but how the effectors get through the mycomembrane is unknown. Importantly, mycobacterial genomes contain several copies of the T7SS gene cluster, so it is likely that numerous effector molecules are exported through T7SSs. Surprisingly, T7SS gene clusters are present in several other species of gram-positive pathogens, including *Staphylococcus aureus*, *Listeria monocytogenes*, and *Streptococcus agalactiae* (group B streptococci), and the T7SS is functional for the transport of virulence factor effectors by *S. aureus*. Why gram-positive bacteria lacking a mycomembrane encode T7SSs is a mystery, but this observation suggests that T7SSs may play multiple roles in several gram-positive species.

Secretion Systems as Vaccine and Therapeutic Targets

Specialized protein secretion systems have emerged as a common strategy used by gram-negative pathogens for establishing infections, not only in terms of assembly of protein adhesins on the surfaces of bacteria, but also in terms of delivering toxins and other effector proteins to the medium or host cell cytosol to manipulate host cell signaling and immune response. It stands to reason, then, that they might serve as good targets for therapeutic interventions, since blocking delivery of the toxic proteins or adhesins would effectively reduce the virulence of the pathogen. Although all of the secretion systems are differ-

SELECTED READINGS

Abdallah, A. M., N. C. G. van Pittius, P. A. DiGiuseppe Champion, J. Cox, J. Luirink, C. M. J. E. Vandenbroucke-Grauls, B. J. Appelmelk, and W. Bitter. 2007. Type VII secretion—mycobacteria show the way. *Nat. Rev. Microbiol.* **5:**883–891.

Backert, S., and T. F. Meyer. 2006. Type IV secretion systems and their effectors in bacterial pathogenesis. *Curr. Opin. Microbiol.* **9:**207–217.

Buist, G., A. N. Ridder, J. Kok, and O. P. Kuipers. 2006. Different subcellular locations of secretome components of gram-positive bacteria. *Microbiology* **152:**2867–2874.

Dautin, N., and H. D. Bernstein. 2007. Protein secretion in gram-negative bacteria via the autotransporter pathway. *Annu. Rev. Microbiol.* **61:**89–112.

Day, J. B., F. Ferracci, and G. V. Plano. 2003. Translocation of YopE and YopN into eukaryotic cells by *Yersinia pestis yopN, tyeA, sycN, yscB,* and *lcrG* deletion mutants measured using a phosphorylatable peptide tag and phosphospecific antibodies. *Mol. Microbiol.* **47:**807–823.

Driessen, A. J. M., and N. Nouwen. 2008. Protein translocation across the bacterial cytoplasmic membrane. *Annu. Rev. Biochem.* **77:**643–667.

Gerlach, R. G., and M. Hensel. 2007. Protein secretion systems and adhesins: the molecular armory of gram-negative pathogens. *Int. J. Med. Microbiol.* **297:**401–415.

Leiman, P. G., M. Basler, U. A. Ramagopal, J. B. Bonanno, J. M. Sauder, S. Pukatzki, S. K. Burley, S. C. Almo, and J. J. Mekalanos. 2009. Type VI secretion apparatus and phage tail-associated protein complexes share a common evolutionary origin. *Proc. Natl. Acad. Sci. USA* **106:**4154–4159.

Madden, J. C., N. Ruiz, and M. Caparon. 2001. Cytolysin-mediated translocation (CMT): a functional equivalent of type III secretion in gram-positive bacteria. *Cell* **104:**143–152.

Mougous, J. D., M. E. Cuff, S. Raunser, A. Shen, M. Zhou, C. A. Gifford, A. L. Goodman, G. Joachimiak, C. L. Ordonez, S. Lory, T. Walz, A. Joachimiak, and J. J. Mekalanos. 2006. A virulence locus of *Pseudomonas aeruginosa* encodes a protein secretion apparatus. *Science* **312:**1526–1530.

Natale, P., T. Brüser, and A. J. M. Driessen. 2008. Sec- and Tat-mediated protein secretion across the bacterial cytoplasmic membrane—distinct translocases and mechanisms. *Biochim. Biophys. Acta* **1778:**1735–1756.

Pukatzki, S., A. T. Ma, D. Sturtevant, B. Krastins, D. Sarracino, W. C. Nelson, J. F. Heidelberg, and J. J. Mekalanos. 2006. Identification of a conserved bacterial protein secretion system in *Vibrio cholerae* using the *Dictyostelium* host model system. *Proc. Natl. Acad. Sci. USA* **103:**1528–1533.

Rosch, J. W., and M. G. Caparon. 2005. The ExPortal: an organelle dedicated to the biogenesis of secreted proteins in *Streptococcus pyogenes. Mol. Microbiol.* **58:**959–968.

Skrzypek, E., C. Cowan, and S. C. Straley. 1998. Targeting of the *Yersinia pestis* YopM protein into HeLa cells

and intracellular trafficking to the nucleus. *Mol. Microbiol.* **30:**1051–1065.

Sory, M.-P., and G. R. Cornelis. 1994. Translocation of a hybrid YopE-adenylate cyclase from *Yersinia enterocolitica* into HeLa cells. *Mol. Microbiol.* **14:**583–594.

Torruellas Garcia, J., F. Ferracci, M. W. Jackson, S. S. Joseph, I. Pattis, L. R. W. Plano, W. Fischer, and G. V. Plano. 2006. Measurement of effector protein injection by Type III and Type IV secretion systems by using a 13-residue phosphorylatable glycogen synthase kinase tag. *Infect. Immun.* **74:**5645–5657.

Trosky, J. E., A. D. B. Liverman, and K. Orth. 2008. *Yersinia* outer proteins: Yops. *Cell. Microbiol.* **10:**557–565.

QUESTIONS

1. *Shigella* expresses a number of virulence factors. Describe in detail how you would go about showing which of the *Shigella*-secreted proteins were actually being translocated into host cells during infection.

2. What are the known secretion systems in gram-positive and gram-negative bacteria? Describe their basic features and how you would distinguish among them.

3. Describe how you would go about showing that putative translocated effector proteins from *Salmonella* were involved in inducing actin rearrangements, changing protein phosphorylations of host proteins, and causing induction of apoptosis or DNA synthesis in host cells.

4. Why do secretion systems differ so much in their strategies? Why don't all of the systems directly inject proteins into mammalian cells?

5. How could a secretion system like some of the type IV systems secrete DNA when these systems mostly secrete proteins? (Hint: the first step in conjugation covalently attaches a protein to one strand of the nicked DNA.) How could you show that this explanation of DNA transfer by a protein secretion system is correct?

6. You suspect that a newly discovered secretion system is important for virulence. How would you test

this hypothesis? Keep in mind that mutations that disrupt an essential secretion system may kill the cell, thus making the bacterium appear less virulent.

7. Could you use a secretion system to secrete a protein it does not normally secrete? This has been considered in some biotechnology applications to avoid having to lyse bacterial cells. What type of secretion system would you use in the case of a gram-negative bacterium, or is there one? What about a gram-positive bacterium? What would you have to do in the way of genetic engineering to make this work? Why might your strategy fail?

8. You want to design a plasmid vector that would enable researchers to have their favorite protein(s) secreted by *E. coli*, even though that protein is not normally a secreted protein. Which of the secretion systems would you choose and why? Describe what this plasmid vector would entail.

9. What are the functions of cytoplasmic and periplasmic chaperones in secretion?

10. How does the autotransporter secretory pathway differ from the other secretory pathways? Is this secretion mechanism considered Sec dependent or Sec independent?

SOLVING PROBLEMS IN BACTERIAL PATHOGENESIS

1. Although most community-acquired methicillin-resistant *S. aureus* (CA-MRSA) bacteria are not as resistant to antibiotics as the hospital-acquired MRSA strains, the recently identified strains that we hear about in the news appear to be more virulent. A group of researchers has recently discovered that bacterial

cultures of a particularly virulent strain of CA-MRSA (called USA300) cause apoptosis in leukocytes and neutrophils, as evidenced by release of cytochrome *c* from mitochondria, followed by cytokine release and cell death. Antibodies generated against bacterial culture medium from the USA300 strain identified two proteins of 33 kDa and 44 kDa on Western blots. To determine the localization of these two proteins, researchers performed sodium dodecyl sulfate-polyacrylamide gel electrophoresis and Western blot analysis of lysates from the bacterial cell pellets and the bacterial culture medium, as well as lysates of human neutrophils treated with the USA300 strain. The results from Western blot analysis are shown in the left blot in Figure 1. The researchers then applied the cell-free bacterial culture medium to the neutrophils and performed subcellular fractionation of the treated neutrophils. The results from Western blot analysis are shown in the right blot in Figure 1.

A. Provide an interpretation (with rationale) of the results shown in Figure 1.

B. Predict (with rationale) possible functions for the two proteins, which have been named LukS and LukF (LukS is the 33-kDa protein, and LukF is the 44-kDa protein). Provide at least one additional (different) experiment that could be performed to confirm your prediction.

C. The researchers speculate that the two proteins, LukS and LukF, are two subunits of an A-B-type toxin. Based on the results shown in Figure 1, do you agree with this interpretation? Provide your rationale. Provide an experiment that would allow you to verify your answer.

2. A group of researchers at the university veterinary diagnostics laboratory isolated a new bacterium from the blood and stool of several horses that became ill and died at a local farm. Prior to death, symptoms included disorientation, loss of motor function, and flaccid paralysis, so the researchers suspected central nervous system involvement and possible production of neurotoxin. Based on 16S rRNA comparison, they found that the new bacterium was related to the gram-positive bacterium *Clostridium botulinum,* and they named it *Clostridium equiniae.* The researchers subsequently developed a mouse model of infection, in which bacteria are injected into the bloodstreams of mice, and the mice are monitored for paralysis and death. Using this animal model, they have conducted signature-tagged mutagenesis to identify genes in-

volved in virulence. From these studies, they have isolated three avirulent mutants (CeMut1 through CeMut3) that have genes encoding putative virulence factors deleted. They have conducted a series of experiments to determine the role of each of these virulence factors in pathogenesis. Interestingly, the researchers subsequently found that CeMut1, while avirulent, still produces and secretes the proteins encoded by the genes deleted in CeMut2 and CeMut3. Sequence comparison with protein databases revealed that the gene deleted in CeMut2 encodes a protein of 50 kDa and has significant sequence similarity to the catalytic N-terminal domain of botulinum neurotoxin (therefore, they named this protein BltA for botulinum-like toxin A part). The gene deleted in CeMut3 encodes a protein of 100 kDa and has significant sequence similarity in its N terminus to cholesterol-dependent cytolysins and in its C terminus to the B subunit of cholera toxin (therefore, they named this protein CptB for cytolysin plus toxin B part). Based on this, they suspect that the two proteins BltA and CptB form an A-B-type neurotoxin and decide to test this idea. They make antibodies against the culture filtrate from wild-type *C. equiniae* and then use the antisera in Western blots to compare the localization of the putative toxin proteins in the wild-type and CeMut1 bacteria and the neuronal cells treated with each bacterium. The results are shown in Figure 2.

A. Provide a detailed interpretation (with rationale) of the results shown in Figure 2. Were the researchers correct in their hypothesis? Provide your rationale.

B. Provide (with rationale) a possible function for the protein encoded by the gene deleted in CeMut1. Provide an experiment that could be performed to confirm your prediction.

C. Draw a clearly labeled schematic diagram depicting a possible model that accounts for all of the experimental observations, i.e., show how BltA and CptB are delivered to host cells and show the mode of action of the putative toxin on neuronal cells. Be

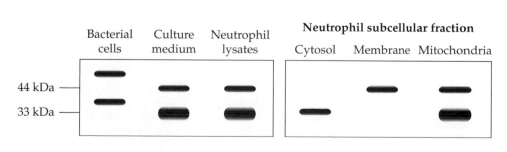

Neutrophil subcellular fraction

| Bacterial cells | Culture medium | Neutrophil lysates | Cytosol | Membrane | Mitochondria |

44 kDa —
33 kDa —

Figure 1 Western blots using antibodies against USA300-secreted proteins.

(continued)

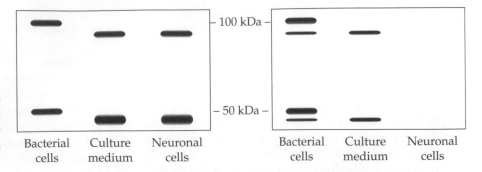

Figure 2 Western blot of cellular fractions from the wild type (left) and CeMut1 (right) using polyclonal antibodies against secreted proteins from *C. equiniae.*

sure your model is consistent with your answers to parts A and B.

3. Aphids (plant-eating insects) possess several types of symbiotic bacteria. Pea aphids often harbor as a symbiont *Hamiltonella defensa*, which is a gram-negative bacterium related to *Salmonella*. It is believed that the presence of *H. defensa* confers substantial resistance to parasitic wasps, which are natural enemies of the aphids that lay their eggs inside the aphids. In aphids harboring *H. defensa*, the wasp larvae die prematurely, allowing the aphid hosts to develop into adults and reproduce. Recent genome-sequencing data have revealed that *H. defensa* possesses two PAIs that are prophages, each encoding a T3SS and carrying putative virulence genes. The first PAI, called APSE-1, encodes a homolog of the catalytic subunit of Shiga toxin (Stx), produced by *Shigella*, which inhibits protein synthesis by disrupting 28S rRNA. The second PAI, called APSE-2, encodes a homolog of the catalytic subunit of the cytolethal distending toxin (CdtB), produced by *Campylobacter jejuni*, which is a DNase with an NLS that targets it to the eukaryotic nucleus, where it interrupts cell cycle progression by arresting the cells in G_2 phase. It is possible that these APSE-encoded toxins might be selectively targeted to destroy larval cells of attacking parasitic wasps. However, neither of these PAIs includes genes that might encode the corresponding binding subunits of the two putative toxins, suggesting that the Stx and CdtB proteins might be delivered to the wasp cells by the T3SS. To further characterize this fascinating symbiotic relationship, you have developed a system for genetically manipulating *H. defensa* that is based on *Salmonella* genetic tools, as well as a system for growing wasp larval cells in tissue culture. Describe how you would experimentally go about demonstrating that Stx and CdtB are indeed delivered by the *H. defensa* T3SS into the wasp cells and that they end up in the expected cellular locations. (Hint: you may need more than one type of experiment to demonstrate this.)

4. A group of researchers has been studying the pathogenesis of a bacterium Q that they isolated from a patient with a severe sore throat. During the course of their studies, they found that rabbit polyclonal antisera generated against filtered, dialyzed, and concentrated culture filtrates from bacterium Q are reactive against two proteins produced by the bacterium, a 70-kDa protein that has hemolysin-like activity, which they call Q bacteriolysin O (QLO), and a 50-kDa protein that has ADP-ribosyltransferase activity, which they call exotoxin Q (ExoQ). N-terminal sequencing of each protein and comparison with the protein databases allowed the researchers to identify highly homologous genes in several streptococcal species. Using the streptococcal gene sequence information, the researchers cloned the genes from bacterium Q. From sequence analysis, they determined that both proteins are probably synthesized with an additional 25 amino acid residues at the N terminus that are not present in the proteins found in the culture filtrates. The researchers also created in-frame deletion mutants of each gene, which they denoted ΔQLO and ΔExoQ. They also have a mutant strain of bacterium Q that is deficient in a protein adhesin, denoted ΔAdhQ. All three mutant strains were avirulent in an animal model of infection. They then performed cytotoxicity assays that they developed to examine the effects of the bacteria on epithelial cell cultures. Following exposure of the epithelial cell cultures to the wild-type bacterium and each of the mutants, either alone or as a coinfection (ΔQLO plus ΔExoQ), the researchers washed the epithelial cells to remove the unattached bacteria and lysed the cells. They centrifuged the lysed-cell extracts to separate them into a soluble cytosolic fraction and a membrane pellet fraction. They resuspended the membrane pellet in buffer containing a mild detergent that solubilizes membranes and proteins but does not lyse bacteria (so that they could then remove the bacteria by centrifugation). They separated the proteins from each of the samples by sodium dodecyl sulfate-polyacrylamide gel electrophoresis, transferred the proteins to a membrane, and detected the proteins by performing Western blot analysis using the antisera against QLO and ExoQ. The Western blots are shown in Figure 3.

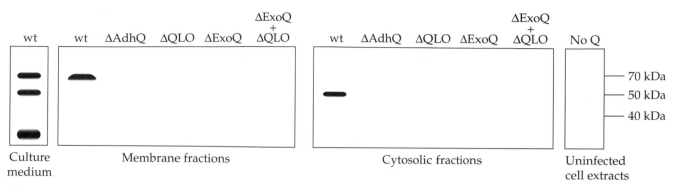

Figure 3 Western blots of cellular fractions and culture media, using antisera against QLO and ExoQ.

Using only the information provided and your knowledge of the virulence mechanisms that we covered in class, address each of the following.

A. Provide an explanation (with rationale) for the difference between the predicted sizes of the QLO and ExoQ proteins and the sizes of the proteins found in the culture medium. How does this occur? Be specific!

B. Provide a detailed explanation (with rationale) and interpretation for the results shown in the Western blots.

C. Provide a diagram depicting a possible model that accounts for the experimental observations and your interpretation of them. Be consistent with your answers to A and B. Be sure to clearly label your diagram.

5. By performing mutagenesis studies, a group of researchers have identified a gene *(expA)* that encodes a protein (ExpA) that is involved in degrading extracellular-matrix components. Based on sequence analysis, they suspect that ExpA is secreted via the general secretory pathway.

A. What feature of the protein sequence led the researchers to this conclusion?

B. Describe how the researchers might demonstrate experimentally that ExpA is indeed secreted via the general secretory pathway.

C. What cytoplasmic accessory proteins are important for secreting ExpA?

D. What function might ExpA have in pathogenesis?

6. You are a graduate student and have performed a transposon mutagenesis screen as part of your dissertation research on a medically important bacterium, *Helpmium graduatii*. Based on previous experimental evidence, this pathogen is thought to deliver at least one cytotoxic effector protein, called HgtA, directly into host cells, presumably via a T3SS. You have iden-

tified three independent transposon mutations. You are interested in studying the proteins that are disrupted in these mutants for your doctoral research. You design a series of experiments to characterize these mutants. Using the results for each mutant provided below, answer the questions for that particular mutant.

Mutant 1

- In mutant 1, the effector protein HgtA is constitutively secreted into the supernatant, whereas in the wild-type bacterium, HgtA is secreted only in the presence of host cells.

- When you tag the wild-type protein that is mutated with a fluorescent label and look at the wild-type bacterium under the microscope, the protein appears to be localized on the outside of the bacterial cytoplasmic membrane but remains associated with the cell envelope.

- Overexpression of the protein reduces, but does not inhibit, secretion of HgtA in the presence of host cells.

A. Is *H. graduatii* a gram-negative or a gram-positive bacterium?

B. Is the protein mutated in mutant 1 a member of a Sec-dependent or Sec-independent pathway?

C. Propose a function for your protein. Provide your rationale.

D. Why would this function be important to the bacterium?

Mutant 2

- Genetic analysis reveals that the transposon inserted into the first gene of a two-gene operon that encodes two proteins involved in T3SS.

- Polyclonal antibodies that bind to the two proteins in the wild-type bacterium do not recognize any proteins in the mutant bacterium.

(continued)

- In the mutant bacterium, the HgtA effector protein is secreted but cannot be translocated into cultured host cells.

- Sequence analysis of the proteins reveals the presence of at least one region of hydrophobicity in each of the proteins.

    A. What type of mutation has the transposon insertion resulted in? How do you know this?

    B. Propose a function for the two proteins in the operon. Provide your rationale.

Mutant 3

- Genetic analysis of the mutant bacterium reveals that the transposon has inserted into a gene that is adjacent to the gene encoding the HgtA effector protein.

- In the mutant bacterium, secretion of HgtA is severely reduced.

- A polyclonal antibody that binds to wild-type protein detects the protein in the cytoplasm of the wild-type bacterial cells.

- The half-life of HgtA is longer in the wild-type bacterium than in the mutant bacterium.

- In the wild-type bacterium, antibodies that bind to HgtA also coimmunoprecipitate the protein, but no other proteins. However, in the mutant bacterium, antibodies that bind HgtA also coimmunoprecipitate a number of other cytoplasmic proteins.

    A. Propose a function for the protein.

    B. Why is this protein important for secretion of HgtA?

    C. Is the function of this particular protein unique? Explain.

14

Virulence Regulation

Here is where the pathogens-as-criminals paradigm, which has played a role in introductions to other chapters in this book, runs into a possible snag, because now it is necessary to raise the question of whether "criminals" can police themselves. If I have criminal propensities, can I really control myself so that I am not as bad as I could be if some restraint is in my best criminal interests? One example you might think of is organized crime syndicates, which might set up legitimate business "fronts" to mask their more nefarious behind-the-scenes activities. Then, to keep the real police from removing their operations, the organized crime leaders might control their own members and "police" each other so that they all might stay below the police radar. Do pathogens have similar control mechanisms to keep under the surveillance of the host immune system? To get an answer of sorts, read on.

Mechanisms of Virulence Gene Regulation

Most pathogens encounter a number of different environments outside and inside the host, and even inside the host there will be many different environmental niches to which pathogens will be exposed. For each disparate environment, they will be subjected to a different set of stressful conditions, which will require appropriate responses for the bacteria to survive and grow. For instance, how do bacteria that normally reside as saprophytes in soil, vegetative, or aqueous habitats implement the transition to life inside an animal's gut and then, in the case of intracellular pathogens, inside a host cell? How do certain extracellular pathogens reside in the very different environments of the nasopharynx, lungs, blood, and brain? To do this, the bacteria must be able to sense a myriad of signals from their environment and, in response, induce or repress a defined set of genes through spatial and temporal regulation.

In this chapter, we will learn about mechanisms by which bacteria switch from one state to another and how they change their phenotypes in response to their environments or to avoid host defenses through regulation, often in a coordinated fashion, of their virulence genes. We will also discuss how bacteria sense signals from their environment and translate that sensory input into modulation of gene transcription and protein expression and function.

Types of Regulation

Gene Rearrangements

We have already explored several mechanisms for regulating virulence gene expression through phase or antigenic variation and gene rearrangements (deletions, insertions, duplications, and inversions) in the chapter on genetic exchange (chapter 7), so here we will focus more on the implications of these types of events for bacterial adaptation to the environment and virulence regulation.

PHASE AND ANTIGENIC VARIATION. Switching between "on" and "off" states of a gene or its promoter is the basis for the common phase variation mechanism for increasing the variability of surface structures of pathogenic bacteria, such as capsules, flagella, pili, and other cell wall and membrane proteins, lipoproteins, and liposaccharides. As we saw in chapter 7, this phase variation can occur through inversion of a gene or its promoter or through slipped-strand base mispairing. For example, nonvirulent mutants of *Bordetella pertussis*, the causative agent of whooping cough, can be obtained from virulent strains at frequencies of about 1×10^{-6}. Virulent revertants arise from the nonvirulent mutants at a similar frequency. These changes are caused by frameshifts in the *bvgS* gene, which encodes a component of a global regulator of virulence (see below). This phase variation is not simply a laboratory artifact. Nonvirulent phase variants of *B. pertussis* are frequently isolated from children who are recovering from a case of whooping cough, whereas the initially infecting strain was fully virulent. Antigenic variation from changes in the gene sequence (mutation) or switching of the gene being transcribed from one copy to another variant (gene conversion) can also lead to considerable diversity in the bacterial components being expressed and presented to the host immune system. This high antigenic variation prevents an effective immune response from being mounted and facilitates adaptation to new host niches. It also has serious implications for vaccine development (see chapter 17).

The frequency at which phase and antigenic variation occurs usually exceeds 1×10^{-5} variants per bacterial cell per generation, which is at least 3 orders of magnitude higher than basal mutation frequencies. Recent studies have uncovered evidence that suggests certain phase or antigenic switches can occur in a co-ordinately regulated way, so that there may be an order to the appearance of certain phenotypes or the turning on of one gene at the same time as the turning off of another (Figure 14–1). For example, consider a bacterium that can express three phase-variable genes encoding surface proteins with similar functions, such as pilins, but that have amino acid sequences that are antigenically different. Selection will favor expression of a version of the pilin gene that is best adapted to the environment, and this variant will proliferate and become the most abundant in the population. But what happens when a new selective pressure arises, such as an antibody response against the first pilin? In the independent-expression model, bacteria expressing the first variant will be removed from the population, thereby allowing one of the other variants to outgrow and take over the population. This process can continue to occur in response to new waves of antibody responses. However, in some cases, this phase variation is a coordinated process. For example, the original selective pressure can turn on the expression of the first variant and other operons containing genes that lead to adaptation. Some of these other genes may induce or directly regulate expression of a second variant, which will then begin to accumulate and can become the predominant species when there is selection against the first variant. Likewise, expression of the second variant can directly or indirectly stimulate expression of genes that favor expression of a third variant. A major difference between the independent and coordinated modes of regulation is that the variants will appear in a random order in the former and a specific sequential and temporal order in the latter.

GENE DUPLICATION AND AMPLIFICATION. Recent evidence from an extensive and ever-growing repertoire of complete genome sequences for closely related and more distantly related bacteria has revealed that genomic rearrangements (deletions, insertions, inversions, and duplications) serve a critical function in generating genomic variability (sometimes referred to as **genomic plasticity**). This genomic variability results in the phenotypic diversity needed for adaptation to selective constraints imposed on the bacteria under different environmental conditions. A large number of extensive, nonrandomly distributed repetitive sequences are scattered throughout most bacterial genomes, where they can serve as sites of recombination. This recombination leads to enhanced diversification of the bacterial species and thereby provides for flexible adaptive responses of a population to various environmental stresses, allowing the more fit variants to be selected for survival.

A Independent

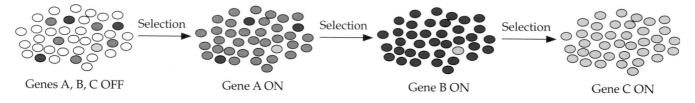

Genes A, B, C OFF   Gene A ON   Gene B ON   Gene C ON

B Coordinated

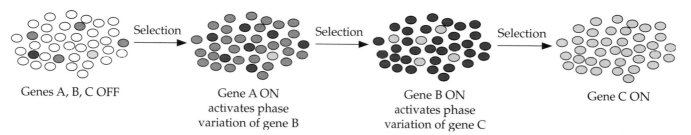

Genes A, B, C OFF   Gene A ON
activates phase
variation of gene B   Gene B ON
activates phase
variation of gene C   Gene C ON

Figure 14–1 Independent or coordinated phase and antigenic variation. For a bacterium with three phase-variable genes that encode proteins with identical or similar functions but with different sequences (i.e., antigenic variants of the gene), the genes are only randomly turned on in some of the bacteria when there is no selective pressure. Once the first selective pressure is introduced, however, those bacteria with the variant best adapted to the environment will predominate. Subsequent switching on or off of the different variants upon introduction of new selective pressures could occur through either separate independent events **(A)** or coordinated events **(B).** Grey, gene A; dark blue, gene B; light blue, gene C.

Consider, for example, *Helicobacter pylori,* whose environmental niche is the harsh, low-pH environment of the stomach. Isolates of *H. pylori* from different individuals are highly diverse, much more so than would be expected for a bacterium that lives in an environment where there are no cocolonizing bacteria of different species through which intergenic recombination or horizontal gene transfer could occur. Some strains of *H. pylori* are **hypermutable** (i.e., have higher rates of mutation), but even this does not explain the high level of diversity observed. Instead, evidence points to the presence of numerous nonrandom, repetitive direct-repeat sequences throughout the *H. pylori* genome, particularly within virulence-associated loci, as the source for the observed genomic plasticity.

Gene duplication and amplification constitute an important mechanism for adaptation to antibiotics and clearly play roles in the evolution of antibiotic resistance (for more discussion of this topic, see chapter 16), particularly in regard to increased production of antibiotic-modifying enzymes, target molecules, and drug efflux pumps. Once an antibiotic resistance gene is acquired, a gene duplication event often occurs, usually through homologous recombination between direct repeats. Further amplification can in-

crease the number of copies of the resistance gene, particularly under strong selective pressure through continued exposure to the antibiotic. Duplication or amplification events can also sometimes lead to enhanced expression of the resistance protein by introducing changes in the promoter region, ribosome-binding site (RBS), or other transcriptional or translational regulatory elements so that the gene is constitutively turned on. More copies of the resistance gene increase the likelihood of a point mutation occurring in one of the copies that might improve the efficiency of the resistance gene and thereby favor selection for the improved resistance gene. Although gene amplification places an increased metabolic or energy burden on the bacteria, the bacteria can often adapt to this stress by acquiring further secondary mutations that compensate for the added fitness cost to the cell (see chapter 16).

Transcriptional Regulation

LEVELS OF REGULATION. The environmental conditions that a pathogen faces during its life cycle, and not just the phase when it is actually causing an infection in the host, determine the types of virulence factors it produces and the extent of their regulation. Common conditions that control the expression of regulated virulence genes in bacterial pathogens are temperature, pH, iron (Fe^{2+}) and other divalent metal cations (Ca^{2+}, Mg^{2+}, and Mn^{2+}), availability of carbon and nitrogen sources (sugars and amino acids), growth phase (cell density), osmolarity, oxygen, carbon dioxide, and light. Those genes that are regulated in response to the host environment and result in survival within the host, subsequent infection, and disease are referred to as virulence genes. For any given condition, more than one gene may need to be regulated either positively or negatively and either at the same time or sequentially in order for the bacterium to respond correctly to the environmental stimulus. This regulation of a defined set of one or more virulence genes can occur at several levels.

In an **operon,** the virulence gene or genes are all transcribed as part of a single mRNA transcript controlled by a single promoter upstream of the genes (see Figure 9–4). The genes in an operon are all regulated and expressed together. In a **regulon,** the virulence genes are in different locations on the chromosome, but the genes all have a promoter that is controlled by the same regulatory signal protein, called a **regulator.** This regulator, which is often a **transcription factor** that binds to specific regions of the virulence gene promoters, controls the expression of genes within multiple operons, as well as multiple single genes. **Coordinated regulation** is the regulation of multiple genes in response to a particular condition or signal. When a single regulator controls a large number of different types of genes that affect a broad range of physiological processes, the regulator is sometimes called a **global regulator.** Coordinate regulation provides a framework for working out circuits of virulence gene control, since different virulence genes or regulons often share a common regulator.

As an example, the protein PrfA positively regulates the transcription of many of the genes involved in virulence of the food-borne pathogen *Listeria monocytogenes* (see Figure 11–13 for the life cycle of *Listeria*). The PrfA regulon is comprised of three multigene operons and several single-gene operons (Figure 14–2). The first two multigene operons consist of the *plcA-prfA* genes and the *mpl-actA-plcB* genes, followed by three open reading frames (ORFs) with unknown functions *(orfX-orfY-orfZ)*. These two operons flank the single-gene *hly* operon, which is divergently transcribed from *plcA-prfA*. All of these genes are located in *Listeria* pathogenicity island 1 (LPI-1). Northern blot analysis indicates that there are multiple promoters in this region that are regulated by PrfA. Some of these promoters are located in the intercistronic

regions between genes that are also cotranscribed (e.g., between *plcA* and *prfA*). These internal promoters allow an additional level of transcriptional control by PrfA and possibly other regulators. The other three operons that are part of the PrfA regulon, one comprised of two genes *(inlA-inlB)* and the other two of one gene each *(inlC and hpt),* were identified at other locations on the chromosome and were shown to also encode proteins involved in virulence.

TRANSCRIPTIONAL ACTIVATORS AND REPRESSORS. A **transcriptional activator** is a protein that can stimulate the expression of genes by binding to specific DNA sequences and recruiting RNA polymerase to the promoter region to enhance transcription of the genes. Some transcriptional activators require a coactivator, a ligand, or a modification, such as phosphorylation, to bind to the DNA or to interact with the RNA polymerase. A **transcriptional repressor** is a protein that acts opposite to an activator and prevents transcription of genes by binding to specific DNA sequences (often called **operators**) and physically blocking RNA polymerase from binding to the promoter region or starting transcription. Just as for activators, some transcriptional repressors require a corepressor, a ligand, or a modification to bind to the DNA. In some cases, a ligand binds to a repressor that is already bound to an operator, thereby releasing the repressor and the block in transcription.

The classic example of regulation of gene transcription by the opposing actions of an activator and a repressor is the *lac* operon of *Escherichia coli*, which is involved in the transport and metabolism of lactose (Figure 14–3). The levels of glucose and lactose in the cell regulate the *lac* operon. When glucose is scarce, the bacterium can use lactose as an energy source by upregulating expression of β-galactosidase (encoded by *lacZ*), which hydrolyzes lactose into glucose and galactose, a membrane-bound permease (encoded by *lacY*) that transports lactose into the cell, and β-galactoside transacetylase (encoded by *lacA*), which transfers acetyl groups from acetyl-coenzyme A (CoA) to β-galactosides. When lactose is not present, the constitutively expressed LacI repressor protein binds to operator sites in the promoter region, blocking the binding of RNA polymerase and thereby preventing transcription of the *lacZYA* genes. Low levels of glucose result in increased levels of cyclic AMP (cAMP), which binds to the cAMP activator protein (CAP) and changes its conformation so that it can bind to its cognate binding site upstream from the promoter and recruit RNA polymerase. When lactose and glucose are available, neither the LacI repressor nor the CAP activator protein is able to bind the DNA, so only basal

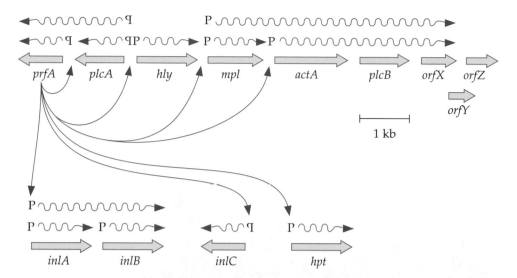

Figure 14–2 Organization and regulatory control of *L. monocytogenes* virulence genes that are members of the PrfA regulon. Shown is the core PrfA regulon involved in *L. monocytogenes* virulence. The gene *prfA* encodes the regulatory protein PrfA, which activates genes in three multigene operons and several single-gene operons, indicated by the arrows. The genes located in LPI-1 are *hly* (encoding LLO, a pore-forming toxin critical for escape from the phagosome and vacuoles), *plcA* (encoding PI-PLC, a phosphatidyl inositol [PI]-specific phospholipase that helps in escape from the phagosome and vacuoles), *plcB* (encoding PC-PLC, a phosphatidyl choline [PC]-specific phospholipase that helps in escape from the vacuoles), *actA* (encoding ActA, a protein that nucleates actin on the bacteria and promotes polymerization and intracellular motility), *mpl* (encoding Mpl, a zinc-dependent metalloprotease that processes PC-PLC precursor to its mature form), and *orfX, -Y,* and *-Z* (encoding several genes of unknown function). The genes in the *inlAB* operon are *inlA* and *inlB* (encoding InlA and InlB, adhesion proteins called internalins that contribute to invasion of host cells). The *inlC* operon contains a single gene, *inlC* (encoding InlC, an internalin-like protein that promotes protrusion formation at apical junctions and cell-to-cell spread). The *hpt* operon also contains a single gene, *hpt* (encoding Hpt, a hexose phosphate transporter required for optimal intracellular growth). The locations of promoters are indicated by P; wavy lines indicate mRNA transcripts.

levels of transcription occur. However, when lactose, but not glucose, is available, LacI binds an isomer of lactose called allolactose, which releases LacI from the operator. CAP binds to its binding site and recruits RNA polymerase, resulting in high levels of *lacZYA* transcription.

Many iron-regulated virulence genes are regulated through the action of iron-binding transcriptional repressors that are homologous to the **ferric uptake repressor (Fur),** also referred to as the Fe utilization regulator of *E. coli.* An example of iron regulation of virulence gene expression is the case of diphtheria toxin (DT), produced by the gram-positive bacterium *Corynebacterium diphtheriae.* The DT gene *(tox)* is carried on a closely related group of temperate (lysogenic) bacteriophages, β-phage and ω-phage. The gene for the regulator of DT gene expression *(dtxR)* is located on the bacterial chromosome and is constitu-

tively expressed (Figure 14–4). DtxR is a Fur-like transcriptional repressor that senses the presence of ferrous ions (Fe^{2+}). When iron levels are high, the Fe^{2+}-DtxR complex binds as a dimer to the operator region of the *tox* gene and represses gene transcription. When iron becomes the growth rate-limiting nutrient, which is the prevailing condition in the host, the ferrous ions dissociate from the repressor, causing it to dissociate from the DNA, thereby allowing transcription of the *tox* gene, as well as other iron uptake genes, such as those encoding a corynebacterial siderophore and other components of the high-affinity iron transport system.

TRANSCRIPTIONAL REGULATION THROUGH COMPLEX REGULATORY NETWORKS. Production of virulence factors by *Vibrio cholerae* is regulated at multiple levels and differs among pathogenic strains, and even the

A

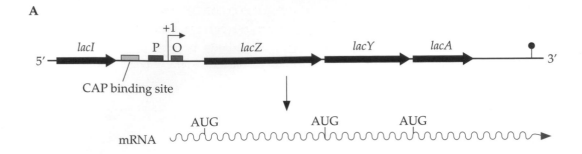

B

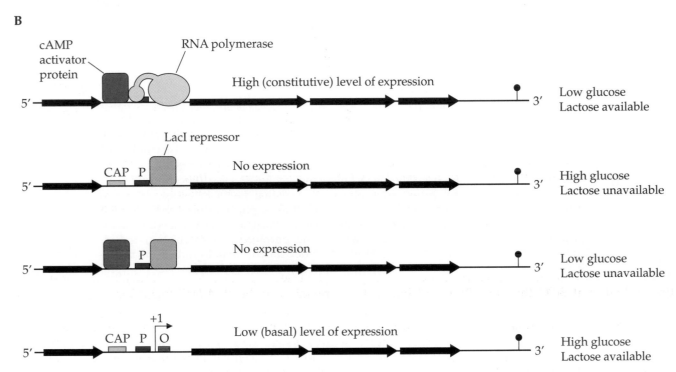

Figure 14–3 Regulation of the *lac* operon paradigm of *E. coli*. **(A)** The *lac* operon of *E. coli*. The *lacZ* gene encodes β-galactosidase, *lacY* encodes a permease, and *lacA* encodes an acetyl transferase. LacI is a repressor of the *lac* operon. **(B)** In the absence of lactose, the LacI repressor protein binds to the operator (O), which overlaps the promoter (P) and prevents transcription of the *lacZYA* genes. For optimal transcription of the *lacZYA* genes, two conditions must be met: there must be sufficient lactose present so that an isomer of lactose (allolactose) can bind to the LacI repressor protein and release it from the promoter region, and there must be low levels of glucose so that CAP binds to the DNA and recruits RNA polymerase to the promoter.

extent of disease (i.e., diarrhea) can be regulated at the level of the host. For instance, regulation of toxin activity occurs through duplication of the *ctxAB* (cholera toxin gene) region of the chromosome, regulation of virulence gene transcription, differential translation of *ctxA* and *ctxB* mRNAs, activation of the A subunit by proteolytic nicking in the host, and host-mediated activation of the catalytic A1 subunit inside the host cell.

Regulation of *V. cholerae* virulence genes at the transcriptional level involves the global regulator ToxT (Figure 14–5A). Transcription of virulence genes, including the operons *ctxAB* (cholera toxin), *tcp* (toxin-coregulated pilin), *acf* (accessory colonization factor), and as many as 20 other virulence genes collectively referred to as *tag* (ToxT-activated genes), are affected by a number of environmental signals, including pH (transcription is higher at pH 6 than at pH 8.5), certain

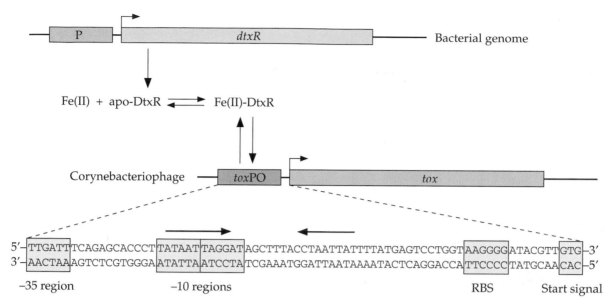

Figure 14–4 Regulation of DT expression. The DT is regulated by the Fur family repressor DtxR, which in the presence of Fe^{2+} binds as a dimer to two adjacent inverted repeats (indicated by arrows) that overlap in the –10 regions of the promoter and the transcriptional start site. Fe(II), Fe^{2+}. (Adapted from A. White, X. Ding, J. C. vanderSpek, J. R. Murphy, and D. Ringe. 1998. Structure of the metal-ion-activated DT repressor/*tox* operator complex. *Nature* **394**:502–506, with permission from Macmillan Publishers, Ltd.)

amino acids, low osmolarity, temperature (transcription is higher at 30°C than at 37°C), and bile salts. Pandemic serotype O1 strains of *V. cholerae* are divided into two biotypes, classical and El Tor. In both biotypes, regulation of virulence gene expression depends on a cascade of regulators that control ToxT expression. In the classical biotype, maximal expression of the ToxT regulon occurs at 30°C and pH 6.5, which are conditions that induce the upstream ToxR regulator (see below), whereas in the El Tor biotype, production of these virulence genes occurs under much more restrictive conditions and is inhibited by higher temperatures, such as 37°C, and the presence of bile salts.

The fact that genes regulated similarly to *ctxAB* are located in different places on the chromosome suggests that these genes are part of the **ToxT regulon.** Besides **ToxT,** the regulatory proteins involved in controlling expression of this regulon are **ToxR, ToxS, TcpP,** and **TcpH.** The genes encoding these proteins were first identified and cloned by taking advantage of the fact that the β-galactosidase activity of a *ctxA-lacZ* fusion was low in *E. coli.* Investigators transformed *E. coli* carrying the *ctxA-lacZ* fusion with cloned DNA fragments from the *V. cholerae* chromosome and looked for clones that increased β-galactosidase activity. The *toxT* regulatory gene was

identified on the basis of its ability to enhance expression of the *ctxA-lacZ* fusions in *E. coli* in the absence of *toxR* (see below). The *toxT* gene encodes a cytoplasmic 32-kDa protein that has amino acid sequence similarity to a family of transcriptional activators. The *toxT* gene is located within the *tcp* gene cluster, between *tcpF* and *tcpJ* on the Tcp pathogenicity island (PAI) that encodes the machinery for Tcp pilin biosynthesis. As noted above, ToxT serves as the global activator of several virulence genes.

The ToxR regulator controls the expression of *toxT* (Figure 14–5A), and the *toxR* and *toxS* genes were also found by the screen for *V. cholerae* genes that increase expression of a *ctxA-lacZ* fusion in *E. coli.* Expression of the *toxRS* operon is itself regulated by temperature at the transcriptional level. Immediately upstream of this operon is a gene *(htpG)* that encodes a heat shock protein (Figure 14–5B). Heat shock proteins are proteins produced at high levels when bacteria are exposed to high temperatures or to other forms of stress (e.g., low pH or changes in osmolarity). The *htpG* gene is transcribed divergently from *toxRS,* and its promoter region overlaps that of *toxRS.* The lower level of *toxRS* expression at 37°C could be due to increased expression of *htpG,* because RNA polymerase binding and transcription from the strong *htpG* promoter would be expected to interfere with RNA polymerase

A

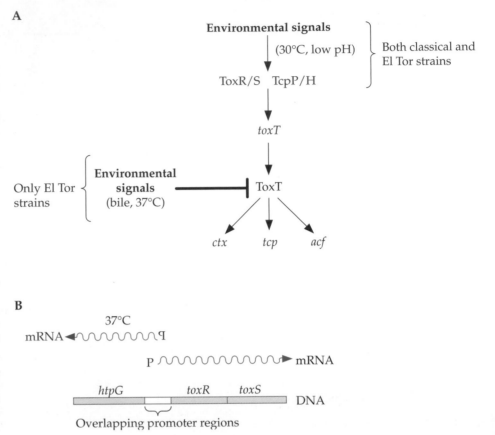

Figure 14–5 Differential regulation of virulence genes in pandemic strains of *V. cholerae* in response to environmental conditions. Expression of virulence genes, including the operons *ctxAB* (cholera toxin), *tcp* (toxin-coregulated pilin), and *acf* (accessory colonization factor), in *V. cholerae* strains is highly regulated by environmental conditions: low pH, some amino acids, low osmolarity, temperature (optimal expression at 30°C), and bile salts. **(A)** El Tor strains are negatively regulated by high temperature (37°C) and bile salts. Classical strains do not show this restricted regulation. **(B)** Regulation of the *toxRS* operon by *htpG*, a heat shock protein gene. Expression of *htpG* is enhanced at temperatures above 30°C and interferes with expression of the *toxRS* operon, which is transcribed in the opposite direction.

transcription from the somewhat weaker *toxRS* promoter.

Despite their names, the ToxRS regulators do not form a classical **two-component regulatory system (TCS)** that signals by phosphoryl group transfers (see below). ToxR is a 32-kDa protein that spans the cytoplasmic membrane, with the amino terminus exposed to the cytoplasm, and this portion contains the DNA-binding domain, which recognizes a 7-bp repeated sequence upstream of the genes it regulates (Figure 14–6). ToxR monomers do not bind DNA and activate transcription; only the dimer form with ToxS is active. ToxS, a 19-kDa transmembrane protein with a periplasmic domain, acts to facilitate formation of the ToxRS dimers.

ToxR also controls the expression of a second regulatory system of the ToxT operon made up of TcpP and TcpH, which are needed in the classical biotype for maximal *toxT* transcription. TcpP and TcpH are membrane proteins with periplasmic and cytoplasmic domains analogous to those of ToxRS. The genes for TcpP and TcpH are on the VPI PAI and are required for full expression of the *tcp* pilin genes. Transcription of *tcpPH* in the classical biotype is regulated by pH and temperature independently of ToxR or ToxT, suggesting that TcpPH can couple environmental signals

with the transcription of *toxT*. Transcription of *toxT* is dependent on two different sets of promoters, one immediately upstream of the *toxT* gene that has two binding sites for ToxR and one binding site for TcpP, and one much further upstream that is activated by ToxT itself. Some genes, such as *tcp* and *acf*, are activated directly by ToxT, but not by ToxR (i.e., ToxR does not bind to their operator regions but instead controls their synthesis indirectly by activating *toxT* expression). Other genes, such as *ctxAB*, are activated directly by ToxR, as well as by ToxT.

TCSs. The transcriptional control of many bacterial virulence and metabolic genes occurs through TCSs (Figure 14–7). With the exception of *Mycoplasma* species, which have relatively small genomes, all eubacteria contain varying numbers of TCSs, depending on the species. For example, *Streptococcus pneumoniae* and *E. coli* contain 13 and nearly 30 TCSs, respectively. A typical TCS is composed of a sensory **histidine kinase** (also called **sensor kinase**) and a cognate **response regulator,** which often act as transcription activators or repressors, depending on the TCS. Most histidine kinases are membrane-bound dimers that have extracellular domains. These extracellular domains sense signals, which can be small molecules, ions, or stress

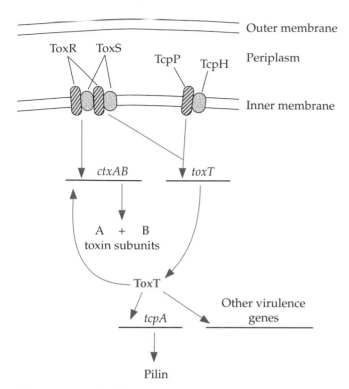

Figure 14–6 ToxT regulatory network in *V. cholerae.* ToxRS, TcpPH, and ToxT constitute a regulatory cascade that controls the transcription of virulence genes in *V. cholerae.*

conditions. The change in conformation of the extracellular domain is transferred via transmembrane domains to the cytoplasmic domains of the histidine kinases. Stimulation by the signal results in increased autophosphorylation of the histidine kinase on specific histidine residues. This region containing the phosphoryl-histidine residue then interacts with a domain of a cognate response regulator, and the phosphoryl group is then transferred into a specific aspartate residue in the response regulator. This phosphorylation changes the conformation of the response regulator. In many cases, binding of phosphorylated response regulators changes the transcription of regulons encoding proteins that can help the cell respond to the original signal or stress condition. Many histidine kinases also have a second phosphatase activity that can remove the phosphoryl group from the response regulator and turn the system back off (Figure 14–7). In this way, the TCS can be finely tuned to respond to changing environmental and stress conditions. The phosphatase activities of histidine kinases also play an important role in preventing aberrant cross talk between different TCSs.

Some TCSs are comprised of a complex phosphorelay between histidine and aspartate residues present in multiple signaling domains within the sensor kinase, as is found in the BvgAS TCS of *B. pertussis* (Figure 14–8). As is the case for most TCSs, the genes encoding the cognate BvgA response regulator and BvgS sensor kinase are cotranscribed as an operon (where *bvg* stands for *Bordetella virulence genes*). Virulence genes, including those encoding adhesins *(fha)* and components of pertussis toxin *(ptxAB),* are scattered over the bacterial chromosome in *B. pertussis.* Some virulence genes, such as the *ptxAB* genes and the genes required for secretion of pertussis toxin *(ptsAB),* are organized in two-gene operons. Expression of the *fha, ptx,* and *pts* operons is highest at 37°C and is suppressed when bacteria are grown at lower temperatures or in media containing high levels of $MgSO_4$ or nicotinic acid. Under conditions where *fha, ptx,* and other genes are actively expressed, some genes, such as the genes encoding siderophore production and respiratory enzymes, are repressed. It is important to note, however, that although $MgSO_4$, nicotinic acid, and temperature modulate virulence gene expression in the laboratory, the actual signals sensed by the BvgS histidine kinase are unknown for bacteria in the human body.

Some virulence genes are activated by the BvgAS system, while others are repressed. The differences between cells grown under conditions that cause the Bvg system to go into action and cells grown under conditions that turn the Bvg system off are sufficiently large that the two phenotypic states have been termed Bvg$^+$ and Bvg$^-$, respectively. There is also an intermediate state, called Bvg^i, which occurs in the presence of intermediate levels of modulating signals. Under Bvg^i conditions, Bvg$^-$ phase adhesins are expressed, but the toxins are not. Also, new surface proteins specific to this phase are synthesized. What do these phases have to do with infection of the human body? The low titer of antibodies against the proteins produced by Bvg$^-$ cells in infected people suggests that the Bvg$^-$ state is not achieved in bacteria in the human body. Since antibodies are made against the proteins produced by Bvg$^+$ cells, this phase is clearly achieved in the human body. Thus, in the human body, the cells are either always in the Bvg$^+$ phase or at most switch from Bvg$^+$ to Bvg^i. Bvg^i might even be involved in survival during transmission and colonization of the nasopharynx. A progression from the Bvg^i state, in which adhesins are produced, to the Bvg$^+$ state, in which toxins are produced, matches the paradigm introduced early in this book for the progression of a bacterial infection from colonization to persistence to breakout. But what about the Bvg$^-$ state? Scientists tend to assume that bacteria do not maintain genes that are never used. Possibly, the Bvg$^-$

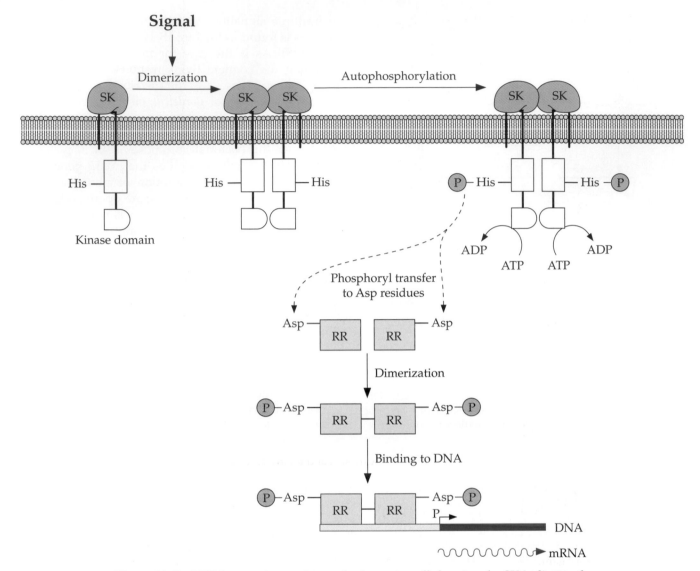

Figure 14–7 TCS for sensing and transducing extracellular signals. SK indicates the sensor kinase, and RR indicates the response regulator. Typical sensor kinase dimers consist of four domains: extracellular sensing domains (marked SK), transmembrane domains, dimerization-histidine phosphorylation domains, and catalytic ATPase domains that catalyze the autophosphorylation reaction on the histidine residues in response to a signal. Sensor kinases also catalyze dephosphorylation of their phosphorylated cognate response regulators, releasing inorganic phosphate ion (P). Response regulators consist of three domains: a receiver domain that contains the aspartate group that is phosphorylated, a linker region, and an effector domain that often contains a version of the helix-turn-helix DNA-binding motif. The phosphorylated response regulator activates or sometimes represses transcription of genes that respond to the signal.

state was important for the ancestral *Bordetella* species that gave rise to the modern species, some of which live outside the animal body. In this case, one would have to posit that the rise of different species has occurred too recently for the Bvg⁻ traits to be lost. A more intriguing possibility is that *B. pertussis* has a niche outside the human body. Such traits as flagella, biosynthetic enzymes, and respiratory enzymes, which are associated with the Bvg⁻ state, could well be needed for survival outside the human body. Conventional wisdom has it that *B. pertussis* does not have such a niche outside the human body, but it is difficult

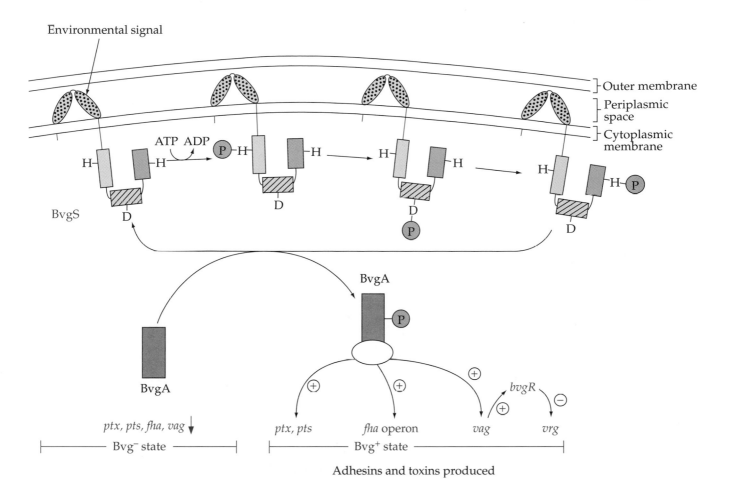

Figure 14–8 The BvgAS phosphorelay two-component system, which serves as a master regulator of virulence genes in *Bordetella*. BvgS has four domains. One is exposed in the periplasmic space and presumably senses signals. The three cytoplasmic domains (shown as rectangular boxes) contain different phosphorylation sites. The first amino acid residue to be phosphorylated is a histidine (H) in the first cytoplasmic domain. Then, the phosphoryl group (P) is transferred to an aspartate residue (D) in the second domain and finally to another histidine residue at the last C-terminal domain before being transferred to BvgA to form phosphorylated BvgA (BvgA~P), the active form of the response regulator. BvgA~P activates the expression of several virulence operons, including the *ptx, pts, fha,* and *vag* operons, while others, such as the *vrg* (*v*irulence *r*epressed *p*rotein) operon, are repressed through upregulation of *bvgR*. When these genes are being expressed, the bacteria are said to be in the Bvg⁺ state. The state characterized by no expression of these genes is called Bvg⁻. There is an intermediate state (Bvgⁱ) in which only the BvgA-controlled adhesin genes are expressed.

to support a negative conclusion, and scientists may simply have failed to identify this new environmental niche for *B. pertussis*.

The BvgS histidine kinase, which senses environmental signals, contains a multistep phosphorelay system that is more complex than the simpler set of reactions catalyzed by most histidine kinases (Figure 14–7). Instead of the single domain that is autophosphorylated and then serves as the phosphoryl group

donor for phosphorylation of the response regulator (BvgA), BvgS has not only one domain of this sort, but also two other domains, one of which is similar to the phosphoryl group receiver domain of the response regulator. In this case, the phosphoryl group is transferred to the aspartate residue of this internal receiver domain and then to a histidine of the third domain, which ultimately transfers the phosphoryl group to the BvgA response regulator. The phospho-

Table 14–1 Sigma factors and their functions in *E. coli* and *M. tuberculosis*

E. coli sigma factor	Regulatory function	*M. tuberculosis* sigma factor	Regulatory function
σ^{70} (RpoD)	Housekeeping genes necessary for cell growth	SigA	Housekeeping genes necessary for cell growth
σ^{54} (RpoN)	Response to nitrogen-limiting conditions	SigB	Response to hypoxia, oxidative stress, cell envelope stress
σ^{38} (RpoS)	Response to starvation and/or stationary phase	SigC	Virulence
σ^{32} (RpoH)	Response to heat shock	SigD	Response to starvation and/or stationary phase; stringent response
σ^{28} (RpoF)	Flagellar gene expression	SigE	Response to surface stress, heat shock, virulence
σ^{24} (RpoE)	Response to extreme cytoplasmic or extreme heat stress	SigF	Biosynthesis of mycobacterial cell envelope
σ^{19} (FecI)	Regulates expression of the *fec* (ferric citrate) gene for iron transport	SigG	SOS response, survival during macrophage infection
		SigH	Response to oxidative stress, heat shock
		SigI	Unknown
		SigJ	Response to oxidative stress
		SigK	Unknown
		SigL	Virulence, biosynthesis of phthiocerol dimycocerosate
		SigM	Long-term in vivo adaptation

rylated BvgA then activates the transcription of the virulence genes. It is thought that these added steps in the signal transduction pathway allow the regulatory system to sense more complex signals or modulate responses to environmental signals more effectively by allowing more points of regulation.

SIGMA FACTORS. Sigma (σ) factors are a subunit of the RNA polymerase holoenzyme. Sigma factors allow RNA polymerase to recognize specific DNA sequences that serve as promoters. Bacteria use different sigma factors to control the expression of different sets of genes under certain environmental conditions. Sigma factors are usually distinguished by their molecular masses. For example, σ^{70} has a molecular mass of 70 kDa and is the "housekeeping" sigma factor that recognizes and contacts the canonical -35 and -10 regions of the majority of promoters in bacteria. Besides promoter recognition specificity for RNA poly-

merase, sigma factors assist with DNA strand separation and then dissociate from, or at least interact less tightly with, the core enzyme during transcription elongation. Sigma factors provide effective mechanisms for simultaneously regulating the expression of large numbers of bacterial genes. The number of sigma factors varies among bacterial species; for example, *S. pneumoniae* has only 2 (σ^{70} and ComX, which induces competence), *E. coli* has at least 7 sigma factors, and *Mycobacterium tuberculosis* appears to have as many as 13 (Table 14–1). Each RNA polymerase contains only one of these sigma factors at a time, and the promoter sequences recognized by each form of RNA polymerase are different, depending on the sigma factor.

There are also **anti-sigma factors** that counteract the functions of some sigma factors. Anti-sigma factors regulate their cognate sigma factors posttranslationally by sequestering them away from the RNA polymerase. When a stress condition is sensed, the anti-sigma factor releases the sigma factor so that it can interact with RNA polymerase and change the transcription pattern to respond to the stress condition. Anti-sigma factors do this by a so-called **partner-switching mechanism,** in which protein-protein interactions are controlled by modifications, such as phosphorylation or redox-sensitive disulfide bond formation. Adding to this complexity is the regulation of the anti-sigma factors themselves by specific anti-anti-sigma factors.

Examples of regulatory networks involving single or multiple sigma factors and other transcriptional regulators include the sigma B-PrfA network of *L. monocytogenes* (Figure 14–9A). As we saw in Figure 14–2, PrfA is a positive regulator that activates σ^{70} RNA polymerase to transcribe many genes critical for pathogenesis, including *hly* and *actA*. However, the regulatory circuit in *Listeria* is actually a bit more complicated than that. The *prfA* gene is not only positively autoregulated by PrfA, it is also transcribed by RNA polymerase, which contains sigma B under some stress conditions. In addition to *prfA,* the sigma B regulon consists of some genes that are activated solely by sigma B-containing RNA polymerase (e.g., *hfq* and *opuCA*) and some that are activated by both sigma B and PrfA mechanisms at two different promoters (e.g., *inlA* and *bsh*). RNA polymerases containing some stress-induced sigma factors can transcribe the genes encoding other sigma factors and thereby cause a cascade of regulation. For example, RNA polymerase containing stress-induced σ^N in *Pseudomonas syringae* transcribes genes involved in polyketide biosynthesis leading to the phytotoxin coronatine and the *hrpL* gene, which encodes another alternative sigma factor called HrpL (Figure 14–9B). RNA polymerase containing HrpL, in turn, transcribes genes involved in a type II secretion system (T2SS). In *M. tuberculosis*, the regulatory cascade is even more complicated, with multiple sigma factors regulating the transcription of other sigma factors, in addition to virulence-associated genes (Figure 14–9C).

cis-Acting RNA Thermosensors

SUPERCOILING DNA. During DNA/RNA synthesis, **supercoiling** occurs as the double-helical DNA/RNA twists around the helical axis in order for the DNA/RNA polymerases to bind and act. Circular DNA relieves some of the stress by introducing a supercoiled shape. DNA topoisomerases (some of which are known as DNA gyrases) can either generate or dissipate the supercoiling to change the DNA topology. Changes in supercoiling in a particular region of DNA

can occur in response to changes in temperature and osmolarity and may play roles in virulence gene expression. For example, DNA topology is thought to differentially regulate gene expression during the developmental cycle of the intracellular pathogen *Chlamydia trachomatis,* in which chromosomal DNA supercoiling was found to vary depending on the stage in the cycle. The promoters of two midcycle-specific genes, *ompA* and *pgk,* were sensitive to alterations in supercoiling, with promoter activity changing by more than eightfold.

In *H. pylori,* flagellar and global gene regulation is thought to be modulated by changes in DNA supercoiling. When the DNA gyrase was inhibited by treatment with the antibiotic novobiocin, the resulting decrease in negatively supercoiled DNA lowered transcription of the *flaA* gene, blocking flagellum production. Genome-wide transcript analysis of *H. pylori* under conditions of reduced supercoiling showed that flagellar, housekeeping, and virulence genes were coordinately regulated by the state of global DNA supercoiling.

There is also a growing amount of data that suggests DNA/RNA structures in certain regions of the chromosome are sensitive to temperature and can serve as sensors of environmental conditions, modulating the expression of genes in those regions. In *Shigella flexneri,* a causative agent of diarrhea in humans, an increase in temperature changes the conformation of the promoter region of the *virF* gene, which encodes the main transcriptional activator of virulence factors. As shown in Figure 14–10, this conformational change in DNA prevents the binding of the global transcriptional repressor **H-NS,** which is a histone-like protein commonly found in gram-negative bacteria that is important for repressing transcription of foreign genes introduced through horizontal gene transfer.

TRANSCRIPTIONAL TERMINATORS AND ANTITERMINATORS. In bacteria, intrinsic (also called rho-independent) **transcriptional terminators** are particular RNA secondary structures, usually hairpin loops with stems of 7 to 10 bp rich in GC content followed by a number of uracil residues, that form in the nascent transcript. When the transcribing RNA polymerase encounters these hairpin structures, the polymerase stalls, which destabilizes the mRNA-DNA-RNA polymerase complex and terminates transcription. **Rho-dependent transcriptional terminators** require rho (ρ) factor, an RNA-binding hexameric helicase complex, to help disrupt the mRNA-DNA-RNA polymerase complex. If transcription is terminated prematurely, **transcriptional anti-**

A

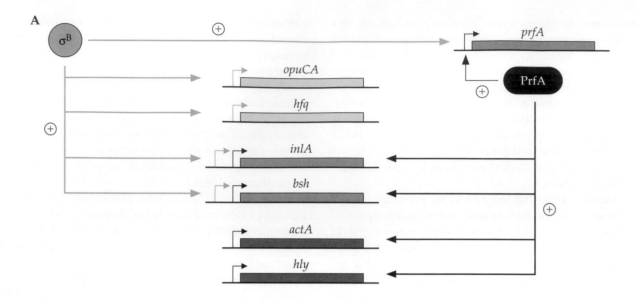

B

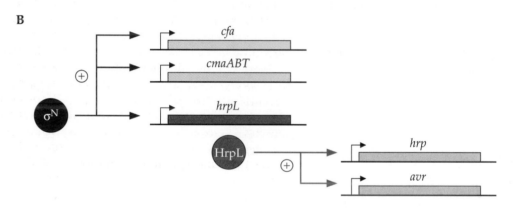

C

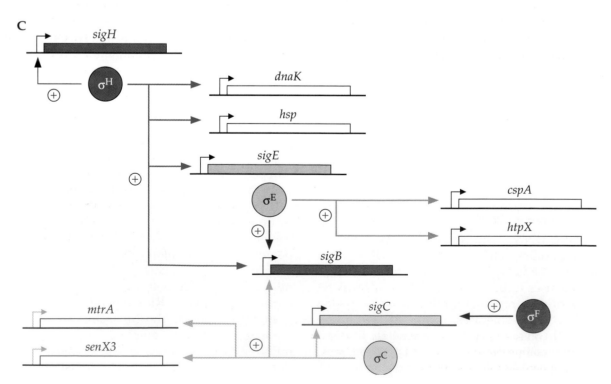

terminators sometimes enable the RNA polymerase to resume transcription by ignoring the termination signal.

The regulation of transcription termination is found in many bacteria, including most major pathogens. These mechanisms often regulate fundamental metabolic and biosynthetic processes in response to the availability of small molecules, such as amino acids and cofactors. Three of the best-studied mechanisms that regulate transcription termination are **attenuation,** which controls amino acid biosynthetic operons in gram-negative bacteria; **T-box riboswitches,** which control amino acid biosynthesis and aminoacyl-tRNA synthetase expression in grampositive bacteria; and **metabolite-sensing riboswitches,** which control the expression of genes involved in biosynthesis of the small ligands that bind to the switch. In addition, bacteriophages often use regulated transcription termination to decide between lytic and lysogenic outcomes of infection. So far, not many virulence factors whose expression is controlled directly by transcription termination mechanisms have been identified, so these important forms of regulation that control the underlying metabolism of many bacterial pathogens are not covered further here.

Translational Regulation

RNA THERMOSENSORS. Certain RNA structures can serve as sensors of temperature changes through the stability of their double-stranded hairpin loops. These **thermostable riboswitches** do not bind small-molecule ligands but instead form stable secondary structures at low temperatures that melt at higher temperatures. These thermostable riboswitches regulate translation by masking RBSs (also called Shine-Dalgarno [SD] sequences) in the secondary structures at low temperatures, thereby preventing translation. At higher temperatures, the RBSs are exposed and free to initiate translation. For instance, translation of IcrF, the primary transcriptional activator of some virulence genes in *Yersinia pestis,* the causative agent of plague, is controlled by temperature changes. The RBS near the 5' end of the *icrF* mRNA is sequestered in a hairpin loop structure. The stability of the hairpin structure decreases with increasing temperature, resulting in more expression of the IcrF protein at 37°C than at lower temperatures. This kind of temperature switch likely operates in many zoonotic pathogens, like *Y. pestis,* that infect both cold- and warm-blooded hosts (fleas and mammals in this case).

In another example, expression of the heat shock sigma factor σ^{32} of *E. coli* is controlled by an RNA thermosensor that resides within the gene *rpoH* (Figure 14–11A). The σ^{32} mRNA forms a secondary structure containing hairpins within the translation-initiation region at low temperatures (below 30°C). This inhibitory structure is disrupted at higher temperatures (above 42°C), resulting in translation of *rpoH*. A similar mechanism is used to control the expression of virulence genes in *L. monocytogenes* (Figure 14–11B). Protein levels of the global transcriptional activator PrfA are high at 37°C but low at 30°C. By contrast, *prfA* is still transcribed at high levels at 30°C. This thermoregulation occurs due to formation of a secondary structure in the untranslated 5' end of the *prfA* mRNA transcript. Base substitution mutations that destabilize this structure increase the level of PrfA protein at low temperatures, thereby enabling the ex-

Figure 14–9 Examples of regulatory networks involving sigma factors and other transcriptional regulators. **(A)** The PrfA-sigma B network of *L. monocytogenes*. The gene *inlA* encodes internalin A; *hly* encodes LLO, involved in escape from the phagolysosome; *actA* encodes ActA, involved in actin tail formation; *bsh* encodes a bile salt hydrolase involved in *Listeria* resistance to bile; *hfq* encodes an RNA-binding protein chaperone important for modulating stability or translation of mRNAs and interacts with numerous small regulatory sRNAs; and *opuCA* encodes a solute transporter. **(B)** The sigma factor cascade regulating expression of the T2SS in *P. syringae*. Sigma N mediates transcription of the *cfa* and *cma* operons (biosynthesis of polyketide components of the phytotoxin coronatine) and *hrpL*, which encodes a sigma factor (HrpL) responsible for transcriptional regulation of the *hrp* and *avr* genes involved in T2SS expression. **(C)** The complex interaction of several sigma factors involved in virulence of *M. tuberculosis*. Multiple sigma factors activate transcription of other sigma factors and virulence-associated genes. The gene *dnaK* encodes the DnaK corepressor for the HspR regulator, *hsp* encodes the heat shock protein Hsp, *cspA* encodes a cold shock protein, *htpX* encodes the membrane-bound zinc-dependent metalloprotease HtpX, and *sigB* encodes the sigma 54 subunit (Table 14–1), which mediates control of other genes.

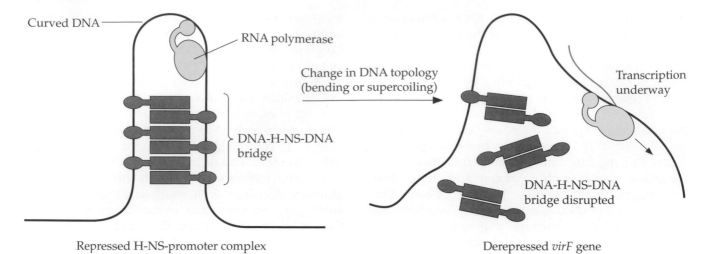

Curved DNA

RNA polymerase

Change in DNA topology
(bending or supercoiling)

Transcription
underway

DNA-H-NS-DNA
bridge

DNA-H-NS-DNA
bridge disrupted

Repressed H-NS-promoter complex

Derepressed *virF* gene

Figure 14–10   Relief of H-NS repression in *S. flexneri* by changes in DNA topology. Temperature-dependent bending or supercoiling of DNA disrupts the H-NS–promoter complex to enable transcription of the *virF* gene.

pression of virulence factors at the lower temperatures. When the DNA encoding the RNA secondary structure was fused with the *gfp* reporter gene, the bacteria became fluorescent at high temperatures but not at low temperatures, suggesting that the thermosensor of *prfA* could function ectopically (i.e., in the absence of the translated region of the *prfA* gene).

sRNA REGULATORY SYSTEMS.   Expression of **small RNA (sRNA)** transcripts that are complementary to mRNAs of target genes can modulate the cellular levels of virulence proteins or proteins with roles in adaptive responses to environmental cues or stresses. Bacterial sRNAs come in two major classes: **Hfq-dependent sRNAs** (the larger class) and **protein activity-modifying sRNAs.** Often, these sRNAs do not encode peptides and exert their regulation by pairing with mRNA targets or proteins; however, some sRNAs that function by binding to mRNA targets and by encoding regulatory peptides have recently been found.

In gram-negative bacteria, sRNAs prevent translation of certain encoded proteins by pairing with the RBS site on the mRNA and preventing the binding of the ribosome. In other cases, sRNAs bind to regions in the mRNA that themselves block the RBS, and this binding stimulates translation because the RBS is no longer sequestered. In some cases, though, binding of an sRNA to an mRNA signals rapid degradation of the transcript. Many of these sRNAs use Hfq protein as an RNA chaperone, which facilitates base pairing between the sRNA and its mRNA target, as well as protecting the sRNA from nuclease degradation. For

example, the DsrA sRNA positively regulates translation of the stationary-phase sigma factor RpoS by binding to and opening a hairpin in the target mRNA that normally inhibits translation of RpoS (Figure 14–12A). DsrA also negatively regulates translation of the H-NS protein by base pairing downstream of the translation initiation codon. The noncoding 85-nucleotide DsrA sRNA is highly conserved among closely related gram-negative bacteria. The Hfq protein both facilitates pairing of DsrA to its target mRNAs and protects DsrA sRNA from cleavage and degradation by RNase E endonuclease. The Hfq RNA chaperone is found in some gram-positive bacteria, where it does not seem to play a dramatic global role in sRNA metabolism. Some gram-positive pathogens, such as *S. pneumoniae,* synthesize sRNAs but do not contain Hfq homologs or known RNA chaperones.

Full virulence in the marine fish pathogen *Vibrio anguillarum* requires an efficient iron uptake system encoded by the *fat* operon on a 65-kb plasmid (Figure 14–12B). To avoid potentially harmful excess iron being taken up into the cell, *V. anguillarum* uses two systems to downregulate the expression of the iron uptake complex. The first is the well-known Fur regulator described above, in which the iron-bound dimer of the Fur repressor protein binds to DNA and thereby blocks transcription of the *fat* operon. RNAα, which mediates the second regulation, is a 650-nucleotide antisense RNA transcript within the iron transporter gene *fatB*. An antisense RNA is different from the sRNAs mentioned above, which are transcribed from regions on the chromosome that are different from their mRNA targets. In addition, pairing

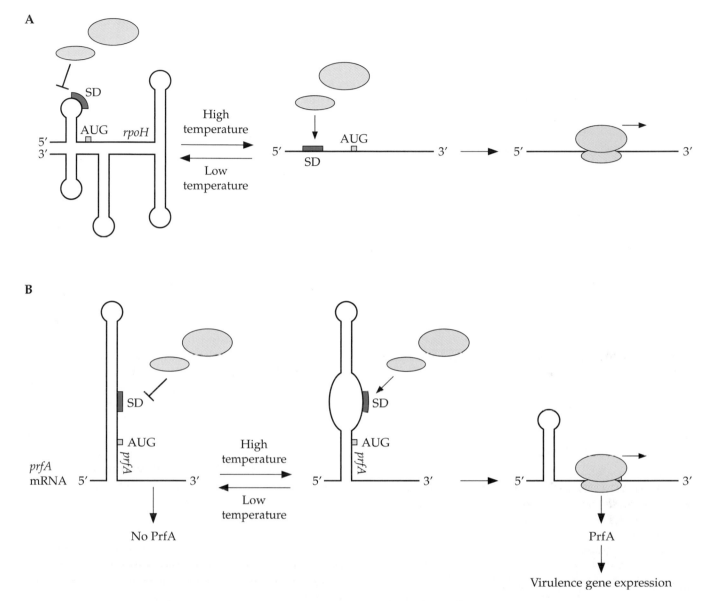

Figure 14–11 RNA thermosensing and regulation of translation. At low temperatures, mRNA forms a hairpin loop structure that prevents binding of the ribosome to the RBS (which is also called the SD region). At higher temperatures, the hairpin loop stem destabilizes and opens the SD site so that the ribosome (light-blue ovals) can bind and translation can ensue. **(A)** Regulation of σ^{32} expression in *E. coli*. **(B)** Regulation of PrfA expression in *L. monocytogenes*.

between sRNAs and their RNA targets usually involves relatively short segments of imperfect pairing, whereas antisense RNAs perfectly pair with their other strand targets. Intracellular levels of RNAα are increased under iron-rich conditions. Transcription of RNAα starts from the Pα promoter and in this instance is positively regulated by the Fur protein. Transcription of *fatDCBA* is initiated from the *fat* promoter and is repressed by the Fur protein. Binding of the RNAα transcript to the *fatDCBA* mRNA transcript

promotes processing upstream of *fatA*, thereby lowering the expression of FatA. Binding of RNAα to the *fatB* mRNA also causes degradation of the *fatB* transcript by modifying its structure to enhance the accessibility of this region of the mRNA to ribonucleases.

Protein activity-modifying sRNAs, as the name implies, act by binding to and modifying the activity of target proteins. As an example, *csrB* and *csrC* transcribe two sRNAs 268 and 245 nucleotides in length,

A

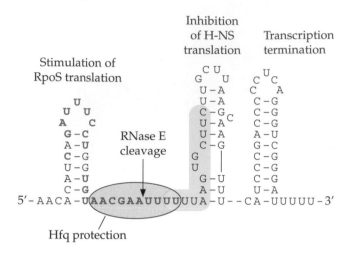

B

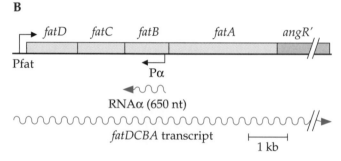

Figure 14–12 Translation regulation by complementary regulatory RNAs. Many regulatory RNAs regulate translation or mRNA stability by base pairing with a complementary sequence in target mRNAs to form duplexes. sRNAs are synthesized from sites distant from the target and perform imperfect base pairing with their targets that is facilitated by RNA chaperone proteins, such as Hfq. Antisense RNAs are synthesized from the opposite strand to the target RNA and involve long base-paired regions. **(A)** DsrA is an sRNA that positively regulates the stationary-phase sigma factor RpoS and negatively regulates the histone-like protein HNS. Hfq binding to the site indicated (grey oval) protects DsrA against RNase E cleavage and facilitates DsrA binding to complementary target sequences in the *rpoS* (bold blue) or *hns* (shaded blue) mRNA in the region of the RBS. **(B)** Antisense-RNAα regulation of the *fat* operon in *V. anguillarum*.

respectively, which bind to and inhibit the translational regulator CsrA, a protein involved in regulation of carbon metabolism in *E. coli* (Figure 14–13A). The *csrB* and *csrC* transcripts appear to have the same function of negatively regulating glycogen biosynthesis, glyconeogenesis, and glycogen catabolism while positively regulating glycolysis. Both sRNA transcripts have repeat sequences (18 in CsrB and 9 in CsrC) that form hairpin loops with a conserved motif sequence of CAGGXXG (where X is any nucleotide). These loops bind and sequester multiple copies of the CsrA protein. Titration of CsrA by CsrB/CsrC opens up the RBSs and allows translation to occur. CsrA homologs found in other bacteria are frequently involved in regulation of gluconeogenesis, biofilm formation, and virulence factor expression.

The pathogen *Erwinia carotovora* causes soft-rot disease in plants. *Erwinia* species produce a number of extracellular proteins, polysaccharides, and small secondary metabolites, including a diffusible quorum-sensing signal, *N*-(3-oxohexanoyl)-L-homoserine lactone (OHL) (quorum sensing will be discussed further below). Production of these substances is greatly increased during late exponential and early stationary growth phases and is regulated by at least three sensory mechanisms. The first involves the alternative stationary-phase sigma factor RpoS. The second involves quorum sensing by OHL, which, in conjunction with the LuxR response regulator homolog, regulates the production of a number of stationary-phase proteins. The third system, depicted in Figure 14–13B, consists of the RNA-binding protein RsmA (for *reg*ulator of *s*econdary *m*etabolism), which in turn regulates, along with *rsmB* RNA, the expression of a number of the extracellular proteins, phytohormones, antibiotics, pigments, polysaccharides, flagella, and secondary metabolites, such as OHL. RsmC is the global regulator that controls the production of RsmA and *rsmB* mRNA. RsmA promotes mRNA instability by binding to the region near the RBSs of target gene transcripts, blocking translation and exposing the now untranslated mRNA to degradation. The primary transcript of the untranslated *rsmB* mRNA is processed at its 5′ end upon binding of RsmA, yielding the functional *rsmB′* sRNA, 259 nucleotides in length. Binding of RsmA to *rsmB′* sRNA depletes the pool of free RsmA, thereby sequestering RsmA and preventing it from promoting the degradation of target gene mRNA.

Quorum Sensing

For a long time, bacteria were thought to exist as individual cells that did not communicate with each other. We now know that bacteria are able to coordinate their activities as a group through small diffusible signaling molecules called **autoinducers,** which are usually produced under positive regulatory feedback control. Bacteria use autoinducers to sense the population densities of not only themselves, but also other bacterial species. Acting as a group (a **quorum)**

A

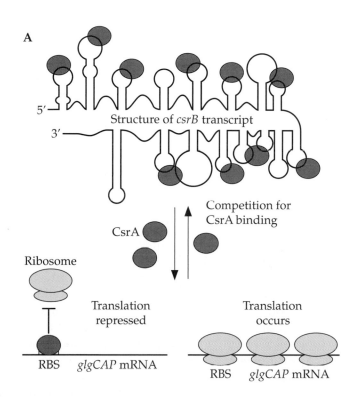

Structure of *csrB* transcript

Competition for CsrA binding

CsrA

Ribosome

Translation repressed

Translation occurs

RBS   *glgCAP* mRNA

RBS   *glgCAP* mRNA

B

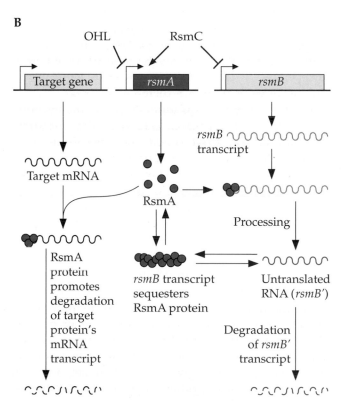

OHL   RsmC

Target gene   *rsmA*   *rsmB*

rsmB transcript

Target mRNA

RsmA

Processing

RsmA protein promotes degradation of target protein's mRNA transcript

rsmB transcript sequesters RsmA protein

Untranslated RNA (*rsmB'*)

Degradation of *rsmB'* transcript

Figure 14–13 Translation regulation by protein activity-modifying sRNAs. Some sRNAs stimulate the translation of target genes indirectly by binding to and sequestering, or modifying the activity of, proteins that act as translation repressors. Inhibition of translation by these repressor proteins often stimulates degradation of the mRNA, which is normally covered by translating ribosomes. **(A)** Regulation of CsrA by CsrB in *E. coli*. **(B)** The RsmA-*rsmB'* regulatory system in *E. carotovora*.

allows the individual bacteria in that group to benefit from the activity of the entire group. This has significant consequences for many pathogens, where the outcome of the interaction between host and bacterium is strongly affected by the bacterial population size. Coupling the production of virulence factors with cell population density ensures that the mammalian host does not have enough time to mount an effective defense against attack by pathogens. Population sensing also tells the bacteria when a host site is becoming saturated and that it may be time to spread to a new site through the production of toxins and flagella. Such a strategy depends on the ability of an individual bacterial cell to sense other members of the same species and, in response to that signal, differentially express specific sets of genes. Such cell-cell communication is called **"quorum sensing"** and involves the direct or indirect activation of a cognate receptor protein by the autoinducer, which results in up- or downregulation of specific genes that are under the control of that regulator. Quorum sensing is a sophisticated strategy for a bacterial population to coordinately regulate the expression of virulence factors (e.g., toxins, adhesins, flagella, capsules, siderophores, antibiotic resistance, biofilm formation, and secretion systems), but it also regulates a number of other cellular processes, including bioluminescence, sporulation, and mating.

The quorum-sensing signaling molecules used by gram-negative bacteria for intraspecies communication differ from those used by gram-positive bacteria (Figure 14–14A, B). Possibly because of the inherent differences in their cell morphologies, the signaling circuitries used by gram-negative and gram-positive bacterial autoinducer systems also differ (Figure 14–15). However, it should be pointed out that as we learn more and more about the different quorum-sensing circuitries in different bacteria, it is clear that the signaling pathways involved are much more complex than what is depicted here. It is beyond the scope of this textbook to cover all of the myriad permutations that bacteria have adapted to relay their messages. Instead, we provide here a generalized mechanism for each, followed by some examples.

Most gram-negative bacteria use **N-acyl homoserine lactones (AHLs)** as intraspecies quorum-sensing signals (Figure 14–14A), although recently other signaling molecules, such as the **α-hydroxyketones (AHKs)** (Figure 14–14D), have been discovered in some bacteria. The AHL and AHK autoinducers are small, relatively hydrophobic, membrane-diffusible molecules that can readily traverse bacterial cell membranes (Figure 14–15A). AHL is made from a fatty acid biosynthetic precursor, acyl-ACP, and S-adenosylmethionine (SAM) (Figure 14–16A and B). AHK is also made from the condensation of a fatty acyl-CoA with another small amino acid molecule, such as 2-amino-butyrate, through a pyridoxal phosphate-dependent reaction, followed by conversion of the amino group into a hydroxyl group (Figure 14–16C). Binding of the AHL or AHK ligand to its respective receptor protein results in dimerization of the regulator, which then binds to the promoter regions of target genes and thereby modulates the up- or downregulation of those genes.

Gram-positive bacteria, in contrast, use small post-translationally modified peptides as intraspecies autoinducers (Figure 14–14B). These **autoinducing peptides (AIPs)** are expressed as larger precursor peptides that are then processed into smaller, sometimes cyclic thiolactone-containing peptides that are transported across the bacterial cell membrane (Figure 14–17). These AIPs are too hydrophilic to cross membranes on their own and so remain in the extracellular medium once transported out of the bacterial cell. Sensing of the AIP occurs through binding of the AIP to its cognate receptor, which is a transmembrane protein located on the surface of the bacterium. The receptor is the histidine kinase part of a TCS (see above). Binding of the AIP to the sensor kinase induces autophosphorylation of the kinase on a conserved histidine residue, from which the phosphoryl group is transferred to an aspartate residue of the response regulator. The phosphorylated response regulator, in turn, binds to the promoter regions of target genes and activates or represses their transcription.

A different type of quorum-sensing signaling molecule is used for interspecies cell-cell communication. This **autoinducer 2 (AI-2)** is a small furanosyl borate diester compound (Figure 14–14C) that is produced in the catabolic degradation pathway of SAM and requires three enzymatic steps (Figure 14–16A), the last one involving LuxS, which converts S-ribosylhomocysteine into 4,5-dihydroxy-2,3-pentanedione (DPD). The DPD product then nonenzymatically cyclizes and converts into AI-2 in the presence of borate (Figure 14–16B). The advantage of using a molecule that is the product of a metabolic process common to all bacteria is that its production is a clear indicator of the presence of other bacteria but is independent of the type of bacteria that produce it. Thus, by detecting the concentration of this common metabolite, any given bacterium not only can sense the presence of other bacteria, but also can assess how many bacteria will be competing with it for colonization and growth. The receptor of AI-2 is a histidine kinase of a TCS.

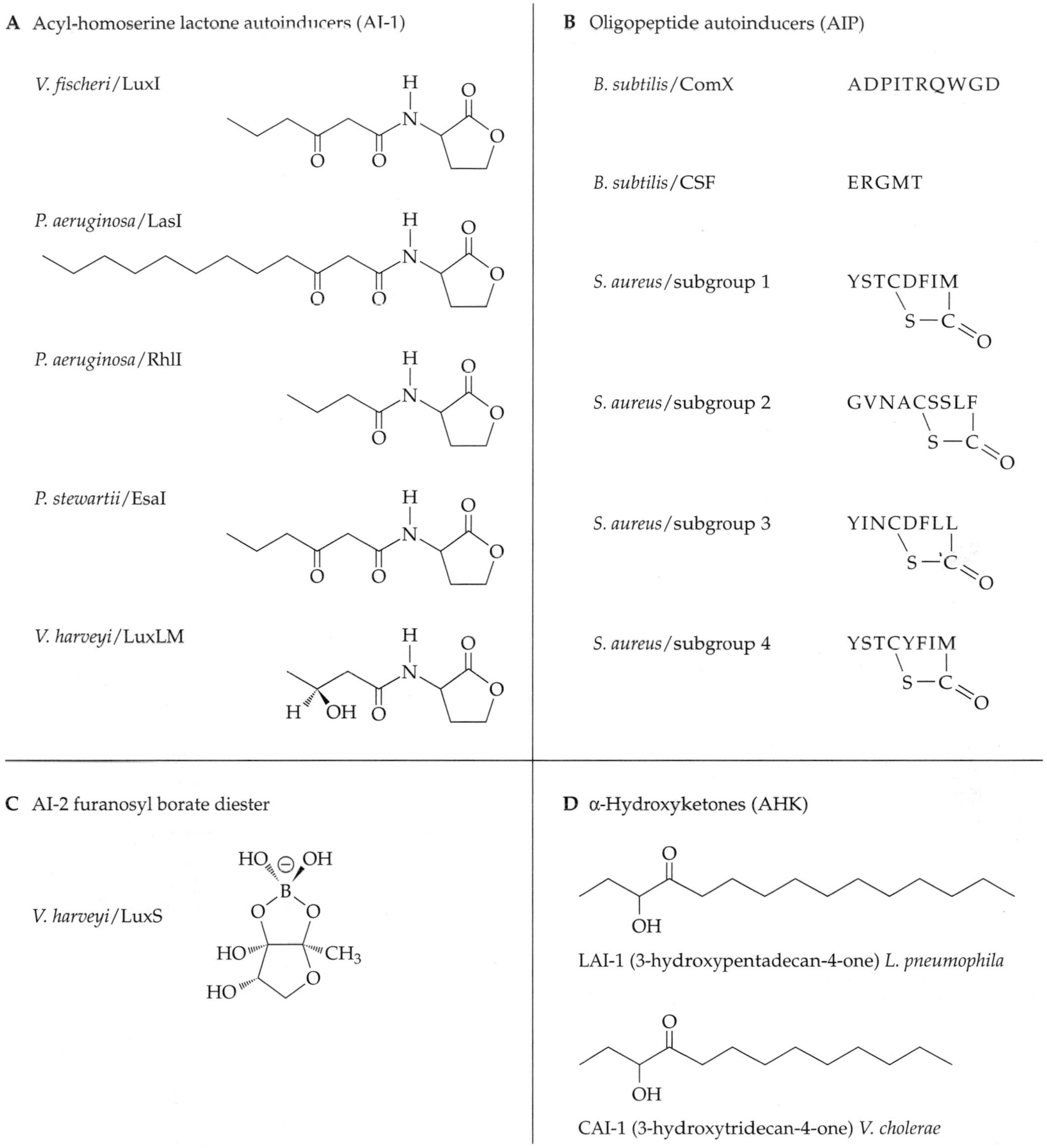

A Acyl-homoserine lactone autoinducers (AI-1)

V. fischeri/LuxI

P. aeruginosa/LasI

P. aeruginosa/RhlI

P. stewartii/EsaI

V. harveyi/LuxLM

B Oligopeptide autoinducers (AIP)

B. subtilis/ComX   ADPITRQWGD

B. subtilis/CSF   ERGMT

S. aureus/subgroup 1   YSTCDFIM

S. aureus/subgroup 2   GVNACSSLF

S. aureus/subgroup 3   YINCDFLL

S. aureus/subgroup 4   YSTCYFIM

C AI-2 furanosyl borate diester

V. harveyi/LuxS

D α-Hydroxyketones (AHK)

LAI-1 (3-hydroxypentadecan-4-one) *L. pneumophila*

CAI-1 (3-hydroxytridecan-4-one) *V. cholerae*

Figure 14–14 Quorum-sensing autoinducers. The quorum-sensing signaling molecules used in gram-negative bacteria differ from those of gram-positive bacteria. Shown are examples of autoinducers from different gram-negative and gram-positive bacteria. Also indicated are the names of corresponding proteins responsible for their biosynthesis. **(A)** Most gram-negative bacteria use AHL autoinducers. **(B)** Gram-positive bacteria use posttranslationally modified oligopeptides as autoinducers. Different *Staphylococcus aureus* subspecies produce different autoinducer peptides (e.g., subgroups 1 to 4) that can interfere with each other (bacterial interference [Figure 14–23]). **(C)** Furanosyl borate diester (AI-2) found in both gram-negative and gram-positive bacteria. **(D)** AHKs of *L. pneumophila* (LAI-1) and *V. cholerae* (CAI-1).

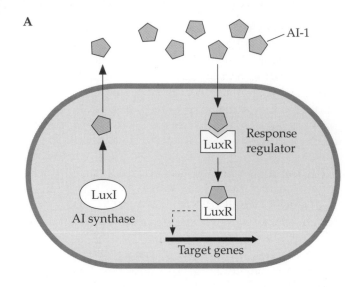

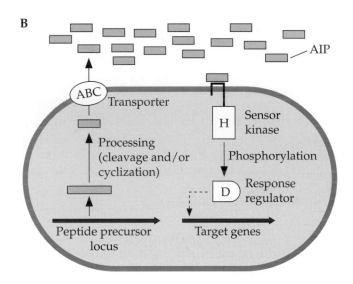

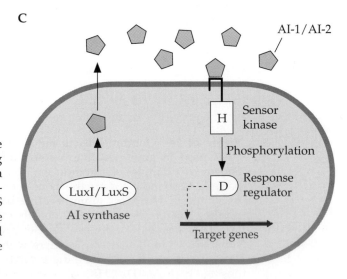

Figure 14–15 Quorum-sensing systems of gram-negative and gram-positive bacteria. **(A)** The AI-1 quorum-sensing system utilized for AHL and AHK signaling molecules in gram-negative bacteria involves a receptor protein response regulator that is not part of a TCS. **(B)** The AIP TCS quorum-sensing system of gram-positive bacteria. **(C)** The hybrid autoinducer TCS quorum-sensing systems utilized for AI-1 and AI-2 signaling molecules in gram-negative bacteria and for AI-2 in gram-positive bacteria.

Quorum Sensing in Gram-Negative Bacteria

THE LUX SYSTEMS OF *VIBRIO FISCHERI*.   Quorum sensing in gram-negative bacteria was first described in the bacterium *Vibrio fischeri*. *V. fischeri* has bioluminescent properties and is found as a symbiont with various marine animals, most notably in the light organs of the bobtail squid, *Euprymna scolopes*. Baby bobtail squid do not have light organs when they hatch, but as they develop, the squid secrete a mucus gel to which *V. fischeri* binds. Then, ciliated cells within the ducts leading to the light organs help the bacteria to migrate up to the light organs, where they colonize and grow. Once the density of the bacteria reaches a certain threshold, the bacteria turn on the Lux system, which produces bioluminescence, and the light organs begin to glow. The bioluminescence results from transcriptional activation of the bacterial Lux operon, *luxCDAB(F)EG*, where *luxA* and *luxB* encode the luciferase protein components and the other *lux* genes encode proteins that form a biosynthetic enzyme complex needed for synthesizing the long-chain aldehyde substrate that is oxidized by the luciferase to produce blue-green light (Figure 14–18A).

In the Lux system, the two proteins LuxI and LuxR are key to the regulation of the bioluminescence. LuxI is produced constitutively at low levels and is responsible for producing the quorum-sensing signal molecule, **AHL** (also called **acyl-HSL**, or **AI-1**, for **autoinducer 1**) (Figure 14–18B). When the cell density of the bacterial culture increases, the AHL reaches a threshold concentration that causes it to bind and activate the LuxR receptor protein through dimerization. The AHL-bound LuxR dimer then binds to the promoter region of the *lux* operon and positively regulates the *luxI* gene, making more LuxI and in turn more AHL, as well as the downstream *luxCDABEG* genes, which leads to production of luciferase and its substrate, which then generate bioluminescence (Figure 14–18C). Binding of activated AHL-LuxR also negatively regulates transcription of the divergent *luxR* gene, reducing the amount of the LuxR regulator.

OTHER AUTOINDUCER SYSTEMS IN *VIBRIO* SPECIES.   *V. fischeri* produces and responds to three different autoinducers: two different AHLs (one produced by LuxI and another produced by AinS), both of which are involved in regulation of bioluminescence gene expression, and AI-2 (produced by LuxS) for interspecies cell-cell communication. The Ain system is a hybrid system (Figure 14–19) in which AinR is a transmembrane sensor kinase that binds to the AHL signal molecule, octanoyl-homoserine lactone (C_8-HSL), produced by AinS and initiates the same signaling cascade as the AI-2 system (LuxPQ). In the absence of autoinducer, the sensor kinases LuxQ and AinR phosphorylate LuxU, which phosphorylates and activates the σ^{54}-dependent activator LuxO, which in turn expresses several regulatory sRNAs that, together with the RNA chaperone Hfq, destabilize *litR* rRNA. In the presence of autoinducer, the sensor kinases are converted into phosphatases, which dephosphorylate LuxU, which in turn represses LuxO and allows translation of the master transcriptional activator LitR, presumably because the sRNA that destabilizes *litR* mRNA is no longer made. LitR upregulates AinS and LuxR. LuxR can bind to the AHL produced by AinS, which in turn leads to low-level expression of LuxI, production of 3-oxohexanoyl-homoserine lactone (3-oxo-C_6-HSL), and bioluminescence. The Ain system induces bioluminescence at bacterial cell densities that precede activation of the Lux system. Once 3-oxo-C_6-HSL accumulates enough, it binds to LuxR and leads to even more induction of LuxIR signaling and increased bioluminescence. The sequential regulation of these systems in *V. fischeri* enables the bacteria to differentiate and respond to at least three conditions: low cell density, when no AHL autoinducer is sensed; intermediate cell density, when only C_8-AHL is sensed; and high cell density, when both C_8-AHL and 3-oxo-C_6-HSL are sensed.

In *V. cholerae*, multiple regulatory pathways control virulence and biofilm formation, which differ between the classical and El Tor pandemic strains. In the El Tor strain, quorum sensing involves two autoinducers, the AHL molecule CAI-1, produced by CqsA, and AI-2, produced by LuxS (Figure 14–20). Under low-cell-density conditions, the LuxPX and CqsS histidine kinases phosphorylate LuxU, when the autoinducers are at low concentrations. LuxU phosphorylates LuxO, which then transcribes sRNAs that destabilize HapR mRNA. Under high-cell-density conditions, the histidine kinases act as phosphatases to dephosphorylate the phosphorylated LuxU, expression of the sRNAs decreases, *hapR* mRNA is stabilized, and expression of HapR is increased. HapR downregulates the levels of the cyclic dinucleotide cyclic-di-GMP (c-di-GMP), which is a small signaling molecule. Decreased c-di-GMP leads to decreased *vps* transcription (biofilm formation) but increased virulence gene transcription. In the classical strain, the VieSAB sensory system, in response to an as-yet-unknown signal, controls c-di-GMP levels to control biofilm formation and virulence in a HapR-independent manner.

Recently, a new class of autoinducers, AHKs, was discovered in *V. cholerae*, *Legionella pneumophila*, and several other environmental bacteria, where they regulate host-pathogen interactions, bacterial virulence, and biofilm formation. Whereas different *Vibrio* spe-

A

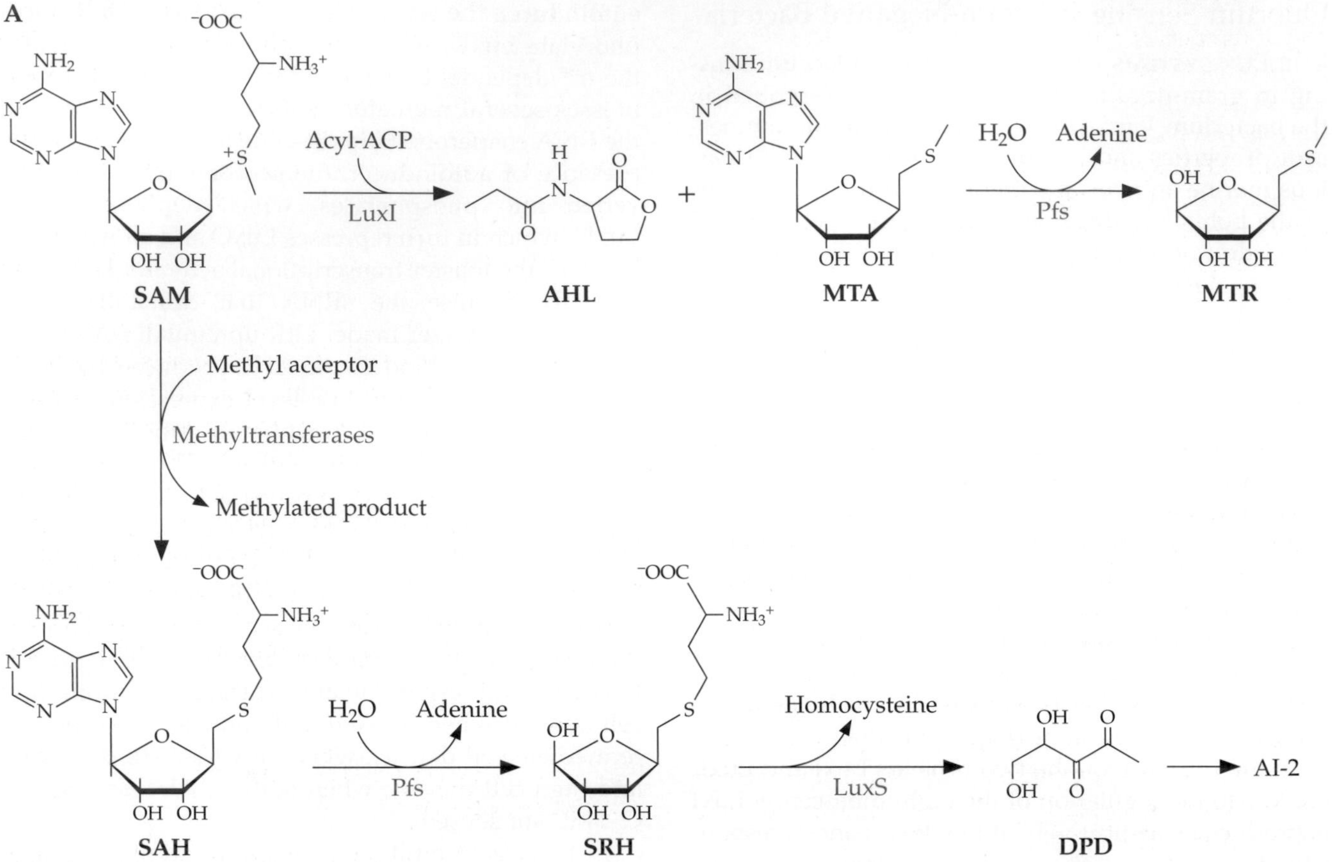

SAM → (Acyl-ACP / LuxI) → AHL + MTA → (H₂O, Adenine / Pfs) → MTR

Methyl acceptor

Methyltransferases

Methylated product

SAH → (H₂O, Adenine / Pfs) → SRH → (Homocysteine / LuxS) → DPD → AI-2

B

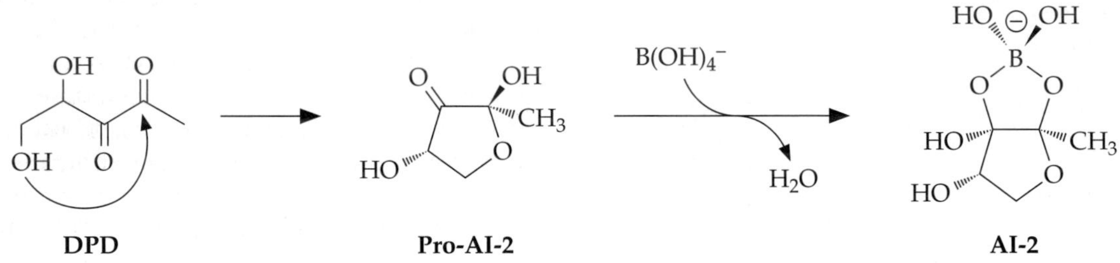

DPD → Pro-AI-2 → (B(OH)₄⁻ / H₂O) → AI-2

C

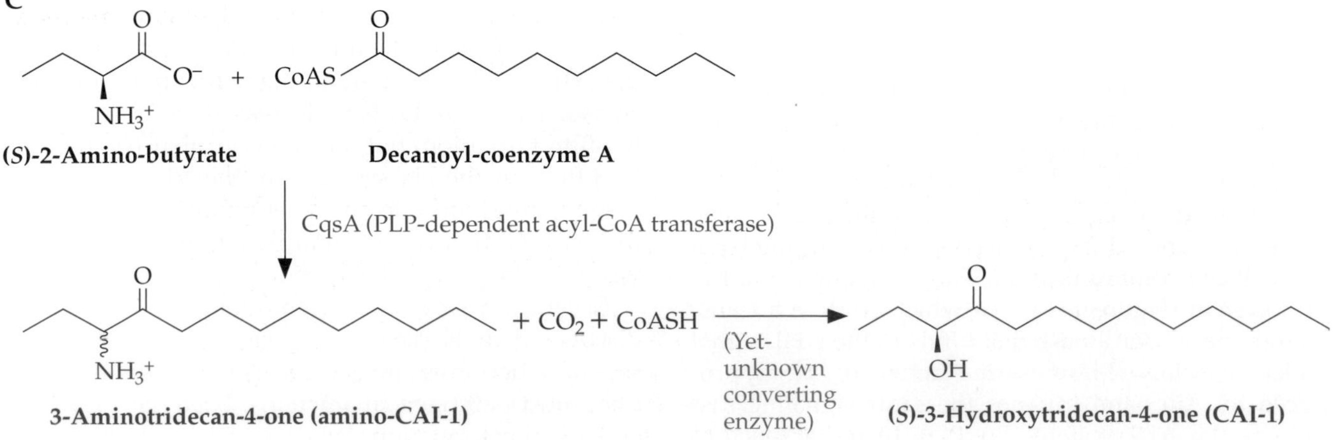

(S)-2-Amino-butyrate + Decanoyl-coenzyme A

CqsA (PLP-dependent acyl-CoA transferase)

3-Aminotridecan-4-one (amino-CAI-1) + CO₂ + CoASH → (Yet-unknown converting enzyme) → (S)-3-Hydroxytridecan-4-one (CAI-1)

cies use AI-2, AHL, and/or AHK signaling for intra- and interspecies communication, *L. pneumophila* employs only AHK-mediated quorum sensing. The AHK signaling systems of *L. pneumophila* and *V. cholerae* have many parallels (Figure 14–21), including homologs of AHK synthases (LqsA and CqsA), sensor kinases (LqsS and CqsS), TCSs (LetAS and VarAS), global repressors of transmissible traits (CsrA), small noncoding RNA regulators (RsmYZ and CsrBCD), and overall signal circuitry.

Quorum Sensing in Gram-Positive Bacteria

THE AGR SYSTEM OF *STAPHYLOCOCCUS AUREUS*. The Agr system in *S. aureus* was one of the first quorum-sensing systems elucidated in gram-positive bacteria (Figure 14–22). The expression of most virulence factors, including toxins and numerous surface proteins, is globally controlled by the *agr* locus, which contains genes for a two-component signaling system *(agrAC)*, for the biosynthesis of the activating ligand *(agrDB)*, and for the regulatory sRNA (RNAIII). The *agrD* gene encodes the 46-amino-acid precursor peptide of the cyclic thiolactone-containing AIP that is processed and transported out of the cell by the transmembrane protein AgrB. The thiol group of this cysteine is used to form the cyclic thiolactone structure of the secreted AIP. The secreted AIP serves as the ligand for the transmembrane receptor AgrC, which is the sensor histidine kinase that transduces the signal through phosphorylation of its cognate response regulator, AgrA. Phosphorylated AgrA activates transcription from the *agr* P2 and P3 promoters, which in turn upregulate the expression of the *agrACDB* genes through the P2 promoter in one direction and expression of the *agr* RNAIII through the P3 promoter in the opposite direction. RNAIII is the global regulator that upregulates genes for virulence factors and secreted proteins while downregulating genes for certain surface proteins.

BACTERIAL INTERFERENCE. The mature AIPs from different staphylococcal species consist of 7 to 9 amino acids with a highly conserved cysteine at a position located 5 residues from the C terminus (Figure 14–23). The *agr* genes show considerable sequence variation but can be divided into four distinct groups based on the specificities of the AIPs and their cognate histidine kinase receptors. Variations in the structures of the AIPs have compensatory variations in the sequences of the corresponding AgrB processor and AgrC receptor proteins at sites involved in protein-ligand interactions. AIPs from different staphylococcal species interfere with the signaling from each other, so that AIPs of one group activate the *agr* response within the same group but inhibit *agr*-mediated regulation in strains producing different AIPs. It is thought that an AIP binds its cognate AgrC receptor in a different manner than it does a receptor from another group. This interesting mechanism of bacterial interference has been exploited for the development of novel therapeutics against staphylococcal infections (Figure 14–24).

New Theory: Antibiotics and Quorum Sensing

A new theory is emerging regarding antibiotics and their roles in natural environments. The idea is that the majority of low-molecular-weight organic compounds made and secreted by microbes (i.e., antibiotics) play roles as cell-signaling molecules in the environment. The proposal is that these molecules modulate interactions among bacteria within a microbial community in the environment. This theory is supported by the findings that many antibiotics produced by one microbe at low concentrations can regulate gene transcription in another microbe. This idea has implications for the role of natural environments in the evolution of antibiotic resistance mechanisms, namely, that some resistance genes may have originated through selection in natural environments as

Figure 14–16 Biosynthesis of the autoinducers AHL, AI-2, and AHK. **(A)** Like AHL, AI-2 is derived from SAM in its normal catabolic degradation pathway. The last enzymatic step involves LuxS, which produces DPD. SAM, *S*-adenosyl-L-methionine; SAH, *S*-adenosyl-L-homocysteine; SRH, *S*-ribosyl-L-homocysteine; MTA, 5′-methylthioadenosine; MTR, 5′-methylthioribose; DPD, 4,5-dihydroxy-2,3-pentanedione. **(B)** Conversion to AI-2 occurs nonenzymatically in the presence of borate. **(C)** AHKs, such as CAI-1 from *V. cholerae*, are derived from reaction of a fatty acyl-CoA with 2-aminobutyrate through the action of a pyridoxal phosphate (PLP)-dependent acyl-CoA transferase, followed by conversion of the amino group into a hydroxyl group by an as-yet-unidentified enzyme. CoASH, coenzyme A.

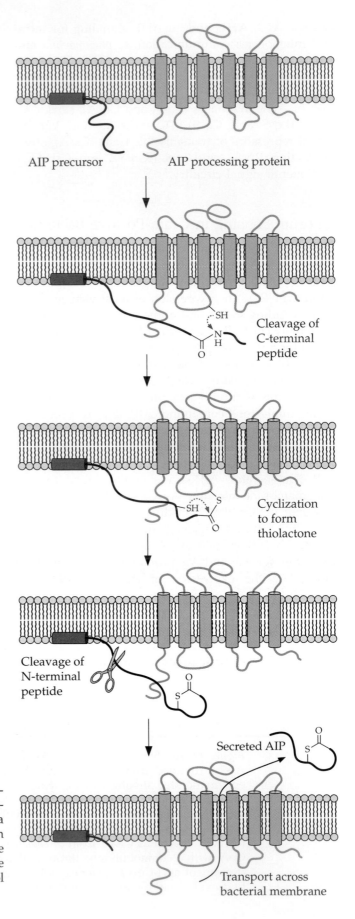

AIP precursor   AIP processing protein

Cleavage of
C-terminal
peptide

Cyclization
to form
thiolactone

Cleavage of
N-terminal
peptide

Secreted AIP

Transport across
bacterial membrane

Figure 14–17 Processing of AIPs. In gram-positive bacteria, the quorum-sensing AIP is derived from a larger precursor peptide that is posttranslationally modified into a smaller peptide by a transmembrane processing protein that mediates the catalytic cleavage and transport of the AIP across the bacterial membrane. During the process, the peptide is often cyclized into a thiolactone using the thiol side chain of a cysteine residue.

A

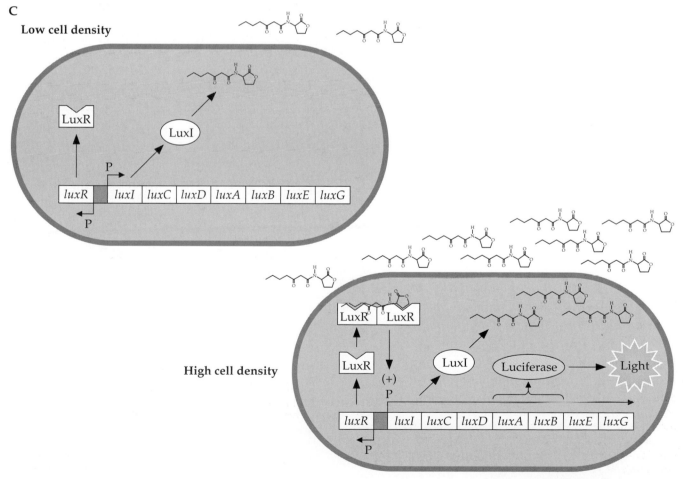

B

Acyl-ACP

SAM

AHL autoinducer

C

Low cell density

High cell density

(see legend for Figure 14–18 on next page)

Figure 14–18 Luciferase-catalyzed bioluminescence. In the Lux system of *V. fischeri*, bioluminescence is generated through the actions of two sets of proteins, one responsible for the production of the luciferase (the *luxCDABEG* operon) and the other responsible for regulating the Lux system through quorum sensing (the *luxIR* operon). **(A)** Luciferase generates blue-green light through the oxidation of reduced flavin mononucleotide ($FMNH_2$) and a long-chain fatty aldehyde (RCHO) in the presence of oxygen. **(B)** LuxI catalyzes the reaction for producing the autoinducer acyl-HSL. **(C)** Density-dependent light production. Acyl-HSL freely diffuses across the bacterial cell membrane to bind to the LuxR regulator. The acyl-HSL–LuxR complex positively regulates the *luxI* and downstream *lux* genes, which results in generation of blue-green light.

Figure 14–19 Quorum-sensing systems of *V. fischeri*. *V. fischeri* has two AHL quorum-sensing systems, the hybrid AinSR system and the LuxIR system, which work together in a sequential manner to regulate motility, colonization, and bioluminescence. PO_4, inorganic phosphate ion. (Adapted from Milton, 2006, with permission from Elsevier.)

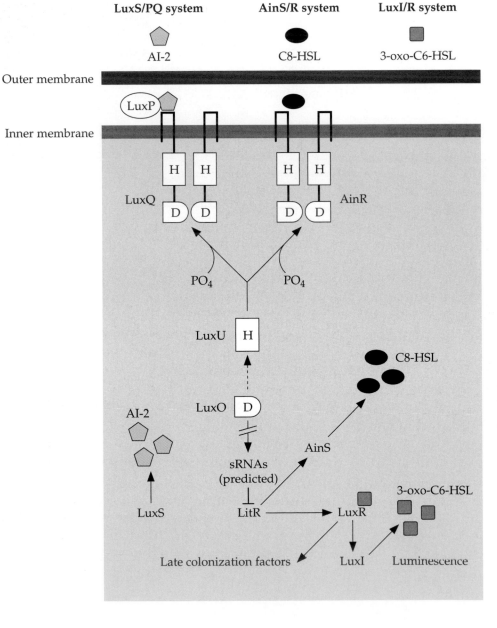

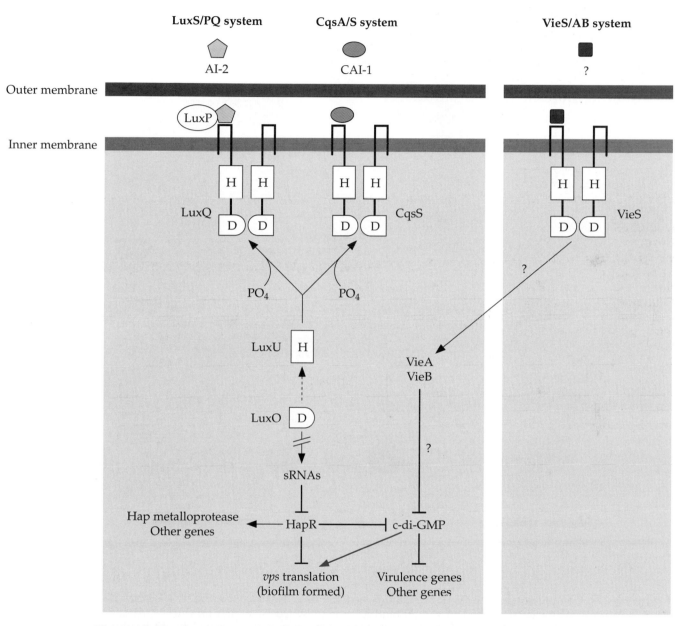

Figure 14–20 Quorum sensing in *V. cholerae*. Multiple regulatory pathways control biofilm formation and virulence gene expression in the classical and El Tor pandemic strains of *V. cholerae*. Under high cell densities, *hapR* mRNA is stabilized and HapR downregulates c-di-GMP levels, thereby blocking biofilm formation and increasing virulence gene expression. Although still largely uncharacterized, the VieSAB sensory system in the classical strain appears to control c-di-GMP levels in a HapR-independent manner. PO_4, inorganic phosphate ion. (Adapted from Milton, 2006, with permission from Elsevier.)

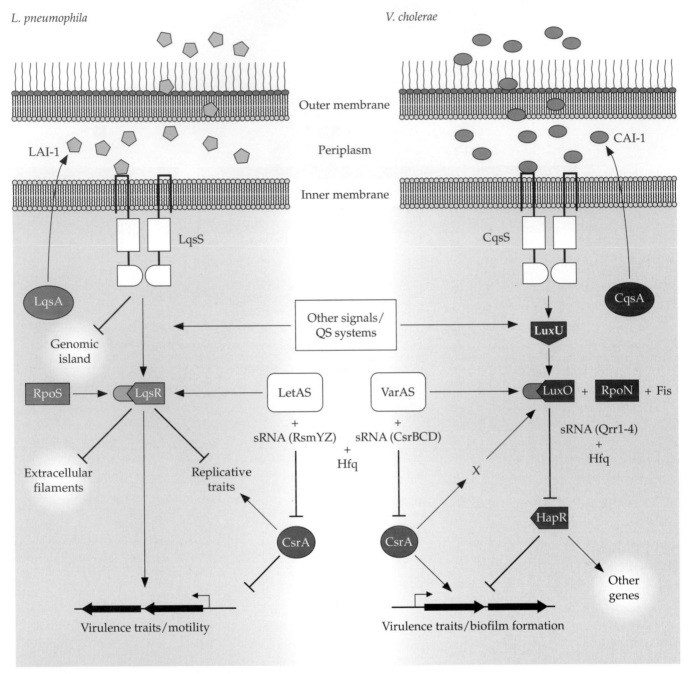

Figure 14–21 Comparison of the AHK regulatory systems in *L. pneumophila* and *V. cholerae*. (Right) In *V. cholerae*, AHK quorum sensing is comprised of the autoinducer synthase CqsA, which produces CAI-1, and the membrane-bound sensor kinase CqsS. At low bacterial cell density, the response regulator LuxO is phosphorylated by the phosphotransferase LuxU. Along with the sigma factor RpoN (σ^{54}) and the small nucleoid protein Fis, phosphorylated LuxO induces expression of sRNAs *qrr1* to *qrr4*, which, with the RNA chaperone Hfq, destabilize the *hapR* mRNA and thereby inhibit expression of the global regulator HapR. At high bacterial cell density, LuxO is dephosphorylated, the *qrr1* to *qrr4* sRNAs are not made, and HapR is produced and represses virulence and biofilm formation. The two-component system VarAS regulates expression of the sRNAs *csrBCD*, which inhibit the global regulator CsrA and thereby also regulate expression of HapR. The intermediate "X" has not yet been identified. (Left) In *L. pneumophila*, AHK quorum sensing is comprised of the autoinducer synthase LqsA, which produces LAI-1, and the membrane-bound sensor kinase LqsS and the response regulator LqsR, which induce expression of virulence factors and motility. The stationary sigma factor RpoS (σ^{38}) and the two-component system LetAS upregulate the expression of the sRNAs *rsmYZ*, which, along with Hfq, sequester the RNA-binding global repressor CsrA. (Adapted from Tiaden et al., 2010, with permission from Elsevier.)

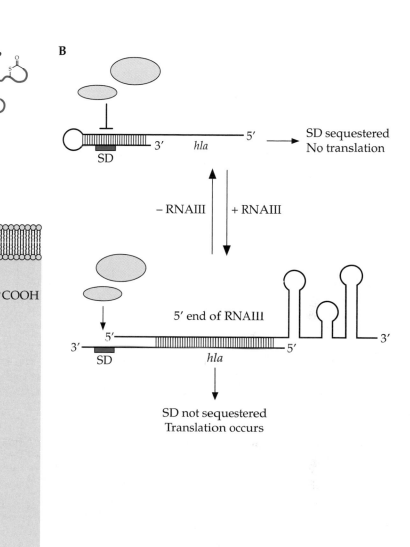

A

Interfering AIP

AIP

AgrB

H₂N

AgrC

COOH

AgrD

P2 P3

agrA   agrC   agrD   agrB   RNAIII

RNAIII

↑ Exotoxins
↓ Surface proteins

B

SD

3′   *hla*   5′

SD sequestered
No translation

− RNAIII   + RNAIII

3′   5′
SD   *hla*   5′

5′ end of RNAIII

3′

SD not sequestered
Translation occurs

Figure 14–22 The Agr quorum-sensing system of staphylococci. **(A)** The Agr quorum-sensing system controls the expression of virulence genes and surface proteins in gram-positive staphylococci. **(B)** RNAIII regulation of virulence genes in *Staphylococcus aureus*. The inhibitory hairpin secondary structure of the mRNA blocks access of the ribosome to the SD region of the *hla* mRNA. Binding of the 5′ end of the RNAIII sRNA transcript perturbs this hairpin structure, which reveals the SD site so that the ribosome can bind and translation can proceed. (Adapted from Novick and Geisinger, 2008, with permission from Annual Reviews, Inc.)

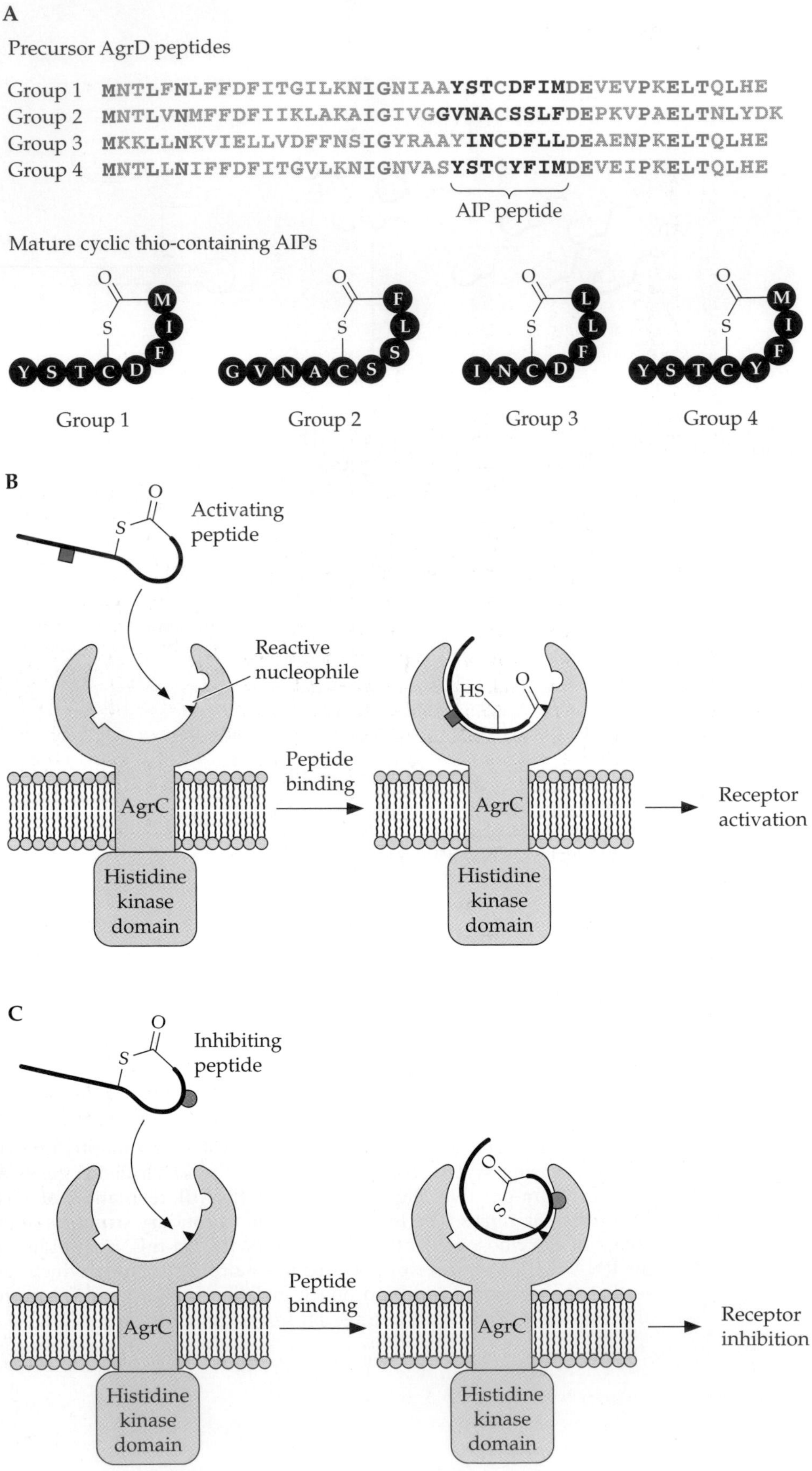
A
Precursor AgrD peptides
Group 1 MNTLFNLFFDFITGILKNIGNIAAYSTCDFIMDEVEVPKELTQLHE
Group 2 MNTLVNMFFDFIIKLAKAIGIVGGVNACSSLFDEPKVPAELTNLYDK
Group 3 MKKLLNKVIELLVDFFNSIGYRAAYINCDFLLDEAENPKELTQLHE
Group 4 MNTLLNIFFDFITGVLKNIGNVASYSTCYFIMDEVEIPKELTQLHE
AIP peptide
Mature cyclic thio-containing AIPs
Group 1
Group 2
Group 3
Group 4
B
Activating peptide
Reactive nucleophile
Peptide binding
AgrC
Histidine kinase domain
Receptor activation
C
Inhibiting peptide
Peptide binding
AgrC
Histidine kinase domain
Receptor inhibition

Figure 14–23 Bacterial interference in staphylococci. **(A)** Sequences of precursor AgrD peptides from *S. aureus* groups 1 to 4. **(B)** Activation of AgrC receptor by cognate AIP ligand. **(C)** Inhibition of AgrC receptor by AIP ligand from a different staphylococcal group. (Panels B and C adapted from P. Mayville, G. Ji, R. Beavis, H. Yang, M. Goger, R. P. Novick, and T. W. Muir. 1999. Structure-activity analysis of synthetic autoinducing thiolactone peptides from *Staphylococcus aureus* responsible for virulence. *Proc. Natl. Acad. Sci. USA* **96**:1218–1223, with permission from the National Academy of Sciences of the United States of America.)

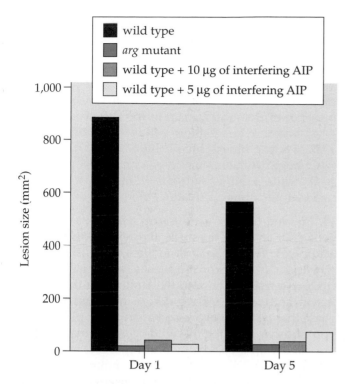

Figure 14–24 Experiment demonstrating control of staphylococcal infection through bacterial interference. Staphylococcal skin abscess formation in mice could be inhibited by treatment with a thiolactone-containing AIP that could compete with the wild-type AIP for binding to the AgrC receptor but could not induce signaling, which effectively blocked virulence in the inoculated wild-type bacteria. A mutant bacterium lacking the *agr* locus (*agr* mutant) was used as a negative control. Shown are the sizes of the skin lesions caused by injection of the wild-type staphylococci or *agr* mutant in the presence or absence of interfering AIP. (Adapted from P. Mayville, G. Ji, R. Beavis, H. Yang, M. Goger, R. P. Novick, and T. W. Muir. 1999. Structure-activity analysis of synthetic autoinducing thiolactone peptides from *Staphylococcus aureus* responsible for virulence. *Proc. Natl. Acad. Sci. USA* **96:**1218–1223, with permission from the National Academy of Sciences of the United States of America.)

an adaptive phenotypic or genotypic response to the presence of other bacteria in a microbial community and were then acquired by medically relevant bacteria through horizontal gene transfer.

SELECTED READINGS

Aras, R. A., J. Kang, A. I. Tschumi, Y. Harasaki, and M. J. Blaser. 2003. Extensive repetitive DNA facilitates prokaryotic genome plasticity. *Proc. Natl. Acad. Sci. USA* **100:**13579–13584.

Babitzke, P., and T. Romeo. 2007. CsrB sRNA family: sequestration of RNA-binding regulatory proteins. *Curr. Opin. Microbiol.* **10:**156–163.

Dorman, C. J., and C. P. Corcoran. 2009. Bacterial DNA topology and infectious disease. *Nucleic Acids Res.* **37:** 672–678.

Freitag, N. E., G. C. Port, and M. D. Miner. 2009. *Listeria monocytogenes*—from saprophyte to intracellular pathogen. *Nat. Rev. Microbiol.* **7:**623–628.

Henkin, T. M. 2008. Riboswitch RNAs: using RNA to sense cellular metabolism. *Genes Dev.* **22:**3383–3390.

Milton, D. L. 2006. Quorum sensing in vibrios: complexity for diversification. *Int. J. Med. Microbiol.* **296:**61–71.

Ng, W. L., and B. L. Bassler. 2009. Bacterial quorum-sensing network architectures. *Annu. Rev Genet.* **43:** 197–222.

Novick, R. P., and E. Geisinger. 2008. Quorum sensing in staphylococci. *Annu. Rev. Genet.* **42:**541–564.

Sandegren, L., and D. I. Andersson. 2009. Bacterial gene amplification: implications for the evolution of antibiotic resistance. *Nat. Rev. Microbiol.* **7:**578–588.

Schauder, S., K. Shokat, M. G. Surette, and B. L. Bassler. 2001. The LuxS family of bacterial autoinducers: biosynthesis of a novel quorum-sensing signal molecule. *Mol. Microbiol.* **41:**463–476.

Stoebel, D. M., A. Free, and C. J. Dorman. 2008. Antisilencing: overcoming H-NS-mediated repression of transcription in gram-negative enteric bacteria. *Microbiology* **154:**2533–2545.

Tiaden, A., T. Spirig, and H. Hilbi. 2010. Bacterial gene regulation by α-hydroxyketone signaling. *Trends Microbiol.* **18:**288–297.

Yanofsky, C. 2007. RNA-based regulation of genes of tryptophan synthesis and degradation in bacteria. *RNA* **13:**1141–1154.

Yim, G., H. H. Wang, and J. Davies. 2007. Antibiotics as signaling molecules. *Philos. Trans. R. Soc. Lond. B Biol. Sci.* **362:**1195–1200.

QUESTIONS

1. Describe the means by which bacteria sense and respond to changes in their environment.

2. Why would having regulated virulence genes be an advantage to a bacterium?

3. How do simple activators differ from two-component regulatory systems?

4. What type of mutation(s) would cause genes regulated by a two-component regulatory system to be expressed constitutively?

5. How do sRNAs regulate virulence genes?

6. Briefly describe the mechanism by which rho-independent termination occurs.

7. Compare the autoinducers of gram-negative and gram-positive bacteria.

8. Osmolarity, temperature, bile salts, and other environmental signals regulate the production of *V. cholerae* virulence factors.

> Name the key transcriptional regulatory systems responsible for toxin production found in *V. cholerae*. Where are they located in the cell?

> What roles do these proteins play in the regulation of cholera toxin and the Tcp pilus?

9. Describe how the regulatory mechanism of quorum sensing might play a role in the pathogenesis of a *Vibrio* infection and how it works using the example of Lux (bacterial luminescence).

10. Describe how the regulatory mechanism of quorum sensing might play a role in the pathogenesis of staphylococcal infection and how it works using the example of the *AgrABCD* locus.

11. Bacteria adapt to changes in their environment and alter regulation of gene expression accordingly. One of the mechanisms of regulation in bacteria involves transduction of an environmental signal. Explain the functions of bacterial sensors and transducers for gene regulation. How do TCSs control gene expression in response to environmental changes? Give examples.

12. The *B. pertussis* Bvg system is essential for virulence.

> **A.** What type of system is this, and what are the representative proteins?

> **B.** How is the Bvg system unique, and how does it become activated?

> **C.** List two virulence factors that Bvg controls (either directly or indirectly).

13. You are using a gene fusion between a very bad gene *(vbg)* and *lacZ* to study the regulation of *vbg* by iron. (When *lacZ* is expressed, cells are blue; when *lacZ* is not expressed, cells are white.) You introduce a transposon into the strain carrying a *vbg-lacZ* fusion and screen for regulatory mutants. You conclude that

vbg is expressed in the presence of iron because you observe that, for the wild-type fusions, colonies are _____ in the presence of iron and _____ in the absence of iron. You conclude that the transposon has inserted into a gene encoding an activator protein because you observe that colonies are _____ in the presence of iron and _____ in the absence of iron.

> **A.** blue ... white ... blue ... blue
> **B.** white ... blue ... blue ... blue
> **C.** blue ... white ... white ... white
> **D.** white ... blue ... white ... white
> **E.** white ... blue ... white ... blue

14. You are using a gene fusion between a virulence gene *(vir)* and *lacZ* to study the regulation of *vir* by acidic pH. (When *lacZ* is expressed, cells are blue; when *lacZ* is not expressed, cells are white.) You introduce a transposon into the strain with the *vir-lacZ* fusion and screen for regulatory mutants. You conclude that *vir* is not expressed at an acidic pH because you observe that, for the wild-type fusions, colonies are _____ on an acidic medium (pH 5.0) and _____ on a neutral medium (pH 7.0). In a mutant strain, you conclude that the transposon has inserted into a gene encoding a repressor protein because you observe that colonies are _____ on the acidic medium and _____ on the neutral medium.

> **A.** blue ... white ... blue ... blue
> **B.** white ... blue ... blue ... blue
> **C.** blue ... white ... white ... white
> **D.** white ... blue ... white ... white
> **E.** white ... blue ... white ... blue

15. Using the following gene, answer the questions below. Make sure you clearly indicate what you are labeling in relation to a particular question.

```
5'TTGACACGAATCCGCATAGCTCCTATAATCAGCCTTA-
3'AACTGTGCTTAGGCGTATCGAGGATATTAGTCGGAAT-
5'CCCTTCCAGGAGGTACCGCTATGCGCGGATCGTATGT-
3'GGGAAGGTCCTCCATGGCGATACGCGCCTAGCATACA-
5'CTAA3'
3'GATT5'
```

> **A.** Label the promoter region of the gene.
> **B.** Label the start of transcription.
> **C.** Label the DNA strand that will serve as the template during transcription.
> **D.** Write out what the mRNA sequence would be for this gene.
> **E.** Label the RBS.
> **F.** Label the translation start codon.
> **G.** Where would be the most likely location for a regulatory sRNA to bind?
> **H.** Where would be the most likely location for a transcriptional repressor to bind?

1. You have purified a protein from a bacterium Y that causes cells in tissue culture to increase their cAMP levels. You subsequently cloned a fragment of DNA from bacterium Y that contained two ORFs, one of which you identified as the gene encoding the protein you isolated, which you called *yctA* (for Y bacterium cytotoxin A). You then determined that expression of the YctA protein occurs only under iron-restrictive conditions (i.e., when iron is absent from the medium).

A. Describe how you would determine if the expression of YctA is regulated by iron at the transcriptional level.

B. Describe how you would determine whether the other ORF, which you have named *yctR*, encodes a possible transcriptional regulator.

C. If YctR is a transcriptional regulator, how could you determine if it is an activator or repressor protein?

2. The PhoPQ TCS of the intracellular pathogen *Salmonella enterica* serovar Typhimurium, encoded by the genes *phoPQ*, is one of the best-studied global regulatory systems and has been shown to regulate the majority of virulence genes involved in intracellular survival in macrophages, resistance to antimicrobial peptides, acid tolerance, and phagosome remodeling to prevent phagolysosome fusion. In *S. enterica* serovar Typhimurium, a large number of virulence genes are regulated by a two-component regulatory system, which senses divalent cations, such as Mg^{2+}, Mn^{2+}, and Ca^{2+}, as well as low-pH environments. PhoQ is a dimeric transmembrane protein that binds divalent cations to form bridges with membranes that maintain a PhoQ-repressed state. Antimicrobial peptides and acidic pH activate PhoQ. PhoP has an aspartate that when phosphorylated binds to DNA, in particular HilA, the master regulator of *Salmonella* PAI 1, and controls expression of *phoP*-activated (*pag*) and *phoP*-repressed (*prg*) genes, of which there are over 200.

A. Which of the two components is the sensor kinase, and which is the response regulator?

B. What is the phenotype (virulence and expression of regulated proteins) that would result from a null mutation in the response regulator gene?

C. What is the phenotype (virulence and expression of regulated proteins) that would result from a response regulator mutation that led to constitutive activation of the response regulator?

D. Predict what the effects would be for each of the following mutations in activity (on or off) of a

lacZ fusion with the given gene under the given conditions. Be sure to state your rationale.

Mutation 1: effect of transposon insertion into *phoP* on *prg-lacZ* when *Salmonella* is in a phagosome.

Mutation 2: effect of transposon insertion into *phoQ* and constitutive expression of *phoP* from a plasmid on *pag-lacZ* when *Salmonella* is in a phagosome.

Mutation 3: effect of transposon insertion into *phoQ* on *prg-lacZ* when *Salmonella* is in the lumen of the gut.

Mutation 4: effect of transposon insertion into *hilA* on *prg-lacZ* when *Salmonella* is in the lumen of the gut.

Mutation 5: effect of transposon insertion into *phoQ* and constitutive expression of *phoP* from a plasmid on *pag-lacZ* when *Salmonella* is in a phagosome.

3. *C. diphtheriae* was the first bacterium for which the symptoms of disease could be explained entirely by the action of a single secreted protein, DT. DT is produced when iron levels are low. Describe in detail how you would determine if the *dtxR* gene encodes a transcriptional activator or repressor of DT gene *(tox)* expression under low-iron conditions.

4. A group of researchers have isolated from the lungs of several cystic fibrosis patients a bacterium Q that produces an unusual polysaccharide capsule called QPS. They can see QPS on the surfaces of the bacteria in electron micrographs from fresh isolates (bacteria just obtained from patients). However, after culturing the bacteria in vitro on agar plates, they find that most of the bacteria no longer have QPS on their surfaces. When the researchers grew the bacteria overnight in culture media, surprisingly, they found that some of the bacteria now had QPS on their surfaces. If they then incubated the cultures for longer periods, they found that more and more of the bacteria produced QPS on their surfaces. They also found that if the media from old cultures were filtered and added to bacteria lacking QPS on their surfaces, within a short time those bacteria also began to express QPS on their surfaces.

A. What possible mechanism could account for these observations?

B. The same researchers subsequently cloned two separate DNA sequences, one containing 2 ORFs, which they named *qpsAB*, and a second containing

(continued)

four ORFs, which they named *qpsCDEF*. The researchers suspected that the first operon might include two biosynthetic genes responsible for making QPS and the second operon might encode the regulatory system controlling expression of QPS. They then proceeded to conduct a series of experiments to test their hypothesis. The results follow.

- Introduction of the two *qps* operons into the gram-positive bacterium *Bacillus subtilis* was able to cause expression of QPS on the surface of the bacterium. Deletion of one or more of the *qps* genes resulted in complete loss of QPS expression, even though each of the remaining genes was expressed.

- Cellular fractionation, followed by sodium dodecyl sulfate-polyacrylamide gel electrophoresis (SDS-PAGE) and Western blot analysis using antibodies against the proteins encoded by *qpsC*, *qpsD*, *qpsF*, and *qpsE*, showed that QpsD and QpsF were membrane associated, QpsC was located in the cytosol, and QpsE was found mostly in the medium.

- The size of the QpsE peptide in the medium (10 residues) was much smaller than its size in the cytosol (45 residues). A synthetic peptide corresponding to the 10-residue peptide could not substitute for native QpsE peptide.

- Deletion of QpsF resulted in QpsE being localized as a 45-residue peptide only to the cytosol and not in the medium.

- Incubation of bacterial membranes with filtered media from old cultures in the presence of [γ-^{32}P]ATP, followed by SDS-PAGE and autoradiography, revealed that both QpsC and QpsD were radiolabeled. QpsD could radiolabel itself, but QpsC could not.

- Incubation of cell-free extracts containing QpsC, QpsD, QpsE, and QpsF with oligonucleotides including the promoter region of *qpsAB* showed that only QpsC was able to bind the DNA.

- A series of QpsD-PhoA fusions at various points in the 350-residue QpsD protein were constructed and tested in *B. subtilis*. The results are summarized in the table below:

QpsD-PhoA proteins: position of fusion (residue)	Alkaline phosphatase activity	Cellular localization of fusion protein
33	+	Medium
60	−	Membrane
105	+	Membrane
142	−	Membrane
160	+	Membrane
176	+	Membrane
200	−	Membrane
256	−	Membrane

Using the above information, provide a detailed model for QPS regulation in bacterium Q that accounts for all of the observed results. You may draw a diagram if that helps, but be sure to provide a clear role for each of the Qps proteins in your model. Based on the above information, predict the membrane topology of the QpsD protein. Draw a diagram of your model.

5. You are a physician scientist studying the mechanisms by which *Neisseria gonorrhoeae* can survive in the human host and are performing a variety of different experiments, including monitoring recurring *N. gonorrhoeae* infections in sexually active males. One of your patients is a sexually promiscuous 22-year-old man who has had gonorrhea five times in the last 2 years. Each time he visits your clinic to donate bacterial samples during an episode, he is given antibiotics, and the infection has cleared up. Throughout the course of the study, you have isolated a number of different clinical strains of *N. gonorrhoeae* from this individual.

A. List two reasons why *N. gonorrhoeae* is capable of causing recurrent infections in this same individual.

B. Could this patient be vaccinated to avoid these recurring infections? Provide your rationale.

C. Immunoblot analysis using a monoclonal antibody against the surface protein OpaC shows that OpaC is not detected in *N. gonorrhoeae* strain 1. However, you can PCR amplify the *opaC* gene using chromosomal DNA from this strain. Based on sequence analysis of the *opaC* gene, you detect differences at the 5′ end of the gene from this strain compared to a related strain that does produce OpaC. What is the specific type of variation that is occurring in regard to OpaC in this strain? Based on the experimental evidence, what is causing OpaC not to be produced in this strain? Explain very briefly how this is most likely occurring.

D. In *N. gonorrhoeae* strains 2 and 3 you have sequenced the pilin gene *pilE* and have denoted the minicassette variable regions in the gene for strain 2 A-B-C-D-E-F and those for strain 3 1-2-3-4-5-6. You isolate chromosomal DNA from strain 2 and add it to a culture of strain 3. After a period of time, you isolate colonies from the culture. Sequencing the *pilE* gene from one of the colonies (denoted *N. gonorrhoeae* strain 4) reveals that you have created a new PilE protein encoded by a gene with the minicassette regions 1-B-C-4-5-6. What has occurred in strain 4 to change the *pilE* minicassette regions to 1-B-C-4-5-6? Will polyclonal antibodies that recognize PilE from strain 2 also recognize PilE from strain 4? Explain.

E. Another strain (denoted *N. gonorrhoeae* strain 5) isolated at your patient's most recent visit appears to be a new strain that does not have pili (Pil⁻ phenotype). Using polyclonal antibodies generated against PilE proteins from strains 2 and 3, you are able to detect a PilE protein in cell extracts from strain 5 that is much larger than the PilE from strain 2 or 3, but this pilin is not detected on the surfaces of the bacteria. What event has occurred in strain 5 to give you the Pil⁻ phenotype? Explain why there is a Pil⁻ phenotype with strain 5 even though you can detect the presence of the PilE protein.

F. On the last visit from your patient, you also collect some blood and find that there are both immunoglobulin M (IgM) and IgG antibodies present in the patient's serum. You further find that the IgG antibodies recognize PilE from all of the strains, whereas the IgM antibodies do not recognize any of the PilE proteins. Interpret these results.

6. You are a senior research scientist working for a large agricultural firm that is interested in novel, nonpesticide ways of controlling plant insects. Your research group is studying pea aphids (plant insects) that possess several types of symbiotic bacteria. Pea aphids often harbor as a symbiont *Hamiltonella defensa*, which is a gram-negative bacterium related to *Salmonella*. It is believed that colonization with *H. defensa* confers substantial resistance to parasitic wasps, which are natural enemies of the aphids that lay their larvae inside the aphids. In aphids harboring *H. defensa*, the wasp larvae die prematurely, allowing the aphid hosts to develop into adults and reproduce. Recent genome-sequencing data have revealed that *H. defensa* possesses two PAIs that are prophages, each encoding a type III secretion system (T3SS) and putative virulence genes. The first PAI, called APSE-1, encodes a homolog of the catalytic subunit of Shiga toxin (encoded by *stx*) produced by *Shigella*, which inhibits protein synthesis by disrupting 28S rRNA. The second PAI, called APSE-2, encodes a homolog of the catalytic subunit of the cytolethal distending toxin (encoded by *cdtB*) produced by *Campylobacter jejuni*, which is a DNase with a nuclear localization sequence that targets it to the eukaryotic nucleus, where it interrupts cell cycle progression by arresting the cells in G_2 phase. It is possible that these APSE-encoded toxins might be selectively targeted to destroy larval cells of attacking parasitic wasps. However, neither of these PAIs includes genes that might be the corresponding binding subunits of the two putative toxins,

suggesting that the binding subunits are novel and not recognizable by sequence comparison or that the Stx- and CdtB-like proteins might instead be delivered to the wasp cells by the T3SS. To further characterize this fascinating symbiotic relationship, you have developed a system for genetically manipulating *H. defensa* that is based on *Salmonella* genetic tools, as well as a system for growing wasp larva cells in tissue culture.

A. Describe how you would go about experimentally demonstrating that the Stx-like and CdtB-like proteins are indeed delivered by the *H. defensa* T3SS into the wasp cells and that they end up in the expected cellular locations. Be sure to include the experimental design, the reagents used, and how you visualized and interpreted the data. (Hint: you may need more than one type of experiment to demonstrate this.)

B. It is thought that binding of *H. defensa* to a surface antigen on the wasp larva cell triggers the expression of the CdtB gene. Describe in detail how you would confirm quantitatively that CdtB expression is regulated at the transcriptional level by the wasp larva cell surface antigen.

C. Describe how you would go about identifying potential regulatory proteins of *cdtB* expression. Be sure to provide the reagents, the screening conditions, how you would visualize/measure the results, and how you would verify the results.

D. Using your strategy, you identify three genes (which you name *cdtR*, *cdtS*, and *cdtT*) that might be regulators of *cdtB* expression. Describe in detail how you would determine if the gene product of *cdtR* is a transcriptional activator or repressor of *cdtB* gene expression.

E. To demonstrate that the gene products of *cdtR* and *cdtS* directly regulate the *cdtB* gene, you purify the corresponding CdtR and CdtS recombinant proteins in *E. coli* and perform an experiment to demonstrate that they can bind to the promoter region of the *cdtB* gene. However, your results show that only CdtR can bind to the DNA. Provide (with rationale) a function for CdtS and at least one additional experiment that could be performed to confirm your hypothesis.

F. Despite numerous attempts, you discover that you cannot visualize the predicted CdtT protein in Western blots of bacterial cells, yet you were able to observe transcriptional regulation of the *cdtB* gene using the reporter system that you made as described above. After further analysis of the *cdtT*

(continued)

gene sequence, you discover that part of the sequence is complementary to the 5′ promoter regions of an *H. defensa* polysaccharide biosynthetic gene and several adhesin-like genes. Provide (with rationale) an explanation for these observations.

7. *Clostridium perfringens* is a gram-positive anaerobic pathogen that causes serious human and animal diseases, including clostridial myonecrosis (i.e., "flesh-eating" disease) and gas gangrene (i.e., "tissue rotting"). *C. perfringens* produces many extracellular enzymes and toxins, including alpha, theta, and kappa toxins, that are encoded by the genes *plc*, *pfoA*, and *colA*, respectively. Being interested in understanding the regulation of these toxin genes, you performed signature-tagged mutagenesis and isolated a number of transposon-insertion mutants that are avirulent in a mouse model of gas gangrene infection. Sequence analysis of these mutants revealed that five of these avirulent mutants, Cpmut1 to Cpmut5, were in genes unrelated to the toxin genes, yet these mutants no longer produced the three toxins. Four of the five putative regulatory genes (*mut2* to *mut5*) were found in a single operon. You then deleted each of these genes and designated the resulting mutants CpΔmut1 through CpΔmut5. You also cloned each of the genes and found that you could express the corresponding proteins in *E. coli*, as well as *C. perfringens*. You were interested in characterizing these mutants and determining the functional identities of the proteins encoded by the putative regulatory genes, so you designed a series of experiments.

  A. In the first experiment, you determined the cellular localizations of the proteins encoded by the genes *mut1* through *mut5*. Describe how you conducted this experiment. Be sure to include the experimental design, the reagents used, and how you visualized and interpreted the data.

  B. In the second experiment, you found that addition of cell-free filtrates from overnight cultures of wild-type *C. perfringens* or any of the mutants CpΔmut2 to CpΔmut5 to freshly inoculated cultures of the gram-negative bacterium *Vibrio harveyi* greatly stimulated the luminescence of the *V. harveyi* cells, whereas overnight culture filtrates of CpΔmut1 had no effect on *V. harveyi* luminescence. You found that overnight culture filtrates from *E. coli* expressing *mut1*, but not *mut2* to *mut5*, could also stimulate *V. harveyi* luminescence. Provide a possible explanation for these results. What is the most likely identity of the *mut1* gene? Be sure to state your rationale. Provide at least one additional experiment that you might perform to confirm your conclusion.

  C. In the third experiment, you find that although CpΔmut2, CpΔmut3, CpΔmut4, and CpΔmut5 are all defective in making the three toxins (alpha, theta, and kappa toxins), when you add overnight culture filtrates of wild-type *C. perfringens*, CpΔmut2, or CpΔmut5 to cultures of CpΔmut3 or CpΔmut4, all three toxins are produced. However, when overnight culture filtrates of CpΔmut3 or CpΔmut4 are added to cultures of CpΔmut2 or CpΔmut5, no toxin is produced. Cellular-fractionation results indicate that both Mut2 and Mut3 are cytoplasmic proteins, while Mut4 and Mut5 are integral membrane proteins. Provide a possible explanation for these results. What is the most likely identity of each of the *mut2* to *mut5* genes? Provide at least one additional experiment for each of the proteins Mut2 to Mut5 that you might perform to confirm your conclusions. (Hint: you should describe at least four separate experiments, one for each of the *mut*-encoded proteins.)

8. You are a graduate student who has been performing a transposon mutagenesis screen on the medically important bacterium *Americana badbugium*. Based on previous experimental evidence, this pathogen encodes a virulence-associated type III secretion apparatus that is turned on when *A. badbugium* is growing at 37°C and turned off when *A. badbugium* is growing at 23°C. Although a cytotoxic effector protein (CepA) of this secretory machinery has been defined, the other proteins required for secretion have not been identified. You have found a transposon mutation that results in an attenuated phenotype in vivo and are interested in characterizing this mutant for your doctoral dissertation. You develop a series of experiments to characterize the protein and obtain various results for the mutants. Use the above information and the results listed under each mutant to answer questions for that particular mutation. The results of these studies are as follows.

- In the wild-type bacterium, CepA is expressed and secreted when the bacterial cells are grown at 37°C and upon bacterial contact with host cells.

- In the mutant bacterium, CepA is not expressed and not secreted.

- You have determined that all structural components of the T3SS are intact and functioning properly.

- You determine that *cepA* mRNA is not detected when the mutant bacteria are growing at 37°C.

- Very low levels of *cepA* mRNA are detected when the mutant bacteria are growing at 23°C.

- You have identified and sequenced the gene that is mutated and have expressed the recombinant protein. You find that the N-terminal domain features several regions of conserved amino acids that are similar to other proteins that function in a similar manner.

- Using transcriptional fusions, you have demonstrated that the recombinant protein, encoded by your gene of interest, can bind to the *cepA* promoter and is required for transcriptional activation of *cepA* when the bacterial cells are growing at 37°C.

- The defect in *cepA* transcriptional activation and secretion at 37°C is complemented by expression of your gene of interest (on the chromosome or on a plasmid) in the mutant bacterium.

    A. What type of protein have you identified?

    B. Propose a function for your gene/protein of interest.

    C. When is your protein/gene activated in *A. badbugium?*

    D. What other protein is required for your protein to function properly, and how does this additional protein function?

9. Chemotactic responses in *E. coli* and *Salmonella* are regulated through the Che signal transduction pathway. Swimming behavior can be affected by mutations in the pathway. For each of the following mutations, indicate the phenotype you would expect the mutant to exhibit. Choose from phenotypes A through C, using each phenotype as many times as necessary.

    A. Runs (swims smoothly) constantly; never changes direction (i.e., never tumbles under any conditions)

    B. Tumbles constantly under normal conditions but can run in response to high concentrations of stimulant

    C. Runs constantly under normal conditions but can tumble in response to high concentrations of repellent

Site of mutation	Phenotype
1. *cheZ*; the chemotaxic regulator that dephosphorylates CheY; mutation results in loss of phosphatase activity	_____
2. *cheY*; the response regulator gene; mutation results in loss of the conserved aspartate (Asp [D])	_____
3. *cheR*; the methyltransferase gene; mutation results in loss of methyltransferase activity	_____
4. *cheA*; the histidine protein kinase (sensor) gene; mutation results in loss of the conserved histidine (His [H])	_____
5. *cheB*; the methylesterase response regulator gene; mutation results in loss of the conserved aspartate (Asp [D])	_____

SPECIAL GLOBAL-PERSPECTIVE PROBLEMS: INTEGRATING CONCEPTS IN PATHOGENESIS

1. A group of researchers at the university veterinary diagnostics laboratory isolated a new bacterium from the blood and stool of several horses that became ill and died at a local farm. Prior to death, symptoms included disorientation, loss of motor function, and flaccid paralysis, so the researchers suspected central nervous system involvement and possible production of neurotoxin. Based on 16S rRNA comparison, they found that the new bacterium was related to the gram-positive bacterium *Clostridium botulinum,* and they subsequently named it *Clostridium equiniae.* They also determined that the bacterium was sensitive to metronidazole but resistant to beta-lactam antibiotics, such as penicillin, and to macrolide antibiotics, such as erythromycin and azithromycin. However, although *C. equiniae* was resistant to azithromycin, the researchers found that when they treated infected mice with metronidazole in combination with azith- romycin, the 50% lethal dose (LD_{50}) value went from 10 for a control without antibiotic treatment to 10^4 for treatment with metronidazole alone to 10^7 for combined treatment of metronidazole plus azithromycin, and the mice recovered from paralysis much sooner with the combined treatment than with metronidazole treatment alone. Interestingly, when they were plating out the bacteria to determine the LD_{50} values, the researchers observed that the colonies isolated from the control mice and mice treated with metronidazole alone were mucoid, but those from mice treated with both antibiotics were not.

    A. What possible mechanism(s) could account for all of these observations (i.e., change in the LD_{50} value, faster recovery from paralysis, and change in the mucoid phenotype)? Be sure to provide your rationale.

(continued)

B. Provide an experiment that could be performed to confirm your hypothesis.

2. A group of researchers are studying a gram-negative respiratory pathogen called bacterium K. During the course of their studies, they identify a secreted protein cytotoxin (which they call Kct). They determine that Kct causes increased phosphorylation of mitogen-activated protein kinase proteins and cellular proliferation in cultured lung cells. They generate rabbit polyclonal antisera against filtered, dialyzed, and concentrated culture filtrates from bacterium K and determine that these antibodies neutralize the toxin and protect animals from disease. They separate the proteins present in bacterial culture filtrates by SDS-PAGE, transfer the proteins to a membrane, and detect the proteins by performing Western blot analysis using the rabbit antisera against Kct. The Western blot analyses reveal that the antibodies are reactive against two proteins produced by the bacterium, one with a mass of 80 kDa and another with a mass of 20 kDa. Both of these proteins are secreted into the medium only when they are grown under in vivo conditions, including temperature (37°C) and high blood/tissue levels of CO_2. Using an in vivo expression technology screening strategy, they subsequently clone three fragments of DNA from bacterium K, each of which contains an operon important for production and secretion of the toxin, and hence virulence.

- The first operon has two ORFs. One (which they call *kctS*) has five membrane-spanning hydrophobic sequences in its N-terminal domain and some sequence homology to protein kinases in its C-terminal domain; the other (which they call *kctR*) has no significant sequence homology to any protein in the database, but it does have a helix-loop-helix secondary-structure motif that is frequently found in DNA-binding proteins.

- The second operon has two ORFs, which they call *kctBA*. They suspect these genes encode the protein toxin they isolated, since the first ORF encodes a protein of 85 kDa that has some sequence homology to the B subunit of anthrax toxin and the second ORF encodes a protein of 25 kDa that has some homology to mammalian protein kinases.

- The third operon has 13 ORFs, which they name *kctC* through *kctO*. Many of the proteins encoded by these genes have membrane-spanning hydrophobic sequences indicative of integral membrane proteins; some also have sequence similarity to proteins involved in biogenesis of type IV pili, and one has an ATP-binding motif.

A. The researchers determine that transposon insertion into *kctS* removes the ability of CO_2 to regulate expression of Kct. Describe (in detail) how the researchers could determine if the expression of Kct under physiological CO_2 conditions is regulated at the transcriptional level by a TCS consisting of KctS and KctR and how they could determine if *kctR* encodes a transcriptional activator or repressor of Kct gene expression under physiological CO_2 conditions. Based on the given information, what is the predicted result?

B. The researchers perform cellular-fractionation studies of toxin production in the presence of lung cells and examine the protein contents of the fractions by using SDS-PAGE and Western blot analyses. Provide a detailed explanation (with rationale) and interpretation for the results shown in the Western blot below. Be sure to also provide an explanation for the difference between the predicted sizes of the A and B subunits of Kct and the sizes of the proteins found in the bacterial culture medium and associated with the mammalian lung cells (i.e., how does this occur?). What evidence could they obtain from sequence analysis that would support your explanation?

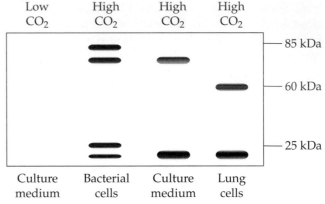

| Low CO_2 | High CO_2 | High CO_2 | High CO_2 | |
| Culture medium | Bacterial cells | Culture medium | Lung cells | |

85 kDa
60 kDa
25 kDa

C. Further studies reveal that Kct action on lung cells requires a low-pH-dependent step, since agents that prevent acidification, such as basic buffers, block toxin effects. Based on the information given above and these results, propose a molecular mechanism by which Kct might cause the observed increased mitogenesis in lung cells. Provide details of each of the steps involved!

D. To determine the role of the third operon *(kctC* to *kctO)* in Kct production, the researchers make in-frame deletions of each of the genes. They then perform fractionation studies of toxin production in the presence of lung cells at high CO_2, followed by SDS-PAGE and Western blot analyses, and find that in each case the Kct A and B subunits are not secreted into the medium but instead localize to the periplasm with molecular masses of 20 kDa and 80 kDa, respectively. Provide a detailed explanation for these results.

3. *Bacillus anthracis,* the causative agent of anthrax disease, is a spore-forming gram-positive bacterium that is touted as a top choice for a biological weapon because of its lethality (anthrax is almost always deadly if not treated early) and because the spores are colorless, tasteless, odorless, and extremely hardy (they can be stored dry for decades without losing viability) and can be easily dispersed (even through the mail). There are three forms of the disease: cutaneous anthrax, acquired through cuts or abrasions of the skin, which is characterized by swelling (edema), intense but painless inflammation, and a necrotic (dying) center that turns black; intestinal anthrax, usually caused by consumption of contaminated meat, which is characterized by acute inflammation of the intestinal tract, abdominal pain, vomiting of blood, and severe diarrhea; and systemic anthrax (or inhalational anthrax), which is characterized by generalized edema and massive pulmonary edema and hypotension (low blood pressure) and is nearly always fatal once symptoms occur, with the victim usually succumbing to toxic shock, heart failure, and/or encephalitis. Epidemic outbreaks of inhalational anthrax are not natural and so can only be attributed to deliberate release of a massive number of spores into the atmosphere by humans. The major virulence factors of anthrax are the A3B7 anthrax toxin complex of the A subunit's lethal factor (LF, a zinc-dependent metalloprotease) and edema factor (EF, a calcium-calmodulin-dependent adenylate cyclase) and the B subunit's protective antigen (PA), which is encoded on a large 184-kb virulence plasmid called pXO1, and an antiphagocytic poly-γ-D-glutamic acid capsule, the biosynthetic genes for which are carried on a large, 97-kb virulence plasmid called pXO2. While the three toxin genes for LF (*lef*), EF (*cya*), and PA (*pagA*) are located on the same plasmid, they are not part of the same operon and appear to have their own promoters. The capsule genes (*capBCADE*) are all part of the same operon.

   A. The anthrax toxin subunits are secreted into the medium by the general secretory pathway. What feature of the sequences of these proteins supports this conclusion? Describe how you might demonstrate experimentally that they are indeed secreted by the general secretory pathway. Be sure to provide an example of what the expected resulting data might look like and your rationale.

   B. Describe how you might demonstrate experimentally that anthrax toxin is indeed an AB toxin, where PA is the B subunit that delivers both LF and EF to the cytosol. Show your expected results.

C. In vivo, the signals that stimulate expression of anthrax virulence genes, *lef, cya,* and *pagA,* are thought to be temperature (on at 37°C) and blood/tissue levels of carbon dioxide (on at 5% CO_2). Transposon insertion into *atxA* removes the ability of CO_2 to stimulate expression of anthrax virulence genes. Describe how you would determine if the expression of the *lef, cya,* and *pagA* genes under physiological conditions is regulated at the transcriptional level and if AtxA is a transcriptional activator or repressor of anthrax virulence gene expression under high-CO_2 conditions. Be sure to provide the reagents and conditions used, how the results will be visualized or measured, and what the expected results would be for these genes. Based on the given information, what is the predicted result?

D. You suspect that there is a global regulator of virulence gene expression that senses physiological conditions. Describe how you would go about finding this putative regulator of both toxin production and capsule biosynthesis. Be sure to provide the reagents and conditions used in your screen, how the results from your screen will be visualized or measured, what the expected results would be, and how you would distinguish between regulators of the toxin genes alone, the capsule biosynthetic genes alone, and regulators of both (i.e., the putative "global" regulator).

E. From your screen, you identify seven putative regulatory genes (denoted *bamut1* to *bamut7*). The *bamut1* gene is immediately downstream of *pagA,* and these two genes appear to be part of the same operon. You suspect that the gene product of *bamut1* might be a regulator of *pagA,* so you rename it *pagR.* Describe how you would determine if the gene product of *pagR* is an activator or repressor of *pagA* and whether it acts as a protein or as an sRNA.

F. Deletion of *pagR* results in increased basal expression of *pagA* at low temperatures and low CO_2 levels, but under physiological conditions, *pagA* expression is just as high as that of the wild type without deletion of *pagR.* Provide an interpretation with rationale for these results.

G. The *bamut2* deletion mutant is no longer stimulated by CO_2 to express *lef, cya, pag,* or *capBCADE,* whereas in the *bamut3* deletion mutant, the expression of these genes is constitutively on. Provide an interpretation with rationale for these results.

(continued)

H. Toxin production is also controlled by the growth phase, i.e., toxin production is highest during exponential phase, but once the bacteria reach stationary phase, toxin production is turned off. What mechanism could account for this observation? Provide your rationale.

I. The *bamut4* to *bamut7* genes are all located on the same operon, and toxin production by the mutants with any one of these genes deleted is no longer regulated by the growth phase. Analysis of the protein sequences of these genes reveals that none of them has any significant sequence homology to known proteins in the databases, but there are some sequence motifs. Protein sequence analysis of *bamut4* reveals that it has four membrane-spanning sequences in its N terminus and an ATP-binding kinase-like motif in its C terminus. The protein encoded by *bamut5* has a helix-loop-helix

secondary-structure motif that is found in DNA-binding proteins. The protein encoded by *bamut6* has an ATP-binding motif and six membrane-spanning hydrophobic sequences. The protein encoded by *bamut7* is a small protein of 50 amino acids with a stretch of 23 hydrophobic residues at the N terminus. Predict the function and cellular location of each of the proteins encoded by *bamut4* to *bamut7* and provide one additional experiment for each that could be performed to confirm your prediction for each.

J. Using the information from the hydrophobicity plot in the figure below, design an experiment that would allow you to determine the membrane topology of the Bamut4 protein. Be as specific and efficient as possible, and provide details of your experimental setup, as well as your expected results. Draw a model of the membrane topology of your protein.

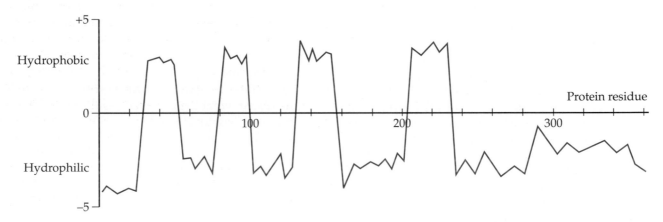

K. Antibiotics will suppress infection only if administered very early after exposure, usually within the first 2 days. After that, there is no effective treatment for unvaccinated victims of inhalation anthrax. By the time symptoms manifest, it is highly likely that death will occur despite the best efforts of modern medical science, i.e., the mortality rate is 50% if treated within the first 3 to 5 days after symptoms appear and 100% thereafter. Ciprofloxacin is the preferred antibiotic, but penicillin,

erythromycin, tetracycline, chloramphenicol, and clindamycin can be used. Recent studies show that patients with inhalation anthrax receiving pleural (lung) fluid drainage treatment and multidrug antibiotic therapy with a combination of ciprofloxacin, clindamycin, and Augmentin (ampicillin plus clavulanic acid) have significantly higher survival rates. Provide a possible explanation for this observation, including the rationale for the inclusion of each component in the treatment regimen.

15

Antimicrobial Compounds

A ntibiotics are indeed "wonder drugs." It is hard now to imagine the era prior to World War II, when most bacterial diseases were life threatening. Before then, bacterial infectious diseases and infections of simple wounds or surgical wounds acquired during operations in hospitals or in childbirth could rapidly spin out of control and result in death. The appearance of antibiotics, first sulfa drugs and then penicillin, heralded a new and exciting era of antibiotic discovery and development that completely revolutionized human and animal health, to the point where there are only a few bacterial diseases that we have not been able to cure. Unfortunately, it looks like this era may now be over. What happens if—for the first time in history—we lose a cure? This scenario may happen in the not-too-distant future for bacterial diseases and will likely result in a drastically negative view of public health care systems, regulatory and funding agencies, and the pharmaceutical industry.

Antimicrobial Compounds: the Safety Net of Modern Medicine

Importance of Antimicrobial Compounds

The nonspecific and specific defenses of the body are remarkably effective, but they are not perfect. Not only have bacterial pathogens developed ways of circumventing them, they also have means of taking advantage of instances where there is disruption of the defenses, such as during surgery or cancer chemotherapy, when the body is more open to infection. Such medical disruptions have great benefits if the risk of subsequent infection can be minimized. One of the greatest advances in human health during the past century was the discovery that our natural defenses could be augmented with externally provided chemical defenses: **disinfectants** (applied to nonliving objects or surfaces), **antiseptics** (applied to living tissues or skin), and **antibiotics** (administered outside and inside the body).

The public has taken antiseptics, disinfectants, and antibiotics so much for granted that it has forgotten how crucial these compounds are to human well-being. Only with the appearance of antibiotic-resistant bacteria have physicians and public health officials begun to recall how essential the effectiveness of antibiotics is to modern medicine. Without antibiotics, nonessential surgical procedures, such as knee and hip replacements, could not be done on a routine basis. Essential surgical procedures, such as bypass surgery or heart valve replacement, would become a lot riskier for the patient, who would be put in the unenviable position of having to decide whether to risk death because of failure to undergo the operation or to risk death due to an overwhelming infection.

Surgical procedures have been practiced in emergency situations for centuries, but the patients often died from ensuing infections. In the 1800s and early 1900s, surgeons discovered how to reduce postsurgical infections dramatically by using sterilizing procedures, such as vaporized phenol (carbolic acid), in the operating room. A surgeon of that era owed his success not only to the level of his skill at wielding the knife, but also to his ability to endure the effects of repeated phenol exposure. Only when antibiotics and disinfectants made phenol a thing of the past did surgery begin to flourish and medical students begin to vie for admission to the specialty.

Unfortunately, just as virulence factors enable bacteria to evade or undermine the natural defenses of the human body, so the adaptability of bacteria has enabled them to devise strategies for evading these chemical defenses. These strategies, called mechanisms of antimicrobial resistance, could undermine many of the advances we have made over the past century. In this chapter, we examine the mechanisms of action of antiseptics and disinfectants and we explain how bacteria become resistant to these agents. We then cover how the major classes of antibiotics work. How bacteria become resistant to antibiotics is the subject of chapter 16.

Killing versus Inhibiting Growth

A key concept in antimicrobial therapy is the difference between bactericidal and bacteriostatic compounds. Antimicrobial compounds that kill bacteria are called **bactericidal.** Others, called **bacteriostatic** compounds, merely stop or slow the growth of bacteria, but the bacteria are not killed. With bacteriostatic agents, the bacteria can frequently recover once the antibiotic is removed (referred to as **tolerance**). In patients with an intact immune system, bacteriostatic compounds can be very effective, because the defenses of the body need only a little help in slowing the growth of bacteria so that they can be eliminated

from the body. In people with defective defenses, however, there is much more reliance on the antimicrobial compound to effect a cure. Bactericidal antibiotics are best for treating infections in such patients.

At one time, the distinction between bactericidal and bacteriostatic antimicrobial compounds seemed very clear-cut. Certain antimicrobials were bacteriostatic, and others were bactericidal. That is, the distinction resided in the properties of the antimicrobial compound. It is now clear that properties of the bacteria can affect this distinction, too. The best example of this is bacterial biofilms. Bacteria that form biofilms change metabolically to a less active state. Such bacteria are able to resist an antibiotic that would be bactericidal if the bacteria were dividing rapidly. In other words, a bactericidal antibiotic is rendered bacteriostatic or completely ineffective by the growth state of the bacteria. This is the reason that bacteria in biofilms that form on plastic implants and catheters are so difficult to eliminate.

The susceptibility of bacteria to antibiotics and other compounds is measured by simple growth tests. In one version, bacteria from freshly grown cultures are diluted into a row of wells of a microtiter plate containing rich growth medium. Different dilutions of the compound to be tested are then added to the wells, and the bacteria are allowed to grow overnight. Bacteria in the control well, which lacks the compound, and in wells containing low concentrations of the compound grow to stationary phase, making the wells appear turbid. Above a certain concentration of the compound, the bacteria fail to grow and the wells appear clear. This concentration of the compound is defined as the **minimal inhibitory concentration (MIC).** The lower the MIC, the more potent the compound is as an inhibitor. For a bacteriostatic compound, the bacteria in the clear wells can be recovered, because their growth has only been inhibited. At low concentrations, even bactericidal compounds may inhibit growth without completely killing the bacteria, but at higher concentrations, all of the bacteria will be killed, and this concentration is referred to as the **minimal bactericidal concentration (MBC).** Thus, the MBC is greater than or equal to the MIC for bactericidal compounds.

This simple dilution susceptibility assay format to determine MICs can be simultaneously performed on different species of bacterial pathogens growing in the same microtiter plate. These comparisons show that different bacterial pathogens are susceptible to a range of concentrations of a given antibiotic. For the most susceptible bacteria, the MIC may be less than 0.01 μg of antibiotic per ml of culture. Some antibiotics, such as vancomycin, inhibit only gram-positive, but not gram-negative, bacteria, because these compounds fail to penetrate the outer membrane of gram-negative bacteria. For these compounds, all of the wells will be turbid in the MIC assays of gram-negative bacteria. Importantly, MIC values provide a measure of resistance, because resistance to an antibiotic leads to an increase in the MIC value for a given bacterial species. The MIC assay can also be used to test whether compounds of interest bind to carrier proteins in blood, such as serum albumin. Tight binding to carrier proteins can be a serious hurdle for a compound to act as an effective antibiotic, because the compound may not be free to attack bacteria in mammalian hosts. When a compound binds tightly to serum albumin, its MIC will increase in assays containing the protein compared to those lacking the protein. Besides the broth format, there are other formats of susceptibility assays in which bacteria are spread on agar plates. The susceptibility of the bacteria is based on the zones of inhibition that are present around discs or strips containing different concentrations of the antibiotic being tested.

A second key concept in antimicrobial therapy is the distribution of the drug in the body. Some antimicrobial compounds can be used only on body surfaces or inanimate objects because they are too toxic for internal use. Antiseptics and disinfectants are examples of this type of compound. Antibiotics are compounds that can be used internally, as well as externally. Antibiotics are not all equal inside the body, because each class of antibiotic has its own special **pharmacokinetics,** which describes the distribution of the antimicrobial compound in the body. For example, some of the antimicrobials used to treat urinary tract infections concentrate primarily in the kidneys and urine and do not disseminate widely to other parts of the body. Others are readily absorbed from the gastrointestinal tract or bloodstream and permeate tissues all over the body. Obviously, the location of the bacteria responsible for causing an infection is key to selection of the appropriate antimicrobial compound. It does not help to treat a meningitis patient with a bactericidal antibiotic if the antibiotic does not cross the blood-brain barrier. Besides the initial distribution, how long an antibiotic remains in a location in the body before being degraded or removed is also important to understanding its efficacy. Considerable research time spent in the development of new antibiotics is focused on determining where and for how long these antibiotics are localized in the body after they are administered.

A third key concept in antimicrobial therapy is side effects. Although the ideal antimicrobial kills or inhibits the growth of bacteria specifically and does not affect mammalian cells (differential toxicity), such an ideal compound is not always available, and even an antibiotic like penicillin, which comes very close to this ideal, can cause allergic reactions in some people. An example of side effects is the ability of some of the aminoglycoside antibiotics to destroy hearing. Obviously, the state of the patient and the seriousness of the infection affect the choice of appropriate therapy. An aminoglycoside that causes deafness if administered long enough would not be considered for treatment of a relatively minor infection but might be the drug of choice for a critically ill patient whose infection can only be treated with this class of antibiotics. Unfortunately, since few new antibiotics against emerging multidrug-resistant bacterial pathogens, such as multidrug-resistant *Mycobacterium tuberculosis* (the causative agent of tuberculosis), are available, older antibiotics that were not widely used earlier because of serious side effects are increasingly being used as treatments of last resort. However, this strategy only makes treatment that much more difficult, because the long-term regimens required to treat tuberculosis, which involve the simultaneous administration of several different antibiotics, are tolerated by patients even less well than before.

All of these factors must be considered in the choice of an effective therapeutic compound. In the remainder of this chapter, the focus will be on the effects of antimicrobials on the bacteria themselves, but these traits of antimicrobials and of the bacteria they target are only part of the treatment story.

Antiseptics and Disinfectants

Mechanisms of Action

Antiseptics and disinfectants, like antibiotics, are chemicals that kill or inhibit the growth of bacteria and other microorganisms. Most antiseptics and disinfectants are bactericidal. Most are also effective against other types of disease-causing microbes, such as viruses, fungi, and protozoa. This broad coverage has a drawback, however, because the chemicals used as antiseptics and disinfectants are often too toxic for internal use in humans and other animals. Accordingly, they are applied only on inanimate surfaces or externally to skin or mucosal surfaces.

"Disinfectant" is the term used to describe antimicrobial compounds applied to inanimate objects and surfaces. They can include substances such as chlorine compounds (e.g., bleach), reactive oxygen compounds (e.g., peroxides), iodine (e.g., Betadine), alcohols (e.g., isopropanol), phenolic compounds (e.g., carbolic acid, Lysol, triclosan [Box 15–1], and hexachlorophene), cat-

BOX 15–1   Triclo-Insanity Hits the United States

An antiseptic compound much in the news in the late 1990s was an antiseptic/disinfectant called triclosan. Despite its name, triclosan actually consists of two phenolic rings, one of which contains two chlorine atoms. During this period, many companies began to market a variety of antibacterial plastic products, ranging from cutting boards to toys, products that are still popular with consumers today. This proved to be a very effective marketing strategy, which sent millions of householders to their local stores to purchase these products that were guaranteed to protect them from deadly bacteria. The active ingredient in these products was triclosan. Triclosan was not actually added to the products to make them safer. Instead, triclosan was impregnated into the plastic to prevent bacterial degradation of the plastics. Alert to the bacteriophobia of the times, however, advertising executives decided that they could sell products by touting the "antibacterial compound" they contained. The ads implied that these products contained a special additive other products did not have when in fact virtually all plastic products contain it. On a scientific level, triclosan turned out to be unusual among disinfectants and antiseptics in that, unlike them, triclosan did have a specific target, an enzyme involved in fatty acid biosynthesis. A structural relative of triclosan (both chlorinated bisphenols) is hexachlorophene, the active ingredient in many deodorants.

The popularity of products containing antibacterial compounds raised yet again the question of whether it is possible to be too clean. The human body evolved to accommodate massive microbial populations and regular contact with a variety of disease-causing microbes. The degree of sanitation and hygiene that has become the rule in modern times—at least in developed countries—has changed drastically, and in a short time, the amount of human exposure to disease-causing microbes has decreased markedly. Some scientists have speculated that this change might be responsible for the rise in the incidence of such conditions as asthma and inflammatory bowel disease, which are caused by a malfunctioning immune system. This rise has occurred primarily in countries that are noted for their high standards of hygiene. The argument is that an immune system that is not adequately challenged during childhood may become unbalanced or overreactive to stimuli, leading to autoimmune reactions. If this view is correct, the incorporation of antibacterial compounds in soaps and other household products only makes the situation worse.

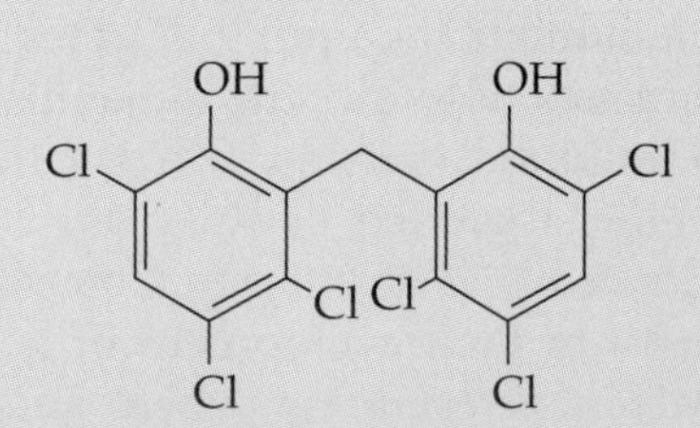

Structures of triclosan (A) and hexachlorophene (B). Triclosan is routinely added to plastics to prevent microbial degradation. Hexachlorophene is an antibacterial compound used in deodorants to prevent odor-producing activities of skin bacteria.

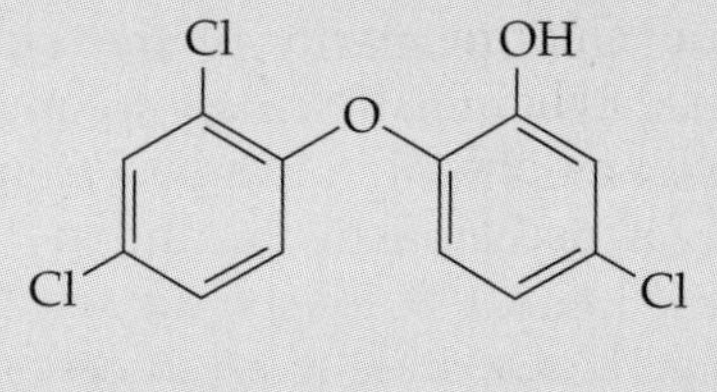

A Triclosan

B Hexachlorophene

Table 15–1 Common disinfectants and their mode of killing

Disinfectant	Mode of killing
Alcohols (ethanol, isopropanol)	Denature proteins
Alkylating agents (formaldehyde, ethylene oxide)	Form epoxide bridges that inactivate proteins
Halides (I^-, Cl^-, NaClO)	Oxidizing agents
Heavy metals (Hg^{2+}, Ag^+)	Bind –SH groups, thus denaturing proteins
Phenols	Denature proteins; disrupt cell membranes by intercalating in them
QACs	Disrupt cell membranes by intercalating in them
UV radiation	Blocks DNA replication and transcription by damaging DNA

ionic surfactants (e.g., **quaternary ammonium compounds [QACs]**), strong oxidizers (e.g., ozone), alkylating agents (e.g., formaldehyde), heavy metals and their salts, strong acids, and strong bases. "Antiseptic" is the term used to describe antimicrobial compounds applied to the skin, e.g., hand-washing preparations used in hospitals and doctors' offices and the waterless hand sanitizers that are now widely available in public locations. Antiseptics often include appropriately diluted or weaker versions of the above-mentioned disinfectants, such as alcohol, hydrogen peroxide, iodine, chlorine, and QACs, or weak acids (e.g., vinegar, salicylic acid, and sorbic acid). Other compounds, such as phenol or heavy metals, are too harsh or toxic for use on skin and are used only as disinfectants.

Antiseptics and disinfectants do best against actively replicating microorganisms. Bacterial spores are generally resistant to them, although the germination of spores can be inhibited. Otherwise, antiseptics and disinfectants are effective against a wide range of fungi, protozoa, viruses, and bacteria, and they tend to attack multiple targets in microbes. For example, **halides,** such as chlorine or sodium hypochlorite (household bleach), and iodine are strong oxidants that inactivate many bacterial proteins. **Hydrogen peroxide** has a similar mechanism of action. Halides and peroxide can also damage microbial DNA. QACs, such as benzalkonium chlorides, which are some of the most widely used antimicrobial compounds, intercalate into phospholipid bilayer membranes, caus-

ing cells to leak vital ions and other small molecules. These compounds also disrupt electron transport chains. Examples of QACs are cetrimide and benzalkonium chloride. Table 15–1 summarizes the mechanisms of action of these and other common disinfectants.

Resistance to Antiseptics and Disinfectants

Resistance to antiseptics and disinfectants is poorly understood in general, but some resistance mechanisms are known. Many antiseptics and disinfectants, especially those that attack membranes, such as QACs, are less effective against gram-negative bacteria than gram-positive bacteria. The reason seems to be that lipopolysaccharide in the outer membrane of gram-negative bacteria prevents hydrophobic molecules from intercalating into the outer membrane, while porins restrict access to the cytoplasmic membrane by limiting diffusion. Some membrane-active antiseptics, however, can breach the barrier posed by the outer membrane. An interesting type of resistance to QACs has been found in staphylococci: a cytoplasmic membrane pump that pumps the QAC out of the cell cytoplasm. Why this would make the bacteria resistant to QACs, which are thought to act mainly by dissolving membranes, is still unclear. Whatever the explanation, these pumps are fairly effective in protecting the bacteria from QACs. Genes encoding QAC efflux pumps have been found on plasmids, as well as on putative mobile elements in the chromosome. The fact that resistance to antiseptics and disinfectants can develop is a disturbing discovery, because these compounds have been considered an important line of defense against microbial infections.

Antibiotics

Characteristics of Antibiotics

Antibiotics ("agents against life") are low- to medium-molecular-weight compounds that kill or inhibit the growth of bacteria and can be ingested by or injected into humans and animals with minimal side effects. Table 15–2 provides a list of commonly used classes of antibiotics and their characteristics. Antibiotics can be either bactericidal or bacteriostatic. In contrast to most disinfectants and antiseptics, antibiotics generally interfere with a specific bacterial enzyme or process, such as the enzyme DNA gyrase, which negatively supercoils the genomic DNA, or the transpeptidase enzymes that cross-link peptidoglycan in the cell wall.

Several characteristics define a good antibiotic. First, the antibiotic must have few or no side effects.

Table 15–2 Antibiotics and their mechanisms of action

Class	Mechanism of action	Resistance mechanisms	Spectrum of activity	Common name(s)
β-Lactams (penicillins, cephalosporins, carbapenems, monobactams)	Inhibit transpeptidation step in peptidoglycan synthesis; bind penicillin-binding proteins; stimulate autolysins	Gram-negative outer membrane; porin mutations; β-lactamase; modify target (alteration of penicillin-binding protein)	Gram-positive and/or gram-negative bacteria (depends on agent)	Penicillin, ampicillin, Cefobid, Augmentin
Glycopeptides	Inhibit transglycosylation and transpeptidation steps in peptidoglycan synthesis by binding D-Ala-D-Ala	Gram-negative outer membrane; modify target (substitute D-Ala-D-lactate for D-Ala-D-Ala)	Most effective against gram-positive bacteria	Vancomycin, teichoplanin, daptomycin
Aminoglycosides	Bind 16S rRNA in 30S subunit of bacterial ribosome	Inactivation of antibiotic by adding groups	Broadly bactericidal	Kanamycin, gentamicin, streptomycin
Tetracyclines	Bind 16S rRNA in 30S subunit of bacterial ribosome; disrupt bacterial membrane	Inactivation of antibiotic (?); ribosome protection; efflux system	Broadly bacteriostatic; some protozoa	Tetracycline, doxycycline
Macrolides/lincosamides	Bind 23S rRNA in 50S subunit of bacterial ribosome	Methylation of target; efflux	Bacteriostatic for most; bactericidal for some gram-positive bacteria	Erythromycin (macrolide); lincomycin, clindamycin (lincosamides)
Streptogramins	Bind 23S rRNA in 50S subunit of bacterial ribosome	Inactivation of antibiotic by removing groups	Bacteriostatic individually; bactericidal in combination; used for multidrug-resistant enterococcal infections	Synercid
Fluoroquinolones	Bind DNA gyrase	Efflux (?); reduced uptake (?); mutation in DNA gyrase	Broadly bactericidal; can enter phagocytes, kill intracellular bacteria	Ciprofloxacin, norfloxacin
Rifampin	Binds β subunit of bacterial RNA polymerase	Mutation in RNA polymerase	Broadly antibacterial; effective against mycobacteria	Rifampin, rifadin
Trimethoprim/sulfonamides	Inhibit enzymes responsible for tetrahydrofolate production	Mutations alter affinity of target enzymes	Broadly antibacterial; some fungi (*Pneumocystis jiroveci*), protozoa	Bactrim, Septra
Metronidazole	Nicks bacterial DNA and interferes with DNA replication	Decreased production of flavodoxin gene (?)	Antibacterial (mainly anaerobes); antiprotozoal	Flagyl
Oxazolidinones	Bind 50S ribosomal subunit	Mutation in 23S rRNA genes	Bacteriostatic; broad spectrum against gram-positive bacteria, mycobacteria	Zyvox

That is, it must be far more toxic for bacteria than for the human body. This is known as the principle of **differential toxicity.** A second desirable characteristic, especially from the physician's perspective, is a **broad spectrum of activity** against many different types of bacteria. This is important because it is usually not possible to determine the identity of the bacterium causing an infection from the symptoms alone, and clinical testing is often not convenient or expeditious. In addition, clinical testing is costly. Patients with bacterial infections often have nonspecific symptoms, such as fever, malaise, and pus formation. Since it takes time to isolate and identify the bacterium responsible for an infection, it is useful to have antibiotics that are effective against the entire range of bacteria capable of producing a particular set of symptoms. Especially in the case of serious, rapidly progressing diseases, such as bacterial pneumonia and septic shock, where there is only a narrow window of efficacy and not much margin for error in selection of an effective antibiotic.

Several other properties besides minimal side effects and maximum spectrum are required to make a good antibiotic. Antibiotics must have the appropriate **bioavailability** and pharmokinetics to get to sites of infections. Bioavailability is a measurement of the fraction of the drug that enters the systemic circulation and reaches the site of action or the rate at which it does so. For the same reasons that a broad spectrum is desirable, excellent uptake, distribution, and half-life of an antibiotic in humans or animals add to the usefulness of the antibiotic. In addition, an antibiotic that can be absorbed from the digestive tract and administered in pill form is far easier to dispense than one that can only be administered intravenously by medical personnel. Finally, cost becomes a serious consideration in antibiotic discovery and use. If an antibiotic is too costly to manufacture, it likely will not be developed by the pharmaceutical industry. Conversely, first-line antibiotics can be prohibitively expensive in developing countries.

Broad-spectrum antibiotics have an important drawback that has caused scientists to take another look at this type of antibiotic. Such antibiotics not only attack the bacterium causing the infection, but can also attack members of the resident microbiota of the body. Disruption of the microbiota can allow other pathogens, which are normally outcompeted by the resident microbiota, to cause infections. A timely example of this problem is *Clostridium difficile,* which is emerging as an extremely serious pathogen. *C. difficile* is normally a minor member of the microbiota of the colon. Disruption of this normal colonic microbiota by antibiotic treatment allows *C. difficile* to overgrow and

produce toxins that seriously damage the colon (see chapters 5 and 18). Because *C. difficile* forms hardy spores, it can be difficult to treat once established, and it can readily spread through the diarrhea caused by the infection. A second example is yeast vaginitis in women who have taken antibiotics that disrupted their vaginal microbiota, thus allowing the yeast to overgrow and elicit an inflammatory response. Because of such experiences, more consideration is now being given to the effects of new antibiotics on the resident microbiota, with the goal of minimizing the impact of the antibiotic on the normal microbiota.

Another problematic feature of broad-spectrum antibiotics is that even if they do not disrupt the resident microbiota significantly, they may select for resistance to the antibiotic. Some bacteria in the resident microbiota are capable of causing serious infections if they escape from the area where they normally reside. Members of the resident microbiota are a significant cause of hospital-acquired infections. An example is vancomycin-resistant *Enterococcus* (VRE) species. Enterococci are common inhabitants of the human colon. The widespread use (and overuse) of vancomycin to treat or prevent infections caused by *Staphylococcus aureus* and other gram-positive pathogens has selected for VRE strains. These VREs are now wreaking havoc in some hospitals. Although enterococci are not highly virulent pathogens, when they do cause infection, they can kill the patient if the infection is not brought under control quickly. Enterococcal infections caused by VRE strains kill nearly 40% of the patients who have them. We will discuss this topic more in chapter 18.

Tests Used To Assess Antibiotics

To identify good antibiotics, numerous assays are performed. Compounds are often first screened at a single fixed concentration to see if they inhibit the growth of bacteria that show sensitivity to a variety of antibiotics, such as *Streptococcus pneumoniae*. These whole-cell growth assays are then followed by MIC testing with a variety of bacterial species. Some compounds inhibit the growth of all species tested, whereas others inhibit only gram-positive or gram-negative species. Rarely, compounds are found that inhibit only one type of bacteria, such as the recently discovered diarylquinolines that inhibit only mycobacteria.

MIC testing can be coupled with determinations of whether compounds are bacteriostatic or bactericidal or bind strongly to carrier proteins in serum, as described above. Compounds with favorable properties are then tested to see if they are cytotoxic to cultured eukaryotic cells or whether they affect eukaryotic-cell respiration. Compounds that pass these hurdles are next tested in rodent models to determine whether the injected compounds reduce bacterial infections at distant sites or cause overt toxicity to the animals. Candidate compounds are then assayed in the bacterium-based **Ames test** to determine whether they can act as mutagens that damage DNA. The compounds may also be tested at this stage for resistance development in several key bacterial pathogens. If resistance develops at a low frequency ($<10^{-7}$ resistant mutants per total CFU), the genomes of resistant mutants are sequenced (see chapter 5) to locate possible targets or processes affected by the compounds. Compounds with favorable properties may then be assayed by a simple surrogate test using cultured mammalian epithelial cells, such as Caco-2 cells, to determine whether the compound is absorbed by and passed through epithelial cells.

Finally, compounds that get this far are tested in two general kinds of systematic animal experiments. Pharmacokinetic analyses, including rates and levels of absorption, distribution, metabolism, and excretion **(ADME)** at different compound doses, are performed to determine what happens to the compounds once inside the animal hosts. Pharmacodynamic analyses are performed in parallel to learn what the compounds do to the animal hosts, including detecting any cytological changes, effects on organ functions, toxicities, and side effects. Only after these types of data are acquired in model systems showing that the candidate compounds are effective and safe for possible use as new antibiotics can applications be filed with government agencies to carry out further testing, eventually in human volunteers. This is a daunting gauntlet that compounds have to pass, and it is little wonder that discovering new antibiotics is a difficult undertaking, despite its necessity.

The Process of Antibiotic Discovery

Natural products. Many antibiotics were obtained from bacteria or fungi that were isolated from soil and other environmental sources. These antibiotics are often referred to as **natural products** to distinguish them from compounds that arise solely by chemical synthesis. Naturally occurring antibiotics are almost all products of secondary metabolism and are often produced in response to environmental stress or competition with other microbes. With the recent dearth of new antibiotics, there has been interest in screening for new natural-product antibiotics from microbial and untapped eukaryotic sources, such as marine organisms. However, screening for antibiotic activity from natural products is challenging, because the source material can vary from batch to batch, and natural-product extracts are usually complex mixtures of chemicals that need to be fractionated to find active compounds. In addition, it should always be kept in mind that antibiotics synthesized by bacteria bring the complication that the producer species must have its own intrinsic mechanism of resistance. Therefore, operons imparting resistance to natural-product antibiotics are already out there in nature and can become a serious problem if they are genetically transferred from the producer bacteria to pathogens treated with the antibiotics.

The fact that bacteria produce antibiotics has raised the question of what the role of antibiotic production is in nature. An obvious explanation is that antibiotics are a kind of "germ warfare," in which the producing species use antimicrobial compounds to discourage microbial competitors. The problem with this widely accepted explanation is that antibiotic production by microbes growing in nature is so low that the levels of antibiotics are undetectable under many conditions. An alternative explanation for the role of antibiotic production in nature is that bacteria use these compounds as signaling molecules (see chapter 14). Although there is no definitive experimental basis yet for either the germ warfare theory or the signaling theory, there is growing evidence that bacteria use antibiotic-like compounds as signals. Recent studies indicate that antibiotics can modulate gene transcription and specific adaptive responses in bacteria in a dose-dependent manner, suggesting that these antibiotics may have a role in cell-to-cell communication.

Historically, the way compounds with desired properties were discovered and developed was by finding a "lead compound" in nature that exhibited the desired antibiotic activity in the tests described above. However, these initial compounds almost always lacked sufficient activity or had unfavorable properties. The lead compound was then chemically modified into **semisynthetic antibiotics** that are more active, more stable, less toxic to hosts, or capable of overriding bacterial resistance mechanisms. This process is referred to as building **structure-activity relationships (SARs)** and remains a keystone of drug discovery. The lead compound is chemically modified in numerous ways, and the biological properties of the modified compound sets are assayed. Some chemical modifications improve antibiotic properties, whereas others have the opposite effect. The SAR is optimized until threshold goals of efficacy, spectrum, specificity, lack of toxicity, and bioavailability are reached. Of course, many compounds cannot be optimized for all of these properties and so fail as antibiotics.

One of the greatest challenges of natural products is their chemical complexity. The traditional semisynthetic route offered few ways to specifically modify the starting lead compounds through chemical synthesis. Therefore, there were a limited number of options for making chemical modifications in SARs, and the whole process still depended on biological production of the starting lead compound. Fortunately, some natural products, such as the penicillins, were simple and could be chemically synthesized completely and modified extensively. Recent advances in organic synthesis methods have made the complete synthesis of complex natural products feasible in reasonably high yields. Therefore, extensive SAR strategies are becoming possible for natural-product antibiotics. In addition, much has been learned about the biosynthetic pathways used by natural producers to synthesize antibiotics, including polyketides, which include the macrolide and tetracycline families of antibiotics. This knowledge is being applied to greatly increase yields of antibiotics by producer strains or by optimized genetically engineered surrogate strains. In addition, these biosynthetic pathways are being genetically manipulated or even mixed among different organisms to produce new antibiotics with different structures and activities. Also, genome sequencing of microbes has led to the unexpected finding that putative clusters of genes for antibiotic biosynthesis are present in many bacteria, in addition to the soil bacteria, such as *Actinomycetes* or *Streptomyces* species, which have traditionally been used as sources of antibiotics. Indeed, other soil species, such as *Flavobacterium* species and even *M. tuberculosis*, contain genes that may encode new types of antibiotics. Natural products from bacterial and eukaryotic sources remain a possibly rich source of new antibiotic compounds.

SYNTHETIC ANTIBIOTICS. Some antibiotics, such as the fluoroquinolones, are completely synthetic. That is, these antibiotics were derived by SAR studies from libraries of starting compounds that were synthesized by organic chemical methods. One might think that with the great chemical diversity out there, many organic molecules would be found that exhibit favorable antibiotic activity. In theory, this might be the case, but in practice, it has been difficult to find new synthetic antibiotics. So far, human ingenuity has not been nearly as effective as the process of evolution, which has had over 3 billion years to get things right and produce effective antibiotics.

There are many approaches being tried to generate new synthetic antibiotics (Figure 15–1). One approach has been **combinatorial chemistry.** Starting with a particular chemical structure or scaffold, a large variety of chemical groups are added in a random fashion to create an array of new chemicals. This bank of compounds is usually then tested for inactivation of a particular bacterial target, such as a purified enzyme that plays an essential role in pathogenic bacteria but lacks strong homologs in mammals. By using a core template structure and randomly screening a large bank of chemical groups to make new compounds, combinatorial chemistry changes the paradigm for discovery by introducing the notion that compounds can be synthesized as mixtures and then subsequently screened for inhibitory activities as mixtures.

When mixtures are produced in chemical reactions, the number of compounds produced increases exponentially. For example, in traditional chemical synthesis, a reaction joins two types of functional groups, A and B, to form a molecule, A-B, but in a combinatorial approach, two of each functional group are included in a single reaction mixture to produce four product compounds: A1 + A2 + B1 + B2 = A1-B1 + A1-B2 + A2-B1 + A2-B2. If three types of each group are included in the same reaction mixture, then nine compounds are produced, and so on. By combining many components in these types of reactions, very large mixtures (or "libraries") can be made. The biggest advantage of this approach is that the number of tests for the desired property can be greatly reduced if the libraries are used for testing instead of individual compounds. That is, it is much easier to perform an antibiotic screen on one mixture of 100 compounds than separately on 100 compounds. The screen can then be dramatically ramped up and robotically automated to assay thousands of libraries, with each library consisting of 100 or more compounds. This process is referred to as **high-throughput screening (HTS).** This way, hundreds of thousands of compounds can be screened in a relatively short time and using considerably fewer chemical reagents and materials. Once it has been determined that a particular library has a desired property, the problem changes to simply identifying which of the 100 compounds in that particular library mixture is the active one. The process of making that determination is called **"deconvolution."**

A drawback of this combinatorial method is that since a purified target, such as an enzyme or rRNA molecule, is used in the screening, it is not guaranteed that a compound that inactivates or inhibits the activity of that target will be able to get to the target in an intact bacterium. To get to an intracellular target, the

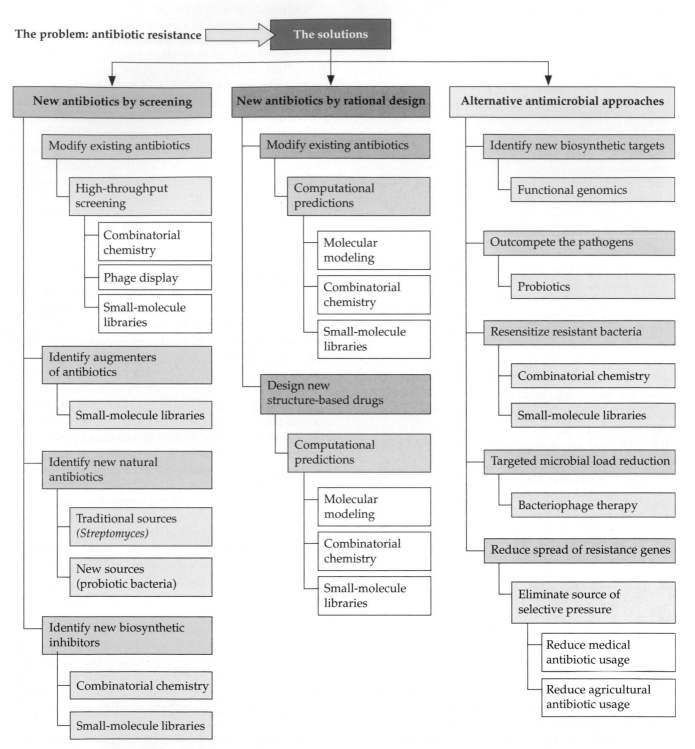

Figure 15–1 The search for new antimicrobials. The growing problem of antibiotic resistance has spurred the need for new antibiotics and alternative strategies to combat infectious bacteria. Solutions being pursued by biotechnology companies include refining and streamlining traditional approaches to discover new or to modify existing antibiotics, as well as creating new designer drugs and exploiting genomics, probiotics, bacteriophages, and other alternative approaches. (Adapted from Wilson and Salyers, 2002, with permission from Elsevier.)

antibiotic has to transit the outer membrane of a gram-negative bacterium and the cytoplasmic membrane of both gram-negative and gram-positive bacteria. Also, the interior chemical milieu of the bacterial cell may not be conducive to antibiotic action. For these reasons, essential proteins on bacterial cell surfaces, such as enzymes involved in peptidoglycan biosynthesis, cell division, or signal transduction, are often chosen as targets. These targets are considered more "druggable," because antibiotics can reach them without traversing the cytoplasmic membrane. Another drawback to combinatorial chemistry is that, so far, it has not worked nearly as well as expected. No one really knows why. Possibly, the reason is that scientists have to choose a particular base molecular structure for chemical modification and a particular target molecule, and we do not yet seem to know the rules for picking the most effective combinations to take forward.

A second new approach is similar, except that scientists use the crystal structure of the target molecule as a guide to model the potential inhibitor into the binding sites of the target protein by using computer "docking" simulations and then synthesize chemicals that will bind to the target molecule and inactivate it. This approach is called **rational drug design** and was widely heralded as the salvation of the pharmaceutical companies when it was first introduced. Rational drug design has some of the same drawbacks as combinatorial chemistry. More discouraging, however, is the fact that the results have been very disappointing. A hybrid approach that combines screening with rational drug design seems to hold more promise for finding synthetic antibiotics. In this approach, chemical libraries are screened for inhibitors of a purified bacterial target protein or RNA structure. The inhibitor is then cocrystallized with the target protein or RNA, and the three-dimensional structure of the inhibited complex is determined by X-ray crystallography. Alternatively, for smaller targets, the nuclear magnetic resonance structures of inhibited complexes can be determined. The three-dimensional structure of the inhibitor in the target is used as a starting point for modeling to further optimize the inhibitor. The modified inhibitors are synthesized and tested for improved inhibition. The tremendous advantage of this hybrid approach is that the modeling is driven by the structures of actual inhibitor complexes instead of a theoretical prediction of what molecules bind to the target. However, even if strong inhibitors are synthesized, they still must be transported into cells for cytoplasmic targets and be amenable to further chemical modification to optimize all of the other properties of good antibiotics described above.

Finding New Targets by Exploiting Genomics

As we shall see below, most of the clinically successful antibiotics act on a very short list of known cellular targets, that is, a limited number of proteins involved in ribosomal function, DNA synthesis, or cell wall biosynthesis. Whole-genome sequencing of hundreds of bacteria opened the prospect of identifying new targets for antibacterial drug discovery. The idea behind this **genomics-based discovery** approach was to compare the sequences of different pathogenic bacteria to find those genes that were highly conserved and consequently more likely to be potential broad-spectrum targets. The set of conserved genes could then be narrowed further to those that were unique to bacteria (or at least with low homology to any human counterpart) and that were essential for bacterial survival (i.e., that would kill the bacteria if inhibited). The resulting targets would then be used in HTS to identify lead compounds for antibiotic development. There are currently many large small-molecule library collections that have been assembled for HTS, usually consisting of up to 200,000 and in some instances over a million compounds.

With all the bacterial genome sequences that have been made available over the past 10 years, you might ask how it is possible that so few new antibiotics have made it to the market. A number of explanations for this meager level of success have been put forward. Interestingly enough, it does not appear to be because of an inability to move potential targets to the HTS stage, the lack of robust assays, or the known challenge of converting a lead compound into a safe bioactive drug. Rather, it appears to be a combination of other factors, the primary one being the lack of sufficient molecular diversity in the nature of the compounds that make up the small-molecule libraries used for HTS.

Most chemical libraries are made up of compounds that have particular readily synthesized structural features and that have pharmacological properties suitable for mammalian drug targets. That is, they are heavily biased toward compounds that follow **Lipinski's "Rule of Five,"** which is a medicinal chemist's empirical rule of thumb used to evaluate whether a potential lead compound has chemical properties that would make it a likely orally bioactive drug in humans without significant toxicity. Lipinski's rule states that a potential drug cannot violate more than one of the following criteria: have no more than 5 hydrogen bond donors, have no more than 10 hydrogen bond acceptors, have a molecular mass under 500 daltons, and have an octanol-water partition coefficient (log P) of less than 5. Note the numbers are all multiples of

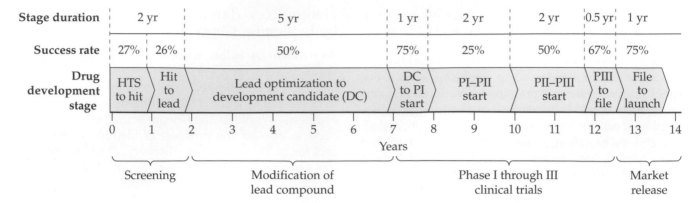

Figure 15–2 Timelines and risks associated with broad-spectrum antibacterial drug discovery. Shown are the timelines and estimated success metrics for the development of a broad-spectrum antibacterial drug based on the probability of success (percentages) metrics reported by GlaxoSmithKline and on clinical success rates based on industry averages reported by the Centers for Medicines Research. The most challenging, risky, and lengthy stage is lead optimization, which requires a sizable investment of time and medicinal-chemistry effort to bring a lead compound to the beginning of phase I (PI) clinical trials. (Adapted from Payne et al., 2007, with permission from Macmillan Publishers Ltd.)

5; hence, the rule's name. The problem is that most known naturally occurring antibiotics do not follow Lipinski's rule, so it looks like more chemical diversity and a better understanding of the physical and biochemical properties that are important for effective antibacterial activity are important if we are to move forward in discovering new antibiotics. The problem is compounded by the fact that most of these libraries are compendia of compounds and their derivatives that were initially screened against mammalian targets involved in other human diseases. Although the numbers of compounds in libraries sound impressive, relatively few of the synthetic compounds have precedents as antibiotics.

Another factor is the lack of annotated bacterial gene targets with validated functions or activities, which precluded them from consideration for assay development in the first place. A surprisingly large percentage of the genes sequenced in each bacterial genome could not be annotated or could be only tentatively annotated based on weak homology to genes in other bacteria. It is impossible to develop a bioassay for a target protein of unknown function, and so by default, a large portion of the genes could not even be tested as potential targets. Researchers also found that the putative targets identified through genomic comparisons as homologs of known genes in other bacteria did not always have the predicted function in the bacterial pathogens of interest. This was particularly a problem when the researchers were interested in more broad-spectrum antibiotics. In addition, it

was found that many bacteria have considerable built-in redundancy, with more than one unrelated gene catalyzing the same reaction or having the same function, so that an inhibitor against one does not necessarily work against the other, and thus, the efficacy or spectrum of the inhibitor is greatly reduced. One of the current challenges of bacterial physiology, genomics, and bioinformatics is to identify these genes of unknown function.

The Economics of Antibiotic Discovery

The above description makes it clear that antibiotic discovery and development are challenging processes with many hurdles. The process is also costly and time-consuming. A company can spend $800 million or more to bring a new drug to market, and the process can take more than 10 years. At any point, complications, such as unfavorable side effects or insufficient efficacy in animals or human volunteers, may be uncovered, and the company loses its investment. Figure 15–2 shows an example of the process for development of a broad-spectrum antibacterial drug and the associated timelines and risks.

Why is bringing a drug from the research laboratory to market so expensive and time-consuming? First, the discovery and characterization processes require the efforts of a large team of highly paid scientists from many scientific disciplines, including microbiologists, medicinal organic chemists, structural biologists, and chemical modelers. At some stage, detailed pharmacokinetic and pharmacodynamic studies need to be performed in at least two animal species to evaluate efficacy and safety. A promising **investigational new drug (IND)** represents **intellectual property** and will have to be patented during these early stages, a process that is both expensive and time-consuming.

But what really contribute to the cost and time taken to bring a drug to market are the eventual human clinical trials. Human clinical trials can occur only after data from extensive preclinical studies of the types described above are evaluated and approved by **institutional review boards** that contain scientists, disease experts, statisticians, physicians, and bioethicists. Planning a clinical trial is an exceedingly complicated process that is beyond the scope of this book, but briefly, the study protocol must establish dosing and administration regimens, patient recruitment and evaluation, criteria of efficacy and safety, and contingency plans for when to stop a failing trial. Large teams of physicians, statisticians, drug experts, and hospital administrators are required to conduct clinical trials, usually carried out at multiple institutions.

For antibiotic IND testing, phase I involves a relatively small number (10 to 100) of healthy adult volunteers to test for potential adverse side effects and safety. Phase II testing involves administering the drug to a moderate-size (30 to 300) population of volunteers with an infection that the IND antibiotic should treat. The efficacy and safety of the antibiotic are closely monitored, as is the health of the volunteers in the event that the trial needs to be stopped. Many drugs fail at phase II because they are not efficacious enough in humans or cause unexpected side effects that were not seen in preclinical animal studies. Phase III trials are the most extensive and expensive of the testing paradigm. Phase III trials involve many more volunteers (usually thousands) with infections that are expected to be treatable by the antibiotic IND. They are usually set up using a **double-blind** protocol in which the identities of the patients receiving the test drug, a comparator drug already on the market, or a placebo are unknown to participants or to hospital staff during the trial. Phase III trials are designed to answer the critical question of whether the IND antibiotic is equal to or better than current treatments and whether side effects occur in a larger patient population. Only in the last stages of clinical testing are children and pregnant women included. Often two or more independent phase III trials are completed for a particular infectious disease (also called an "indication"), and separate phase II and III trials need to be carried out for each infectious disease that will eventually be on the drug approval label.

After the clinical trials are completed and the data are analyzed, the company presents its data to the Food and Drug Administration (FDA) for approval, a process that can take months or years. The FDA is charged with ensuring that any drug released for general use has passed a highly rigorous and thorough evaluation. Importantly, the FDA also requires that new drugs match or exceed those currently on the market in some way. Sometimes, restrictions have to be put on the patient populations or types of uses. An antibiotic that is very effective but has some toxic side effects may be acceptable if used as a last-ditch treatment for critically ill patients but not for routine use or for children or pregnant women. The final stage, if FDA approval is obtained, is manufacturing and marketing. Sometimes, if a new IND antibiotic appears to be performing well in clinical trials, it will be used for critically ill patients on a compassionate-use basis even before clinical trials are completed.

As new antibiotics became harder to find and more expensive to develop and test, pharmaceutical companies became less enthusiastic about them. No antibiotic is going to be as profitable as simple synthetic drugs, such as Prozac or Viagra, which are taken repeatedly over long periods. The upshot was that, starting in the 1970s when, to make matters worse, there was a glut of antibiotics on the market, the pharmaceutical companies began reducing or shutting down their antibiotic discovery programs based on marketing decisions. There was a small resurgence of antibiotic drug discovery in the late 1990s when high-throughput screening, routine protein structure determination, and genomic information became available. However, at this writing, the situation is dire, and only one or two of the larger pharmaceutical companies have retained antibiotic discovery groups large enough to have any chance of success. These developments have also negatively impacted antibiotic discovery by small biotechnology companies, which depend on attracting funding from larger pharmaceutical companies for their lead compounds or the biotechnology firm itself. Several promising biotechnology companies focused on antibiotic discovery have recently folded.

What about academic antibiotic discovery? New study panels have been established to review grant proposals on antibiotic drug discovery and resistance development and to foster high-throughput screening. However, at this writing, some government agencies are funding less than 10% of the investigator-initiated proposals submitted. This is an extremely low overall success rate, and in many ways, these grant study panels have become as conservative as the pharmaceutical companies by funding research on known lead compounds rather than trying to find new kinds of lead compounds or supporting high-risk projects. In addition, many academic studies identify inhibitors of bacterial growth or the activities of purified targets, but they often do not move to the next stages of establishing these inhibitors as true drug candidates by performing the assays for cytotoxicity, serum binding, mechanism of action in the bacterium, and bioavailability in mammals described above. Besides this serious gap in approach, it is hard to assemble and maintain the multidisciplinary teams needed for drug discovery in academic institutions, especially using the short-term funding provided by many research grants. Finally, protection of intellectual property and rapid reporting of scientific advances in peer-reviewed journals, which are the lifeblood of companies and academic institutions, respectively, are seriously at odds. For these reasons, academic antibiotic discovery has not been able to fill the void left by pharmaceutical companies leaving antibiotic discovery and development.

This recent history has profound implications for public health. By the end of the 1980s, the flood of new antibiotics had slowed to a trickle (Table 15–3). Moreover, bacteria are becoming increasingly resistant to antibiotics and are making some of the former antibiotic stars into has-beens. New antibiotics to cope with these resistant bacteria are badly needed for an increasing number of diseases, but at the same time, the major players in the pharmaceutical industry have largely decimated their antibacterial research efforts. To add to this impending crisis, this chapter makes it clear that antibiotic discovery and development are complicated processes that take considerable time, effort, and resources (Figure 15–2). At this stage, even an abrupt resurgence in meaningful, concerted support of antibiotic discovery by the government, foundations, and industry would take years to have an impact before new antibiotics became available for use by the public.

Mechanisms of Antibiotic Action

Targets of Antibiotic Action

A drug that attacks a vital eukaryotic-cell target will have serious side effects. Thus, the ideal antibiotic should act on a bacterial target that either is not present in eukaryotic cells or is different enough from the same molecule or process in eukaryotic cells that there is little or no cross-reactivity. Although there are a number of targets that potentially satisfy this criterion, the currently used antibiotics tend to focus on a very limited set of targets. The most clinically relevant targets to date have been cell wall biosynthesis, protein synthesis, DNA synthesis, and folic acid synthesis. There are two main reasons why pharmaceutical companies have focused on these targets. First, in earlier screens of antibiotics, compounds that inhibited one of these processes emerged as the ones that were most effective and relatively nontoxic. Second, the expense of developing and testing a new antibiotic has become so high that administrators who approve funding for new antibiotic development feel more comfortable with drug classes and targets with which the pharmaceutical industry has had successful past experience. Reaching out for new chemical classes of antibiotics that hit new targets runs the risk that the new antibiotics could fail at the clinical-trial phase because of unexpected side effects and unexpected pharmacological properties. Hence, considerable effort has been expended using SAR and new chemical methods to improve antibiotic classes that already exist (the so-called "me too, me better" approach).

However, even modification of known classes of antibiotics has its limitations. For example, the antibiotic mupirocin (Figure 15–3) is the only approved antibiotic that inhibits an aminoacyl-tRNA synthetase. Mupirocin is a polyketide-derived antibiotic originally isolated from *Pseudomonas fluorescens*. It inhibits the isoleucyl-tRNA synthetase that attaches the amino acid isoleucine to its cognate tRNA^{Ile}. The isoleucyl-tRNA^{Ile} is subsequently used in translation, so mupirocin acts by ultimately blocking translation and inducing the stringent response. Mupirocin is bacteriostatic at low concentrations and bactericidal at

Table 15–3 Evolution of antibiotic resistance

Antibiotic	Yr deployed	Yr resistance observed
Sulfonamides	1930s	1940s
Penicillin	1943	1946
Streptomycin	1943	1959
Chloramphenicol	1947	1959
Tetracycline	1948	1953
Erythromycin	1952	1988
Vancomycin	1956	1988
Methicillin	1960	1961
Ampicillin	1961	1973
Cephalosporins	1960s	1960s
Ciprofloxacin	1987	1990
Linezolid	1999	2003
Daptomycin	2003	2005

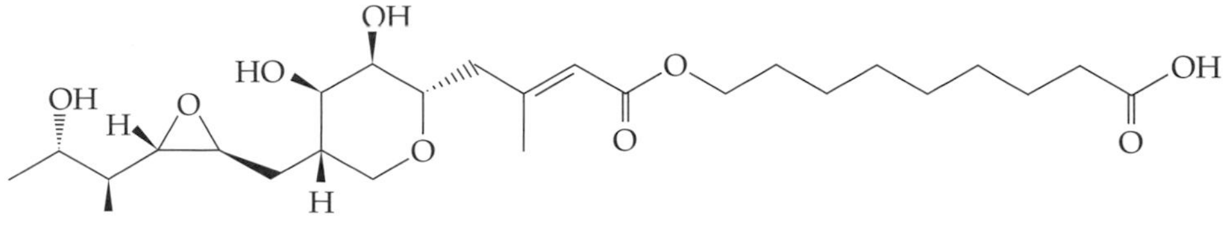

Figure 15–3 Mupirocin, a polyketide-derived antibiotic produced by certain *Pseudomonas* species that targets bacterial tRNA synthetases.

high concentrations. It does not effectively inhibit the human isoleucyl-tRNA synthetase. Unfortunately, mupirocin can only be used topically in ointments on the skin and in the nose, because it is rapidly metabolized and excreted by humans. Mupirocin has been used successfully to eliminate gram-positive bacteria, including methicillin-resistant *S. aureus* (MRSA), from the noses of carriers. A hospital care worker who is a carrier of MRSA is a danger to patients who have surgical wounds, because *S. aureus* is a common cause of postsurgical infections. It is likely that the company that developed mupirocin performed extensive SAR to try to overcome its unfavorable pharmacokinetics in humans. However, these data are often considered proprietary and are seldom published. Similarly, it is difficult to know what SAR has been performed on other established antibiotics to see if they might be improved.

Cell Wall Synthesis Inhibitors

β-Lactam antibiotics. The **β-lactam antibiotics** get their name from the four-member β-lactam ring they all have in common. This group of antibiotics now includes four main types: **penicillins, cephalosporins, carbapenems,** and **monobactams** (Figure 15–4). β-Lactam antibiotics have been among the most useful of all antibiotics. The main toxicity problem with these antibiotics is an allergic reaction that occurs due to formation of a β-lactam–serum protein conjugate, which evokes an immune response. Allergy to penicillins may result in allergy to cephalosporins, as well, and vice versa. Fortunately, the monobactams are different enough in structure from penicillins and cephalosporins that they can often be used on people who are allergic to penicillin.

β-Lactam antibiotics kill bacteria by inhibiting the last step in peptidoglycan synthesis, the transpeptidation reaction that cross-links the peptide side chains of the polysaccharide peptidoglycan backbone (Figure 15–5). β-Lactam antibiotics also bind to and inhibit the actions of other inner-membrane proteins that may have a role in peptidoglycan synthesis. Transpeptidase and these other proteins are sometimes called **"penicillin-binding proteins."** β-Lactam anti-

biotics work by forming a covalent bond with a reactive serine residue in the active site of the transpeptidase (Figure 15–5). This reaction opens the β-lactam ring, which is sometimes referred to as a **"warhead,"** because it is the structure that inactivates the enzyme. Notably, the transpeptidase active site prevents water from freeing the serine residue, and the resulting covalent complexes are unusually stable. Inhibition of cross-bridge formation by β-lactam binding also triggers endogenous enzymes called **autolysins** that degrade the peptidoglycan. Normally, these enzymes function in turnover of peptidoglycan that allows orderly growth and division of the bacteria. The action of β-lactam antibiotics apparently removes controls that normally keep these enzymes in check and stimulates their attack on peptidoglycan. Since the peptidoglycan cell wall prevents the bacteria from bursting due to the high osmotic strength of the cytoplasmic contents relative to the external medium, weakening of the cell wall leads to bacterial lysis, particularly during the exponential phase of cell growth. β-Lactam antibiotics are normally bactericidal. Occasionally, if bacteria are in a high-osmolarity compartment of the body (kidney) or if environmental conditions, such as pH, prevent activation of the autolysins, bacteria can sometimes escape the killing effects of β-lactam antibiotics, but such cases are uncommon.

Different β-lactams differ in their spectra of activity. Some are effective against both gram-positive and gram-negative bacteria, whereas others are much more effective against gram-positive than gram-negative bacteria, or vice versa. Many thousands of β-lactam derivatives have been synthesized and characterized in SAR studies. Despite all of the time and money invested to date in studying β-lactam antibiotics, it is still not possible to predict with certainty what changes in the basic β-lactam structure will produce a more effective antibiotic. β-Lactams also differ widely in their toxicity, stability in the human body, rate of clearance from blood, whether they can be taken orally, and their ability to penetrate the blood-brain barrier. However, we seem to be approaching a limit in the extent to which we can change β-lactam antibiotics to improve them as antibiotics that evade resistance mechanisms.

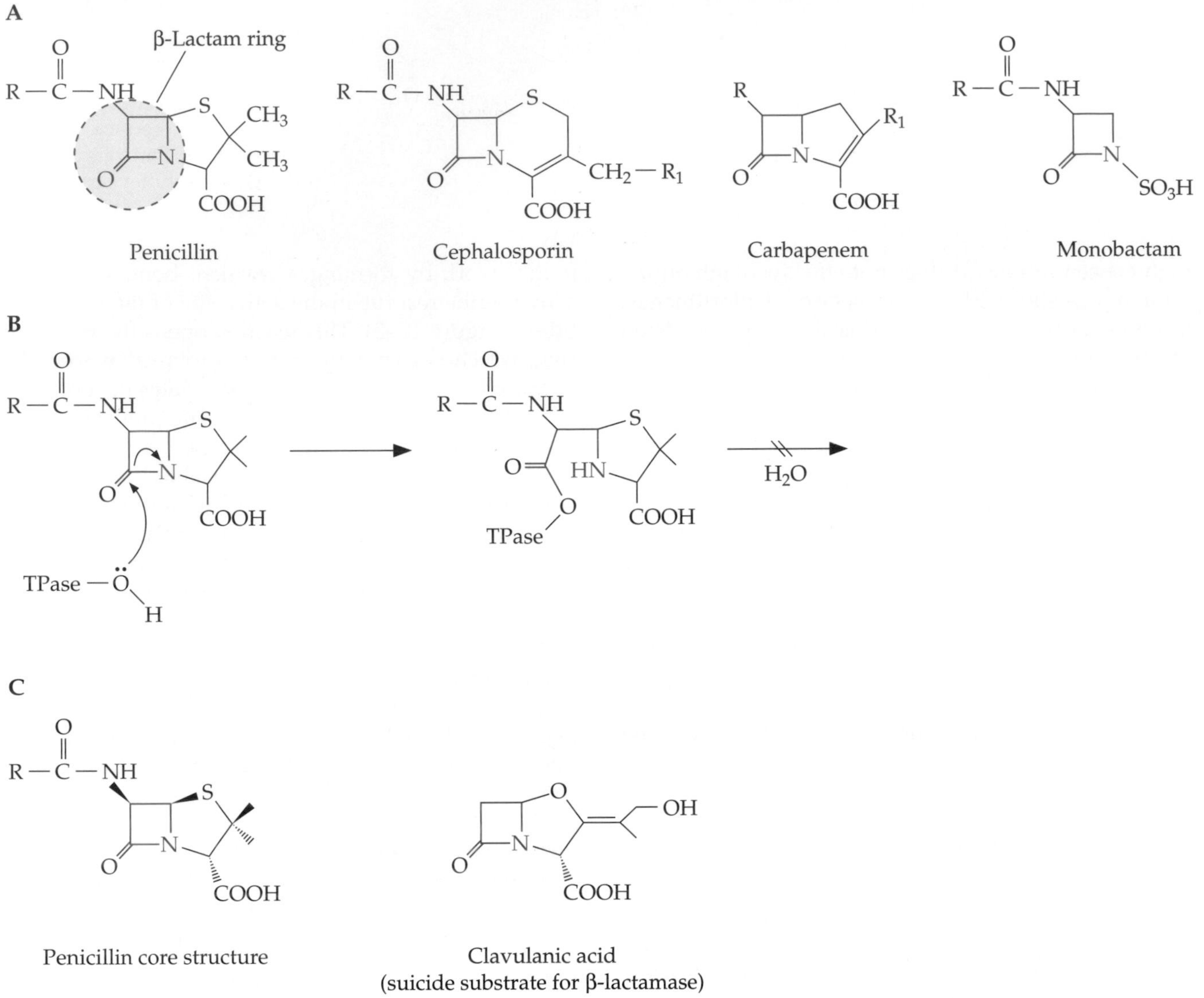

A

Penicillin   Cephalosporin   Carbapenem   Monobactam

B

C

Penicillin core structure   Clavulanic acid
(suicide substrate for β-lactamase)

Figure 15–4 Structures of β-lactam antibiotics and mechanism-based inactivators of β-lactamases as augmenters to enhance antibiotic efficacy. **(A)** Core structures of representative β-lactam antibiotics and the mechanism by which β-lactam antibiotics bind covalently to and inhibit transpeptidase (penicillin-binding protein) enzymes (TPase). **(B)** A reactive serine in the TPase active site attacks and opens the β-lactam ring, and water is not allowed into the active site to release the covalently bound TPase enzyme. (Adapted from Walsh, 2003.) **(C)** Clavulanic acid and other related suicide substrate inhibitors of β-lactams are often combined with the active β-lactam antibiotics to prevent their degradation by β-lactamases.

GLYCOPEPTIDES. Another group of peptidoglycan synthesis inhibitors is the **glycopeptides** and/or **lipopeptides,** which include **vancomycin, daptomycin,** and **teichoplanin** (Figure 15–6). These antibiotics, especially vancomycin and, more recently, daptomycin, have become extremely important medically because they are the last drugs that are effective against some gram-positive pathogens, such as *S. aureus* and *Entero-* *coccus* species. Glycopeptide antibiotics bind to the D-Ala-D-Ala portion of the UDP-muramyl-pentapeptide after it is transferred out of the cell cytoplasm. Binding appears to inhibit **transglycosylation** and **transpeptidation,** the two final steps in peptidoglycan synthesis. Vancomycin and teichoplanin are used primarily to treat infections caused by gram-positive bacteria and are not very effective

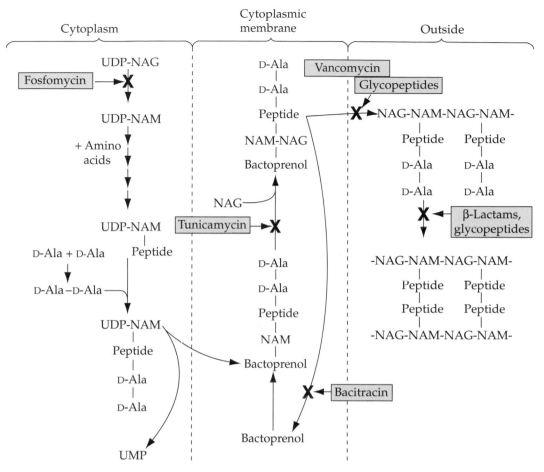

Figure 15–5 Steps in the synthesis of peptidoglycan and the effects of different classes of antibiotics. Fosfomycin blocks the conversion of UDP-NAG to UDP-NAM. Bacitracin inhibits the dephosphorylation and recycling of bactoprenol (the cell membrane lipid anchor). β-Lactams and glycopeptides, such as vancomycin, block the transpeptidation step (cross-linking of the peptidoglycan layer). Because the enzymes that carry out transpeptidation also mediate transglycosylation, these antibiotics tend to block both steps.

against gram-negative bacteria because they cannot penetrate the outer membrane. Despite their relatively narrow spectrum, these antibiotics are clinically important. For example, vancomycin has become particularly important for treatment of infections due to strains of *S. aureus* that are resistant to virtually all other antibiotics. Daptomycin was originally discovered and developed by researchers at Eli Lilly in the 1980s. It showed promise in phase I/II clinical trials in the 1980s for the treatment of infections by gram-positive bacteria, but it was found to have adverse side effects on skeletal muscle and was dropped by the company. Cubist Pharmaceuticals subsequently acquired the rights to the compound in 1997, and the drug was reintroduced on the market in 2003 with a different treatment regimen that reduced toxicity to acceptable levels.

OTHER ANTIBIOTICS THAT INHIBIT CELL WALL SYNTHESIS. **Fosfomycin** (Figure 15–7) and **bacitracin** (Figure 15–8) are two antibiotics that inhibit peptidoglycan synthesis by inhibiting earlier steps in the pathway than those inhibited by β-lactams and glycopeptides. Fosfomycin inhibits the conversion of UDP-NAG (*N*-acetylglucosamine) to UDP-NAM (*N*-acetylmuramic acid) (Figure 15–5) by acting as a **mechanism-based inhibitor (suicide substrate).** The epoxy ring of the compound covalently reacts with a cysteine side chain in the enzyme's active site, which irreversibly inactivates the enzyme. Fosfomycin has limited clinical utility but is used to treat urinary tract infections and MRSA. Bacitracin interferes with the recycling of bactoprenol (Figure 15–5). It is very effective against gram-positive bacterial cell walls. Bacitracin is used primarily in topical ointments available

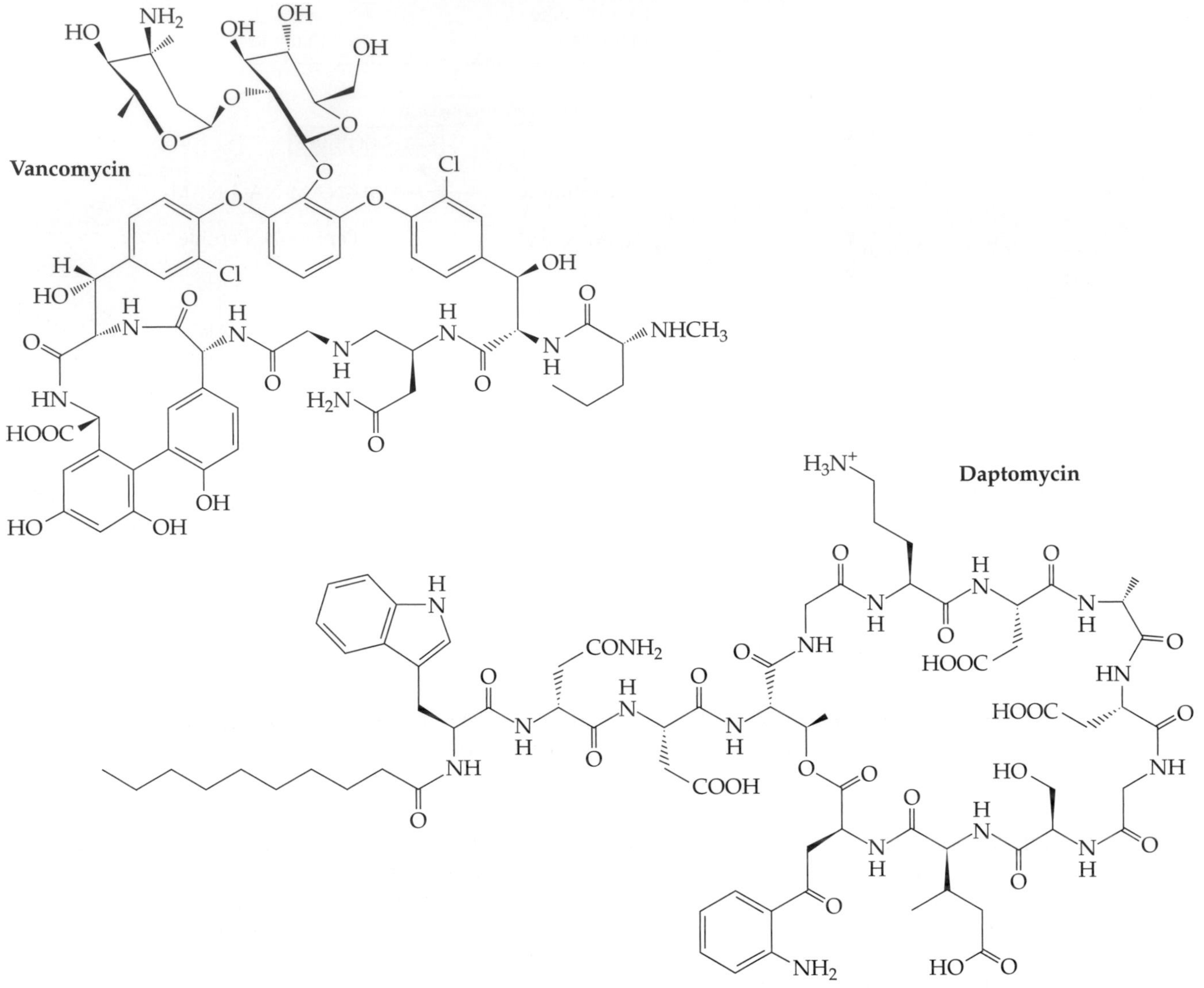

Figure 15–6 Structures of the cyclic-peptide-derived antibiotics vancomycin, daptomycin, and teicoplanin. Vancomycin is a cyclic glycopeptide antibiotic produced by *Amycolatopsis orientalis*. Daptomycin is a lipopeptide antibiotic produced by *Streptomyces roseosporus*.

over the counter for treating a variety of skin infections, as well as preventing wound infections, but it is too toxic for internal use.

Protein Synthesis Inhibitors

One of the most significant discoveries in structural molecular biology occurred recently when the three-dimensional structures of ribosomes were determined. These elegant structures revealed that the rRNA molecules catalyzed peptide bond formation instead of the many ribosomal proteins. Besides having profound implications for evolution and the mechanisms of translation, these structures have led to deep understanding of how antibiotics block bacterial translation. Most translation inhibitors are large natural products with side groups that can form multiple hydrogen bonds and salt bridges. It turns out that these antibiotics inhibit translation by forming spatially specific hydrogen bonds to the bases and phosphate backbones of the 16S and 23S rRNAs in the ribosome functional sites. In addition, ribosome structures have revealed differences between prokaryotic and eukaryotic ribosomes that impart differential selectivity. This knowledge is driving new efforts to modify older antibiotics to improve their binding and to discover chemicals that bind to bacterial rRNAs in new ways.

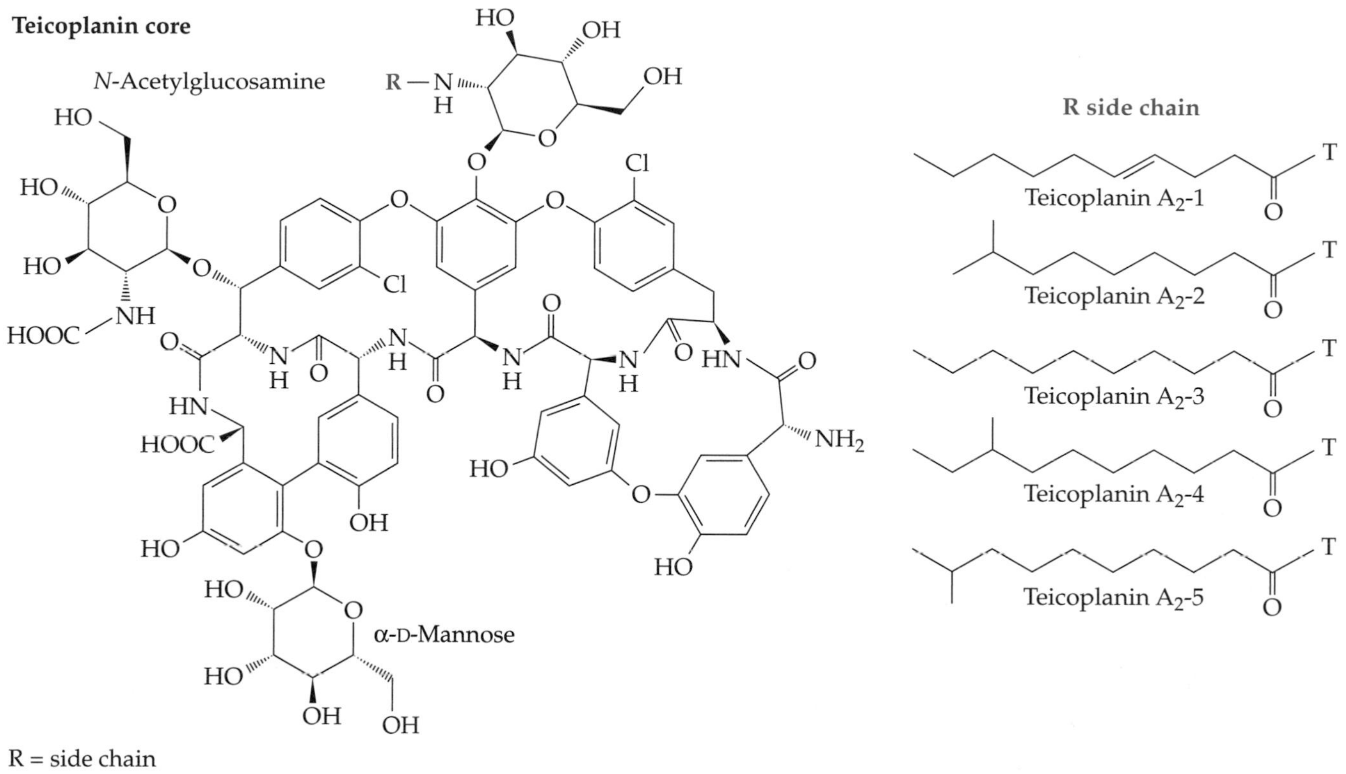

Teicoplanin core

R = side chain

Figure 15–6 *(continued)* Teicoplanin is produced by *Actinoplanes teichomyceticus* and has a cyclic glycopeptide core with different lipid side chains.

AMINOGLYCOSIDES. **Aminoglycosides** are trisaccharides with amino groups that act by binding to specific sites in the 16S rRNA in the 30S subunit of the bacterial ribosome, thereby blocking protein synthesis (Figure 15–9). Binding of aminoglycosides, such as **kanamycin** and **gentamicin** (Figure 15–10), to the bacterial ribosome does not prevent the 30S subunit from binding mRNA and placing the initial tRNA^{fMet} in the P site at the AUG start codon of the mRNA to introduce the first amino acid, formyl methionine (fMet), but it does block the 50S subunit from joining the 30S subunit to form the active ribosome. Aminoglycosides are bactericidal, because protein synthesis is essential for continued viability of a bacterium. Aminoglyco-

sides are effective against a number of pathogenic bacteria, but they have serious side effects that limit their use. Prolonged use of aminoglycosides can lead to hearing loss and to impairment of kidney function.

TETRACYCLINES. **Tetracyclines,** as the name suggests, are compounds consisting of four fused cyclic six-member rings (Figure 15–11). Tetracyclines also target the bacterial ribosome and bind to the 16S rRNA in the 30S subunit. The effect of binding is to distort the A site and prevent the alignment of aminoacylated tRNA with the codon on mRNA. **Tigecycline** is a tetracycline derivative that has a bulky glycyl-glycine side chain. It has recently been approved for use against a number of both gram-negative and gram-positive pathogens. Although most tetracyclines unquestionably act by interfering with protein synthesis, some members of the tetracycline family (e.g., **chelocardin**) appear to act instead by disrupting the bacterial membrane, not by stopping protein synthesis. At present, most of these atypical tetracyclines are of academic interest only, because they have so far proven to be too toxic for use in humans. There is hope, however, that some nontoxic derivatives can be found.

Figure 15–7 Structure of fosfomycin, a metabolite from *Streptomyces* species that contains a reactive epoxy ring and an unusual carbon-phosphorus bond.

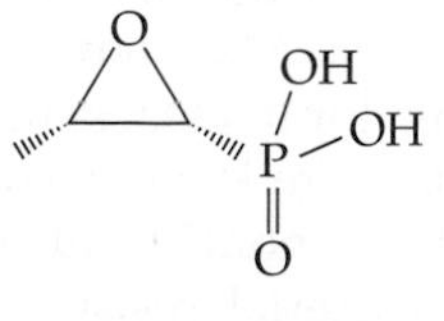

Fosfomycin

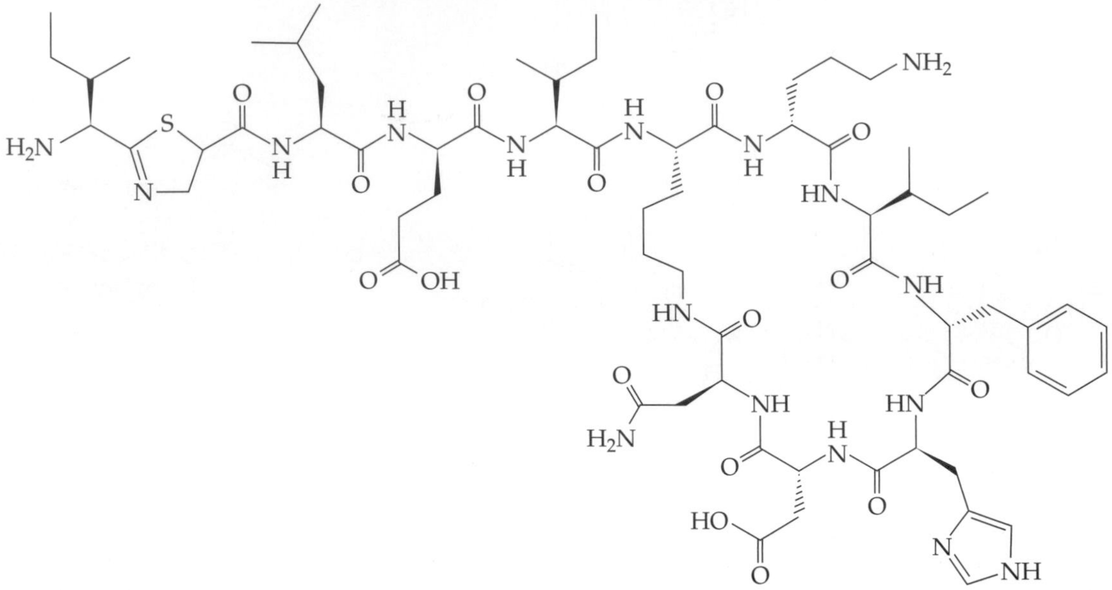

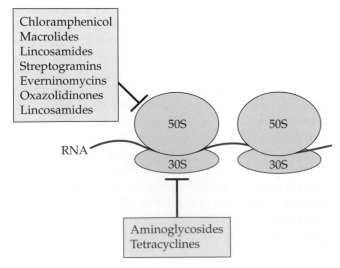

Bacitracin

Figure 15–8 Structure of bacitracin, a cyclic polypeptide produced by certain *Bacillus* species.

Tetracyclines are generally bacteriostatic. Although tetracycline can cause discoloration of teeth if given to young children or pregnant women and sometimes causes nausea or phototoxicity, it has been one of the least toxic antibiotics ever produced. This, together with the fact that it can be given orally, has led to overuse of tetracycline in clinical practice. It has also been used in the treatment of acne and as a feed additive to promote growth of livestock. Not too surprisingly, tetracycline resistance is now so widespread

that the utility of the tetracycline family has been considerably diminished. Nonetheless, tetracyclines still have some important uses, e.g., in the treatment of Lyme disease and some sexually transmitted bacterial diseases, such as chlamydia and gonorrhea. Dermatologists still use tetracycline extensively to treat acne and rosacea (excess reddening of the skin of the face). Scientists would like to find new members of the tetracycline family, because this family of antibiotics has many good properties. It has been so widely used (and abused) precisely because it is not only effective, but also nontoxic.

Figure 15–9 The bacterial ribosome as a target for antibiotics that interfere with protein synthesis.

Chloramphenicol
Macrolides
Lincosamides
Streptogramins
Everninomycins
Oxazolidinones
Lincosamides

50S 50S

RNA

30S 30S

Aminoglycosides
Tetracyclines

MACROLIDES. The **macrolide** family, exemplified by **erythromycin** (Figure 15–12), is another group of polyketide-derived antibiotics that have relatively few side effects. Macrolides inhibit bacterial protein synthesis by binding the 23S rRNA in the 50S ribosomal subunit (Figure 15–9). Binding of macrolides blocks the exit tunnel used by nascent peptides to exit the ribosome. This is a particularly striking example of an antibiotic "throwing a monkey wrench" into the ribosome molecular machine. Macrolides are bacteriostatic for most bacteria but are bactericidal for some gram-positive bacteria. Macrolides, like tetracyclines, have also been used for livestock, primarily to prevent shipping sickness. There is some concern, as with the tetracyclines, that this nonclinical use is contributing to the spread of bacterial resistance to this class of antibiotics.

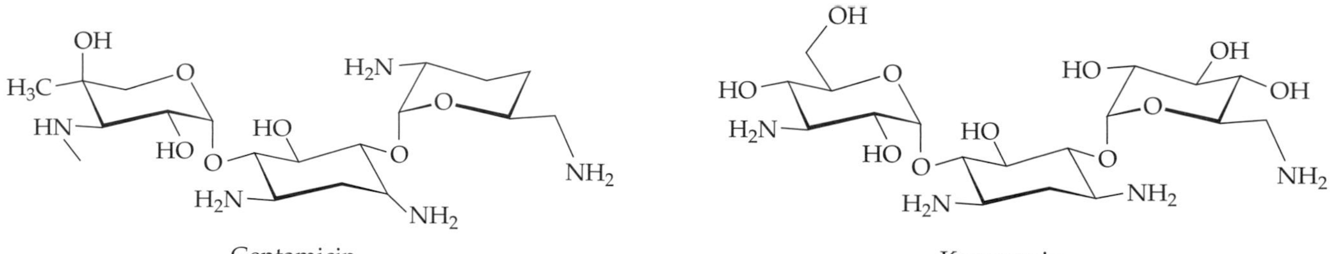

Figure 15–10 Structures of gentamicin and kanamycin and of aminoglycosides. Gentamicin is produced by gram-positive *Micromonospora* species. Kanamycin is produced by *Streptomyces kanamyceticus*.

Azithromycin (Figure 15–12) has been an exciting newer-generation macrolide. Azithromycin is one of the great triumphs of SAR to improve the antibiotic properties of an existing class of antibiotics. Azithromycin is so effective in treating sexually transmitted bacterial diseases, such as chlamydial disease and gonorrhea, that it needs only to be administered orally in one or two closely spaced doses because of its stability, long half-life, and excellent bioavailability in humans. This solves a problem that has plagued clinics that treat people with sexually transmitted diseases. The patient is given tetracycline or another antibiotic and instructed to take it daily for a week. If the patient abuses drugs or alcohol or is homeless, the full regimen may not be taken. Also, the patient may not abstain from sexual contact until the antibiotic regimen has been completed. A patient who can be treated in a supervised manner at a clinic will take the entire course of antibiotic, will be cured more quickly, and will thus be less likely to transmit the disease to others.

Unfortunately, azithromycin has one rather serious drawback. It is very expensive—too expensive for many cash-strapped inner city clinics to use. It is also too expensive for many developing countries, where

Figure 15–11 Structure of tetracycline. Tetracycline is a polyketide-derived antibiotic produced by *Streptomyces* species. There are also semisynthetic derivatives available.

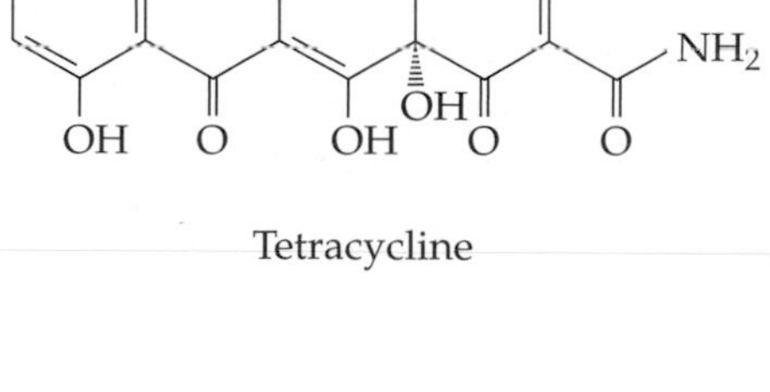

Tetracycline

resistance to tetracycline and other commonly used antibiotics in bacteria that cause sexually transmitted diseases is beginning to be a serious problem. This illustrates yet another problem that has been only partly dealt with. Pharmaceutical companies need to make enough return from new antibiotics to pay for the expense of their development. This need is at odds with the need of vulnerable impoverished populations for cheap antibiotics.

LINCOSAMIDES. **Lincosamides**, such as **lincomycin** and **clindamycin** (Figure 15–13), are synthesized by a condensation between a modified sugar and an amino acid derivative, but they have the same mechanism of action as the macrolides and bind the 23S rRNA near the same site as the macrolides. Perhaps the most widely used lincosamide is clindamycin. Clindamycin has been used extensively to treat infections caused by obligately anaerobic bacteria, such as *Bacteroides* species. Clindamycin is also one of the few antibacterial compounds that is effective against disease-causing protozoa, such as *Giardia,* the cause of a persistent and sometimes life-threatening diarrhea. Clindamycin's strength—its effectiveness against obligate anaerobes—also proved to be its downfall. Patients who took clindamycin experienced a decrease in numbers of the obligate anaerobes that comprise the major populations of the colonic microbiota. In some of these patients, this disruption of the normal microbiota allowed the pathogen *C. difficile* to overgrow and produce powerful toxins that could kill the patient (see chapter 18).

Antibiotics That Target DNA Synthesis

QUINOLONES AND FLUOROQUINOLONES. **Quinolones** are synthetic compounds that inhibit bacterial DNA synthesis and replication (Figure 15–14). For a long time, the only quinolone available, **nalidixic acid** (Figure 15–15), was very effective in treating urinary

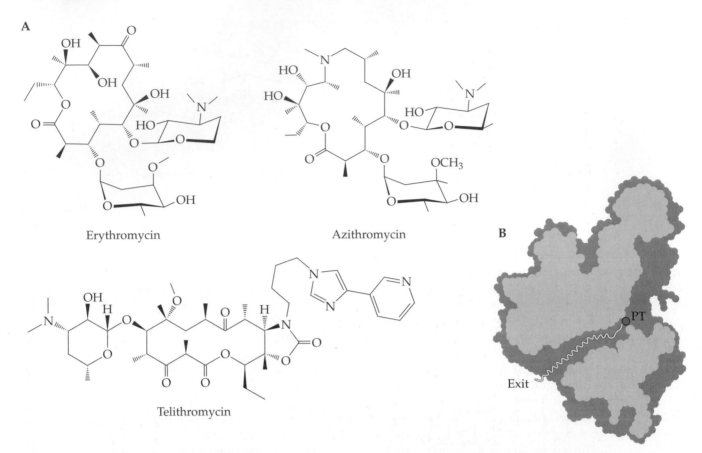

A

Erythromycin

Azithromycin

Telithromycin

B

PT

Exit

Figure 15–12 Polyketide-derived macrolide antibiotics that target the ribosome. **(A)** Structures of erythromycin, azithromycin, and telithromycin. **(B)** Exit tunnel for nascent polypeptide chains in the 50S bacterial ribosome. PT, peptidyltransferase active site where peptide bonds are formed. The blue dot marks the site of binding of macrolides and lincosamide antibiotic to 23S rRNA at the front of the tunnel. This binding blocks exit of the polypeptide chains and halts translation elongation. (Panel B adapted from Walsh, 2003.)

tract infections due to gram-negative bacteria but was not particularly useful for many other types of infections. Then, scientists added a fluorine group to the molecule, and a new class of antibiotics useful for humans, the **fluoroquinolones,** was born. This new group of antibiotics caused considerable excitement in clinical circles because of their impressive antibacterial activity and good pharmacological properties.

Fluoroquinolones, such as **norfloxacin** and **ciprofloxacin** (Figure 15–15), are broad-spectrum bactericidal antibiotics that inhibit bacterial DNA replication by binding to and inhibiting the activity of DNA gyrase, which is a type II topoisomerase that introduces negative supercoils (i.e., relaxes positive supercoils) into DNA. During DNA replication, recombination, and transcription, DNA strands must be broken and re-

Figure 15–13 Structures of lincomycin and clindamycin, examples of lincosamide antibiotics.

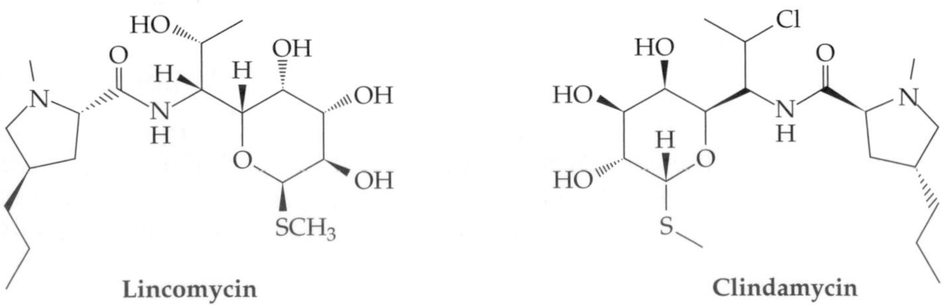

Lincomycin

Clindamycin

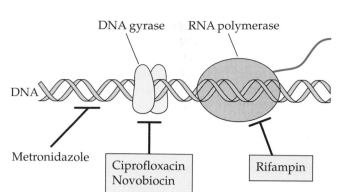

Figure 15–14 Antibiotics that inhibit DNA and RNA synthesis.

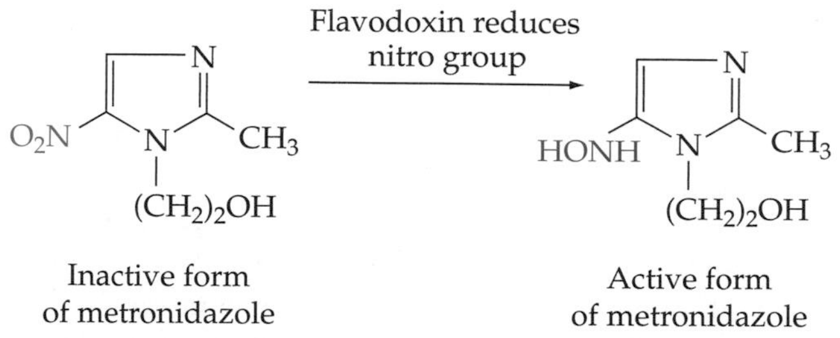

Figure 15–16 Structure and activation of metronidazole. Metronidazole is first activated by flavodoxin to a form that can attack and cause strand breaks in DNA. Exactly how the activated form of metronidazole breaks the phosphodiester backbone of a DNA strand is still not clear.

formed because of supercoiling. Fluoroquinolones block this essential process after the DNA gyrase has formed covalent links to and nicked each DNA strand, thereby creating fatal double-strand breaks in the bacterial chromosome.

Ciprofloxacin, commonly known as **Cipro,** gained considerable recognition during the anthrax attacks of 2001, when it was dispensed to a large population of people who had potentially been exposed to the anthrax spores through the U.S. mail system (more on this in chapter 20). At the time, Cipro was the only official FDA-approved antibiotic for the treatment of anthrax infection. Fluoroquinolones have poor activity against streptococci and anaerobes, which comprise a majority of the resident microbiota of the mouth, colon, and vaginal tract. Thus, they are less likely than other antibiotics to disrupt the normal microbiota. Also, fluoroquinolones penetrate macrophages and polymorphonuclear leukocytes better than most other antibiotics and are thus useful for curing infections caused by bacteria that survive in phagocytes.

The Achilles heel of the fluoroquinolones is that a single mutation in DNA gyrase makes bacteria resistant to them. Overuse of fluoroquinolones in human medicine has had the predictable effect of giving rise

to fluoroquinolone-resistant bacteria. Fluoroquinolones have even been approved for use in the water provided to chickens being raised in intensive chicken farming operations to prevent infections from sweeping through the bird population. The FDA's approval of this use of fluoroquinolones has been controversial. No one knows how much of a contribution this extensive new use of fluoroquinolones will make to resistance of bacteria, such as *Salmonella* and *Campylobacter,* which are found in the avian intestine and can cause human disease. A strain of *Salmonella enterica* serovar Typhimurium (strain DT104) that has caused a number of outbreaks in humans in recent years is resistant to fluoroquinolones and several other antibiotic classes.

METRONIDAZOLE. Metronidazole, like the fluoroquinolones, interferes with DNA synthesis, but it does so not by inhibiting an enzyme, but rather by making breaks in the DNA (Figure 15–14). Metronidazole, unlike other antibiotics, must first be activated by a housekeeping protein, flavodoxin, before it can attack DNA (Figure 15–16). Metronidazole has become important recently because it is one of the drugs used to eliminate *Helicobacter pylori,* which causes stomach ul-

Figure 15–15 Structures of nalidixic acid, norfloxacin, and ciprofloxacin, examples of fluoroquinolone antibiotics.

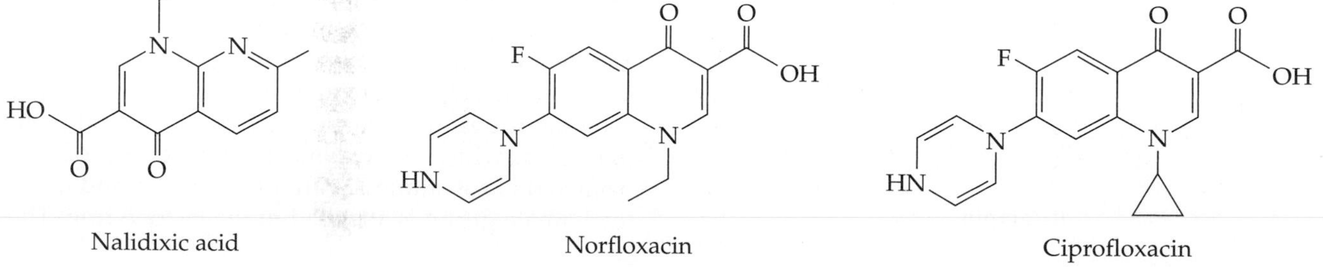

Nalidixic acid

Norfloxacin

Ciprofloxacin

cers, and because it is important for treating infections caused by anaerobic pathogens that are resistant to most antibiotics. The ferredoxins and flavodoxins found in microaerophiles and obligate anaerobes, but not aerobes, are capable of reducing the nitro group of metronidazole to a form that interacts with DNA and causes nicks in the DNA strands. The activated form of metronidazole could be considered to be a mutagen. At one time, this feature of metronidazole raised questions about its safety, but since human cells do not covert metronidazole to a mutagen, these concerns were short-lived. Metronidazole has proved to be an antibiotic that has few side effects. Since such antibiotics are increasingly thin on the ground, metronidazole is looking better every day.

Metronidazole, like the lincosamide clindamycin, is not specific for bacteria. It is also effective against certain eukaryotic pathogens, such as *Trichomonas vaginalis* (vaginitis) and *Giardia intestinalis* (diarrhea). These protozoa have in common with anaerobic bacteria the capability to activate metronidazole to its active, DNA-damaging form.

Antibiotics That Inhibit RNA Synthesis

Rifampin (Figure 15–17) is a semisynthetic antibiotic produced by *Amycolatopsis rifamycinica* that inhibits the activity of bacterial DNA-dependent RNA polymerase by binding to its β subunit and thereby preventing transcription into RNA (Figure 15–14). By analogy to the macrolides, rifampin binds to an exit tunnel in RNA polymerase, thereby blocking the emergence of nascent RNA molecules, but an important difference to keep in mind is that the mouth of the tunnel in ribosomes blocked by macrolides is made of 23S RNA, whereas the tunnel in RNA polymerase blocked by rifampin is made up of protein chains. Although rifampin has always had its clinical

Figure 15–17 Structure of rifampin.

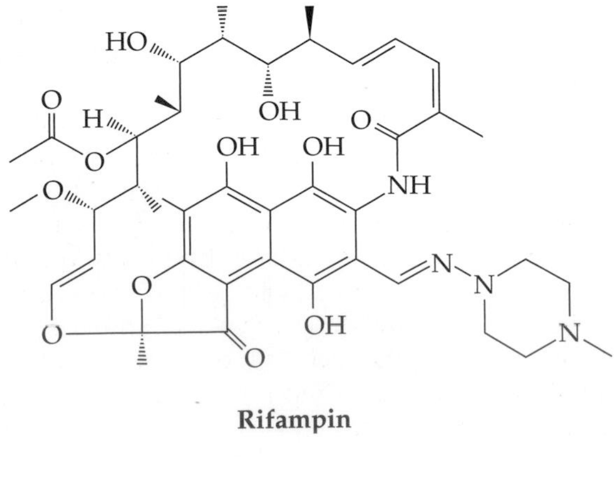

Rifampin

uses, e.g., prophylactic use during outbreaks of *Neisseria meningitidis* meningitis and as part of the drug combinations used to treat tuberculosis, it has not received much attention until recently. The appearance of strains of *M. tuberculosis* that are resistant to isoniazid, one of the front-line antibiotics used to treat the disease, caused renewed interest in rifampin. Unfortunately, resistance to rifampin can arise fairly easily as a result of single mutations in RNA polymerase, so rifampin is most often used in combination with other antibiotics. Nonetheless, rifampin is one of the few drugs available for treating isoniazid-resistant tuberculosis. Rifampin has also been used recently to treat some other types of bacterial infections caused by strains that are resistant to the more commonly used antibiotics.

Inhibitors of Tetrahydrofolate Biosynthesis

The sulfa drugs were the first systemic antibiotics to become available in the 1930s. **Trimethoprim** and **sulfonamides** are inhibitors of enzymes in the bacterial pathway for the production of **tetrahydrofolic acid** (Figure 15–18). Tetrahydrofolic acid is an essential cofactor for 1-carbon transfer reactions that occur in pathways for synthesis of nucleic acids and fMet. Mammalian cells require preformed folic acid, because they do not make their own tetrahydrofolate. Thus, inhibitors of the tetrahydrofolic acid pathway do not affect them. Sulfonamides are structurally similar to *p*-aminobenzoic acid, a substrate for the first enzyme in the pathway, and they competitively inhibit that step. Trimethoprim is structurally similar to dihydrofolic acid and acts as a competitive inhibitor of dihydrofolate reductase, an enzyme that catalyzes the last step in the pathway. Trimethoprim and sulfonamides, such as sulfamethoxazole, are bacteriostatic and are not usually used separately. However, a combination of the two drugs seems to act synergistically, resulting in bactericidal activity; decreases the chances of resistance development; and is effective against bacterial and fungal respiratory infections.

The Newest Antibiotics

A few new antibiotics have been introduced during the past several years. Two are members of known classes of antibiotics. As mentioned earlier, tigecycline is a tetracycline derivative with a glycyl-glycyl group added that helps overcome resistance to tetracycline. The **ketolides** are new derivatives of macrolide antibiotics in which the cladinose sugar group of erythromycin is substituted with a keto group and a cyclic carbamate group is attached to the lactone ring. These

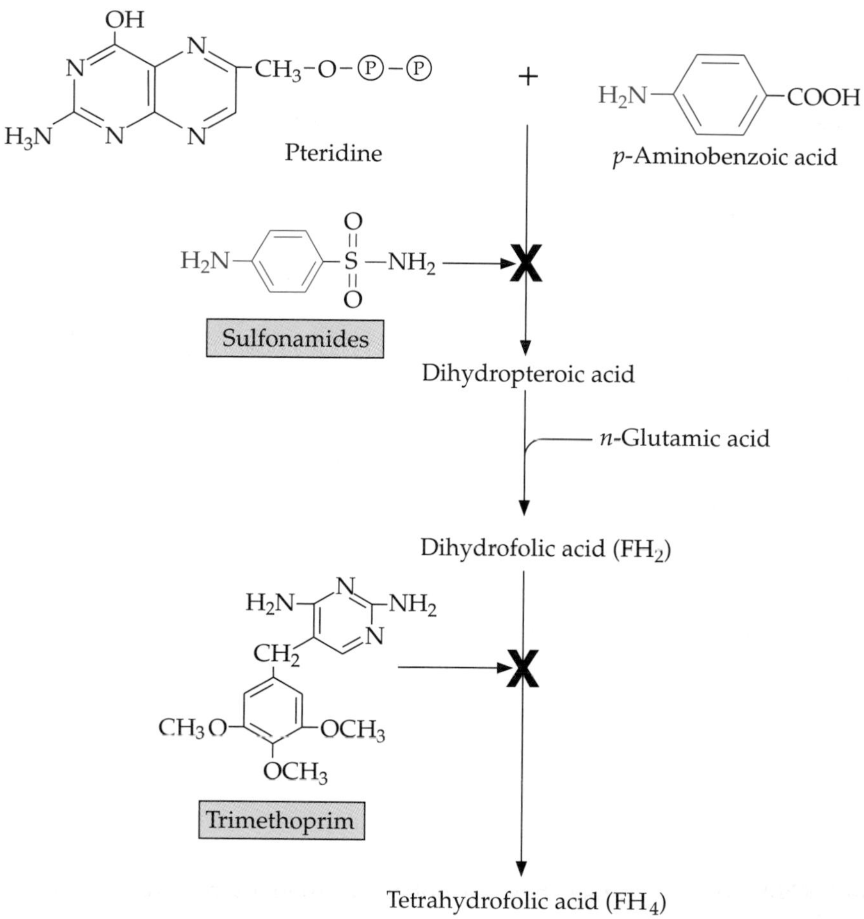

OH

Pteridine

p-Aminobenzoic acid

Sulfonamides

Dihydropteroic acid

n-Glutamic acid

Dihydrofolic acid (FH$_2$)

Trimethoprim

Tetrahydrofolic acid (FH$_4$)

Figure 15–18 Structures and actions of trimethoprim and sulfonamides. Sulfonamides resemble p-aminobenzoic acid in structure and are thus able to competitively inhibit the action of the first enzyme in the tetrahydrofolate pathway. Trimethoprim resembles the substrate of a later enzyme and competitively inhibits this step in synthesis.

modifications provide ketolides with a much broader spectrum than other macrolides. Since they bind to two sites in bacterial 23S rRNA, ketolides are effective against macrolide-resistant bacteria. The only ketolide currently on the market is **telithromycin** (Figure 15–12). However, serious side effects, including liver failure, have limited the use of this antibiotic.

A completely new class of antibiotics, the first new class in decades, has been given the tongue-twisting name **oxazolidinones.** The name under which the first member of this class, **linezolid** (Figure 15–19), is being sold is easier to pronounce: Zyvox. The oxazoli-

Figure 15–19 Structure of the oxazolidinone linezolid (Zyvox).

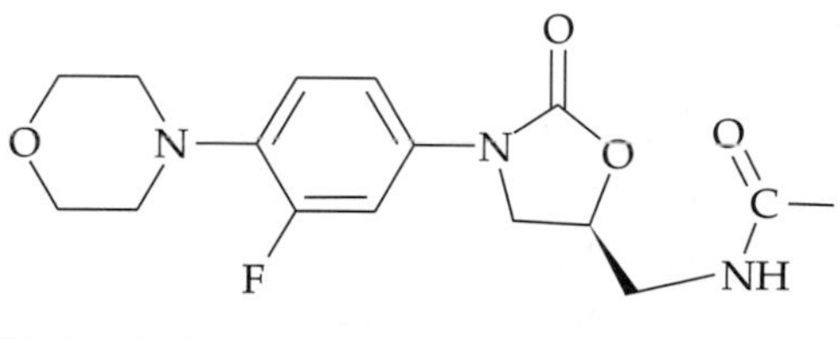

Linezolid

dinones are completely synthetic and bind the large ribosomal subunit to which the macrolides, lincosamides, and streptogramins bind (Figure 15–9). However, the oxazolidinones seem to act at an earlier step in translation than these other antibiotics. They seem to bind to the peptidyltransferase site and block formation of the first peptide bond during protein translation. Linezolid is being used primarily to treat infections by gram-positive bacteria, such as VRE, that are resistant to other antibiotics. The oxazolidinones have high oral bioavailability and few side effects. Unfortunately for a new drug introduced in 2000, resistance has already started to appear, including one class due to base changes in the 23S rRNA.

Among the more promising new IND antibiotics reported recently are diarylquinoline compounds (Figure 15–20) that specifically inhibit *Mycobacteria*, including drug-resistant *M. tuberculosis*. The lead compound in this group was identified by medium-throughput whole-cell screening for inhibition of bacterial growth. The lead compound inhibited only the growth of *Mycobacteria* and did not inhibit other bacterial species. This compound fit Lipinski's rules for a good drug candidate and showed excellent phar-

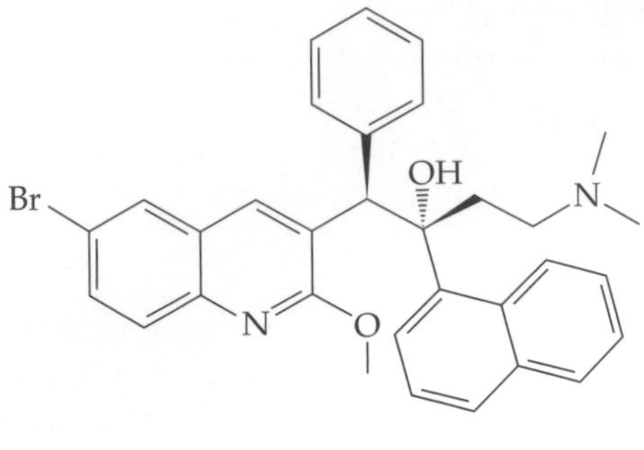

Diarylquinoline R207910

Figure 15–20 Structure of diarylquinoline R207910, which blocks the growth of mycobacteria by inhibiting the ATP synthase.

macokinetic and pharmacodynamic profiles in animal models. Resistance studies indicated that the lead compound inhibits a unique target, the essential membrane-bound mycobacterial ATP synthase that synthesizes ATP. All bacteria contain a homologous ATP synthase, and it was unprecedented that these compounds specifically inhibited the enzyme in *Mycobacteria*. Efficacy studies in animal models of tuberculosis suggested that the diarylquinoline was effective alone and in combination with other drugs in combating infection. We will need to see whether this promising new kind of antibiotic emerges from human trials to add to the arsenal of antituberculosis drugs. The fact that resistance can develop by simple point mutations in genes encoding the ATP synthase means that diarylquinolines will need to be used in combination with other antituberculosis drugs.

The Newest Antibiotic Targets

Some recent antibiotic discovery strategies have moved away from targeting essential bacterial housekeeping functions. Instead, various pathways involved in virulence are being targeted for inhibition, such as quorum sensing, toxin delivery, virulence regulation, and bacterial adhesion. **Quorum sensing** provides regulatory control over specific groups of genes,

including those encoding biofilm formation and various virulence factors, such as toxins, secretion systems, and adhesins. Inhibition of these bacterial cell-cell communication systems may result in attenuation of virulence (bacteriostatic rather than bactericidal) and may thus represent attractive new targets for antibiotic discovery.

Recent studies suggest that targeting quorum sensing may be particularly relevant for bacteria, such as MRSA and *Pseudomonas aeruginosa*, that use quorum sensing to regulate the expression of virulence genes and that are becoming increasingly resistant to other antibiotics. Indeed, a recent structure-based screen of a small-molecule library that targeted the quorum-sensing regulator of MRSA turned up a small-molecule inhibitor that occurs as a natural product called **hamamelitannin** (Figure 15–21) in the bark of witch hazel. The compound also controlled MRSA infection.

A very exciting recent finding is that azithromycin (Figure 15–12) exhibits strong quorum-sensing-antagonistic activity against virulence factor production, biofilm formation, and oxidative-stress response in *P. aeruginosa*. Even though *P. aeruginosa* infections, especially chronic infections of the lung in cystic fibrosis patients, are normally quite refractory to killing by macrolides, azithromycin has been reported to positively influence the clinical outcome in patients with lung infections.

Another success story involved high-throughput phenotypic screening of compound libraries for direct inhibition of virulence, which resulted in the discovery of a small molecule, **virstatin** (Figure 15–22), that inhibited ToxT, a known transcriptional regulator of cholera toxin and pilus production in *Vibrio cholerae*. Virstatin blocked toxin production and protected mice from intestinal colonization by the bacteria.

One should keep in mind that these new classes of targets, while interesting, may result in inhibitors that are largely bacteriostatic and that have very narrow spectra. Bacteriostatic antibiotics are less desirable in patients with compromised immune systems. A narrow spectrum means that physicians need to be certain that a condition is solely caused by the bacterium

Figure 15–21 Structure of hamamelitannin, a natural product from witch hazel that blocks quorum sensing in methicillin-resistant staphylococci.

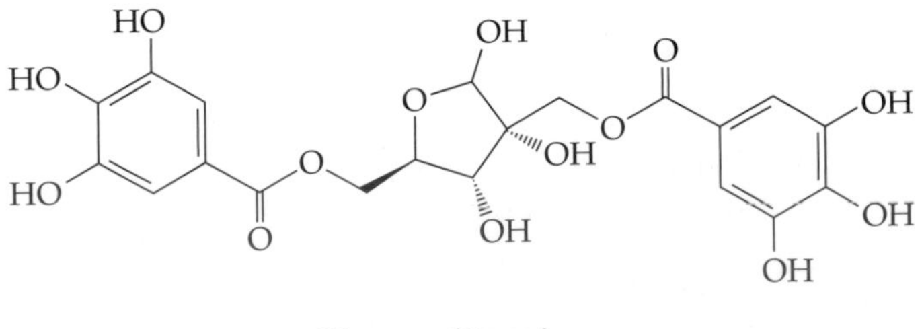

Hamamelitannin

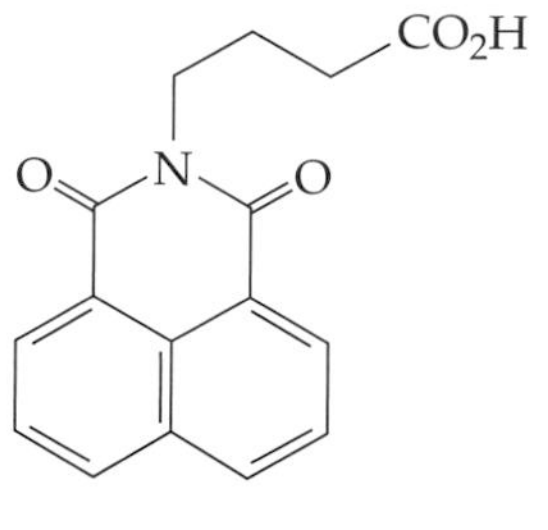

Virstatin

Figure 15–22 Structure of virstatin, a small-molecule inhibitor of ToxT, a transcriptional regulator of *V. cholerae* virulence.

that will respond to the drug, and this is not always possible. Finally, FDA approval depends on showing that these new drugs are at least as efficacious and safe as potent drugs that are already approved.

Strategies for Enhancing Antibiotic Efficacy

There has been considerable success in screening for and using drug design approaches to find **augmenters** (i.e., helpers) of existing antibiotics that make the existing antibiotics more potent or stable. One of the most successful examples of this approach is **clavulanic acid,** an irreversible mechanism-based (suicide substrate) inhibitor of one class of β-lactamases. Clavulanic acid was initially discovered as a natural product that lacks antibiotic activity itself but strongly inhibits one class of β-lactamases. Synthetic clavulanic acid is often included in drug formulations with other β-lactam antibiotics to prevent their degradation by β-lactamases (the products of antibiotic resistance genes). Clavulanic acid combined with the penicillin derivative amoxicillin is marketed as the drug Augmentin (Figure 15–4). Unfortunately, resistance caused by a second kind of β-lactamase is not inhibited by clavulanic acid, and no augmenter has yet been found that inhibits this class of β-lactamases.

Screening and combinatorial methods have also been applied to combating vancomycin resistance. Vancomycin interferes with bacterial cell wall biosynthesis by binding to the D-Ala-D-Ala linkage of peptidoglycan precursors. Bacteria become resistant to vancomycin by modifying this D-Ala-D-Ala linkage. Resistance genes encode enzymes that substitute D-Ala-D-Ala with D-Ala-D-lactate in the cell wall precursors. The presence of D-Ala-D-lactate in the cell wall and its precursors reduces the affinity of vancomycin for the peptidoglycan. In this case, library screening included compounds that are not themselves antibiotics but that were able to resensitize the

resistant bacteria to the antibiotic. One screen identified small compounds that selectively cleaved the modified peptidoglycan D-Ala-D-lactate linkage. Because these D-Ala-D-lactate linkages were removed, the bacteria once again became sensitive to vancomycin.

Streptogramins are usually overlooked in most textbooks because they have not been used until very recently to treat human infections. The appearance of **Synercid** on the market changed that picture in a radical way. Synercid is a combination of two streptogramins, **dalfopristin** and **quinupristin** (Figure 15–23). The reason the mixture is needed is that each of the drugs alone is bacteriostatic, but together they are bactericidal. Synercid has the same mechanism of action as the macrolides and lincosamides; it binds to the same portion of the 23S rRNA in the 50S subunit, thereby blocking the peptide exit tunnel and blocking elongation of the peptide chain (Figure 15–9).

Synercid was developed in response to the growing resistance of *Enterococcus* species and *S. aureus* to vancomycin, the last-ditch treatment of infections caused by these two gram-positive opportunists. Streptogramins would seem to be an ideal solution to the question of how to develop a new antibiotic that would be effective against the multiresistant bacterial threats looming in our future. Since streptogramins have not been used to treat human infections in the past, resistance would not be a problem, at least not at first, right? Well, not quite.

It turns out that streptogramins have been used widely in agriculture for a number of years, under the names virginiamycin and pristinamycin, to treat and prevent disease in animals being raised under crowded conditions. Not surprisingly, resistance to streptogramins in nature has emerged, and the CDC is now monitoring the emergence of streptogramin-resistant strains. The question remains of how separate the consequences of agricultural use of antibiotics actually are from the consequences of human use of antibiotics. Another complicating factor, which is covered in more detail in chapter 16, is that since the macrolides, lincosamides, and streptogramins bind to overlapping sites on the bacterial ribosome, some of the mechanisms bacteria develop to resist the actions of these antibiotics confer resistance to all three classes. This means that use of one class can select for resistance to the other classes in this triumvirate.

The Continuing Challenge

The few new antibiotics that are trickling through the pipelines present a quandary for physicians and the pharmaceutical companies. There is understandable

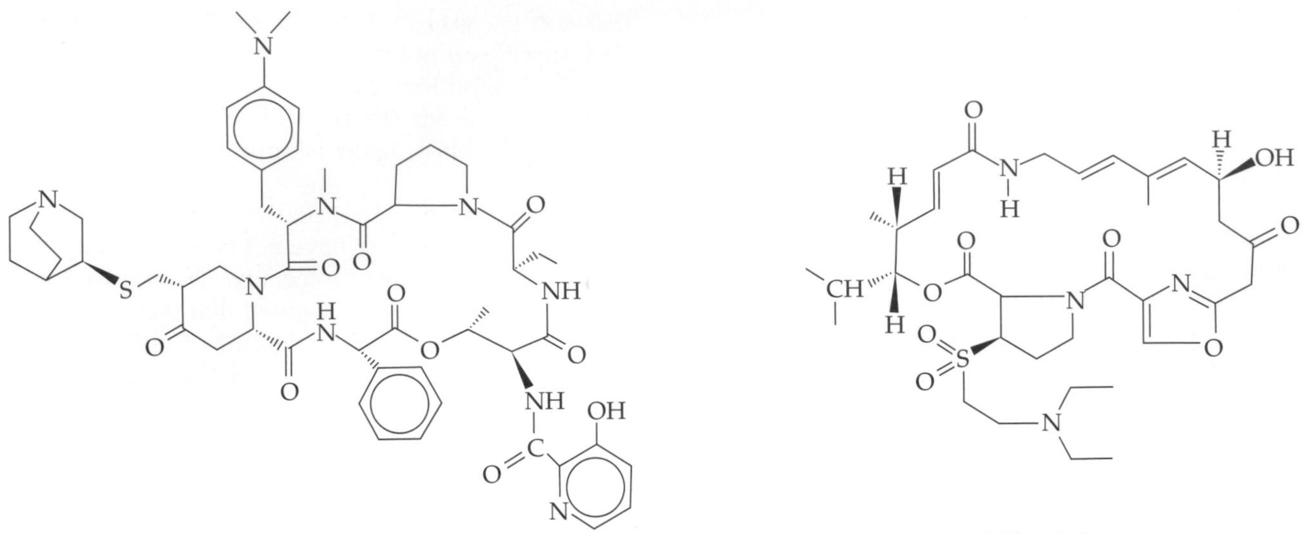

Quinupristin

Dalfopristin

Figure 15–23 Structures of the two streptogramins in Synercid, quinupristin and dalfopristin.

pressure to reserve these new drugs for the cases in which nothing else works, yet it is difficult to ask physicians, who have to make accurate guesses as to what antibiotic to use on their very sick patients, to forgo the latest antibiotics, especially if multidrug-resistant bacteria have become a problem in their hospital. It is also difficult to ask the pharmaceutical companies to restrict the sales of an antibiotic that has taken hundreds of millions of dollars and years to develop and test. Fear of not being able to recoup research and development costs and make a sizable profit has been one of the arguments used by pharmaceutical companies to justify their retreat from antibiotic research. Clearly, how to preserve new antibiotic drugs is an ethical and economic challenge.

There is also the challenge of how to control the use of the older antibiotics that still work against many bacterial infections. Physicians have been accustomed to making their own decisions as to what antibiotic to prescribe, and they are already chafing at the restrictions on medical practice being imposed by the HMOs. There are prudent-use guidelines that have the imprimatur of the American Medical Association, but they are often ignored. Two of the most widely flouted guidelines are the one that discourages the use of antibiotics to treat sore throats and flu and the one that discourages the use of antibiotics to treat ear infections in children until they have persisted for 24 to 48 h. Physicians like to please their patients and have an economic, as well as psychological, incentive for doing so. What does a physician do when a patient threatens to go elsewhere? Also, complaints lodged by

patients against physicians can get the physician in trouble with hospital administrators. Clearly the problem of enforcing prudent antibiotic use goes beyond scientific considerations and spills over into economic and political areas. Perhaps you will see the solution to these problems materialize and a new era of prudent use of antibiotics, in which they are revered and conserved as the vital resource they are, ushered in. We devoutly hope that you and your children will not instead witness the disastrous consequences of failure to contain abuse and overuse of antibiotics, coupled with the fact that the well of new antibiotics has been dry for a long time. Will we return to the situation of the 1920s, when any bacterial infection was life threatening?

SELECTED READINGS

Andries, K., P. Verhasselt, J. Guillemont, H. W. H. Göhlmann, J.-M. Neefs, H. Winkler, J. Van Gestel, P. Timmerman, M. Zhu, E. Lee, P. Williams, D. de Chaffoy, E. Huitric, S. Hoffner, E. Cambau, C. Truffot-Pernot, N. Lounis, and V. Jarlier. 2005. A diarylquinoline drug active on the ATP synthase of *Mycobacterium tuberculosis*. *Science* **307:**223–227.

Bjorland, J., T. Steinum, M. Sunde, S. Waage, and E. Heir. 2003. Novel plasmid-borne gene *qacJ* mediates resistance to quaternary ammonium compounds in equine *Staphylococcus aureus, Staphylococcus simulans,* and *Staphylococcus intermedius. Antimicrob. Agents Chemother.* **47:**3046–3052.

Chan, P. F., D. J. Holmes, and D. J. Payne. 2004. Finding the gems using genomic discovery: antibacterial drug discovery strategies—the successes and the challenges. *Drug Discov. Today Ther. Strateg.* **1**:519–527.

Chiosis, G., and I. G. Boneca. 2001. Selective cleavage of D-Ala-D-Lac by small molecules: re-sensitizing resistant bacteria to vancomycin. *Science* **293**:1484–1487.

Clatworthy, A. E., E. Pierson, and D. T. Hung. 2007. Targeting virulence: a new paradigm for antimicrobial therapy. *Nat. Chem. Biol.* **3**:541–548.

Fajardo, A., and J. L. Martinez. 2008. Antibiotics as signals that trigger specific bacterial responses. *Curr. Opin. Microbiol.* **11**:161–167.

Hoffmann, N., B. Lee, M. Hentzer, T. B. Rasmussen, Z. Song, H. K. Johansen, M. Givskov, and N. Hoiby. 2007. Azithromycin blocks quorum sensing and alginate polymer formation and increases sensitivity to serum and stationary-growth-phase killing of *Pseudomonas aeruginosa* and attenuates chronic *P. aeruginosa* lung infection in Cftr⁻/⁻ mice. *Antimicrob. Agents Chemother.* **51**:3677–3687.

Hung, D. T., E. A. Shakhnovich, E. Pierson, and J. J. Mekalanos. 2005. Small-molecule inhibitor of *Vibrio cholerae* virulence and intestinal colonization. *Science* **310**:670–674.

Kiran, M. D., N. V. Adikesavan, O. Cirioni, A. Giacometti, C. Silvestri, G. Scalise, R. Ghiselli, V. Saba, F. Orlando, M. Shoham, and M. Balaban. 2008. Discovery of a quorum sensing inhibitor of drug resistant staphylococcal infections by structure-based virtual screening. *Mol. Pharmacol.* **73**:1578–1586.

Lynch, S. V., and J. P. Wiener-Kronish. 2008. Novel strategies to combat bacterial virulence. *Curr. Opin. Crit. Care* **14**:593–599.

Mascaretti, O. A. 2003. Bacteria versus antibacterial agents; an integrated approach. ASM Press, Washington, DC.

Payne, D. J., M. N. Gwynn, D. J. Holmes, and D. L. Pompliano. 2007. Drugs for bad bugs: confronting the challenges of antibacterial discovery. *Nat. Rev. Drug Discov.* **6**:29–40.

Poole, K. 2005. Efflux-mediated antimicrobial resistance. *J. Antimicrob. Chemother.* **56**:20–51.

Reese, R. E., R. F. Betts, and B. Gumustop. 2000. Handbook of antibiotics, 3rd ed. Lippincott, Williams, and Wilkins, Philadelphia, PA.

Salyers, A. A., and D. D. Whitt. 2004. *Revenge of the Microbes.* ASM Press, Washington, DC.

Yim, G., H. H. Wang, and J. Davies. 2007. Antibiotics as signaling molecules. *Phil. Trans. R. Soc. Lond. B. Biol. Sci.* **362**:1195–1200.

Walsh, C. 2003. *Antibiotics. Actions, Origin, Resistance.* ASM Press, Washington, DC.

Wax, R. G., K. Lewis, A. A. Salyers, and H. Taber (ed.). 2007. *Bacterial Resistance to Antimicrobials*, 2nd ed. CRC Press, Boca Raton, FL.

Wilson, B. A., and A. A. Salyers. 2002. Ecology and physiology of infectious bacteria—implications for biotechnology. *Curr. Opin. Biotechnol.* **13**:267–274.

QUESTIONS

1. How do antiseptics, disinfectants, and antibiotics differ from each other?

2. Why might it be harder for a bacterium to become resistant to a disinfectant than to an antibiotic?

3. Many of the antibiotics described in this chapter have similar types of action: they bind to an important bacterial target and inhibit its activity. Are there exceptions to this rule?

4. The pharmaceutical industry has been trying to develop antibiotics by solving the crystal structure of a protein target and designing molecules that fit into the active site of this target (rational drug design). So far, this strategy has not been nearly as successful as expected. Why did scientists think this would be superior to the old ways of finding antibiotics, and what are some of the reasons this approach might fail?

5. Many of the aminoglycoside antibiotics are much less effective against *Escherichia coli* under anaerobic conditions than under aerobic conditions. Assume that the rates of protein synthesis are about the same in both cases and that the antibiotic inhibits protein synthesis in vitro. How would you explain the reduced effectiveness under anaerobic conditions?

6. Although in vitro tests of susceptibility to antibiotics are widely used and considered useful, there have been cases in which the in vitro tests have not predicted how well the drug would work when administered to humans. This works both ways. Some antibiotics that are effective in vitro fail when administered to humans, and some antibiotics that test resistant in vitro actually work when used therapeutically. Give some explanations for both of these outcomes.

7. In the quest for new antibiotics, what are the pharmaceutical companies trying to do? What criteria would you use to judge whether to develop a new antibiotic? What is the attraction of sticking with known antibiotics and trying to modify their structures rather than looking for new antibiotic targets?

8. For the following antibiotics, briefly describe how each kills or inhibits the growth of bacteria: β-lactams, glycopeptides, aminoglycosides, tetracyclines, quinolones, and macrolides.

9. An ideal antibiotic must have certain characteristics in order to be effective in human infections. Name four characteristics of a good antibiotic.

10. Indicate whether the following statements are true or false.

Aminoglycosides are bacteriostatic antibiotics.

Quinolones affect the replication of DNA.

Pharmaceutical companies synthesize antibiotics in the laboratory.

Antibiotics are administered only for short periods of time because they have side effects on the patient.

Broad-spectrum antibiotics are often used when the pathogen has not been identified.

Broad-spectrum antibiotics mainly kill gram-positive bacteria.

Tetracyclines cause discoloration of teeth in children.

Rifampin inhibits the activity of bacterial RNA polymerase.

Streptogramins inhibit bacterial DNA replication.

Trimethoprim inhibits tetrahydrofolate biosynthesis.

Ketolides are new derivatives of erythromycin antibiotics.

The β-lactam ring in penicillin can be cleaved to inactivate the antibiotic.

The β-lactam ring in penicillin can be modified to increase the spectrum of antibiotic activity.

Taking antibiotics can make you more susceptible to other infections.

Isoniazid, one of the primary antibiotics used against tuberculosis, inhibits mycolic acid biosynthesis.

SOLVING PROBLEMS IN BACTERIAL PATHOGENESIS

1. A group of researchers at the university veterinary diagnostics laboratory isolated a new bacterium from the blood and stool of several horses that became ill and died at a local farm. Prior to death, symptoms included disorientation, loss of motor function, and flaccid paralysis, so the researchers suspected central nervous system involvement and possible production of neurotoxin. Based on 16S rRNA comparison, they found that the new bacterium was related to the gram-positive bacterium *Clostridium botulinum*, and they subsequently named it *Clostridium equiniae*. They also determined that the bacterium was sensitive to metronidazole but resistant to β-lactam antibiotics, such as penicillin, or to macrolide antibiotics, such as erythromycin and azithromycin. However, although *C. equiniae* was resistant to azithromycin, the researchers found that when they treated infected mice with metronidazole in combination with azithromycin, the 50% lethal dose (LD_{50}) value went from 10 for a control without antibiotic treatment to 10^4 for treatment with metronidazole alone to 10^7 for combined treatment with azithromycin, and the mice recovered from paralysis much sooner with the combined treatment than with metronidazole treatment alone. Interestingly, when they were plating out the bacteria to determine the LD_{50} values, the researchers observed that the colonies isolated from the control mice and mice treated with metronidazole alone were mucoid, but those from mice treated with both antibiotics were not.

A. What possible mechanism(s) could account for all of these observations (i.e., change in LD_{50} value, faster recovery from paralysis, and change in mucoid phenotype)? Be sure to provide your rationale.

B. Provide an experiment that could be performed to confirm your hypothesis.

16

How Bacteria Become Resistant to Antibiotics

A s resistance soon emerges in any war, so too did resistance to antibiotics appear in clinical isolates shortly after antibiotics were introduced onto the battlefield in our war against bacterial diseases. At first, research on bacterial resistance to antibiotics focused primarily on basic science issues, such as the mechanism of resistance and regulation of resistance gene expression. Recently, however, we have begun to appreciate the economic consequences of increasing bacterial resistance. Hospital administrators and government regulators who have to confront health care costs have made a troubling discovery: not only does resistance make it harder to cure disease, but also, it is expensive! Lawsuits are now being brought against hospitals where patients have acquired drug-resistant nosocomial infections. Concern about resistant bacteria has also moved from the hospital to the farm; antibiotic-resistant bacteria have become the center of the controversy over the use of antibiotics in agriculture. To make matters worse, whereas most isolates were initially resistant only to single classes of antibiotics, we are facing an even direr problem of multidrug resistance. Clearly, concern about antibiotic resistance has come out of the ivory tower and into the real world.

The Dawning of Awareness

The 1990s was the decade when the public first began to take real notice of antibiotic-resistant bacteria. Prior to this, physicians had tended to downplay the importance of antibiotic-resistant strains of disease-causing bacteria, because in virtually all cases there were still antibiotics—sometimes many antibiotics—that continued to work. However, a shift in attitude had already begun to appear. Whereas once having to turn to a second antibiotic was considered a treatment failure, gradually the definition of treatment failure shifted to failure to find a successful antibiotic after trying several. As long as the physician finally hit upon an effective antibiotic, the treatment was considered a success. Patients did not always agree: those with serious systemic infections sometimes had irreversible damage to important organs or suffered a stroke because the infection was not brought under control quickly enough.

347

Arguably, the first groups to become publicly concerned about antibiotic-resistant bacteria were the officials of health insurance companies and HMOs. Infections caused by resistant bacteria were proving to be expensive (Box 16–1). Resistant bacteria were also costing state governments money. It cost New York City nearly a billion dollars to bring the **multidrug-resistant (MDR)** tuberculosis outbreak of the mid-1990s under control. About this same time, the New York Chamber of Commerce approached the state legislators to ask what was being done about antibiotic resistance. Even in the absence of a high-profile epidemic, businesses were losing money because of days lost from work and higher health care costs, and there

appeared to be no end to this problem. Congress held legislative hearings about antibiotic-resistant bacteria and their potential impact on human health.

Predictably, the media fell on the issue with their usual gusto. Headlines such as "The End of Antibiotics" and "Return to the Pre-Antibiotic Era" began to appear in the news magazines. Not surprisingly, the content of most of these articles was sensational and frightening. Even the normally staid journal *Science* had a cover on an issue about the antibiotic resistance problem that made the mainstream media seem conservative by comparison. On the cover of that issue of *Science* was a diptych. The left panel was a painting by Bruegel that depicted the skeletons pil-

---

BOX 16–1   The Cot of Antibiotic Resistance—a Telling Example

During the 1990s, *S. aureus* became one of the most common causes of hospital-acquired infections in the United States and other developed countries. Hospital-acquired infections, especially postsurgical infections, are much more common than they should be. It has been estimated that in the United States alone, about 2 million hospital patients per year will acquire such an infection. *S. aureus* has become more and more resistant to a variety of antibiotics, with the MRSA strains currently the most troublesome. Actually, the acronym "MRSA" would be more accurately rendered as MDR *S. aureus*, because these strains are usually resistant to several antibiotics in addition to methicillin. The only drug currently able to control MRSA infections is vancomycin, and isolated reports of MRSA strains with reduced susceptibility to vancomycin have begun to appear.

In a recent study, Rubin and colleagues attempted to estimate the costs of MRSA infections in New York City in 1995. They found that about 21% of all *S. aureus* infections acquired in hospitals or in the community were caused by MRSA strains. In the case of community-acquired infections, the additional cost per patient to treat these infections was about $2,500. Frequently, these patients had to be hospitalized. For hospital-acquired infections, the additional cost was higher, $3,700 per patient, probably because the patients involved were sicker than the community patients and were thus less able to control the disease.

The higher cost of treating MRSA infections was due to a variety of factors. First, vancomycin is more expensive than the drugs normally used to treat *S. aureus* infections. Second, it is often necessary to isolate patients to keep them from infecting other patients. Third, patients with MRSA infection stayed longer in the hospital. The increased financial cost was not the only toll taken by MRSA. The death rate for patients with MRSA infections was a shocking 21%, about 2.5 times higher than the death rate due to infections caused by susceptible *S. aureus* strains. The resistant strains are not necessarily more virulent than the susceptible ones, but their resistance makes it harder to control them with antibiotics. Thus, the increased death rate due to MRSA strains can be attributed largely to antibiotic resistance.

Not included in the economic estimates were costs to the patient. Longer hospital stays mean more days lost at work and more disruption of family life. The patients who died obviously paid the highest price, but even patients who survived could leave the hospital with irreversible damage to vital organs, such as the brain, lung, and kidney. Keep in mind that MRSA strains are still treatable with vancomycin. Imagine the carnage if vancomycin-resistant MRSA strains appear.

Source: R. J. Rubin, C. A. Harrington, A. Poon, K. Dietrich, J. A. Greene, and A. Moiduddin. 1999. The economic impact of *Staphylococcus aureus* infection in New York City hospitals. *Emerg. Infect. Dis.* **5:**9–17.

ing up during the plague years and being taken away to wherever skeletons go. A modern painter painted the right panel, which showed an inner city scene in which skeletons consorted with the living (but clearly not long for this world) and fires loomed in the background.

The media soon discovered agricultural use of antibiotics, and the possible impact of this use on human health began to be the subject of news articles and TV programs. Environmental advocacy groups started to consider antibiotic use and antibiotic resistance issues on which they needed to take a stand. The Humane Society got involved because it realized that severely restricting the use of antibiotics in agriculture might force improved hygiene and reduced crowding, changes that would improve the quality of life for the animals being raised.

Pharmaceutical companies responded to the steady increase in resistance by performing a highly effective "me-too, me-better" strategy to improve existing antibiotics, especially the β-lactam and macrolide classes. Several generations of new β-lactam antibiotics appeared over the years that stayed one step ahead of resistance for a time. In addition, several new classes of antibiotics appeared. However, eventually the chemical options to modify known antibiotics to outmaneuver resistance became limiting around the same time that the discovery of new classes of antibiotics nearly ceased. These setbacks, along with drastic changes in business models, contributed to the decisions of many pharmaceutical companies to curtail antibiotic discovery and development, which has only exacerbated the resistance problem.

Ironically, the one group that seemed to be left out of the growing awareness of the problem was the scientific community. Many scientists felt that the intensive research effort mounted in the 1980s to define the mechanisms of resistance and the transmissible elements, such as plasmids and transposons, carrying antibiotic resistance genes had uncovered all that was worth knowing about antibiotic-resistant bacteria. The funding agencies, following the lead of the scientific community, de-emphasized support for research in the area. The field of antibiotic resistance research came to be viewed as somewhat old-fashioned, and the number of scientists continuing to work in this area declined to a perilously low level.

Regrettably, a daunting number of important questions remained unanswered. When the gram-positive cocci began to reassume prominence as the most serious causes of human infections, it became evident that virtually nothing was known about their mechanisms of resistance or mechanisms of transfer of resistance genes. Moreover, many of the questions that

began to dominate the debate over antibiotic use patterns and possible preventive strategies in human medicine or agriculture turned out to be questions about the ecology and evolution of antibiotic-resistant bacteria and their genes. This area had been almost untouched even in the heyday of research on antibiotic-resistant bacteria, and now, with the realization that horizontal gene transfer is so prevalent among environmental and pathogenic bacteria, the need to understand antibiotic resistance mechanisms is reaching a critically urgent state.

How Did We Get to Where We Are?

Many social and medical factors have contributed to the development of antibiotic resistance. One of the major contributors is the tremendous genetic plasticity of bacteria. As discussed throughout this book and later in this chapter, bacteria have many mechanisms to acquire mutations and to exchange genetic material. They exist in phenomenally high populations and as members of complicated microbial communities. Resistance provides strong selective pressure, especially against bacteriostatic antibiotics, and resistance mechanisms that emerge have spread rapidly among bacterial species, but there are other causes for the spread of antibiotic resistance. About 50% of antibiotic use in the United States has been estimated to be inappropriate, in that antibiotics have often been prescribed for viral infections or at the wrong doses or durations for bacterial infections. Earlier, new antibiotics were often overprescribed, although this situation has changed because there are so few new antibiotics available.

One of the greatest sources of resistance continues to be the use of antibiotics to enhance the growth of livestock in crowded feedlots. At the time of writing, the U.S. Congress has yet to pass the Preservation of Antibiotics for Medical Treatment Act, which would greatly limit the use of antibiotics as animal growth supplements. Crowding, homelessness, poor nutrition and sanitation, and inadequate routine medical care remain problems that promote the spread of antibiotic resistance in developed, as well as developing, countries. In developed countries, certain social organizations, such as day care centers and hospitals, remain sources of bacteria harboring antibiotic resistance genes that can spread to visitors and family members. International travel has become commonplace, so resistant bacteria can rapidly be spread worldwide. Another source of antibiotic resistance is immunosuppression due to greater life expectancy; diseases, such as HIV infection; invasive medical procedures, such as organ transplants; and the use of implanted medical devices that support biofilm growth. At the same time, public health infrastructures have eroded in the United States and other developed countries. These formidable contributing factors are beginning to be addressed, especially with the increasing appearance of community-acquired and hospital-acquired (**nosocomial**) bacterial infections caused by MDR bacteria.

Mechanisms of Antibiotic Resistance

Overview of Resistance Mechanisms

Mechanisms of antibiotic resistance can be grouped into four main categories. One is restricted access of the antibiotic to its target. For example, the outer membrane of gram-negative bacteria can serve as an effective barrier against certain antibiotics, and this resistance can be enhanced by changes in the outer membrane properties that allow the bacteria to avoid taking up antibiotics. Another way to restrict access is to prevent the antibiotic from accumulating to high inhibitory concentrations in the cell by increasing active efflux (pumping out) of the drug from the bacteria. Efflux pumps are ubiquitous in gram-negative and gram-positive bacteria. A second category of resistance mechanisms is enzymes that inactivate or chemically modify the antibiotic, either by hydrolyzing it or by adding chemical groups to some important part of the antibiotic that interferes with binding of the drug to its target. A third category is modification of the antibiotic target. In this type of resistance, the bacteria accumulate mutations in a gene encoding a target protein or rRNA or acquire proteins that modify the target so that the target protein still works but no longer binds the antibiotic strongly enough to cause inhibition. In the fourth category, failure to activate the antibiotic, mutations that decrease the expression of an enzyme that activates the antibiotic can occur. If the bacteria do not activate the antibiotic, it is harmless.

An interesting feature of many resistance mechanisms is that the proteins that mediate them are often related to bacterial housekeeping proteins. For example, some enzymes that inactivate penicillin are related to and may have evolved from the transpeptidase enzymes that carry out the cross-linking of peptidoglycan (see chapter 15), the enzymes that are inactivated by penicillin. Apparently, bacteria sometimes adapt the target of an antibiotic to become an offensive weapon against antibiotics. Although the different types of resistance mechanisms are considered individually here, it is important to realize that bacteria can combine more than one mechanism of resistance to increase their defensive shield against an antibiotic. In addition, MDR bacteria contain separate mechanisms that impart resistance to several different classes of antibiotics.

Limiting Access of the Antibiotic

OUTER MEMBRANE PORINS. Antibiotics must first reach their target in order to have an effect. β-Lactam antibiotics must transit the gram-negative outer membrane to reach the cytoplasmic membrane, where the penicillin-binding proteins are located. Other types of antibiotics that have targets in the bacterial cytoplasm must be transported across the cytoplasmic membrane. In gram-negative bacteria, the outer membrane can function as a barrier to antibiotic entry. The reason vancomycin, which is very effective against gram-positive bacteria, is not effective against most gram-negative bacteria is that it is too bulky to diffuse through the outer membrane porin proteins. Porin proteins form beta-barrel structures in the outer membrane that allow the selective diffusion of small molecules into the periplasm. The genomes of gram-negative bacteria encode many different porins with a variety of permeability limits, and changing stress conditions regulate the expression of porin genes. For example, *Pseudomonas aeruginosa*, an organism that causes eye infections associated with improper contact lens use, bacteremia in burn patients, and lung infections in cystic fibrosis patients, has a large genome that includes over 5,500 genes, over 70 of which encode porin proteins from three different structural families. Clearly, *P. aeruginosa* has considerable capacity to modulate the uptake of molecules by its outer membrane.

Mutations in genes encoding porins can increase the permeability of the outer membrane to bulky compounds and thereby confer sensitivity to vancomycin and other bulky antibiotics on gram-negative bacteria. Conversely, bacteria can accumulate mutations that further restrict the diffusion of antibiotics through the outer membrane and increase resistance, and since some types of porins are relatively nonselective, a single porin mutation can confer resistance to more than one type of antibiotic. For a long time, the importance of mutations in porins as a mechanism of resistance was somewhat underappreciated because this type of resistance usually confers increases in resistance of only 5- to 10-fold. In contrast, other resistance mechanisms can confer greater than 50- to 100-fold increases in resistance. However, in clinical settings where the highest concentration of antibiotic achievable at the site of infection is sometimes less than 5 times higher than the level required to kill or inhibit growth of the bacteria, a 10-fold increase in resistance can be as disastrous as a 100-fold increase.

REDUCED UPTAKE ACROSS THE CYTOPLASMIC MEMBRANE. An obvious way for bacteria to resist the action of an antibiotic that has a target in the bacterial cytoplasm (e.g., the ribosome or DNA gyrase) is to fail to transport the antibiotic across the cytoplasmic membrane, yet this does not seem to be a common mechanism of resistance. In some cases, the lack of such a mechanism of resistance is understandable. Tetracycline, for example, diffuses readily through membranes because it is a hydrophobic compound. Penicillin and other β-lactam antibiotics do not need to reach the cytoplasm, since they act on targets on the extracellular surfaces of bacteria. Nonetheless, there are other antibiotics, such as the aminoglycosides, that use specific transporters to enter bacterial cells. However, resistance does not readily appear by accumulation of mutations in a transporter gene. There are several possible reasons for this observation, such as the possibility that this transporter might be essential or that there may be multiple redundant transporters that can take up this class of antibiotic. An interesting observation is that some bacteria become much more resistant to aminoglycosides when they are growing under anaerobic conditions. *Escherichia coli* becomes almost 10 times more resistant to aminoglycosides when growing anaerobically, and many anaerobes are totally resistant to aminoglycosides. In both cases, the resistance appears to be due to drastically reduced uptake of the antibiotic.

ACTIVE EFFLUX OF THE ANTIBIOTIC. Efflux pumps are membrane proteins that use energy to pump small molecules out of the bacterial cytoplasm. If this small molecule is an antibiotic, resistance results, because the antibiotic is prevented from reaching a high enough concentration in the cytoplasm to be effective. Bacteria contain multiple efflux pumps, many of which can contribute to reducing cytoplasmic concentrations of antibiotics. For example, the genomes of *E. coli* and *P. aeruginosa* each encode over 30 different efflux pumps. These pumps normally play roles in maintaining homeostasis by pumping metabolites and toxic substances out of the bacterial cell. Some efflux pumps are highly specific for a metal or compound (e.g., the TetA tetracycline efflux pump), whereas others pump out many compounds (e.g., tetracyclines and macrolides), resulting in resistance to multiple antibiotics. Some efflux pumps move horizontally between different bacterial pathogens on mobile elements or are part of pathogenicity islands.

Efflux pumps fall into two structural classes. The majority are **antiporters** that use the uptake of protons

(H⁺) as the source of energy to pump the antibiotics and other small molecules from the cytoplasm. Some efflux pumps are **ABC transporters,** which are multisubunit complexes that use ATP hydrolysis to power the pump. The structures of efflux pumps of gram-positive bacteria are usually relatively simple, because these pumps only need to remove the antibiotic to the outside of the cell. In contrast, efflux pumps of gram-negative bacteria often consist of proteins in the inner membrane, periplasm, and outer membrane that channel the antibiotic outside of the cell.

The first efflux mechanism to be characterized extensively mediates resistance to tetracyclines. The resistance protein is an antiporter pump (called TetA, TetB, etc., in different bacteria) located in the cytoplasmic membrane. These pumps catalyze energy-dependent transport of tetracycline out of the bacterium. Since tetracycline is removed as rapidly as it is taken up, the intracellular concentration of tetracycline remains too low to inhibit protein synthesis. Efflux pumps for every class of antibiotic have now been discovered, and they cause serious clinical problems by imparting resistance to β-lactams, macrolides, fluoroquinolones, and streptogramin, as well as tetracycline antibiotics, especially in *Staphylococcus* species. They consequently represent an extremely serious form of antibiotic resistance that has been hard to overcome.

Enzymatic Inactivation of the Antibiotic

Examples of resistance due to enzymatic inactivation of the antibiotic are the enzymatic hydrolysis of the β-lactam ring in penicillin-related antibiotics through the action of β-lactamases and the covalent modification of chloramphenicol through acetylation of one or two of its hydroxyl groups, which sterically prevent them from binding to their targets.

β-LACTAMASES. A major mechanism of resistance to β-lactam antibiotics, especially among gram-negative bacteria, is the production of **β-lactamases,** enzymes that cleave the β-lactam ring and render the antibiotic inactive (Figure 16–1). The "serine class" of β-lactamases forms covalent bonds between an active-site serine residue and the β-lactam ring (Figure 16–1A). This is analogous to the covalent bond formed between critical active-site serine residues in transpeptidases (penicillin-binding proteins) (see chapter 15). However, unlike the transpeptidases, which preserve this covalent bond, β-lactamases allow water molecules to attack it, thereby converting the antibiotic into an inactivated form with an opened β-lactam ring and freeing the β-lactamase for another round of

A

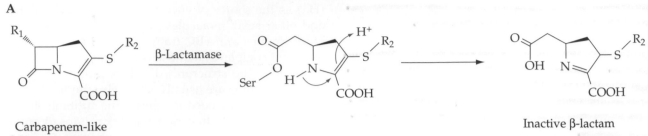

Carbapenem-like
β-lactam antibiotics

Inactive β-lactam

B

Clavulanic acid

Covalently modified
and inactivated
β-lactamase

C

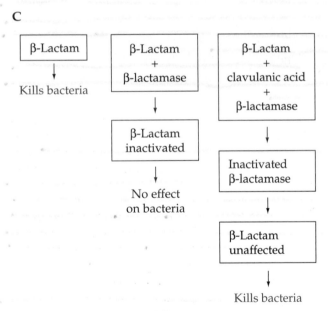

Figure 16–1 Modes of action and inhibitors of β-lactamase. **(A and B)** Modes of action of β-lactamase on β-lactam antibiotics **(A)** and clavulanic acid **(B)**, a suicide substrate inhibitor of β-lactamase. **(C)** Clavulanic acid inactivates the β-lactamase so that β-lactam antibiotics can then kill the bacteria.

catalysis. β-Lactamases are secreted into the periplasmic space by gram-negative bacteria and into the extracellular fluid by gram-positive bacteria. Because the gram-negative bacteria confine their β-lactamases to the periplasm and have porins that restrict the entry of β-lactams into this region, they can achieve the same level of resistance with a lower level of enzyme than gram-positive bacteria. Unlike porin mutations, which confer resistance to many different antibiotics, β-lactamases are much more specific and are usually active against only a subset of β-lactam antibiotics. In fact, the main reason for the large number of β-lactam antibiotics now on the market is the need for new β-lactam antibiotics that are not cleaved by existing β-lactamases. So far, the appearance of each new β-lactam on the market has been followed not long after

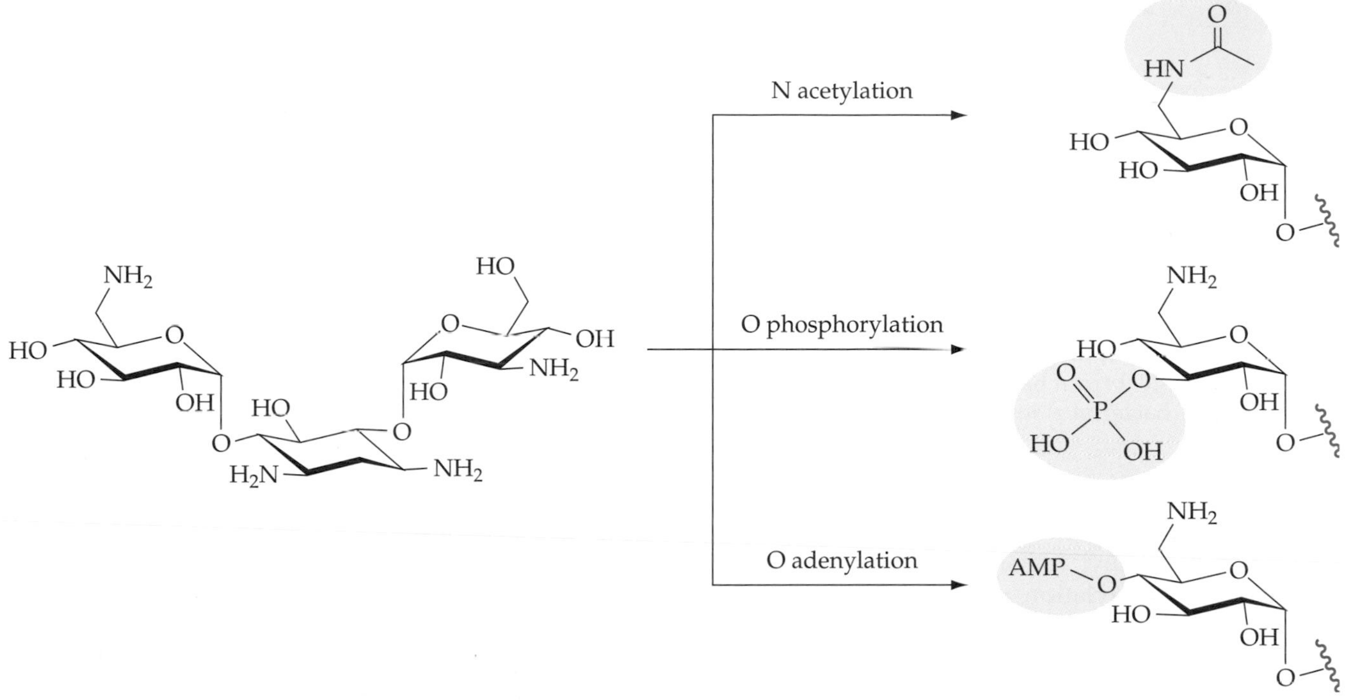

Figure 16–2 Actions of aminoglycoside-inactivating enzymes. Shown are examples of three enzymatic routes to aminoglycoside modification and deactivation: acetylation of amino groups by acetyl-coenzyme A, phosphorylation of hydroxyl groups by ATP, and adenylation of hydroxyl groups by ATP.

by the first report of a new β-lactamase that inactivates it.

Besides changing the β-lactam antibiotic, another strategy for countering β-lactamases is to mix the β-lactam antibiotic with a mechanism-based β-lactamase inhibitor, such as **clavulanic acid** (Figure 16–1) or **sulbactam**. Clavulanic acid does not kill bacteria; instead, it prevents the β-lactamase from inactivating the antibiotic, which can then proceed to kill the bacteria. These β-lactamase inhibitors have expanded the spectrum and have enabled once again the use of some old β-lactams, such as ampicillin, which were in danger of becoming obsolete. However, predictably, β-lactamases have appeared that are resistant to both clavulanic acid and sulbactam inhibition. In one mechanism, a chromosomal gene encoding a β-lactamase was duplicated many times to give a much higher level of β-lactamase production. Apparently, the excess β-lactamase was able to bind enough clavulanic acid to allow the remaining β-lactamase to inactivate the antibiotic. A far more serious challenge to the β-lactam class of antibiotics is the appearance of **zinc β-lactamases**. These metalloenzymes use a catalytic mechanism that does not involve active-site serine residues. Many classes of β-lactam antibiotics that are not cleaved by serine β-

lactamases are avidly cleaved by zinc β-lactamases. To make matters worse, the current β-lactamase inhibitors, such as clavulanic acid, inhibit only the serine β-lactamases and are completely ineffective against the metallo-β-lactamases.

AMINOGLYCOSIDE-MODIFYING ENZYMES. The main mechanism of aminoglycoside resistance is inactivation of the antibiotic. In contrast to β-lactamases, which cleave a C–N bond in the antibiotic and destroy the β-lactam "warhead," aminoglycoside-modifying enzymes inactivate the antibiotic by adding groups (phosphoryl, adenylyl, or acetyl groups) to the –OH and –NH₂ groups of these antibiotics (Figure 16–2). These modifications interfere with the hydrogen-bonding network that the antibiotics use to bind tightly to the 16S rRNA and to inhibit translation. In some gram-negative species, resistance also results from inhibition of aminoglycoside uptake.

CHLORAMPHENICOL AND STREPTOGRAMIN ACETYLTRANSFERASES. A common mechanism of resistance to chloramphenicol is acquisition of an enzyme that adds an acetyl group to the chloramphenicol (Figure 16–3). The enzyme is called **chloramphenicol acetyltransferase,** because it transfers an acetyl group from

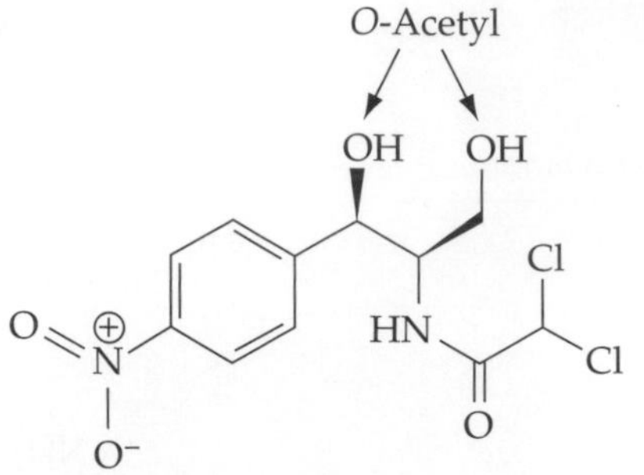

Figure 16–3 Action of chloramphenicol acetyltransfer-ase enzyme. The addition of acetyl groups to chloram-phenicol prevents it from binding to the 23S rRNA in the 50S subunit of bacterial ribosomes.

S-adenosyl-L-methionine, a compound used in many housekeeping methyl transfer reactions, to chloram-phenicol. This acetylation prevents tight binding of chloramphenicol to the 23S rRNA peptidyltransferase site. Although this form of resistance is common in many bacteria, its clinical impact is somewhat limited, because chloramphenicol use is restricted to an anti-biotic of last resort due to serious potential side ef-fects, including aplastic anemia and other blood dis-orders.

Likewise, acetyltransferases that modify and inac-tivate streptogramins have appeared. Again, the acet-ylation weakens the binding of the streptogramins to their targets in 23S rRNA. Synercid, which is a mix-ture of two streptogramins, had been in use to treat human infections for only a relatively short time when resistance began to be seen. Acetyltransferases that modify streptogramins are encoded by *vat* and *sat* genes of staphylococcal and enterococcal strains. In addition, efflux by an ABC transporter pump has been found in some clinical isolates of staphylococci.

OXIDATION OF TETRACYCLINE. A novel enzyme that uses chemical modification to inactivate tetracycline has been discovered. The reaction requires oxygen and NADPH and thus works only in aerobically growing bacteria. This form of resistance, encoded by **tetX**, is not nearly as prevalent as that caused by ef-flux. In this regard, it is also important to note that, except for the nasopharynx and lungs, most body sites are relatively low in free oxygen, since oxygen is tightly bound to hemoglobin. An odd feature of this resistance is that the gene was found originally in an obligate anaerobe *(Bacteroides fragilis)*, despite the fact that the resistance mechanism cannot work in this type of organism. This finding highlights the potential for further surprises from nature.

Modification or Protection of the Antibiotic Target

Besides modifying or destroying the antibiotic, bac-teria can become resistant to several different classes of antibiotics by modifying the target of the antibiotic. Target modifications can be divided into two general classes. The first class of target modifications is the accumulation of spontaneous mutations in the target that interfere with antibiotic binding. Antibiotic stress is a powerful selection condition, especially for bac-teriostatic antibiotics that do not kill bacteria outright. A classical example of this mechanism is the effect that mutational changes in residue A2058 in bacterial 23S rRNA have on sensitivity to macrolides, such as erythromycin. Residue A2058 is involved in hydrogen bond formation with macrolide antibiotics. When res-idue A2058 is mutated to G2058, the ribosome 23S rRNA binds the macrolide less tightly and resistance results. It turns out that the base corresponding to A2058 is a G residue in eukaryotic 28S rRNA, and this difference partially explains the specificity of macro-lides as prokaryotic translation inhibitors that do not harm humans. The second class of target modifica-tions is chemical additions or changes to the targets, such as addition of methyl groups, which impede an-tibiotic binding but still allow target function. Several examples of these two mechanisms are discussed be-low.

RESISTANCE TO β-LACTAMS. Alteration of the target of the antibiotic is a second mechanism of resistance to β-lactam antibiotics. In this case, the binding spec-ificity of the **penicillin-binding proteins** is altered so that they no longer bind the β-lactam antibiotic. This type of resistance is particularly common among gram-positive bacteria and is currently a type of β-lactam resistance that is causing problems clinically. β-Lactamase inhibitors can counter resistance due to β-lactamase, but this fix does not work for resistance due to alteration in the penicillin-binding proteins. Probably the best-characterized resistance gene of this type is *mecA,* a gene encoding resistance to methicillin that is found in *Staphylococcus aureus.* This resistance gene encodes a β-lactam-binding protein (also called penicillin-binding protein 2' or PBP2'), which is not inhibited as readily by methicillin as are the bacte-rium's normal β-lactam-binding proteins. Apparently, this new protein replaces the normal transpeptidase and allows peptidoglycan cross-linking to occur in the presence of the β-lactam antibiotic. Another clinically important example is the development of β-lactam re-sistance by *Streptococcus pneumoniae* isolates. To date, genes encoding β-lactamases have not made their way to *S. pneumoniae,* and all of the β-lactam resis-

A Susceptible bacteria

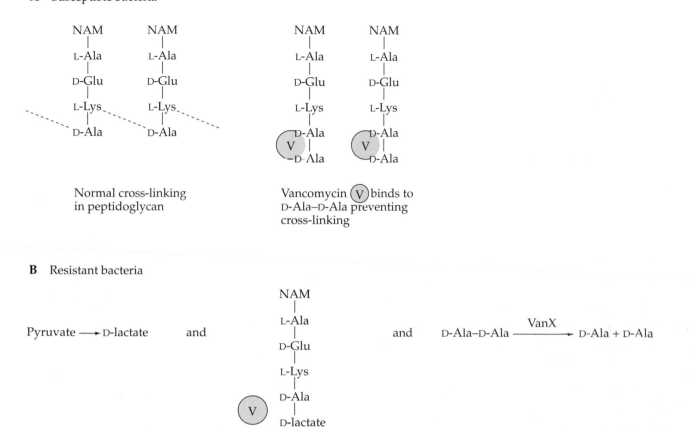

Figure 16–4 Mechanism of vancomycin resistance. **(A)** Action of vancomycin in susceptible bacteria. Vancomycin binds to D-Ala-D-Ala, preventing cross-linking of the peptidoglycan. **(B)** Mechanisms that prevent vancomycin binding in resistant bacteria. Three enzymes are involved, one (VanH) that catalyzes conversion of pyruvate to D-lactate, a second (VanA or VanB) that catalyzes the synthesis of D-Ala-D-lactate, and a third (VanX) that cleaves D-Ala-D-Ala that is synthesized by the normal pathway.

tance is due to mutations in the chromosomal copies of the penicillin-binding proteins. In fact, the combinations of mutations in the genes encoding penicillin-binding proteins in *S. pneumoniae* likely arose by horizontal gene transfer during natural transformation. Thus, these genes that impart resistance appear as "mosaics" of DNA segments that can be found in isolates of different *Streptococcus* species.

RESISTANCE TO GLYCOPEPTIDE ANTIBIOTICS. Vancomycin prevents cross-linking of peptidoglycan by binding to the D-Ala-D-Ala at the ends of muropeptides. For this reason, it initially seemed that resis-

tance to vancomycin would not readily develop because replacing the D-Ala-D-Ala dipeptide with another group that does not bind vancomycin but still functions in cross-linking is a very tall order. Nevertheless, this eventually did happen after vancomycin use as an antibiotic became widespread. Vancomycin-resistant *Enterococcus* (VRE) isolates, which act as opportunistic pathogens, were the first clinically important bacteria to appear that become resistant to vancomycin by replacing D-Ala-D-Ala in muropeptides with D-Ala-D-lactate, which does not bind vancomycin. There are three essential enzymes needed for this resistance phenotype (Figure 16–4). One is a ligase encoded by either *vanA* or *vanB* that makes D-Ala-D-lactate from D-Ala and D-lactate. A second gene, *vanH*, encodes a lactate dehydrogenase that makes D-lactate from pyruvate. These two enzymes make it possible for the bacteria to make the substitute part of the murodipeptide. However, as long as the bacteria still produce the original D-Ala-D-Ala, they will remain susceptible to vancomycin. This is where the third gene, *vanX*, comes into the picture. VanX is an enzyme that cleaves D-Ala-D-Ala but not D-Ala-D-lactate. The mechanism of vancomycin resistance is amazingly complex and shows how resourceful bacteria can be when it comes to protecting themselves from antibiotics.

The origin of the vancomycin resistance genes is still a mystery. They could have come from vancomycin-producing bacteria, such as *Amycolatopsis coloradensis*, although the currently circulating resistance genes show only about 50 to 60% amino acid identity with the corresponding genes of *A. coloradensis*. Another possible source is bacteria, such as *Lactobacillus*, that are naturally vancomycin resistant because they do not use the D-Ala-D-Ala dipeptide in their cell wall peptides. Gram-negative bacteria are also naturally resistant to vancomycin and other glycopeptides because the glycopeptides are very bulky molecules that do not diffuse through the outer membrane porins of gram-negative bacteria. Intermediate resistance to vancomycin has now spread to *Staphylococcus* species, and there is considerable apprehension that vancomycin resistance will continue to spread to other serious gram-positive pathogens.

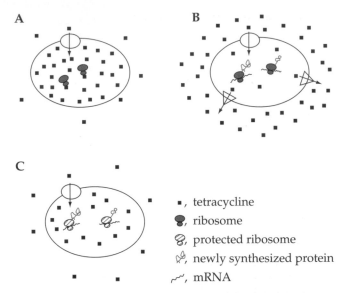

Figure 16–5 Mechanisms of tetracycline resistance. **(A)** Sensitive bacterial cell. Tetracycline is taken up by diffusion and possibly a transporter (open ellipse). The intracellular concentration becomes higher than the extracellular concentration. Tetracycline binds to ribosomes and stops protein synthesis. **(B)** Tetracycline efflux pump. A cytoplasmic membrane protein pump (open triangles) pumps tetracycline out of the cell as fast as the transporter takes it up. The intracellular concentration of tetracycline remains too low for effective binding to ribosomes. **(C)** Ribosome modification and protection. Tetracycline accumulation within the cell is similar to that in a sensitive cell, but the ribosome is protected through modification or mutation (cross-hatching), so that tetracycline no longer binds to it.

RESISTANCE TO TETRACYCLINES. Besides drug efflux, another clinically important type of resistance to tetracycline, called **ribosome protection,** is conferred by a cytoplasmic protein, called TetM, TetO, or TetQ in different bacteria, that protects ribosomes from tetracycline inhibition (Figure 16–5). When the protein is present in the bacterial cytoplasm, tetracycline no longer binds to the ribosome. This mechanism does not involve covalent modification of the ribosome, similar to macrolide resistance, which is discussed next. An interesting feature of this type of resistance protein is that it has GTPase activity and shares high amino acid homology in its amino-terminal region with bacterial elongation factors involved in protein synthesis. One model is that the GTPase of these resistance proteins perturbs a helix in the 16S rRNA involved in tetracycline binding. Therefore, the proteins again cause resistance by reducing the affinity of the target for the antibiotic. Although tetracycline efflux has been studied for decades, the ribosome protection type of resistance was discovered later and appears to

be quite widespread among a number of different groups of bacteria, including gram-positive bacteria, mycoplasmas, and some gram-negative genera, such as *Neisseria*, *Haemophilus*, and *Bacteroides*.

RESISTANCE TO MACROLIDES, STREPTOGRAMINS, AND LINCOSAMIDES. RNA methylases, called ErmA, ErmB, ErmF, or ErmG in different bacteria, impart simultaneous resistance to several antibiotics that bind to 23S rRNA. These RNA methylases add one or two methyl groups to the A2058 adenine in 23S rRNA. This is the same adenine residue mentioned above that can mutate spontaneously to a guanine residue and cause resistance. The methylation of A2058 imparts widespread resistance to macrolides, streptogramins, and lincosamides, which otherwise block the ribosome exit tunnel (see chapter 15). We can surmise the mechanism for this multiple resistance from a common theme that has emerged in this chapter. The A2058 base in 23S rRNA forms important hydrogen bonds with groups in each of these antibiotic classes.

Methylation of A2058 prevents hydrogen bond formation, the antibiotics fail to bind tightly to the ribosome exit channel, the tunnel remains unblocked by the antibiotic, and resistance results. This type of resistance has been found mainly in gram-positive cocci and in the *Bacteroides* group. Some gram-negative bacteria, such as *E. coli* strains, tend to be naturally resistant to macrolides, probably because their porins do not admit the antibiotic into the periplasm.

RESISTANCE TO QUINOLONES, RIFAMPIN, AND STREPTOMYCIN. Resistance to quinolones commonly involves amino acid changes that alter the way these antibiotics interact with the A or B subunit of DNA gyrase. DNA gyrase is an essential enzyme, but mutations that impart resistance allow sufficient function of the gyrase for nearly normal growth. Similarly, resistance to rifampin is caused by mutations that result in amino acid changes in the β-subunit of RNA polymerase. These amino acid changes reduce the affinity of the antibiotic for the RNA exit channel in RNA polymerase. Finally, we have stressed the amazing interactions between rRNA and antibiotics. However, it should be kept in mind that ribosomes are complicated machines that also contain proteins. Mutations in some of these proteins can alter antibiotic binding or ribosome function in such a way that resistance results. A classical example of this form of resistance is the streptomycin resistance that results from specific amino acid changes in the S12 protein (encoded by the *rpsL* gene) in the 30S ribosomal subunit.

RESISTANCE TO TRIMETHOPRIM AND SULFONAMIDES. Resistance to trimethoprim and sulfonamides arises from mutations in the folate pathway biosynthetic enzymes inhibited by these antibiotics. The mutant forms of the enzymes no longer bind the antibiotic with a higher affinity than their natural substrate. Mutations conferring resistance to sulfonamides or to trimethoprim occur rather frequently, but simultaneous double mutations that confer resistance to both types of antibiotic occur only rarely. For this reason, a combination of trimethoprim and one of the sulfonamides is currently used for antibacterial therapy.

Failure To Activate an Antibiotic

Metronidazole, which is often used to treat dental plaque caused by *Porphyromonas gingivalis* and gastric ulcers caused by *Helicobacter pylori*, must be activated before it can attack bacterial DNA. Acquisition of resistance to metronidazole by *H. pylori*, the cause of ulcers, is an ominous development for ulcer sufferers. Resistance to metronidazole is poorly understood, but in at least some cases, mutations that decrease the expression of the activation enzyme flavodoxin, which is required to convert metronidazole into its active form, can occur.

Isoniazid is one of the mainstays of antituberculosis therapy. It must be activated by a catalase (KatG) of mycobacteria. The activated form of isoniazid then covalently attaches to an NADH molecule at the active site of an acyl carrier protein reductase called InhA. InhA acts on long-chain fatty acids and catalyzes a step in the biosynthesis of mycolic acid, which is part of the mycobacterial cell wall. Inhibition of InhA blocks cell wall biosynthesis. One known mechanism of resistance to isoniazid is a mutation that inactivates KatG.

Regulation of Resistance Genes

REPRESSORS. Since bacteria need resistance genes only when they encounter antibiotics, a relatively uncommon occurrence in their lives, it makes sense that many antibiotic resistance genes are regulated. The first type of regulation described for a resistance mechanism was repression of the genes encoding tetracycline efflux pumps. In *E. coli*, the amount of the TetB pump is regulated by classical repression control mediated by the TetR repressor (Figure 16–6A). When tetracycline is absent from the cytoplasm, TetR binds to an operator that blocks high levels of transcription of the *tetB* gene. The *tetR* gene, encoding TetR, is divergently transcribed from *tetB*, and *tetR* transcription is autoregulated by this repression mechanism. When tetracycline is present, it enters the cell, complexes with Mg^{2+} ions, and then binds to TetR, which causes release from the operator DNA, thereby allowing high-level expression of the TetB pump. A new class of tetracycline antibiotics, called glycylcyclines, has recently been developed to circumvent this resistance mechanism and restore the effectiveness of tetracycline antibiotics. Glycylcyclines still inhibit translation by binding to the 16S rRNA, but they bind more effectively than other tetracyclines and they are not good substrates for the tetracycline efflux pumps.

Repression mechanisms regulate many other resistance genes. In methicillin-resistant *S. aureus* (MRSA), there is an interesting variation of the repression mechanism (Figure 16–6B). In the absence of β-lactam antibiotics, a repressor called BlaI inhibits the transcription of the *blaZ* gene, which encodes a β-lactamase. BlaI also autoregulates its own transcription by turning off the *blaI* gene and other genes in the signal transduction pathway. When cells encounter β-lactam antibiotics, a serine residue in a surface

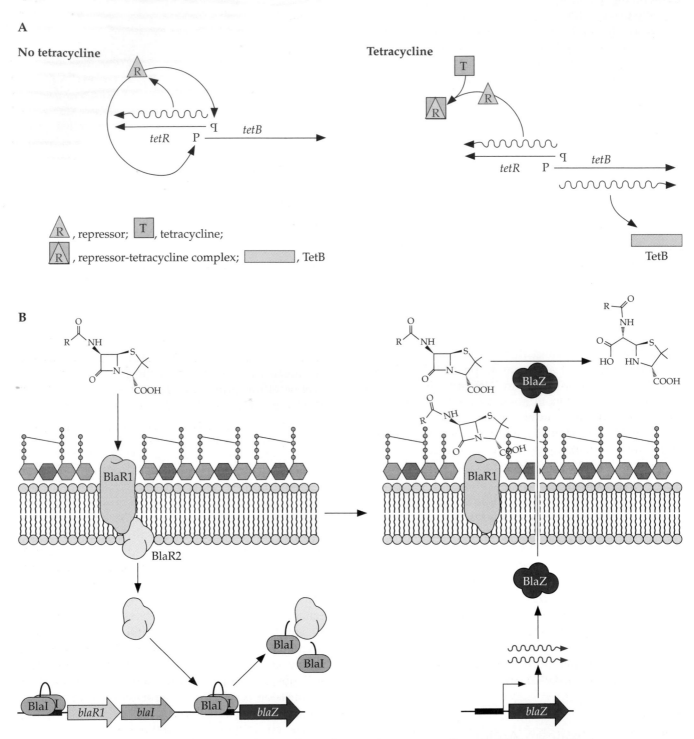

Figure 16–6 Repressor-mediated regulation of genes encoding the TetB tetracycline efflux pump in *E. coli* **(A)** and the BlaZ β-lactamase gene in MRSA **(B).** Wavey arrows, mRNA; P, promoter; backward P, promoter in opposite direction. (Panel B adapted from Walsh, 2003.)

protein called BlaR1 forms a covalent bond with the β-lactam ring analogous to the one that forms in transpeptidases or serine β-lactamases (see above). This binding signals the BlaR1 protein, which extends all the way through the membrane, to release a dormant protease called BlaR2 into the cytoplasm. The release activates BlaR2 for cleavage of its substrate, BlaI. As intact BlaI repressor disappears, the *blaZ* gene is transcribed, which allows production of the BlaZ β-lactamase. BlaZ is exported to the cell surface, where it inactivates the β-lactam antibiotic. This multiple-component repression system is much more elaborate than the classical repression of expression of the TetB efflux pump, where tetracycline itself relieves the repression directly. Nevertheless, it accomplishes a similar goal and allows MRSA to express the BlaZ β-lactamase only when it needs it. A similar protease-dependent repression system regulates expression of the alternate PBP2a protein by MRSA, which is more resistant to β-lactam antibiotics than the ones expressed in the absence of antibiotic (see above).

TRANSLATIONAL ATTENUATION. Another type of regulation of resistance genes, first described for *erm* RNA methylase genes of gram-positive bacteria, is a form of **translational attenuation** (Figure 16–7). The mRNA for the resistance gene starts nearly 100 bp upstream of the start codon for the gene. This 100-bp **leader region** in the transcript encodes a short peptide **(leader peptide)**. In the absence of the macrolide, ribosomes move rapidly along the mRNA and the leader peptide is efficiently translated. Under these conditions, two **stem-loop structures** form in the mRNA in such a way that the second stem-loop structure masks the ribosome-binding site and start codon of the *erm* gene, thereby preventing translation of the *erm* gene product, RNA methylase. When erythromycin is present, the bacterial ribosomes cannot translocate and thus do not move along the mRNA. Stalling during translation of the *erm* leader peptide allows formation of an alternative RNA stem-loop structure so that the ribosome-binding site and start codon are exposed, allowing *erm* gene translation and resistance. Methylase-modified ribosomes do not stall during translation of the *erm* leader peptide, allowing autoregulation to reduce *erm* gene translation according to need. Conversely, low-level formation of the alternative stem-loop structure in the absence of macrolides allows basal translation to occur so that small amounts of the methylase are synthesized to keep some ribosomes functional until the antibiotic is again encountered.

No erythromycin

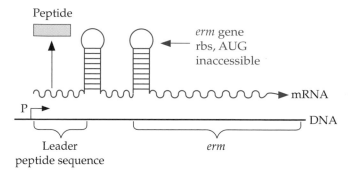

Erythromycin

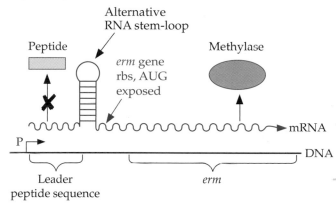

Figure 16–7 Regulation of erythromycin resistance genes by translational attenuation.

ACTIVATORS. Many resistance genes are regulated by **transcriptional activators.** In this mechanism, the antibiotic generates a signal molecule that binds to the activator protein. This complex then binds to the promoter region of the resistance gene and increases transcription of the gene. An excellent example of this mechanism is activation of the vancomycin resistance genes from *Enterococcus* species by the VanRS two-component regulatory system (Figure 16–8A). The VanS histidine kinase senses cell wall damage caused by vancomycin. The signal sensed by the VanS extracellular domain may be a cell wall fragment but does not seem to be vancomycin itself. Binding of this signal is transduced by VanS across the cell membrane and leads to autophosphorylation of a specific histidine residue in a cytoplasmic domain of VanS. This phosphoryl group is then transferred to a specific aspartate residue in the VanR response regulator, which alters its conformation to an activated state. Phosphorylated VanR then binds to the promoter regions upstream of the *vanRS* operon (autoregulation) and the *vanHAX* operon (encoding the muropeptide modifi-

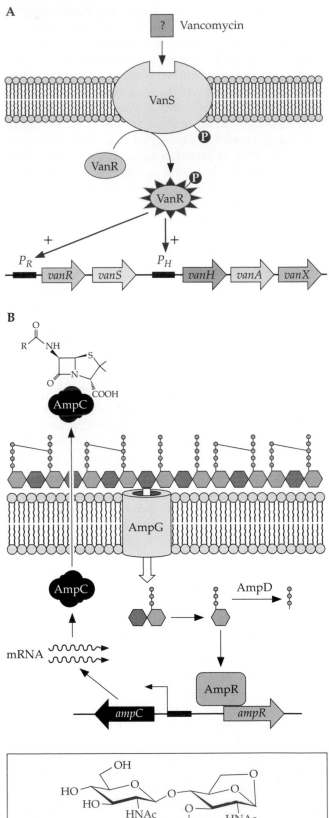

Figure 16–8 Regulation of resistance gene expression by transcriptional activation. **(A)** Activation of vancomycin resistance by the VanRS two-component regulatory system in *Enterococcus* species. **(B)** Activation of β-lactam antibiotic resistance by the AmpR positive regulator in *E. coli*. (Adapted from Walsh, 2003.)

cation enzymes described above) and activates their transcription.

A different mechanism regulates activation of *ampC*-encoded β-lactamase expression in *E. coli* and some other gram-negative bacteria. Inhibition of transpeptidation by β-lactam antibiotics induces cell wall autolysis that leads to accumulation of peptidoglycan fragments, including an anhydrodisaccharide tripeptide (Figure 16–8B, box). This peptidoglycan fragment is transported into the cytoplasm by the AmpG protein, and the *N*-acetylglucosamine sugar (solid shading in box) is enzymatically removed, leaving the anhydromuramyl tripeptide. This molecule binds to the AmpR protein, which then changes conformation, and the complex binds upstream and activates transcription of the *ampC* gene. The AmpC β-lactamase is exported to the periplasm of the cell, where it degrades the β-lactam antibiotic. This system is fine tuned by the AmpD protein, which is an amidase that inactivates the signal molecule by cleaving the sugar from the tripeptide. The AmpD activity ensures that high concentrations of the anhydromuramyl tripeptide signal accumulate only when there is significant damage to the cell wall in response to β-lactam antibiotics.

INSERTION SEQUENCES AND PROMOTER MUTATIONS. Although transcriptional-regulation mechanisms are usually reversible, so that expression of a gene can be turned off as well as on, there is a type of increased transcription that might be considered regulatory, although it results in a permanent alteration. Mutations in a promoter region or insertion of a transposon upstream of a resistance gene can increase the expression of the resistance gene and thus the level of resistance. For example, a noninducible version of the *ampC* gene in *Enterobacter* was originally expressed at such a low level that it did not make the bacteria resistant to ampicillin. Gradually, with continued selection by antibiotic use, mutations accumulated in the promoter region that increased β-lactamase gene expression to the point that it has become a serious contributor to antibiotic resistance. In this case, expression of the mutant promoter is constitutive, and the β-lactamase enzyme is always produced.

Other resistance genes that were originally silent when they entered a new host could later acquire an insertion sequence in their promoter regions that would cause the genes to be expressed. Many insertion sequences have promoters that point outward from their ends. Thus, when they insert into a region upstream of an open reading frame, they provide a promoter that controls expression of the gene. An ex-

ample of this is a plasmid-borne erythromycin resistance gene in *Bacteroides* species, *ermF*, which is expressed because it has acquired a promoter from an adjacent insertion sequence. Since there are many such examples of promoter mutations or insertion of an insertion sequence that activates the expression of a gene, scientists now consider that any resistance gene that enters a bacterial strain, whether it is expressed initially or not, is of concern because it can with time be activated through mutation.

Multiple Resistance and Genetic Linkage

The first resistance mechanisms to be described conferred resistance to a single class of antibiotics. For example, β-lactamases confer resistance to members of the penicillin-cephalosporin family but not to protein synthesis inhibitors. Two exceptions to this rule have appeared in recent years. The first is the **multidrug efflux pumps,** pumps that excrete antibiotics of more than one type. Recent structural work has been aimed at addressing how such pumps are nonspecific for more than one antibiotic class, whereas they do not export small molecules essential to the bacterium. The second exception is the macrolide-streptogramin-lincosamide type of erythromycin resistance, which makes a bacterium resistant to three different classes of antibiotics. This is possible because these three types of antibiotics bind to overlapping sites near the mouth of the exit tunnel of bacterial ribosomes.

A distinct but similar problem is the development of multidrug resistance due to genetically linked resistance genes. These genetic linkages can develop when two or more resistance genes are picked up by a plasmid. Two mechanisms by which resistance genes can move onto plasmids are transposons and integrons, both of which are described below.

Both the multidrug resistance mechanisms and the development of genetically linked resistance gene clusters create troubling problems. In bacterial strains where either occurs, selection by one class of antibiotic can hold in place resistance genes that confer resistance to unrelated antibiotics. Therefore, if a plasmid contains both a tetracycline resistance gene and a macrolide resistance gene, exposure of that strain to tetracycline selects not only for maintenance of the tetracycline resistance gene, but for maintenance of the erythromycin resistance gene, as well.

Many physicians have assumed that use of a particular class of antibiotic only selects for resistance to that particular class of antibiotic. If this were true, cessation of use of a type of antibiotic should allow resistance to that antibiotic to decrease or disappear. The multidrug resistance genes and the linkages of genes ensure that this desirable outcome will not occur in some cases due to cross-selection. To make matters worse, there are some cases in which disinfectant resistance genes have proved to be linked genetically to antibiotic resistance genes. In such cases, disinfectant use could select for the maintenance of the antibiotic resistance genes. Whereas we had all hoped that disinfectants and antiseptics would help protect us from antibiotic-resistant bacteria, they may in some cases have exactly the opposite effect. Moreover, cases of multiple resistance and genetic linkage are becoming more common. Perhaps the best example of this alarming trend is the escalating spread of MRSA strains, which can be acquired in the community as well as in hospitals. MRSA strains, which were initially identified for their resistance to the β-lactam methicillin, are now often resistant to multiple antibiotics besides β-lactams, including macrolides, tetracycline, aminoglycosides, and antiseptics.

Antibiotic Tolerance

The antibiotics that inhibit cell wall synthesis are bactericidal because the bacterium participates in its own destruction; bacterial enzymes that normally participate in cell wall turnover (lytic enzymes or autolysins) degrade the peptidoglycan, leaving the bacterium without the protection of its cell wall. Bacteria that can prevent their autolysins from destroying their peptidoglycan or that are located in an area where they can survive without a cell wall can avoid killing by the antibiotic. This type of response to antibiotics is called **tolerance.** A resistant bacterium continues to grow in the presence of the antibiotic. Tolerance differs from resistance because a tolerant bacterium just stops growing when the antibiotic is present; it is not killed, however, so it has a chance to recover when levels of the antibiotic fall. Unlike resistance, tolerance is not due to mutations or acquisition of additional genes and is reversible.

Tolerance is particularly significant in the case of bacterial biofilms, which are associated with the majority of infections of catheters, orthopedic devices, heart valves, the urinary tract, and the lungs of cystic fibrosis patients. Interestingly, it has been found that a small percentage (about 1%) of dormant bacterial cells within a biofilm, called **persisters,** contribute to the high level of tolerance observed. These persister cells appear to be nongrowing cells in a bacterial population that survive antibiotic treatment. The molecular basis for the presence of persister cells and how they lead to tolerance are not well understood. It seems likely that there may be multiple mechanisms that can cause persisters to appear at different stages

of growth. Recent data have accumulated showing that bacterial populations are not uniform with respect to the expression of certain genes. Noise in regulatory circuits, formally called **stochastic processes,** can lead to the appearance of subpopulations with different phenotypes and environmental responses. One idea is that persisters represent just such a subpopulation in which certain metabolic stress genes, such as inhibitors called toxins, from so-called **toxin-antitoxin modules,** or signal molecules, such as the **alarmone (p)ppGpp,** are induced and inhibit macromolecular synthesis processes susceptible to antibiotics, such as translation and cell wall biosynthesis. This inhibition shuts down the growth of the persister subpopulation and allows it to weather the presence of antibiotics.

A different type of tolerance has been seen in the case of *E. coli* strains that cause kidney infections. Since the kidney filters blood and excretes wastes in urine, the concentration of salts in the kidney is higher than in other tissues. Thus, when *E. coli* loses its cell wall because penicillin stops cell wall synthesis and autolysins degrade the existing cell wall, the osmotic strength of the fluid in which the bacteria are bathed is high enough to keep the bacteria from lysing due to internal turgor pressure. Forms of *E. coli* and other bacteria that lack a cell wall have been called **L forms.**

Horizontal Gene Transfer of Resistance Genes

Bacteria can become resistant to antibiotics by mutation of existing genes, but this process is very expensive for the bacteria. In the process of testing many different mutations, many bacteria die. Sometimes a single mutation is sufficient to confer resistance. This is the case with fluoroquinolone resistance. Single mutations in the bacterial DNA gyrase can make the enzyme unable to bind or respond to the antibiotic, thereby conferring resistance to the fluoroquinolone on the bacteria. In such cases, mutation to resistance is a feasible option for the bacterium. By analogy, bacteria readily accumulate spontaneous mutations in the gene encoding the β-subunit of RNA polymerase that confers resistance to rifampin. However, in other cases, such as resistance to penicillins or tetracyclines, several mutations are needed, and the development of resistance by accumulation of spontaneous mutations can take a long time, usually requiring repeated selection at sublethal concentrations of the antibiotic. In some cases, such as resistance to vancomycin, resistance does not develop by spontaneous mutation.

A much easier, quicker, and safer way for a bacterium to become resistant to an antibiotic is to acquire the resistance gene or genes from some other bacterium through horizontal gene transfer (covered in detail in chapter 7). The potential power of this rapid mechanism for evolution of resistance has been demonstrated in a number of studies, which showed that *Salmonella* and *E. coli* can invade and be taken up by plants, such as lettuce, and that the bacteria can exchange DNA, not only on, but also within, the plant cells. Couple this with the knowledge that genetic exchange among bacteria is frequent within animal guts and within insect midguts and can even occur within animal epithelial cells, and the implications of these findings are far-reaching and quite alarming. If water contaminated with bacteria containing antibiotic resistance genes is used for irrigation of crops, it is possible that resistance can spread to other bacteria. Moreover, if animals or humans then eat the crops, those bacteria present in the food could exchange DNA with our resident gut microbiota and confer resistance on them and any potential pathogens that might come along.

Scientists believe that the most common mode of acquiring resistance genes, especially when the donor is a member of a different species, is by **conjugation**. Although bacteria can acquire new genes by bacteriophage **transduction** or by **transformation** (uptake of DNA from the external environment), these types of transfer tend to occur mainly between members of the same species or members of very closely related species. The reason is that in phage transduction, the bacteria have to have the right phage receptor on their surfaces, a trait that is usually restricted to a closely related group of bacteria. In the case of natural transformation, in which linear single-stranded fragments of DNA are taken up by a specialized set of proteins, the DNA must integrate into the genome by homologous recombination. Thus, bacteria must be close enough to each other genetically for homologous recombination to be possible. Conjugation has no such limitations.

Narrow-host-range resistance gene transfer can be important clinically. For example, transformation may be transferring the mutant penicillin-binding proteins that make *S. pneumoniae* resistant to penicillin. However, the spread of resistance genes, especially between members of different species, can be a much more serious general threat. Accordingly, attention has tended to focus on transfer of resistance genes by conjugation. Conjugation is the direct cell-to-cell transfer of DNA through a protein complex that transits the membranes of two bacteria. As detailed in chapter 7, there are two types of conjugative elements: plasmids and conjugative transposons.

The best-studied type of conjugative element is the conjugative plasmid (Figure 16–9). Not all plasmids

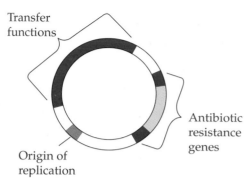

Figure 16–9 Examples of functions associated with conjugative plasmids.

are capable of self-transfer. Plasmids that transfer themselves by conjugation must carry a number of genes encoding proteins needed for the conjugation process itself (*tra* genes). Thus, **self-transmissible plasmids** are usually at least 25 kbp in size. Some plasmids that cannot transfer themselves can still be mobilized by self-transmissible plasmids. These are called **mobilizable plasmids,** and they can be much smaller than self-transmissible plasmids, because they need only one or two genes (*mob* **genes**) to take advantage of the transfer machinery provided by the other plasmids.

Clinical isolates resistant to many different types of antibiotics are being seen with increasing frequency. This multidrug resistance can arise in two ways. First, some types of resistance mechanisms, such as mutations in gram-negative porins or some types of antibiotic efflux, confer resistance to more than one type of antibiotic. Second, mobile genetic elements, such as plasmids, can acquire multiple resistance genes. Thus, acquisition of a single plasmid can make the recipient resistant to multiple drugs. As discussed in chapter 7, multiple virulence determinants can also be transferred by plasmids, often on the same plasmid as the resistance genes. Many examples of plasmids carrying multiple resistance genes have been reported, but until recently, it was not clear how such plasmids arose.

A plasmid can pick up more than one resistance gene if the resistance genes are carried on **transposons,** DNA segments that can insert into a chromosome or plasmid independently of homologous recombination. Transposons are flanked by DNA segments known as **insertion sequences,** which encode the enzyme that catalyzes transposition (**transposase**) and provide the ends recognized by transposase when it cuts and pastes the DNA during an insertion event (Figure 16–10). Insertion sequences have structural similarities that make it possible to recognize them from their DNA sequences even when

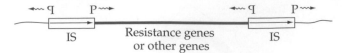

Figure 16–10 Structure of a transposon. A transposon is composed of two insertion sequences (IS) and intervening DNA, which can carry antibiotic resistance genes or genes conferring other traits. The ISs and the DNA they flank move as a single unit.

transposition activity cannot be demonstrated. Between the insertion sequences are resistance genes or other genes not involved in transposition, which are also carried by the transposon. Some multiresistance plasmids may have arisen by acquiring sequential transposon insertions. However, most multiresistance plasmids appear to have arisen by a different mechanism.

Another type of integrating element, called an **integron**, has now been discovered and is probably responsible for the evolution of many of the plasmids that carry multiple resistance genes. Integrons are usually transposons, but they have an extra feature. They contain an **integrase** gene and an **attachment (att) site**, in addition to insertion sequences. The integrase integrates circular DNA segments containing a promoterless single open reading frame (gene cassettes) into the *att* site (Figure 16–11). In effect, integrons create operons by sequential integration of the gene cassettes. The *att* site is a promoter provided by

Figure 16–11 Integration of two gene cassettes, carrying promoterless resistance genes X and Y, into an integron. The integron supplies the promoter (P) and an integrase gene (*int*). The gene cassettes insert site specifically and direction specifically into the integron *att* site. Although two genes are shown here, some integrons accumulate many genes.

the integron that allows the operon genes to be expressed.

A second type of conjugative element is the **conjugative transposon**. Conjugative transposons are usually located in the bacterial chromosome and can transfer themselves from the chromosome of the donor to the chromosome of the recipient. They can also integrate into plasmids. Their mechanism of transfer is different from those of other known gene transfer elements. They excise themselves from the donor genome by a process of nearly precise excision (Figure 16–12). The transfer intermediate is a covalently closed circle that does not replicate but transfers similarly to a plasmid. In the recipient, they integrate into

Figure 16–12 Transfer of a conjugative transposon. The transposon, which is integrated into the genome of the donor cell, excises itself to form a circular intermediate. The intermediate form transfers by conjugation into the recipient, where it integrates into the recipient's genome.

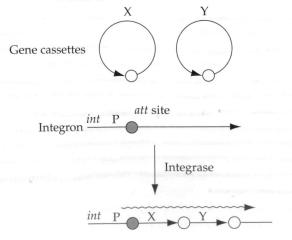

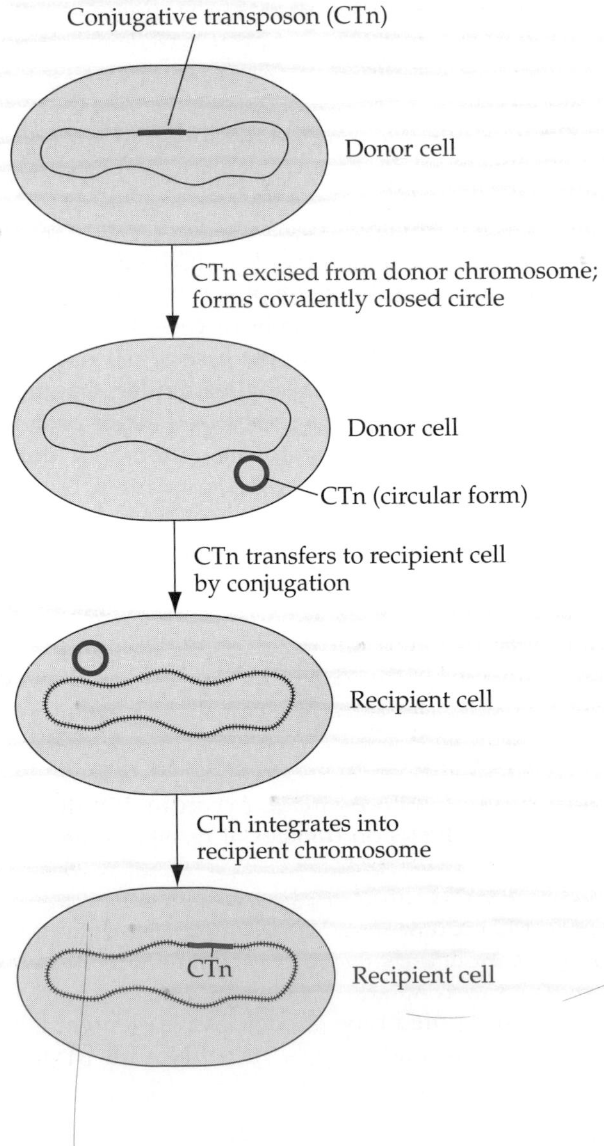

BOX 16–2   How Common Is Horizontal Resistance Gene Transfer in Nature?

Plasmids or conjugative transposons that have been isolated recently from bacteria in nature generally transfer in the laboratory at very low frequencies. In a mating between a donor and a recipient, only one recipient in a million will acquire the plasmid or conjugative transposon. If this is the frequency of transfer under optimal laboratory conditions, would not transfer in nature, where conditions are presumed to be less ideal, be a very rare event?

To answer this question, scientists focused on *Bacteroides* species in the human colon. *Bacteroides* is one of the numerically major genera in the human colon, accounting for about 25% of all colon isolates. *Bacteroides* species are known to have plasmids and conjugative transposons, both of which are capable of transferring antibiotic resistance genes. If the colon is perforated during surgery or other abdominal trauma, *Bacteroides* can cause life-threatening infections of tissue and blood. This group of bacteria has become resistant to most antibiotics.

The scientists doing this study had access to a collection of *Bacteroides* strains that had been isolated before 1970 and to another collection of strains that had been isolated in the late 1990s. They found that before 1970, about 25% of the strains were resistant to tetracycline and

none were resistant to erythromycin. By the late 1990s, over 80% of the *Bacteroides* strains were resistant to tetracycline and nearly one-third were resistant to erythromycin. By using DNA hybridization and DNA sequence analysis, the scientists were able to determine that the big increase in tetracycline resistance was due to a single resistance gene, *tetQ*, which had been spread to many different *Bacteroides* species. The increase in erythromycin resistance was due primarily to the spread of two genes, *ermF* and *ermG*. The DNA sequences of the *tetQ* and *erm* genes found in different species were over 95% identical, a finding that suggested the genes had been transferred by horizontal transfer and did not arise by independent mutational events. Finally, the investigators were able to show that *tetQ*, and probably the *erm* genes as well, were carried on conjugative transposons. The results of this study suggest that horizontal gene transfer by conjugation occurs very readily in the human colon.

Source: N. B. Shoemaker, H. Vlamakis, K. Hayes, and A. A. Salyers. 2001. Evidence for extensive resistance gene transfer among *Bacteroides* and other genera in the human colon. *Appl. Environ. Microbiol.* **67**:561–568.

the chromosome by a mechanism that does not duplicate the target site. Thus, they differ in a number of ways from standard transposons, phages, and plasmids.

Conjugative transposons are probably responsible for at least as much resistance gene transfer as plasmids, especially among gram-positive bacteria. Also, they have a broader host range than most plasmids. Conjugative transposons can transfer not only among species within the gram-positive group or within the gram-negative group, but between gram-positive and gram-negative bacteria. Conjugative transposons were overlooked for a long time, because they are located in the chromosome and thus cannot be isolated as easily as plasmids. Now that scientists are looking for them, they are finding conjugative transposons in many bacteria, including such familiar genera as *Salmonella*, *Vibrio*, *Rhizobium*, and *Agrobacterium*. Indeed,

a recent study of resistance gene transfers among *Bacteroides* species in the human colon has revealed that conjugal transfers of resistance genes occur more frequently than expected (Box 16–2).

Propagating and Maintaining Antibiotic Resistance through Selective Pressure and Changes in Fitness

Resistance can be genetically and biochemically complex and depends on the frequency at which the pathogen undergoes mutation and intra- or interspecies genetic exchange (horizontal gene transfer). However, another confounding factor is the intensity of the selective pressure imposed on the pathogen by use of the drug. An inevitable side effect of the use of antibiotics is the emergence and dissemination of resistant bacteria. The question is whether removing the selective pressure can eliminate antibiotic resistance.

Both vancomycin and third-generation cephalosporin use have been strongly correlated with a sharp rise in VRE infections. A patient with VRE infection could be at increased risk of serious postsurgical infection that is difficult to treat. Of particular concern nowadays is the potential for spread of vancomycin resistance to MRSA, which in a hospital setting is very problematic. Most health care personnel believe that restricting antibiotic use will help to limit the spread and development of more VREs, so it is now becoming common practice, particularly for hospitals, to campaign against overprescription of antibiotics by physicians; to implement more stringent hygiene practices, such as frequent hand washing; and to strictly isolate patients that have VRE infections. The goal of this strategy is to reduce the transfer of the resistance genes by removing the selective pressure through reduction of medical usage of broad-spectrum antibiotics and through restriction of transmission. The problem with this approach is that it has been quite ineffective in halting the spread. Contrary to popular belief, the resistance genes never really go away once the selective pressure is removed.

There is no longer any doubt that subtherapeutic doses of antibiotics as growth stimulants for animals can select for resistant bacterial strains. Avoparcin, a vancomycin derivative, has never been approved for use in the United States, but it was used as a feed additive in Western Europe from 1986 until 1995. It was banned in 1995 due to the discovery that it was contributing to the selection of VanA-type VRE in animal husbandry. It was hoped that banning its use would reduce the potential for human exposure to this new VRE, but that has not been the case. There are noticeable differences in the epidemiology of VRE between Europe and the United States. European VRE are usually still susceptible to other antibiotics, most likely due to the overuse of avoparcin and subsequent emergence and spread of resistant strains that resulted specifically from its overuse. In contrast, VRE in the United States are resistant to many antibiotics, not just one. This is presumably due to hospital-acquired resistances from the overuse of various antibiotics in a hospital setting.

Why does resistance remain in bacterial populations, even when antibiotic use diminishes? This phenomenon results partly from the tremendous bacterial genetic plasticity, which is responsible for the development and spread of resistance in the first place. You may have wondered what happens to bacterial growth and physiology when bacteria pick up resistance mutations or genes. In general, resistance phenotypes initially make bacteria less "fit" than their nonresistant, sensitive parent strains. This makes sense, because many resistance mechanisms change the structures of the macromolecular targets of the antibiotics. These changes often reduce the functions of these important cellular machines, and the resistant strain grows and competes less well than the parent strain when the antibiotic is absent. However, bacteria have two ways to cope with this fitness problem. As discussed above, many resistance mechanisms are tightly regulated so that the resistance genes are turned on strongly only when they are needed in the presence of an antibiotic, but bacteria also have other mechanisms to cope with constitutively expressed resistance. One common mechanism is to evolve their way out of the problem. For example, resistance mutations in rRNA, ribosomal proteins, DNA gyrase, or RNA polymerase often result in reduced fitness that decreases relative growth. Therefore, there is a window of time when the resistant bacteria may be outcompeted by the sensitive bacteria in the absence of antibiotic. However, optimal growth is a powerful genetic selection in itself, and the less fit resistant bacteria begin to accumulate secondary mutations that restore growth to the resistant bacteria. These are not simple revertants but, rather, are compensatory mutations in the same gene (intragenic) or in other genes (intergenic) that restore growth. Thus, the window in which sensitive bacteria can outcompete resistant bacteria can rapidly close, especially in large populations of bacteria, such as those in sizable geographic locations. The outcome is the worst of both situations: entrenched resistant bacteria that are now as fit (or sometimes even more fit) than the starting sensitive bacteria.

Will We Return to the Preantibiotic Era?

A scenario portrayed in some of the more lurid news stories about antibiotic-resistant bacteria is the return to a world much like that of the 1800s, when there were no antibiotics and people commonly died of diseases like pneumonia and wound infections. Surgery would once more become an intervention of last resort, with a high mortality rate due to untreatable postsurgical infections. Is this likely to happen? Well, not entirely. First, it is highly unlikely that all antibiotics will become ineffective against all bacteria. There are some bacteria, such as *Streptococcus pyogenes*, a dreaded cause of wound infections, that have remained susceptible to most antibiotics. Why this continued susceptibility is seen in some—although, unfortunately, not many—bacteria is not clear, but it gives room for hope that there will still be some treatable diseases.

A second consideration is that we have learned a lot about bacterial infections since the 1800s. Passive immunization is currently approved to treat infections caused by a number of toxin-producing bacteria. In this form of immunotherapy, serum or immunoglobulin fractions are isolated from humans or horses that have been inoculated with a specific antigen. These antibody fractions can be injected into a person and provide limited protection against infections in which the antigens play critical roles. It is feasible that passive immunization could be extended to treat infections caused by highly drug-resistant bacteria, such as *S. aureus*. For certain high-risk groups, such as the elderly and the very young, vaccines have offered significant protection against some bacterial pathogens, such as the capsule and conjugated-capsule vaccines directed against *S. pneumoniae*. A third consideration is that advances continue to be made in surgical procedures that make them less invasive. Laser surgery, which does not create huge surgical wounds, is not nearly as conducive to the development of postsurgical infections as cut-and-stitch surgery. Good hygienic practices, such as hand washing and sterilizing surfaces, will continue to prevent many infections, if these practices are observed rigorously. As long as we do not lose disinfectants, a great deal can be done to prevent infections from occurring in the first place. Also, bacteria, inventive little devils that they are, are not likely to become resistant to autoclaving. There may be other preventive measures that allow people to protect themselves from developing the predisposing conditions that increase the risk of infection. People with uncontrolled diabetes are one such high-risk group for bacterial infections. Improvements in control of diabetes do reduce this risk. Similarly, new methods for bolstering the flagging immune systems of the elderly may make them less vulnerable to disease.

The much-maligned pharmaceutical industry should not be overlooked. Although current market forces discourage antibiotic discovery and development, the pharmaceutical industry has had a brilliant past record of finding and marketing new antibiotics. Finding ways to speed up the approval of new antibiotics may be needed in some cases. However, this could mean that the public will have to accept a higher level of risk of side effects if the clinical trials that cost so much time and money are abbreviated. Faced with the alternative of dying, however, people may find that side effects are acceptable after all. AIDS and cancer patients have already made this psychological transition from insisting on completely risk-free treatments to a willingness to take chances on new therapies.

Finally, technological and scientific advances are dramatically and quickly expanding our understanding of the mechanisms by which antibiotics work and how resistance develops and evolves. Although first passes at applying combinatorial chemistry, structure-based design, genomics, and robotic screening were not as successful in antibiotic discovery as initially hoped, these approaches are very much in their infancy and still have huge potential to discover new antibiotics and modify existing ones. Many older natural products with antibiotic activity that were passed over might now be revisited using new microbiological and chemical approaches. Antibiotic combinations also hold new promise to treat emerging and re-emerging diseases. This brings us back to a theme that was introduced in chapter 4 on host defenses: the human mind is one of the most important defenses against disease. If scientists and the public put their minds (and resources) to conquering the worsening resistance problem instead of ignoring it, as they have until recently, surprising and wonderful things could happen.

One of the biggest casualties of bacterial diseases that become incurable may be confidence in the medical establishment. The public is disgruntled because scientists have not come up with better cures for cancer or for HIV infection, but at least we were never in a position in the past to cure these diseases. How will the public react if a point comes when parents have to watch children die of infections that were once curable? No one knows what the psychological fallout of lost cures will be. Moreover, as parts of the developing world, which were largely left out of the antibiotic and vaccine revolutions, become more prosperous, how will people in those countries feel about those in developed countries who squandered the miracle drugs people in developing countries are now in a position to afford?

It does not take a rocket scientist to figure out that part of the solution is not to lose antibiotics, but instead to conserve this precious resource by curbing the reckless abuse and overuse of antibiotics (especially those related to precious "drugs of last resort") by the medical profession and agricultural industry. We do not need paintings of skeletons, such as those that graced the cover of *Science*, to evoke images of a coming plague. Instead, we need images of responsible behavior by patients, physicians, and industries dedicated to the preservation of antibiotics for future generations. We also need a periodic supply of new antibiotics and vaccines to set back the clock in the relentless race against the evolution and spread of bacterial resistance.

SELECTED READINGS

Andersson, D. I., and D. Hughes. 2010. Antibiotic resistance and its cost: is it possible to reverse resistance? *Nat. Rev. Microbiol.* **8:**260–271.

Arias, C. A., and B. E. Murray. 2009. Antibiotic-resistant bugs in the 21st century—a clinical super-challenge. *N. Engl. J. Med.* **360:**439–443.

Cosgrove, S. E., and Y. Carmilli. 2006. The impact of antimicrobial resistance on health and economic outcomes. *Clin. Infect. Dis.* **36:**1433–1437.

Ferguson, G. C., J. A. Heinemann, and M. A. Kennedy. 2002. Gene transfer between *Salmonella enterica* serovar Typhimurium inside epithelial cells. *J. Bacteriol.* **184:**2235–2242.

Franz, E., and A. H. van Bruggen. 2008. Ecology of *E. coli* O157:H7 and *Salmonella enterica* in the primary vegetable production chain. *Crit. Rev. Microbiol.* **34:**143–161.

Hinnebusch, B. J., M.-L. Rosso, T. G. Schwan, and E. Carniel. 2002. High-frequency conjugative transfer of antibiotic resistance genes to *Yersinia pestis* in the flea midgut. *Mol. Microbiol.* **2:**349–354.

Jayaraman, R. 2008. Bacterial persistence: some new insights into an old phenomenon. *J. Biosci.* **33:**795–805.

Mascaretti, O. A. 2003. *Bacteria versus Antibacterial Agents: an Integrated Approach.* ASM Press, Washington, DC.

Matthew, A. G., R. Cissell, and S. Liamthong. 2007. Antibiotic resistance in bacteria associated with food animals: a United States perspective of livestock production. *Foodborne Pathog. Dis.* **4:**115–133.

Reese, R. E., R. F. Betts, and B. Gumustop. 2000. *Handbook of Antibiotics,* 3rd ed. Lippincott, Williams, and Wilkins, Philadelphia, PA.

Salyers, A. A., and C. F. Amabile-Cuevas. 1997. Why are antibiotic resistance genes so resistant to elimination? *Antimicrob. Agents Chemother.* **41:**2321–2325.

Salyers, A. A., and D. D. Whitt. 2004. *Revenge of the Microbes.* ASM Press, Washington, DC.

Smith, P. A., and F. E. Romesberg. 2007. Combating bacteria and drug resistance by inhibiting mechanisms of persistence and adaptation. *Nat. Chem. Biol.* **3:**549–556.

Smith, R. D., M. Yago, M. Millar, and J. Coast. 2006. A macro-economic approach to contain antimicrobial resistance: a case study of methicillin-resistant *Staphylococcus aureus. Appl. Health Econ. Health Policy* **5:**55–65.

Walsh, C. 2003. *Antibiotics: Actions, Origin, Resistance.* ASM Press, Washington, DC.

Wax, R. G., K. Lewis, A. Salyers, and H. Taber (ed.). 2007. *Bacterial Resistance to Antimicrobials,* 2nd ed. CRC Press, Boca Raton, FL.

Wright, G. D. 2003. Mechanisms of resistance to antibiotics. *Curr. Opin. Chem. Biol.* **7:**563–569.

QUESTIONS

1. Explain how mutant porins could help to make β-lactamases more effective. Could mutant porins team up with other types of resistance mechanisms to increase the effectiveness of the resistance mechanism?

2. Efflux pumps seem very energy inefficient. Why do bacteria not simply fail to take up antibiotics into the cytoplasm?

3. Efflux pumps only reduce the level of an intracellular antibiotic, they do not eliminate the antibiotic completely. Why does this make the bacteria resistant to the antibiotic?

4. Some target modifications require only one or a few mutations (e.g., resistance to fluoroquinolones), yet others, such as protection of the ribosome by methylation of rRNA, require new enzymes. Explain the difference.

5. Why does mutation to resistance occur within a short time in some cases (e.g., erythromycin or penicillin) and only over a period of decades in other cases (e.g., vancomycin)?

6. In the case of gram-negative bacteria, efflux pumps are generally coupled with outer membrane proteins. Why is this necessary?

7. Can you think of a type of resistance mechanism that is possible in theory although it has not yet been found?

8. Why can transfer of DNA by conjugation cross genus lines, whereas transfer of DNA by natural transformation or phage transduction is usually limited to a few closely related organisms?

9. Some integrons that contain disinfectant resistance genes or mercury resistance genes, as well as antibiotic resistance genes, have been found. What is the potential significance of this association?

10. Give a possible set of bacterial factors (or lack thereof) and what properties they have that would explain an increase in the 50% inhibitory concentration for tetracycline by 1,000-fold in an organism.

11. You are characterizing a mutant of *S. aureus*. The mutation affects the cell wall so that the cells make only a thin layer of peptidoglycan (making up about 20% of the cell wall). What would you expect the result to be for treatment with each of the following antibiotics: penicillin, tetracycline, erythromycin, and vancomycin?

12. Indicate which of the following statements are true or false:

_____Antibiotic resistance is a consequence of the mutagenic action of the antibiotic on the bacterium.

_____Antibiotic resistance is clinically important because it results in increased toxicity.

_____Antibiotic resistance is clinically important because it results in increased morbidity and mortality.

_____Antibiotic resistance is clinically important because it results in increased costs.

_____Misuse of antibiotics has resulted in greater numbers of antibiotic-resistant organisms.

_____The patient's normal microbiota can serve as a reservoir for antibiotic resistance genes.

13. Which of the following is the easiest way for bacteria to develop antibiotic resistance?
A. Generate a mutation to alter the target protein.
B. Add a functional group to the antibiotic.
C. Acquire a conjugative plasmid harboring an antibiotic resistance gene.
D. Mutate the gene that regulates the antibiotic resistance genes.
E. Mutate the efflux pump so that it pumps out the antibiotics faster than they can accumulate in the bacterial cell.

14. Bacteria can become resistant to fluoroquinolones by producing a mutation in what protein?

15. Fill in the blank spaces with the most appropriate words.
Some members of enterococci are resistant to _____, a last-resort antibiotic that binds to the _____ dipeptide of *N*-acetyl muramic acid within the peptidoglycan. The enterococci develop resistance by replacing this dipeptide with _____, which is not bound by the antibiotic. A novel way to overcome this resistance is to _____.
The most common way antibiotic resistance genes are spread is by _____.
The main source of antibiotic resistance genes is believed to be _____ between bacteria found in the _____.

16. How might a mutant porin in a gram-negative bacterium allow the bacterium to become less susceptible to antibiotics? How could this mutant porin help to make β-lactamases and efflux pumps more effective?

17. Enzymes, such as acetyltransferases, phosphotransferases, or adenyltransferases, would inactivate which of the following antibiotics: methicillin, streptogramin, chloramphenicol, ciprofloxacin, tetracycline, vancomycin, erythromycin, and penicillin.

18. Production of a new penicillin-binding protein would reduce the affinity for which of the following antibiotics: methicillin, streptogramin, chloramphenicol, ciprofloxacin, tetracycline, vancomycin, erythromycin, or penicillin.

SOLVING PROBLEMS IN BACTERIAL PATHOGENESIS

1. Recent headlines have highlighted the spread of MRSA infection in the United States. According to the CDC, MRSA is responsible for over 90,000 serious infections and over 18,000 hospital stay-related deaths per year in the United States. These MRSA strains are responsible for many serious skin and soft tissue infections, as well as pneumonia. One major problem with MRSA is that occasionally the skin infection can spread to other organs of the body with more severe, life-threatening symptoms, including necrotizing fasciitis (hence the name "flesh-eating" bacteria) and necrotizing pneumonia (tissue destruction), followed by sepsis and toxic shock, and then death in up to 50% of cases. A striking finding about these infections is that they occur even in young immunocompetent patients who were previously healthy. MRSA is resistant to several commonly prescribed antibiotics that are usually effective against gram-positive bacteria (methicillin, penicillin, and cephalosporins), and an infection with a MRSA strain can be deadly if left untreated. MRSA is subcategorized as community acquired (CA-MRSA) or hospital acquired (HA-MRSA), depending on how the infection is usually acquired. Most CA-MRSA strains are still sensitive to many antibiotics, such as trimethoprim, tetracycline, and clindamycin, but HA-MRSA strains are often resistant to these drugs while still sensitive to vancomycin and linezolid.

(continued)

A. Provide a common mechanism that accounts for the observed resistance of MRSA to methicillin, penicillin, and cephalosporin. Provide two different strategies that could be used to overcome this particular resistance.

B. For each of the antibiotics trimethoprim, tetracycline, and clindamycin, provide a possible mechanism to account for the observed resistance of HA-MRSA to the antibiotic. Provide a strategy that could be used to treat patients infected with HA-MRSA resistant to these antibiotics.

C. Why would HA-MRSA strains that are resistant to methicillin, penicillin, cephalosporins, trimethoprim, tetracycline, and clindamycin still show sensitivity to vancomycin and linezolid?

D. Although CA-MRSA is resistant to clindamycin, treatment with clindamycin in combination with rifampin results in an increase in the 50% lethal dose (LD_{50}) value from 10 without antibiotic treatment to 10^4 with rifampin to 10^8 for combined treatment with clindamycin and rifampin. In addition, treatment with clindamycin enhanced opsonization of CA-MRSA by macrophages. What possible mechanism(s) could account for these observations (i.e., change in the LD_{50} value and enhanced opsonization)? Provide your rationale. Provide an experiment that could be performed to confirm your hypothesis.

SPECIAL GLOBAL-PERSPECTIVE PROBLEM: INTEGRATING CONCEPTS IN PATHOGENESIS

A researcher ordered a colony of rabbits for his studies. Shortly after their arrival at the animal care facility, however, the university veterinarian discovered that most of the animals in the colony had contracted a bacterial infection. Clinical signs were generally limited to chronic infection of the lungs in the healthy adult animals, but young animals succumbed to systemic infection, including brain lesions and death. The veterinarian and researcher enlisted your help to study the cause of this disease. Subsequently, you isolated a new virulent gram-positive bacterium related to *Listeria monocytogenes*, which you named *Listeria leporine*. Using a signature-tagged mutagenesis approach based on existing reagents already developed for *L. monocytogenes*, you identified eight genes encoding putative virulence factors, which you named *llp1* through *llp8*. You then created mutant strains with in-frame deletions in each of these eight genes. When administered to rabbits through a breath nebulizer, the wild-type bacterium has a 50% infective dose (ID_{50}) value of 10 for lung colonization in young and adult animals, whereas it has an LD_{50} value of 10 for systemic infection with brain lesions in young animals but an LD_{50} value of 10^6 for adult animals. Mutant $\Delta llp4$, $\Delta llp7$, and $\Delta llp8$ strains have LD_{50} values in young animals similar to those in the wild type but have ID_{50} values for lung colonization of 10^7. On the other hand, the $\Delta llp1$, $\Delta llp2$, $\Delta llp3$, $\Delta llp5$, and $\Delta llp6$ strains have LD_{50} values in young animals of $>10^9$ but have ID_{50} values for lung colonization similar to those of the wild type. To help in your studies, you have generated antibodies against each of the proteins based on synthetic peptides of the antigen using multiple-antigen peptide conjugation technology, where multiple copies of the synthetic peptides are attached to lysine groups on a multivalent core resin

that is used to immunize the animals. You use these antibodies to visualize your proteins in Western blots. You have also developed an in vitro cell culture infection model using rabbit lung cells and have determined that *L. leporine* invades the rabbit cells and enters a phagosome but then escapes from the phagolysosome to replicate in the cytosol. Similar to *L. monocytogenes*, *L. leporine* also appears to move about in cells using actin filaments that gather at one end of the bacterium (forming what looks like a comet tail) as a means for propulsion.

1. You determine that cationic peptides, such as those released into the phagosome upon fusion with a lysosome, stimulate expression of the *llp1*, *llp2*, and *llp3* genes that are deleted in the $\Delta llp1$, $\Delta llp2$, and $\Delta llp3$ strains, respectively, but inhibit the expression of the *llp4* gene that is deleted in the $\Delta llp4$ strain. Describe how you could determine if the expression of the *llp1* to *llp4* virulence factors in the presence of cationic peptides is regulated at the transcriptional level. Be sure to include your expected results.

2. The $\Delta llp5$ and $\Delta llp6$ strains are no longer stimulated by cationic peptides to express the *llp1*, *llp2*, and *llp3* virulence factors, whereas *llp4* is expressed. Cellular fractionation revealed that Llp5 is cytoplasmic and Llp6 is membrane bound. Sequence analysis showed that Llp6 has four hydrophobic helices and an ATP-binding motif in its C terminus.

    A. Predict (with rationale) a function for the proteins deleted in the $\Delta llp5$ and $\Delta llp6$ strains.

    B. Provide (with rationale) an experiment to confirm your prediction for each protein.

3. Despite numerous attempts, you discover that you cannot visualize the Llp4 protein in Western blots of bacterial cells, yet you were able to observe transcriptional regulation of the *llp4* gene by cationic peptides. After further analysis of the *llp4* sequence, you discover that part of the sequence is complementary to the 5′ promoter regions of several polysaccharide biosynthetic genes and two adhesin-like genes. Provide (with rationale) a possible explanation for these observations.

4. You find that the Δ*llp1* and Δ*llp2* mutants can no longer escape from the phagolysosome, whereas the Δ*llp3* mutant can still escape into the cytosol. However, the Δ*llp1*, Δ*llp2*, and Δ*llp3* strains are all no longer able to make actin tails. Sequence comparison with protein databases revealed that the gene *llp1* encodes a 100-kDa protein with homology to pore-forming cytolysins, while *llp2* encodes a 50-kDa protein with homology to phospholipases and *llp3* encodes a 20-kDa protein with no homology to any known protein. All three proteins have ~20 hydrophobic amino acids at their N termini. You perform a series of cellular fractionations to determine the localization of the proteins. In the first fractionation, you examine wild-type bacterial cells cultured in medium with or without cationic peptides. In the second fractionation, you examine lung cells cultured with bacteria and then lysed with a detergent that does not lyse bacterial cells, followed by centrifugation to pellet the bacteria (the pellet) and collection of the supernatant fraction (the lysate). You then run sodium dodecyl sulfate-polyacrylamide gel electrophoresis to separate the proteins, followed by Western blot analysis.

A. Provide a detailed interpretation (with rationale) of the results shown in the figure.

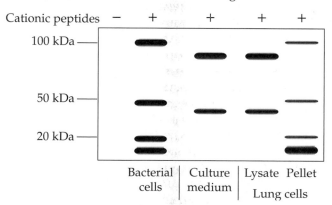

B. Predict (with rationale) a possible function for each of the Llp1, Llp2, and Llp3 proteins.

C. Provide an experiment that could be performed for each protein to confirm your predictions.

5. Curious about the possible source of the contamination in the rabbit housing facility, you find that wa-

ter samples from several of the water bowls in the animal cages near an air-handling vent inlet are indeed contaminated with the bacteria, and you then find that there is a biofilm of the bacteria coating the moist inside of the vent. When you examine the bacteria from fresh isolates obtained from the vents, you notice that, like the bacteria from fresh lung isolates but unlike the bacteria obtained from blood or brain lesions, the colonies are mucoid with a thick polysaccharide capsule. Using the rabbit lung inhalation model, you infect the animals with wild-type and mutant bacteria for 1 week. You find that wild-type bacteria and Δ*llp1*, Δ*llp2*, Δ*llp3*, Δ*llp5*, and Δ*llp6* mutants can all be isolated easily from the lungs of the animals with the mucoid phenotype, but Δ*llp4*, Δ*llp7*, and Δ*llp8* mutants cannot be isolated from the lungs after 1 week but instead can be isolated from brain lesions of young animals that die and are not mucoid.

A. Although the wild-type bacteria lost their mucoid phenotype after being plated on agar overnight, you find that culturing the wild-type bacteria in broth for a few days restores the mucoid character. You also find that if the medium from old cultures is added to the bacteria lacking polysaccharide on their surfaces, within a short time, those bacteria begin to express polysaccharide on their surfaces. Provide a mechanism for both the in vitro and in vivo (in lungs and in brain) observations regarding capsule.

B. You found that the genes deleted in the Δ*llp7* and Δ*llp8* strains are located within the same operon comprised of four genes, which you have tentatively named *llp7*, *llp8*, *llp9*, and *llp10*. Sequence analysis revealed that *llp7* encodes a 100-amino-acid peptide with an N-terminal hydrophobic region, *llp8* encodes a 400-amino-acid protein with five hydrophobic regions and a C-terminal nucleotide-binding motif, *llp9* encodes a 300-amino-acid protein with three hydrophobic regions and a C-terminal nucleotide-binding motif, and *llp10* encodes a 200-amino-acid protein with a DNA-binding motif. You make additional Δ*llp9* and Δ*llp10* deletion mutants. You observe that old medium from the wild-type, Δ*llp4*, Δ*llp8*, or Δ*llp10* strain added to nonmucoid cultures of the Δ*llp7* or Δ*llp9* strain restores the mucoid phenotype, but when old medium from Δ*llp7* or Δ*llp9* culture is added to nonmucoid cultures of the wild-type, Δ*llp4*, Δ*llp8*, or Δ*llp10* strain, the mucoid character is not restored. Provide a possible explanation for these observations. Draw a clearly labeled schematic diagram depicting the location of each gene

(continued)

product (for *llp4*, *llp7*, *llp8*, *llp9*, and *llp10*) and its possible role in regulation of the mucoid phenotype.

C. You find that although the Δ*llp4*, Δ*llp7*, and Δ*llp8* mutant strains have LD_{50} values in young animals similar to those of wild-type bacteria, they are now sensitive to treatment with erythromycin, azithromycin, tetracycline, kanamycin, and chloramphenicol, and these antibiotics can be used to treat infection, whereas wild-type bacteria and the other mutant strains are still resistant to the antibiotics and cause chronic lung infections. Provide (with rationale) a possible explanation for this observation.

17

Vaccination—an Underappreciated Component of the Modern Medical Armamentarium

When Edward Jenner (1749–1823) introduced the first vaccine, a cowpox vaccine that protected millions of people from the dreaded disease smallpox and thereby enabled the eradication of the disease, you would think that the vaccine would have been greeted with jubilation and gratitude—and it was by most people. However, there was also considerable ambivalence, as indicated by a famous print by the English satirist James Gillray (1756–1815) that depicted people who had been vaccinated as having small cow heads sprouting from various areas of their bodies. In subsequent years, many more vaccines have appeared and have had a major beneficial public health impact, yet even today, there are still some pockets of ambivalence, such as parents who refuse to vaccinate their children because they believe, against overwhelming evidence to the contrary, that certain vaccines cause neurological diseases, such as autism. As the old saying goes, the more things change, the more they remain the same.

Vaccines: a Major Health Care Bargain

The concept of **acquired immunity** has been around for a long time. In his account of the plague of Athens during the Peloponnesian War (ca. 430 B.C.), the ancient Athenian historian Thucydides clearly recognized that those individuals who recovered from the plague never developed the disease again; that is, they had acquired immunity to the plague. Two millennia later, in the 1790s, Edward Jenner demonstrated the concept of acquiring protective immunity against smallpox by using material that he had taken from the sores of milkmaids with cowpox, and hence, he dubbed the process of immunization by inoculation with cowpox **vaccination** (*vacca*: Latin for cow). A century later, Louis Pasteur (1822–1895) extended the concept of vaccination to include gaining protective immunity through inoculation with other attenuated infectious agents, such as other viruses (rabies virus) and bacteria (fowl cholera and anthrax).

Today, **vaccines** are defined as nontoxic or greatly attenuated **antigens** that are injected, ingested, or inhaled to induce the specific protective host immune defenses without having to go through the more serious and harmful disease process. We all recognize that vaccines have changed the course of history and that we have benefited enormously from their use as powerful means of protecting against certain highly contagious or dangerous infectious diseases. Indeed, one of the greatest achievements of the past century was the eradication of smallpox through a massive worldwide campaign that was conducted in the 1960s and 1970s by the World Health Organization, with the last known case reported in 1979.

Vaccines are not only much cheaper than the cost of treatment of infectious diseases; they also reduce suffering and long-term consequences of the diseases. Patients who are diagnosed and treated for an infection that is under way have already suffered symptoms serious enough to cause them to seek medical attention and may even have sustained irreversible damage. In contrast, vaccines prevent the infection from becoming established in the first place. Importantly, in cases where there is no effective therapy for an infection, vaccines may be the only way to protect patients from serious disease. Vaccines are clearly a major health care bargain.

If vaccines are so important, why are there not more of them? Unfortunately, many of the barriers to vaccine development are financial, legal, or political rather than scientific. In the market economies of the developed countries, there have been financial disincentives to develop or produce new vaccines, especially vaccines that would have their greatest impact in the developing world. Vaccines are not nearly as profitable as cholesterol-lowering drugs or blood pressure medications that have to be taken daily for long periods. In recent years, development and testing of vaccines have focused more on diseases that are widespread, such as otitis media in children. These vaccines have made important recent contributions in controlling several childhood diseases, at least in developed countries, and they have proven relatively profitable for their pharmaceutical company producers.

In addition, there have been legal reasons for industry's recent lack of enthusiasm for vaccine development. There have been numerous lawsuits based on alleged rare side effects of vaccines, which can be quite expensive for companies, especially when they involve children. Also, juries may not be adequately informed of the fact that many of the alleged side effects of vaccines are not supported by scientific studies. Vaccines have proven to be extremely safe, but there is no question that each new vaccine carries a risk that rare, serious side effects may occur. In an era where most people in developed countries have not experienced the horrors of an outbreak of smallpox,

polio, whooping cough, or diphtheria; have not watched a child die of bacterial meningitis; or have not witnessed the anguish of a mother who gives birth to a malformed child because she contracted rubella during pregnancy, it is easy to take these benefits for granted and to focus instead on the rare side effects that are the inevitable cost of such benefits.

There are also formidable technical barriers to development of vaccines against some of the diseases considered most important today: AIDS, malaria, gonorrhea, tuberculosis (TB), and pneumonia. To understand some of the technical problems that have frustrated development of vaccines against these infectious diseases, it is first useful to understand how successful vaccines work, because the successful vaccines reveal traits of infectious microorganisms that make them amenable to development of a vaccine. Microorganisms that do not have these traits pose a much greater vaccine challenge. It is also encouraging to recall that despite the rather poor showing in recent years, some highly effective vaccines have been developed in the past. Moreover, several nonprofit research organizations funded by private charitable foundations are spearheading the use and development of cost-effective vaccines in developing countries.

The Antivaccination Movement

There is a small but vocal group of individuals who are refusing to have their children vaccinated. Moreover, these individuals are now working to make refusal of vaccinations possible for people who cannot claim religious grounds for their objections. Why would parents willingly expose their children to such horrible diseases as diphtheria and whooping cough, which used to kill many thousands of children? It certainly is not because these diseases have disappeared. If you need to be convinced that they are still around, consider the "involunteer experiment" done accidentally in the United Kingdom during the 1970s and 1980s. In response to sensational and highly inaccurate news accounts of alleged serious side effects of the vaccine against whooping cough, parents of young children began to refuse to have them receive the vaccine, and vaccination rates dropped from around 80% to 30% (Figure 17–1). This vaccine is the **P** (for pertussis) part of the three-component **DTP vaccine** routinely administered to infants and toddlers at that time. Parents demanded—and their children received instead—a **DT** formulation that lacked the pertussis component of the vaccine. The effects of this miscalculation by gullible parents were tragic but, unfortunately, entirely predictable. A series of epi-

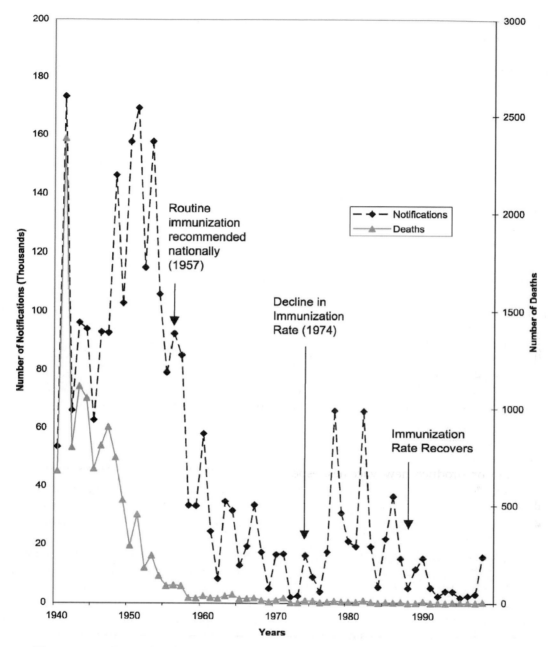

Figure 17–1 Example of the impact of the antivaccination movement. Public backlash against the pertussis vaccine resulted in a large drop, from around 80% to 30%, in vaccination rates in Great Britain during the mid-1970s, which in turn was followed by a series of whooping cough epidemics. (Reprinted from J. P. Baker. 2003. The pertussis vaccine controversy in Great Britain, 1974–1986. *Vaccine* **21:**4003–4010, with permission from Elsevier.)

demics of whooping cough subsequently occurred in the United Kingdom, in which about 100,000 children were infected and 36 died. Those children who survived went through a horrible experience. The disease produces a cough so violent and repetitive that children feel as if they are suffocating. Permanent brain damage occurred in some of those who survived.

Given this, why would any parent refuse to have a child vaccinated? Most of these parents are well intentioned and believe allegations that vaccines cause all manner of evils, in particular, neurological damage. The reason for this association is easy to understand. Most neurological problems in children appear in the first few years of life, the time when virtually all children in developed countries are receiving their vaccinations. However, as we stressed ad tedium in chapter 6, association is not proof of cause and effect. In particular, in cases where over 80% of the population experiences a particular treatment, that treatment could be associated with almost everything. As for the argument that there are fewer cases of neurological disorders in developing countries where vaccinations are not available, the sad fact is that most children in such countries who develop neurological problems do not live long enough to be counted.

In an attempt to introduce some medical rationality into the debate over vaccine safety, the Institute of Medicine, an independent research agency sponsored by the National Academy of Sciences, undertook a painstaking review of the purported side effects of vaccines in 1994. Their conclusion was that there is little or no evidence for most of the alleged side effects of vaccines, and where there might be a connection, the side effects are rare and not of the most serious type (Table 17–1). This and other studies have also emphasized the benefit figures, which are almost never included in sensational press articles about vaccines. For vaccines recommended in the United States, the reduction in incidence of the targeted disease was at least 96% and in most cases was 99% or higher. Reductions of nearly 100% in the case of diseases such as polio have made it possible to attempt to eradicate the disease. This has already been accomplished with smallpox, and if it were not for the bioterrorists eyeing the few remaining stocks of smallpox virus in scientists' freezers, we could safely forget smallpox, a disease that terrorized Europe, scourged Asia, and wiped out whole populations of Native Americans.

Those who refuse to have their children vaccinated are, quite consciously in many cases, taking advantage of a phenomenon known as the **herd immunity effect.** Herd immunity arises from the fact that an unvaccinated person in a population that is mostly vaccinated will be protected from disease because there are not enough susceptible people in the population to allow infectious disease transmission and an outbreak to materialize. There is a serious defect in this reasoning, however, given the increase in international travel. A child who has not been immunized against viral diseases, such as measles and mumps, and who has been protected from measles and mumps during the childhood years due to the herd effect would be ill advised to travel as an adult to areas of the world where measles is still endemic. For reasons that are still not well understood, these diseases have a much more serious effect if they strike teens or young adults than if they occur in children. Parents who refuse to have their children vaccinated are setting those children up for a dangerous future.

The Success Stories

Vaccines against smallpox, a viral disease, and diphtheria, a bacterial disease, have virtually eradicated these diseases in countries with vaccination programs in effect. A good vaccine can provide lifelong immunity to an infectious disease by stimulating long-lived effector and memory T cells and B cells, as described in chapter 4. Some of the vaccine success stories are summarized in Table 17–2.

Subunit Vaccines

Most successful vaccines used today against bacterial pathogens work by inducing humoral immunity against bacterial antigens that are relatively invariant, e.g., toxins or surface proteins. The antibacterial vaccines listed in Table 17–3 are administered routinely to all children in developed countries. A vaccine that illustrates some features associated with vaccines that have few side effects is the trivalent DTaP vaccine. The **D** (for diphtheria) and **T** (for tetanus) in DTaP are detoxified forms of the bacterial toxins (called **toxoids**). Vaccines like DTaP, which consist of one or more purified proteins, are called **subunit vaccines.** Because they are soluble and administered extracellularly, they primarily elicit an antibody rather than a cell-mediated response. This is effective in the cases of diphtheria and tetanus because the toxins circulating in the bloodstream are the causes of the symptoms of these diseases. In both diseases, bacteria grow at a localized site in the body (the throat in the case of diphtheria or a wound in the case of tetanus). The protein toxin released by the bacteria enters the bloodstream and damages essential organs. Vaccination with DT elicits an antibody response that neutralizes diphtheria and tetanus toxins circulating in the body. Even if the bacteria that cause diphtheria and tetanus repeatedly colonize a person, no symptoms will develop because an immediate antibody response neutralizes circulating toxin. Bacteria that produce toxins, which are responsible for causing the symptoms of the disease, are thus among the easiest and most efficacious vaccine targets.

Diphtheria and tetanus toxoids can be administered to infants, and side effects are rare. This is not true of the P part of the old DTP vaccine. Although the pertussis vaccine has been very successful in preventing whooping cough, a disease that was once a common killer of children, it has some troublesome side effects. In about 20% of infants given DTP, side effects that range from generalized discomfort, which makes the infant fussy for a couple of days, to convulsions are experienced. The pertussis vaccine may also cause hearing loss or irreversible brain damage

Table 17–1 Summary of conclusions based on evidence regarding the possible association between specific adverse effects and receipt of childhood vaccines, by determination of causality—Institute of Medicine, 1994[a]

DT/Td/tetanus toxoid[b]	Measles vaccine[c]	Mumps vaccine[c]	OPV/IPV[d]	Hepatitis B vaccine	*Haemophilus influenzae* type b (Hib) vaccine
1. No evidence was available to establish a causal relationship					
None	None	Neuropathy Residual seizure disorder	Transverse myelitis (IPV) Thrombocytopenia (IPV) Anaphylaxis (IPV)	None	None
2. Inadequate evidence to accept or reject a causal relationship					
Residual seizure disorder other than infantile spasms Demyelinating diseases of the central nervous system Mononeuropathy Arthritis Erythema multiforme	Encephalopathy Subacute sclerosing panencephalitis Residual seizure disorder Sensorineural deafness (MMR) Optic neuritis Transverse myelitis Guillain-Barré syndrome Thrombocytopenia Insulin-dependent diabetes mellitus	Encephalopathy Aseptic meningitis Sensorineural deafness (MMR) Insulin-dependent diabetes mellitus Sterility Thromobocytopenia Anaphylaxis[e]	Transverse myelitis (OPV) Guillain-Barré syndrome (IPV) Death from SIDS[f]	Guillain-Barré syndrome Demyelinating diseases of the central nervous system Arthritis Death from SIDS[f]	Guillain-Barré syndrome Transverse myelitis Thrombocytopenia Anaphylaxis Death from SIDS[f]
3. Evidence favored rejection of a causal relationship					
Encephalopathy[g] Infantile spasms (DT only)[h] Death from SIDS (DT only)[h,i]	None	None	None	None	Early-onset Hib disease (conjugate vaccines)
4. Evidence favored acceptance of a causal relationship					
Guillain-Barré syndrome[j,k] Brachial neuritis[j]	Anaphylaxis[e]	None	Guillain-Barré syndrome (OPV)[k]	None	Early-onset Hib disease in children aged ≥18 mo whose first Hib vaccination was with unconjugated vaccine
5. Evidence established a causal relationship					
Anaphylaxis[j]	Thrombocytopenia (MMR) Anaphylaxis (MMR)[e] Death from measles vaccine strain viral infection[f,l]	None	Poliomyelitis in recipient or contact (OPV) Death from polio vaccine strain viral infection[f,l]	Anaphylaxis	None

[a] Reprinted from CDC. 1996. Update: vaccine side effects, adverse reactions, contraindications, and precautions—recommendations of the Advisory Committee on Immunization Practices (ACIP). *MMWR Morb. Mortal. Wkly. Rep.* **45**(RR-12):1–35. This table is an adaptation of a table published previously by the Institute of Medicine (IOM) (K. R. Stratton, C. J. Howe, and R. B. Johnston [ed.]. 1994. *Adverse Events Associated with Childhood Vaccines: Evidence Bearing on Causality.* Institute of Medicine, National Academy Press, Washington, DC). IOM is an independent research organization chartered by the National Academy of Sciences. The National Childhood Vaccine Injury Act of 1986 mandated that IOM review scientific and other evidence (e.g., epidemiologic studies, case series, individual case reports, and testimonials) regarding the possible adverse consequences of vaccines administered to children. IOM constituted an expert committee to review and summarize all available information; this committee created five categories of causality to describe the relationships between vaccines and specific adverse events.

(continued)

Table 17–1 Summary of conclusions based on evidence regarding the possible association between specific adverse effects and receipt of childhood vaccines, by determination of causality—Institute of Medicine, 1994[a] (*continued*)

[b] DT, diphtheria and tetanus toxoids for pediatric use; Td, diphtheria and tetanus toxoids for adult use.

[c] If the data derived from studies of a monovalent preparation, then the causal relationship also extended to multivalent preparations. If the data derived exclusively from studies of the measles-mumps-rubella (MMR) vaccine, the vaccine is specified parenthetically. In the absence of data concerning the monovalent preparation, the causal relationship determined for the multivalent preparations did not extend to the monovalent components.

[d] For some adverse events, the IOM committee was charged with assessing the causal relationship between the adverse event and only oral poliovirus vaccine (OPV) (i.e., for poliomyelitis) or only inactivated poliovirus vaccine (IPV) (i.e., for anaphylaxis and thrombocytopenia). If the conclusions for the two vaccines differed for the other adverse events, the vaccine to which the adverse event applied is specified parenthetically.

[e] The evidence used to establish a causal relationship for anaphylaxis applies to MMR vaccine. The evidence regarding monovalent measles vaccine favored acceptance of a causal relationship, but this evidence was less convincing than that for MMR vaccine because of either incomplete documentation of symptoms or the possible attenuation of symptoms by medical intervention.

[f] This table lists weight-of-evidence determinations only for deaths that were classified as sudden infant death syndrome (SIDS) and deaths that were a consequence of vaccine strain viral infection. However, if the evidence favored the acceptance of (or established) a causal relationship between the vaccine and a possibly fatal adverse event, then the evidence also favored the acceptance of (or established) a causal relationship between the vaccine and death from the adverse event. Direct evidence regarding death in association with a vaccine-associated adverse event was limited to (i) Td and Guillain-Barré syndrome, (ii) tetanus toxoid and anaphylaxis, and (iii) OPV and poliomyelitis.

[g] The evidence derived from studies of DT. If the evidence favored rejection of a causal relationship between DT and encephalopathy, then the evidence also favored rejection of a causal relationship between Td and tetanus toxoid and encephalopathy.

[h] Infantile spasms and SIDS occur only in an age group that is administered DT but not Td or tetanus toxoid.

[i] The evidence derived primarily from studies of DTP, although the evidence also favored rejection of a causal relationship between DT and SIDS.

[j] The evidence derived from studies of tetanus toxoid. If the evidence favored acceptance of (or established) a causal relationship between tetanus toxoid and an adverse event, then the evidence also favored acceptance of (or established) a causal relationship between DT and Td and the adverse event.

[k] This conclusion differs from the information contained in the Advisory Committee on Immunization Practices recommendations because of new information that became available after IOM published this table.

[l] Deaths occurred primarily among persons known to be immunocompromised.

in some children, although this side effect is extremely rare. The reason the old form of the pertussis vaccine has so many side effects compared to the diphtheria and tetanus toxoids is that it consists of whole killed *Bordetella pertussis* bacterial cells, not just the pertussis toxin toxoid, and thus, it is a complex mixture of antigens, some of which cause adverse effects.

New, more defined versions of the pertussis vaccine are now available. These vaccines usually consist of the **pertussis toxoid** plus one or more *B. pertussis* cell surface components, e.g., **filamentous hemagglutinin (FHA)**, a 69-kilodalton outer membrane protein called **pertactin (Pn)**, or **fimbriae (Fim)** type 2 or 3. The new subunit vaccines are called **acellular pertussis (aP) vaccines** to distinguish them from the old whole-cell vaccine. The DTaP vaccines have fewer side effects than the DTP vaccine, although their efficacy is somewhat reduced compared to the whole-cell vaccine.

The pertussis vaccine story illustrates the general principle that the simplest vaccines, which consist of only one or a few well-defined antigens, are the ones least likely to cause side effects. Also, at least in theory, the simple vaccines should make it easier to target the desirable type of immune response. As already mentioned in the case of the DT-based vaccine, subunit vaccines elicit an antibody rather than a cell-mediated response. Whooping cough is a more complex disease than tetanus or diphtheria, with more than one virulence factor responsible for disease symptoms. Since *B. pertussis* replicates extracellularly in the lung airways, a cell-mediated response is not essential in this case, as it would be if the bacteria

multiplied intracellularly. However, antibodies that neutralize the pertussis toxin and also bind to surface antigens and opsonize the bacteria are needed to control the infection.

The DTaP vaccine also illustrates a common problem with subunit vaccines. Prolonged exposure to antigens is needed to generate a good memory T- and B-cell response. Since subunit vaccine components do not replicate in the body, they must be administered repeatedly to elicit a maximally effective memory response. The DTaP vaccine has to be administered in a series of five injections. Although the vaccine regimen produces very long-term immunity to diphtheria and tetanus, a tetanus booster is usually administered to persons with the sort of wounds that are associated with tetanus to make sure that antibody levels are high enough to be protective. A possible solution to this problem, which is currently under intensive study, is to have the vaccine proteins produced by a self-replicating microbe, to exogenously express and secrete the antigens in the host cell, or to encapsulate it so that it is released slowly over time (see below).

To afford protection when it is most needed, a vaccine must be administered in a timely manner (Figure 17–2). The fact that subunit vaccines must be administered several times creates a potential public health problem, because it increases the likelihood that a child will not receive the full set of vaccinations before being exposed to a disease and thus will not be adequately protected. The problem of children who do not receive the full recommended course of vaccine administration is still a formidable one, even in developed countries (Box 17–1).

Table 17–2 Baseline 20th century annual morbidity (cases of disease) before the vaccine became available and 1998 morbidity from nine diseases with vaccines recommended before 1990 for universal use in children[a]

Disease	Baseline 20th century annual morbidity	1998 morbidity	% Decrease
Smallpox	48,164[b]	0	100
Diphtheria	175,885[c]	1	100[d]
Pertussis	147,271[e]	6,279	95.7
Tetanus	1,314[f]	34	97.4
Poliomyelitis (paralytic)	16,316[g]	0[h]	100
Measles	503,282[i]	89	100
Mumps	152,209[j]	606	99.6
Rubella	47,745[k]	345	99.3
Congenital rubella syndrome	823[l]	5	99.4
Haemophilus influenzae type b	20,000[m]	54[n]	99.7

[a]Reprinted from *MMWR Morb. Mortal. Wkly. Rep.* **48:**243–247, 1999.
[b]Average annual number of cases during 1900 to 1904.
[c]Average annual number of reported cases during 1920 to 1922, 3 years before vaccine development.
[d]Rounded to nearest tenth.
[e]Average annual number of reported cases during 1922 to 1925, 4 years before vaccine development.
[f]Estimated number of cases based on reported number of deaths during 1922 to 1926, assuming a case-fatality rate of 90%.
[g]Average annual number of reported cases during 1951 to 1954, 4 years before vaccine licensure.
[h]Excludes one case of vaccine-associated polio reported in 1998.
[i]Average annual number of reported cases during 1958 to 1962, 5 years before vaccine licensure.
[j]Number of reported cases in 1968, the first year reporting began and the first year after vaccine licensure.
[k]Average annual number of reported cases during 1966 to 1968, 3 years before vaccine licensure.
[l]Estimated number of cases based on seroprevalence data in the population and on the risk that women infected during a childbearing year would have a fetus with congenital rubella syndrome.
[m]Estimated number of cases from population-based surveillance studies before vaccine licensure in 1985.
[n]Excludes 71 cases of *Haemophilus influenzae* disease of unknown serotype.

Conjugate Vaccines

Another successful antibacterial vaccine is the one that protects against **childhood meningitis** caused by the gram-negative bacterium *Haemophilus influenzae* type b (hence the vaccine name, **Hib**). *H. influenzae* type b used to be the most common cause of fatal meningitis in children and also caused epiglottitis, a rapidly progressing disease that can close the airway and lead to suffocation. *H. influenzae* type b also causes pneumonia, particularly in the elderly. Unlike

the bacteria that cause diphtheria, tetanus, and whooping cough, *H. influenzae* does not produce a protein toxin. Instead, *H. influenzae* coats its surface with a **polysaccharide capsule** that prevents C3b binding and thus allows the bacteria to avoid ingestion by phagocytes. Uncontrolled growth of the bacteria triggers an inflammatory response, which is the cause of local tissue damage and can develop into septic shock. Inflammation of the lining that separates the brain and spinal cord from the rest of the body (meninges) allows the bacteria and blood components to enter the spinal fluid, where the cytokines released by macrophages and other cells cause increased pressure on the brain that can lead to brain damage. Antibodies that bind to capsular polysaccharides allow phagocytes to opsonize and kill the bacteria and thus prevent this destructive (and frequently fatal) process from getting under way.

Since *H. influenzae* causes meningitis primarily in children under the age of 5 years, it is necessary to immunize young children as early as possible. The problem was that capsular polysaccharides, which are recognized primarily through T-cell-independent mechanisms, do not reliably elicit protective, long-lasting antibody responses (see chapter 4). In addition, young children are not able to mount an effective T-cell-independent antibody response to polysaccharides. To solve this problem, the *H. influenzae* type b capsular polysaccharide antigens were attached covalently to a protein antigen to produce a **conjugate vaccine.** This way, the immune system processes the vaccine antigen through the major histocompatibility complex type II (MHC-II) pathway as if it were a protein, producing a strong, long-lasting humoral response against the protein and the polysaccharide capsule. This vaccine has been highly effective and has already dramatically reduced the incidence of this type of meningitis.

A similar success story has recently occurred for invasive diseases caused by *Streptococcus pneumoniae*, which is a major cause of otitis media, bacteremia, and meningitis in infants. A 7-valent conjugate vaccine based on different capsule polysaccharides has reduced the frequency of pneumococcal invasive diseases in developed countries, and there have been attempts to make this vaccine available in developing countries. In addition, a 13-valent pneumococcal vaccine that will protect against additional common serotype strains of *S. pneumoniae* was approved by the U.S. Food and Drug Administration in February 2010.

The Less-than-Success Stories

The list of successful vaccines is rather pathetic if viewed from a historical perspective. DTP has been around for years, as has the polio vaccine. New vac-

Table 17–3   Antibacterial vaccines commonly recommended to be administered to all children or adolescents in the United States

Component name	Composition	Disease prevented	Administration
D	Diphtheria toxoid	Diphtheria	Injected as part of DTaP trivalent vaccine
T	Tetanus toxoid	Tetanus	
aP	Pertussis toxoid, adhesion proteins (acellular vaccine)	Pertussis (whooping cough)	
Hib	Polysaccharide-protein (conjugated vaccine) (*Haemophilus influenzae* type b)	Neonatal and childhood meningitis	Injected
PCV7	Heptavalent conjugated vaccine containing polysaccharide capsules from 7 types of *Streptococcus pneumoniae* linked to diphtheria toxoid	Meningitis, pneumonia, otitis media	Injected
MCV4	Meningococcal tetravalent capsular polysaccharide vaccine conjugated to diphtheria toxoid	Meningitis (meningococcemia) caused by infection with *Neisseria meningitidis*	Injected

cines, such as those against *H. influenzae* type b (Hib), hepatitis B virus (HBV), and *S. pneumoniae,* have been few and far between despite dramatically increased funding for infectious disease research and technological breakthroughs that should have accelerated vaccine development. Numerous diseases, such as TB, viral and bacterial genital tract infections, and intestinal infections, have so far not yielded to the efforts of scientists to create effective vaccines to prevent them.

Clearly, not all vaccines are successful. Vaccines fail primarily for three reasons. One is failure to elicit the anticipated protective immune response. Three of the great vaccine disappointments of the late 20th century have been the failure, despite considerable funding support and Herculean efforts, to develop effective vaccines against the three bacterial diseases salmonellosis, cholera, and TB. Despite major advances in our understanding of how the bacteria that cause these diseases interact with the human body, there must still be important missing pieces in the picture, because numerous candidate vaccines have failed to show efficacy in clinical trials. So-far-unsuccessful attempts to develop vaccines effective against human immunodeficiency virus (HIV) infection and malaria can be added to this list. Here, too, it is frustrating and perplexing that expenditure of huge amounts of grant money and involvement of hundreds of scientists have not produced a better outcome. What these examples show is that throwing money at a problem does not always lead to a solution and that high scientific sophistication does not ensure success in developing effective prevention strategies. Clearly, scien-

tists still have a lot to learn about the host-pathogen interaction.

In some cases, the lack of efficacy is understandable. For example, the *Neisseria* bacterium that causes gonorrhea is constantly changing the amino acid composition of its surface proteins, the ones targeted by opsonizing antibodies (see chapter 7). This antigenic variability makes it very difficult to elicit the type of immunological response that will clear the bacteria from an infected person and protect against subsequent exposure. Similarly, HIV not only rapidly changes the surface proteins it uses to enter cells from the outside, but also can move from cell to cell, a mode of spread that protects it from antibodies designed to prevent its attachment to cell surface receptors.

A second reason for vaccine failure is side effects. This was the downfall of a vaccine against a common type of infant diarrhea caused by rotavirus. The vaccine was very effective in preventing diarrhea, but a few infants developed intestinal blockage, a side effect that caused the vaccine to be abruptly withdrawn from use during clinical trials. In developed countries, rotavirus diarrhea can be treated effectively by replacing fluids lost during the diarrheal phase. Thus, in this case, the protection afforded by the vaccine was not sufficient to justify the risk. On the other hand, in developing countries, where emergency medical treatment is much more limited in availability, parents may find the vaccine more appealing than parents in developed countries because the infant death rate due to diarrheal disease is so much higher.

Recommended immunization schedule for persons aged 0 through 6 years —United States • 2010

Vaccine	Birth	1 month	2 months	4 months	6 months	12 months	15 months	18 months	19–23 months	2–3 years	4–6 years
Hepatitis B	HepB	HepB			HepB						
Rotavirus			RV	RV	RV						
Diphtheria, tetanus, pertussis			DTaP	DTaP	DTaP		DTaP				DTaP
Haemophilus influenzae type b			Hib	Hib	Hib	Hib					
Pneumococcal			PCV	PCV	PCV	PCV				PPSV	
Inactivated poliovirus			IPV	IPV	IPV						IPV
Influenza					Influenza (yearly)						
Measles, mumps, rubella						MMR					MMR
Varicella						Varicella					Varicella
Hepatitis A						HepA (2 doses)				HepA series	
Meningococcal										MCV	

░ Range of recommended ages for all children except certain high-risk groups

▓ Range of recommended ages for certain high-risk groups

Recommended immunization schedule for persons aged 7 through 18 years —United States • 2010

Vaccine	7–10 years	11–12 years	13–18 years
Tetanus, diphtheria, pertussis		Tdap	Tdap
Human papillomavirus		HPV (3 doses)	HPV series
Meningococcal	MCV	MCV	MCV
Influenza	Influenza (yearly)		
Pneumococcal	PPSV		
Hepatitis A	HepA series		
Hepatitis B	HepB series		
Inactivated poliovirus	IPV series		
Measles, mumps, rubella	MMR series ·		
Varicella	Varicella series		

░ Range of recommended ages for all children except certain high-risk groups

▓ Range of recommended ages for catch-up immunization

▓ Range of recommended ages for certain high-risk groups

Figure 17–2 Timing of childhood vaccinations. The bars represent the optimal time ranges for administering vaccines and illustrate the complexity of the vaccination program for a child. The recommendations are adjusted each year to reflect new or improved vaccines.

BOX 17–1   Having an Effective Vaccine Is Not Enough: Missed Opportunities for Vaccination

In developed countries, the delivery of vaccines to the general public has been relatively effective. In the early 1990s, U.S. public health workers were not satisfied with the level of coverage. They established an official goal for the year 2010 for at least 90% of all young children to have been vaccinated with DTaP, measles-mumps-rubella vaccine (MMR), polio vaccine, and Hib and at least 70% to have been vaccinated with HBV. A study of 2008 vaccination levels in the United States showed that among families surveyed, approximately 76% of the children had received the recommended number of DTaP, polio vaccine, Hib, MMR, and HBV administrations by the recommended age. While many states have exceeded the goals, there is great state-to-state variation (from 82% to 59%), which is worrisome. At the year 2010 mark, it was clear that the goals for several target areas had not been met, and those objectives that were not reached for the 2010 goal were carried forward into the next 10-year period. Some new target goals were added, while others required resetting and refocusing with new objectives or modifications before the launch of Healthy People 2020. (For an update, see http://www.healthypeople.gov/hp2020/default.asp.)

One strategy to increase the percentage of children vaccinated is to require that vaccinations be completed before a child enters elementary school. This has been effective, because the percentage of children entering school who had received the full vaccine series was over 95% for all vaccines in 75% of states. One objective of the U.S. Department of Health and Human Service's Healthy People 2010 project was to achieve a 90% vaccination rate by 2010. This was not satisfactory, however, because infants and young children are the ones at highest risk for the severe forms of many of these diseases. The study of current vaccine usage also revealed that the percent coverage was even worse among low-income urban populations than among other populations. The study also revealed that many of the children who had not been vaccinated could have been vaccinated in a timely manner, because they had passed through the health care system as infants and young children. These target areas are now the focus of the new Healthy People 2020 project period.

Why were these opportunities missed? For one thing, not all health care providers were routinely checking the vaccination status of all young patients during each visit. This was especially true in cases where the child was brought in for treatment of an illness rather than for a routine checkup. For another, many health care providers seemed to think that illness precluded vaccination. There are some contraindications, but most childhood illnesses, particularly mild ones, do not preclude vaccination. Finally, advantage was not being taken of the fact that the vaccines listed in Table 17–3 can be administered simultaneously. Several recommendations were made to reduce these missed opportunities.

1. Accurate vaccination records should be maintained and should be easily accessible to health care providers.
2. Vaccination status should be checked for every infant and young child during every visit to a health care provider.
3. Health care providers should be made aware that mild illness does not preclude vaccination.
4. Needed vaccines should be administered simultaneously.

Once the problem of undervaccination was recognized, the Centers for Disease Control and Prevention launched an advertising campaign to alert parents and physicians to the problem. Despite this campaign, which definitely helped raise public awareness, reports continue to come in of undervaccination of certain populations. Further investigations of the reason for the continued undervaccination rate in certain settings revealed that since physicians were not reimbursed adequately for vaccinations, even with the vaccine being provided free, they were routinely referring parents to vaccination clinics, thus necessitating another clinic visit and another wait in line. Even parents who are

(continued)

BOX 17–1   Having an Effective Vaccine Is Not Enough (*continued*)

devoted to their children have limits, especially when they see the pediatrician acting as if vaccination were low on the scale of things to worry about.

Sources: Centers for Disease Control and Prevention. 2009. National, state and local area

vaccination coverage levels among children aged 19–35 months—United States 2008. *MMWR Morbid. Mortal. Wkly. Rep.* **58:**921–926; Centers for Disease Control and Prevention. 2006. Vaccination coverage among children in kindergarten—United States, 2006–2007 school year. *MMWR Morbid. Mortal. Wkly. Rep.* **56:**819–821.

A third reason for vaccine failure is that the vaccine unexpectedly makes the disease worse. This is unarguably the worst nightmare for those who work long hours to develop a vaccine. The classic example of this type of vaccine failure is the respiratory syncytial virus vaccine. Respiratory syncytial virus attacks infants in their first months of life and is a major cause of hospital admission of infants under the age of 3 months. A vaccine that showed promising results in animals was administered to infants in the early stages of clinical trials. These trials were abandoned abruptly when it became apparent that vaccinated infants were getting a more severe form of respiratory disease than unvaccinated infants. Why this happened is still not clear, but this example is another indication of how little scientists understand the immune response, despite major breakthroughs in this area. More recently, a very promising vaccine against strains of *Chlamydia trachomatis*, an intracellular bacterial pathogen that causes genital infections in the developed world but causes blindness (trachoma) in certain areas of the developing world, made the eye infections in vaccinated children worse than those in unvaccinated children.

Working on vaccine development takes the kind of courage exhibited by the earlier microbiologists who walked bravely into cholera epidemics or, in more recent times, volunteered to study HIV and multidrug-resistant TB strains. In the case of vaccine development, however, a special sort of courage is needed—the courage not only to face the possible failure of a vaccine to protect, but also to face the possibility that it could make the disease worse. Just as the earlier discoveries that vaccination could protect people from such a devastating infectious disease as smallpox revolutionized preventive medicine, future discoveries that explain why some vaccines cause side effects or even make the symptoms of a disease worse will revolutionize vaccine development and provide a safer and surer source of new vaccines in the future.

What Makes an Ideal Vaccine?

Early work developing vaccines was largely based on empirical data and a lot of trial and error. Recent advances in the development of new-generation vaccines are based on rational design approaches. These advances have been made possible by our improved understanding of microbial and host factors required for virulence and the nature of the immune response to specific bacterial infections. Thus, modern vaccinologists have a battery of questions that they must first answer in order to carry out the task of designing an effective vaccine. The first two questions that must be considered are where the body first encounters the pathogen and what parts of the body the pathogen encounters during infection. Answering these questions will determine what type of immune response will be most effective at preventing infection, mucosal versus systemic or both (the first one at early times, and the other at later times). The next question that must be considered is what virulence factors the pathogen produces. Answering this question will determine whether a humoral (antibody) response or a cell-mediated (cytotoxic T-lymphocyte [CTL]) response will be most effective and whether a single antigen or multiple antigens will need to be targeted (see chapter 4). Once target virulence factors have been identified as potential vaccine antigens, then it is necessary to determine the extent of antigenic variation that might be found in these virulence factors. Finally, it will be necessary to determine how best to present the antigens to the immune system for an optimal protective response.

It is important to note that not all antigens (defined as a substance that interacts with an antibody) are capable of eliciting an immune response. An **immunogen** is an antigen that elicits an immune response, but alas, not all immunogens elicit protective immune responses. The part of the antigen that is recognized by the antibody **(antigenic determinant)** is called an **ep-**

itope. The epitope may be buried or not exposed in the native antigenic substance so that it is inaccessible to the antibody during infection. Thus, not all immunogenic epitopes elicit a protective immune response. To develop an effective vaccine, it is helpful to determine which epitope(s) of an antigen is protective and which will elicit the desired immune response, such as Th1 versus Th2, where Th1 responses are effective against intracellular pathogens and Th2 responses control extracellular pathogens (see chapter 4). A complication that sometimes arises is that the strongest and most protective immunogenic epitope is not against a single linear sequence of an antigen (called a **continuous epitope**) but rather against a structural conformation or against multiple segments of the antigen that have come together to form a topological surface (called a **discontinuous epitope**) (Figure 17–3). In such cases, it is very difficult to replicate the epitope in recombinant vaccines without using the antigen's native structure.

Identifying a protective immunogen is just the first step that may lead to actually generating an effective vaccine. To move to the next step in vaccine development, it is important to consider what the requirements of an "ideal" vaccine might be. An effective vaccine must be safe, with minimal side effects, yet it must be able to elicit a strong, protective, and long-lasting immune response. It should be heat stable, dryness stable, and have a long shelf life, all of which have been very challenging requirements to satisfy. It must be low cost with minimal infrastructural needs and minimal maintenance programs required. A vaccine that requires numerous boosters, that cannot be administered in combination with or at the same time as other vaccines, and that requires injection (as opposed to oral or nasal inoculation) is not low cost or easily administered, which makes it less amenable to application in developing countries that lack the resources and facilities to implement effective vaccine programs. Therefore, we will consider ways in which vaccinologists are attempting to improve existing vaccines and to develop new vaccines.

Approaches to Enhancing Immunogenicity

Eliciting the Correct Type of Immune Response

Vaccines against extracellular versus intracellular bacterial pathogens need to elicit different types of immunity, including CD4$^+$-mediated antibody responses and CD8$^+$-mediated CTL responses. In the case of many extracellular pathogens, especially those that produce protein toxins, protective immunity can be readily achieved through vaccines that result in stimulation of a strong humoral response, and as noted above, these types of vaccines have been the most successful to date. Vaccines against intracellular pathogens, where cellular immunity is critical for protection, have been much more challenging to develop. Unlike extracellular pathogens, intracellular pathogens in general have multiple virulence factors that contribute to pathogenesis, and because they are intracellular, these potential antigens are not always easy to identify and often are not available in vivo to the antibodies that might be generated against them. Although there are a few exceptions, as in the case of *Listeria monocytogenes*, where a CD8$^+$ CTL response to a single protein antigen (listeriolysin O) is sufficient to provide protective immunity, as a rule, it is necessary to immunize with a mixture of antigens to elicit a protective response against most intracellular pathogens.

Adjuvants

From the foregoing description, it might appear that subunit vaccines, like D, T, and aP, consist solely of the purified protein antigens. Although proteins given alone can produce an immune response, they work much better if they are given in conjunction with additional components, called **adjuvants,** which stimulate, modify, or augment the immune system and thereby enhance the host response to the vaccine. In addition, adjuvants can be used to direct the immune system toward a desired Th1 or Th2 response. In the

Figure 17–3 Continuous versus discontinuous epitopes. The part of an antigen that is recognized by an antibody is called an epitope. If the antibody binds to a part of the antigen that is not disrupted or has no tertiary structure, the epitope is considered to be continuous. If the antibody recognizes a surface topology that is made of multiple segments, such as more than one peptide sequence, or that contains tertiary structure (i.e., it is not linear), then the epitope is referred to as a discontinuous epitope.

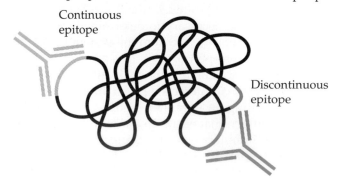

Continuous epitope

Discontinuous epitope

case of peptide vaccines, adjuvants are essential to elicit a strong immune response, because peptides given alone not only do not elicit much of an immune response, but also can actually cause the body to ignore the epitope through a process known as **induced tolerance** (or **anergy**). This anergy is a consequence of antigen-induced cell death of activated, antigen-specific T cells. Adjuvants, through an as-yet-unknown mechanism, can increase the half-life of activated T cells and counter this effect.

FREUND'S ADJUVANT. First developed in the 1940s, **Freund's complete adjuvant** is composed of inactivated and dried *Mycobacterium tuberculosis,* the causative agent of TB, emulsified in mineral oil or paraffin. It is highly effective in stimulating cell-mediated immunity, but because it stimulates production of the proinflammatory cytokine tumor necrosis factor, its use, even in animals, is now restricted due to the associated pain and tissue damage. **Freund's incomplete adjuvant** is the same adjuvant, but without the mycobacterial components that cause the strong inflammatory response, and is frequently used in veterinary vaccines.

ALUMINUM SALTS. Although a number of new adjuvants are currently under development and some are at the clinical trial stage, the only approved adjuvant currently licensed for human use in the United States is **alum,** which is composed of particulate aluminum salts, such as $Al(OH)_3$ or $AlPO_4$. In other countries, however, a number of other vaccine adjuvants have already been approved for human use (see below). For a long time it was thought that alum, which has a very long half-life at the site of injection, enhanced the immune response of antigens by allowing gradual release of the adsorbed antigen and thereby increasing exposure of antigen-presenting cells (APCs) to the antigen. This view has now been challenged by evidence that radiolabeled antigen is released very rapidly from the alum complex and disseminates away from the site of injection. Instead, recent evidence suggests that alum acts by promoting antigen uptake into dendritic cells (DCs) or other APCs and by activating innate immune reactions and cytokine release, but the exact mechanism of its action is still not clear. The primary advantage of using alum is its long history of safety for human use. Although alum is a good adjuvant for toxoid-based vaccines that rely on generation of Th2 humoral immunity, its major limitation is that it is a poor inducer of Th1 cellular immune responses. In addition, there is a limit to the total amount of alum that can be administered per dose, which curtails its use for potential multicomponent vaccine formulations.

TRITERPENOID-BASED ADJUVANTS. Plant-derived triterpene glycosides **(saponins)** have gained popularity as vaccine adjuvants because they can elicit strong Th1 and CTL immune responses. The most widely used saponin-based adjuvants are **QuilA** and its derivative **QS-21** (Figure 17–4), which are mixtures of saponins that are prepared in oil-in-water emulsions. Unfortunately, the high toxicity and undesirable hemolytic side effects, as well as instability in water, have so far restricted their use to veterinary vaccines. **Squalene,** on the other hand, is a triterpenoid that is a precursor of cholesterol and other steroid hormones. Since it is a natural component found in humans and is also biodegradable, it has been used to develop the first new adjuvant to gain approval for vaccine use in humans since alum. In 1997, the adjuvant **MF59** was licensed for human use. MF59 is an oil-in-water emulsion of squalene and two nonionic surfactants (Tween 80 and Span 85) as stabilizers. It has been included in an influenza vaccine that is commercially available in 23 countries worldwide. Postmarketing surveillance data have confirmed that it is safe and that it elicits higher antigen-specific antibody titers than alum and also promotes innate immune responses. MF59 has now been evaluated in clinical trials for a number of other antiviral vaccines and so far has shown no serious adverse effects and in general is well tolerated.

VIROSOMES. Influenza virus-derived **virosomes** are another market-approved, efficient antigen carrier and adjuvant for vaccines. Virosomes are vesicles (nanoparticles) composed of a lipid membrane with membrane-bound viral proteins, mostly hemagglutinin and neuraminidase. Virosomes retain the host cell receptor-binding and pH-dependent membrane fusion activities of the virus but lack the viral genetic material. These properties enable them to deliver their encapsulated antigens to DCs and other APCs and thereby confer their T-cell-mediated immunostimulatory properties (Figure 17–5). Another advantageous property of virosomes is the possibility of incorporation of other lipophilic adjuvants to provide additional immune-enhancing functions, and studies are under way to explore this possibility.

MICROSPHERES. A particularly interesting type of adjuvant, which may solve the problem of having to administer DTaP so many times, is **microspheres (liposomes** or **microcapsules)** made of resorbable suture material or some other inert substance that gradually breaks down in the body. For example, microparticles of biodegradable polymers, such as poly(D,L-lactide-co-glycolide), entrap the protein or polysaccharide antigen, and the encapsulated antigens are then released

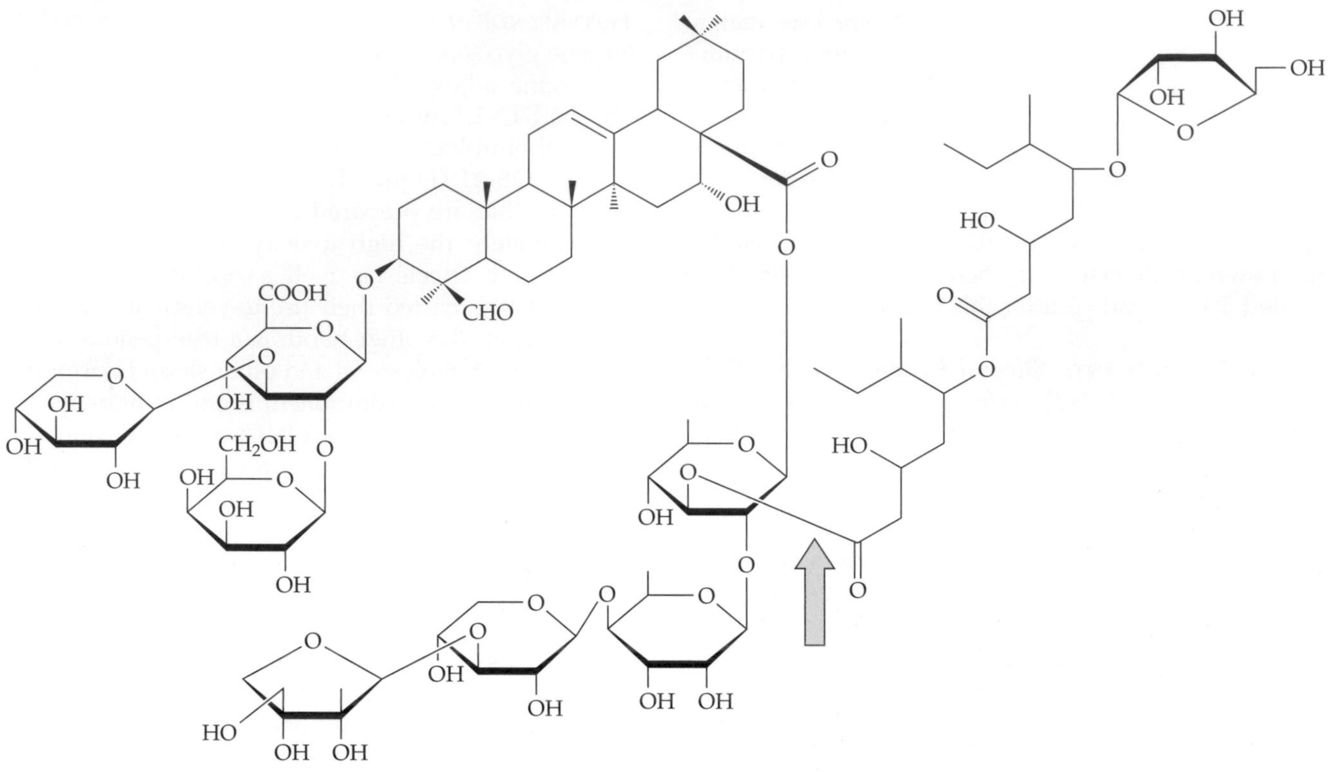

Figure 17–4 Chemical structure of the saponin-based adjuvant QS-21. The arrow points to the location of the acyl side chain group that is responsible for stimulation of the Th1 and CTL responses and, unfortunately, also for its toxicity.

slowly over a period of time to increase exposure time to the immune system. By administering a mixture of microspheres of different sizes, it is hoped that the vaccine will be released over a long enough period so that a memory response will be elicited with only one or two injections.

New Strategies

Programming Adaptive Immunity

TARGETING DCs AND INNATE IMMUNE SENSING. Although adjuvants have long been known to shape the immune response, it is only recently that the underlying mechanisms by which they do so have begun to be unraveled. A key event that triggers the immune response is when the immune system "senses" the vaccine antigen or microbe. The activation of DCs and other APCs translates into the release of cytokines and activation of antigen-specific T cells and B cells, which consequently modulate the strength, quality, and persistence of the adaptive immune response. Critical to this process is triggering the innate immune responses of DCs and APCs through Toll-like receptors (TLRs), as well as C-type lectins and NOD proteins. Trigger-

ing distinct TLRs activates DCs to elicit different cytokine profiles that differentiate Th1 from Th2 responses (Figure 17–6).

Different TLR agonists are currently being tested as adjuvants to tailor the immune response so that optimal protection is induced. For example, the TLR4-specific agonist lipopolysaccharide (in the form of detoxified monophosphoryl lipid A [MPL] obtained from *Salmonella*) is being combined with other adjuvants, such as alum or the saponin QS-21, for vaccine formulations. Recently, one such adjuvant, labeled AS04, which consists of MPL adsorbed on alum, has already been approved for human vaccines. Other TLR agonists, such as flagellin, poly(IC) double-stranded RNA, and bacterial DNA (in the form of CpG oligonucleotides that are underrepresented in humans), which stimulate TLR5, TLR3, and TLR9, respectively, are currently being tested in preclinical trials.

Heat shock proteins and antigens with RGD tags (Arg-Gly-Asp peptide motifs that act as ligands for cell surface integrins), which specifically bind to and activate DCs, are also being explored as potential adjuvant components. Activation of DCs increases their ability to process and present antigen and to attract

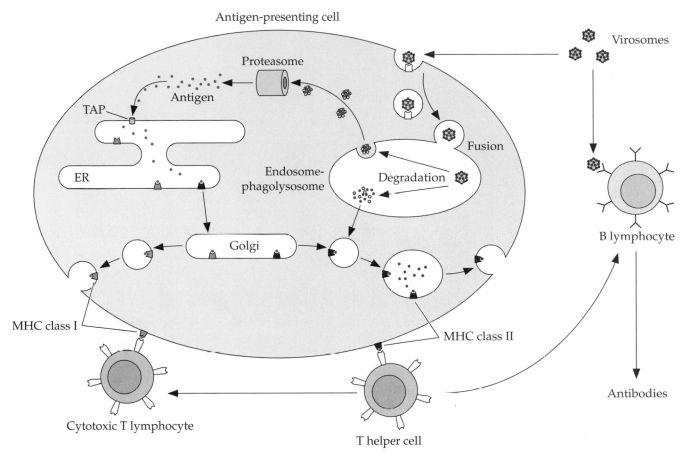

Figure 17–5 Virosome interactions with immune cells. Through multiple binding interactions of hemagglutinin on the virosome surface with APCs, the virosomes are taken up and the encapsulated antigens are processed through both the MHC-I and MHC-II pathways, activating T helper cells and stimulating both Th1 and Th2 responses. (Adapted from Wilschut, 2009, with permission from Elsevier.)

and activate T cells through cytokine secretion. Because of this, recent vaccine and adjuvant improvement strategies include using **cytokine technology** to modulate the immune responses. For example, interleukin 12 (IL-12) has strong adjuvant activity when administered with antigen because it stimulates a Th1-specific response. In chronic infections, such as those caused by the intracellular parasite *Leishmania*, the severity of the disease is determined by whether Th1 or Th2 mediates the immune response. The ability to manipulate the Th1 versus Th2 response could enhance protection and reduce the pathology of the disease.

EPITOPE-BASED SYNTHETIC-VACCINE STRATEGIES. Recombinant DNA technology revolutionized the development of vaccines by allowing the production and testing of recombinant nontoxic antigens, such as site-directed, catalytically inactive toxins, instead of

toxoids (chemically treated or heat-inactivated toxins). The tremendous leap in understanding of the mechanisms underlying protective immunity that we have achieved in recent years has also led to the development of a new class of recombinant vaccines. This newfound knowledge, coupled with the power of molecular biological techniques, has enabled protein engineers to change the conformation and structure of the antigen. Genetic manipulation can enhance DC-binding properties, stimulate desired immune responses, and enhance antigen presentation through the design of **epitope-based targeting signals** into the peptide sequence. For example, as mentioned above, addition of the RGD peptide (an integrin-binding motif) to the recombinant vaccine protein allowed the targeting of the antigen to DCs. However, there are a number of other immune-directing features that can be added. Inclusion of known MHC-I-restricted epitopes that activate the desired T helper cells in the

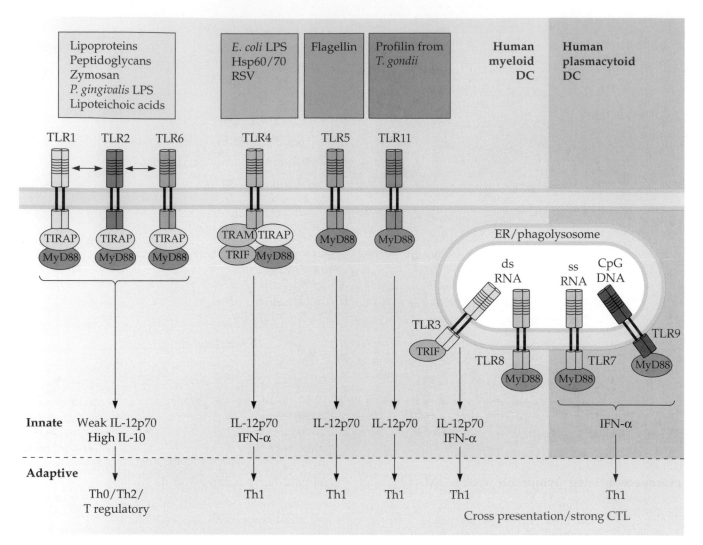

Figure 17–6 Triggering distinct TLRs on DCs elicits different cytokine profiles and different immune responses. Microbially derived ligands bind distinct TLRs to stimulate different innate immune responses, which cause the selective release of certain cytokines that in turn stimulate different adaptive immune responses. LPS, lipopolysaccharide; Hsp 60/70, heat shock protein 60/70; RSV, respiratory syncytial virus; *T. gondii, Toxoplasma gondii*; ER, endoplasmic reticulum; IFN-α, alpha interferon; dsRNA, double-stranded RNA; ssRNA, single-stranded RNA.

recombinant vaccine protein allows modulation of the appropriate cellular immune response. Attaching a TLR-specific ligand to the antigen allows stimulation of appropriate innate immune responses. Incorporation of lysosomal protease cleavage sites between known epitopes in the protein antigen allowed enhanced antigen processing, especially through the MHC-II pathway. Based on these concepts, it has been possible to design an entirely synthetic vaccine (Figure 17–7). However, keep in mind that the multiple chemical steps needed to manufacture synthetic vaccines can add to their cost, which can impact their attractiveness to pharmaceutical companies.

Targeting Mucosal Immunity

VACCINE VECTOR TECHNOLOGY. One strategy for eliciting the correct immune responses against bacterial pathogens is to use molecular techniques to make a live, avirulent bacterial vaccine strain that produces the protein antigen (or portions of it) on its surface. For example, since many pathogens gain entry into the body through mucosal surfaces, targeting mucosal immunity is a desirable approach for the development of vaccines against such pathogens. Vaccines based on killed whole cells or on isolated protein or polysaccharide antigens do not generally induce strong CD8+

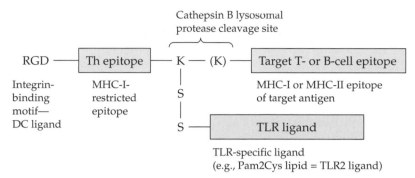

Figure 17–7 Schematic representation of an entirely synthetic epitope-based vaccine candidate. Such a vaccine would contain an RGD motif, which would target the antigen for DC binding and uptake. It would consist of at least two epitopes, one strong known Th-specific epitope at the N terminus and one or more B- or T-cell epitopes from the target antigen, which would elicit the desired CD4⁺ or CD8⁺ response, respectively. A lysosomal protease cleavage site (e.g., containing lysines for cathepsin B cleavage), which allows optimal antigen processing, separates the epitopes. A TLR-specific ligand is also included by linking the ligand (such as a lipid adjuvant) via two serine residues to the ε-amino group of one of the lysine residues of the peptide containing the epitope.

T-cell responses, so live but attenuated mutants are the most effective strategy against pathogens that require cellular immunity for protection.

Avirulent bacterial vectors, such as attenuated, antigen-expressing *Salmonella* strains that bind to and invade M cells in the gut, where they interact with **mucosa-associated lymphoid tissue (MALT)** and stimulate secretory immunoglobulin A (sIgA), as well as cellular immune responses, are being tested as vaccines targeting mucosal immunity against gut pathogens. *Salmonella* has another advantageous feature that makes it a desirable candidate as a vaccine vector. It has type 3 secretion systems that could be used to deliver immunogenic antigens or epitopes directly into host immune cells. Other avirulent or attenuated strains of pathogens being explored as antigen delivery vectors are *Vibrio cholerae*, *L. monocytogenes*, *Shigella* spp., *Mycobacterium bovis*, and *Yersinia enterocolitica*. Bacteria found in the normal intestinal microbiota that are being explored include *Streptococcus gordonii*, *Lactobacillus* spp., and *Lactococcus* spp.

The anti-TB vaccine *M. bovis* BCG (for bacillus Calmette-Guérin) is an attenuated strain of *M. bovis* that causes a TB-like disease in humans and cattle that results in subsequent immunity to TB. Controversy over BCG's efficacy (ranging from 0% to 80% in field trials, now thought to be due to large gene deletions and other polymorphisms in some of the vaccine strain preparations that were used in the trials) and the problem that people vaccinated with it test positive in the TB skin test caused the disuse of BCG in the United States. However, because of the emergence of multidrug-resistant TB since the 1990s, BCG is now administered to at-risk health care workers. BCG is also being investigated as a vaccine vehicle because of its extensive safety record and its ability to elicit a strong immune response. Efforts are under way to engineer BCG to display recombinant vaccine antigens against other pathogens on its surface.

A number of problems have been encountered with the generation of effective live vaccines. One problem has been getting sufficient amounts of the recombinant foreign antigens or epitopes to be expressed on the bacterium's surface, yet without overburdening the bacterium's metabolism so that it cannot grow adequately. Another major difficulty is generating mutant strains of the bacteria that are avirulent or attenuated enough to be nontoxic and safe yet are still able to survive long enough in the body to generate a strong immune response. This requires a delicate balance that is extremely challenging to achieve.

The approach of using live attenuated bacterial vaccines has been greatly complicated recently with the realization that there are an abundance of bacteriophages in the environment and the body and that these bacteriophages often carry virulence factors, which could result in converting the avirulent vaccine strain into a potentially deadly pathogen. One recent solution to this problem is the use of **psoralen** (Figure 17–8) plus UV treatment, which cross-links DNA. This blocks DNA synthesis, replication, and repair in the bacteria, but not protein synthesis or metabolism. Such cells do not proliferate, but they remain alive and can still produce antigens.

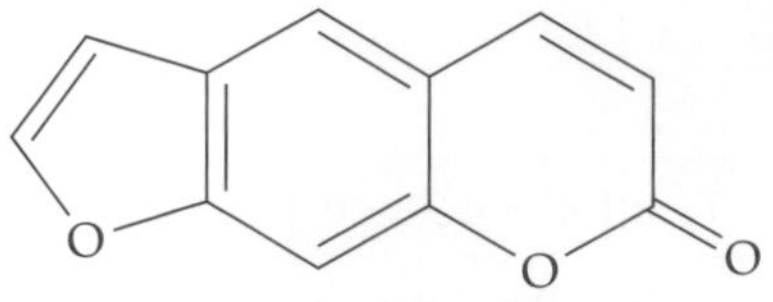

Figure 17–8 Chemical structure of psoralen, a photoactivatable DNA cross-linker.

TOXINS TO STIMULATE MUCOSAL IMMUNITY. Another powerful adjuvant comes from an unexpected source: the bacterium that causes cholera *(V. cholerae). V. cholerae* produces a toxic protein complex called **cholera toxin,** which causes massive dehydrating watery diarrhea. For reasons that are still not completely apparent, cholera toxin is a very effective stimulator of the antibody-specific immune response. Do not worry, we are not asking you to ingest large doses of cholera toxin. Instead, cholera toxin is being used to stimulate immune responses to vaccines administered through the skin. There is currently also a lot of interest in how to create vaccines that do not require needle injections but can be administered intranasally or by skin patches (see below).

Researchers are beginning to explore the use of catalytically inactive toxins to deliver epitopes in candidate vaccines that require stimulation of both humoral and CTL responses. Diphtheria toxin was the prototype of such "tools," referred to as **immunotoxins,** but researchers have now expanded the immunotoxin repertoire to include other toxins, such as cholera toxin and anthrax toxin (Figure 17–9). Since the B subunit of cholera toxin binds to M cells of the gut, linking an antigen to the B subunit will target the antigen for direct delivery to the MALT.

INJECTION-FREE VACCINES. There are some new vaccine developments that stimulate mucosal immunity via MALT that are very exciting. One such development is rather pedestrian in the sense that it applies to vaccine administration, but it is an advance that will be hailed as revolutionary by physicians, parents, and children alike: the shot-free vaccination. Children, like adults, do not take kindly to being stuck with a needle, however thin and "painless." Production of a syringe by the doctor or nurse produces a child's version of sticker shock and is guaranteed to raise howls of indignation and terror. The words "this won't hurt a bit" rank right up there with "the check is in the mail" in the list of definitely-not-to-be-believed statements.

Three recent developments may signal the end of the dreaded syringe. One is **intranasal inoculation** of the vaccine. Nasal inoculation activates the **bronchial-associated lymphoid tissue (BALT),** the respiratory tract equivalent of the intestinal **gastrointestinal-associated lymphoid tissue (GALT).** Nasal vaccines against influenza virus have come into widespread use recently. One major advantage of targeting BALT over GALT is that nasal inoculation does not require protection of the antigen from degradation in the harsh environment of the stomach (Table 17–4). Also, as we learned in chapter 3, mucosal immunity generated at one mucosal site in the body can provide protective responses at other mucosal sites. Nasal influenza vaccines contain live attenuated viruses, which cannot replicate in most areas of the body because it is too warm, but they can replicate in the nose, which is a few degrees cooler than the rest of the body. Production of virus for these vaccines is currently a bottleneck, as exemplified by delays encountered in the distribution of vaccine against H1N1 "swine flu," because the viruses are currently propagated in chicken eggs. Improvement in the technology to produce large amounts of virus for live attenuated influenza vaccines, as well as to inactivate the virus for the more traditional "flu shot," will undoubtedly continue to improve to allow rapid deployment to prevent pandemics.

A second development is a **patch** containing many tiny prongs. The patch is applied to the skin surface to stimulate **skin-associated lymphoid tissue (SALT),** but the prongs do not penetrate far enough to encounter the epidermal neurons, so that application of the patch is painless. Coupled with the use of cholera toxin as an adjuvant (noted above), administration of a vaccine using the patch could stimulate antibody-specific immune responses to various antigens via SALT. This response is often accompanied by a general mucosal immune response, a very desirable feature, and since the skin cannot develop cholera diarrhea, cholera toxin as an adjuvant should be safe.

EDIBLE VACCINES. Still another interesting approach is the development of **edible vaccines.** The ability to genetically engineer plants has made it possible to produce plants that synthesize vaccine proteins. The original, and somewhat naïve, idea was that vaccination could occur by the simple act of ingesting the plant. This did not take into account that dosing would be variable and that regulatory agencies will not approve vaccines with variable dosing, but the concept was intriguing, and the approach was explored nonetheless. Currently, the edible vaccines are being produced by a large number of plant species. Several of these vaccines have been through small

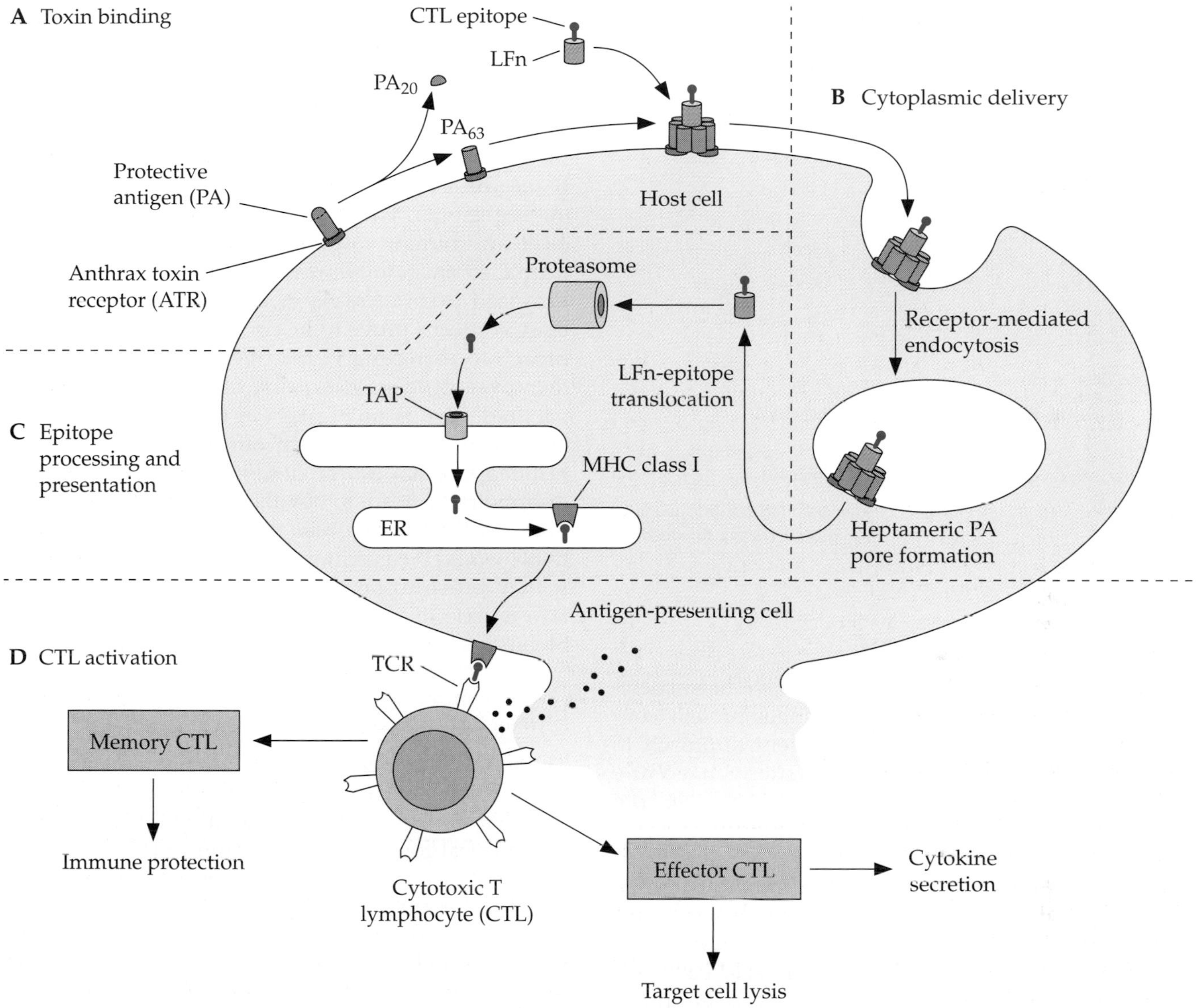

A Toxin binding

CTL epitope

LFn

PA_{20}

PA_{63}

B Cytoplasmic delivery

Protective
antigen (PA)

Host cell

Anthrax toxin
receptor (ATR)

Proteasome

Receptor-mediated
endocytosis

C Epitope
processing and
presentation

TAP

LFn-epitope
translocation

MHC class I

ER

Heptameric PA
pore formation

Antigen-presenting cell

D CTL activation

TCR

Memory CTL

Immune protection

Cytotoxic T
lymphocyte (CTL)

Effector CTL

Cytokine
secretion

Target cell lysis

Figure 17–9 Delivery of vaccine antigens as fusions with bacterial toxins. Shown is an example of the use of anthrax toxin for vaccine delivery. Coupling a CTL-specific epitope to a catalytically inactive form of lethal factor (LF) of anthrax toxin allows the uptake, processing, and presentation of the epitope in APCs. (A) Toxin binds to APC. (B) Toxin is endocytosed, and epitope is translocated into cytoplasm. (C) Epitope is processed and presented on cell surface through the MHC-I complex. (D) Activation of CTL response. TCR, T-cell receptor; ER, endoplasmic reticulum; TAP, antigen-binding cassette transporter associated with antigen processing. (Adapted from C. A. Shaw and M. N. Starnbach. 2003. Using modified bacterial toxins to deliver vaccine antigens. *ASM News* **69:**384–389.)

trials in humans to test safety. So far, they have been shown to be safe, but alas, they have not produced a robust immune response. One concern that has been voiced regarding the development of edible vaccines using commonly eaten fruits or vegetables is the possibility that if the vaccine is presented together with the food in which it is produced, the gut's immune system may view the vaccine as "food" and thus become tolerant. This would decrease rather than increase future response to invaders. Another concern sometimes voiced is the possibility that food allergies might arise as a result of introducing foreign antigens.

Table 17–4 Pros and cons of oral and nasal vaccines

Vaccine property	Oral	Nasal
Delivery	Easy	Simple device
Patient compliance	High	Moderate
Antigen digestion/degradation	Extensive	Limited
Antigen uptake	Poor	Good
Immunogenicity	Poor	Modest to good
Immunological tolerance	Possible	Possible
Antigen dose required	High	Medium
Adjuvant required	Powerful mucosal	Mucosal
Targeting required	Mandatory	Helpful
Safety	Safe	Safe (so far, but needs more testing to confirm)

DNA Vaccine Technology

All of the vaccines discussed above have been bacteria, proteins, polysaccharides, or sugar-protein conjugate antigens. A completely different approach to vaccination that has received a lot of attention recently is to inject DNA that encodes a vaccine protein into human muscle cells, where it is expressed under a mammalian promoter. These vaccines are called **DNA vaccines.** The DNA is first adsorbed to gold particles, which are injected with an air gun into muscle tissue. Incredibly enough, foreign antigens are produced and displayed on the surfaces of the muscle cells. Although display of the foreign antigen lasts only a month or two, this is long enough to evoke a robust response. Because the display is localized, side effects associated with an autoimmune attack on the body's own cells are minimized. Also, the DNA is in differentiated cells, not in the germ line, so it will not be passed on to subsequent progeny. So far, this approach has been tested mainly in animals, but the results have caused considerable excitement, and human trials are already under way. This approach, if successful, would have three important advantages. First, unlike live vaccines or subunit vaccines with most adjuvants, DNA on gold particles can be stored dry and need not be refrigerated. Second, it is much less expensive to prepare pure DNA than to prepare pure protein. Third, DNA vaccines do not require adjuvants.

Just as DNA vaccines appeared to be ready to revolutionize vaccination procedures, some troubling second thoughts began to be expressed. These second thoughts centered on a rediscovery of an old wheel: DNA is immunostimulatory. Scientists working on the disease lupus have long suspected that this autoimmune disease was caused by the body's immune reaction to its own DNA. Articles have begun to appear that question the safety of injecting DNA into human tissue. Basically, the claim is that certain DNA sequence motifs, which are found in bacterial DNA, elicit an immune response that might lead in some people to an autoimmune response, which could in turn lead to an increase in such diseases as lupus. If these concerns prove to be correct, another trendy approach to correcting human genetic disorders—gene therapy—could go down the tubes along with DNA vaccines. This is no big loss in the case of gene therapy, since intense research efforts in this area have resulted in what can be described most charitably as disappointing results, but the loss of DNA vaccines, which actually have real promise as well as megahype, would be a serious setback for vaccine research. It may prove to be the case that if DNA is injected into muscle tissue, not enough of it will get into the bloodstream to trigger an immune response.

Storage of Vaccines

There is a major need for development of new vaccine formulations that are more stable with longer shelf life and that do not require refrigeration. This is particularly critical for vaccines that are needed in developing countries, where storage conditions are less than optimal. Recent efforts in this arena include the use of sugar-based **glassification,** which is a **lyophilization (freeze-drying)** process that includes vacuum drying with or without freezing. Glassification provides enhanced stability at room temperature. Other processes include CO_2-assisted nebulization with a bubble dryer that allows particles to be made in the 3- to 5-μm size range, which is desirable for lung delivery, without destroying the biological activity of the vaccine.

Immunization Programs

The administration of an effective immunization program requires a large national infrastructure with the capacity to facilitate coherent planning, coordination, financing, public information, and government-industry cooperation for the proper testing, production, distribution, use, and monitoring of vaccines. In the United States, this process requires the proper coordination and functioning of over 20 federal agencies, not including state departments of health, vaccine and biotechnology companies, professional medical facilities, voluntary organizations, and public and private health care officials. A clear example of what can happen when this complex infrastructure fails is the diphtheria epidemic that swept through the former Soviet Union following its collapse in the early 1990s.

A different type of vaccine problem, and one that at least in theory is more amenable to solution, is that of simplifying the current confusing mixture of separate inoculations of different vaccines at different times. It would help a lot to be able to administer several vaccines simultaneously in a single dose, preferably by the oral, skin, or nasal route, and to eliminate the need for so many booster shots. Using live viral or bacterial vaccine strains that present antigens of more than one pathogen is a possible solution. The U.S. Food and Drug Administration approved one such vaccine, Pediarix, in 2002. It combines DTaP, the vaccine against HBV, and the polio vaccine into one shot. Use of Pediarix reduces the number of shots in the first 6 months of life from 10 to 3. Orally administered forms of DTaP, Hib, and HBV are being developed, and work on the administration of such vaccines by inhalation has begun. The ability to administer a variety of different vaccines in a single swallow or a single inhalation has obvious appeal.

Passive Immunization

The administration of antibodies obtained from immune animals or humans is an old solution to the problem of how to treat a patient who is not immune to a particular disease. This is called **passive immunization.** This type of therapy offers short-term protection to people who have recently been exposed or soon may be exposed to a disease. Three types of passive immunization products are used. The first is standard human immunoglobulin derived from pooled human sera that have been treated to kill viruses. It contains primarily IgG against a broad spectrum of antigens and has been used routinely to protect travelers from hepatitis A virus. It is now also a way to treat immunocompromised patients, especially those infected with common viral infections, such as cytomegalovirus infections. In this case, preparations of immunoglobulin made from human sera are used because antibodies to cytomegalovirus antigens are commonly found in the sera of most people.

The second type is specific human immunoglobulin, which is derived from pooled human serum known to contain a high concentration of a specific antibody. This is a much more expensive reagent than the standard human immunoglobulin described above. It is often used to treat patients recently exposed to HBV.

The third type, which has been used for over 100 years, is animal-derived serum antibodies or antitoxins. An animal (usually a horse) is immunized with a specific antigen, and then the antibodies are harvested. Antitoxin derived from horse serum is often used for cases of diphtheria and botulism to neutralize the toxins. Major problems with this type of product are serum sickness in about 10% of patients and occasionally death due to anaphylactic shock. Because of the increased mortality rate due to serum sickness in infant botulism (caused by botulinum neurotoxins), special human immunoglobulin (sold as BabyBIG) generated from pooled adult sera (containing high titers of antibodies against mainly neurotoxin types A and B) from persons immunized with botulinum toxoid is available (at a rather steep price of about $45,300 per vial) for young children.

Passive immunization has been receiving much more attention recently, as physicians have finally admitted that major human killers of the preantibiotic era, such as *S. pneumoniae*, the most common cause of bacterial pneumonia, have become so resistant to antibiotics that patients with multidrug-resistant disease may no longer be treatable with antibiotics. Passive immunization then becomes even more important as a firewall against multiple antimicrobial-resistant microbes. Another treatment option is to provide antibodies that bind to the antiphagocytic capsule of the bacterium and enhance phagocytosis, thus destroying its ability to evade the nonspecific host defenses. More and more often, passive immunization is being invoked as a possible clinical response to multidrug-resistant bacteria.

Passive immunization is also gaining importance as the number of immunocompromised people has increased. This growing group includes not only people with HIV infections, but also people undergoing cancer chemotherapy or organ transplantation and those living longer. In these people, who are unable to mount an effective immune response, passive immunization may well prove to be the only option for life-threatening infections. As the cost of vaccine development escalates and companies become more and more reluctant to pursue the production of vaccines, passive immunization may well be the last line of defense against any disease for which there are no vaccines available.

SELECTED READINGS

Glenting, J., and S. Wessels. 2005. Ensuring safety of DNA vaccines. *Microb. Cell Factories* **4:**e26.

Jackson, D. C., Y. F. Lau, T. Le, A. Suhrbier, G. Deliyannis, C. Cheers, C. Smith, W. Zeng, and L. E.

Brown. 2004. A totally synthetic vaccine of generic structure that targets Toll-like receptor 2 on dendritic cells and promotes antibody or cytotoxic T cell responses. *Proc. Natl. Acad. Sci. USA* **101:**15440–15445.

Leung, A. S., V. Tran, Z. Wu, X. Yu, D. C. Alexander, G. F. Gao, B. Zhu, and J. Liu. 2008. Novel genome polymorphisms in BCG vaccine strains and impact on efficacy. *BMC Genomics* **9:**e413.

Mestecky, J., H. Nguyen, C. Czerkinsky, and H. Kiyono. 2008. Oral immunization: an update. *Curr. Opin. Gastroenterol.* **24:**713–719.

Schalk, J. A., F. R. Mooi, G. A. Berbers, L. A. van Aerts, H. Ovelgönne, and T. G. Kimman. 2006. Preclinical and clinical safety studies on DNA vaccines. *Hum. Vaccin.* **2:**45–53.

Schultze, V., V. D'Agosto, A. Wack, D. Novicki, J. Zorn, and R. Hennig. 2008. Safety of MF59™ adjuvant. *Vaccine* **26:**3209–3222.

Shortman, K., M. H. Lahoud, and I. Caminschi. 2009. Improving vaccines by targeting antigens to dendritic cells. *Exp. Mol. Med.* **41:**61–66.

Sun, H.-X., Y. Xie, and Y.-P. Ye. 2009. Advances in saponin-based adjuvants. *Vaccine* **27:**1787–1796.

Titball, R. W. 2008. Vaccines against intracellular bacterial pathogens. *Drug Discov. Today* **13:**596–600.

Tritto, E., F. Mosca, and E. De Gregorio. 2009. Mechanism of action of licensed vaccine adjuvants. *Vaccine* **27:**3331–3334.

Ulmer, J. B., B. Wahren, and M. A. Liu. 2006. Gene-based vaccines: recent technical and clinical advances. *Trends Mol. Med.* **12:**215–222.

Vermij, P. 2004. Edible vaccines not ready for main course. *Nat. Med.* **10:**881.

Wack, A., and R. Rappuoli. 2005. Vaccinology at the beginning of the 21st century. *Curr. Opin. Immunol.* **17:**411–418.

Wilschut, J. 2009. Influenza vaccines: the virosome concept. *Immunol. Lett.* **122:**118–121.

Wilson, C. B., and E. K. Marcuse. 2001. Vaccine safety-vaccine benefits: science and the public's perception. *Nat. Rev. Immunol.* **1:**160–167.

QUESTIONS

1. In the former Soviet Union, as it began to fall apart in the early 1990s, there were great unrest and population dislocations. Suddenly, the incidence of diphtheria began to rise to alarming proportions. This might be understandable if all the cases were in young children who might have fallen through the cracks of a compromised health system (which had actually been of high quality, at least in some areas, during earlier days), but many of the cases were in adults. It is still not clear what happened, but speculate on how the diphtheria vaccination program in the Soviet Union might have failed.

2. In 2004, there was an influenza vaccine crisis in the United States, when many lots of vaccine had to be recalled. The problem was bacterial contamination. Why would this be a problem with an injectable vaccine? Would it have been a problem with an oral vaccine? If so, under what conditions? There was only one major manufacturer of the vaccine, a fact that made the shortage worse. How could this shortage of manufacturers have happened?

3. Perhaps DNA vaccines would work better if the same approach taken with some attempts at gene therapy—to clone the bacterial DNA into a viral vector that would integrate more efficiently into the human chromosome than naked DNA—was used. In the case of gene therapy using viral vectors, cancer has been a potential problem in some cases. How could this happen?

4. In this chapter, it was asserted that failures in vaccines could occur for three reasons. Actually there are more. Can you think of some of them?

5. Most vaccines are equally effective in most populations. An exception to this rule was the TB vaccine BCG. When administered to college students in the United States during the mid-1950s, it was very effective (about 80%) in preventing TB. In Bangladesh, the efficacy of the same vaccine was close to 0%. What are some of the reasons this could happen? Suppose that the vaccine was of high quality and the cold chain (refrigeration conditions) had been maintained but the disparity still occurred. Try to explain this. Hint (and this is highly speculative): there are bacteria related to *M. tuberculosis* that are found widely in the Indian subcontinent. How could this fact have affected the outcome? Remember, the percent effectiveness is the effectiveness above prevailing rates of TB.

6. How does passive immunization differ from active immunization? Why does passive immunization not confer long-term immunity? Under what circumstances might passive immunization be preferable to vaccination for treating infections?

7. What are the advantages and disadvantages of live vaccines? Of DNA vaccines?

8. Consider a bacterium that is ingested via contaminated water and locally infects the small intestine. What type of vaccine would be most useful against this bacterium? What type of immune response would this vaccine elicit?

9. How can recombinant DNA technology be used to enhance immunity and improve a vaccine?

10. What is a potential problem with the use of diphtheria toxin as an immunotoxin? Can the use of anthrax toxin overcome this problem?

11. Some scientists are looking into the strategy of using passive immunization to control bacteria that are resistant to many antibiotics. How would such a strategy work? (Recall that passive immunization is the injection of preformed antibodies into a patient.)

12. Do vaccines necessarily work better if they target virulence factors than if they target molecules not involved in virulence? Explain why or why not.

13. Why might an IgG response to *Salmonella enterica* serovar Typhi be ineffective?

SOLVING PROBLEMS IN BACTERIAL PATHOGENESIS

1. A group of researchers have isolated a new bacterium, W, from the lymph nodes of several patients who returned from a camping trip and presented in the emergency room with high fever, rash, and swollen lymph nodes. The bacterium produces a unique surface protein (called OmpW) and an unusual polysaccharide capsule (called WPS) that the researchers can see on the surfaces of the bacteria in electron micrographs from fresh isolates (bacteria just obtained from the patients). However, after culturing the bacteria in vitro on agar plates, they find that most of the bacteria still have OmpW but no longer have WPS on their surfaces. The researchers also find that bacterium W binds to and invades phagocytic cells and inhibits their function.

   A. The researchers propose that OmpW and WPS might make excellent targets for development of a vaccine against bacterium W. What led the researchers to propose this?

   B. The researchers find that a vaccine made of WPS alone does not evoke long-lasting immune responses. How does it elicit an immune response? What strategy could the researchers use to generate long-lasting immunity to bacterium W using WPS as part of the vaccine?

2. One of the alternative structures for the outer core of the lipooligosaccharide (LOS) found in clinical isolates of *Neisseria meningitidis* and *Neisseria gonorrhoeae* is a Galα[1-4]Galβ[1-4]Glc group, which has the same structure as the exposed group of globotriaosyl ceramide, Galα[1-4]Galβ[1-4]Glc-ceramide, a common glycolipid component of mammalian cell membranes. The loci including the biosynthetic genes that encode the outer core of LOS in *Neisseria* contain five conserved glycosyl transferase genes, *lgtA* to *-E*, arranged in an operon. The genes *lgtA*, *lgtC*, and *lgtD* contain poly(G) tracts. The *lgtC* and *lgtE* genes encode the glycosyl transferases that connect the α-galactose and β-galactose, respectively, to the glucose of the outer core to form the Galα[1-4]Galβ[1-4]Glc group.

   A. What would be the advantage for *Neisseria* in having the Galα[1-4]Galβ[1-4]Glc group as part of its LOS? Be specific.

   B. One of the researchers in a vaccine development group at a biotechnology company proposes that the Galα[1-4]Galβ[1-4]Glc group might make an excellent target for vaccine development against *Neisseria*. What might have led the researcher to propose this? Provide two possible reasons.

   C. Another researcher in the same vaccine development group challenged this proposal, stating that it was a poor choice for a vaccine candidate. What might have led the second researcher to say this? Provide two possible reasons.

3. Lyme disease is the most common vector-borne disease in the United States. It is caused by the gram-negative spirochete *Borrelia burgdorferi*, which is transmitted through the bites of infected ticks. The ability of the bacteria to survive in the tick and in mammals depends on the differential expression or repression of particular genes. For example, the bacteria express an outer surface protein A (OspA) in the midgut of nymphal and adult ticks but downregulate OspA and upregulate another outer surface protein (OspC) in response to increased temperature after the tick begins to feed on blood and the bacteria enter the host. Both OspA and OspC have become primary targets for development of a vaccine against Lyme disease, and indeed, several commercial vaccines based on these proteins have been made and have been shown to have high efficacy in providing protective immunity in clinical trials.

   A. Provide at least three reasons why OspA and OspC might be good vaccine candidates. Would OspA or OspC make a better vaccine candidate? State your rationale.

   B. What type of immune response is primarily responsible for controlling *Borrelia* infection in

(continued)

healthy individuals who have been vaccinated against OspA or OspC? Be specific.

C. Describe how the immune response in an individual vaccinated with a vaccine based on OspC would function to control infection upon subsequent exposure to *Borrelia*. How would this differ from the response in an individual vaccinated against OspA?

D. Describe how *Borrelia* infection would be controlled during the first few days after exposure in healthy individuals who have not been vaccinated. Be specific. Provide the specific pathway(s).

E. Recent studies have determined that the immunodominant epitope in OspC responsible for protective immunity in humans is located in the highly conserved C terminus of OspC. How is this OspC epitope presented to immune cells when it is used as a vaccine? Provide the specific pathway(s).

F. One consequence of disseminated *B. burgdorferi* infection in ~10% of infected individuals is a condition described as treatment-resistant Lyme arthritis (TRLA), which persists in patients even after antibiotic treatment. Recent studies have revealed that TRLA is correlated with a particular T-cell epitope associated with rheumatoid arthritis that has sequence similarity to a region of OspA spanning residues 165 to 173, which in turn is within a region of the protein that is important for providing protective immunity in humans. In clinical trials testing the efficacy of OspA as a vaccine candidate, this finding caused a serious problem. What was that problem? How could you correct it to improve the vaccine's safety yet still retain its efficacy?

G. Studies aimed at designing diagnostic tests for Lyme disease have determined that animals, such as mice, hamsters, and dogs, do not react to OspC in the same way humans do. Indeed, animal models failed to produce immune responses to the dominant C-terminal epitope of OspC that is immunoprotective in humans. Consequently, the use of animal models to evaluate the efficacy and reliability of diagnostic tests for Lyme disease in humans has come into question recently. However, as a researcher, you are seeking to find additional diagnostic or vaccine targets against Lyme disease in humans. You would like to take advantage of the differential expression of proteins that occurs in *B. burgdorferi* when it is in the tick versus when it is in the human to identify such targets. Describe how you might accomplish this with no animal or human infection model available. Provide details of the experimental design and rationale for your choice of experimental approach.

4. Both diphtheria toxin and anthrax toxin have been used as prototypes for the development of novel vaccines and therapeutics.

A. Describe how these toxins might be used as vaccines to deliver other antigens. What is the advantage of using these toxins as vaccine delivery vehicles?

B. Diphtheria toxin was the first toxin to be exploited as an immunotoxin. Many other toxins have since been used as immunotoxins for specific purposes. Name at least two problems with the use of diphtheria toxin as an immunotoxin. What strategies could be employed to overcome these limitations? Name at least two.

C. Describe how you might design an immunotoxin based on diphtheria toxin that might work to cure a tumor that overexpresses on the surfaces of the tumor cells the ATR (anthrax toxin receptor) protein, which has an extracellular von Willebrandt factor A domain that binds to anthrax protective antigen.

5. The alarming spread of methicillin-resistant *Staphylococcus aureus* (MRSA) infection in the United States has prompted a number of research efforts to develop a vaccine. MRSA is subcategorized as community-acquired (CA) MRSA or hospital-acquired (HA) MRSA, depending on how the infection is usually acquired. Although most CA-MRSA bacteria are not as resistant to antibiotics as the HA-MRSA strains, the recent strains that we are hearing about in the news appear to be more virulent. A group of researchers have recently discovered that bacterial cultures of a particularly virulent strain of CA-MRSA (called USA300) cause apoptosis in leukocytes and neutrophils, as evidenced by release of cytochrome *c* from mitochondria, followed by cytokine release and cell death. Most MRSA strains produce type CP5 or CP8 polysaccharide capsules, surface-exposed protein A, and a number of toxins and superantigens.

A. Antibodies generated against bacterial culture medium from the USA300 strain identified two proteins of 33 and 44 kDa on Western blots. To determine the localization of these two proteins, researchers performed sodium dodecyl sulfate-polyacrylamide gel electrophoresis and Western blot analysis of lysates from the bacterial cell pellets and the bacterial culture medium, as well as lysates of human neutrophils treated with the USA300 strain. The results from Western blot analysis are shown in the left panel of Figure 1. The researchers then applied the cell-free bacterial culture medium to the neutrophils and performed subcellular fractionation of the treated neutrophils. The results from Western blot analysis are shown in the right panel of Figure 1. Provide an interpretation (with rationale) of the results shown in Figure 1. Predict (with rationale) possible functions for the two proteins, which have been named LukS and LukF, where LukS is the 33-kDa protein and LukF is the 44-kDa protein. Provide at least one additional (different) experiment that could be performed to confirm your prediction. The researchers speculate that the two proteins, LukS and LukF, are two subunits of an AB-type toxin. Based on the results shown in Figure 1, do you agree with this interpretation? Provide your rationale. Provide an experiment that would allow you to verify your answer.

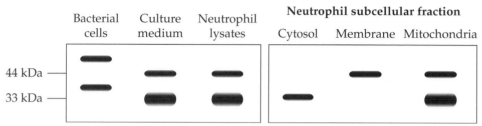

Figure 1 Western blots using antibodies against US300 secreted proteins.

B. Provide a mechanism by which protein A contributes to virulence (i.e., what is its function?).

C. Provide a mechanism by which the superantigens might contribute to MRSA virulence. What is a potential long-term consequence of infection with MRSA strains that produce superantigens?

D. MRSA strains producing CP5 show a significantly higher bacteremia level than strains producing CP8. A group of researchers propose that development of a two-component vaccine made up of LukS or LukF in combination with CP5 or CP8 would be effective protection against CA-MRSA. What led the researchers to propose this? Provide at least two possible reasons. Which two of the possible vaccine targets (LukS, LukF, CP5, and CP8) would be most effective at generating protective immunity as part of a two-component vaccine against CA-MRSA? Be sure to provide your rationale.

SPECIAL GLOBAL-PERSPECTIVE PROBLEM: INTEGRATING CONCEPTS IN PATHOGENESIS

Dental plaque is a biofilm consisting of a complex community of over 700 different bacterial species. Epidemiological evidence suggests that a population shift toward certain gram-negative anaerobes is responsible for the initiation and progression of periodontal diseases. *Tannerella forsythia*, a gram-negative, filamentous, nonmotile, anaerobic bacterium, is considered one of the pathogens implicated in contributing to advanced forms of periodontal disease in humans and is strongly associated with cases of severe periodontitis. It is found coaggregated in periodontal pockets with other putative periodontal pathogens, such as *Porphyromonas gingivalis* and *Fusobacterium nucleatum*. Infection with *T. forsythia* induces alveolar bone resorption in a mouse infection model involving inoculation under the gums of mice and then measuring for loss of dental bone. The organism is naturally resistant to gentamicin and erythromycin but is sensitive to tetracycline, chloramphenicol, and kanamycin. Genome sequencing of the bacterium is under way but has not been completed and is still unavailable. Due to the fastidious (very picky) nature of this bacterium for growth (there are no auxotrophic

mutants, and its growth on agar plates requires blood- or hemin-supplemented amino acid-rich culture medium and incubation in an anaerobic chamber) and the difficulties in genetically manipulating it (there are no known transposons available, and the one *Escherichia coli* shuttle vector reported thus far has very low conjugation frequencies and very poor transformation efficiencies), virulence factors that contribute to the role of *T. forsythia* in the progression of periodontal disease have been difficult to identify.

1. Considering all of the above information, which of Koch's postulates, if any, have been satisfied so far for *T. forsythia* involvement in periodontal disease? Be sure to state your rationale. Provide at least two additional modern molecular experiments (different from those already described) that could be performed to help satisfy Koch's postulates. (Be specific!)

2. Considering all of the above information, describe a strategy that researchers, who have a limited budget, might use to identify virulence factors of *T. forsythia* that are associated with periodontal disease. Be sure to provide a rationale for your choice of strategy, the appropriate reagents that they will need to use for it (hint: a clearly labeled, detailed diagram or schematic of the overall strategy and/or tools to be used should be provided, along with a rationale for each of its features), the overall experimental design, and how the researchers might determine the identities of the putative virulence factors once they are found by this strategy.

3. Using your proposed screening strategy, the researchers have identified a number of putative virulence factors that they believe are involved in *T. forsythia* pathogenesis. Among these are several that have homology with proteins of known function:

> BspA, a fibronectin-binding cell surface-associated protein
>
> PrtH, a secreted trypsin-like protease
>
> SiaHI, a secreted sialidase
>
> HmuR, a hemin-binding outer membrane protein
>
> TfsA, a large, 230-kDa surface (S-layer) glycoprotein with homology to *P. gingivalis* hemagglutinin

A. What is the possible role of each of these factors in *T. forsythia* pathogenesis? (Be specific!)

B. What do these researchers have to do next to verify that these putative virulence factors are indeed involved in *T. forsythia* pathogenesis? Describe an experiment (which does not involve human subjects) that could be performed to demonstrate this. What are the expected results (in terms of data) from such an experiment if these proteins are indeed virulence factors (i.e., provide a table, graph, plot, or gel illustrating this; you may make up example values or data points if that helps)?

4. Another group of researchers found that BspA and TfsA, in addition to having hemagglutination properties, are also involved in mediating bacterial binding and invasion of human epidermal cells.

A. What type(s) of protective adaptive immune response does wild-type *T. forsythia* expressing these two proteins elicit? Be sure to provide your rationale.

B. Describe in detail how these two antigens are presented to the immune cells in an unimmunized individual. (Hint: a detailed, clearly labeled schematic diagram might be helpful.)

C. The researchers propose that BspA and TfsA might make excellent candidates for development of a vaccine against *T. forsythia*-associated periodontal disease. What led the researchers to propose this? Provide two different possible reasons.

5. The production of hemagglutinins is a well-established virulence strategy for a number of bacterial pathogens. Thus, it was not surprising that *P. gingivalis,* which is another putative dental plaque pathogen, also expresses several hemagglutinins on its cell surface. It was found that one of the hemagglutinin genes *(hagA)* of *P. gingivalis* contains four large contiguous direct repeats varying from 1,318 to 1,368 bp in length, which together encode a protein of 283 kDa. The repeat unit (denoted *HArep*), which contains the hemagglutinin domain, was also found to be present in the other hemagglutinin genes in *P. gingivalis* and in the *tfsA* gene (which encodes TfsA) from *T. forsythia*. The beginning amino acid sequence encoded by the first repeat *(HArep1)* in *hagA* is PNPNPGTTT, while that of the other three repeats *(HArep2* to *-4)* is GTPNPNPNPGTTT. The amino acid sequence at the C terminus of the fourth repeat *(HArep4)* is GTPNPNPNP.

A. Provide an explanation that could account for the presence of this *HArep* repeat unit four times in *hagA* and also in the other hemagglutinin genes from *P. gingivalis* and in TfsA from *T. forsythia* (from a molecular evolutionary point of view).

B. Provide a possible mechanism for how this occurs. (Hint: a detailed, clearly labeled diagram or schematic might be helpful.)

6. Another group of researchers isolated from detergent extracts of *T. forsythia* bacterial cells a lipoprotein (which they named TfLP) that, when added to human oral fibroblasts or epithelial or monocytic cells, induced proinflammatory cytokine production (IL-1, IL-6, and tumor necrosis factor alpha) and NF-κB-mediated apoptotic cell death. When this group of researchers submitted their manuscript reporting their findings to a scientific journal, one of the journal's reviewers rejected the manuscript because the researchers failed to include an important control experiment, stating that the cytokine production could have been due to a contaminating factor in the TfLP-containing bacterial extracts.

A. What factor did the reviewer think was a contaminant in the lipoprotein preparation, and why was the possibility of this factor being present in the TfLP-containing extracts a concern? Be sure to provide your rationale.

B. If the researchers were to remove this factor or ensure that it is not present in the TfLP-containing preparations, provide two possible mechanisms by which the TfLP lipoprotein might induce cytokine production and apoptotic cell death.

C. Describe how you might experimentally distinguish between these two possibilities.

18

The Gram-Positive Opportunistic Pathogens

Some scientists have viewed opportunistic pathogens, such as *Staphylococcus aureus* and *Staphylococcus epidermidis,* as less important than "real" pathogens, such as *Yersinia pestis* and *Mycobacterium tuberculosis,* because the opportunists usually do not infect healthy people but need an opportunity in the form of an impaired host defense. We do not agree with this assessment. In fact, we have it on the highest medical authority that someone who dies of a *Staphylococcus* infection is just as dead as someone who dies of plague or tuberculosis, and in the developed world, despite all of the advances in modern medicine (and sometimes because of them), the number of people being killed by "opportunists" is a lot greater than the number killed by "real" pathogens.

What Is an Opportunist?

The term "opportunistic pathogen" is used to describe a microorganism that rarely causes infections in healthy people. That is, some impairment of host defenses such as burns, cuts, surgery, or immunocompromising conditions (having cancer, taking immunosuppressive drugs, or coinfection with AIDS or cystic fibrosis), must preexist for the microbe to take advantage of and cause disease. A problem with this definition is that there are exceptions that raise questions about its accuracy. One obvious case is *Staphylococcus aureus,* the cause of a variety of serious diseases ranging from skin and soft tissue infections to endocarditis and toxic shock syndrome. Although many cases of *S. aureus* infections are seen in hospital patients or patients with conditions that impair their defenses, there are numerous cases in the community involving people with no obvious underlying conditions. This trend is showing an alarming increase and applies to other opportunistic pathogens, such as community-acquired (CA) *Clostridium difficile,* which causes inflammatory colon diseases. In addition, some opportunistic pathogens seem to be acquiring increased virulence properties, such as strains of *Streptococcus pyogenes,* which can cause benign, albeit still serious, diseases, such as pharyngitis ("strep throat"), but which have received considerable press in recent years because of some of the "scary" strains that have "flesh-eating" properties.

399

What these cases may illustrate is that within the average "healthy" population there is a wide variation in susceptibility to different infectious diseases. Looking at things from the "true-pathogen" side, it is worth noting that even in the case of pathogens such as *Vibrio cholerae* (the cause of cholera), *Salmonella enterica* serovar Typhi (the cause of typhoid fever), and *Mycobacterium tuberculosis* (the cause of tuberculosis), there are many cases in which infected people are asymptomatic. The picture of host-pathogen interactions is evolving toward a view that there are probably no hard and fast distinctions, such as "opportunists" or "frank pathogens," but rather a continuum that includes a wide range of outcomes. Continua associated with different diseases overlap, with some skewing toward disease in seemingly healthy individuals and others toward people with clear underlying conditions. Nonetheless, the term "opportunist" has proven useful as a general concept and will be used here and in chapter 19 discussing gram-negative opportunistic pathogens.

Shared Characteristics

All of the gram-positive opportunists featured in this chapter have in common that they are normal commensal inhabitants of the human body. Some of these bacteria colonize everyone all of the time, whereas others colonize only a portion of the adult human population at any one time. In chapter 19, examples of soil- or even insect-borne bacteria will also be given. In the case of the gram-positive opportunists, however, those of human origin predominate in terms of their numbers and the seriousness of the infections they cause. Bacteria that normally reside on the human body clearly have an advantage, because they are always present in large numbers at the site of a potential breach of host defenses. Another unfortunate trait that is shared by many of the gram-positive opportunists is resistance to a variety of antibiotics. Thus, when they manage to cause an infection, the infection can be very difficult to treat. As was explained in earlier chapters, it is important to bring infections, especially bloodstream infections, under control as soon as possible. Both gram-positive and gram-negative pathogens trigger sepsis, and the longer sepsis persists, the more likely the infected person is to suffer long-term damage to essential organs or death.

From the brief introduction above, it is clear that many of these opportunistic gram-positive pathogens cause a spectrum of different diseases ranging in severity from mild to fatal. How could this be? The answer that has emerged from recent genomic, tran-scriptomic, and proteomic comparisons is that each of these species is actually composed of many subgroups that contain profound differences in their genomes and virulence properties. It turns out that each species really has a "pangenome," or "total composite genome," consisting of more genes than are found in the genome of any given isolate. Each isolate of each species contains the same set of core genes, which may account for about 80% of the genome. However, the other 20% of the genome is made up of various pathogenicity islands, lysogenic bacteriophages, plasmids, and transposons that carry distinctive sets of virulence factors. The makeup of these "accessory" genes determines whether that subgroup of a given species acts as a well-behaved commensal that is happy colonizing its host but not causing disease or as a pathogen that has tendencies to cause different forms of invasive diseases. Thus, because of pangenomes, each pathogenic species is really greater than the sum of its genomic parts, leading to astounding genetic diversity.

Why Commensal Bacteria Act as Pathogens

Before considering the representative gram-positive pathogens in this chapter, it is worth considering why commensal bacteria cause invasive diseases in the first place. Some opportunists do seem to be "accidental" pathogens. For example, *Staphylococcus epidermidis* is ubiquitous on the human skin and on other animals. In this niche, it spreads easily and does not cause problems. However, it still must contend with the innate immune system on the skin, such as antibacterial peptides. In response, the formation of biofilms and the expression of other virulence traits on implanted medical devices may just be a manifestation of its normal defenses against the human immune system. At any rate, *S. epidermidis* gains access through wounds or by contamination during medical procedures, and it has been argued that commensal bacteria with easy routes of transmission between hosts, such as *S. epidermidis*, have not had to evolve mechanisms of virulence that allow them to invade. *C. difficile* is another "accidental" opportunist whose recent emergence has depended on the widespread use of antibiotics that disrupt the normal microbiota in the colon (i.e., they cause a microbial community shift).

Streptococcus pneumoniae, which is a leading cause of CA pneumonia and other invasive infections, is harder to classify as a strict opportunist. All serotypes of *S. pneumoniae* reside almost exclusively in humans (although there seems to be some recent spread to pets), with one exception that is also found in horses. Therefore, carriage from person to person through contact and colonization of the nasopharynx is essential for the persistence of *S. pneumoniae*—but then, do the invasive diseases caused by *S. pneumoniae* play a role in bacterial transmission, or are they just accidental states that exaggerate mechanisms used by *S. pneumoniae* during colonization? It is hard to understand how otitis media (earache), meningitis, and bacteremia could contribute to pneumococcus transmission and spread to other humans. It is less clear whether this argument strictly applies to pneumonia, since there are numerous examples of the spread of *S. pneumoniae* resulting in small-scale outbreaks of pneumonia among people in cramped settings, such as day care centers, jails, and military bases, or even an outbreak in a large city. Moreover, several virulence factors that attenuate invasive disease in animal models of *S. pneumoniae* infection do not seem to have strong effects in some colonization models. However, it is important not to overinterpret results from animal models, which cannot fully reflect human infection states, because numerous proteins used by *S. pneumoniae* to evade human innate immunity do not bind to or interact with homologous proteins in other animals, such as mice (e.g., a protease that specifically cleaves human secretory immunoglobulin A [sIgA] or a surface protein that binds to human factor H). In addition, human commensal bacteria that can act as pathogens, such as *S. pneumoniae*, have probably resided in their human hosts for hundreds of thousands of years, and it is difficult to extrapolate how human lifestyles influenced the evolution and spread of these bacteria.

On the other end of the spectrum are *S. aureus* and *S. pyogenes*. Like *S. pneumoniae*, these bacteria colonize about one-third of the adult human population, but unlike *S. pneumoniae*, which produces a fairly limited repertory of toxins and tissue-destructive proteins and compounds, *S. aureus* and *S. pyogenes* are extraordinarily toxic bacteria, producing an array of toxins and superantigens that can lead to septic shock. Why, then, are these commensal species so ready for battle with their hosts? It has been argued that *S. aureus* encodes more virulence factors than *S. epidermidis* because the pathway of transmission of *S. aureus* leading to colonization of the anterior nares is more complex and difficult than colonization of the skin by the ubiquitous *S. epidermidis*. Transmission through direct contact and secretions that leads to nasal colonization does not seem to be that different for *S. aureus* and *S. pneumoniae*, yet *S. aureus* has many more virulence mechanisms than *S. pneumoniae*, and it cannot simply be because *S. aureus* is not confined to human hosts. Like *S. pneumoniae*, *S. pyogenes* is a strict human commensal, but *S. pyogenes* is also much more toxic than

S. pneumoniae. In fact, it has been proposed that strains of *S. pyogenes* that colonize saliva and the mouth may have devolved through mutation from virulent antecedent strains. Based on these examples, it could be concluded that there are relatively few clear-cut examples of truly accidental or opportunistic pathogens, and numerous complex processes appear to drive the evolution of commensalism and the development and roles of virulence properties of bacterial pathogens.

Examples of Notable Gram-Positive Opportunists

S. aureus: a Toxically Loaded Commensal

CHARACTERISTICS AND HABITAT. *S. aureus* is a common inhabitant of part of the human nose (the anterior nares, which contain the nostrils) and the upper respiratory tract. It colonizes the noses of at least one-third of people, often asymptomatically. Many people are persistent carriers of *S. aureus* and are always colonized, whereas about an equal number are colonized transiently. Its name (Latin: *aureus* = "golden") comes from the golden color of the colonies it forms on rich medium. A classical defining characteristic of *S. aureus* that separates it from *S. epidermidis* is the presence of an enzyme called coagulase (which catalyzes the conversion of fibrinogen to fibrin, resulting in blood clotting) that is easily detected by a simple clinical test. *S. aureus* is also known for its durability in the environment. Although the main human site for *S. aureus* colonization is the nose, *S. aureus* bacteria shed onto the skin and into the environment can last for long periods and serve as a reservoir of infection.

The propensity of *S. aureus* to colonize the nose has led to shedding of the bacterium by hospital staff. This is especially serious with multidrug-resistant *S. aureus* strains, such as **methicillin-resistant *S. aureus* (MRSA)**. Bloodstream and tissue infections with MRSA are the greatest current causes for concern. Understanding where *S. aureus* lurks has also led to a strategy for limiting the spread of these bacteria. The antibiotic mupirocin is too toxic for internal use but can be used topically. It is effective even against MRSA, so the application of mupirocin to the noses of colonized health care workers can lead to decolonization of these workers, almost literally defusing a ticking infectious-disease time bomb. *S. aureus* survives on articles of clothing, including neckties. Male doctors are even being asked to abandon their neckties as a safety measure, although hospital gowns and even gloves are still the main vectors of transmission to patients.

A disturbing recent development is the emergence of CA-MRSA strains that cause severe necrotizing pneumonia and contagious skin infections. As the name implies, strains of CA-MRSA are not confined to hospitals and are acquired from contact in schools and other community settings. Nasal carriers of methicillin-sensitive *S. aureus* and MRSA have a higher risk of infection than noncarriers and serve as important reservoirs for infection. In contrast, emerging CA-MRSA seems to be spread in the absence of nasal colonization, probably from colonization sites on the skin. The emergence of CA-MRSA is likely due to changes in the expression and functions of some of the bacterial virulence factors discussed below.

TOXINS.   We start this section by describing the many toxins encoded by *S. aureus* strains. Different *S. aureus* strains can cause a variety of toxin-induced diseases, including skin infections, such as impetigo and boils, and they can colonize plastic implants, leading to endocarditis. Enterotoxin-producing *S. aureus* strains are a major cause of food-borne disease, with the enterotoxin stimulating the vagus nerve in the stomach to induce vomiting and abdominal pain. Another type of *S. aureus* exotoxin causes toxic shock syndrome. Toxic shock syndrome is an interesting example of an opportunistic infection, because cases were first seen when a tampon that was absorbent enough to be left in the vagina for long periods was introduced on the market. Strains of *S. aureus* that produced toxic shock syndrome toxin (TSST) found these tampons to be a site conducive to growth and produced the toxin, which entered the bloodstream. A fever, rash, and sloughing skin were early symptoms of the disease, but the most dangerous results of the disease were shock and death induced by the toxin acting as a superantigen (see chapter 12).

Superantigens are proteins that force an association between major histocompatibility complexes on antigen-presenting cells and the T-cell receptor that would not normally occur (Figure 18–1; see chapter 12). This association can occur in the absence of presented antigens or in the presence of nonspecific antigens and is tight enough to trigger cytokine release by both cell types. If many such complexes form, enough cytokine release can occur to trigger the shock process. *S. aureus* produces two classes of superantigens, **toxic shock syndrome toxin 1 (TSST-1)** and **staphylococcal enterotoxin (SE).** There are now seven types of SEs: **SEA, SEB, SEC1, SEC2, SEC3, SED,** and **SEE.**

TSST-1, encoded by the *tst* gene, is the toxin responsible for the symptoms of toxic shock syndrome that were first associated with superabsorbent tampons. Not only did the tampons provide a special environment where *S. aureus* could grow, they also contained air pockets that provided the oxygen necessary for expression of the *tst* gene. Cases of toxic shock syndrome caused by *S. aureus* virtually disappeared once the tampon was removed from the market. TSST-1 in the bloodstream can trigger a massive release of cytokines that cause shock and death. In animals, TSST-1 makes the animal hypersusceptible to lipopolysaccharide (LPS), which may enter the bloodstream regularly in small amounts due to lysis of gram-negative bacteria in the gut microbiota. Whether this is true of humans is not known. LPS might not even be the most important molecule whose action TSST-1 potentiates. One might expect lipoteichoic acid (LTA) of gram-positive bacteria to be potentiated in the same way, but this has not been tested. An unanswered question is, what is the true function of TSST-1? Production of a toxin in the vaginal tract that has its pathogenic effect only when it leaves the site and enters the bloodstream is clearly not doing anything of obvious benefit to the bacteria. A possible role could be to block mucosal immunity. TSST-1 may interfere with the T cells found in sites where the mucosal immune system is located and prevent efficient development of an effective sIgA response.

SEs are responsible for the symptoms of a staphylococcal disease that are more common than toxic shock syndrome and a lot less lethal—food-borne disease. SEs, encoded by *ent* (for enterotoxin genes), are produced by bacteria in contaminated food. The toxin is ingested with the food. In the stomach, the SE stimulates the vagus nerve endings, which control the vomiting reflex. Projectile vomiting and abdominal pain are the hallmarks of *S. aureus* food poisoning. This type of food poisoning is usually not fatal, although people who have suffered it report wanting to die during the symptomatic period. Also, there have been cases in which pilots of small planes or people in similarly vulnerable situations have begun to experience the symptoms and come close to having a fatal accident.

The SE-mediated disease has a rapid onset, usually just a few hours after eating contaminated food, and symptoms subside in a day or two, as expected from the fact that this is an intoxication, not a bacterial infection. Once again, there is the mystery of why *S. aureus* produces SEs, since they confer no obvious benefit on the bacteria, which are rapidly eliminated from the digestive tract. The picture gets stranger when one considers that both TSST-1 and SEs are single-chain polypeptides with some sequence similarity, raising the question of whether they are members of a protein superfamily. The *entA* and *entB* genes

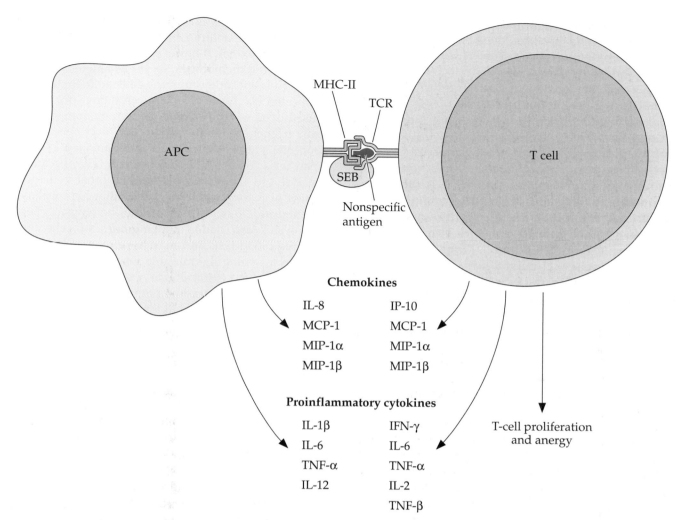

Figure 18–1 Actions of superantigen toxins in inducing massive cytokine production leading to sepsis and shock. Staphylococcal enterotoxin B (SEB) is shown in this example. MIP, macrophage inflammatory protein; MCP, monocyte chemoattractant protein; IFN, interferon; IP, interferon-inducible cytokine; TNF, tumor necrosis factor.

are carried on bacteriophage, another similarity to *tst,* but other SE genes (e.g., *entC* and *entD*) are carried on plasmids. The prevalence of superantigen genes in *S. aureus* gets back to the enigma raised earlier. If the virulence of opportunistic commensal pathogens is an aberration of defenses used in colonization, then why are superantigen genes prevalent in *S. aureus?* It seems likely that the maintenance of these genes and the virulence they cause may contribute to transmission in humans or other hosts, either now or at some time in the past.

Although the role of the superantigens in the biology of *S. aureus* is unclear, other toxic exoenzymes make a clearer contribution to the survival and spread of the bacteria in the human body. An example of an exoenzyme that may promote the spread of the bacteria is a group of proteases called **exfoliative toxins**

(ETs). As mentioned above, some *S. aureus* strains cause skin conditions, such as **scalded-skin syndrome** in infants, a disease that gets its name from the fact that the infant develops the sort of red, peeling appearance that suggests a bad scald. These same strains have also been implicated in **bullous impetigo,** a type of skin infection that produces blister-type lesions. In both cases, the epidermis is separated from the underlying tissue. The ETs are responsible for this exfoliation, or separation of skin layers.

The best-studied ET is **ETA.** The target for the protease activity of ETA has now been identified. It is a protein belonging to the cadherin family found on the surfaces of epidermal cells, called **desmoglein 1 (Dsg-1).** Dsg-1 is produced only in the skin, a fact that would explain the localized effect of ETA. Its role is to maintain keratinocyte cell-cell adhesion. Cleavage of Dsg-1 leads to separation of skin keratinocytes, a result that would produce the sort of separation of layers of epidermal tissue seen in scalded-skin syndrome and bullous impetigo.

Another exoenzyme that contributes to the spread of bacteria is **staphylokinase (Sak).** Sak dissolves clots. To understand how Sak does this, consider the process by which the body normally breaks down clots as part of the wound-healing process. Clots are made up of platelets held together by a fibrin mesh. One of the components of this mesh is the protein plasminogen. Normally, during dissolution of a clot, endothelial cells secrete a serine protease called **tissue plasminogen activator (tPA),** which cleaves plasminogen into a form called plasmin. Plasmin itself acts as a serine protease that degrades the fibrin mesh, dissolving the clot (Figure 18–2). This is a carefully controlled process that is confined to the clot. Free plasmin is rapidly degraded in blood. Sak forms a 1:1 complex with plasminogen, and this complex then cleaves other plasminogen molecules into plasmin in an uncontrolled way. Sak activity destroys the extracellular matrix and fibrin fibers that hold cells together, thus allowing the bacteria to move through tissue. It also helps the bacteria escape from abscesses, walled-off regions of dead tissue that provide nutrients for bacteria within them but that also confine bacteria to the site.

Figure 18–2 Normal fibrinolytic system involving conversion of plasminogen to plasmin and degradative cleavage of fibrin to release clots. Normally, activators and inhibitors in the blood carefully control the cascade process. Sak leads to uncontrolled cleavage of plasminogen, causing fibrin degradation. Other pathogens also manipulate the fibrinolytic system to dissolve clots that allow bacterial spread. PAI, plasminogen activator inhibitor. (Adapted from D. Collen. 1998. Staphylokinase: a potent, uniquely fibrin-selective thrombolytic agent. *Nat. Med.* **4:**279–284, with permission from Macmillan Publishers Ltd.)

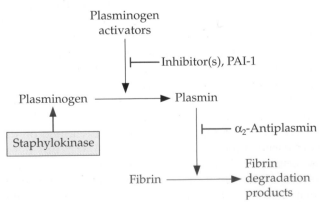

Another function of staphylococcal exotoxins is to kill or limit the ability of immune cells to attack the bacteria. **α-Toxin,** a β-barrel exotoxin that forms pores in human cell membranes, serves as a defense against neutrophils and other immune cells. α-Toxin has been implicated in some of the necrotizing pneumonia caused by CA-MRSA. Some membrane damage by α-toxin triggers cytokine production and sets off the apoptotic pathways that kill the damaged cells. α-Toxin used to be called α-hemolysin (Hla) because it can lyse red blood cells as well as other cell types. Strains of *S. aureus* also produce other hemolysins—β-hemolysin, δ-hemolysin, and γ-hemolysin. These hemolysins can damage membranes of cells other than red blood cells and may have roles similar to α-toxin in lessening the immune response.

Yet another extracellular exotoxin, **leukocidin,** damages mammalian cell membranes. Its name arises from the fact that it can kill leukocytes (of which neutrophils are one type). Leukocidin consists of two protein components, LukS and LukF, that assemble to form large pores in membranes (Figure 18–3). It has been shown that all three proteins responsible for γ-hemolysin activity (HylA, HylB, and HylC) have amino acid sequences nearly identical to those of leukocidin components S and F. Leukocidin is associated with virulence in CA-MRSA, but its role in disease caused by this emerging pathogen has been controversial.

OTHER VIRULENCE FACTORS. *S. aureus* strains produce a variety of other factors involved in colonization and virulence. One important class of factors mediate attachment of *S. aureus* cells to certain blood proteins, such as IgG and fibrinogen, or to extracellular matrix proteins, such as fibronectin and collagen. The surface adhesins that bind to extracellular matrix proteins have been given the mind-boggling acronym **MSCRAMMs,** which stands for **"microbial surface components recognizing adhesive matrix molecules."** Several MSCRAMMs are involved in colonization and virulence and allow the bacteria to adhere to and persist in the host. These MSCRAMMs include fibronectin-binding protein, collagen-binding protein, laminin-binding protein, vitronectin-binding protein, and elastin-binding protein (Figure 18–4; see Figure 11–14). Close proximity of *S. aureus* to the extracellular matrix provides a rich mixture of host proteins and polysaccharides, such as hyaluronic acid. *S. aureus* produces hyaluronidase and surface proteases that digest and dissolve these components of the extracellular matrix, providing food for the bacteria, exposing additional attachment sites for bacterial adherence, and increasing the chances for the bacteria to spread

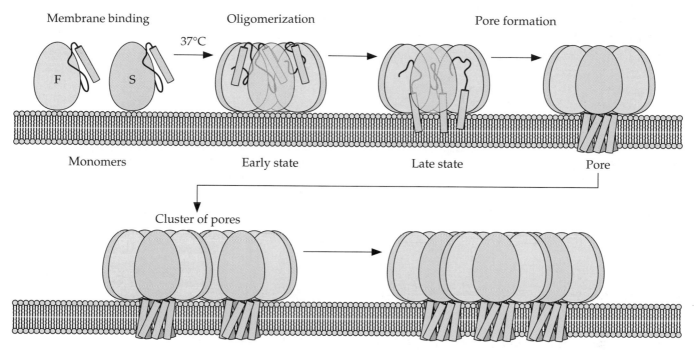

Membrane binding　　　　　Oligomerization　　　　　　Pore formation

37°C

F　　S

Monomers　　　　　　　Early state　　　　　　Late state　　　　　Pore

Cluster of pores

Multiple pores fuse to form larger holes

Figure 18–3　Steps in pore formation by the two-component toxin leukocidin. Monomers of LukS (S) and LukF (F) bind to GM1 gangliosides and assemble on the surfaces of host cells. Upon formation of a heptamer, the stem-loop undergoes a conformational change and inserts into the plasma membrane to form the pore. Multiple pores can cluster to form larger holes.

throughout tissues. The clumping factor MSCRAMM, which binds to fibrinogen in blood clots and can cause clumping of human plasma, has an additional role in protecting *S. aureus* from the immune system. Clumping factor binds to serum factor I, which stimulates the cleavage of complement C3b fragment on the *S. aureus* surface to its inactive (iC3b) form. This reduction in bound intact C3b opsonin reduces phagocytosis of *S. aureus* by human neutrophils.

Most MSCRAMMs and other enzymes that digest host proteins and polysaccharides, including the sugars on glycoproteins, are covalently attached to the bacterial peptidoglycan cell wall by sortases (Figure 18–5A; see Figure 11–8). In *S. aureus*, the muropeptide chains in the peptidoglycan are linked by pentaglycine cross-bridges (Figure 18–5B). These pentaglycine bridges are used to connect the muropeptide chains, an action that is catalyzed by penicillin-binding transpeptidases, and to covalently attach secreted proteins containing the sortase recognition motif, LPXTG (see Figure 11–8). In fact, many other gram-positive bacteria contain several classes of sortases that recognize slightly different recognition motifs or signals (Figure 18–5A). *S. aureus* contains a general sortase that attaches numerous MSCRAMMs and surface proteins

used for phage recognition, a class used to attach proteins involved in heme uptake and iron acquisition, and a class used to attach pili. Gram-positive species that form spores (such as *C. difficile*, discussed below) also contain a specialized sortase involved in spore coat assembly. The characteristic sortase recognition motifs can be easily picked out in putative proteins predicted by genomic sequences of gram-positive bacteria. These likely surface proteins can then be further tested for functions such as the effects of mutants on adherence to cell lines in culture or virulence in animal models of infection. Since sortase-attached proteins play such critical roles in colonization and virulence, the sortase enzyme has become a possible target for the development of new antibiotics.

Given the exposed nature of the surface adhesins and their putative roles in virulence, it might seem likely that they would be good targets for the immune system, yet people or animals infected with *S. aureus* do not become immune to reinfection, even though they make antibodies to *S. aureus* surface proteins. One explanation for this is suggested by the presence of antibodies to the **fibronectin-binding protein (FnBP)**. These antibodies do not block binding of the staphylococcal protein FnBP to fibronectin. Rather,

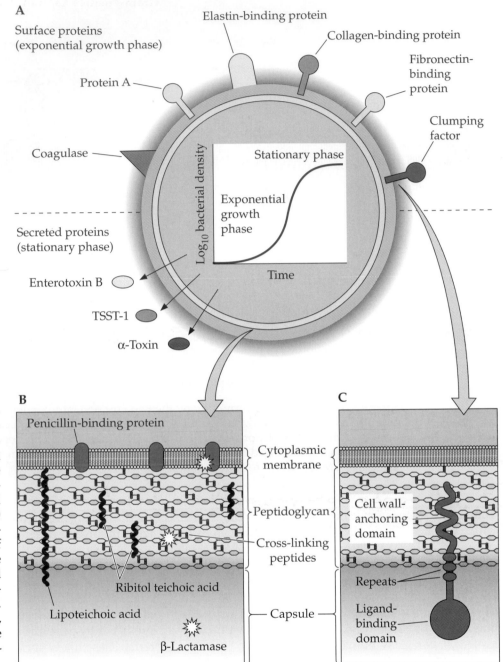

A

Surface proteins
(exponential growth phase)

Protein A

Coagulase

Secreted proteins
(stationary phase)

Enterotoxin B

TSST-1

α-Toxin

Elastin-binding protein

Collagen-binding protein

Fibronectin-
binding
protein

Clumping
factor

Stationary phase

Exponential
growth
phase

Log_{10} bacterial density

Time

B

Penicillin-binding protein

Cytoplasmic
membrane

Peptidoglycan

Cross-linking
peptides

Ribitol teichoic acid

Lipoteichoic acid

Capsule

β-Lactamase

C

Cell wall-
anchoring
domain

Repeats

Ligand-
binding
domain

Figure 18–4 Virulence factors of *S. aureus*. **(A)** Surface-expressed and secreted proteins as a function of growth phase. **(B)** Cross section of the cell envelope showing locations of various cell wall and membrane components. **(C)** Cross section of the cell envelope showing locations of the protein parts of the clumping factor. (Adapted from F. D. Lowy. 1998. *Staphylococcus aureus* infections. *N. Engl. J. Med.* **339:**520–532, with permission from the Massachusetts Medical Society.)

Figure 18–5 Four different classes of gram-positive bacterial sortases based on sequence homology and cellular function. **(A)** Class A enzymes (SrtA, or housekeeping sortases) are responsible for the cell wall anchoring of proteins that are involved in bacterial adhesion, immune evasion, or internalization or function as receptors for phage binding. Class B enzymes, SrtB, anchor proteins that are specifically involved in iron acquisition to the cell wall envelope. Class C enzymes, SrtC, assemble pili on the surfaces of gram-positive bacteria, whereas class D sortases anchor proteins to cell wall peptidoglycan as bacilli or streptomyces cells engage in sporulation, a developmental program that generates dissimilarly sized daughter and mother cells. Recognition motifs for each class of sortase are highlighted. CM, cell membrane; CW, cell wall; OM, outer membrane; SC, spore coat; ES, exosporium. **(B)** Cell wall of *S. aureus*. The repeating disaccharide *N*-acetylmuramic acid-(1-4)-*N*-acetylglucosamine (GlcNAc-MurNAc) is amide linked to the alanine of the pentapeptide [L-Ala-D-iGln-L-Lys-(Gly5)-D-Ala]. Its pentaglycine cross-bridge is linked to the carboxyl group of D-Ala at position 4 of a neighboring cell wall tetrapeptide. The amino group of pentaglycine cross-bridges is also the site of sortase-mediated anchoring of surface proteins (arrow). iGln, isoglutamine. (Adapted from Maresso and Schneewind, 2008, with permission from the American Society for Pharmacology and Experimental Therapeutics.)

A Sortase classes

SrtA
Adhesion
Immune evasion
Internalization
Phage recognition

SrtB
Iron acquisition

SrtC
Pilus formation

SrtD
Spore formation

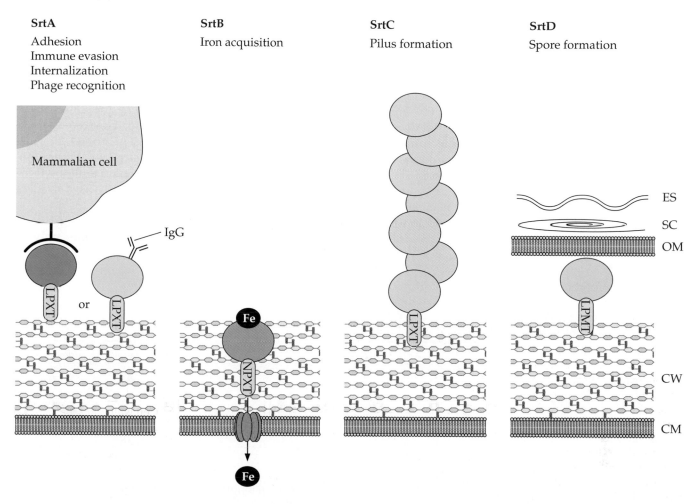

B

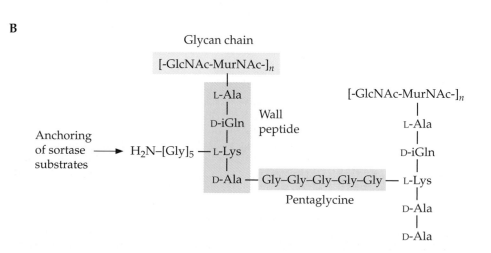

they bind to FnBP after it has attached to fibronectin. This sort of misdirecting of the antibody response could be a protective mechanism for the bacteria.

Sortase-attached MSCRAMMs also play roles in the molecular mimicry that camouflages the bacteria so they are not recognized by the immune system. **Protein A** binds the Fc portion of the IgG antibodies (Figure 18–6A; see Figure 11–14), so that this portion of the antibody cannot bind receptors on phagocytic cells. That is, the antibody bound to protein A is in the wrong orientation for opsonizing the bacteria. Moreover, human antibody molecules that do not induce an immune response now coat the surface of *S. aureus*. This is another type of misdirection that helps the bacteria to evade the immune response.

The IgG-binding property of protein A, the IgG-binding protein, has proven useful to scientists in a variety of fields. If a scientist is interested in whether a certain protein, called X in this example, interacts with another protein, called Y, in a prokaryotic or eukaryotic cell, one approach is to mix cell extracts with antibodies to protein X. Then, agarose beads coated with *S. aureus* protein A are added. The protein A traps the Fc portion of the antibodies bound to protein X and any other protein bound to protein X, including protein Y (Figure 18–6B). The beads are then allowed to settle or are briefly centrifuged at low speed and are washed to remove all unbound protein. If protein Y is bound to protein X, it will be trapped on the bead, along with protein X. The bound proteins can be eluted from the beads by harsher conditions, separated by gel electrophoresis, and identified by mass spectrometry. This process is referred to as **immunoprecipitation** (also sometimes called a pulldown assay), and variations of it are commonly used to identify protein-protein-interacting partners.

However, not all surface adhesins of *S. aureus* are proteins. At least one is a polysaccharide, **poly-*N*-**

Figure 18–6 Binding of IgG by protein A. **(A)** Protein A binds to IgG through the Fc region rather than the antigen-binding sites in the normal way. **(B)** Use of protein A-conjugated agarose beads to harvest IgG that is bound to other proteins through the antigen-binding sites.

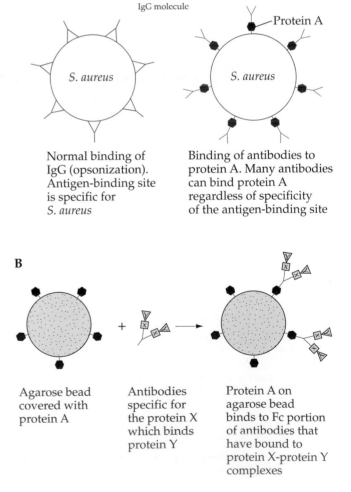

succinyl-β-1,6-glucosamine (PNSG) (Figure 18–7). PNSG has attracted attention because it is produced in vivo during an infection. Indeed, at least one study has found that antibodies against it are protective. All strains of *S. aureus* tested thus far produce this surface polysaccharide. It is also expressed by *S. epidermidis* strains, a fact that has made it attractive as a possible vaccine candidate. The genes *(icaA, icaB, icaC,* and *icaD)* encoding the enzymes that synthesize this polysaccharide are found in the *ica* operon (for *intercellular adhesin* locus). The PNSG adhesin allows bacteria to adhere to each other and may also promote adherence of the bacteria to other molecules, such as extracellular matrix components. PNSG may also provide protection against IgG antibodies by creating a physical barrier over the cell surface and cationic **antimicrobial peptides (AMPs).** The utility of this antigen as a vaccine target, however, remains to be proven.

At least three-fourths of *S. aureus* strains have a polysaccharide microcapsule (Figure 18–7). This poly-γ-glutamic acid (PGA) capsule is distinct from the PNSG involved in adherence and biofilm formation. In fact, the capsule interferes with binding to some cell types. This capsule is thought to limit phagocytosis. Recent studies have shown that a conjugated form of the capsular polysaccharide may have promise as a vaccine that could provide hope for preventing MRSA infections. About two-thirds of MRSA strains have one of these two capsule serotypes, a feature that makes the conjugated-vaccine approach more attractive.

S. aureus has not generally been considered to be an intracellular pathogen, because in pathology specimens it is always seen outside of cells. However, *S. aureus* strains can adhere to, invade, and grow within tissue culture cells, including endothelial cells. Invasion of cells might contribute to the ability of *S. aureus* to enter the bloodstream. Although bacteria do not have to be invasive to enter the bloodstream, since just the ability to produce local inflammation can create enough damage to endothelial cells to allow access to a blood vessel, the ability to invade and pass through endothelial cells could facilitate transit from tissues to blood. Thus, invasion of cells may play a role in dissemination of the bacteria in the body.

ANTIBIOTIC RESISTANCE. It is impossible to discuss *S. aureus* without going more deeply into the previously mentioned problem of antibiotic resistance. The first report of penicillin resistance in *S. aureus* was published about a year after the first use of penicillin

Figure 18–7 Factors that contribute to colonization and pathogenesis of *S. epidermidis*. In animal models, only the roles of PNSG, PGA capsule, and the MSCRAMM SdrG in infection have been determined. Other roles are based on in vitro experiments and environmental challenges during colonization and infection. (Adapted from Otto, 2009, with permission from Macmillan Publishers Ltd.)

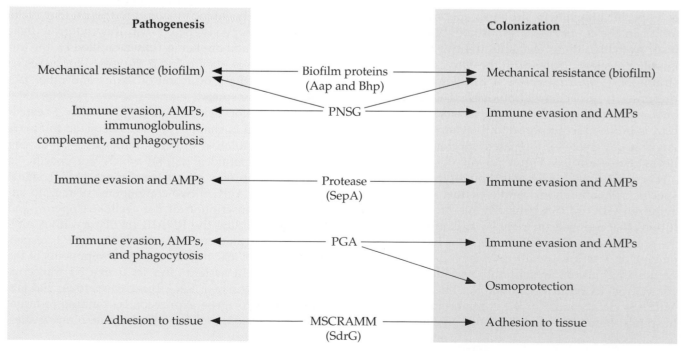

to treat human infections (see chapter 15 for the mechanism by which penicillin kills bacteria). It is likely that penicillin-resistant *S. aureus* strains were around even earlier. Ever since, strains of *S. aureus* have become resistant to more and more antibiotics, until some strains are only treatable with vancomycin or, more recently, two new antibiotics: a combination of two streptogramins (Synercid) and another class of protein synthesis inhibitors, the linezolids (e.g., Zyvox). The streptogramins have been known as protein synthesis inhibitors that prevent translocation of the ribosome (see chapter 15). Linezolids constitute a completely new class of protein synthesis inhibitor that seems to bind to 23S rRNA and interfere with the translation initiation process. Unfortunately, *S. aureus* resistance to linezolid was reported only about 1 year after the drug was first introduced in 2000.

MRSA strains use a form of regulated resistance to β-lactam antibiotics, such as methicillin. β-Lactam antibiotics induce the expression of a novel penicillin-binding protein **(PBP2a)** encoded by the *mecA* gene (see chapter 16). PBP2a still carries out the transpeptidation reaction required for peptidoglycan cross-linking but is insensitive to β-lactam antibiotics. *mecA* and the genes that regulate its expression reside on a mobile genetic element called a **staphylococcal cassette chromosome, *mec* (SCC*mec*)**. This element contains an insertion sequence linked to the *mecA* gene (IS*431mec*). SCC*mec* also includes genes for a site-specific recombinase complex *(ccr)* and contains recognition sequences for the recombinase enzyme, which catalyzes the insertion and excision of the element from the chromosome. SCC*mec* elements insert at a specific attachment site in an open reading frame *(orfX)* in the *S. aureus* chromosome, but the mechanism by which these elements transfer between cells is still not clear. Instead of the *mecA* system, some strains of *S. aureus* express a β-lactamase **(BlaZ)** in response to β-lactam antibiotics. The β-lactamase breaks down the antibiotic. Expression of *mecA* and *blaZ* is induced in response to β-lactam antibiotics by an unusual protease-mediated mechanism that relieves repression (see Figure 16–6).

For some time, MRSA was thought to be a hospital-specific phenomenon, and victims of **hospital-acquired (HA)** MRSA generally had underlying conditions that reduced their ability to combat infections. As noted above, CA-MRSA infections have started to emerge as a serious health problem. CA-MRSA often occurs in children, which was a group not usually considered to be at high risk for MRSA infections. MRSA is clearly loose in the community as well as in hospitals. Before the discovery of penicillin, *S. aureus*

was a common cause of death in surgical patients and soldiers with battle wounds. The arrival of virtually untreatable *S. aureus* strains might lead hospitals to curtail surgical procedures that are not needed to respond to medical emergencies. People have become used to having access to a range of surgical options, such as bypass surgery or knee surgery, which improve the quality of life but are not essential. How people would respond to reduced access to such surgical options is not something health officials like to think about.

REGULATION OF VIRULENCE. Many of the *S. aureus* virulence genes, especially those encoding surface adhesins and exoenzymes, are regulated by a quorum-sensing system. When cells are in the early stages of growth, analogous to exponential growth, the adhesin genes are preferentially expressed (Figure 18–7). Once the bacteria enter late exponential phase and reach a high population density, adhesin production is decreased and exoenzyme production is increased. This progression makes sense because the exoenzymes from many bacteria that are all localized in the same place have a much greater effect than exoenzymes from a few isolated bacteria. *S. aureus* wound infections are often characterized by large pus-filled lesions. This is the kind of damage that is produced by multiple bacteria acting in concert. These regulatory patterns fit the general paradigm of bacterial pathogenesis proceeding by a progression from colonization to persistence to spread (see Figure 11–1). During colonization, the bacteria adhere and grow. As the bacteria are attacked by the first stages of the immune response and begin to run out of nutrients, they make a transition to persistence. When they outgrow their environment and are being fully assaulted by the immune responses, they turn off the genes for adhesins and increase the production of toxins and other factors that allow spread.

The Agr quorum-sensing system used by *S. aureus* was described in chapter 14 and is depicted in Figure 14–22. Accumulation of cyclic thiolactone-containing autoinducer peptides (AIPs), which are specific to each subgroup of *S. aureus* (see Figure 14–23), signal through the AgrCA two-component system to increase the transcription of the *agrBDCA* operon and the gene encoding the RNAIII regulatory RNA molecule from promoters designated P2 and P3, respectively. The *agrBDCA* genes encode components of the two-component system, and its increased transcription is autoregulated. RNAIII increases toxin and late virulence factor gene expression by binding to target mRNA molecules and freeing ribosome-binding sites, thereby allowing increased translation. Conversely, RNAIII binds to mRNAs encoding adhesins and other colonization factors and blocks translation, thereby decreasing expression and destabilizing these mRNAs. Besides the Agr system, *S. aureus* virulence gene expression is regulated by another transcriptional regulatory system, designated SarA, which also stimulates transcription of the *agrBDCA* operon and the RNAIII gene. However, the mechanism of this activation remains unclear. The *sarA* gene is transcribed from multiple promoters by RNA polymerase containing the standard σ^{70} subunit or the stress-response σ^B subunit (see chapter 14). The availability of σ^B is controlled by a complicated anti-sigma factor and an anti-anti-sigma factor regulatory mechanism that responds to a variety of environmental stress conditions, but the important general point is that σ^B-dependent expression of *sarA* allows colonization and virulence gene expression to respond to cellular stress conditions in addition to culture density. Finally, several global transcriptional regulators couple virulence factor expression, biofilm formation, and antibiotic resistance to the host central metabolism. These regulators include CcpA (catabolite regulation), CodY (branched-chain amino acid supply and GTP), GlnR (nitrogen supply), Fur (iron supply), Rex (redox state), and MgrA/PerR/SarZ (oxidative state). The critical roles that physiology and metabolic regulation play in virulence are only beginning to be studied and understood.

S. epidermidis: an Accidental Pathogen

LOCATION AND CHARACTERISTICS. *S. epidermidis*, as its name suggests, is found mainly on human skin. Unlike *S. aureus*, which colonizes only a fraction of humans at any given time, virtually everyone's skin is colonized by *S. epidermidis*, and it is likely that *S. epidermidis* colonization may play a role in protecting the skin from unwanted pathogenic species, such as *S. aureus*. Unlike *S. aureus*, *S. epidermidis* strains lack the enzyme coagulase that converts fibrinogen to fibrin. This property has been used in traditional clinical microbiological tests to distinguish between the two species. In addition, colonies of *S. epidermidis* are chalky white, unlike the golden colonies of *S. aureus*. They are sadly familiar to many microbiologists as irritating contaminants of supposedly pure cultures of other bacteria, such as *Escherichia coli*. Given its location site, it is perhaps not surprising that *S. epidermidis* is such a common contaminant of laboratory cultures. If anything, *S. epidermidis* is even hardier than *S. aureus* and can survive for long periods in the external environment. Hardiness, existing ubiquitously in nature, and the ability to form robust biofilms (Figure 18–8) are properties that have allowed *S. epidermidis* to become an accidental pathogen that infects implanted medical devices in humans.

The common use of implanted medical devices has turned bloodstream infections caused by *S. epidermidis* into a serious and costly problem. Although plastics are a comparatively new human invention, both *S.*

Figure 18–8 Steps in biofilm formation in *S. epidermidis*. (Adapted from Otto, 2009, with permission from Macmillan Publishers Ltd.)

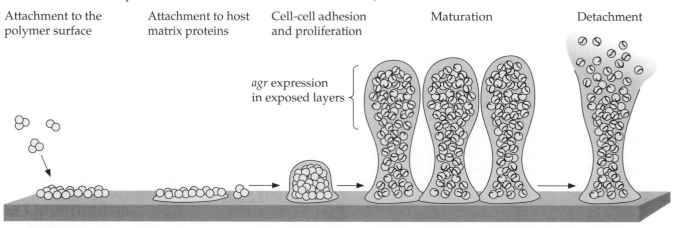

Attachment to the polymer surface	Attachment to host matrix proteins	Cell-cell adhesion and proliferation	Maturation	Detachment

agr expression in exposed layers

| Polymer surface: hydrophobicity, AtlE, Aae, and teichoic acids | Host matrix proteins: SdrF, SdrG, SdrH, Ebp, AtlE, and Aae | PNAG, teichoic acids, Bap, and Aap | PSMs? Proteases? | |

aureus and, especially, *S. epidermidis* have taken to plastic medical devices, such as indwelling venous catheters and medical prostheses, like the proverbial duck to water. *S. epidermidis* binds to plastic catheters and forms biofilms that extend into tissue and ultimately shed bacteria into the bloodstream, leading to sepsis. *S. epidermidis* is introduced into the human body largely through contamination from the skin. This contamination can occur on medical devices that are about to be implanted; during procedures that involve implanted devices, such as dialysis; or during surgery if gloves become compromised by small tears, especially during cardiac surgery, when there are many sharp segments of bone and metal sutures. Some conditions, such as heart valve damage from rheumatic heart disease, can provide sites for bacterial colonization even in the absence of implants.

Once the bacteria have established a biofilm on the damaged valve or valve implant, it is very difficult to eliminate them using antibiotics, because sessile bacteria in biofilms are resistant to antibiotics and *S. epidermidis* strains often carry resistance genes to a number of antibiotics. In the case of contaminated plastic heart valves, it is often necessary to undertake an operation to remove the valve, then treat the patient with antibiotics, and finally perform another operation to introduce a new, hopefully sterile valve. A strategy for preventing infections by plastic-loving bacteria like *S. epidermidis* is to impregnate catheters and valve implants with antibiotics or other bactericidal compounds, such as colloidal silver. This approach seems to be helping by preventing biofilm formation, but it is not a panacea. Prevention of contamination in the first place through good hygienic practices, such as hand washing, proper gloves, and careful scrutiny of indwelling catheters, is key to combating infection with *S. epidermidis*.

Virulence factors. Unlike *S. aureus*, *S. epidermidis* is a mild-mannered commensal bacterium that encodes only a single type of peptide that causes inflammation and could be considered a weak toxin. There is seemingly no endless list of superantigens and other nasty agents that can do severe harm to the human host. Recent progress in characterizing the factors used by *S. epidermidis* to colonize the skin strongly supports the model that these same factors function aberrantly when *S. epidermidis* finds itself as a contaminant on an implanted medical device inside the body (Figure 18–7). To colonize the skin, *S. epidermidis* needs to adhere and resist mechanical disruption, to deal with high salt concentrations and osmotic pressure, and to fend off the innate immune defenses of the skin, such as AMPs, and those provided by the

skin-associated lymphoid tissue (see chapter 2). These same mechanisms allow *S. epidermidis* to propagate and persist on mechanical devices and to evade the host immune system.

Osmoprotection of *S. epidermidis* occurs by several mechanisms. *S. epidermidis* encodes eight sodium ion/proton exchangers and six transport systems that take up osmoprotectants. In addition, *S. epidermidis* also produces an exopeptide capsule-like polymer that consists of PGA, which is induced by high salt concentrations and seems to provide osmoprotection. PGA, which is not synthesized by *S. aureus*, also contributes to biofilm formation and protects the bacteria against AMPs and phagocytosis by neutrophils. *S. epidermidis* biofilms seem to be mixtures of adhesive proteins, exopolysaccharides, and released nucleic acids. The formation of these biofilms involves the participation of numerous MSCRAMMs linked by sortase to the *S. epidermidis* surface (Figure 18–8).

The bacteria first attach to surfaces through hydrophobic interactions using abundant surface proteins that function as cell wall hydrolases. Teichoic acids (TAs) attached to the peptidoglycan may also interact with surfaces. Biofilm formation also involves MSCRAMMs that bind to collagen (SdrF), fibrinogen (SdrG), and other extracellular matrix proteins. Intercellular aggregation then starts to occur, mediated by exopolysaccharide adhesions, such as poly-*N*-acetylglucosamine (PNAG) homopolymer and biofilm-associated adhesion proteins (Bap and Aap). Like PGA, PNAG also protects *S. epidermidis* from AMPs and phagocytosis. At some point after the biofilm has matured into its final structure, it begins to detach and release planktonic cells. It is noteworthy that biofilm detachment is regulated by the Agr quorum-sensing system (see Figure 14–22). Agr activation leads to the production of proteases and short amphipathic **phenol-soluble modulins (PSMs),** which is one of the few types of molecules produced by *S. epidermidis* that can be considered a weak toxin. PSMs can induce proinflammatory cytokines and sometimes have cytolytic functions that may act on neutrophils. Various types of PSMs are also produced by *S. aureus* strains and have been linked to some of the virulence properties of CA-MRSA.

Many colonization factors of *S. epidermidis* are involved in destroying AMPs produced by the skin (Figure 18–7). This is such an important defense mechanism that *S. epidermidis* uses a specific regulatory system to respond to AMPs. Cationic AMPs are sensed by the ApsS histidine kinase (sensor kinase), which then phosphorylates the AspR response regulator (Figure 18–9). Phosphorylated AspR turns on genes that encode Dlt system proteins and the MprF

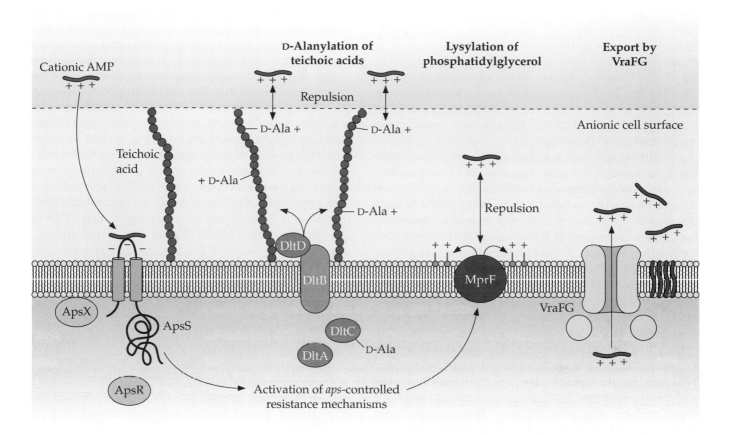

Figure 18–9   The AMP sensor Aps regulatory system in *S. epidermidis*. Cationic AMPs attach to the negatively charged bacterial surface and membrane by electrostatic interactions, a prerequisite for AMP antimicrobial activity that is often based on pore formation in the bacterial cytoplasmic membrane. The *S. epidermidis* ApsS AMP sensor has one short extracellular loop with a high density of negatively charged amino acid residues that interact with cationic AMPs. Transduction of this interaction signal through ApsS and the essential accessory ApsX, which has an unknown function, triggers the expression of key AMP resistance mechanisms. The D-alanylation of TAs, which is carried out by the products of the *dlt* operon, and lysylation of phosphatidylglycerol, which is catalyzed by the MprF enzyme, result in the decreased negative charge of the cell surface and membrane, respectively, leading to decreased attraction or repulsion of cationic AMPs. The VraF and VraG ABC transporter also promotes resistance to AMPs and probably functions as an AMP exporter. Gray shading represents negative charge, and blue shading represents positive charge. (Adapted from Otto, 2009, with permission from Macmillan Publishers Ltd.)

protein, which add D-alanine groups to surface TAs and lysine groups to phospholipids in the cell membrane. These modifications increase the positive charge of the cell surface, which repels cationic AMPs from reaching the bacterial cell membrane. In addition, *S. epidermidis* produces efflux pumps, such as the VraFG ABC transporter, that pump AMPs out of the cytoplasmic membrane, and several secreted proteases, such as SepA, that probably play roles in degrading AMPs.

COMPARISONS BETWEEN S. EPIDERMIDIS AND S. AUREUS.   It is instructive to compare some shared properties of *S. aureus* and *S. epidermidis*. Both species use the Agr quorum-sensing system and produce species-specific AIPs (see Figures 14–22 and 14–23). Notably, induction of the Agr system in *S. aureus* leads to production of a full range of toxins, whereas induction in *S. epidermidis* leads to production of PSMs and no powerful toxins. The *S. epidermidis* AIPs inhibit the Agr systems of most subgroups of *S. aureus*, whereas the AIP from only one rare subgroup of *S. aureus* inhibits the Agr system of *S. epidermidis*. Thus, *S. epidermidis* seems to win the interference battle against *S. aureus,* and it has been speculated that this property may contribute to *S. epidermidis* predominating in skin colonization over *S. aureus*. In fact, *S. epidermidis* may play such an important role in keeping *S. aureus* in check that its elimination from the skin could cause microbiota shift diseases.

Another interesting comparison between these two species concerns the prevalence of SCC*mec* cassettes in *S. epidermidis* compared to *S. aureus*. Nearly every clinical isolate of *S. epidermidis* isolated in the United States and some other countries is methicillin resistant and carries the SCC*mec* cassette. Furthermore, methicillin-resistant *S. epidermidis* almost always carries plasmids or other elements imparting resistance to many other antibiotics, including tetracyclines, aminoglycosides, and macrolides. Thus, methicillin-resistant *S. epidermidis* is more endemic in hospitals than MRSA and is very hard to treat, because it forms antibiotic-resistant biofilms on implanted devices. At this stage, only vancomycin can be used with reasonable success to treat catheters infected with *S. epidermidis,* but strains with intermediate resistance to vancomycin have begun to appear. Taking these data together, it seems that *S. epidermidis* acts as a dangerous reservoir for the accumulation of antibiotic resistance genes in *Staphylococcus* spp., and it has been suggested that transfer of SCC*mec* from an *S. epidermidis* isolate contributed to the emergence of CA-MRSA strains. Interestingly, this route of genetic transfer may be unidirectional from *S. epidermidis* to *S. aureus*. This would explain why *S. epidermidis* has not picked up toxin genes from *S. aureus*. The basis for this unidirectional transfer may be the presence of a molecular defense mechanism (called **CRISPR,** for **clustered regularly interspaced short palindromic repeats**) against genetic exchange that is present in *S. epidermidis* but absent from *S. aureus*. CRISPR are direct repeats of 24 to 48 bp that are found scattered throughout the chromosome in many bacteria and that function as a quasi-immune system to confer resistance to exogenously introduced genetic elements, such as plasmids or bacteriophages, by targeting them for degradation.

S. pneumoniae: a Commensal Nicknamed "the Captain of All the Men of Death"

LOCATION AND CHARACTERISTICS. *S. pneumoniae* (also called pneumococcus) is arguably the most common cause of deaths due to infectious disease in the world. Annually, more than 1.2 million infants die worldwide of *S. pneumoniae*-related diseases, and the number of at-risk adults who die each year probably exceeds the number of infants. Thus, well over 2 million people perish each year from pneumococcal invasive diseases, a staggering number that may be increasing. *S. pneumoniae* has played important roles in the history of molecular biology and epidemics. Oswald Avery (1877–1955), Michael Heidelberger (1888–1991), and their coworkers were the first to discover that the *S. pneumoniae* capsule was an exopolysaccharide and immunogenic. Avery, along with his colleagues Colin MacLeod (1909–1973) and Maclyn McCarty (1911–2005), later used the natural transformation of *S. pneumoniae* to demonstrate that DNA is the genetic material (see Figure 7–6). Therefore, *S. pneumoniae* was one of the first model bacterial systems used to study fundamental molecular genetic processes. Part of Avery's intense interest in *S. pneumoniae* stemmed from the fact that pneumococcal pneumonia is a common lethal secondary infection of influenza. More than half of the people who died in the great 1918 influenza epidemic died of invasive pneumococcal disease.

S. pneumoniae is a gram-positive, ovococcus-shaped bacterium that is usually seen as a diplococcus or as short chains of cells. The deadliest diseases caused by *S. pneumoniae* are pneumonia, bacteremia, and meningitis, but *S. pneumoniae* is also a common cause of the less serious childhood disease earache (otitis media). *S. pneumoniae* colonizes the nasopharynx, which extends from the base of the skull to the upper surface of the soft palate, in 10 to 20% of adults and 40% of young children. The colonization frequency may exceed 60% for infants in day care. Similar to *S. aureus,* there are numerous subgroups of *S. pneumoniae* strains that are usually grouped by the composition of their exopolysaccharide capsules. There are 91 different serotype strains with different capsules, but these subgroups do not simply differ in the sugar compositions of their capsules, and recent genomic comparisons have shown that the genomes of different serotype strains may differ by as much as 10%, mainly at numerous sites of diversity scattered around the bacterial chromosome. Many of these sites contain pathogenic islands that include genes that mediate invasive diseases. There is even genetic diversity within pneumococcal serotype strains. In addition, *S. pneumoniae* genomes carry a large number of complete and defective insertion sequence transposons, including conjugative transposons. Since *S. pneumoniae* is naturally transformable, there is considerable genetic plasticity among *S. pneumoniae* subgroups and even between related species, such as *Streptococcus mitis*. *S. pneumoniae* is a prime example of a bacterium with a dynamic pangenome and the capacity for rapid genetic change.

The duration of colonization depends strongly on the serotype subgroup and usually lasts for weeks in adults and months in children. In most cases, colonization is asymptomatic or may result in a mild runny nose. Colonization is the major reservoir for transmission of *S. pneumoniae* among humans, who are its only known host (with the exception of horses for one particular serotype). Interestingly, most people are usually colonized by only one serotype of *S. pneumoniae* at a time. This intraspecies competition is partly mediated by the production of bacteriocins called pneumocins. Quorum sensing through a two-component system regulates the production of these small, serotype-specific antibacterial peptides. Each serotype strain that secrets a specific pneumocin also produces proteins that impart immunity to that pneumocin, thereby conferring self-protection.

Colonization has also been thought to be a prerequisite for invasive diseases. Major risk factors include age (younger than 2 years or older than 65 years); debilitation due to conditions such as poverty or alcoholism; untreated chronic diseases, such as diabetes; immunosuppression due to immune system defects or infections (e.g., human immunodeficiency virus [HIV]); and antecedent viral respiratory infections, especially influenza. As the worldwide incidence of diabetes, HIV infection, and influenza continues to increase, so does the incidence of serious pneumococcal infections.

The risk of invasive disease also depends on the serotype. Some serotypes are benign commensals that are not linked to invasive diseases, whereas other serotypes can often lead to invasive disease. As with *S. aureus,* a disturbing recent realization is that some colonizing serotype strains can progress to invasive diseases in healthy individuals without obvious risk factors. As might be expected, the type of invasive disease is also correlated with different serotype subgroups. With the exception of pneumonia, invasive pneumococcal diseases, such as otitis media, bacteremia, and meningitis, do not lead to transmission among hosts. Pneumococcal pneumonia is not considered contagious enough in the hospital setting to require isolation. On the other hand, large and, more commonly, small community outbreaks of pneumococcal pneumonia have been reported, consistent with transmission through coughing.

VACCINES. Once *S. pneumoniae* reaches the lungs, it encounters a major defense of the lung, the alveolar macrophages. Later, as infection progresses, neutrophils also enter the area. The capsule is the main defense of *S. pneumoniae* against phagocytosis by the cells that protect the lung. The oldest and most widely used method for prevention of bacterial pneumonia due to *S. pneumoniae* is a vaccine that consists of the 23 most common antigenic capsular types. This 23-valent capsular vaccine has proven to be very safe and has been routinely given to the elderly, who are at greatest risk of contracting pneumonia. Unfortunately, its efficacy, especially among the elderly, is only around 60%. In addition, the 23-valent pneumococcal vaccine is useless in infants, who lack T-cell-independent responses to generate antibodies to long-chain polysaccharides.

Given the fact that influenza is frequently a precursor to secondary *S. pneumoniae* lung and invasive infections, a second preventive strategy has long been in place—urging, or even requiring, health care workers who care for the elderly to have an anti-influenza vaccination. Similarly, people caring for elderly residents of nursing homes have been urged to accept an influenza vaccination in order to build a shield of protection (so-called **"herd immunity"**) around the elderly residents. During 2009, the issue of requiring influenza vaccination of health care workers resurfaced in connection with the swine flu (H1N1) pandemic. Unexpectedly, given the willingness of health care workers in past years to accept an influenza vaccine in order to protect their patients, a small number of vocal health care workers made the H1N1 vaccine a personal-freedom issue. This was a debatable action, because the H1N1 vaccine is relatively safe, and over half of the people who died in the influenza pandemic died of secondary pneumococcal invasive disease.

To circumvent the problem with the simple capsule vaccines in young children, a protein conjugated-capsule vaccine was introduced in 2000 in developed countries. Recall that covalently linking carbohydrates to protein segments can induce a major histocompatibility complex type II-mediated T-cell-dependent response to the carbohydrates, even in infants. However, conjugated vaccines are expensive to manufacture, and the capsule types covered had to be limited to the seven most prevalent serotypes recovered in clinical isolates from developed countries at the time. **PCV-7 (pneumococcal conjugated seven-valent vaccine)** has been highly successful in the United States and other developed countries (Figure 18–10). Invasive pneumococcal diseases have dropped about 70 to 80% for infants under 2 years of age, and most infants in the United States now routinely receive this vaccine, which has been a "blockbuster" product for the pharmaceutical company that produced it. Herd immunity has also likely protected unvaccinated infants in crowded day care settings. Unfortunately, the serotypes that cause invasive pneumococcal diseases in developed countries only partly

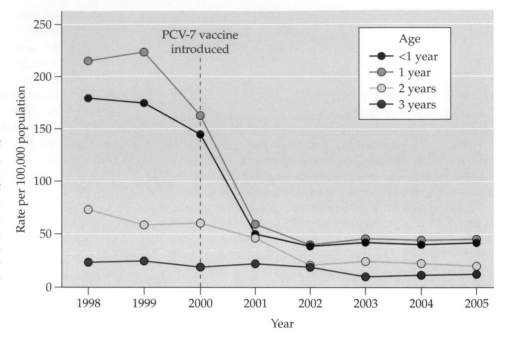

Figure 18–10 Incidence of pneumococcal disease in children in the United States before and after introduction of the PCV-7 vaccine. (Adapted from Centers for Disease Control and Prevention. 2008. Invasive pneumococcal disease in children 5 years after conjugate vaccine introduction—eight states, 1998–2005. *MMWR Morb. Mortal. Wkly. Rep.* **57**:144–148.)

overlap those in developing countries, and conjugated vaccines are often too expensive for use in the developing world. Alarmingly, capsule serotype strains not covered by PCV-7 have already started to appear frequently as the cause of invasive pneumococcal diseases in vaccinated infants, and some of these emerging strains are already resistant to multiple antibiotics. To stay ahead of this onslaught, a new conjugated 13-valent vaccine covering six more capsule serotypes than PCV-7, including those that have commonly emerged as a result of the previous vaccine, was recently approved. This development should hold infant pneumococcal infections at bay for several years, at least in developed countries, until some of the other 78 capsule serotype strains slip past PCV-13. Meanwhile, a search continues for pneumococcal surface proteins that can serve as vaccine candidates for eliciting a protective immune response. One challenge has been that many pneumococcal surface proteins show variability in different serotype strains. Recently, two highly conserved surface proteins involved in cell division (PcsB) and cell signaling (serine-threonine kinase, StkP) were identified as strong antigens in humans that may act as new vaccine candidates.

COLONIZATION AND VIRULENCE FACTORS. Before discussing some aspects of pneumococcus colonization and infection, it is instructive to consider the genetic composition of *S. mitis,* which is the closest relative of *S. pneumoniae. S. mitis* is a commensal inhabitant of hard surfaces in the oral cavity and is a member of the oral microbiota, but unlike *S. pneu-*

moniae, S. mitis is a well-behaved commensal that does not cause widespread invasive diseases. Recent genomic analyses suggest that *S. pneumoniae* evolved from an ancient *S. mitis* progenitor. Consistent with this idea, the *S. pneumoniae* genome contains a fairly large number of truncated genes that are still intact in *S. mitis.* Loss of these functions may contribute to the effectiveness of *S. pneumoniae* in colonizing the nasopharynx and possibly occupying other niches during invasive infections. Conversely, *S. pneumoniae* has acquired several key genes that are not present in *S. mitis,* including genes that encode numerous sugar transporters and several accessory factor genes, including operons that mediate biosynthesis of the all-important capsule, the pneumolysin toxin, and hyaluronate lyase (Hyl), which degrades the anionic polysaccharide, hyaluronic acid, in the extracellular matrix. Acquisition of these accessory factor genes has allowed *S. pneumoniae* to colonize the nasopharynx niche, and they are critical to invasive pneumococcal diseases.

Colonization of the nasopharynx by *S. pneumoniae* starts with adherence to nasal surfaces. Isolates that effectively colonize seem to express less capsule material and form more transparent-looking colonies on some media than those isolated later in invasive disease, which seem to express more capsule material and to form more opaque-looking colonies. It has been proposed that switching between transparent and opaque colony phenotypes is a form of phase variation, but no genetic mechanism has been established for this phenomenon, and the tendency to switch colony appearances depends on the serotype and can be influenced by the expression of numerous genes. Part of this phenomenon may be downregulation of the genes encoding proteins that mediate capsule exopolysaccharide biosynthesis. In general, pneumococcal cells that produce less capsule material adhere significantly better to eukaryotic cells in culture and produce certain forms of biofilms more effectively. Nevertheless, a low level of capsule expression does seem to play a role in preventing entrapment in nasal mucus, so that the nonmotile *S. pneumoniae* cells reach the epithelial surfaces of cells lining the nasopharynx.

Several mechanisms allow *S. pneumoniae* cells to adhere to the apical surfaces of epithelial cells (Figure 18–11). Some serotypes produce pili that are covalently attached to peptidoglycan by a special sortase (Figure 18–5A). When present, these pili act as adhesins to epithelial cells. The TAs and LTAs are decorated with phosphocholine (ChoP) groups, which serve two purposes in *S. pneumoniae*. ChoP groups serve as sites of binding for a group of 10 to 15 secreted choline-binding proteins (CBPs) that play important roles in pneumococcal colonization and virulence. This mechanism of attaching surface proteins to the gram-positive surface is characteristic of *S. pneumoniae* and its relatives, such as *S. mitis*. ChoP groups on TAs and LTAs also bind to the platelet-activating factor receptor (PAFR) on the surfaces of epithelial cells. PAFR normally binds the cellular signaling molecule, PAF, which contains a ChoP group. Thus, the pneumococcal TAs and LTAs use mimicry to make PAFR a surface receptor for adhesins. Besides protein adhesins, *S. pneumoniae* cells bind to carbohydrate groups on the eukaryotic cell surface, including *N*-acetylglucosamine-β-(1,3) or (1,4)-galactose, which is attached to the surfaces of glycosphingolipids in epithelial cell membranes. *S. pneumoniae* also encodes several exoglycosidases that remove sugar residues from host glycoproteins. Besides providing the bacteria with a source of food, these hydrolytic enzymes, such as neuraminidases (NanA) that release sialic acid, may expose new binding sites for adherence of *S. pneumoniae* cells. Again, exoglycosidases are covalently linked to the peptidoglycan by sortases.

S. pneumoniae is primarily an extracellular pathogen that causes disease by inducing severe inflammatory responses. However, *S. pneumoniae* has ways to breach epithelial cell layers and reach host basal surfaces and the bloodstream. One of the CBPs (CbpA) on the pneumococcal surface binds to a segment called secretory component in the polymeric Ig receptors that transport sIgA from the basal surfaces to the apical surfaces of epithelial cells, where it is then released into the lumen of the gut. Binding of the

bacterium via CbpA to this receptor when it is exposed on the apical side seems to be one way for *S. pneumoniae* cells to transcytose (i.e., hitch a ride) when the receptor returns to the basal surface for reloading. Another more straightforward way for *S. pneumoniae* to breach epithelial cell layers is to slip through openings made to allow neutrophils to enter the nasal cavity to combat the bacterial infection. This process of occult bacteremia (i.e., the presence of bacteria in the bloodstream of febrile patients with no apparent site of infection) can allow *S. pneumoniae* cells to reach the basement membrane layer and to seed the bloodstream. After *S. pneumoniae* has reached the extracellular matrix, several other surface proteins act as adhesins. Proteins PavA and PavB bind to fibronectin. **Hyaluronate lyase (Hyl)** dissolves hyaluronic acid, thereby promoting spread and possibly providing food for the bacteria. Abundant glycolytic enzymes, such as enolase and glyceraldehyde phosphate dehydrogenase, possibly released by autolysis, bind to the pneumococcal surface and aid in binding to plasminogen.

At some point during colonization, *S. pneumoniae* cells release the soluble cytolytic toxin pneumolysin, which has several functions. Pneumolysin does not seem to be transported directly out of cells and is likely released by the activity of cell wall hydrolases, such as the amidase LytA, which causes cell lysis, especially when cells reach stationary phase. Autolysis also releases inflammatory substances, such as peptidoglycan and LTA, which interact with Toll-like receptor 2 (TLR2) of the innate immune system (see chapter 3) and induce the release of proinflammatory cytokines and chemokines. Once released, pneumolysin assembles in the cholesterol-containing membranes of the host cells and produces large donut-shaped transmembrane pores. As the concentration of pneumolysin builds up in the membranes, the beating of cilia is inhibited, cytokine and chemokine release is further stimulated, and CD4$^+$ T cells and chemotaxis of immune cells are activated.

S. pneumoniae also secrets copious amounts of hydrogen peroxide, which can stimulate cytokine production by surrounding host cells and possibly inhibit the growth of competing microbial species for colonization of the nasopharynx. This endogenous hydrogen peroxide is mostly produced by the enzyme **pyruvate oxidase (SpxB),** which is involved in synthesizing additional ATP from pyruvate in this energy-challenged bacterium, which lacks an electron transport system and a citric acid cycle. Several genes protect *S. pneumoniae* from this endogenously produced hydrogen peroxide, including those encoding the PsaBCA transporter, which takes up manganese

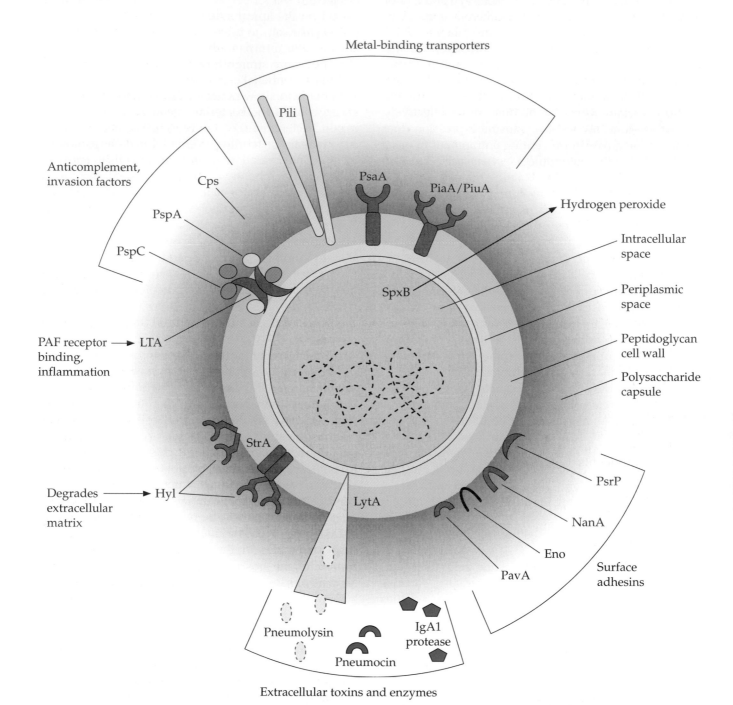

Figure 18–11 Virulence factors of *S. pneumoniae*. PsaA, pneumococcal surface antigen A; PiaA/PiuA, pneumococcal iron acquisition and uptake; PsrP, pneumococcal serine-rich repeat protein; NanA, neuraminidase; Eno, enolase; PavA, pneumococcal adhesion and virulence; LytA, autolysin; StrA, sortase A; Hyl, hyaluronate lyase; PspC, pneumococcal surface protein C; PspA, pneumococcal surface protein A; Cps, polysaccharide capsule; SpxB, pyruvate oxidase. (Adapted from van der Poll and Opal, 2009, with permission from Elsevier.)

ions. In this transporter, PsaA is the extracellular lipoprotein receptor protein for manganese. On the other hand, *S. pneumoniae* encodes three different ABC transporters (Pia, Piu, and Pit) that take up iron, and the iron content per unit of cell mass is about the same as that in *E. coli*, which does not produce high endogenous levels of hydrogen peroxide. A full understanding of how *S. pneumoniae* survives and protects its DNA from endogenous hydrogen peroxide concentrations that approximate those produced by oxidative bursts in macrophages is lacking. In addition, the question of how *S. pneumoniae* competes with the myriad of other bacterial species in the nasopharynx and throat is only beginning to be studied. Experiments in a mouse model of colonization demonstrated that the host plays a role in determining the outcome of competition between two bacterial species. One species may stimulate innate immune responses by the host cell that do not eliminate it but tend to eliminate a competing bacterial species.

Within 3 days of colonization, the host strikes back with an influx of neutrophils to the lumen of the nasal cavity. *S. pneumoniae* strains that do not produce pneumolysin actually persist longer in the nasopharynx than the parent strain. This suggests that the bacteria intentionally induce inflammation, perhaps to further transmission even at the expense of being cleared more rapidly. Several other elements of the host defenses are stimulated by interactions with pneumococcal components (called PAMPs [pathogen-associated molecular patterns] [see chapter 3]) during colonization (Figure 18–12). Binding to PAFR not only allows adherence, but also sets off a G protein-signaling cascade in the host cells. Besides its cytolytic activity, pneumolysin binds directly to TLR4 and induces an innate immune response. TLR2 stimulation leads to expression by CD4$^+$ T cells of the specific cytokine interleukin 17 (IL-17), which recruits neutrophils, monocytes, and macrophages to the mucosal surface to clear pneumococcus from the nasopharynx. This antibody-independent involvement of CD4$^+$ T cells in controlling pneumococcal colonization provides an explanation for why people infected by HIV, which reduces CD4$^+$ T-cell numbers, are highly susceptible to persistent pneumococcal colonization and invasive disease. Eventually, the mucosal immune system starts producing sIgA antibodies directed against pneumococcal surface epitopes to further eliminate the bacteria. Another host defense molecule, called C-reactive protein, binds to ChoP in the pneumococcal TAs and LTAs. This complex then binds the C1 component and activates the classical complement pathway (see chapter 3).

S. pneumoniae resists this onslaught from the host, at least temporarily. The capsule inhibits opsonization by sterically inhibiting the binding of bound complement factors and Ig molecules to receptors on neutrophils and macrophages. The capsule reduces the deposition of complement factors on the bacterial surface and the capture of pneumococcal cells in extracellular traps produced by neutrophils. CBP PspA protrudes through the capsule and blocks the binding of complement component C3b to factor B. PspA also binds host lactoferrin, which also protects the cells. Likewise, CBP PspC is multifunctional. Besides binding to the polymeric Ig receptor, it binds factor H, which prevents formation of C3b and activation of the alternative complement cascade. In addition, *S. pneumoniae* releases a protease that specifically cleaves human sIgA. The bacteria can also ward off host cationic AMPs by decorating its LTA with D-alanine residues (Figure 18–8). The pneumococcus protects its peptidoglycan from host lysozyme cleavage by modifying it through deacetylation of *N*-acetylglucosamine residues.

The same factors that play roles in colonization are important for invasive disease. Foremost among these is the capsule. Mutants that lack capsules are avirulent in mouse models of infection, even at extremely high bacterial doses. Pneumolysin is well established as a critical factor for invasive diseases. The contributions of other factors are more graded and seem to depend on the serotype of the strain and the mouse model of infection used. Several bacterial factors that strongly influence virulence do not affect colonization, but this may only reflect the limitation of the animal models available for studying colonization. *S. pneumoniae* bacteria that enter the lung are first met by the alveolar macrophages. If the infection is not brought under control, neutrophils are recruited and play a major role in phagocytosis. The alveolar macrophages then switch roles to clearing out apoptotic neutrophils.

Stimulation of the host innate immune system leads to a substantial increase in the production of proinflammatory cytokines and chemokines (Figure 18–13). Macrophages use several additional innate immunity signaling pathways besides the ones mentioned above. A receptor called MARCO (macrophage receptor with collagenous structure) is expressed on the surfaces of alveolar macrophages (Figure 18–12). This scavenger receptor binds to *S. pneumoniae* cells and aids their internalization, and mice deficient in the receptor are highly susceptible to fatal pneumococcal infections. In addition, pneumococcal cells internalized in endosomes release their chromosomal DNA, which sets off an innate immune response mediated by TLR9. Fragments of digested peptidoglycan that are released into the cytoplasm further stimulate the inflammatory response via the NOD (nucleotide-

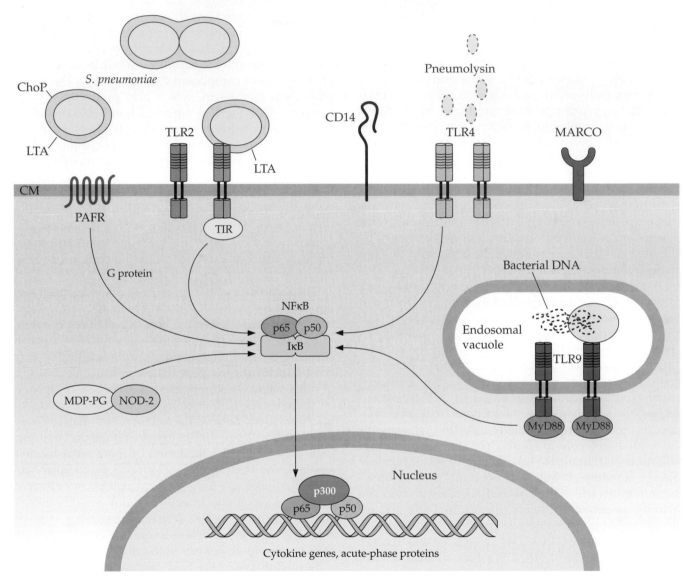

Figure 18–12 Pattern recognition signaling receptors and pathways in pneumococcal infection. *S. pneumoniae* is recognized as a pathogen in the lung by several TLRs, including TLR2 (with pneumococcal LTA as its major ligand), TLR4 (which recognizes pneumolysin), and TLR9 (within endosomes; interacts with bacterial DNA). MARCO expressed by alveolar macrophages contributes to the innate immune response in the lungs. PAFR is shown as a pattern recognition receptor because it recognizes pneumococcal phosphorylcholine and LTA, thereby contributing to tissue invasion. CD14 probably further helps *S. pneumoniae* invade from the airways into the blood. Within the cytoplasm, the muramyl dipeptide component of pneumococcal peptidoglycan (MDP-PG) is recognized by NOD-2 and can activate host defense and inflammation. TIR, Toll–IL-1 receptor domain; MyD88, myeloid differentiation primary response protein 88. IκB, inhibitor κB. (Adapted from van der Poll and Opal, 2009, with permission from Elsevier.)

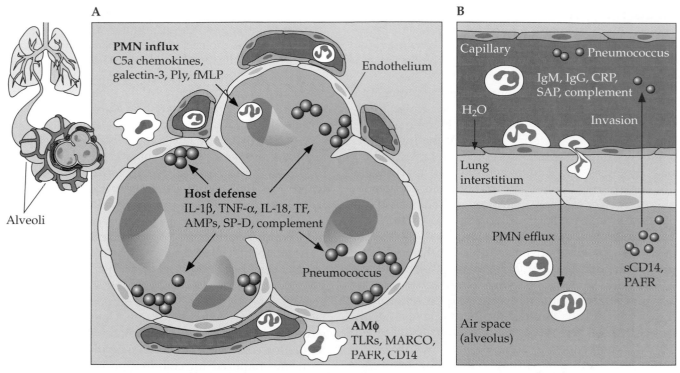

Figure 18–13 Steps in invasive *S. pneumoniae* infection. **(A)** Pneumococci that enter lower airways are recognized by pattern recognition receptors, including TLRs (on epithelial cells and alveolar macrophages) and MARCO (on alveolar macrophages). At low infectious doses, epithelial cells and alveolar macrophages can clear *S. pneumoniae* without help from recruited neutrophils, in part by the release of protective inflammatory mediators, such as IL-1, tumor necrosis factor alpha (TNF-α), IL-18, complement products, surfactant protein D (SP-D), and AMPs. These mediators continue to have a role after infection with a high infectious dose, whereby PMNs are recruited by chemoattraction through various mediators, including C5a and galectin 3, and pneumococcal products, such as pneumolysin (Ply) and formyl-methionine-leucine-phenylalanine (fMLP). **(B)** If alveolar defense mechanisms are overwhelmed by the multiplication of pneumococci, invasion of *S. pneumoniae* into the bloodstream takes place, helped by PAFR and CD14. In the bloodstream, several host proteins contribute to host defense, including natural IgM antibodies, complement components, C-reactive protein (CRP), and serum amyloid peptide (SAP). TF, tissue factor. AMϕ, alveolar macrophage. (Adapted from van der Poll and Opal, 2009, with permission from Elsevier.)

binding oligomerization domain) receptors. Other surface proteins, such as C-type (SIGNR1) lectins, on macrophages in the spleen, also combat pneumococcal infection. These lectins mediate the uptake of capsule and *S. pneumoniae* cells and the presentation of capsule antigens to B cells.

As alveolar macrophages become overwhelmed, signals are released to recruit an influx of neutrophils. These signals include C5a complement fragment, α-chemokines, a soluble adhesin called galectin, and pneumolysin (Figure 18–13). Some *S. pneumoniae* cells may escape to the bloodstream by slipping out of openings that allow neutrophil influx. Escape of *S.*

pneumoniae from the blood is facilitated by PAFR binding and a soluble recognition protein called cluster of differentiation 14 (CD14), which acts as a coreceptor, along with TLR2, for bacterial cell wall components. Some of these mechanisms contribute to the susceptibility of influenza patients to secondary pneumococcal pneumonia. Viral infection temporarily destroys the ciliated cells that guard the airway. Viral infections also increase the production of PAFR on lung and endothelial cells. The ability of pneumococci to adhere to these cell types increases the chances of the bacteria attaching to lung and endothelial cells. Influenza also may impair some of the signaling pathways mediated by the TLRs, such as TLR9, that are important for phagocytosis capacity. Cytokines produced in response to influenza virus may also downregulate MARCO scavenger receptors on alveolar macrophages.

Many of the symptoms of pneumococcal pneumonia (fever and lung damage) can be accounted for by the intense and largely unrestrained inflammatory response caused by bacteria growing in the lung. Damage to endothelial cells allows blood to enter the lung and produces a common symptom of pneumonia, bloody sputum. Peptidoglycan fragments activate the alternative complement pathway and elicit IL-1 production by macrophages. LTA from the bacterial cell wall elicits cytokine production and is an effective activator of the alternative pathway. Most people produce antibodies to bacterial cell wall antigens, but not to capsular antigens, possibly because the capsular antigens are less antigenic. Antibodies to cell wall antigens can diffuse through the porous matrix of the capsule to the bacterial surface, where they bind and activate complement by the classical pathway. This contributes further to produce C5a, the neutrophil chemotactic factor, but does not aid phagocytic uptake of bacteria, because C3b bound to the cell wall cannot make contact with phagocyte receptors due to the physical interference of the capsule. The net result is a continually expanding inflammatory response that causes considerable tissue damage but does not clear the bacteria. The fluid that accumulates in the lungs as pneumonia develops is due to leakage from blood vessels as neutrophils move through vessel walls and from activated complement components that increase vascular permeability. Damage to the area also disrupts gas exchange so that the patient literally suffocates. Ever-decreasing amounts of oxygen do not inhibit pneumococcal growth, because *S. pneumoniae* is metabolically an aerotolerant anaerobe.

REGULATION. Colonization and virulence factor expression in *S. pneumoniae* do not seem to be regulated by a single master regulatory pathway, such as the Agr system in *S. aureus*. Moreover, the simplified paradigm that adhesin gene expression is followed by virulence gene expression as pathogen cultures progress from exponential to stationary phase does not easily apply to *S, pneumoniae*, partly because the laboratory cultures begin to undergo autolysis shortly after reaching stationary phase. Instead, specific regulators and several global regulatory systems, such as catabolite repression, stringent-response, and two-component systems, control the genes that mediate pneumococcus colonization and virulence.

The single largest regulon (>100 genes) in *S. pneumoniae* mediates the development of natural competence. Competence induction is a complicated process that partly involves the accumulation of a secreted peptide that resembles a quorum-sensing autoinducer. This competence-stimulatory peptide is sensed by an autoregulatory two-component system (ComDE) that stimulates the expression of the single alternative sigma factor (ComX) encoded by *S. pneumoniae*. RNA polymerase containing ComX stimulates the transcription of the many late competence genes required for DNA uptake and recombination. An interesting aspect of pneumococcal competence is fratricide, where cells that become competent produce cell wall hydrolases that kill noncompetent cells in the population. Relatively soon after the competence system is induced, it is turned off by a mechanism that is not well understood. Besides providing a mechanism for great genetic plasticity, it has been speculated that the induction of pneumococcal competence can act as a general stress response to conditions that damage DNA or disrupt cellular function. Considerably more work needs to be done to understand the higher-level regulatory networks that coordinate the expression of the genes required for colonization and virulence in *S. pneumoniae*.

ANTIBIOTIC RESISTANCE. When the penicillin-resistant pneumococci first appeared, some scientists assumed this was yet another example of resistance due to inactivation of the drug by a β-lactamase. However, in *S. pneumoniae,* mutant PBPs exclusively mediate resistance to β-lactam antibiotics, and a β-lactamase has yet to be discovered in clinical isolates. Interestingly, the PBPs in *S. pneumoniae* and its *S. mitis* relatives have mosaic amino acid sequences that seem to have arisen by reshuffling of gene segments through recombination during transformation. To explain why the discovery of these resistant PBPs was so discouraging, it is necessary to review the heartening progress that had been made at the time in dealing with β-lactamase-producing bacteria. The idea was simple but effective. By combining an old β-lactam antibiotic that had been rendered obsolete by bacterial β-lactamases with a β-lactamase inhibitor, the old antibiotic became magically effective once again (see chapter 16).

There was still more bad news to come. Besides the appearance of mosaic PBPs that imparted resistance to β-lactam antibiotics, *S. pneumoniae* clinical isolates were acquiring resistance to other classes of antibiotics, such as erythromycin, tetracyclines, and chloramphenicol (see chapter 16 for the mechanisms). These resistance genes have been moving rapidly among the gram-positive bacteria on conjugative transposons (see chapter 7). The origin of these conjugative transposons, which were first discovered in gram-positive cocci, is still unclear, but they are increasingly widespread.

A troubling feature of these types of resistance mechanisms is that they often confer resistance to several different kinds of antibiotics. For example, erythromycin resistance imparts additional resistance to other macrolides, such as streptogramins and lincosamides (see chapter 16). Multidrug resistance in *S. pneumoniae* clinical isolates is increasing rapidly. This alarming trend is probably being exacerbated by the increase in the number of adults who are at risk for persistent colonization and invasive diseases (see above). The genetic plasticity and pangenome of the different serotypes of *S. pneumoniae* will likely contribute to the development of ever-more virulent and antibiotic-resistant strains. For quite some time, clinical isolates of *S. pneumoniae* that are highly resistant to β-lactam antibiotics are often also resistant to erythromycin, clarithromycin, trimethoprim-sulfamethoxazole, and tetracyclines (see chapter 16). Currently, high-level resistant strains of *S. pneumoniae* remain sensitive to later-generation quinolones, such as levofloxacin and vancomycin, whose resistance cassettes have not yet moved into *S. pneumoniae*, but again, we are running out of effective, inexpensive antibiotics to treat a major bacterial pathogen.

C. difficile ("*C. diff.*"): a True Opportunist

DISEASE AND CHARACTERISTICS. The discovery of antibiotics revolutionized medicine by providing cures for many infectious diseases that had previously caused untold suffering and death. Understandably, antibiotics were viewed as miracle drugs, and because of this, it was difficult at first for physicians and the public at large to accept the fact that antibiotic use might have some negative aspects. The fact that some antibiotics had toxic side effects was recognized early on. The connection between overuse of antibiotics and the emergence of resistant strains was made much later and is still not being addressed aggressively enough by regulatory agencies. Even harder to accept was the idea that antibiotics being used successfully to treat one type of bacterial infection might actually cause another type of bacterial disease in the same patient. This happens when the antibiotics depress the resident microbiota of various body sites, thus allowing pathogens that had been kept in check by the microbiota to overgrow. **Pseudomembranous colitis,** one example of this type of disease, is probably the best documented. Another example is yeast vaginitis, a disease that develops in some women who take antibiotics that affect the normal vaginal microbiota.

Pseudomembranous colitis is a disease characterized by severe ulceration of the colon. It was first described nearly a century ago, before the advent of antibiotics. It was a rare disease until around 1970, when outbreaks began to occur in hospitals, particularly among elderly patients. Because pseudomembranous colitis is often fatal and can kill within a few days, these outbreaks caused alarm. The reason for the sudden increase in pseudomembranous colitis cases turned out to be the widespread use of antibiotics such as clindamycin, cephalosporins, and ampicillin, which inhibit the growth of the predominant genera of colonic bacteria (Figure 18–14). This gave *C. difficile*, the causative agent that is normally present in only about 5% of the population and then in very low numbers, the chance to overgrow the colon environment and cause disease. Some antitumor drugs have the same effect, although the reason for this is not clear, since antitumor drugs are not overtly antibacterial. Thus, once again, changes in human practices created a niche that a microbe could exploit. At first, clindamycin was the antibiotic most frequently associated with the disease. Certainly, clindamycin has probably caused more cases per unit of antibiotic used than any other antibiotic, but other antibiotics, such as ampicillin and cephalosporins, have caused more cases of the disease because they are used much more widely than clindamycin. A recent disturbing development has been the appearance of hypervirulent epidemic strains of *C. difficile* that are also resistant to fluoroquinolones.

C. difficile is a gram-positive, anaerobic, rod-shaped, motile bacterium. Like other clostridia, *C. difficile* is a spore former and is notable for its ability to produce exotoxins. *C. difficile* causes a spectrum of diseases, called CDIs (*C. difficile* infections), which range from a mild diarrhea to pseudomembranous colitis. Now that diagnostic tests for detecting toxin in feces are available and the types of patients who are at highest risk for pseudomembranous colitis can be identified, aggressive treatment of patients who show early signs of developing full-blown disease has lessened its incidence. Also, restrictions on the use of antibiotics known to cause pseudomembranous colitis have helped to reduce the incidence of the disease even further. However, it has recently been found that *C. difficile* is widespread in the environment and can cause diseases in animals. Furthermore, CDIs caused by hypervirulent strains are no longer confined to hospitals and are increasingly being acquired in the community. Some possible sources of CA CDIs are soil, water, animals used for food, meats, and produce, although the actual sources are not well understood. Because of community outbreaks, *C. difficile* has increasingly been called "*C. diff.*" in the popular press.

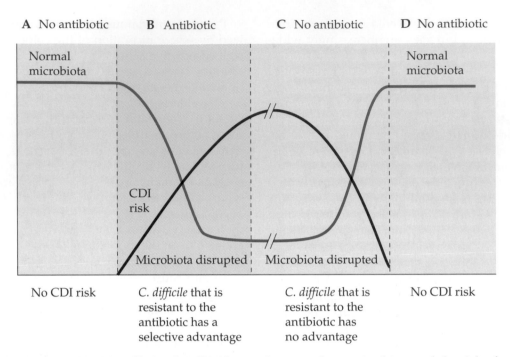

A No antibiotic   **B** Antibiotic   **C** No antibiotic   **D** No antibiotic

No CDI risk   *C. difficile* that is resistant to the antibiotic has a selective advantage   *C. difficile* that is resistant to the antibiotic has no advantage   No CDI risk

Figure 18–14   The effects of antibiotics on the normal gut microbiota and the risk of CDI. **(A)** Patients are resistant to CDI if antibiotics do not disrupt their normal gut microbiota. **(B)** Once antibiotic treatment starts, infection with a *C. difficile* strain that is resistant to the antibiotic is more likely while the antibiotic is being administered owing to the presence of the antibiotic in the gut. **(C)** When the antibiotic treatment stops, the levels of the antibiotic in the gut diminish rapidly, but the microbiota remains disturbed for a variable period of time (indicated by the break in the graph), depending on the antibiotic given. **(D)** During this time, patients can be infected with either resistant or susceptible *C. difficile*. Finally, after the microbiota recovers, resistance to *C. difficile* colonization is restored. (Adapted from Rupnik et al., 2009, with permission from Macmillan Publishers Ltd.)

An interesting feature of pseudomembranous colitis is that many strains of *C. difficile* are susceptible to the antibiotics that precipitate episodes of the disease. How, then, could *C. difficile* overgrow if the antibiotic is present? The reason seems to be that as antibiotic treatment comes to an end and levels of antibiotic in the intestine fall, *C. difficile* is able to resume growing more quickly than most other colonic bacteria and repopulates the colon before the normal microbiota can become reestablished. Also, although many strains of *C. difficile* are scored as "susceptible" using the standard cutoff between resistant and susceptible strains, they are less susceptible than many of the major groups of colon bacteria. A small difference in antibiotic susceptibility as measured by in vitro tests could give bacteria a big edge in the colon, and *C. difficile* is somewhat more resistant to many antibiotics than the other colonic species even though it is technically susceptible as defined by inability to grow in concentrations of the antibiotic achievable in the colon during therapy. Also, as noted above, many hypervir-

ulent strains seem to take advantage of the fact that they are highly resistant to fluoroquinolones.

It is important to note that even in people who are colonized with *C. difficile* and who are treated with the antibiotics associated with pseudomembranous colitis, only a subset develop the disease (Figure 18–15). Some people become colonized with nontoxigenic strains of *C. difficile* that do not produce the clostridial toxins, such as TcdA, described below, while others colonized by toxigenic strains seem to mount a sufficient humoral response against the toxins to remain asymptomatic. However, a small number of people do not produce a sufficient antibody response and consequently progress to symptomatic CDI. In addition, it appears that certain people are more likely to develop CDIs because of differences in the composition of their colonic microbiota or because they are physiologically more susceptible to the effects of the clostridial toxins that mediate the disease.

The steps in the development of CDIs and pseudomembranous colitis are outlined in Figure 18–16.

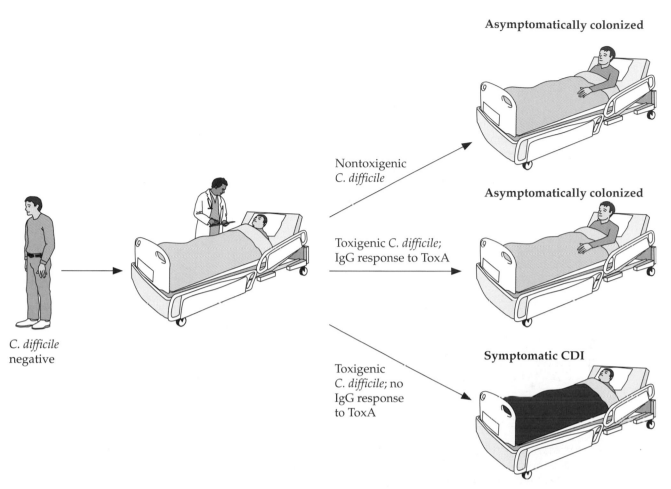

Figure 18–15 Model for the acquisition of CDI. Patients are exposed to *C. difficile* spores through contact with the hospital environment or health care workers. After taking an antibiotic, they develop CDI if they acquire a toxigenic *C. difficile* strain and fail to mount a serum IgG antibody response to the toxin TcdA; if they can mount an antibody response, they become asymptomatically colonized with *C. difficile*. If they acquire a nontoxigenic *C. difficile* strain, they also become asymptomatically colonized. Colonized patients have been shown to be protected from CDI. (Adapted from Rupnik et al., 2009, with permission from Macmillan Publishers Ltd.)

Antibiotics or other drugs used to treat another infection afflicting the patient can cause a reduction in the concentrations of bacterial genera that normally predominate in the colon. When this happens, an important protective barrier to colonization of the site by pathogens is lost, and *C. difficile* is apparently poised to take advantage of this opportunity. Normally, less than 5% of people in the healthy population harbor *C. difficile* in their intestinal tracts, but the percentage of people colonized by *C. difficile* can become as high as 20% in a hospital setting, where the bacteria are spread from one patient to another. In some settings, such as nursing homes, over 50% of the residents may be asymptomatically colonized by *C. difficile*. The risk of colonization is linked directly to the length of time

spent in the hospital, particularly in the case of a hospital that has experienced outbreaks of *C. difficile* disease.

C. difficile is a strict anaerobe and dies rapidly outside the colon. Thus, at first glance, it might appear unlikely that *C. difficile* would be transmissible from one patient to another. However, *C. difficile* is a spore former, and spores not only persist in the environment for many months, they survive passage through the stomach if they are ingested. The spread of spores from a colonized person to the surrounding environment and thence to other patients is facilitated by the fact that an early symptom of pseudomembranous colitis is diarrhea, a notorious source of aerosols and fecal contamination. Members of the hospital staff

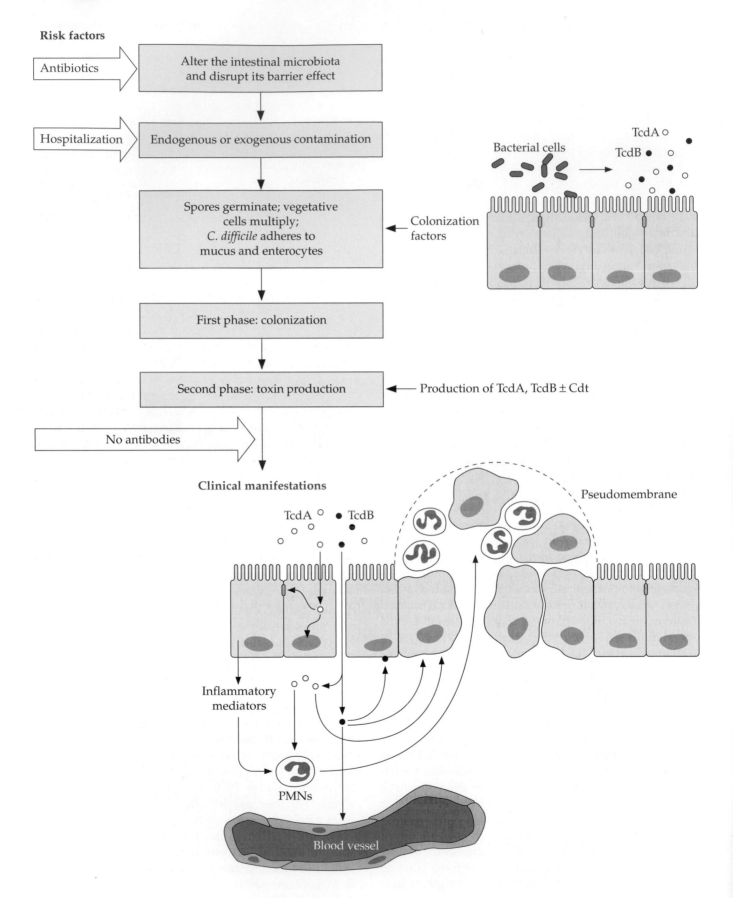

who care for patients colonized with *C. difficile* also transmit the spores to other patients on their hands. Even apparently innocuous practices, such as reuse of handles on electronic thermometers, have been found to spread *C. difficile* spores. Colonization of an otherwise healthy person with *C. difficile* causes no apparent symptoms as long as the clostridia are present in low concentrations. Only when *C. difficile* is able to grow to high enough concentrations to allow appreciable amounts of toxin to be produced does symptomatic disease occur.

VIRULENCE FACTORS. At first, in attempts to predict patients at highest risk for developing pseudomembranous colitis, attention focused on patients who entered the hospital colonized with *C. difficile*. However, careful studies of these patients revealed that they were, if anything, less likely to develop pseudomembranous colitis than patients who acquired the bacteria in the hospital. The current model for development of *C. difficile* disease is that an uncolonized person enters the hospital, is exposed to antibiotics, and becomes colonized with *C. difficile* (Figures 18–15 and 18–16). Development of symptomatic disease now appears more likely in such cases than in previously colonized people. Why this is so is not at all clear. Possibly *C. difficile* strains that survive in the hospital environment are more resistant to antibiotics than strains colonizing people in the community, or there may be some partial immunity to *C. difficile* disease in cases where people have carried the bacteria asymptomatically for a long time.

The first step in virulence is adherence to mucin and cells that line the colon. Several surface proteins have been proposed as adhesins, including surface layer (S-layer) proteins, such as SlpA; other cell wall-binding proteins; fibronectin-binding adhesin; and flagellar components. However, the clostridial toxins that are produced following colonization cause CDIs. Toxins produced by *C. difficile* damage the colonic mucosa. Accumulations of fibrin, mucin, and dead host cells form a yellowish layer on the surface of the colon (Figure 18–16). Damage by the toxins provokes an inflammatory response that recruits neutrophils to the area and to the lesions. Initially, numerous separate lesions appear, covered by patches of debris. Eventually, these scattered lesions coalesce, and the sheet-like layer of debris covers a larger area. This layer of debris **(pseudomembrane)** distinguishes the disease from other types of intestinal inflammation **(colitis)** and thus gives the disease its name. Untreated pseudomembranous colitis can be fatal, with symptoms similar to those seen in cases of septic shock.

Currently, the treatment of choice is to stop administering the antibiotic that precipitated the disease and to begin to administer antibiotics known to be effective against *C. difficile*, such as vancomycin or metronidazole. If this treatment is successful and the patient recovers, there is still the chance of a relapse once the therapy is discontinued. Relapses occur in as many as 10 to 20% of cases, and multiple relapses in the same person are not uncommon. Presumably, relapses are due to failure to clear *C. difficile* and to restore a stable, nonpathogenic microbiota.

C. difficile strains that cause pseudomembranous colitis produce two large AB-type protein toxins, **toxin A (TcdA)** and **toxin B (TcdB)** (Figure 18–16). Some strains also produce a third AB toxin, called **bi-**

Figure 18–16 Steps in the development of CDI and pseudomembranous colitis. *C. difficile* colonizes the intestine (colon) after disruption of the normal intestinal microbiota by antibiotic treatment. After endogenous or exogenous contamination by *C. difficile* spores, the spores germinate and vegetative forms multiply and then adhere to the mucus layer to colonize. The bacteria then produce the main virulence factors, the two large clostridial toxins A and B (TcdA and TcdB), which enter host cells and inactivate small Rho GTPases involved in cytoskeletal function through glucosylation. Some strains also produce another AB toxin (CdtAB) that ADP-ribosylates actin. TcdA binds to the apical side of the cell and, after internalization, causes cytoskeletal changes that result in disruption of tight junctions and loosening of the epithelial barrier, in cell death, or in the production of inflammatory mediators that attract neutrophils. The disruption of tight junctions enables both TcdA and TcdB to cross the epithelium. TcdB binds preferentially to the basolateral cell membrane. TcdB also enters the bloodstream and appears to have a tropism for cardiac tissue. Both toxins induce the release of various immunomodulatory cytokines from epithelial cells, phagocytes, and mast cells, resulting in inflammation, recruitment of neutrophils, and formation of the pseudomembrane. (Adapted from Deneve et al., 2009, with permission from Elsevier, and from Rupnik et al., 2009, with permission from Macmillan Publishers Ltd.)

nary toxin CDT, which modifies actin. TcdA and TcdB are thought to be responsible for the symptoms of pseudomembranous colitis, with TcdB being sufficient to cause the symptoms of CDIs. These toxins have a number of interesting properties. TcdA and TcdB are two of the largest single-polypeptide bacterial exotoxins known. Molecular masses reported in the literature vary considerably due to the tendency of these proteins to aggregate but are now considered to be 308 kDa and 269 kDa, respectively, based both on biochemical data and on the sequences of the genes encoding the toxins. The toxins have no recognizable signal sequence and do not appear to be activated by proteolytic nicking. They act by modifying host cell membrane G proteins that control many cellular activities, including actin polymerization.

From the beginning, TcdA and TcdB have frustrated and sometimes misled scientists working on them. As mentioned above, they were difficult to size and purify because of their tendency to aggregate. Until the genes encoding TcdA (*tcdA*) and TcdB (*tcdB*) were cloned, an added difficulty in the way of purifying one or the other of these toxins was that they have very similar properties. Another feature of the toxins that initially misled scientists into thinking they had different mechanisms was their different behaviors in standard toxin assays. TcdA acted like an enterotoxin; it caused fluid release when injected into a ligated rabbit ileal loop. The fluid was not the watery type of fluid seen with enterotoxins like cholera toxin, however, but had a more viscous, bloody appearance. This is due to the fact that TcdA, unlike cholera toxin and similar enterotoxins, does not cause a fluid imbalance by disrupting ion pumps but causes fluid accumulation by damaging mucosal cells so extensively that they can no longer control water movement.

TcdB was cytotoxic for tissue culture cells, whereas TcdA was at least 1,000-fold less cytotoxic in assays of this type. TcdB collapsed the actin cytoskeleton of the tissue culture cell, resulting in rounded-up cells, some of which still had long point-like projections. TcdA appeared not to have this activity. It was thus a surprise to find that both toxins not only have the same mechanism of action, but also are very similar in amino acid sequence (45% identity). The explanation for the apparent difference between the two toxins in modes of action proved to be that TcdB is much more active than TcdA and thus kills cells more quickly. TcdA acts more slowly and produces milder symptoms, thus explaining the enterotoxin-like behavior in the rabbit ileal loop model.

A model for the contributions of TcdA and TcdB to development of CDI and mucosal damage is shown in Figure 18–16. TcdA activates neurons and attracts and activates neutrophils (polymorphonuclear cells [PMNs]) by stimulating intestinal mucosal cells to produce cytokines and other proinflammatory proteins. Activation of the enteric neurons affects the motility of the intestinal contents and thus may contribute to diarrhea. Neutrophils moving into the area create an inflammatory response that further damages cells, contributing to mucosal-cell destruction. Neutrophils migrating between mucosal cells disrupt the tight junctions that normally prevent fluids from flowing across the mucosal membrane. This not only allows water to leak from tissue into the lumen of the intestine, but also allows TcdB to cross the mucosal membrane by diffusion. TcdB damages the tissue underlying the mucosal membrane, further contributing to damage of the intestinal wall. If damage becomes too extensive, LPS or bacteria from the colon can breach the intestinal wall, enter the bloodstream, and cause septic shock. Also, the extensive tissue damage caused by the toxins elicits a strong inflammatory response that may contribute to septic shock.

Both toxins glucosylate (i.e., transfer a glucose residue from UDP-glucose to the target) a threonine residue on the G protein targets, Rho, Rac, and Cdc42, which are located in the host cell cytosol (Figure 18–17). TcdA has two additional G protein targets, the small Rap GTPases. All of these targets are important regulatory proteins of mammalian cells, and if they malfunction, many cellular processes are affected. Among other things, they control the polymerization and depolymerization of actin. This process is normally dynamic, because the actin cytoskeleton not only helps determine the shape of the cell, but also allows a cell to make pseudopods for phagocytosis or to change shape, as PMNs do when they migrate between endothelial cells or epithelial cells. Thus, the cytoskeleton is constantly turning over. Impairment of polymerization, depolymerization, or both upsets the natural balance of these processes in the cell.

Small G proteins cycle between an inactive state, in which they bind GDP, and an active state, in which they bind GTP. The GTP-bound form mediates the effects of the G protein. The ratio of the GDP-bound form and the GTP-bound form is determined by signals that stimulate the exchange of GDP with GTP (Figure 18–17A). The toxins act preferentially on the GDP-bound form, because in that form, the G protein takes on a configuration that exposes the threonine that is then glucosylated by the toxin. In the GTP-bound form, the conformation of the G protein changes enough to bury the threonine residue in the protein, where it is inaccessible to toxin activity. Glucosylation does not completely inactivate the G protein, but it does reduce its GTPase activity by increas-

A Normal function of G protein

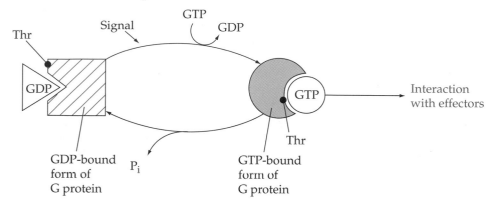

B Effects of glucosylation of G protein
by TcdA or TcdB

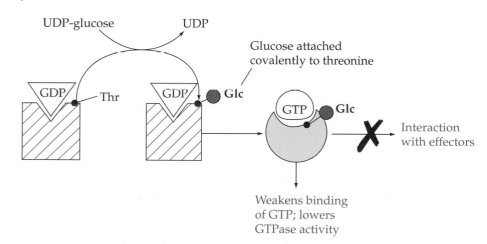

Figure 18–17 Effect of glucosylation by TcdA or TcdB on G protein function. **(A)** Normal cycling of G proteins between the GDP-bound form and the GTP-bound form. The threonine (Thr) residue that will be glucosylated by the toxins is shown. It is exposed on the surface of the GDP-bound form but buried in the protein in the GTP-bound form. P_i, phosphate ion. **(B)** Glucosylation of the threonine residue of the GDP-bound form does not prevent it from converting to the GTP-bound form, but the glucosylated GTP-bound form has a lower affinity for GTP and a lower GTPase activity. These changes disrupt the normal functions of the G protein, causing numerous changes in cell physiology.

ing the conversion of GTP to GDP and thereby inactivating the G protein. This creates an imbalance that disrupts the control of processes normally controlled tightly by the G protein. The modification also interferes with the interaction between the G protein and its downstream effectors.

The structure of the *C. difficile* toxins is illustrated in Figure 18–18 and follows the usual single-chain toxin structural plan. There is a binding domain that binds to the host cell receptor, a translocation domain that mediates translocation of the enzymatic portion into the cell cytoplasm, and the catalytic domain that glucosylates the G proteins. A trisaccharide is part of the TcdA receptor in animal cells, but since this compound is not present in humans, the TcdA receptor is likely a host surface glycoprotein called Gp96. This receptor is found on the apical side of colon cells (Figure 18–16). In contrast, TcdB binds to an unknown receptor on the basolateral side of the epithelial cell layer. Hence, one role of TcdA toxicity is likely to disrupt the tight junctions to allow egress of TcdB to the basolateral membranes of the host cells. After receptor binding, the toxins are endocytosed into the cells, where an autoproteolytic function in the translocation domain cleaves off the catalytic domain, which is then transferred into the cytoplasm through a toxin-mediated pore. Differences in the identities and distributions of the receptors for TcdA and TcdB undoubtedly contribute, along with different levels of autoproteolytic cleavage, transfer, and catalytic activity, to the different functions of these toxins in human disease.

The *tcdA* and *tcdB* genes are located on a pathogenicity island called **PaLoc** (for **pathogenicity locus**) (Figure 18–18). Also in the same vicinity are the regulatory genes, *tcdC* and *tcdR*. The small open reading frame between *tcdB* and *tcdA* encodes a protein with sequence similarity to bacteriophage release proteins

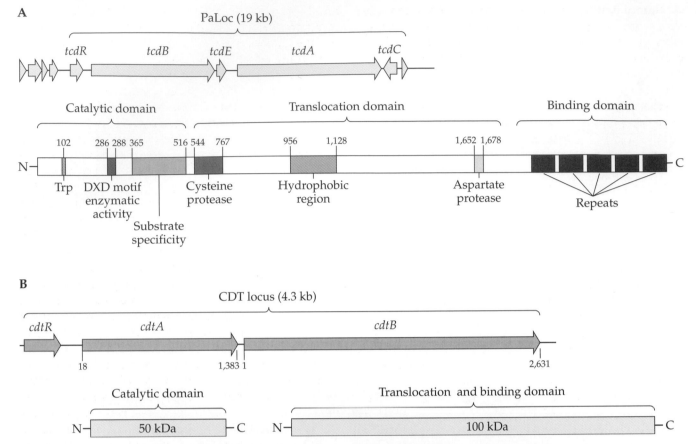

A

Figure 18–18 Toxins produced by *C. difficile*. **(A)** Toxigenic strains of *C. difficile* that cause CDI produce two main virulence factors, the two large clostridial toxins A and B (TcdA and TcdB), which enter host cells and inactivate small Rho GTPases involved in cytoskeletal function through glucosylation. The genes for the toxins are in the PaLoc, which is comprised of five genes. Both toxins are single-chain proteins, and several functional domains and motifs have been identified. TcdB is shown in detail below the PaLoc. **(B)** Some strains also produce another AB toxin (CdtAB) that ADP-ribosylates actin and potentiates the effects of TcdA and TcdB. The three genes are in a separate region of the chromosome (CdtLoc). The binary toxin is composed of two unlinked proteins, CdtB and CdtA. CdtB has a binding function, and CdtA is the enzymatic component. (Adapted from Rupnik et al., 2009, with permission from Macmillan Publishers Ltd.)

called **holins.** The PaLoc is present in all toxigenic strains of *C. difficile* and is replaced by a short spacer in nontoxigenic strains.

Toxin levels are highest in the medium during late exponential and stationary growth phases. The TcdR protein is an alternative sigma factor that mediates transcription of the *tcdBEA* operon. TcdC is an anti-sigma factor for TcdR that antagonizes TcdR function during exponential growth. The growth phase is not the only signal sensed by the toxin regulatory genes. Stresses of various types, including limitation of carbon, nitrogen, and oxygen, also increase toxin production. These stresses may be linked, because one effect of oxygen on an obligate anaerobe would be to

reduce metabolic activities, causing a greater reliance on exogenous amino acids and other nutrients. Recently, the global repressor protein CodY, which senses the metabolic state by responding to intracellular concentrations of branched-chain amino acids and GTP, was shown to regulate toxin gene expression.

Strains of *C. difficile* vary considerably in their levels of TcdA and TcdB production, and many strains produce little or no toxin. Toxigenic strains almost always produce both TcdA and TcdB, and the levels of the two toxins covary. This is not too surprising, given the fact that the toxin genes are cotranscribed. The levels of these toxins can be quite high. Under in vitro conditions that give maximum toxin production, the toxins can account for up to 5% of the total protein when the bacteria are growing in pure culture, which is an enormous expenditure of energy. Epidemic hypervirulent strains that have recently emerged produce at least 15-fold more TcdA and TcdB than previous strains. Toxin production in these strains seems to be completely deregulated, and the toxins are produced throughout exponential and stationary phases. This massive increase in toxin production results from a frameshift mutation that inactivates the TcdC antisigma factor that normally negatively regulates toxin production during exponential phase.

Considerably less is known about CDT binary toxin (Figure 18–18B), which is present in fewer than 10% of *C. difficile* strains. This AB-type toxin is comprised of a catalytic subunit (CdtA) and a receptor-binding/translocation subunit (CdtB). The CDT toxin is encoded by the CDT locus located in a different region in the chromosome from the PaLoc. The CDT locus also encodes a positive regulator (CdtR) of the CdtAB toxin genes. CdtR likely acts as a response regulator of a two-component system, but no cognate histidine kinase (sensor kinase) gene is included in the CDT locus. The CdtB protein seems to target the CdtAB complex to colon epithelial cells, where the CdtA is internalized and then ADP-ribosylates actin molecules, leading to cytoskeleton disorganization. One function of the CDT toxin may be to potentiate the toxicity of TcdA and TcdB, thereby causing more severe diseases with hypervirulent strains.

Successful treatment of CDI depends on early diagnosis. Patients receiving antibiotics known to cause pseudomembranous colitis are watched carefully for signs of diarrhea. In many U.S. hospitals, nearly every patient with antibiotic-associated diarrhea is tested for the presence of TcdA and TcdB in the feces. If pseudomembranous colitis is suspected, the patient is given vancomycin or metronidazole, two antibiotics that are still effective against *C. difficile*. Timely administration of oral vancomycin or metronidazole is usually successful in preventing the development of full-blown pseudomembranous colitis. As mentioned above, relapses can occur once vancomycin therapy ceases.

One strategy to treat recurrent pseudomembranous colitis is to restore the resident microbiota by giving the patient enemas with dilute feces taken from family members. Although an enema does not reach very high into the colon and thus might not seem to be an effective strategy for repopulating the entire colon, this treatment has been effective at preventing relapses. Another approach has been to replenish the colonic microbiota with probiotic nonpathogenic bacterial and yeast species, such as *Saccharomyces boulardii* or bifidobacteria. To date, convincing data are lacking for the effectiveness of probiotic administration in preventing or treating CDIs.

So far, the most effective efforts to prevent pseudomembranous colitis have been restrictions on the use of the antibiotics that are most likely to cause it, coupled with increased hygiene practices to minimize contamination. However, one approach will not work without the other. At one time, some dermatologists gave clindamycin to acne patients, because acne was exacerbated, if not caused, by another anaerobe, *Propionibacterium acnes*. It is now agreed that giving oral clindamycin to people who are not closely supervised in a hospital setting is dangerous. Careful observation of patients taking clindamycin, cephalosporins, ampicillin, or fluoroquinolones is critical in hospitals, particularly following an outbreak of pseudomembranous colitis. The fact that spores can be spread from patient to patient has increased awareness of the importance of hand washing by hospital staff caring for patients who have pseudomembranous colitis or who develop diarrhea after administration of antibiotics. The fact that carriage by people in the community (<5%) is significantly lower than carriage by hospital patients, especially in intensive-care units (as high as 20%), strongly supports the hypothesis that *C. difficile* can be spread from patient to patient in the hospital environment. Careful cleaning of rooms where pseudomembranous colitis patients have stayed can reduce, albeit not eliminate, the spores in the environment.

That's Not All, Folks

In this chapter, we have considered some issues concerning what defines a gram-positive commensal bacterium that can also act as an opportunistic pathogen. Four examples of gram-positive opportunists were chosen that included an accidental pathogen (*S. epidermidis*), a commensal bacterium that kills millions of people each year (*S. pneumoniae*), an increasingly antibiotic-resistant commensal bacterium that can produce an amazing arsenal of powerful toxins (*S. aureus*), and an opportunist that causes microbial community shift diseases (*C. difficile*). These examples were chosen to provide a framework for thinking about the diverse factors and mechanisms used by gram-positive opportunists during colonization of their normal resident sites and when they cause diseases. We have also tried to introduce some concepts from current evolutionary theory about the relationships between modes of transmission between human hosts and the need to produce virulence factors.

Naturally, we have presented only a small sampling of important gram-positive opportunistic pathogens. We refer the reader to reviews cited at the end of this chapter about three other important pathogens that we did not cover here. *S. pyogenes* (group A streptococcus), which is a commensal found in the throats and saliva of about 30% of people, causes a range of serious diseases from pharyngitis (the well-known "strep throat") and endocarditis to necrotizing fasciitis (the much-feared "flesh-eating" disease). Analogous to *S. aureus, S. pyogenes* comprises a diverse group of bacteria that have an astounding capacity to regulate and produce adhesins and toxins. In fact, we used this regulation as the paradigm for thinking about the sequential expression of genes as bacterial pathogenesis progresses from colonization to persistence to spread (see Figure 11–1). Rather astoundingly, *S. pyogenes* is one of the few commensal species that has not acquired multidrug resistance and so far remains sensitive to β-lactam antibiotics.

Analogous to *S. pneumoniae, Streptococcus agalactiae* (group B streptococcus) is a common commensal bacterium with a high potential to cause a variety of invasive diseases. *S. agalactiae* is a common member of the vaginal microbiota of about 30% of women and can spread to newborn infants causing life-threatening diseases. *S. agalactiae* is also emerging as a serious antibiotic-resistant pathogen of adults, especially those at risk for invasive pneumococcal diseases. Similar to *S. pneumoniae, S. agalactiae* is equipped with a set of colonization and virulence factors, including a sialic acid-rich capsular polysaccharide, pore-forming toxins, a hyaluronate lyase, adhesins that bind to the extracellular matrix, D-alanine decoration of LTA, and peptidases and proteases that interfere with the innate immune response.

Finally, analogous to *S. epidermidis, Enterococcus faecium* has emerged as a serious cause of HA (also known as **nosocomial**) bacteremia, urinary tract infections, wound infections, and catheter-associated infections. This ubiquitous commensal bacterium found in the gastrointestinal tracts of humans and livestock, including poultry, was once thought to be benign. However, in the past 20 years, *E. faecium* has rapidly acquired antibiotic resistance mechanisms, including resistance to vancomycin (see chapter 16), which make it extremely difficult to treat once infection has occurred. In fact, hospital isolates of *E. faecium* seem to be evolving rapidly and have acquired accessory genes that are not present in gut isolates. *E. faecium* seems to exhibit the disturbing trend noted numerous times in this chapter of diversifying its genome to become more resistant to antibiotics and more virulent to human hosts.

SELECTED READINGS

Bartlett, J. G. 2009. *Clostridium difficile* infection: historic review. *Anaerobe* **15:**227–229.

Dean, N. 2010. Methicillin-resistant *Staphylococcus aureus* in community-acquired and healthcare-associated pneumonia: incidence, diagnosis, and treatment options. *Hosp. Pract. (Minneapolis)* **38:**7–15.

Denapaite, D., R. Bruckner, M. Nuhn, P. Reichmann, B. Henrich, P. Maurer, Y. Schahle, P. Selbmann, W. Zimmerman, R. Wambutt, and R. Hakenbeck. 2010. The genome of *Streptococcus mitis* B6—what is a commensal? *PLoS One* **5:**e9426.

Deneve, C., C. Janoir, I. Poilane, C. Fantinato, and A. Collignon. 2009. New trends in *Clostridium difficile* virulence and pathogenesis. *Int. J. Antimicrob. Agents* **33:** S24–S28.

Gordon, R. J., and F. D. Lowy. 2008. Pathogenesis of methicillin-resistant *Staphylococcus aureus* infection. *Clin. Infect. Dis.* **46:**S350–S359.

Henriques-Normark, B., and S. Normark. 2010. Commensal pathogens, with a focus on *Streptococcus pneumoniae,* and interactions with the human host. *Exp. Cell Res.* **316:**1408–1414.

Hookman, P., and J. S. Barkin. 2009. *Clostridium difficile* associated infection, diarrhea and colitis. *World J. Gastroenterol.* **15:**1554–1580.

Jank, T., T. Giesemann, and K. Aktories. 2007. Rhoglucosylating *Clostridium difficile* toxins A and B: new insights into structure and function. *Glycobiology* **17:** 15R–22R.

Kadioglu, A., J. N. Weiser, J. C. Paton, and P. W. Andrew. 2008. The role of *Streptococcus pneumoniae* virulence factors in host respiratory colonization and disease. *Nat. Rev. Microbiol.* **6:**288–301.

Langley, R., D. Patel, N. Jackson, F. Clow, and J. D. Fraser. 2010. Staphylococcal superantigen super-domains in immune evasion. *Crit. Rev. Immunol.* **30:**149–165.

Larkin, E. A., R. J. Carman, T. Krakauer, and B. G. Stiles. 2009. *Staphylococcus aureus:* the toxic presence of a pathogen extraordinaire. *Curr. Med. Chem.* **16:** 4003–4019.

Lynch, J. P., III, and G. G. Zhanel. 2010. *Streptococcus pneumoniae:* epidemiology and risk factors, evolution of antimicrobial resistance, and impact of vaccines. *Curr. Opin. Pulm. Med.* **16:**217–225.

Maisey, H. C., K. S. Doran, and V. Nizet. 2008. Recent advances in understanding the molecular basis of Group B *Streptococcus* virulence. *Expert Rev. Mol. Med.* **10:**e27.

Maresso, A. W., and O. Schneewind. 2008. Sortase as a target of anti-infective therapy. *Pharmacol. Rev.* **60:**128–141.

Massey, R. C., M. J. Horsburgh, G. Lina, M. Höök, and M. Recker. 2006. The evolution and maintenance of virulence in *Staphylococcus aureus:* a role for host-to-host transmission? *Nat. Rev. Microbiol.* **4:**953–958.

Olsen, R. J., and J. M. Musser. 2010. Molecular pathogenesis of necrotizing fasciitis. *Annu. Rev. Pathol. Mech. Dis.* **5:**1–31.

Otto, M. 2009. *Staphylococcus epidermidis*—the 'accidental' pathogen. *Nat. Rev. Microbiol.* **7:**555–567.

Plata, K., A. E. Rosato, and G. Węgrzyn. 2009. *Staphylococcus aureus* as an infectious agent: overview of biochemistry and molecular genetics of its pathogenicity. *Acta Biochim. Polon.* **56:**597–612.

Rajagopal, L. 2009. Understanding the regulation of Group B Streptococcal virulence factors. *Future Microbiol.* **4:**201–221.

Ramos-Montañez S., H.-C. T. Tsui, K. J. Wayne, J. L. Morris, L. E. Peters, F. Zhang, K. M. Kazmierczak, L.-T. Sham, and M. E. Winkler. 2008. Polymorphism and regulation of the *spxB* (pyruvate oxidase) virulence factor gene by a CBS-HotDog domain protein (SpxR) in serotype 2 *Streptococcus pneumoniae*. *Mol. Microbiol.* **67:**729–746.

Rupnik, M., M. H. Wilcox, and D. N. Gerding. 2009. *Clostridium difficile* infection: new developments in epidemiology and pathogenesis. *Nat. Rev. Microbiol.* **7:** 526–536.

Somerville, G. A., and R. A. Proctor. 2009. At the crossroads of bacterial metabolism and virulence factor synthesis in staphylococci. *Microbiol. Mol. Biol. Rev.* **73:** 233–248.

van der Poll, T., and S. M. Opal. 2009. Pathogenesis, treatment, and prevention of pneumococcal pneumonia. *Lancet* **374:**1543–1556.

Willems, R. J., and W. van Schaik. 2009. Transition of *Enterococcus faecium* from commensal organism to nosocomial pathogen. *Future Microbiol.* **4:**1125–1135.

Weiser, J. N. 2010. The pneumococcus: why a commensal misbehaves. *J. Mol. Med.* **88:**97–102.

QUESTIONS

1. Describe how understanding the mechanisms involved in bacterial adhesion to host cells might lead to development of new preventive therapies or measures.

2. Critique the following statement: members of the same bacterial species are usually found at the same anatomical sites of the host body.

3. What components of the gram-positive cell wall are responsible for septic shock?

4. Superantigens are able to elicit toxic effects throughout the body. Describe how superantigens are able to initiate these effects and how this might be an advantage to the infecting bacteria.

5. On one particularly busy evening, you forget to put a casserole in the refrigerator after dinner. Even though the food has not been refrigerated, it still "looks" okay, and you decide to eat the casserole the next day.
  A. If the casserole is contaminated with *S. aureus*, will reheating protect you from food poisoning? Provide your rationale.

  B. If the casserole is contaminated with *Clostridium botulinum*, will reheating protect you from food poisoning? Provide your rationale.

6. SE and TSST act as _____, cross-linking cells of the innate immune system, and are potent nonspecific activators of _____.

7. *S. pneumoniae* depends on _____ and _____ to evoke the inflammatory response for tissue damage and entry into the bloodstream.

8. *S. aureus* uses a _____ system to regulate its virulence factors: _____ are expressed during early growth phases and then downregulated as _____ are upregulated when high density is reached.

9. You are characterizing a mutant of *S. pyogenes*. The mutation affects the cell wall so that the cells make only a thin layer of peptidoglycan (making up about 20% of the cell wall). What would you expect the Gram stain reaction to be?

SOLVING PROBLEMS IN BACTERIAL PATHOGENESIS

1. MRSA is resistant to several commonly prescribed antibiotics that are usually effective against gram-positive bacteria (methicillin, penicillin, and cephalosporins), and an infection with a MRSA strain can be deadly if left untreated. MRSA is subcategorized as CA-MRSA or HA-MRSA, depending on how the infection is usually acquired. Most CA-MRSA strains are still sensitive to many antibiotics, such as trimethoprim, tetracycline, and clindamycin, but HA-MRSA strains are often resistant to these drugs, while they are still sensitive to vancomycin and linezolid. Although most CA-MRSA bacteria are not as resistant to antibiotics as the HA-MRSA strains, the recent

(continued)

strains that we are hearing about in the news appear to be more virulent. Most MRSA strains produce type CP5 or CP8 polysaccharide capsules, surface-exposed protein A, and a number of toxins, superantigens, and proteases. MRSA strains producing CP5 show significantly higher bacteremia levels than strains producing CP8. A group of researchers have recently discovered that bacterial cultures of a particularly virulent strain of CA-MRSA (called USA300) cause apoptosis in leukocytes and neutrophils, as evidenced by release of cytochrome *c* from mitochondria, followed by cytokine release and cell death.

A. Antibodies generated against bacterial culture medium from the USA300 strain identified two proteins of 33 and 44 kDa on Western blots. To determine the localization of these two proteins, researchers performed sodium dodecyl sulfate-polyacrylamide gel electrophoresis and Western blot analysis of lysates from the bacterial cell pellets and the bacterial culture medium, as well as lysates of human neutrophils treated with the USA300 strain. The results from Western blot analysis are shown in the left panel of Figure 1. The researchers then applied the cell-free bacterial culture medium to the neutrophils and performed subcellular fractionation of the treated neutrophils. The results from Western blot analysis are shown in the right panel of Figure 1. Provide an interpretation (with rationale) of the results shown in Figure 1.

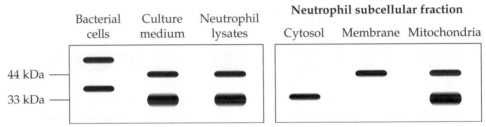

Figure 1 Western blots using antibodies against USA300 secreted proteins.

B. Predict (with rationale) possible functions for the two proteins, which have been named LukS and LukF, where LukS is the 33-kDa protein and LukF is the 44-kDa protein. Provide at least one additional (different) experiment that could be performed to confirm your prediction.

C. The researchers speculate that the two proteins, LukS and LukF, are two subunits of an AB-type toxin. Based on the results shown in Figure 1, do you agree with this interpretation? Provide your rationale. Provide an experiment that would allow you to verify your answer.

D. The researchers suspect that LukF might bind to specific protein receptors on neutrophils. Describe an experiment they might perform to isolate and identify this receptor using the antibodies that recognize LukF.

E. The researchers suspect that LukS and LukF are responsible for the necrotizing lung and tissue damage observed during severe infection with CA-MRSA. What four criteria must the researchers satisfy to establish this connection? Provide at least two additional modern molecular experiments (different from those already described) that could be performed to help satisfy these criteria for this connection.

F. Using Western blot analysis, the researchers discover that the two proteins are not made until late

exponential or stationary phase. When they take medium from overnight bacterial cultures and add it to fresh bacterial cultures, they find that they can stimulate production of the proteins much sooner. Provide a possible explanation for this observation.

G. When the researchers purify the proteins from culture medium and perform N-terminal sequencing of the proteins, they discover that the proteins are encoded by a two-gene operon, which they name *lukSF*. Describe a strategy that the researchers could use to identify potential regulatory factors that affect the expression of the two proteins encoded by *lukSF*. Be sure to provide the reagents and screening conditions, how they would visualize/measure the results, and how they would verify the results.

H. A group of researchers propose that development of a two-component vaccine made up of LukS or LukF in combination with CP5 or CP8 would be effective protection against CA-MRSA. What led the researchers to propose this? Provide at least two possible reasons. Which two of the four possible vaccine targets (LukS, LukF, CP5, and CP8) would be most effective at generating protective immunity as part of a two-component vaccine against CA-MRSA? Be sure to provide your rationale. Describe how these two components plus at least one additional component could be used to make an "ideal" multicomponent vaccine. Be sure to provide a rationale for your choice of a third component and the overall vaccine design.

I. Provide a mechanism by which protein A contributes to virulence (i.e., what is its function?).

J. Provide a mechanism by which the superantigens contribute to MRSA virulence. What is a potential long-term consequence of infection with MRSA strains that produce superantigens?

K. Describe how you might isolate and identify those genes that are unique to CA-MRSA (versus HA-MRSA) using the genomic subtractive hybridization method. (You may consider using a schematic diagram.) How could you determine which of these unique genes encode antigens that generate protective antibodies in patients?

2. In your capacity as an expert working for the Centers for Disease Control and Prevention, you were asked to investigate a disease outbreak at several hospitals in the Washington, DC, area where the patients died after undergoing surgery. You subsequently identified the pathogen as being closely related to *Clostridium perfringens*, which is a gram-positive anaerobic bacterium that can cause serious human and animal diseases, including clostridial myonecrosis (i.e., "flesh-eating" disease) and gas gangrene (i.e., "tissue-rotting" disease). *C. perfringens* strains are not intracellular pathogens but are known to produce many extracellular enzymes and toxins, one of which is perfringolysin O (PFO). Being interested in understanding the regulation of these secreted toxins, you performed transposon mutagenesis using a mouse model of gas gangrene infection and isolated a number of avirulent mutants, five of which had mutations in genes unrelated to the toxin genes, yet these mutants no longer produced the toxins, including PFO. Four of the disrupted genes (*mut1* to *mut4*) were found in a single operon, while the fifth gene (*mut5*) was found upstream of the other genes but transcribed in the opposite direction. You then made in-frame deletions of each of these genes and designated the resulting mutants CpMut1 to CpMut5. You also cloned each of the genes and found that you could express only four of the corresponding proteins (Mut2 to Mut5) in another gram-positive bacterium, *B. subtilis*; however, although the *mut1* gene was transcribed, as assessed by Northern blotting, you were unable to detect any expressed protein corresponding to the *mut1* gene. You were interested in characterizing these mutants and determining the functional identities of the proteins, so you designed a series of experiments.

A. In the first experiment, you determined the cellular localization of the proteins encoded by the genes *mut2* through *mut5*. Describe how you conducted this experiment. Be sure to include the experimental design, generation of the reagents used, and how you visualized and interpreted the data.

B. In the second experiment, you found that the cellular-fractionation results indicated that both CpMut2 and CpMut3 are cytoplasmic proteins, while CpMut4 and CpMut5 are integral membrane proteins. In addition, you found that although all of the mutants were defective in making the toxins, as evidenced by using a reporter gene downstream of the *pfo* gene, when you added cell-free filtrates from overnight cultures of wild-type *C. perfringens* or the mutant CpMut2 or CpMut5 to cultures of CpMut3 or CpMut4, all of the toxins were produced. However, when overnight culture filtrates of CpMut3 or CpMut4 were added to cultures of CpMut2 or CpMut5, no toxin was produced. Provide an explanation for these results. What is the most likely functional identity of each of the proteins encoded by the *mut2* to *mut5* genes? Provide at least one additional experiment for each of the four proteins that you could perform to confirm your conclusions. (Hint: you should describe at least four separate, different experiments, one for each of the proteins. Be sure to provide the expected results.)

C. For the third experiment, you are now interested in determining the role of the *mut1* gene in virulence. Based on the above information, what is the most likely function of the gene? Be sure to provide your rationale. Design an experiment to confirm your hypothesis.

3. You work for a small biotechnology company interested in developing new vaccines against emerging pathogens. You read about the new pathogen described above in problem 2 and thought it would be a great idea to develop a vaccine against it.

A. What type(s) of immune response would the vaccine need to elicit in order to provide protective immunity against the pathogen? Be sure to provide your rationale.

B. You propose that CpMut5 and PFO would make excellent targets for the development of a two-component recombinant vaccine. Name two scientific arguments that you could provide to your colleagues in support of your proposal.

C. To test the effectiveness of these proteins as antigens, you use a vaccine formulation comprised of formalin-treated, heat-inactivated recombinant CpMut5 and PFO mixed with alum as an adjuvant. You find that after immunization with two boost-

(continued)

ers, you get a strong humoral immune response with high titers of antibodies against both CpMut5 and PFO. However, in challenge studies using the mouse infection models, you find that there is very little protection against the wild-type *Clostridium* strain. Provide an explanation for this result.

D. Considering the above result, provide an alternative strategy (with rationale) that is based on CpMut5 and PFO as antigens to design an optimal, safe recombinant vaccine that does not rely on alum as an adjuvant but will still lead to robust, long-lasting protective immunity against the pathogen.

19

The Gram-Negative Opportunistic Pathogens

The opportunists included in this chapter were placed here on the basis of an arbitrary characteristic: their cell wall structure. In fact, gram-negative bacteria seem to be just as adroit at taking advantage of various breaches in the defenses of the host as the gram-positive bacteria. Perhaps microbiologists should replace the old saying "you can't judge a book by its cover" with "you can't judge an opportunist by its cell wall." Instead, special traits of individual organisms seem to be critical to their ability to rise to the occasion, opportunistically speaking.

Jumping Over the (Cell) Wall: Gram-Negative Bacteria Can Be Opportunistic Pathogens Too

The list of gram-negative opportunists now has a large number of entries. Rather than try to cover every bacterium on this increasingly extensive list, examples have been chosen that illustrate some of the different ways in which such bacteria can cause opportunistic infection. Many of these bacteria come from the human microbiota, a fact that is not surprising, given that the location and constant presence of these bacteria make them best able to take advantage of breaches in the host. Three examples of this type of opportunist that exemplify very different ways in which such bacteria cause disease are *Escherichia coli, Bacteroides fragilis*, and *Porphyromonas gingivalis*. Opportunistic gram-negative pathogens can also come from soil. Three examples described in this chapter are *Pseudomonas aeruginosa, Burkholderia cenocepacia*, and *Acinetobacter baumannii*. Gram-negative opportunists can even be transmitted by arthropods, although calling such organisms "opportunists" is still a bit controversial. *Ehrlichia* will serve as an example of this type of pathogen.

A common trait of opportunists is that they are able to take advantage of certain breaches in human defenses to cause infections that they would be unable to cause in the body of a healthy individual. An overview of the opportunists covered in this chapter and the types of opportunities they exploit is provided in Table 19–1. A striking feature of these opportunists is that they do not

Table 19–1   Examples of breaches in human defenses that are used by opportunists to cause infections

Organism	Disease	Breach in defenses
E. coli K1	Infant meningitis	Infants exposed to colonized vaginal tract during birth
B. fragilis	Internal abscesses, bacteremia, sepsis	Lesions in the colon caused by abdominal surgery or other trauma
P. gingivalis	Periodontal disease, internal infections	A shift in the microbiota of the gums, giving rise to inflammation and bone loss
P. aeruginosa	Lung infections	Cystic fibrosis causing aberrant mucus in the lungs
B. cenocepacia	Lung infections	Cystic fibrosis, ventilator pneumonia
A. baumannii	Wound infections	Traumatic war wounds in Iraq veterans
E. chaffeensis	Bacteremia	Arthropod bite, which injects the bacteria into the bloodstream

simply take advantage of any breach in host defenses. For example, none of them is associated with infections involving indwelling catheters. Rather, they seem to be adapted to respond to a limited range of opportunities. To a large extent, this is due to their location relative to a breach in defenses, but it is likely that other characteristics of the specific breaches they favor are responsible for their preferences.

Another trait common to many of them is resistance to a variety of antibiotics. Of the bacteria covered in this chapter, resistance to multiple antibiotics is particularly striking in *B. fragilis, P. aeruginosa, B. cenocepacia,* and *A. baumannii.* The oral anaerobes, such as *P. gingivalis,* are moving quickly to become members of this dangerous club. This trait makes the opportunists that share it particularly difficult to treat successfully. Even if a patient is eventually cured of the infection, delays engendered by having to find the right antibiotic can leave the patient with permanent damage.

The Dark Side of Some Normal Inhabitants of the Human Body

The Ever-Changing Face of *E. coli*

Virotyping.   *E. coli* is associated with many different types of disease manifestations: diarrhea and dysentery in the gastrointestinal tract; bladder and kidney infections and **hemolytic uremic syndrome (HUS)** in the urinary tract; and systemic effects, meningitis and septicemia. Habitat and genetic exchange are the primary reasons why *E. coli* strains have the ability to generate so many different forms of disease symptoms and outcomes. In general, different strains cause different diseases, because the strains have acquired different sets of virulence factors. We saw this pattern for gram-positive opportunists in chapter 18. A few examples for *E. coli* are illustrated in Figure

19–1. Because of the wide range of differences that are found in *E. coli* isolates, attempts to classify *E. coli* strains were first based on the traditional approach of **serotyping** using surface antigens: O antigen of lipopolysaccharide (LPS) to determine the serogroup (~170 identified to date), H flagellar antigen to determine the serotype (~55 identified to date), and K capsule type (used if the strain has a capsule). However, it soon became clear that many clinical isolates were not adequately distinguished using this serotyping classification, since the clinical strains often had the same serotype but very different virulence properties. Consequently, many researchers have switched to using **virotyping** as a means for classification, which is based on profiling the types of virulence factors present, such as invasion and dissemination properties, adherence properties, toxins produced, effects on host cells, and symptoms produced.

For instance, consider *Shigella* spp. (including ***Shigella dysenteriae,*** the causative agent of **dysentery**) and strains of **enteroinvasive *E. coli* (EIEC),** both of which can cause bloody diarrhea. If one looks just at their core "housekeeping" genes, then genetically *Shigella* and *E. coli* are nearly identical, and indeed, many researchers might classify *Shigella* as a strain of *E. coli.* In the early 1980s, Philippe Sansonetti's laboratory found a large (220-kb), highly homologous plasmid that was present in all virulent strains of *Shigella* and EIEC and that was responsible for the invasive phenotype. *Shigella* and EIEC are also able to spread from cell to cell through actin-based motility (actin tail formation), similar to what is observed for *Listeria monocytogenes.* Transfer of a wild-type invasion plasmid into avirulent, plasmid-cured *Shigella* strains or into *E. coli* strains lacking plasmids reconstituted the invasive phenotype and converted the avirulent strains into virulent invasive strains. In addition to having invasive properties, *Shigella* strains produce **Shiga toxin (ST),** which kills cells by depurinating (removing an adenine base from the ribose backbone) 28S rRNA at

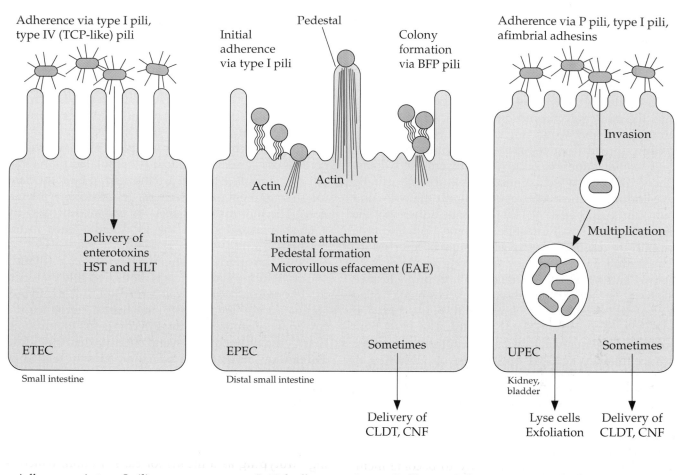

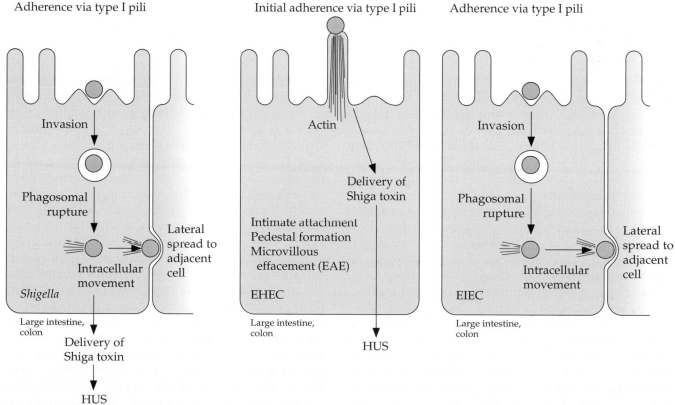

Figure 19–1 Examples of different virotypes of *E. coli.*

a specific position and thereby blocking protein synthesis. This results in the characteristic bloody diarrhea that is observed in dysentery. ST can cause acute, irreversible kidney failure, known as HUS, particularly in children and the elderly. The gene encoding ST is located on a lambdoid bacteriophage integrated into the bacterial chromosome. EIEC strains do not produce ST and so do not cause HUS, but they do invade, spread from cell to cell, and cause diarrhea.

A similar situation occurred with **enterotoxigenic *E. coli* (ETEC),** which is responsible for the usually self-limiting but severe, acute infant diarrhea and "traveler's diarrhea." ETEC bacteria adhere to the mucosa of the small intestine via type I pili and type IV pili yet are noninvasive and cause little or no inflammation. However, they do produce one or both of two enterotoxins. One is a small, peptide-derived **heat-stable toxin (HST),** which survives being boiled for up to 30 minutes, is resistant to acidic pH, and comes in two forms (STIa and STIb). The HSTs bind and activate guanylate cyclase receptors to increase intracellular cyclic GMP (cGMP) levels, resulting in activation of cGMP-dependent protein kinase, which stimulates chloride ion channels and triggers diarrhea. The other is a larger, multisubunit, A_1B_5-type **heat-labile toxin (HLT),** which comes in two forms (HLT-1 and HLT-2) and is related to cholera toxin (CT), which ADP ribosylates the heterotrimeric Gs protein that regulates adenylyl cyclases to increase intracellular cAMP levels, resulting in activation of cAMP-dependent protein kinase, which then stimulates chloride ion channels, thus leading to diarrhea.

Strains of **enteropathogenic *E. coli* (EPEC)** also cause severe, acute infant diarrhea and "traveler's diarrhea," but the pathology is quite different from that of ETEC or EIEC. EPEC strains do not usually have HST or HLT, but sometimes they do have other toxins, such as **cytolethal distending toxin (CLDT),** which causes cell cycle arrest in G_2 phase that leads to distension (enlargement) of the cell, and **cytotoxic necrotizing factors,** which modify and activate small GTPases through deamidation or transglutamination to cause cytoskeletal rearrangements. The most notable characteristic of EPEC strains, however, is that they cause a phenomenon known as **enterocyte attachment and effacement (EAE),** which causes an altered ultrastructure of the apical surfaces of mucosal cells. The attachment and interaction with the host cells result in a characteristic pedestal formation: actin rearrangement, enlargement of short microvilli, and elongation of the microvilli where the bacteria are bound, while the microvilli disappear from areas where the bacteria are not bound. All EPEC genes necessary for the EAE phenotype are found on the chro-

mosome within a 35.6-kb pathogenicity island termed the **locus of enterocyte effacement.** The locus of enterocyte effacement includes genes encoding components of a type 3 secretion system (T3SS) apparatus and secreted proteins (Esp effectors) involved in the disruption of the host cell cytoskeleton and rearrangement to form the pedestals where the bacteria are attached. Another characteristic of EPEC strains is the production of **bundle-forming pili (Bfp),** which are type IV pili similar to the toxin-coregulated pilin (Tcp) pili of *Vibrio cholerae* and are thought to be important for self-aggregation of bacteria (microcolony formation). The biosynthetic genes for Bfp are encoded by a 69-kb plasmid called the EPEC adherence factor plasmid.

Now, consider **enterohemorrhagic *E. coli* (EHEC) strains,** which include the infamous, recently emerged food-borne pathogen *E. coli* serotype O157:H7. EHEC strains do not produce Bfp, but they do produce the EAE phenomenon, like the EPEC strains. In addition, EHEC strains produce several enterotoxins, including **Shiga-like toxins (SLT),** which are closely related to and have the same reaction mechanism as Shiga toxin from toxigenic *Shigella* strains. Again, the SLT is encoded by a gene located on a lysogenic bacteriophage. Hence, EHEC strains have a mixed phenotype, causing EAE like EPEC strains and HUS like *Shigella* strains, but they are not invasive and do not spread from cell to cell like *Shigella* or EIEC.

Yet another virotype of *E. coli* comprises **uropathogenic *E. coli* (UPEC) strains,** which constitute one of the most common causes of **urinary tract infections (UTIs),** particularly in women. UPEC strains often produce hemolysins that can cause lysis of host cells. The unique feature of these strains is their propensity for colonization of the urinary tract (bladder) as opposed to the intestinal tract. UPEC strains produce type I pili, which, through the tip protein adhesin (FimH), bind to mannose-containing glycoproteins **(uroplakins)** located on the luminal surfaces of bladder epithelial cells and mediate invasion, and P pili, which, through the tip protein adhesin (PapG), bind specifically to a globobiose (Gal-α1,4-Gal) disaccharide linked to a ceramide lipid (the P blood group antigen) that is found specifically on the membrane surfaces of host erythrocytes and uroepithelial cells. About 1% of the human population lacks the receptor for P pili, and its presence dictates a person's susceptibility to UTIs caused by UPEC. Although urine flow and a variety of host factors can impede contact of the UPEC with superficial facet cells of the bladder, once contact is made, attachment and/or invasion can result, leading to replication inside the cellular vacuoles and eventually to lysis and **exfoliation** (sloughing) of the cells. Some of the bacteria infect the underlying epithelial cells, where they can persist without being detected by immune surveillance mechanisms.

As illustrated here (and summarized in Figure 19–1), by changing their virulence proteins, *E. coli* strains can dramatically change their virulence and pathological profiles. Horizontal gene transfer, through plasmid conjugation and phage exchange, has already generated a myriad of known phenotypes, and these mixed strains exemplify the ease with which the bacteria can convert from one virotype to another. Clearly, the potential for new virotypes to emerge is high, and *E. coli* subgroups that cause new variations of these diseases, depending on the combinations of virulence factor genes, undoubtedly will appear in the future.

E. COLI NEONATAL MENINGITIS. We turn now to two very serious types of invasive infections that are on the rise, namely, septicemia and meningitis (inflammation of the meninges), caused by **neonatal-meningitis-causing *E. coli* (NMEC)**. If the colon wall is breached by surgery, rupture of the appendix, or other abdominal trauma, *E. coli* enters blood and tissue, where it causes a massive inflammation leading to septic shock. This is the primary form of *E. coli* sepsis seen in adults.

In infants, the presentation of an invasive *E. coli* infection is most commonly associated with meningitis. In fact, next to group B streptococcal infections, *E. coli* is the most common cause of neonatal meningitis (with each species causing 30 to 35% of all cases). During birth, an infant can be exposed to NMEC in a variety of ways. The progression of neonatal meningitis is thought to begin with colonization of the gastrointestinal tract, followed by translocation across the intestinal mucosa. In most cases, this would not have serious consequences, but in the case of some NMEC strains, bacteremia develops, followed by invasion of the cerebrospinal fluid. In adults, the primary risk to survivors of delayed treatment of sepsis is damage to major organs caused by advanced sepsis. However, in the case of neonatal meningitis, the consequences of delayed treatment are more likely to be neurological complications, such as cerebral palsy, learning disabilities, and hearing loss.

Given the different types of progression of *E. coli* invasive disease in adults and in neonates, it is not surprising that the major virulence factors of the causative strains are different. In adult cases, LPS appears to be the main virulence factor. Not all forms of *E. coli* LPS are equally dangerous. LPS molecules with six acyl groups on their lipid A molecule (hexa-acyl LPS) seem to be particularly effective at stimulating the en-

dothelial cell response that results in massive cytokine release, triggering sepsis. An important feature of this type of LPS is that it is poorly recognized by a type of Toll-like receptor (TLR) complex called MD-2 (LPS-binding protein)-TLR4, a form of TLR that normally recognizes and helps to eliminate commensal *E. coli* strains but is not effective against this modified LPS.

In the case of NMEC strains, a type of polysialic acid capsule called K1 seems to be the most important virulence factor. This capsule can be variably O acetylated, a trait that may contribute to virulence. The K1 capsule is poorly immunogenic because sialic acid is a molecule that is widely found on human cells. It confers serum resistance and also has some role in aiding the bacteria to cross the blood-brain barrier. In addition, the inefficiency of the neonates' alternative complement system, as well as a less developed adaptive immunity, hampers their defense against encapsulated bacteria. For reasons that are not clear but that may have something to do with their somewhat suppressed immune systems, pregnant women also appear to be at higher risk of colonization with the K1 capsular strains of NMEC. Colonization of the mother plays an important role in transmission to the neonate.

B. fragilis Internal Abscesses

B. fragilis is a gram-negative, obligately anaerobic, rod-shaped bacterium. Although a subset of *B. fragilis* strains produce an enterotoxin that allows them to cause diarrhea, the main types of infection caused by this species are abscesses and bacteremia. Like *E. coli* disease in adults, the cause of *B. fragilis* infections is trauma to the abdominal area, or primarily surgery, which leads to breaches in the colon wall that allow the bacteria to enter blood and tissue. The genus *Bacteroides* is one of the two major groups of bacteria in the colonic microbiota, the other being a poorly understood group of gram-positive obligate anaerobes that are seldom associated with disease. Although *B. fragilis* is not present in numbers as high as those of other *Bacteroides* spp., it is the one most commonly associated with *Bacteroides* infections.

B. fragilis infections could be considered an emerging disease in the sense that such infections were unknown until an effective antibiotic therapy against invasive *E. coli* infections was developed. *E. coli* septicemia kills so much more quickly that it tended to obscure the much more slowly developing *B. fragilis* infection. The antibiotic mixture that effectively prevented most cases of *E. coli* septicemia was a mixture of an aminoglycoside and a β-lactam antibiotic. *B. fragilis* strains proved to be resistant to this combination.

Another reason *B. fragilis* infections were not formally recognized until the 1970s was that most physicians believed that obligate anaerobes could not cause infection in what they viewed as the aerobic milieu of human tissue and blood. What they failed to take into account is that regions of dead tissue become highly anoxic due to loss of the blood supply. Such lesions are present in most adults and can occur in any organ of the body, including the lungs. *B. fragilis* was able to colonize such dead tissue and form abscesses there that could leak bacteria into the bloodstream. Since most of the oxygen in blood is not free but is bound by hemoglobin, a relatively oxygen-resistant bacterium like *B. fragilis*, which cannot grow in the laboratory if oxygen is present but which can survive oxygen exposure for long periods, survives in blood well enough to cause bacteremia.

The virulence factors of *B. fragilis* have not been defined as clearly and extensively as those of *E. coli*, but some progress has been made toward understanding how *B. fragilis* causes disease. The development of a genetic system for *Bacteroides* spp., which started in the mid-1980s, has helped considerably. As has already been mentioned, the ability of *B. fragilis* to survive in the human body is due in part to its ability to colonize regions of damaged tissue. The production of polysaccharide-degrading enzymes and proteases probably contributes to its ability to grow in the tissue. The major virulence factor, however, is clearly an unusual capsule. This capsule consists of three polysaccharides, PS A, PS B, and PS C. These polysaccharides have a large number of positive and negative charges. PS A and PS B have been shown to cause abscesses in animals even when no bacteria are present. Moreover, unlike most carbohydrate antigens, which are T cell independent, PS A and PS B actually stimulate T-cell activity.

A clinically important feature of *B. fragilis* and other *Bacteroides* spp. that can cause human infections is their increasing resistance to a variety of antibiotics. They are now almost universally resistant to all but the newest tetracyclines. The macrolide antibiotics (clindamycin and erythromycin) that have been used to treat them are losing their efficacy due to the horizontal spread of genes conferring resistance to these antibiotics (see chapter 16). Spread of these resistance genes is occurring actively in the colon due to integrated mobile elements called conjugative transposons, preparing the bacteria to resist treatment should an infection occur.

P. gingivalis and Periodontal Disease

PERIODONTAL DISEASE. *P. gingivalis* is related to *Bacteroides* spp., but whereas the main location of *Bacteroides* spp. is in the colon, *P. gingivalis* is located in the mouth. Like *B. fragilis*, *P. gingivalis* is an obligate an-

aerobe. You might think at first that the mouth is an aerobic environment, and it is for bacteria that stick to the exposed surfaces of teeth or the tissue of the mouth lining. However, the periodontal pocket, the region between the teeth and gums (the gingiva), is quite anoxic, and this is where *P. gingivalis* resides.

P. gingivalis is, as the name suggests, involved in the inflammatory gum disease gingivitis. Although it is normally present in low numbers, some change in conditions, as yet not identified, allows *P. gingivalis* and other gram-negative anaerobes to increase in numbers. The result is inflammation, which causes a deepening of the periodontal pocket, bleeding gums, resorption of bone, and even progression to tooth loss. In fact, periodontal disease is the main cause of tooth loss in adults. Periodontal disease is mainly a disease of adults that is associated with poor oral hygiene, but there is a form of periodontal disease, called juvenile periodontitis, that affects younger people. There is probably also a human genetic factor, because some populations have a higher incidence of periodontal disease than others.

It is important to realize that *P. gingivalis* is not the only gram-negative anaerobe involved in periodontal disease. A number of other, unrelated species (*Tannerella forsythia, Fusobacterium nucleatum, Treponema denticola, Aggregatibacter actinomycetemcomitans, Prevotella intermedia,* and *Eikenella corrodens*) that are members of the oral microbial community have also been associated with periodontal disease, but *P. gingivalis* has so far received the most attention. Also, the virulence factors associated with the other pathogens are similar to those of *P. gingivalis*.

P. gingivalis is the most intensively studied of the oral anaerobes with respect to virulence factors. Already in 1999, a review article on *P. gingivalis* virulence factors that was 70 pages long was published. Two major types of virulence factors appear to be important for the type of tissue damage that occurs during periodontal disease: fimbriae, which mediate attachment to the tooth surface below the gum, and proteases, which damage tissue. Enzymes such as chondroitin sulfatase and hyaluronidase, which degrade host tissue polysaccharides, may also make contributions to tissue destruction.

The proteases not only damage tissue directly and provide food for the bacteria, but they can degrade complement components C3 and C5. The bacteria convert C5 into C5a and C5b. C5a is an attractant for neutrophils, and extensive degradation of C5 may explain the massive influx of neutrophils that occurs during gingivitis. Degradation of C3 may impair its ability to opsonize the bacteria so that they can no longer be ingested and killed by the neutrophils. Proteases of bacteria internalized in gingival epithelial cells (see below) degrade actin fibers and other proteins that maintain cell integrity. This type of activity probably contributes to disruption of the epithelial barrier that normally protects underlying tissue.

Most cases of periodontal disease are still treated by surgery that peels back the gum tissue so that plaque can be scraped from the roots of the teeth. This type of treatment is not pleasant and is very expensive; thus, new laser-based surgery techniques are starting to gain popularity. Not surprisingly, there is also hope that antibiotics, particularly in multidrug combinations, might be used to prevent or even treat periodontal disease. A problem with using antibiotic treatment is that the bacteria form biofilms on the tooth surface, and biofilm-embedded bacteria are generally more resistant to antibiotics than planktonic bacteria.

CORONARY ARTERY DISEASE: INVOLVEMENT OF ORAL BACTERIA? *P. gingivalis* can occasionally cause internal abscesses, but the most serious internal condition with which *P. gingivalis* is associated is cardiovascular disease. Whether *P. gingivalis* actually contributes to causing cardiovascular disease or merely colonizes atherosclerotic plaques in blood vessels is controversial, but the association between periodontal disease and cardiovascular disease is now widely accepted. This possible effect of *P. gingivalis* is exciting, because it opens up the possibility that antibiotics could be used to prevent or treat heart disease.

Virulence factors that might be involved with bacterial invasion of atherosclerotic plaques are not well defined. *P. gingivalis* can invade human coronary artery endothelial cells in culture. Even though these cells are not normally phagocytic, there is a process called **autophagy,** widespread in eukaryotic cells, that allows the cell to degrade damaged or surplus organelles, such as mitochondria. A vesicle forms and fuses with lysosomes, allowing the ingested organelle to be degraded. *P. gingivalis* apparently enters the cell via such vesicles (autophagosomes) but prevents fusion with the lysosome and thus is able to replicate within the eukaryotic cell. Another potential indirect involvement of an oral anaerobe in adverse health outcomes, this time *F. nucleatum,* is described in Box 19–1.

Normal Soil Inhabitants Weigh In as Opportunists

P. aeruginosa Infections

P. aeruginosa has been called the consummate opportunist. *P. aeruginosa* is a soil microorganism with an impressive capacity to utilize a variety of carbon compounds as carbon and energy sources. It has been said

---

BOX 19–1 *F. nucleatum*—a Cause of Preterm Birth?

Preterm birth, defined as birth before 37 weeks of gestation, is well known as a condition that is associated with developmental defects and other lifelong consequences. There are undoubtedly many causes of preterm birth, but now an unexpected one has surfaced—gum disease. An association between periodontitis and preterm birth has been noted for some time. Oral anaerobes, such as *F. nucleatum,* a species with a proven capacity to cause inflammation and tissue damage, was isolated from the placentas and uterine fluid of women who had experienced preterm birth. However, this was still just guilt by association: the bacteria could have been enabled to cause infection by some other condition. Finally, however, there is a mouse model in which *F. nucleatum* is clearly the cause of preterm birth. Granted, animal models can be misleading, but this advance is a step toward making a cause-and-effect connection. The ultimate proof of this connection would be an antibiotic intervention that prevents preterm birth.

Preterm birth has also been associated with bacterial vaginosis, a shift in the vaginal microbiota from predominantly gram-positive to predominantly gram-negative bacteria. In this case, an antibiotic intervention that restored the gram-positive dominance in the vaginal tract did have a protective effect. Although the view of what factors influence preterm birth is still murky, and although it seems clear that there are multiple possible causes of preterm birth, it seems likely that at least some of the predisposing factors are bacterial.

Source: H. Liu, R. W. Redline, and Y. W. Han. 2007. *Fusobacterium nucleatum* induces fetal death in mice via stimulation of TLR4-mediated placental inflammatory response. *J. Immunol.* **179:**2501–2508.

---

that if you can draw a chemical structure containing carbon, oxygen, and hydrogen, some strain of *P. aeruginosa* can catabolize it. Due to its versatility in carbon source utilization, *P. aeruginosa* has been widely used for bioremediation of such toxic compounds as toluene and trichloroacetate. Unfortunately, *P. aeruginosa* is also capable of causing serious disease in humans. Although *P. aeruginosa* can cause a variety of infections, ranging from wound and bloodstream infections to UTIs, it is best known for its ability to infect burned tissue and to cause lung infections in people with cystic fibrosis.

BURN INFECTIONS. The predilection of *P. aeruginosa* for burned tissue is easy to understand. Burned tissues seep plasma, which is full of nutrients. *P. aeruginosa* is likely to be on the scene because it is ubiquitous in the environment and can quickly take advantage of such damaged tissue. Since *P. aeruginosa* infections are the most likely cause of the death of patients who have survived the initial trauma of severe burns, physicians are on the watch for early signs of *P. aeruginosa* infections in these patients.

It is important to treat *P. aeruginosa* infections as early as possible for two reasons. First, *P. aeruginosa* is notorious for its resistance to a variety of antibiotics. The formation of a biofilm on the burned tissue makes it even more resistant to antibiotics. Second, what actually kill patients are bacteria that leak into the bloodstream, so it is important to treat the infection early and effectively in order to prevent the movement of bacteria into the bloodstream. Other bacteria, such as *Staphylococcus aureus* (see chapter 18), cause burn infections, but for some reason, *P. aeruginosa* is the leading cause of such infections.

INFECTIONS IN CYSTIC FIBROSIS PATIENTS. Another classic *P. aeruginosa* infection is infection of the lungs of cystic fibrosis patients. Cystic fibrosis is caused by a disorder of chloride channels that leads, among other things, to thicker mucus in the lungs and airways. This thick mucus can obstruct the airway, causing coughing and wheezing. It also seems to interfere with the ciliated-cell defense of the airway that propels bacteria trapped in mucus out of the airway. This makes it easier for bacteria to gain access to the lungs.

Within the first few years of life, about 80% of cystic fibrosis patients acquire *P. aeruginosa* in their lungs. Oddly enough, although most cystic fibrosis patients are colonized early in life, lung damage and reduced lung capacity do not occur abruptly. Rather, the effects of the infection develop slowly over a period that can extend to many years. Treatment of the infection with antibiotics slows the progression of the damage but does not clear the bacteria from the lung. Gradually, changes in the physiological traits of the bacteria, such as production of an alginate capsule, and increasing resistance to antibiotics combine to impair the lungs to a great enough extent that the patient dies. Cystic fibrosis patients once died early in life, but improve-

ments in treatment of the infection have now extended the life spans of such patients into their 30s and 40s. Thus, although the problem of *P. aeruginosa* infections in cystic fibrosis patients has not been solved, great improvements have been made.

VIRULENCE FACTORS. Of the opportunists covered in this chapter, by far the most is known about virulence factors involved in tissue damage caused by *P. aeruginosa*. These virulence factors can be classified into several different categories (Table 19–2). There are nonpilus adhesins and a flagellum on the cell surface. There are also a number of tissue-damaging toxins and enzymes, such as proteases; hemolysins; exotoxins A, S, T, U, and Y; and the nonprotein redox-active toxic pigment pyocyanin. The proteases, hemolysins, and exotoxin U (ExoU, a phospholipase) have relatively nonspecific activities that damage the surfaces and membranes of eukaryotic cells.

Exotoxin A (ExoA) is an external toxin that binds mammalian cells and translocates into the cytoplasm. It has an A subunit that ADP ribosylates elongation factor 2, an essential factor in protein synthesis. Exotoxins S and T (ExoS and ExoT) are injected directly into the mammalian cell via a T3SS (see chapter 13). These toxins are both bifunctional, with a GTPase-activating activity and ADP-ribosyltransferase activity. Although these two toxins act against different intracellular targets, they both have the same effect—to disrupt the mammalian cell cytoskeleton and trigger an apoptosis-like cell death. ExoU (a phospholipase) and ExoY (an adenylyl cyclase with unknown function) are also injected by the same T3SS.

A different type of cytotoxin is the pigment pyocyanin, which gives colonies of *P. aeruginosa* their blue-green color on agar medium. Large amounts of pyocyanin are produced in the lungs of cystic fibrosis patients. Pyocyanin interferes with the normal redox cycle in mammalian cells, actually leading to production of reactive forms of oxygen that can damage the mammalian cell. Pyocyanin may also be a factor in the tissue necrosis that occurs during infections of burned areas. The production of the blue pigment by the bacteria is seen in burn patients as blue pus. Also, this pigment fluoresces under long-wave UV light. In fact, shining a Woods lamp on a burned area to detect this fluorescence is an old but useful way to discover early signs of a *P. aeruginosa* burn infection.

Another nonprotein toxin of *P. aeruginosa* is LPS, which acts, as in other gram-negative bacteria, by activating cytokine release and the complement cascade. Still other nonprotein toxins produced by *P. aeruginosa* are rhamnolipids, which form a barrier around the bacteria, a phenomenon that has been given the name

Table 19–2 Notable virulence factors of *P. aeruginosa*

Virulence factor	Functional role in pathogenesis
Flagellum (polar)	Motility, dissemination, initiation of the innate immune response
Type IV pili (polar)	Adherence, twitching motility
Siderophores	Iron acquisition
Phospholipases	Hydrolysis of phospholipids in host membranes, tissue damage, phosphate acquisition
Elastase, alkaline protease, and other proteases	Proteolytic degradation of elastin, collagen, immunoglobulins, complement proteins, immune evasion, and nutrient acquisition; tissue damage
LPS	Endotoxic shock, sepsis, serum resistance
Alginate	Adherence, protection from dehydration, antiphagocytic, protection of biofilm from host immune system
Hemolysins	Pore-forming toxins that damage host cell membranes
Pyocyanin	Small-molecule toxic blue-green pigment; redox-active compound that generates reactive oxygen species; disrupts action of cilia and phagocytic function
ExoA	AB toxin that ADP-ribosylates EF-2; inhibits host cell protein synthesis; causes cell death
ExoS	Bifunctional T3SS effector protein with GTPase-activating activity that inactivates small Rho GTPases (Rho, Rac, and Cdc43) to disrupt the actin cytoskeleton and ADP-ribosyltransferase activity that modifies several host signaling proteins, including small Ras GTPases, cell rounding
ExoT	Bifunctional T3SS effector protein with GTPase-activating activity that inactivates small Rho GTPases (Rho, Rac, and Cdc43) to disrupt the actin cytoskeleton and ADP-ribosyltransferase activity that modifies Crk proteins involved in focal adhesion and inhibits phagocytosis, cell rounding
ExoU	T3SS effector protein that is a phospholipase A2; disrupts host cell membranes; cytotoxicity; induces inflammatory response in host
ExoY	T3SS effector protein that has adenylyl cyclase activity; increases intracellular cAMP levels; cytotoxicity; role in pathogenesis unclear
MDR efflux system (MexEF, OprMN, and MexT)	Antibiotic efflux pump (MexEF-OprMN) and its regulator (MexT); confers antibiotic resistance
Quorum sensing—LasR (3-oxo-C_{12}-HSL), RhlR (C_4-HSL)	Key role in controlling virulence factor production, biofilm formation, swarming motility, and expression of antibiotic efflux pumps

shielding. These lipids are toxic for polymorphonuclear leukocytes and so may protect the bacteria from them. Finally, quorum sensing, in the form of the cell-to-cell signaling systems (Las and Rhl) that control the expression of many of the virulence factors, aids the formation of biofilms.

Thus, *P. aeruginosa* has a highly complex repertoire of activities that allow it to be a successful pathogen in certain compromised hosts. Precisely how each of these factors plays into the special ability of *P. aeruginosa* to colonize and damage the lungs of cystic fibrosis patients is still under study. Also unclear is the sequence of action of these factors, which presumably do not act all at once during the long progressive development of infection. Another very interesting question is what these virulence factors and toxins do when *P. aeruginosa* is in the environment or interacting with other hosts.

RESISTANCE TO ANTIBIOTICS. *P. aeruginosa* is not only resistant to a number of antibiotics when it first infects cystic fibrosis patients, but it becomes more resistant as the lung infection progresses. The bacteria often employ a variety of different resistance mechanisms in concert, such as varying porins that prevent entry, efflux systems that pump the antibiotic out of the cell, and antibiotic-inactivating enzymes. Of particular note is an efflux system (MexEF-OprMN) that can expel a variety of antibiotics. It is interesting that a regulator of this system (MexT) has been shown to regulate other *P. aeruginosa* genes that are involved in virulence, such as the genes that encode the T3SS. A direct connection between antibiotic resistance and virulence has been suspected in the cases of many bacterial pathogens, but evidence for such a connection has been sparse.

B. cenocepacia—P. aeruginosa's Evil Twin

B. CENOCEPACIA PNEUMONIA. Many of the traits of *B. cenocepacia* are similar to those of *P. aeruginosa*. (In fact, *B. cenocepacia* was originally called *Pseudomonas cepacia* and then *Burkholderia cepacia*, and although *B. cenocepacia* is the current name, *B. cepacia* still predominates in the literature.) For example, *B. cenocepacia* is a relatively common cause of pneumonia in cystic fibrosis patients. Moreover, it resembles *P. aeruginosa* in the variety of carbon sources it can use. Finally, species of both *Pseudomonas* and *Burkholderia* have been used as bioremediation or protective agents in the recent past. Although *B. cenocepacia* has been and continues to be a serious infectious-disease problem in cystic fibrosis and patients in intensive-care units, an unusual type of opportunistic infection caused by *B. cenocepacia* is of interest because it illustrates how bacteria manage to take advantage of an apparatus seen widely in intensive-care wards, the ventilator tube.

Patients in intensive-care wards can have their breathing assisted by respirators. A tube is inserted deep into the airway to deliver air to the patient. Since the tube bypasses the defenses of the upper airway, such as mucus and ciliated cells, patients on respirators are particularly prone to develop ventilator-associated pneumonia. Also, since these patients are often comatose or semicomatose, they are prone to the buildup of bacterial biofilms on their teeth that can increase the risk of bacteria being inhaled into the lungs, where they can cause infections. To prevent such infections, hospital staff members use mouthwash to clean the mouths and teeth of patients on ventilators. Apparently, batches of mouthwash have been contaminated with *B. cenocepacia*, so hospital staff attempting to prevent infection can actually increase the risk of infection by *B. cenocepacia*.

The virulence factors of *B. cenocepacia* are currently under study. An impressive variety of animal models are available, ranging from a rat model to insect and zebrafish models. *B. cenocepacia* seems to have many of the virulence factors commonly seen in *P. aeruginosa*, including siderophore production, LPS, proteases, pili, and biofilm formation. The resistance of *B. cenocepacia* to a variety of antibiotics is a serious problem and appears to be due in large part to efflux pumps that eject many types of antibiotics from the cell.

A. baumannii—an Emerging Threat from the Iraq War

A. baumannii has long been known as a ubiquitous soil organism, but it was not previously thought to be capable of causing significant human disease. This view of *A. baumannii* changed during the Iraq war, when soldiers with traumatic wounds began to show signs of lethal *A. baumannii* infections. Many whose wounds were not obviously infected became colonized and were thus at risk of subsequent infection. Wounded soldiers from Iraq brought along the Iraq strains of *A. baumannii* when they returned to the United States, and *A. baumannii* has rapidly become the latest disease-causing opportunist to check into Walter Reed hospital and spread from there to other hospitals around the country. These and similar incidents with other United Nations soldiers have helped to spread this newest scourge around the world. Since virtually nothing is known about the virulence factors of *A. baumannii*, it is not clear whether the Iraq strain was unusually virulent or whether it just happened to be in the right place at the right time, when a new group of susceptible hosts presented themselves.

Concern about *A. baumannii* has extended beyond its ability to cause serious wound infections. *A. baumannii* has proven to be one of the multidrug-resistant pathogens that are resistant to so many antibiotics that they have been called **"panresistant"** and are now categorized as one of the **"superbugs,"** along with methicillin-resistant *Staphylococcus aureus* and community-acquired methicillin-resistant *S. aureus* (see chapter 18) and *Mycobacterium tuberculosis*, so called because they are not only **multidrug resistant,** but **extensively drug resistant.** There are still some antibiotics that work, but their number is small and dwindling, making treatment of *A. baumannii* infections or clearance of carriers no simple undertaking. Added to this is the inherent resistance of the bacterium to disinfectants and various environmental conditions, which hampers containment and decontamination and necessitates stringent infection control procedures, such as aggressive sterilization and patient isolation (implementation of the so-called 1 patient = 1 nurse/doctor rule) in hospitals during incidents.

BOX 19–2   The Broader View—Oral Anaerobes

Breaking out of the one microbe-one disease paradigm is hard to do and is frightening to contemplate because it makes formulation of hypotheses and analysis of the data so much more difficult. In a study of periodontal disease using a rat model, investigators tested various combinations of three oral anaerobes: *Prevotella gingivalis, T. forsythia,* and *T. denticola.* The results showed that a combination of the three pathogens was significantly more effective in triggering bone resorption and chronic periodontitis than the individual pathogens alone. This type of multiorganism study is not completely novel, but the increase in sophistication of animal models and the types of measurement available give it new impetus.

A more ambitious broader view asks the question, what bacteria are actually released into the blood during inflammation of the gum? It has been clear for some time that the bacteria released into the blood must be a complex mixture, but in virtually all descriptions of the effects of bacterial release on such processes as abscess formation, atherosclerosis, and infection of the placenta, the picture is one of a single organism entering and traveling through the bloodstream. A group of scientists from the Forsyth Institute have now surveyed 98 isolates from the blood of patients with bacteremia after oral disease. Not surprisingly, these investigators found many different species of oral bacteria. This diversity is probably an underestimate, because the scientists surveyed only cultivated organisms. It would be interesting to look at the diversity of released bacteria using culture-independent methods.

Both of the approaches described here represent the beginning of an attempt to gain a more realistic view of the microbiological situation in oral disease and its sequelae. The real challenge, however, will be to expand our ability to analyze and comprehend the real-world situation, which is a complex picture. In fact, the whole area of mixed infections is bound to yield surprises. Considerable work is now ongoing on biofilms containing two or more bacterial species. As noted in chapter 18, some bacteria seem to stimulate innate immune responses that do not clear them completely but eliminate their bacterial competitors, at least in animal models of infections. Several bacterial pathogens produce hydrogen peroxide, which can damage host cells as well as competing bacteria, and many bacteria use bacteriocins to kill competitors of the same species.

Sources: F. K. Bahrani-Mougeot, B. J. Paster, S. Coleman, J. Ashar, S. Barbuto, and P. B. Lockhart. 2008. Diverse and novel oral bacterial species in blood following dental procedures. *J. Clin. Microbiol.* **46:**2129–2132; L. Kesavalu, S. Sathishkumar, V. Bakthavatchalu, C. Matthews, D. Dawson, M. Steffen, and J. L. Ebersole. 2007. Rat model of polymicrobial infection, immunity, and alveolar bone resorption in periodontal disease. *Infect. Immun.* **75:**1704–1712.

Don't Forget the Arthropods

Ehrlichia spp.

Arthropod-borne infections are well known. Major current diseases, such as malaria and Lyme disease, and diseases of the past, such as plague, have received a lot of attention over the years. All of the well-studied arthropod-borne pathogens are opportunists in the sense that they get into the human body via arthropod bites, but these pathogens are not usually considered opportunists because they cause infections in otherwise healthy people. An example of an arthropod-borne pathogen that could be considered to be an opportunist is ***Ehrlichia chaffeensis.*** *E. chaffeensis* is an obligate intracellular pathogen most closely related to *Rickettsia* spp. (causing Rocky Mountain spotted fever and typhus). It invades monocytes and causes a febrile disease called **human monocytotrophic ehrlichiosis (HME).** *E. chaffeensis* is transmitted by ticks, principally the Lone Star tick. The geographical distribution of ehrlichiosis cases, which occur primarily in the Southeast, is probably due to limitations in the range of the Lone Star tick.

The reason for calling *E. chaffeensis* an opportunist is that HME is seen almost exclusively in older people and people with compromised immune systems. In fact, the early cases of HME occurred in golf-centered

BOX 19–3 The Broader View—*P. aeruginosa* Variation

Most accounts of infections read as if a single strain with a given repertoire of characteristics is responsible for the infection. A recent study of the diversity of strain phenotypes of *P. aeruginosa* in an actual infection produced a surprising finding that challenges the one strain-one phenotype paradigm. The study looked at the phenotypes of 9 to 12 isolates each from the lungs of eight different adult cystic fibrosis patients. The characterization of the isolates focused on quorum-sensing regulators, which have been thought to play a key role in biofilm formation in the cystic fibrosis lung and thus in the pathogenesis of *P. aeruginosa* infections in cystic fibrosis patients. The investigators found a wide range of quorum-sensing phenotypes, from completely inactive alleles of some regulatory genes to completely active ones. This type of variation was seen not only between different patients, but also within the same patient.

The authors stated, "Conclusions about the properties of *P. aeruginosa* QS [quorum-sensing] populations in individual CF [cystic fibrosis] infections cannot be drawn from the characterization of one or a few selected isolates." Perhaps more striking was the implication that there was not a strong selection for maintenance of the functions of regulatory proteins responsible for the control of different quorum-sensing response systems. These results suggest that a broader view of the complexity of different phenotypes that can be sustained in the same lung environment is needed. It will be interesting to see if the same type of functional variation is found in other "essential" virulence factors, such as adhesins, exotoxins, and proteases.

Source: C. N. Wilder, G. Allada, and M. Schuster. 2009. Instantaneous within-patient diversity of *Pseudomonas aeruginosa* quorum-sensing populations from cystic fibrosis lung infections. *Infect. Immun.* **77:**5631–5639.

---

communities in Florida and Georgia, where most of the residents were elderly people who spent time on the golf course, where they could easily come into contact with ticks—especially if they spent a lot of time in the rough. Since then, however, HME has been seen in a wider range of people but still seems to have a predilection for people with waning ability to fight off bacterial invaders.

E. chaffeensis is maintained in the environment in nonhuman mammals, such as deer. In this respect, its life cycle resembles that of ***Borrelia burgdorferi,*** the cause of **Lyme disease.** In both cases, humans seem to be an accidental host, not a host that provides an important reservoir for the bacteria. In humans, *E. chaffeensis* invades monocytes or macrophages, somehow prevents phagosome-lysosome fusion, and multiplies within the phagosome. Ultimately, the bacteria are released and go on to infect other mammalian cells. *E. chaffeensis* is a tiny bacterium with a gram-negative-type cell wall that is not easily visible in a Gram-stained preparation.

The discovery of *E. chaffeensis,* a possible arthropod-borne opportunist, raises the question of whether there are other such arthropod-borne oppor-

tunists. Possibly, the pathogenic potential of the arthropod-borne bacterial pathogens ranges over a continuum as wide as that seen in free-living bacteria. It will be interesting to see if other cases of arthropod-borne opportunists—if you accept the characterization of *E. chaffeensis* as an opportunist—will be discovered in the future, now that scientists have been alerted to their existence. Another area of future research is how these pathogens interact with each other during coinfections of the arthropods or their mammalian prey.

The Broader View

Pathogenesis research has been dominated by the study of single organisms as the sole causes of diseases. Moreover, studies of infections caused by single organisms have generally focused on specific pathways, which may not provide a full picture of the infection process. Although this view is understandable because it simplifies things in a gratifying way, it is likely to be incorrect in many cases. Recently, some investigators have been taking a broader view. Some examples of this broader view are illustrated in Box

19–2 for oral bacteria and Box 19–3 for *P. aeruginosa*. These examples are not necessarily unique to the organisms covered in this chapter, but they illustrate new developments in pathogenesis research that are taking advantage of new technologies and new insights. Some of these approaches were introduced in chapters 5 and 9.

SELECTED READINGS

Amano, A. 2007. Disruption of epithelial barrier and impairment of cellular function by *Porphyromonas gingivalis*. *Front. Biosci.* **12:**3965–3974.

Bergfeld, A. K., H. Claus, U. Vogel, and M. Muhlenhoff. 2007. Biochemical characterization of the polysialic acid-specific O-acetyltransferase NeuO of *Escherichia coli* K1. *J. Biol. Chem.* **282:**22217–22227.

Centers for Disease Control and Prevention. 1998. Nosocomial *Burkholderia cepacia* infection and colonization associated with intrinsically contaminated mouthwash—Arizona, 1998. *MMWR Morb. Mortal. Wkly. Rep.* **47:**926–928.

Cornelis, P. 2009. Shielding, a new pathogen defense mechanism against polymorphonuclear neutrophilic leukocytes (PMNs). *Microbiology* **155:**3474–3475.

Demmer, R. T., and M. Desvarieux. 2006. Periodontal infections and cardiovascular disease. The heart of the matter. *J. Am. Dent. Assoc.* **137**(Suppl. 2):14S–20S.

Eley, B. M., and S. W. Cox. 2000. Proteolytic and hydrolytic enzymes from putative periodontal pathogens: characterization, molecular genetics, effects on host defenses and tissues and detection in gingival crevice fluid. *Periodontology* **31:**105–124.

Hauser, A. R. 2009. The type III secretion system of *Pseudomonas aeruginosa*: infection by injection. *Nat. Rev. Microbiol.* **7:**654–665.

Holt, S. C., L. Kesavalu, S. Walker, and C. A. Genco. 1999. Virulence factors of *P. gingivalis*. *Periodontol. 2000* **20:**168–238.

Tian, Z.-X., E. Fargier, M. M. Aogain, C. Adams, Y.-P. Wang, and F. O'Gara. 2009. Transcriptome profiling defines a novel regulon modulated by the LysR-type transcriptional regulator MexT in *Pseudomonas aeruginosa*. *Nucleic Acids Res.* **37:**7546–7559.

Uehlinger, S., S. Schwager, S. P. Bernier, K. Riedel, D. T. Nguyen, P. A. Sokol, and L. Eberl. 2009. Identification of specific and universal virulence factors in *Burkholderia cenocepacia* strains by using multiple infection hosts. *Infect. Immun.* **77:**4102–4110.

QUESTIONS

1. Speculate about how and why different opportunists seem to be specifically adapted to take advantage of a limited range of opportunities.

2. How would you formulate Koch's postulates for opportunistic pathogens?

3. In the case of the *E. coli* K strains, why would the number of acyl chains on *E. coli* LPS affect the function of a TLR, given that LPS does not bind the TLR directly?

4. Why is there no vaccine against *E. coli* neonatal meningitis?

5. Would it be appropriate to develop a vaccine against *B. fragilis* or *P. gingivalis*? Why or why not? If not, what would be the most effective types of prevention?

6. Speculate about why *P. aeruginosa* and, to a lesser extent, *B. cenocepacia* seem to outdo organisms found in the human body when it comes to infecting the lungs of a cystic fibrosis patient. So far, no one has a definitive answer to this question, but you might have some good ideas.

7. Why might *P. aeruginosa* produce so many virulence factors? Why is cell-to-cell signaling, which controls the production of many of these factors, so important? Could these signaling systems become a target for antibiotics?

8. Do you consider *E. chaffeensis* an "opportunist"? Give arguments for and against this designation. If you accept *E. chaffeensis* as an opportunist, what implications does this have for other possible arthropod-borne microbes?

9. EHEC and EPEC strains are often confused with each other and misidentified. For each of the following, describe at least two of the main similarities or differences between these strains in regard to:

   A. Interactions with host cells
   B. Production of toxins
   C. Symptoms of infection

SOLVING PROBLEMS IN BACTERIAL PATHOGENESIS

1. You have just isolated a new bacterium from a recent hospital outbreak that was based on 16S rRNA comparison. It is closely related to *A. baumannii*. *A. baumannii* is a gram-negative, opportunistic, nosocomial (hospital-acquired) pathogen that is able to colonize patients in intensive-care units, causing pneumonia, UTIs, septicemia, or meningitis, depending on the route of infection. Epidemic strains of *A. baumannii* are often multidrug resistant. You have named your bacterium *Acinetobacter newbii* and have also determined that it is resistant to aminoglycosides, β-lactams, chloramphenicol, tetracyclines, and the dye ethidium bromide. Using an in vivo expression technology (IVET) approach in a swine lung model of infection, you have identified four potential operons that may encode virulence factors.

A. You find that three of the operons identified appear to be part of a regulon. What specific evidence might have led you to this conclusion?

B. One operon consists of two genes in tandem, which you have named *antBA*. Sequence comparison with protein databases reveals that *antA* encodes a protein of 25 kDa that has significant sequence similarity to the catalytic N-terminal 200 amino acid residues of diphtheria toxin, while *antB* encodes a protein of 10 kDa that has significant sequence similarity to the B subunit of CT. You suspect that the Ant protein is an AB-type toxin and decide to test this using a swine lung tissue cell model, and indeed, Ant protein is cytotoxic to the cells. Further studies reveal that Ant action on lung cells requires a low-pH-dependent step, since agents that prevent acidification, such as basic buffers, block the toxin effects. You make antibodies against the culture filtrate from the *A. newbii* cells and use the antisera in Western blots of the cell lysates from the bacterial cells, the media, and the intoxicated lung cells after separation of the proteins by sodium dodecyl sulfate-polyacrylamide gel electrophoresis. The results are shown in Figure 1. Provide a detailed explanation (with rationale) for and interpretation of the results shown in the Western blot in Figure 1. Was your conclusion that Ant is an AB-type toxin correct? Draw a diagram depicting a possible model that accounts for all of the experimental observations, i.e., predict the structure of the toxin and show its mode of action. Be sure to justify your model, i.e., it must be consistent with your answer to part A.

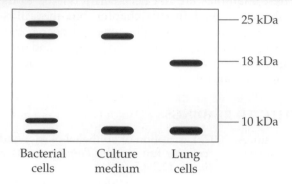

Figure 1 Western blot.

C. In further experiments to determine what regulates the production of Ant, you find that both AntA and AntB are secreted into the medium only when the bacteria are grown under in vivo conditions, including normal body temperature (37°C) and high blood/tissue levels of CO_2. A second operon contains a single open reading frame that has no homology to any protein in the database, but when you knock the gene out with a transposon insertion, you find that the Ant toxin is no longer made in vivo, and you suspect that this gene, which you have now named *antR*, encodes a regulator of *antAB*. Describe how you would confirm quantitatively that Ant expression is regulated at the transcriptional level by physiological CO_2 conditions. Predict your results. Describe how you would determine if AntR is a transcriptional activator or repressor of *antAB* gene expression. Based on the information given, what is the predicted result?

D. The third operon consists of five genes, which you have named *anrCDEFG*. The *anrFG* and *anrCDE* genes are adjacent but separated by an overlapping promoter region, in which they are transcribed in opposite directions. You discover that transposon insertion into any of the *anr* genes, except *anrG*, results in loss of antibiotic resistance, whereas transposon insertion into *anrG* results in increased antibiotic resistance. Analysis of the protein sequence of AnrF reveals that it has four membrane-spanning hydrophobic sequences in its N-terminal domain and some sequence homology to protein kinases in its C-terminal domain. Protein sequence analysis of the other proteins reveals that none of them have any significant sequence homology to known proteins in the database, but AnrG does have a helix-loop-helix secondary-structure motif that is frequently found in DNA-binding proteins, AnrC has an ATP-binding motif and 12 membrane-spanning hydrophobic sequences, AnrD also has multiple stretches of hydrophobic residues, and AnrE has no strongly hydrophobic regions. You make antibodies against each of these proteins. In order to determine the cellular localization of the proteins, you fractionate the bacteria into cytoplasmic, inner membrane, periplasmic, and outer membrane fractions and find that AnrG is located in the cytoplasm, AnrF and AnrC are in the inner membrane, Anr E is in the periplasm, and AnrD is in the outer membrane. Draw a diagram of the operon. Based on the experimental results, draw a schematic diagram of a model of the bacterium and the locations of the Anr proteins that accounts for all of the observations. Provide an interpretation of the results that accounts for all of the observations.

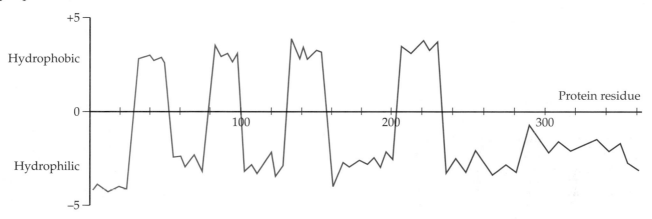

Figure 2 Hydrophobicity plot.

E. Using the information from the hydrophobicity plot in Figure 2, design an experiment that would allow you to determine the membrane topology of the AnrF protein. Be as specific and efficient as possible, and provide details of your experimental setup. Be sure to provide your expected results and a model for the membrane topology of your protein.

F. The fourth operon consists of two genes, which you have named *anqHI*, separated by an overlapping promoter region, with the two genes transcribed in opposite directions. You make antibodies against these proteins, and using Western blots of bacterial cellular fractions, you determine that both AnqH and AnqI are soluble cytoplasmic proteins. When you knock out *anqH* or *anqI* by transposon insertion, you find that *A. newbii* is no longer antibiotic resistant and no longer makes any of the Ant or Anr proteins. When you add the bacterial culture filtrates from wild-type *A. newbii* or from the *anqH* deletion mutant to the *anqI* deletion mutant cultures, you find that antibiotic resistance and toxin production are restored, but adding the culture filtrates from the *anqI* deletion mutant to the *anqH* deletion mutant cultures does not restore antibiotic resistance or toxin production. What possible mechanism could account for these observations? Give details and be specific! Be sure to predict the functions of AnqH and AnqI. Describe or draw a diagram showing how the substance found in the culture filtrates from the wild type and the *anqH* deletion mutant might be generated. Be sure to include your best guess as to its structure.

2. You are a scientist working for the FDA, and you have been given the task of designing a virotyping kit that would allow you to readily identify pathogenic strains of EHEC, EIEC, EPEC, ETEC, and *Shigella* based on probes that recognize virulence factors. Probes to which specific virulence factors would you choose to use in your strain-virotyping kit? Be sure to pick *sufficient* and *unique* factors that would allow each of the strains to be *clearly* and *definitively* identified from the others. Give your rationale!

3. Working for the EPA, you have isolated several *E. coli* strains contaminating a pond near a sewage treatment facility. Using an antibody-based diagnostic kit, you have obtained the results shown below (Figure 3) on your dot blot (similar to a Western blot, but the protein sample is directly applied as a spot to the membrane without sodium dodecyl sulfate-polyacrylamide gel electrophoresis being run first). Based on these results, predict the identity of each of the unknown isolates. Be sure to *briefly* state your rationale. CNF, cytotoxic necrotizing factor.

(continued)

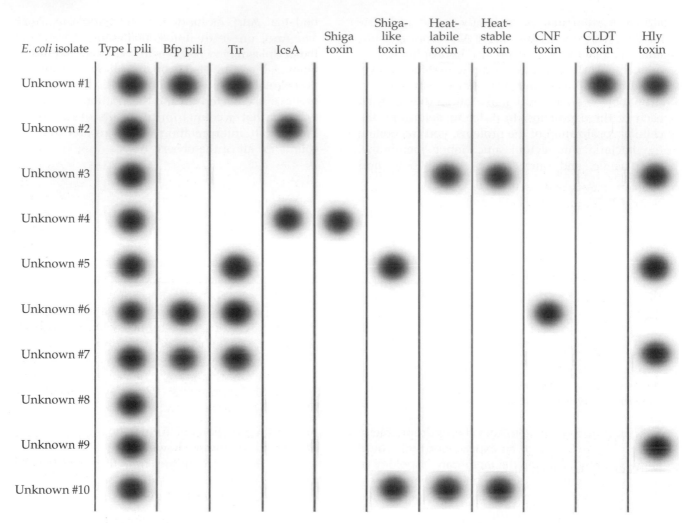

Figure 3 Dot blot.

SPECIAL GLOBAL-PERSPECTIVE PROBLEM: INTEGRATING CONCEPTS IN PATHOGENESIS

1. You are a microbiology graduate student with aspirations of one day completing your doctoral work in bacterial pathogenesis. Your research thesis advisor has ample funds and has allowed you to pursue your own project from scratch. While you are contemplating what bacterium you would like to work on, a group of your friends return from spending holiday break at a resort in southern Florida. Shortly after returning home, five of these friends, who had enjoyed the spa's Jacuzzi facilities, presented to the student health center with an acute generalized eruption of follicular lesions (skin rash with pustules over most of the body). The doctor prescribed some antibiotic ointment containing a combination of gentamicin and Augmentin (a combination of the β-lactams amoxicillin and clavulanic acid). After 1 week of topical treatment, three of the five friends healed without sequelae. However, two of them had sustained some

abrasions at a volleyball game played during the vacation. Despite treatment with the antibiotic ointment, their wounds became further infected, with symptoms of pain, swelling, and redness followed 2 days later by fever and then admission to the hospital with symptoms of septicemia and disseminated intravascular coagulation (DIC). Intravenous administration of gentamicin, along with some surgical debridement of the wound area, rescued these two friends from a potentially dire outcome. Bacterial cultures from the blood of the two friends revealed that the septicemia was caused by a gram-negative rod-shaped bacterium, which looked similar to the organism observed in Gram stains from the follicular lesions. Having a desire to prevent this from happening again (and to fulfill a vendetta on behalf of your friends against this pathogen), you decide to examine the bacterium in more depth, so you obtain a culture of the blood isolate from the diagnostics laboratory. You isolate the genomic DNA and sequence the 16S rRNA gene, and determine that the bacterium is distantly related to *Pseudomonas* and *Aeromonas* spp., which are largely found in aquatic environments but can cause opportunistic infections. Below are a series of experiments you then undertake to study the pathogenesis of this new bacterium, which you have named *Graduamonas exhaustmium*.

A. A combination of gentamicin and amoxicillin-clavulanic acid was used for the follicular lesions, but only gentamicin was used for the septicemia.

 i. Why was amoxicillin-clavulanic acid not used intravenously in this case, yet gentamicin was?

 ii. Why is amoxicillin-clavulanic acid formulated as a combination drug?

 iii. Why was surgical debridement necessary?

 iv. What caused the symptoms of DIC in this case? Be specific.

B. Although you do not have a culture of the bacterium from the follicular lesions, you suspect that the bacterium observed in the Gram stains is the same *G. exhaustmium* isolated from the blood samples.

 i. What four criteria must be satisfied to show that *G. exhaustmium* is indeed responsible for causing the follicular lesions?

 ii. Describe how you would go about demonstrating this experimentally. Provide a reasonable model of infection. Be specific for this case.

 iii. Provide one additional, different molecular experimental approach that could be performed to help satisfy these criteria for this case.

C. Using your experimental model of infection, you wish to identify potential virulence factors. *Briefly* describe how you might go about doing this. Provide your rationale for your choice of approach. Be sure to include how you would verify that they are indeed virulence factors.

D. From your screening, you isolate 10 mutants with deletions in genes that encode putative virulence factors (denoted GeMut1 to GeMut10). Sequence analysis of the genes reveals that Mut1 has an N-terminal stretch of 15 hydrophobic residues but otherwise no homology motifs to any known proteins. Mut2 has four membrane-spanning heli-

ces plus some homology to flagellar-motor protein. Mut3 has homology to Rho GTPase-activating proteins. The gene for Mut4 is in the same operon as *mut3* and is a short gene with some DNA sequence complementary to the 5' region of *mut3*. Mut5 has several membrane-spanning hydrophobic helices and some homology to the protein SecY. Mut6 has an N-terminal stretch of 15 hydrophobic residues, followed by several membrane-spanning hydrophobic helices and a C-terminal domain with homology to cysteine proteases. The genes for Mut7 to Mut10 are all part of the same operon. Mut7 has several membrane-spanning hydrophobic motifs and a C-terminal domain with homology to protein kinases. Mut8 has homology to LuxI. Mut9 has a C-terminal domain with homology to DNA-binding proteins. Mut10 has an N-terminal stretch of 12 hydrophobic residues and a C-terminal domain with homology to glycosyltransferases. To gain some insight into the roles of these virulence factors in pathogenesis, you decide to conduct additional experiments. First, you perform a gentamicin protection assay and a plaque assay using human cultured fibroblast-like cells. The results are shown in the table below.

Bacterium	Gentamicin assay (no. of colonies)		Plaque assay (no. of plaques observed)
	Without gentamicin	With gentamicin	
Wild type and GeMut2 to GeMut10	10^9	10	None
GeMut1	10	10	None

 i. Interpret these results, with your rationale.

 ii. Predict the function of the virulence factor encoded by the gene *mut1*.

 iii. Provide one additional, different experiment that could be performed to confirm your prediction.

E. In the next experiment, you grow the bacteria in medium and then filter the bacterial culture and treat cultured host fibroblasts with either the bacteria or the filtered bacterium-free medium. The results are shown in Figure 1 below.

(continued)

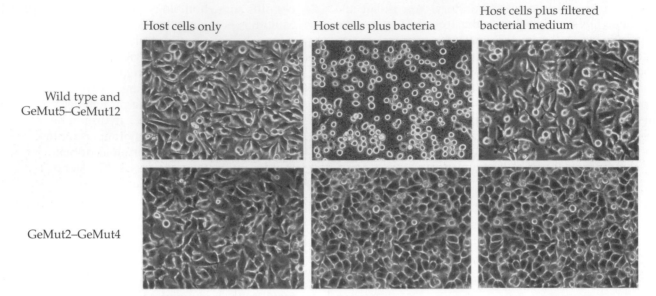

Host cells only   Host cells plus bacteria   Host cells plus filtered bacterial medium

Wild type and GeMut5–GeMut12

GeMut2–GeMut4

Figure 1   Effects on cell monolayers.

i. Interpret these results, with your rationale.

ii. Predict the functions of the virulence factors encoded by the genes *mut2* to *mut4*.

iii. Provide one additional, different experiment that could be performed to confirm your predictions for each.

F. You find that the wild-type bacterium elicits a very strong host humoral immune response, yet wild-type bacteria appear to be able to evade this response and cause septicemia, even in hosts with prior exposure to the bacterium. You find that GeMut5 and GeMut6, while able to colonize and form follicular lesions, are no longer able to cause septicemia and are quickly cleared from the bloodstream in hosts with prior exposure to the bacterium.

i. Describe how *G. exhaustmium* antigens, such as Mut1, are presented to the host immune system to result in a strong humoral response.

ii. Predict, with your rationale, possible functions for Mut5 and Mut6.

iii. Provide one additional, different experiment that could be performed to confirm your prediction.

G. You find that the wild-type bacterium and GeMut2 to GeMut6 form mucoid colonies on agar plates and stain purple with crystal violet dye when grown in test tubes (as shown in Figure 2 below). The mutants GeMut1 and GeMut7 to GeMut10 no longer have this property. However, if you add bacterial culture filtrates from GeMut7 and GeMut9 to cultures of GeMut8, but not

GeMut1 or GeMut10, GeMut8 stains with crystal violet.

GeMut1 and GeMut7–GeMut10   Wild type and GeMut2–GeMut6

Band of crystal violet dye

Figure 2   Crystal violet assay.

i. Interpret these results, with your rationale.

ii. Predict the functions of the virulence factors encoded by the genes *mut7* to *mut10*.

iii. Provide one additional, different experiment that could be performed to confirm your predictions for each.

2. There have been a number of recent high-profile incidents that illustrate the havoc, alarm, and economic consequences that can result from widespread distribution of contaminated food due to accidental introduction of harmful microbes during food processing. What previously was seen only sporadically, such as at family gatherings or community picnics, moved abruptly into the public eye with the large-

Strain no.	16S rRNA phylotype (16S rRNA gene BLASTed against DNA database)	Gentamicin assay (CFU/ml)			Toxic shock assay (high levels of inflammatory cytokines [TNF and IL-1] in blood)[a]	Plaque assay (phenotype observed after 24 h)	Host cell morphology assay (microscopic image of Gram-stained intestinal epithelial cells incubated for 8 h with bacteria)	Acid tolerance assay (no. of live bacteria after 3 h at pH 2 [CFU/ml])
		Without gentamicin	With gentamicin	With gentamicin and mannose				
1	*L. monocytogenes*	10^9	10^8	10^8	No	Large plaques		10
2	*E. coli*	10^9	10^2	10^2	Yes	All clear		1
3	*E. coli*	10^9	10^8	10^2	No	Small plaques		10
4	*E. coli*	10^9	10^2	10^2	Yes	All clear		10^{12}
5	*Bacillus* sp.	10^9	10^8	10^8	Yes	Large plaques		10^{12}
6	*Helicobacter pylori*	10^9	10^2	10^2	Yes	No plaques		10^{10}

[a]TNF, tumor necrosis factor; IL-1, interleukin-1.

scale problem of food-borne pathogens. These dangerous microbes have been associated with hundreds of multistate or multination outbreaks of disease due to contaminated food. In the late summer of 2006, contamination of prepackaged fresh spinach with *E. coli* O157:H7, a toxin-producing bacterium that causes dysentery-like diarrhea and can cause kidney failure and death, especially in children and the elderly, led to 204 cases of illness across 26 states, with 104 hospitalizations, 31 kidney failures, and 3 deaths. The outbreak, which was traced back to spinach obtained from a few fields in California, shook consumer confidence and cost the industry an estimated $150 million in economic losses. The incident also prompted fear that a deliberate act of bioterrorism could result in even worse consequences. In response to the growing need for better food surveillance, the FDA and U.S. Department of Agriculture have implemented several joint surveillance and risk assessment programs for prevention and detection of food-borne diseases. As part of these food safety programs, a group of researchers at one of the testing laboratories have isolated six different strains of bacteria associated with two recent outbreaks of severe diarrheal disease, the source of which was traced back to two large food-processing plants, one in the Midwest (strains 1 to 3) and the other in New England (strains 4 to 6). All of the strains had 50% lethal-dose values of ~10 in a piglet model of intestinal diarrhea, in which bloody diarrheal stool was observed. Being interested in understanding the mechanisms of pathogenesis of these new strains, the researchers subjected each of the strains to a series of experiments to identify the strains and to determine what virulence factors they might have.

A. Based on the results shown in the table on the previous page, the researchers ask you, as an expert consultant in bacterial pathogenesis, to predict the possible virulence factor(s) that must be present (or absent) in strains 1 to 6. Be sure to state your rationale. (Be complete, but base your answer only on the information provided from the assays shown in the table.)

B. Because of the urgency involved in identifying the nature of the virulence properties of these strains and because a lengthy mutagenesis library screening study of each is not practical on such short notice, the researchers ask you to provide at least one additional experiment (not involving signature-tagged mutagenesis, IVET/recombinase-based IVET, genomic subtractive hybridization followed by selective capture of transcribed sequences, in vivo-induced antigen technology, RNA/DNA chips, genome sequencing, or X-ray crystallography) for each strain that they could perform to confirm your predictions. Be sure to state how the data will be visualized or measured, what reagent(s) and instrument(s) are needed, and what the expected results will be for each experiment. (For strains predicted to have multiple virulence factors, you may need more than one type of experiment to verify all of your predictions.)

C. In your expert opinion as a microbiologist and researcher in the field of bacterial pathogenesis, is there plausible cause for the researchers to call the Centers for Disease Control and Prevention and the Department of Homeland Security regarding the possibility that either of these incidences could be an act of bioterrorism? Be sure to state your rationale.

20

Biosecurity: the Changing Roles of Microbiologists in an Age of Bioterrorism and Emerging Diseases

A ny military strategist will tell you that the best defense against a threat is information; hence, intelligence gathering is one of the most critical components of any combat operation. Similarly, the best defense against an infectious disease threat is information—information that is readily accessible and understandable by all parties and that enables those parties to assess the risk, formulate response options, and then react calmly and rationally to the threat.

When Microbiologists Are Called to the Front Line

The concept of biological warfare or bioterrorism, i.e., the deliberate use of biological agents (microorganisms or their toxic products) to further nefarious goals, is not new. Historically, there are numerous accounts of the use of toxic, diseased, or contaminated items being used to sicken, weaken, or kill an enemy. In ancient times, armies were known to poison enemy water wells with rye ergot or purgative herbs (hellebore or skunk cabbage) or diseased animal and human corpses. Archers would dip their arrows in poison, manure, or diseased blood. In medieval times, armies would catapult over castle walls manure, horses that had died from diseases such as glanders or anthrax, or even bodies of humans who had died from plague or smallpox. Enemies would lace wine with the blood of leprosy patients or with foxglove, hemlock, or other poisons. In more recent history, blankets and linens contaminated with smallpox were given to Native Americans during wartime, such as those purportedly given to the Delaware and Shawnee Indians by British troops during the siege of Fort Pitt in the summer of 1763. British forces during the American Revolutionary War gave smallpox to besieged civilians in Boston, and in 1776, Benedict Arnold was accused of, among other things, deliberately spreading smallpox throughout the Continental army.

In modern times, biological warfare, along with chemical warfare, escalated to new and terrifying heights, becoming ever-more sophisticated and grander with each world war. During World War I, biological weapons were mostly directed against animals, with anthrax spores from *Bacillus anthracis* or water contaminated with *Burkholderia mallei,* the cause of glanders, used to infect horses and mules, which hampered troop and artillery movements and supply convoys. The horrific ramifications evidenced by the use of these chemical and biological agents during World War I led to the Geneva Protocol in 1925, which banned any further use (but alas not research or development) of such bioweapons. However, this resolution was all too quickly set aside during World War II with the establishment of extensive bioweapons programs that rapidly switched from defensive to offensive measures and sometimes involved crimes against humanity. By the end of the war, most developed countries had full-blown, active bioweapons programs that continued to expand during the Cold War years.

It was not until 1969 that the United States officially terminated its offensive bioweapons program, destroying all stockpiles and switching back to defensive measures. The first international disarmament treaty banning the development, production, and stockpiling of biological and toxin weapons was initiated in 1972 and ratified by 22 governments in 1975 at the **Biological and Toxins Weapons Convention** (abbreviated as **BWC** or **BTWC**), and over 162 countries and states are now parties. Unfortunately, the lack of a formal verification regime for monitoring compliance limited the effectiveness of the convention, and indeed, it became clear in the 1980s and 1990s that some state-sponsored bioweapons programs, in particular those of Iraq and the Soviet Union, had remained active and offensive.

The 1980s and 1990s also saw several smaller-scale incidents involving intentional release of biological agents by non-state-sponsored entities. In 1984, the Rajneeshee religious cult contaminated salad bars in Oregon with *Salmonella enterica* serovar Typhimurium in an attempt to influence the outcome of a local election. This incident resulted in hundreds of cases of gastroenteritis, but fortunately no deaths. Japan's apocalyptic Aum Shinrikyo cult tried numerous times in the early 1990s to spray anthrax spores around Tokyo, but no one was harmed, presumably because the strain of *B. anthracis* used was avirulent. In 1995, the group turned instead to the release of sarin gas in a Tokyo subway, which killed 12 commuters, seriously injured 54, and affected thousands.

Despite these precedents, the use of the U.S. Postal Service as a vehicle to spread anthrax spores in a deliberate act of bioterrorism, particularly in the wake of the 11 September 2001 attacks on the World Trade Center in New York and the Pentagon Building in Washington, DC, came as a major surprise to the general public and government and health officials, whose mind-set about the use of bioweapons was still mired in more large-scale terms. Amazingly, of the tens of thousands of people who were potentially exposed, there were only 22 documented cases of anthrax attributed to the mailings, thanks to the rapid containment and prophylactic measures taken once the threat was realized. Moreover, only 5 of the 11 victims who actually developed symptomatic inhalation anthrax died. Although tragic, this death toll was far below the 100% death rate that almost certainly would have resulted without rapid medical intervention using today's medical advances.

The 2001 anthrax attacks marked a major paradigm shift in our way of thinking about how and from where potential bioterror acts might occur and what our responses currently are and will need to be to prevent another such attack in the future (Table 20–1). No longer is a bioweapon considered only an implement of a state-sponsored war conducted on a grand scale, as a so-called **"weapon of mass destruction" (WMD),** but rather, it is just as likely to be used on a smaller scale by groups or individuals to spread terror and disrupt normal routine operations, more as a **"weapon of mass disruption"** (still a WMD). As a consequence of the pivotal events of 2001, defense against bioterror agents, collectively dubbed **"select agents"** (Table 20–2), emerged as a major global health priority in the United States and elsewhere. The United States alone has spent many billions of dollars on this effort over the past 10 years.

In this new era, where biological agents, such as anthrax spore-containing powders and botulinum neurotoxin-laced milk, could be introduced as stealth agents to incite terror in the general public and cause massive disruption of daily life, microbiologists have become an important and integral part of the homeland defense network. From the outset, microbiologists have been working closely with the military, the Centers for Disease Control and Prevention (CDC), the FBI, and other law enforcement agencies, as well as the recently instituted Department of Homeland Security (DHS), to detect, identify, trace, monitor, decontaminate, treat, and prevent potential bioterror events and to inform and educate authorities and the public.

Tracking Down a Bioterrorist

Bioterrorism poses unique challenges for security due to the intrinsic differences between an act of releasing a biological agent and, say, building and detonating a bomb or even releasing a chemical, such as a nerve gas. The explosive ingredients used to build a bomb

Table 20–1 Sources of biological threats

Natural biological threats	Biological threats originated by people
Persistent, ever-present infectious diseases	Accidental release of infectious agents
High-impact infectious diseases (major health, ecological, and economic consequences)	Unintentional release through mishap or malfunction
New and reemerging infectious diseases	Anthropogenic impact on environment or ecology that shifts natural habitats or exposes new sources
Multidrug-resistant pathogens	Nonmalicious but intentional release or introduction that results in an unintended outcome
Human-to-human transmission	Sloppiness or negligent hygiene during food preparation or processing
Vector-borne, epizoonotic transmission	
Foreign animal, zoonotic transmission	Deliberate (the "terror effect")
Food-borne transmission	State-sponsored bioweapons: WMD, or "weapons of mass destruction"
Water-borne transmission	Bioterrorism acts by groups or individuals: WMD, or "weapons of mass disruption"
Increasing populations of people with weakened immune systems due to infectious diseases, chronic diseases, and aging	
Invasive alien species	
Plants	
Animals	
Insects	

are obvious, readily identifiable, and unmistakable in their intent. There is only one purpose for which a person might build a bomb, and that is to detonate it and cause destruction and potentially harm people or damage things nearby. The nature and uses of a chemical substance are likewise readily identifiable, and the potential toxicity of the substance is usually well known or predictable. It is also obvious that the presence of any sizable quantity of the toxic substance, especially in an unexpected location, is likely to have but one purpose—again, to cause damage and potential harm. The same cannot always be said about a biological agent.

Unintentional or Deliberate?

The presence of a biological agent or the isolation of a biological agent from an infected individual does not necessarily mean that the biological agent was deliberately introduced to cause harm. Disease could be caused by natural exposure to a biological agent or through accidental, nonmalicious exposure. Biological agents are often used for beneficial biomedical research to help develop therapeutics or for educational purposes to learn about biological processes. Hence, biological agents are often referred to as **"dual-use agents."** An example is the clostridial botulinum neurotoxin, which is not only a highly potent paralyzing toxin, but also a cosmetic for wrinkles (sold commercially as BoTox) and a therapeutic agent used biomedically for nerve disorders, such as dystonia (facial tics [involuntary muscle spasms]).

In some cases, deliberate release of a biological agent is clear-cut. An outbreak of smallpox, caused by the variola virus, nowadays can only be attributed to a deliberate release, since human smallpox has been eradicated globally since 1979 and there are only two known stocks of smallpox still existing, currently in high-level biocontainment facilities at the CDC and in Koltsovo, Russia. Likewise, epidemic outbreaks of inhalation anthrax, which are very rarely seen in developed countries, are not natural, and so can likely be attributed to deliberate or accidental release of a significant number of spores into the atmosphere. For example, epidemiologists and other experts around the world were convinced that the deadly 1979 anthrax incident in Sverdlovsk (today known as Yekaterinburg), Russia, was due to a massive release of anthrax spores from a Soviet military bioweapons facility. About 100 people (the exact number is unknown because records were purportedly destroyed) and numerous animals downwind of the facility were killed. Since anthrax is not contagious (i.e., it is not spread from individual to individual) and many people and animals in the same area were affected, the outbreak had to be caused by a massive release of spores. Although Soviet officials denied it at the time, insisting that it was a natural outbreak, it was later confirmed, after the dissolution of the Soviet Union in 1992, that the outbreak was indeed caused by the accidental (man-made) release from the Sverdlovsk facility. On the other hand, an epidemic outbreak of plague (caused by *Yersinia pestis*) or monkey pox (similar to but not the same as human smallpox), both of

Table 20–2 U.S. Department of Health and Human Services (HHS)/CDC and USDA select agents and toxins (as of May 2010)

Group	Members			
	Toxins	Bacteria	Viruses	Fungi
HHS/CDC select agents and toxins	Abrin	Botulinum neurotoxin-producing species of *Clostridium*	Cercopithecine herpesvirus 1 (herpes B virus)	*Coccidioides posadasii*
	Botulinum neurotoxins	*Coxiella burnetii*	Crimean-Congo hemorrhagic fever virus	*Coccidioides immitis*
	Clostridium perfringens epsilon toxin	*Francisella tularensis*	Eastern equine encephalitis virus	
	Conotoxins	*Rickettsia prowazekii*	Ebola virus	
	Diacetoxyscirpenol	*Rickettsia rickettsii*	Lassa fever virus	
	Ricin	*Y. pestis*	Marburg virus	
	Saxitoxin		Monkeypox virus	
	Shiga-like ribosome-inactivating proteins		Reconstructed replication-competent forms of the 1918 pandemic influenza virus containing any portion of the coding regions of all eight gene segments (reconstructed 1918 influenza virus)	
	Shiga toxin		South American hemorrhagic fever viruses	
	Staphylococcal enterotoxins		Flexal	
	T-2 toxin		Guanarito	
	Tetrodotoxin		Junin	
			Machupo	
			Sabia	
			Tick-borne encephalitis complex viruses (flaviviruses)	
			Central European tick-borne encephalitis virus	
			Far Eastern tick-borne encephalitis virus	
			Kyasanur Forest disease virus	
			Omsk hemorrhagic fever virus	
			Russian spring and summer encephalitis virus	
			Variola major virus (smallpox virus)	
			Variola minor virus (alastrim)	
Overlap select agents and toxins		*B. anthracis*	Hendra virus	
		Brucella abortus	Nipah virus	
		Brucella melitensis	Rift Valley fever virus	
		Brucella suis	Venezuelan equine encephalitis virus	
		Burkholderia mallei (formerly *Pseudomonas mallei*)		
		Burkholderia pseudomallei (formerly *Pseudomonas pseudomallei*)		

USDA select agents
and toxins

Ehrlichia ruminantium
(Heartwater)
Mycoplasma capricolum subsp.
capripneumoniae (contagious
caprine pleuropneumonia)
Mycoplasma mycoides subsp.
mycoides small-colony (*Mmm* SC)
(contagious bovine
pleuropneumonia)

African horse sickness virus
African swine fever virus
Akabane virus
Avian influenza virus (highly pathogenic)
Bluetongue virus (exotic)
Bovine spongiform encephalopathy agent
Camel poxvirus
Classical swine fever virus
Foot-and-mouth disease virus
Goat pox virus
Japanese encephalitis virus
Lumpy skin disease virus
Malignant catarrhal fever virus (alcelaphine
herpesvirus type 1)
Menangle virus
Peste des petits ruminants virus
Rinderpest virus
Sheep pox virus
Swine vesicular disease virus
Vesicular stomatitis virus (exotic): Indiana subtypes
VSV-IN2, VSV-IN3
Virulent Newcastle disease virus 1

USDA plant protection
and quarantine select
agents and toxins

Ralstonia solanacearum race 3,
biovar 2
Rathayibacter toxicus
Xanthomonas oryzae
Xylella fastidiosa (citrus
variegated chlorosis strain)

Peronosclerospora
philippinensis
Peronosclerospora sacchari
Phoma glycinicola
(formerly *Pyrenochaeta*
glycines)
Sclerophthora rayssiae
var. *zeae*
Synchytrium endobioticum

which are contagious and still have natural reservoirs throughout the world, can be caused by exposure to infected animals or individuals instead of by a deliberate, malicious act.

Timing

The devastating effects of detonating a bomb are experienced immediately. Similarly, the toxic effects of a chemical weapon are often immediately evident. This gives the perpetrators very little time to escape and helps the authorities to trace the source of the material and to apprehend the responsible parties. In contrast, acts of bioterrorism usually proceed much more slowly. After exposure to the biological agent, the microorganism has to grow and spread in its targets. Then, there is an incubation period during which the disease develops before symptoms manifest and an accurate diagnosis can be made. In most cases, it is days or even weeks before the nefarious nature of the circumstances is realized. By that time, the perpetrators have long since made their escape and covered their trail. To complicate things, many of the symptoms, especially in the initial stages of an infectious attack, are indistinguishable from those caused by common natural infectious diseases, such as the common cold or flu. Therefore, a bioterrorism attack may not be readily recognized. This happened in the initial phases of the 2001 anthrax attacks, where some of the physicians were not prepared to recognize the specific symptoms of inhalation anthrax and so thought the patients simply had the flu and sent them back home. It was not until after the first two confirmed anthrax cases were made public that physicians began checking incoming patients with suspicious symptoms for anthrax exposure.

Tracing the Source

The identity of the biological agent used in a potential bioterror act is not necessarily helpful in determining the possible origin of the material. For instance, most biological weapon manufacturers and biomedical research laboratories use the same potent Ames strain of *Bacillus anthracis*. Indeed, whole-genome sequencing of the bacteria isolated from the different letters used in the 2001 mailings revealed that the strains were completely identical to each other, as well as to Ames strains from other sources. Moreover, the methods and formulations used for suspending and mixing the *B. anthracis* spores into a fine powdery substance (usually a silica-based matrix) to optimize their aerosol dispersal properties are also common among

bioweapon laboratories and provide little information about the original source of the material.

Lessons Learned

The threat of bioterrorism and biowarfare is now a fact of modern life, and the events of 2001 only served to confirm this and to bring out a number of glaring deficiencies in our current knowledge about microbial pathogenesis (in terms of both the microorganisms and host immunity), response measures, science education, information dissemination, and public policies regarding such acts. The public response to the spread of anthrax spores through the U.S. postal service was particularly illuminating. Whereas the postal workers, who were at highest risk, handled the situation with remarkable calm and courage, many in the general public, including policy makers, reacted just as the terrorists intended—with confusion and fear. Of course, it did not help that in the early stages of the crisis, individuals (including scientists and other experts) with a penchant for horrific scenarios dominated the media reports, while at the same time, politicians and other authorities issued contradictory statements, thereby causing confusion. Consequently, there was considerable hype and propagation of misinformation (for an example, see Box 20–1).

Contributing to this turmoil was the lack of an official, organized network of national and local scientific experts who could readily be called upon to provide accurate, unbiased, yet comprehensive and understandable information to the media and public. Clearly evident was a significant deficiency in scientific literacy, which led to misinterpretation or misunderstanding of the biological threat and nonproductive overreaction to the situation. The media, policy makers, government officials, and the general public were, and to a large extent still are, grossly undereducated in understanding science topics and associated biomedical issues, the nature of scientific endeavor, and how to cope with imposed health threats, whether intentional or not. For example, many people at the time had no idea what a spore is (this is clearly not the case anymore), and in fact, many people, including some officials, referred to the anthrax spores as the anthrax virus. Use of the term "anthrax virus" immediately caused increased fear by giving the deadly anthrax spores the properties of a virus, i.e., a biological agent that is potentially lethal, highly contagious, rapidly spread from person to person, and not treatable with antibiotics, none of which is true for anthrax (while it is deadly, it is not contagious, and it can be treated with antibiotics in the early stages of infection).

BOX 20–1 The Consequences of Propagation of Misinformation: the Ames Strain of *B. anthracis* Was Not from Ames, IA!

Because of its high virulence, researchers use the Ames strain of *B. anthracis* as a "gold standard" positive control for developing and testing the efficacy of vaccines and therapeutics against anthrax. When the Ames strain was officially linked to the anthrax strain used in the 2001 anthrax mailings, everyone assumed that the strain came from Ames, IA.

In fact, the Ames strain was isolated from a diseased cow that died in Texas in 1981. The strain was acquired by researchers at the U.S. Army Medical Research Institute of Infectious Diseases (USAMRIID) in Fort Detrick, MD, as part of a national search for novel types of anthrax to use in testing vaccines. The army researchers at the time mistakenly believed the strain came from Ames, IA, because the label on the standard USDA shipping container bore the return address of the USDA's National Veterinary Services Laboratories, located in

Ames. When the USAMRIID scientists published a paper describing their research involving the strain, they mislabeled the specimen as the Ames strain, and the name stuck. However, during investigations after the anthrax attacks, university and USDA officials in Iowa could find no record of anthrax outbreaks among Iowa cattle in the years prior to 1981 or of anthrax strains being sent to the army.

The unfortunate consequence of this misinformation was that when administration officials at Iowa State University at Ames learned of the possible connection, they ordered destruction of their entire collection of *B. anthracis* stocks, a considerable loss of valuable scientific resource material.

Source: Joby Warrick, *Washington Post,* 29 January 2002.

It was also apparent that public officials were not consulting with unbiased expert scientists familiar with the properties of anthrax spores or they would have realized the impact that a few letters with spore-containing powder could have on postal workers and the general public. Anthrax spores are only a few microns in diameter, whereas the pores in the paper used to make envelopes are about 100 microns in diameter and so could easily allow dispersal of the envelope's contents, especially when smashed through the letter-sorting instruments at a mail facility.

Another issue that came up in the aftermath of the 2001 anthrax attacks was how far off previous predictions made by military strategists were about the circumstances and consequences of such an attack. Most pre-2001 scenarios projected large-scale exposures with catastrophic outcomes, with hundreds to thousands of casualties, total collapse of overwhelmed health care infrastructures, and widespread ensuing pandemonium. What actually happened was far from that. The 2001 attack showed that a highly disruptive, limited anthrax attack was possible. Once it was recognized that the biological agent was anthrax and that transmission was through the postal system, the situation was relatively quickly brought under control with minimal casualties and no col-

lapse, but rather a bolstering of the health care infrastructure. However, there was pandemonium, although briefer and of a different sort than what was projected. There was indeed a media frenzy, generalized public fear, and, in many cases, knee-jerk responses by public officials and policy makers. However, following a short initial period of confusion, more organized, clear, and accurate information was disseminated, and the situation dramatically improved (see Box 20–2 for an insightful anecdotal account of crisis management at a local level).

Over the past 10 years, drastic changes have occurred in how such crises are handled. One of the most valuable lessons learned was the need to provide clear and accurate information and to make everyone knowledgeable and aware of the scientific, medical, and security issues involved in preventing similar situations from occurring again and what to do if they should occur. It is hard to imagine nowadays that anyone would not notice and report a person spraying powder in an urban neighborhood or a crowded venue or would fail to question a colleague acting secretively in a laboratory that works with infectious agents. In fact, most people have become quite sensitive to suspicious behavior, and there is an overall heightened awareness of conditions that might

BOX 20–2 Talking to Postal Workers—and Others

Although the University of Illinois at Urbana-Champaign, where three of the four authors of this text are employed, is located in the Midwest and proved not to be a target of the bioterrorists, local postal workers, health care workers, and law enforcement officials were nonetheless concerned about what risk a "suspicious powder" might pose to them. Immediately after the first anthrax cases were publicized, these people, as first responders, were in the front lines against any potential biothreats that might surface, but they were not prepared to deal with that threat. What the postal workers and other first responders wanted, and indeed were entitled to, was information—information that they were not receiving. This was not due to their supervisors withholding that information from them, but because no one had the expertise or experience to deal with such a situation, there were no guidelines in place to assist them.

To remedy this, the postal workers and other first responders turned to a reliable information source that they knew they could count on for help, namely, faculty members at the University of Illinois. When they contacted the Department of Microbiology asking for experts who could talk to them about anthrax, two of the authors (Abigail Salyers and Brenda Wilson) responded to this request. In short order, Salyers visited with postal workers at many local post offices, while Wilson attended meetings with area mailroom personnel, secretaries, fire fighters, and law enforcement workers.

At each information gathering, the professors provided a brief explanation of what a bacterium is, what a spore is, what the disease anthrax is, and why antibiotics work in the initial stages of the disease but not in later stages. This was followed by a question-and-answer session open to the audience. What was most impressive about these sessions was the serious and intense nature of the interactions. Clearly, the workers had been absorbing and assimilating a lot of information from various media sources because the questions asked showed very critical and logical thinking about various aspects of the crisis.

Meanwhile, Dixie Whitt (another author) was working with university medical students, who were also energized to learn more about the medical aspects of anthrax, as well as other potential bioterror agents. At the same time, Malcolm Winkler, the fourth Midwest author of this book, was working on antibiotic discovery and *Streptococcus pneumoniae* physiology and pathogenesis in the now-defunct Division of Infectious Diseases Research at Eli Lilly and Company. After returning to academia at Indiana University—Bloomington in the spring of 2003, Winkler expanded the sections on select agents that were covered in undergraduate bacterial pathogenesis courses. Since 2003, Indiana University—Bloomington has considerably increased its Microbiology Division in the Biology Department, including the establishment of the first CDC-approved select-agent laboratory in Indiana for work on biodefense projects. The process of setting up this laboratory involved increased communication with the community about the types of work that would take place and the security and safeguards of such a facility.

The involvement of university faculty extended beyond what is mentioned above. Brenda Wilson, who was teaching a course on bacterial pathogenesis at the time and was faculty advisor for the undergraduate Microbiology Club (a student chapter of the American Society for Microbiology), suggested that since no one else seemed to be providing information for the university staff and the community, perhaps the club members could fill that void. These undergraduates jumped into action and organized and convened the first of what turned out to be a series of information panels on anthrax open to the public. The university administration was quite impressed with the outcome of the first student-run panel, which received a lot of local press coverage and was attended by hundreds of grateful university and community members seeking more information—so much so that they quickly sponsored a number of subsequent panel sessions, including one that was posted for the public on the university website. Indeed, this experience paved the way

(continued)

BOX 20–2   Talking to Postal Workers—and Others (continued)

to a noticeable change in local university and public health policy, so that in subsequent potential crisis situations, such as the SARS pandemic of 2003 and the more recent avian and swine flu pandemics, the university administration, along with local health officials, has been very responsive to sponsoring public information forums. Also, as has happened at most academic institutions, biodefense and biosecurity have since been added as key topics in the microbiology and biotechnology curricula.

Could it be that while stumbling around trying to figure out how to help during the

crisis we may have hit upon the best response to such bioterror incidents in the future? Whether the biothreat was the result of a deliberate act or natural causes, clearly an information blitz by local experts who make the effort to communicate with the affected parties in understandable terms instead of jargon was the key to alleviating much of the fear and confusion. As a result, the public workers gained more confidence in their ability to respond to the crisis in a calm and rational manner.

---

be favorable for potential bioterror acts. For example, there is much more scrutiny and higher reporting requirements for requests to obtain and work with any infectious agent, whether it is a potential bioterrorism agent or not (Table 20–2). Editors of journal articles must now scrutinize whether submitted articles pose any potential threat that needs to be examined further. Institutional biosafety committees at universities and companies are charged with reviewing and approving in advance all proposed research with any pathogen or toxin. Moreover, detailed policies, medical infrastructure, emergency response and crisis management plans, and communications networks are now in place and stand ready to be implemented when needed.

The "Top Four" Bioterror Agents

Biowarfare specialists, and presumably bioterrorists, have chosen a limited number of preferred biological agents for development as bioweapons (the so-called "select agents," listed in Table 20–2). Of these, we will consider four examples that stand out among the others: *B. anthracis* spores, plague-causing *Yersinia pestis*, smallpox virus, and *Clostridium botulinum* neurotoxin serotype A. We will briefly explore some of the reasons why these four examples might be among the top choices for use as bioterror agents, as well as some of the limitations that might currently make them less of a threat than previously thought.

B. anthracis Spores

Before 2001, anthrax spores had often been touted as the top choice in bioweapons for biowarfare, or "germ warfare." The events of 2001 only confirmed this

claim, especially when the spores used were found to be of extremely high quality and potency. Anthrax spores are one of the easiest of the bioterror agents to manufacture, manipulate, and develop as a bioweapon. They can be produced in large quantities with a basic knowledge of microbiology. Naturally produced spores usually tend to clump together and are not very infectious, because the clumps are normally too big to pass easily through air passages and enter the lung. However, they can be further refined. With just a little more effort and resources, the spores can be coated with silica or other matrices, and the coated spores can then be ground into a fine powder for aerosol dispersal. The resulting spores can be spread readily by high-tech delivery vehicles, such as missiles, aerial bombs, crop dusters, or sprayers, or even by low-tech methods, such as the mail.

Anthrax spores also have other properties that make them ideal for use as a bioweapon. They are very hardy, are highly heat resistant, and can remain viable in the soil, even with exposure to extreme environmental conditions, for many years. This means that they can be stockpiled in dry form for decades without losing their viability or potency. It also means that they are difficult to destroy, making exposed areas hard to decontaminate. As a case in point, it took over 3 months and more than $23 million to decontaminate the Hart Senate Building after an envelope mailed to a senator was opened in October 2001 and was found to hold anthrax spore-containing powder.

Anthrax spores are hard to detect—any unknown powder that might be discovered most likely consists of the matrix and may not even contain any spores. Indeed, there were numerous "false alarms" reported during the period of the anthrax attacks that upon closer inspection turned out to be just talcum powder, chalk, or powdered sugar. There is also no immediate indication of exposure. There is no obvious cloud, except for massive exposures of fine powder; no color; no smell; and no taste. There is a delay period before the onset of symptoms. Add to this the fact that the initial symptoms are fairly nonspecific and even resemble the common cold or flu and the chances diminish that a doctor will accurately diagnose and treat the disease within the critical window when treatment could still be effective. Unfortunately, once symptoms manifest, the likelihood of survival plummets drastically, even with antibiotic treatment. Inhalation anthrax is almost always deadly if not treated early.

The least serious form of anthrax is cutaneous (skin) anthrax (commonly known as "wool sorter's disease," because people working in tanneries with wool and animal hides frequently contract it). In cutaneous anthrax, the spores enter the body through cuts, abrasions, or other wounds, germinate at the infection site, and cause edema (swelling) and cell death, which causes the formation of a characteristic black-centered lesion (called an **"eschar"**). The death rate for untreated cutaneous anthrax is about 20%, but it is readily treatable with antibiotics. All of the 2001 victims who contracted cutaneous anthrax, including a 7-month-old child, were treated successfully with antibiotics (ciprofloxacin) and survived.

The second form of anthrax is intestinal anthrax, caused by consumption of contaminated, undercooked meat. The spores then germinate in the intestine and cause acute intestinal inflammation, nausea, abdominal pain, and severe bloody diarrhea. Intestinal anthrax results in death in about 60% of untreated cases due to the disease progressing to the systemic form. This natural form of the disease still occurs in certain parts of the world where anthrax spores are endemic.

The most lethal forms of anthrax are the systemic and inhalation forms, with the victim usually succumbing to a rapid onset of massive hypotension, generalized (especially pulmonary) edema, massive bacteremia (with blood bacterial concentrations reaching $>10^8$ per ml of blood), encephalitis, and acute, fatal toxic shock. While the bacteremia can be treated with antibiotics in the early stages, there is a point of no return when the disease switches over to a toxin-mediated disease (see chapter 12 for details on the actions of anthrax edema and lethal toxins); by then, there is little hope for survival, since there are no known antidotes available against the toxin effects. Death rapidly ensues. Administering antitoxin, in the form of antibodies against the edema and lethal toxins, has some benefit for treatment of cutaneous anthrax, but it has not been very successful for inhalation anthrax, presumably because by the time symptoms appear, the disease has progressed too far for the antitoxin to reverse the inflammation and toxic shock.

With all these advantages of using anthrax spores as a bioweapon, the only notable weakness is that, while it is lethal for the individual who received the initial dose of inhaled spores, it is not transmissible from person to person. While not ideal for the goals of a bioterrorist, this is good for us, because unlike a highly contagious disease like smallpox (see below), spread of the illness among a population can be brought rapidly under control, and once the disease is identified, protective measures, such as use of breathing masks and administration of prophylactic antibiotics, can be imposed. The other thing to remember about anthrax is that the spores have to enter the body in order to germinate and cause disease. This means they can be washed off the body and other surfaces. It is amazing what a Band-Aid can do to prevent cutaneous anthrax due to a cut or abrasion.

To date, the FDA has approved a single vaccine against anthrax. The efficacy, safety, and production standards of the anthrax vaccine came under some controversy, but after further testing and improvement of production quality by the manufacturer, the vaccine is now considered to be acceptably safe, except for pregnant women, and is routinely administered to the members of the U.S. armed forces. One problem with this vaccine is the need for yearly boosters to maintain efficacy. Other anthrax vaccines are under development, and perhaps one day anthrax will be eliminated as a bioterrorism threat by vaccination of general populations.

Smallpox

Smallpox is another disease that has been touted as a good bioterror agent, mainly because of its extremely contagious and lethal nature; the long incubation time before characteristic symptoms appear; and its viral nature, which makes it difficult to treat (i.e., it cannot be treated with antibiotics). One bioterror scenario often put forth is the distribution of contaminated material, like the blankets and bed linens given to the Native Americans mentioned above, to the public through commercial retailers. Because of the long incubation period, potentially hundreds of households might be infected before the disease is recognized and measures for containment and treatment are begun.

Another scenario often cited is one in which a suicidal bioterrorist might serve as an incubator by self-inoculation and then would go about spreading the disease through contact with others. Since there is a long lag period before symptoms manifest, the bioterrorist could in theory come into contact with tens to hundreds of people during that period (imagine the person going to a packed stadium for a ballgame, to a large concert, or even to a crowded nightclub). Since smallpox has been eradicated and no one has been vaccinated since the 1970s, the highly contagious, deadly disease could spread rapidly through the younger members of a population. In 1947, years before smallpox was eradicated, such an incident actually did happen accidentally in the United States when a U.S. citizen who had traveled to Mexico and contracted smallpox reentered the United States and traveled by bus to New York City, where he then circulated for a number of days before being hospitalized and diagnosed with smallpox and eventually succumbing to the disease.

Contrary to what you might expect, there were no massive outbreaks, and thousands of New Yorkers did not contract the disease and die. Instead, health officials and local authorities swiftly mobilized, isolating the infected man and all the people with whom he had come in contact and quarantining and vaccinating them. They then set about vaccinating millions of New Yorkers over the course of about 2 weeks, effectively containing the outbreak within 3 weeks. Consequently, the numbers of actual smallpox cases (12) and deaths (2) were relatively small, especially considering what could have happened without the massive isolation and vaccination campaign that was mobilized.

Key to the successful containment of this outbreak, as well as others that have occurred and any that might occur in the future, was the ability to vaccinate everyone who had been exposed but had not yet manifested disease. This was possible because of the long incubation period during which vaccination is still effective in preventing disease. If an exposed person is vaccinated within a window of 3 to 4 days after exposure, the disease is either completely prevented or greatly ameliorated. The critical factor then becomes ensuring that all individuals who may have been exposed to the infected person are indeed identified, quarantined, and vaccinated. Today, this is quite feasible through modern, sophisticated contact-tracing capabilities and epidemiological resources that authorities have at their disposal. It is also possible to get information out quickly through the media and the Internet using new informal media channels, such as FaceBook and Twitter.

Y. pestis

The flea-borne bacterium Y. pestis owes its fear factor primarily to our collective memory of the horrific plagues (the "black death") of the Middle Ages, which decimated entire populations at the time. The main aspect that dampens the effectiveness of using plague as a bioterror weapon is the requirement for the flea as a vehicle. Prevention of transmission by controlling the insect vector is a very effective defense strategy for bubonic (flea-borne) plague. Nevertheless, during World War II, the notorious Unit 731 of the Japanese imperial army experimented with several bioterrorism agents, including Vibrio cholerae and Y. pestis, on unsuspecting Chinese populations. Fleas infected with Y. pestis were disseminated by spraying and in bombs dropped from airplanes. How successful these efforts were is not clear, since records are sketchy, but the purported number of victims was quite high.

Another aspect that diminishes plague as an effective bioterror agent is its sensitivity to antibiotics. For instance, there were periodic outbreaks of bubonic plague in Vietnam during the Vietnam War, which were readily controlled with administration of antibiotics. Although still endemic in certain parts of the world, Y. pestis is a rare cause of disease today in the United States and other developed countries, infecting perhaps a few hunters or hikers each year who inadvertently come in contact with animals carrying infected fleas. Those individuals who do become infected are usually diagnosed and treated effectively with antibiotics.

The current fear most often cited about use of Y. pestis as a bioterror agent is the possibility that the terrorists might render the bacterium resistant to multiple antibiotics and deliver it via aerosol spray. Despite major efforts to understand pneumonic (pulmonary) plague, which is spread directly from person to person via inhalation without using the flea as a vehicle, the fact remains that scientists still do not know how to simulate this most lethal and contagious form of plague well or how to effectively treat it, particularly if it is resistant to multiple antibiotics. Still in our favor is the fact that the fitness toll, which would have to be overcome to develop a strain that is resistant to all available antibiotics, diminishes the likelihood that such a resistant strain could be developed. Nonetheless, there have been reports of naturally occurring multidrug-resistant Y. pestis, so it is conceivable that a viable organism with more than a few resistance genes could be constructed. However, even if such a strain were developed and used, there are still alternative treatment modalities available, including passive immunization and vaccination, and as we saw in chapters 15 and 16, we are learning to deal with the escalating threat of multidrug-resistant microbes of both the natural and man-made varieties. Unfortunately, only one vaccine against Y. pestis has been developed, and it is no longer approved in the United

States because of poor efficacy and issues with side effects. Similar to the case with *B. anthracis,* an effective, safe vaccine that could be used in the general population would largely eliminate *Y. pestis* as a bioterror threat. Several new plague vaccines are currently being tested in clinical trials.

Botulinum Neurotoxin

Although botulinum neurotoxin is more properly defined as a chemical terror agent, since it is a protein rather than a bacterium or virus, it is often included in the bioterrorist's repertoire of select agents. The main fear here is that it might be introduced into water or food supplies. Add the fact that it is widely available for biomedical and cosmetic applications (see chapter 12), and the likelihood that it could be acquired and used to poison vulnerable water and food supplies is not too farfetched.

Many experts discount water supplies as a viable route of delivery because of the dilution factor, which certainly is true for the less stable purified form of the toxin. However, what many do not realize is that the native clostridial neurotoxins are produced naturally from the bacteria as complexes with other chaperone proteins that make them much more heat resistant and stable under acid conditions (such as those encountered during passage through the stomach). For example, the juices from a jar of spoiled string beans contaminated with botulinum neurotoxin dumped into a juice dispenser or a small water supply could have devastating consequences. Even so, there is little doubt that the more vulnerable, centralized milk or food supply chains are generally viewed as being of greater concern. Indeed, there already are numerous incidents of natural or accidental food-borne botulism reported annually due to undercooked or unpasteurized food and juices. Of course, an obvious preventive measure that could be taken is to better educate people regarding the importance of thoroughly cooking their food, an approach that has already had some, albeit not complete, success with other food-borne illnesses (see below).

An advantage of using botulinum neurotoxin as a bioweapon is the difficulty of detection and diagnosis before the onset of disease symptoms. The extreme potency of the toxin makes the limits of detection challenging to reach. Unlike a bacterium or virus that could be propagated through growth or whose DNA/RNA could be amplified by PCR methods to increase the likelihood of detection, botulinum neurotoxins can act at subpicomolar levels, do not replicate, and are proteins (not DNA or RNA). In addition, symptoms can take anywhere from a few hours to several days

to appear. These factors make detection, diagnosis, monitoring, and tracing of the source very difficult.

The biggest concern about an outbreak of botulism is actually logistical in nature, namely, the lack of sufficient numbers of free ventilators in local hospitals. As detailed in chapter 12, once a person has been intoxicated with botulinum neurotoxin and symptoms of paralysis have manifested, no reversal of the paralysis is possible, since there are no postexposure antidotes available. This means that even if a botulism victim receives antitoxin to prevent any further intoxication, that person is still paralyzed and could remain so for up to 4 to 6 months, depending on the toxin serotype used and the total dose received before antitoxin is administered. The only treatment available for the paralyzed patient is supportive, primarily in the form of a mechanical ventilator to breathe. By current practice, most hospitals run at about 95% capacity with regard to the availability of respiratory ventilators, since they are expensive instruments and are also needed for many other medical purposes. Thus, if a moderate to large outbreak of botulism should occur, it would greatly tax the existing infrastructural capacity of nearby hospitals to handle such an acute, long-lasting demand—truly a logistical nightmare.

What If Bioterrorists Came Up with Something Completely New?

Could an inventive, scientifically trained bioterrorist develop, steal, or otherwise procure a biological agent with such new or unusual properties that it could not be initially identified or treated? Fortunately, the chances for such a scenario have recently become less likely, and we now are more ready than ever to meet such a threat head on. The highly sophisticated genomic, analytical, and detection technologies that are currently available (see chapters 5 and 9) have already proven time and again to be able to rapidly and accurately identify the microorganisms and toxins responsible for various disease outbreaks. It is now quite feasible to obtain within days the complete genome sequence of any unknown bacterium, fungus, virus, or protozoal parasite isolated from an encounter. The surveillance and response networks now in place to monitor work with select agents and other pathogens are highly sophisticated and well established.

Clear examples of the effective mobilization of this sort of effort occurred during the severe acute respiratory syndrome (SARS) pandemic of 2003, which was caused by a previously unknown virus, and the more recent avian (H_5N_1) and swine (H_1N_1) flu pandemics, which involved unusual variants of known

Table 20–3 The human cost of selected biological threats of concern

Type of casualties	Biological agent	No. of associated incidents	No. of associated deaths
Bioterrorism	Ricin assassination of Georgi Markov in 1978		1
	Salmonella enterica serovar Typhimurium in salads in 1984 (Rajneeshee)	751 cases	0
	Sarin gas in 1995 (Aum Shinrikyo)	>2,000 hospitalized	12 total
	Anthrax mailings in 2001	22 cases total	5 total
Natural	Malaria worldwide	300–500 million cases per yr	2 million per yr
	Food-borne illnesses in United States	76 million cases/325 hospitalizations per yr	5,000–10,000 per yr
	Tuberculosis worldwide	8–10 million cases per yr	2 million per yr
	Pneumococcus (*S. pneumoniae*)-related illnesses worldwide	Millions of cases per yr	2 million per yr
	Human immunodeficiency virus/AIDS worldwide	>60 million cases total	>20 million total (500,000 in United States)
	Hepatitis C in United States	170 million total	10,000 total
	Septic/toxic shock worldwide	20 million cases per yr (100,000 cases per yr in United States)	10 million per yr (50,000 per yr in United States)
	SARS virus worldwide in 2002–2003	8,096 confirmed cases total	774 total confirmed
	H5N1 (avian) influenza in 2003–2007	349 confirmed cases total	216 total confirmed
	H1N1 (swine) influenza in 2009 in United States	43–89 million cases/274,000 hospitalizations total	12,470 total confirmed
	Seasonal (mostly H1N1 and H3N2) influenza (non-2009) in United States	15–60 million cases/110,000 hospitalizations per yr	36,000–42,000 per yr

viruses. In each case, the viral culprit was rapidly identified, sequenced, and compared to other viruses in the databases, and new PCR-based or enzyme-linked immunosorbent assay-based diagnostics and vaccines were quickly developed for it. In addition to the sequencing technologies, for bacterial threats, there are also high-throughput and multiplex technologies in place to rapidly screen for potential antibiotic therapeutic drugs and identify resistances. There is little doubt that the power of genomics and all of its associated enabling capabilities could be quickly ramped up for any newly encountered biothreat in the future. Nevertheless, constant vigilance and the biosecurity measures discussed below are essential in this new era of bioterrorism.

Biosecurity in an Ever-Changing World

While the anthrax attacks of 2001 made biodefense a major health priority in the United States, the recent pandemics of SARS and avian and swine flu that followed in their wake, the growing number of food-borne illnesses, the alarming spread of antibiotic-resistant pathogens, and the persistence or emergence (or reemergence) of other deadly infectious diseases (Table 20–1) have now converged to make **biosecurity,** which includes biodefense, the most important global health priority. The human cost alone is staggering (Table 20–3) and cannot be ignored.

Biosecurity is a much more comprehensive concept than biodefense alone and entails putting into place a complete set of preventive and containment measures intended to eliminate or reduce a wide spectrum of potential biological risks and threats, ranging from natural or accidental exposures to infectious agents or toxins to deliberate acts of bioterrorism or biowarfare. Many diverse factors impinge on global biosecurity, impacting the severity and urgency of the biothreat event and our ability to quickly and effectively respond to and recover from it. Biosecurity is a challenging yet critical topic of concern that impacts many facets of our society today, and one should have no doubt that microbiologists have played and will continue to play a starring role in this battle.

As pointed out earlier, distinguishing a deliberately introduced infectious disease from a naturally occurring or newly emerging infectious disease is sometimes very difficult. Building a firewall is inherently more challenging for biological agents due to their "dual-use" nature. Thus, the best biosecurity strategy encompasses both intentional and unintentional disease outbreaks. In this regard, while warning and prevention are much preferable to coping with the consequences of an attack, the fact remains that it is possible for a new biothreat to emerge despite our best preventive measures. This means that it is critical that a greater emphasis be placed on bolstering public awareness through education, in addition to improving health care and response policies. For instance, one of the most remarkably effective ways to prevent disease transmission is to practice good hygiene by frequently washing the hands, covering the mouth when coughing, and avoiding direct contact (kissing or handshaking) with others when one of the parties is clearly ill. The massive advertising campaign launched in response to the flu threat was very effective in educating the public and mitigating the spread of disease.

Food Safety and Biosecurity

Each year, over 76 million people in the United States (i.e., 25% of the population) suffer from a food-borne illness. Although most of these individuals have self-limiting diarrhea or nausea and are never diagnosed, there are about 325,000 hospitalizations and 5,000 to 10,000 deaths. In the past few decades, there has been a tremendous scale-up in food production worldwide, from vast herds of cattle, large confined feedlots, and huge slaughterhouses to the many hundreds of distribution centers and supermarkets. Trade barriers have come down, and free trade permits food to be shipped around the world in a matter of a few hours to a few days. This globalization of food distribution also enables the rapid spread of a potential contaminated food source. Undoubtedly, these conditions have contributed to the emergence and prevalence of food-borne illnesses as a major global health concern. Add to this the fact that over 50% of our food dollars are now spent on prepared food outside of the home, and it is not hard to imagine the potential devastating consequences a global biothreat might pose in this situation.

The complexity of the modern food preparation and distribution process makes epidemiological tracking of the sources of contamination challenging, although there have been noticeable advances. Several federal agencies have set up networks, such as the

Food-Borne Diseases Active Surveillance Network (FoodNet) and the Food Emergency Response Network (FERN), to address this critical need. The United States Food and Drug Administration has sponsored a **hazard analysis and critical control point** program that monitors food distribution at critical control points where contamination is most likely to occur. The U.S. Department of Agriculture (USDA) has established a similar assessment method for identifying the most vulnerable target sites for introduction of contamination within the food-processing system, called **CARVER+Shock** (for *c*riticality, *a*ccessibility, *r*ecuperability, *v*ulnerability, *e*ffect, *r*ecognizability, and *shock* effect). However, despite these recent efforts to enhance surveillance, monitoring, reporting, containment, and recovery from food-borne illnesses, the increasing number of high-profile incidents, which have largely been caused by unintentional contamination, amply illustrates the widespread havoc, alarm, and economic consequences that can result from an outbreak of contaminated food. The greatest problem with current efforts to ensure food safety is that they are focused on too expansive a target: the entire food-processing chain, from field to table. There are simply too many spots along the way from the farm to the dinner table for a contaminated food sample to make it undetected to the consumer, sometime with a lethal outcome, and when such a slip occurs, consumer confidence in the food industry rapidly plummets, with substantial economic impact.

The real question that emerges from this situation is, exactly what level of food safety do consumers want the agricultural and food industries to achieve and the government to enforce? In recent years, considerable effort by the industry and policy makers alike has been spent on grappling with this urgent question. So far, most proposed measures have been aimed at further enhancing the infrastructure already in place, such as those mentioned above, but these measures cannot possibly ensure the desired near-complete protection from an incident.

The Case for Food Irradiation

The proposed use of **food irradiation technologies** to enhance food defense capabilities has emerged as a topic of intense current interest. The premise is that food irradiation technology, which can be applied at the very last stage of the food-processing chain, could achieve the nearly 100% free from contamination status of the food product that the consumer is demanding. The problem with its implementation is that, for most consumers, the idea of food irradiation is "scary" because it invokes fear of radiation poisoning and loss of the food's nutritional quality.

Table 20–4 History of food irradiation technology use

Yr approved	Food	Dose (kGy) permitted	Purpose
1963	Wheat flour	0.2–0.5	Control of mold
1964	White potatoes	0.05–0.15	Inhibit sprouting
1986	Pork	0.3–1.0	Kill *Trichia* parasites
1986	Fruits and vegetables	1.0	Insect (fruit fly) control, increase shelf life
1986	Herbs and spices	30.0	Sterilization
1990 (FDA)	Poultry	3.0	Bacterial pathogen reduction
1992 (USDA)	Poultry	1.5–3.0	Bacterial pathogen reduction
1997 (FDA)	Red meat	4.5	Bacterial pathogen reduction
1999 (USDA)	Red meat	4.5	Bacterial pathogen reduction
2008 (FDA)	Lettuce and spinach	4.0	Bacterial pathogen reduction, increased shelf life

Food irradiation technology, also called **"cold pasteurization"** or "cold sterilization," can eliminate disease-causing microbes from foods by treating the food with approved levels of ionizing (high-energy) radiation. There are three types of irradiation used: gamma rays, electron beams, and X rays. **Gamma rays** are high-energy photons produced by radioactive material (usually cobalt-60 or cesium-137) that can penetrate food and other material. Contrary to popular belief, they do not make the target material radioactive. Currently, this method is used to sterilize medical, dental, and household products, especially those used for patients who are immunocompromised, such as those undergoing cancer treatment. **Electron beams** are streams of high-energy electrons produced by an electron gun that can penetrate material up to about a 3-cm depth. There is no radioactive material involved in generating electron beams. This technology has been used over the past 15 years or so to sterilize medical devices and other household products and is gaining some application for food irradiation. **X rays** are generated when the electron beam is directed at a thin plate of gold or other metal. The X rays produced by this technology are more powerful than those used for medical machines to give chest X rays. Again, there is no radioactive material involved, and the beams can penetrate foods up to 15 in. in depth. However, this technology is still being developed and is not currently being used commercially.

In each case, the electron beams generated strike the electrons present in the target material, converting a fraction of their energy into kinetic energy of the secondary electrons. All the primary and secondary electrons undergo additional collisions until all their energy is dissipated by ionization. These high-energy electrons damage the DNA of organisms present in the target material, and the organisms are unable to replicate. Thus, the irradiation technology can effectively kill bacteria, parasites, and insects present in food but is much less effective against viruses and does not work against proteins, such as toxins or prions. The size of the DNA in an organism is a factor in how readily it is killed by the irradiation process (Box 20–3).

The interesting thing about irradiation technology is that, despite public fear, it is already being widely used by consumers. Most medical devices, Band-Aids and other wound care products, cosmetics, diapers, and infant care products, as well as feminine products, are already routinely treated with irradiation. Many food products have also been irradiated to kill insects, to prevent sprout growth, and to prevent spoilage (Table 20–4). Irradiated food generally has a longer shelf life, does not spoil as readily, and in some cases has enhanced antioxidant capacity.

The most often cited concerns regarding food irradiation technology are its potential harmful effects on food quality and the consumer. Extensive longitudinal feeding studies (since the 1950s) have revealed no significant ill effects on animals (mice, rats, dogs, etc.) or humans (hospital patients and astronauts). As the high-energy beam is absorbed, food warms up, with a dose of ~10 kGy warming the food by ~2.5°C. This can result in some treated foods tasting slightly different or, in the case of leafy vegetables, wilting a little. If the food still has living cells (e.g., seeds, shellfish, potatoes, and eggs), then those cells will be killed, just as the microbes are, and this may cause the food to taste a little different. At levels approved for use on foods, some vitamins (e.g., thiamine) are slightly reduced, but most fatty acid, amino acid, and other vitamin content is not significantly altered. Irradiated foods, especially meat containing fat, can have slightly higher amounts of 2-alkyl cyclobuta-

BOX 20–3 How Do You Measure Food Irradiation?

The dose of irradiation is measured in Grays (Gy) (1 Gy = 1 J/kg), i.e., the amount of energy transferred to the food, microbe, or other substance.

- One chest X ray = 0.5 mGy
- One abdominal plus pelvic computed tomography scan = 30 mGy
- Dose to kill 10^4 to 10^6 *Salmonella* bacteria in a piece of chicken, ~4.5 kGy; in juice, ~3 kGy
- Dose to kill 10^4 to 10^6 *Listeria* bacteria in pork, ~2 kGy at 0 to 10°C, ~3 kGy at 16°C

The killing effect of irradiation on microbes is measured as the D value. One irradiation D value will kill 90% of the microbes present, 2 irradiation D values will kill 99% of the microbes present, and 3 D values will kill 99.9% of the microbes. For example, it takes 0.3 kGy to kill 90% of *Escherichia coli* cells, so the D value for *E. coli* is 0.3 kGy; to kill 99% of the *E. coli* cells (2 D values), 0.6 kGy is required.

The size of the DNA of an organism is a factor in how readily that organism is killed by the irradiation.

- Parasites and insects are rapidly killed at D values of <0.1 kGy.
- Bacteria are killed at D values of 0.3 to 0.7 kGy.
- Bacterial spores are killed at D values of ~2.8 kGy. (Note: during the 2001 anthrax attacks, mail was decontaminated with ~30 kGy.)
- Viruses are killed at D values of >10 kGy.
- Prions and toxins (peptides and proteins) are not affected by irradiation.

(Note: NASA routinely sterilizes food for astronauts with 42 kGy of irradiation.)

---

nones (derivatives of triglycerides), which could cause DNA damage, but these compounds are apparently not readily metabolized by animals and are usually excreted. Overall, despite these minor potential concerns, the conclusion of most scientists and government food safety agencies is that food irradiation technologies are generally safe and highly effective in decontaminating food, so be prepared to see more and more use of this technology.

The Future of Biosecurity

The key to an effective biosecurity strategy will be a better understanding of the driving forces that are important for the transmission and perpetuation of infectious diseases and then implementation and management of effective preventive, treatment, and containment measures. To develop better predictive models for disease outbreaks and spread, policy makers will need to increase support for research efforts to understand microbial pathogenesis and the driving forces behind pathogen evolution and ecology. We also need more support for research efforts to develop therapeutics and alternative treatment mo-

dalities for multidrug-resistant microbes, to treat post-exposure intoxication, and to identify and treat new diseases of unknown origin. For this to occur, it will be necessary to strengthen local, national, and international capacities to prevent and control disease outbreaks. Unfortunately, as of this writing, funding for the National Institutes of Health in the United States, one of the major drivers for this type of research, has been largely stagnant for several years now, and the chances of obtaining an investigator-initiated research grant are hovering at historical lows. Moreover, while global disease surveillance systems and international cooperation have improved tremendously over the past decade, they are still not nearly where they need to be in terms of resources and integration. Finally, science education will be a critical deciding factor in enhancing the confidence of policy makers and the public in science and scientists, which will move this endeavor forward for greater preparedness against natural and man-made biothreats. Considering the real biothreats that are already upon us, some of which were considered in detail in chapters 18 and 19, there is real urgency for us to move in this direction rapidly with a long-range vision.

SELECTED READINGS

Arnon, S. S., R. Schechter, T. V. Inglesby, D. A. Henderson, J. G. Bartlett, M. S. Ascher, E. Eitzen, A. D. Fine, J. Hauer, M. Layton, S. Lillibridge, M. T. Osterholm, T. O'Toole, G. Parker, T. M. Perl, P. K. Russell, D. L. Swerdlow, and K. Tonat for the Working Group on Civilian Biodefense. 2001. Botulinum toxin as a biological weapon: medical and public health management. *JAMA* **285**:1059–1070.

Barenblatt, D. 2004. *A Plague upon Humanity: the Hidden History of Japan's Biological Warfare Program.* Harper Collins, New York, NY.

Henderson, D. A., T. V. Inglesby, J. G. Bartlett, M. S. Ascher, E. Eitzen, P. B. Jahrling, J. Hauer, M. Layton, J. McDade, M. T. Osterholm, T. O'Toole, G. Parker, T. M. Perl, P. K. Russell, and K. Tonat for the Working Group on Civilian Biodefense. 1999. Smallpox as a biological weapon: medical and public health management. *JAMA* **281**:2127–2137.

Inglesby, T. V., D. T. Dennis, D. A Henderson, J. G. Bartlett, M. S. Ascher, E. Eitzen, A. D. Fine, A. M. Friedlander, J. Hauer, J. F. Koerner, M. Layton, J. McDade, M. T. Osterholm, T. O'Toole, G. Parker, T. M. Perl, P. K. Russell, M. Schoch-Spana, and K. Tonat for the Working Group on Civilian Biodefense. 2000. Plague as a biological weapon: medical and public health management. *JAMA* **283**:2281–2290.

Inglesby, T. V., D. A Henderson, J. G. Bartlett, M. S. Ascher, E. Eitzen, A. M. Friedlander, J. Hauer, J. McDade, M. T. Osterholm, T. O'Toole, G. Parker, T. M. Perl, P. K. Russell, and K. Tonat for the Working Group on Civilian Biodefense. 1999. Anthrax as a biological weapon: medical and public health management. *JAMA* **281**:1735–1745.

Osterholm, M. T., and A. P. Norgan. 2004. The role of irradiation in food safety. *N. Engl. J. Med.* **350**:1898–1901.

U.S. Department of Health and Human Services, Food and Drug Administration. 2008. Irradiation in the production, processing, and handling of food. Final rule. *Federal Register* **73**:49593–49603.

Wilson, B. A. 2008. Global biosecurity in a complex, dynamic world. *Complexity* **14**:71–88.

QUESTIONS

1. Why is anthrax a "weapon of choice" for bioterrorists? What might make it better than smallpox or plague?

2. What is the major concern about the use of botulinum neurotoxin as a bioweapon?

3. What are some of the ways that we can minimize the detrimental impact of bioterrorism on society and science?

4. How do you think the policies put in place since the 2001 anthrax event have affected our relationships with other countries? Do you think these policies have affected the leadership role of the United States in science and technology? If so, how?

5. How do you think microbiologists and other scientists can improve public policy and governmental infrastructure to best respond to potential biothreats?

6. What is the difference between biodefense and biosecurity?

7. Should local authorities or citizens stockpile antibiotics to protect themselves against a potential bioterror act? Explain your answer.

8. Should citizens buy gas masks to protect themselves and their families from biological agents? Why?

9. Which of the following biological agents is not contagious?

A. Smallpox
B. Anthrax
C. Pneumonic plague
D. Bubonic plague

10. Which of the following is a true statement about smallpox?

A. A victim with smallpox is contagious.
B. Cases of smallpox are still found occasionally in Africa.
C. The recommended treatment for smallpox is ciprofloxacin or doxycycline.

11. Which bioterror agents persist in the environment? Explain your answer. What steps would be necessary to clean up or live in these contaminated environments?

12. Suppose you are doing basic research on an attenuated version of a select agent. What concerns should you have about genetically manipulating these strains to contain antibiotic resistance markers?

13. How informed do you think the public is about key issues of biosecurity, including biodefense? What measures do you think should be taken to increase this knowledge?

14. What roles do science education and science literacy play in biosecurity? Do you think that this is important?

SOLVING PROBLEMS IN BACTERIAL PATHOGENESIS

1. You are working as an emergency doctor at a community health clinic near a convention center. Over the course of the afternoon on the second day of a conference being held in the convention center, approximately 200 of the attendees at the conference come into your clinic with complaints of double vision, dry mouth, difficulty swallowing, drooping eyelids, and overall weakness.

  A. What is the most likely biological agent (plague, anthrax, smallpox, or botulinum toxin) causing these symptoms?

  B. Considering the nature of this outbreak, what are the first steps you take as a first responder to this emergency? What is your primary concern in dealing with this crisis?

  C. By the end of the day, your worst fears are that this may be an act of bioterrorism. What are the next steps that you should take?

2. You work for the DHS and have just received a call from a group of researchers at the USDA to consult on a case of unusual fatal food poisoning in Wisconsin due to contaminated cheese. Based on 16S rRNA gene analysis, the researchers identified the pathogen as *Listeria monocytogenes*. This new, highly virulent strain appears to cause symptoms of listeriosis even in previously healthy, immunocompetent individuals, including fever, headache, muscle ache, nausea, and vomiting. However, over 1 to 3 days, these are followed by the onset of anasarca (severe edema, with widespread swelling and accumulation of fluid in all of the tissues and cavities of the body at the same time), stiff neck, disorientation, convulsions, and, in most cases, death within 1 to 2 days of onset. Autopsy of individuals who died showed that bacteria were present throughout the body. The researchers subsequently determined that in tissue culture this highly virulent *Listeria* strain could invade epithelial and endothelial cells and could spread from cell to cell. They also found that similar disease symptoms occurred in mice fed cheese inoculated with the bacteria. The researchers also discovered that injection of medium from cell cultures into mice resulted in massive edema. In your expert opinion as a microbiologist and researcher in the field of bacterial pathogenesis, is there plausible cause for the DHS to be concerned that the incident described above could be a potential act of bioterrorism? Be sure to state your rationale.

3. Again in your capacity working for the DHS, you have just received an urgent call from the CDC to consult on another, unrelated case of an unusual disease outbreak that researchers at the CDC have been investigating. The researchers isolated a new gram-positive bacterium related to *Listeria* from an outbreak of food poisoning in Wisconsin due to contaminated cheese that appears to cause painful gastritis and, in about half of exposed individuals, sudden onset of bleeding ulcers, followed by death from toxic shock within 2 to 3 days. Upon biopsy of infected individuals, it was found that the bacteria were growing on the surfaces of epithelial cells lining the gastric pit of the stomach. Autopsy of individuals who died showed that bacteria were present only in the stomach and not in any of the other body organs. The researchers at the CDC have subsequently determined that, unlike *L. monocytogenes*, this bacterium does not invade epithelial cells or spread from cell to cell. When the researchers grew the bacteria in laboratory medium overnight and then introduced the filtered culture medium onto the surface of the stomach epithelial layer in experimental rabbits, they found that bleeding stomach ulcers were formed and the rabbits died from toxic shock within 1 to 2 days. When they analyzed the filtered culture medium by sodium dodecyl sulfate-polyacrylamide gel electrophoresis, they found that five protein bands were present on the gel. The researchers cut out each of these bands from the gel and subjected the proteins to protease digestion, followed by N-terminal sequencing of the cleaved peptide fragments. When they compared the sequences of the cleaved peptides from each protein with the known protein databases, they found that protein 1 had sequence homology to *Helicobacter pylori* urease, protein 2 had sequence homology to the mucinases of *H. pylori* and *V. cholerae*, protein 3 had sequence homology to the catalytic A subunit of Shiga toxin, protein 4 had sequence homology to anthrax lethal factor, and protein 5 had sequence homology to the protective antigen of anthrax toxin. In your expert opinion as a microbiologist and researcher in the field of bacterial pathogenesis, is there plausible cause for the DHS to be concerned that either one or both of these incidences could be a potential act of bioterrorism? Be sure to state your rationale.

4. As part of the routine surveillance of several reservoirs supplying drinking water to a major metropolitan area in Washington, DC, researchers at the Environmental Protection Agency have identified four new *V. cholerae* strains associated with severe diarrheal disease outbreaks in the area. Being interested in understanding the pathogenesis of these new strains, the CDC researchers subjected each of these strains to a series of experiments to determine whether they contained virulence factors involved in adhesion, invasion, cell-to-cell spread, biofilm formation, or cholera

Strain no.	Gentamicin assay (no. of colonies)			Plaque assay (phenotype observed)	Agglutination of red blood cells (phenotype observed)		Rabbit ileal-loop assay injecting bacteria (distension [volume/loop length after 24 h] [ml/cm])
	Without gentamicin	With gentamicin	With mannose and gentamicin		Without mannose	With mannose	
1	10^9	10^8	10^8	Large plaques	No clumping	No clumping	0.5
2	10^9	10^2	10^2	Large plaques	Clumping	No clumping	0.6
3	10^9	10^8	10^2	Small plaques	No clumping	No clumping	1.5
4	10^2	1	0	No plaques	Clumping	No clumping	1.7

toxin production. The results are shown in the table above.

A. Based on these results, the researchers ask you to predict the possible virulence factor(s) present in strains 1 to 4. Be sure to state your rationale.

B. In your expert opinion as a microbiologist and researcher in the field of bacterial pathogenesis, is there plausible cause for the DHS to be concerned that any of the above incidences could be a potential act of bioterrorism? Be sure to state your rationale.

Glossary

Note: Some terms included in this glossary have meanings in fields other than bacterial pathogenesis. This glossary focuses on the way the term is used in bacterial pathogenesis.

A-B toxin (also AB-type toxin) a bacterial toxin in which the B part (domain or subunit) of the toxin responsible for binding to target cells is separate from the A part (domain or subunit) that mediates enzymatic activity or other toxic function

ABC transporter a type of integral membrane pump that consists of multiple subunits; uses ATP hydrolysis to power the pump

Abscess a localized collection of pus

Accession number a tracking number assigned to a nucleotide or protein sequence or other data deposited in a database

Accessory factor a periplasmic connecting protein that holds together the two pore-forming components of the outer membrane of the type 1 secretory system

Accessory gene regulator (Agr) the regulatory system controlling production of surface adhesins and exoproteins in *Staphylococcus aureus*

Accessory Sec system a specialized transport system found in some gram-positive bacteria

Acellular pertussis vaccine (aP) the subunit form of pertussis vaccine; consists of pertussis toxoid with an additional component (usually an adhesin, but that differs depending on the manufacturer)

Acellular vaccine a vaccine consisting of purified, nontoxic proteins, not whole cells

Acetylcholine (ACh) a neurotransmitter molecule whose release is inhibited by botulinum neurotoxins

ACh see **acetylcholine**

Acid tolerance response conditioning of bacteria to tolerate low pH, in which bacteria grown at pH 6 for a generation can survive at pH 3; regulated by Fur

Acquired immune response see **adaptive immune response**

Acquired immunity see **adaptive immunity**

ActA a surface protein of *Listeria monocytogenes* responsible for actin nucleation, involved in actin tail formation that enables the bacteria to propel through the host cell and spread from cell to cell

Actin a major protein component of the host cell cytoskeleton

Actin-based motility the mechanism by which bacteria interact with host components to form actin tails, thus becoming able to spread from cell to cell

Actinomyces a group of bacteria that produces many antibiotics

Activated macrophages macrophages with an increased killing capacity due to generation of reactive oxygen intermediates and other toxin compounds

Active immunity immunity that develops as a result of an infection or immunization

Acute inflammatory response see **acute-phase response**

Acute-phase proteins proteins synthesized during the acute phase of an immune response; examples are complement proteins and C-reactive protein

Acute-phase response production of a group of serum proteins in response to infection; a form of induced innate (nonspecific) immunity

Acute respiratory distress syndrome (ARDS) accumulation of fluid and PMNs in the lung; leads to insufficient gas exchange and damage to the lung

Acyl-HSL see *N*-acyl homoserine lactones

Adaptive immune response see **adaptive immunity**

Adaptive immunity host defenses produced in response to invasion by specific bacteria or other infectious agents; involves antibodies, T cells, B cells, and activated macrophages

ADCC see **antibody-dependent cell-mediated cytotoxicity**

Adenylate cyclase a protein toxin produced by *Bordetella pertussis*; activated by intracellular calcium-bound calmodulin to catalyze the conversion of ATP to cyclic AMP

Adhesins microbial surface components that bind to the host cell surface receptors

Adhesion frequency the ratio of cell-associated colony-forming units (CFUs) to total CFUs at the end of an experiment

Adjuvant a substance added to a vaccine that enhances antigenic stimulus of the host's immune response

ADME pharmacokinetic analyses of the absorption, distribution, metabolism, and excretion of different drug doses in a host

ADP-ribosylation transfer of the ADP-ribosyl group from NAD to a host cell protein by a bacterial protein toxin

ADP-ribosyl transferase an enzyme that catalyzes ADP-ribosylation

Aerobic condition where oxygen is required for growth

Aerosol a fine mist that can be inhaled

Aerotaxis movement of bacteria toward higher oxygen concentrations

Aerotolerant able to survive in an aerobic environment (describes anaerobes)

Affinity strength of interaction between a ligand and its receptor, such as an epitope and its cognate antigen-binding site on an antibody

Afimbrial adhesins surface proteins of bacteria important for adhesion (attachment) but not organized in pilus-like structures

Agglutination clumping of cells by specific antibody

agr **genes** see **accessory gene regulator**

AHKs see **α-hydroxyketones**

AHL see *N*-acyl homoserine lactones

AI-1 see **autoinducer 1**

AI-2 see **autoinducer 2**

AIP see **autoinducing peptide**

Alarmone a small regulatory molecule of the host cell that is synthesized in response to a stress condition; see **(p)ppGpp**

Alkaline phosphatase (PhoA) an enzyme commonly used as a reporter group

α-Hemolytic characterized by partial hemolysis of red blood cells on blood agar plates; the area around bacterial colonies on blood agar plates is not clear but has greenish discoloration and is used to identify certain bacteria

α-Hydroxyketones (AHKs) recently discovered quorum-sensing molecules found in gram-negative bacteria

Alpha toxin (α-toxin) (i) exoprotein produced by *Staphylococcus aureus* that forms pores in human cell membranes; (ii) exotoxin produced by *Clostridium perfringens* that hydrolyzes lecithin in human cell membranes

Alternative pathway part of the innate immune response in which the complement cascade is activated by components on the bacterial cell surface

Alum an aluminum salt used as an adjuvant component of a vaccine; the only vaccine adjuvant the FDA approved for use in the United States

Alveolar macrophage a macrophage fixed in the alveoli of the lung

Aminoglycosides a family of antibiotics that bind to the 30S ribosomal subunit and prevent it from binding mRNA; includes kanamycin and gentamicin

Amplicon PCR product of amplification of DNA segments

AMPs antimicrobial peptides that act by disturbing membranes

Anaerobic growth under conditions lacking oxygen; anoxic

Anergy loss of reactivity to an allergen

Animal model a species of animal that develops a disease similar to that in humans infected with the same organism

Anthrax toxin a multisubunit protein toxin produced and secreted by *Bacillus anthracis*; comprised of a complex between up to three subunits of one or both of the catalytic subunits, lethal factor (LF) and edema factor (EF), and a heptameric complex of the binding subunit, protective antigen (PA)

Antibiotic a low-molecular-weight compound that can inhibit growth of or kill microorganisms; administered inside or outside the body

Antibiotic resistance the ability of bacteria to grow in the presence of antibiotics that would normally inhibit their growth or kill them

Antibiotic-resistant bacteria bacteria that are not killed or inhibited by antibiotics

Antibiotic tolerance the ability of bacteria to survive antibiotic treatment although they are unable to divide; removal of the antibiotic restores bacterial growth

Antibody an immunoglobulin molecule produced by B cells that interacts with an antigen

Antibody-dependent cell-mediated cytotoxicity (ADCC) infected host cells that are coated with IgG and then killed by phagocytes or natural killer cells, which bombard them with toxic compounds

Antigen a substance that can interact with a specific antibody (see **immunogen**)

Antigen-binding region a portion of the Fab region of an antibody that binds to a specific antigenic determinant (epitope)

Antigen-binding sites see **antigen-binding region**

Antigenic determinant a site on an antigen that binds to the antigen-binding site of an antibody; an epitope

Antigenic variation the ability of some bacteria to change the amino acid composition of their adhesins or other surface proteins to avoid the host immune system

Antigen-presenting cell (APC) an immune cell, such as a macrophage or dendritic cell, that engulfs a microbe or its products, degrades the proteins, and presents the resulting peptides on its surface to the immune system

Anti-inflammatory cytokines peptides or proteins that regulate the immune response by inhibiting the production of proinflammatory cytokines

Antimicrobial peptides a diverse class of peptides produced by the host that kill bacteria in various ways, including forming channels or holes in bacterial membranes, interfering with bacterial metabolism, or targeting cytoplasmic components; often cationic, but some are anionic

Antiporter a type of efflux pump that uses uptake of protons as the source of energy to pump antibiotics out of the cytoplasm

Antiseptic an antimicrobial compound applied to surfaces, such as the skin

Antiserum serum containing antibodies against a specific antigen

Anti-sigma factors protein factors that counteract the functions of specific sigma factors by sequestering them away from RNA polymerase

Antitoxin (i) an antibody specific for a toxin that neutralizes its activity; (ii) a member of a toxin-antitoxin complex of a bacterial regulatory system that blocks the toxin's action

aP see **acellular pertussis vaccine**

APC see **antigen-presenting cell**

Apical surface the top side of an epithelial cell facing the outside of the body; portion of a cell exposed to the lumen

Apoptosis programmed cell death; occurs in normal body cells; characterized by condensation of chromatin at the boundary of the nucleus; stimulated by some bacteria and toxins

ARDS see **acute respiratory distress syndrome**

AS04 an adjuvant containing monophosphoryl lipid A and alum; approved for use in humans

Aseptic free of microorganisms

Aspiration process in which fluids are introduced into or removed from body cavities

Asymptomatic carrier a person colonized by disease-causing bacteria but who does not have disease symptoms

ATP-binding cassette (ABC) transporter (i) inner membrane-spanning component of the type 1 secretion system; (ii) a separate cytoplasmic protein that binds to membrane-spanning proteins of an ABC transporter

att **site** see **attachment site**

Attaching and effacing (also attachment and effacement) a distortion of microvilli on the apical side of an epithelial cell due to extensive rearrangement of host cell actin by EPEC strains

Attachment site the site on DNA integrons at which integrase integrates circular DNA segments; the insertion site on DNA at which phage attach and integrate into the chromosome (Campbell integration)

Attenuation (i) genetic regulation that involves RNA secondary structure; (ii) decrease in virulence of microorganisms, often through mutation, used in a vaccine

Augmenters compounds used in combination with current antibiotics to increase their efficacy or stability, thereby enhancing antimicrobial activity

Autoimmune disease a disease that occurs when the immune system recognizes a host molecule as foreign; often induced by aberrant immune response to microbial infection

Autoinducer 1 (AI-1) *N*-3-oxohexanoyl-L-homoserine lactone; quorum-sensing signal molecule produced by gram-negative bacteria

Autoinducer 2 (AI-2) a small furanosyl borate diester compound used for interspecies cell-cell communication

Autoinducers small diffusible signaling molecules used by a group of bacteria to coordinate their activities

Autoinducing peptides (AIPs) small quorum-sensing signaling peptides (often containing cyclic thiolactones) used for intraspecies communication in gram-positive bacteria

Autolysins enzymes, produced by bacteria, that digest peptidoglycan and can cause lysis of bacteria; used by phage to induce bacterial cell lysis

Autophagy a process by which components of mammalian cytoplasm or organelles are surrounded by endoplasmic reticulum and slated for destruction or recycling

Autotransporter a protein exported via a type 5 secretion system; delivered to the periplasm by the general secretory pathway and then transports itself across the outer membrane

Avidity a combination of affinity and valence; a measure of the strength of binding of antigen to antibody

Azithromycin a member of the macrolide family of antibiotics

B cell see **B lymphocyte**

B lymphocyte (B cell) the cell type that produces antibodies; see **plasma cell**

Ba a component that results from cleavage of factor B of the complement system

Bacillus a rod-shaped bacterium

Bacillus anthracis a toxin-producing gram-positive bacterium responsible for the disease anthrax

Bacitracin an antibiotic that interferes with the recycling of bactoprenol (lipid 55) during peptidoglycan synthesis; used topically in creams because it is too toxic for internal use

Bacteremia bacteria present in the bloodstream

Bacteria single-cell microbes lacking nuclei that, along with archaea, were first forms of life on Earth

Bacterial luciferase (Lux) see **luciferase**

Bacterial vaginosis a disease that is characterized by a shift of the populations in the vaginal microbiota

Bactericidal a substance that kills bacteria

Bacteriocin a pore-forming toxin produced by one species of bacteria that targets another species or subspecies of bacteria

Bacteriophage a virus that infects bacteria

Bacteriostatic a substance that inhibits the growth of bacteria but does not kill them

BALT see **bronchial-associated lymphoid tissue**

Bands immature forms of PMNs

Barcode a DNA sequence introduced by PCR or transposition that is used to mark and track a DNA segment or bacterium

Basal lamina a layer of extracellular matrix to which epithelial cells attach

Basal surface the portion of a mucosal cell that is in contact with the extracellular matrix

Basement membrane a thin sheet of fibers that underlies the epithelial cell layer lining the cavities and surfaces of organs and skin or the endothelial cell layer lining the interior surfaces of blood vessels

Basolateral membrane or surface the side and bottom portions of polarized epithelial cell membranes in a confluent monolayer; portions in contact with the extracellular matrix

Basophil a granulocyte that makes up a portion of the circulating white blood cells; contains toxic granules that are released when IgE molecules bound to the surface are cross-linked

Bb a component that results from cleavage of factor B of the complement system; complexes with activated C3 and produces C3b; C3b binds Bb, a C3 convertase

BBB see **blood-brain barrier**

B-cell receptor (BCR) a receptor on the surface of B cells that, when bound to antigens, sends activation signals to the B cell to proliferate and produce antibodies

BCR see **B-cell receptor**

β-Galactosidase (LacZ) an enzyme commonly used as a reporter group using chromogenic substrates; catalyzes the hydrolysis of β-galactosides into monosaccharides

β-Glucuronidase (GUS) an enzyme commonly used as a reporter group using chromogenic substrates; catalyzes the hydrolysis of β-glucuronic acid from the nonreducing ends of glycosaminoglycans

β-Hemolytic characterized by complete hemolysis of red blood cells on blood agar plates; the area around bacterial colonies on blood agar plates is clear, which is used to identify certain bacteria

β-Integrins see **integrins**

β-Lactam antibiotics antibiotics that contain a β-lactam ring and act by inhibiting peptidoglycan synthesis; includes penicillins, cephalosporins, carbapenems, and monobactams

β-Lactamase an enzyme that cleaves the β-lactam ring of β-lactam antibiotics and thus inactivates them

β-Phage temperate bacteriophage that carries the gene encoding diphtheria toxin

β-Toxin an exotoxin produced by *Clostridium perfringens*

Bfp see **bundle-forming pili**

Bile salts steroids with detergent-like properties that are produced by the liver and released into the intestine; they disrupt bacterial membranes

Binary toxin (CDT) a heterodimeric subunit toxin produced by *Clostridium difficile* that catalyzes the ADP-ribosylation of host cell actin; may potentiate action of TcdA and TcdB toxins

Bioavailability measurement of the rate or fraction of an administered dose of a drug that enters the circulation and reaches the target site in a mammalian host unchanged; considered to be one of the pharmacokinetic properties of drugs

Biofilm multilayered bacterial populations embedded in a polysaccharide, protein, or DNA matrix that is attached to some surface (plastic, mucosal membrane, teeth, or implanted medical device)

Bioinformatics a field that involves design, development, management, and utilization of life science databases

Biological and Toxins Weapons Convention (BTWC or BWC) a convention held in 1975 for the purpose of ratifying the first international disarmament treaty banning the development, production, and stockpiling of biological and toxin weapons

Biomarker a protein that serves as an indicator of a normal biological process or a pathogenic state in a mammalian host

Biophotonic imaging a technique for following the course of an infection in a mammalian host by monitoring the expression of the luciferase operon that has been incorporated into the chromosome of the bacterial strain being tested

Biosecurity a set of preventive measures designed to reduce the risk of transmission of infectious diseases, quarantine pests, invasive alien species, or living modified organisms

Bioterrorism the deliberate use of microbes or toxins to disrupt normal societal functions and to injure or kill people or livestock

BlaZ β-lactamase produced by *Staphylococcus aureus* in response to β-lactam antibiotics

Blood-brain barrier (BBB) the membrane (meninges) covering of the brain and spinal cord that prevents substances in blood from entering

BoNT see **botulinum neurotoxin**

***Bordetella* tracheal cytotoxin (TCT)** a peptidoglycan fragment of *Bordetella pertussis* membrane that kills ciliated cells, stimulates the release of IL-1, and triggers the coughing response in the disease whooping cough

Borrelia burgdorferi a tick-borne spirochete bacterial species that causes Lyme disease

Botulinum neurotoxin (BoNT) a protein neurotoxin produced by *Clostridium botulinum* that enters motoneurons and catalyzes the zinc-dependent proteolytic cleavage of specific SNARE proteins involved in the release of the neurotransmitter acetylcholine, resulting in flaccid paralysis; there are seven distinct serotypes (A, B, C, D, E, F, and G); also related to tetanus toxin (TeNT) from *Clostridium tetani*

Bradykinin a vasoactive peptide of 9 amino acids that enhances extravasation and mediates a proinflammatory response due to increased vasodilation and contraction of smooth muscle

Broad spectrum of activity description of an antibiotic that is effective against many different types of bacteria

Bronchial-associated lymphoid tissue (BALT) the respiratory tract equivalent of the gastrointestinal-associated lymphoid tissue

BTWC see **Biological and Toxins Weapons Convention**

Bullous impetigo a type of skin infection characterized by painless blister-like lesions; caused by toxins produced by *Staphylococcus aureus*

Bundle-forming pili (Bfp) the type IV pili of EPEC thought to be important in self-aggregation of bacteria (microcolony formation)

BWC (or BTWC) see **Biological and Toxins Weapons Convention**

C1 a complement component important in the classical pathway; activated by bacterium-antibody complexes

C1a a complement component that cross-links Fc regions of IgG or IgM; part of the classical pathway; cleaves C4 and C2

C2 a complement component activated by C1 in the classical pathway

C2a a complement component resulting from cleavage of C2; binds C4b to form C3 convertase

C2b a complement component resulting from cleavage of C2

C3 a complement component that is cleaved into C3a and C3b by C3 convertase

C3 convertase a complex of complement components, comprised of either C3bBb or C2aC4b, that convert C3 to C3a and C3b

C3a a complement component that results from the proteolytic cleavage of C3; acts as a vasodilator

C3b a complement component that results from the proteolytic cleavage of C3; binds to bacterial surfaces and potentiates their opsonization; forms part of C3 convertase

C4 a complement component that is cleaved by activated C1 of the classical pathway or components of the MBL pathway

C4a a complement component that results from cleavage of C4; acts as an anaphylatoxin to mediate local inflammation at the site of infection

C4b a complement component that results from cleavage of C4; binds to C2a to form C3 convertase

C5 a complement component that is cleaved to C5a and C5b by C5 convertase

C5 convertase a complement component comprised of a complex between C3 convertase and C3b that converts C5 to C5a plus C5b

C5a a complement component that results from proteolytic cleavage of C5; acts as a chemoattractant for PMNs

C5b a complement component that results from proteolytic cleavage of C5; C5b recruits C6, C7, C8, and C9 to form the membrane attack complex (MAC) that pokes holes in membranes of gram-negative bacteria

Calcium-calmodulin-dependent adenylate cyclase (Cya) a protein toxin produced by *Bordetella pertussis* that converts ATP into cyclic AMP; only active inside host cells, since it depends on calcium-calmodulin for activity

Calcium mobilization   release of calcium from intracellular stores triggered by interaction of a cell with bacteria or bacterial toxins

Calmodulin   a mammalian protein that binds calcium; required to activate adenylate cyclase

Capsule   an extracellular network, often of exopolysaccharides, but sometimes peptides, that covers the cell surfaces of some bacteria and usually interferes with phagocytosis

Carbapenem   a class of β-lactam antibiotics

Cardiotoxin   a toxin that targets heart cells

Carrier   an apparently healthy person who harbors pathogenic microorganisms

CARVER+Shock (criticality, accessibility, recuperability, vulnerability, effect, recognizability, and shock effect)   an assessment method for identifying the most vulnerable target sites for introduction of contamination within the food-processing system

CAT   see **chloramphenicol acetyltransferase**

cat **gene**   a gene that encodes chloramphenicol acetyltransferase (CAT), which imparts resistance to chloramphenicol and is used as a reporter for transcriptional regulation

Catechol   a class of siderophore

Cathelicidin   an antimicrobial peptide produced by host immune cells

CD1   a protein molecule on the surfaces of antigen-presenting cells that presents lipid or glycolipid antigens to the immune system

CD4   a transmembrane glycoprotein receptor expressed on the surfaces of helper T cells, regulatory T cells, monocytes, macrophages, and dendritic cells that acts as a coreceptor to assist the T-cell receptor (TCR) to activate its cognate T cell following interaction with antigen-bound MHC-II complex on an antigen-presenting cell

CD4⁺ T cells   see **T helper (Th) cells**

CD8   a transmembrane glycoprotein receptor expressed on cytotoxic T cells, natural killer cells, and dendritic cells that acts as a coreceptor to assist the T-cell receptor (TCR) to activate its cognate T cell following interaction with antigen-bound MHC-I complex on a cell

CD8⁺ T cells   see **cytotoxic T cells**

CD14   a pattern recognition receptor on monocytes and macrophages that binds LPS–LPS-binding protein complex and acts as a coreceptor (along with TLR-4 and MD-2) for detection of bacterial LPS by the innate immune system

CD55   see **decay-accelerating factor**

Cdc42   a small G protein (GTPase) involved in regulation of the mammalian cell cycle that is targeted by toxins A and B of *Clostridium difficile*

cDNA   complementary DNA biochemically made from RNA by reverse transcription using a reverse transcriptase enzyme

CDT   see **binary toxin**

CdtA   one of the subunits of the ternary complex of cytolethal distending toxin (CLDT or CDT) from *Escherichia coli*, *Campylobacter*, and several other gram-negative pathogens that, along with CdtC, mediates the receptor binding and translocation of the catalytic subunit, CdtB

CdtB   one of the subunits of the ternary complex of cytolethal distending toxin (CLDT or CDT) from *Escherichia coli*, *Campylobacter*, and several other gram-negative pathogens that mediates host cell DNA damage to cause cell cycle arrest in G_2 phase

CdtC   one of the subunits of the ternary complex of cytolethal distending toxin (CLDT or CDT) from *Escherichia coli*, *Campylobacter*, and several other gram-negative pathogens that, along with CdtA, mediates the receptor binding and translocation of the catalytic subunit, CdtB

CdtR   a positive regulator of the CDT toxin genes

Cell-mediated immunity (CMI)   adaptive immunity due to activation of antigen-specific cytotoxic T cells, macrophages, and natural killer cells; does not involve antibodies or complement

Cell-to-cell spread   a situation where bacteria growing inside one host cell burst out and spread to adjacent host cells; may lead to production of a plaque; characteristic of actin-based motility (via actin tail formation)

Cephalosporins   a class of β-lactam antibiotics

CF   see **cystic fibrosis**

CFTR   see **cystic fibrosis transmembrane conductance regulator**

CFU   see **colony-forming unit**

Chaperone   a protein that aids in the folding of proteins after translation or of RNA molecules after transcription; a protein that binds and stabilizes another protein or brings another protein to a particular site in the cell; a protein that potentiates pairing between small RNAs and mRNAs

Chaperone-usher system   a complex protein system in the periplasm of gram-negative bacteria that plays a role in the assembly of extracellular pili; requires a number of auxiliary proteins

Chelocardin   an antibiotic in the tetracycline family that acts by disrupting the function of the cytoplasmic membrane

Chemokines   small glycopeptides (8 to 10 kDa) produced by many human cell types; they organize activities of cells of innate and adaptive defenses

Chemotaxis   the attraction and movement of bacteria to a particular chemical substance

Chemotherapy   treatment of disease with drugs

Chloramphenicol acetyltransferase (CAT)   an enzyme that inactivates chloramphenicol and imparts antibiotic resistance; often used as a reporter for gene transcription

Cholera toxin (CT)   a multisubunit AB_5-type exotoxin complex produced by *Vibrio cholerae* that ADP-ribosylates and constitutively activates the heterotrimeric Gs protein involved in regulation of mammalian adenylate cyclase; results in elevated intracellular cyclic AMP levels

Chromogenic substrate   a substrate that changes color on plates or in liquid medium if a particular enzyme is present (e.g., X-Gal)

CI   see **competitive index**

Cilia   surface structures of eukaryotic epithelial cells that move mucus over surfaces

Ciliated columnar cells   mucosal cells that have cilia on their apical surfaces; commonly found lining the respiratory tract and fallopian tubes

Cipro   see **ciprofloxacin**

Ciprofloxacin   a member of the fluoroquinolone antibiotic family

Class I major histocompatibility complex (MHC-I)   a major histocompatibility type that when complexed with an epitope triggers activation and proliferation of cytotoxic T cells and natural killer cells; found on most cells in the body

Class II major histocompatibility complex (MHC-II)   a major histocompatibility type that when complexed with an epitope leads to activation and proliferation of helper T cells; found only on a few immune cell types (e.g., B cells and APCs)

Classical complement pathway   a complement pathway that is activated by antigen-antibody complexes

Classical strain   the O1 strain of *Vibrio cholerae* first associated with pandemics

Clavulanic acid   a suicide substrate inhibitor of serine-type β-lactamases that is used in combination therapy with β-lactam antibiotics to enhance their stability by preventing their degradation

Clindamycin   a member of the lincosamide antibiotic family

CLDT   see **cytolethal distending toxin**

CMI   see **cell-mediated immunity**

CMT   see **cytolysin-mediated translocation**

CNFs   see **cytotoxic necrotizing factors**

CNS   see **coagulase-negative staphylococci**

Coagulase-negative staphylococci (CNS)   strains of staphylococci, such as *Staphylococcus epidermidis* and *Staphylococcus saprophyticus*, that do not produce coagulase

Coccus   a sphere-shaped bacterium

Cold pasteurization   see **food irradiation technology**

Colitis   inflammation of the colon

Collectin pathway   see **mannose-binding lectin (MBL)**

Collectins   a family of calcium-binding lectins that are soluble pattern recognition receptors involved in innate immune response; includes mannose-binding lectin

Colonization   the ability of a bacterium to remain at a particular site in the host and multiply there; can be asymptomatic

Colonization factor antigens I and II   different types of pili produced by ETEC

Colony   a discrete mass of cells derived from a single bacterial cell or a chain of bacterial cells

Colony-forming unit (CFU)   a measure of the number of viable bacteria in a sample; each colony on an agar plate arises from a single bacterium or a chain of bacteria

Columnar cells   tall, thin epithelial cells

Combinatorial chemistry   an approach to generate new synthetic antibiotics and drugs; a large variety of chemical groups are added to a scaffold, and the resulting array is tested for activity as a mixture

Commensal microbiota   see **resident microbiota**

Competent   a state in which bacteria are capable of taking up extracellular DNA from the medium and incorporating homologous sequences into their chromosomes by recombination

Competitive index (CI)   [output ratio (CFU mutant/CFU wild type)]/[input ratio (CFU mutant/CFU wild type)]; a means for determining whether a mutation gives a bacterium a competitive advantage in an animal model of infection or mixed-growth experiment

Complement   see **complement system**

Complement activation   proteolytic cleavage of complement components to produce activated proteins that attract phagocytes, cause lysis of gram-negative bacteria, and opsonize bacteria

Complement system   a group of plasma proteins that mediate innate immunity and the inflammatory response when activated

Complementation test   a test to determine whether two different mutants with the same phenotype are due to mutations in the same gene; done by expressing a wild-type copy of the gene from a second location in the chromosome or from a plasmid

Conjugate vaccine   a vaccine comprised of a polysaccharide antigen covalently linked to a carrier protein; forces the polysaccharide to be processed as a protein-like antigen through the T-cell-dependent pathway

Conjugation   the transfer of DNA directly from one bacterial cell to another; mediated by transfer (*tra*) genes and a specific pilus

Conjugative plasmid   a plasmid that carries genes encoding proteins needed for the conjugation process

Conjugative transposons   transposons that can transfer themselves from the genome of the donor to the genome of the recipient

Constant region (Fc)   the portion of an antibody molecule that binds complement component C1 and receptors on phagocytes and other immune cells; see **Fc**

Continuous epitope   a single linear sequence of an antigen recognized by an antibody

Coordinated regulation   regulation of multiple genes in response to a particular signal; see **operon** and **regulon**

CpG oligonucleotides   see **CpG-rich DNA**

CpG-rich DNA   a sequence of DNA present in bacteria but not mammals; acts in stimulation of innate immunity by a toll-like receptor

C-reactive protein (CRP)   an acute-phase protein produced by the liver that appears in blood serum in response to inflammation

CRISPR (clustered regularly interspaced short palindromic repeats)   direct repeats scattered throughout the chromosomes of many bacteria; function as a quasi-immune system against exogenous genetic elements

CRP   see **C-reactive protein**

CT   see **cholera toxin**

CTLs   see **cytotoxic T cells**

C-type lectin   a type of lectin that requires calcium for binding; involved in recognition of cells infected by viruses

Cuboidal cells   cube-shaped epithelial cells

Culture   microorganisms growing in liquid or on solid medium

Cya   see **calcium-calmodulin-dependent adenylate cyclase**

Cystic fibrosis (CF)   a disease caused by a defect in chloride secretion due to a mutation in the gene for a chloride transmembrane conductance regulator protein (CFTR) that regulates production of sweat, digestive juices, and mucus; characterized by production of thick mucin in the lungs that causes difficulty in breathing

Cystic fibrosis transmembrane conductance regulator (CFTR)   an ABC transporter-like chloride ion channel protein that transports chloride ions across epithelial cell membranes

Cytokine technology   a technique for modulating the immune system

Cytokines   signaling proteins (8 to 30 kDa) produced by some mammalian cells in response to stimuli; mediators of inflammation and septic shock

Cytolethal distending toxin (CLDT, or CDT)   a heterodimeric subunit AB toxin (comprised of CdtA, CdtB, and CdtC subunits) produced by EPEC, *Campylobacter*, and several other pathogenic bacteria that causes mammalian cell cycle arrest in G_2 phase; leads to enlargement and distension of host cells

Cytolysin-mediated translocation (CMT)   a proposed mechanism by which gram-positive bacterial effector proteins are secreted by the general secretory system and delivered directly into host cells through a pore in the eukaryotic membrane created by a secreted pore-forming toxin

Cytoskeleton   a complex array of proteins that gives shape to eukaryotic cells

Cytotoxic necrotizing factors (CNFs)   a family of AB-type protein toxins produced by some EPEC and *Yersinia* strains that catalyze the deamidation and activation of small Rho GTPases, which causes cytoskeletal rearrangements

Cytotoxic T cells $CD8^+$ T cells (with CD8 antigen on their surfaces) that kill host cells displaying foreign antigens on their surfaces through the MHC-I complex

Cytotoxic T lymphocytes (CTLs)   see **cytotoxic T cells**

Cytotoxin   a toxin that kills mammalian cells

D   the diphtheria toxoid component of the DTaP vaccine

DAF   see **decay-accelerating factor**

Dalfopristin   a streptogramin antibiotic

Daptomycin   a glycopeptide antibiotic; commercially called Cubicin

Databases   centralized repositories for sequencing, transcriptome, and proteomic information; most are web based and freely available online (e.g., NCBI)

DCs   see **dendritic cells**

DDBJ   DNA Databank of Japan; Japan's nucleotide database

Debilitation   loss of health

Debridement   the surgical removal of dead tissue

Decay-accelerating factor (DAF, also CD55)   a membrane glycoprotein that regulates the complement system on the mammalian cell surface by preventing assembly of or accelerating the disassembly of the C3bBb complex and blocking formation of the membrane attack complex; the site where *Helicobacter pylori* binds to gastric epithelium

Deconvolution   determination of which compound in a large group is the one with the desired activity; used in combinatorial chemistry screening for antibiotics or inhibitors

Defensins   a specific type of antimicrobial peptide comprised of cysteine-rich cationic peptides and used by mammalian host cells to kill bacteria

Degradative enzymes   lysosomal proteins, e.g., proteases and lysozyme, that destroy bacterial surface components

Delta toxin   a membrane pore-forming exotoxin (hemolysin) produced by *Clostridium perfringens*

Denaturing/thermal gradient gel electrophoresis (DGGE, TGGE)   a PCR-based method for profiling a microbial community; amplified regions of DNA (PCR amplicons) are separated by electrophoresis on denaturing or thermal polyacrylamide gels and analyzed for their characteristic banding patterns

Dendritic cells (DCs)   antigen-presenting cells (APCs) that process invading bacteria and activate host immune defenses

Derivative toxin   the toxic portion of botulinum toxin (see **progenitor toxin**)

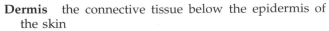

Dermis the connective tissue below the epidermis of the skin

Desmoglein (Dsg) a family of cadherin proteins on the surfaces of epithelial cells that form desmosomes, which join cells to one another; maintains keratinocyte cell-cell adhesion

Desmosomes protein structures that hold epithelial cells together

Desquamation shedding of dead cells of the epidermis

DGGE see **denaturing/thermal gradient gel electrophoresis**

Diapedesis see **transmigration**

Diarrhea abnormal frequency of bowel movements and fluidity of stool

DIC see **disseminated intravascular coagulation**

Differential toxicity a desirable trait of an antibiotic in which it is more toxic to bacterial cells than to the human body

Diphthamide a post-translationally modified histidine found in mammalian translation elongation factor EF-2 that is ADP-ribosylated by diphtheria toxin and *Pseudomonas* exotoxin A (ExoA)

Diphtheria toxin (DT) an AB protein toxin produced by *Corynebacterium diphtheriae* that binds and enters cells and blocks protein synthesis by ADP-ribosylation of mammalian elongation factor 2 (EF-2)

Diphtheria toxin regulation protein (DtxR) a Fur-like protein of *Corynebacterium diphtheriae* that mediates iron regulation of diphtheria toxin transcription

Diphthine a deamidated version of diphthamide found in *Archaea*

Diplococci bacterial cocci arranged as pairs of cells, e.g., pneumococci

Discontinuous epitope the nonlinear portions of an antigen recognized by an antibody that are adjacent to each other due to the structural conformation of the antigen

Disease damage to the host, such as that caused by an infection, which often manifests symptoms

Disease transmission the means by which an infectious agent spreads from one host to another

Disinfectant an antimicrobial compound applied to inanimate objects; often too toxic for internal application

Disseminated intravascular coagulation (DIC) formation of numerous small blood clots that obstruct peripheral blood vessels; a symptom of septic shock

DNA chip the array of DNA on a microarray (usually a glass slide or membrane)

DNA microarray an orderly arrangement of DNA oligonucleotides corresponding to genes of interest

DNA sequence database a resource containing sequences of nucleotides in specific DNA samples, especially genomes

DNA uptake sequence (DUS) short DNA sequences scattered throughout the genomes of some naturally competent bacteria that serve as recognition sites for binding and uptake of that DNA by certain bacteria

DNA vaccine a vaccine consisting of DNA encoding a vaccine protein target that is injected into muscle cells, where the protein is synthesized and stimulates an immune response

DNase deoxyribonuclease, an enzyme that degrades DNA

Donor cell a cell from which DNA is donated during genetic exchange

Dsg see **desmoglein**

DT see **diphtheria toxin**

DT vaccine a divalent vaccine against diphtheria and tetanus comprised of the diphtheria toxoid and tetanus toxoid

DTP (or DTaP) vaccine a trivalent vaccine against diphtheria, tetanus, and pertussis comprised of the diphtheria toxoid, tetanus toxoid, and either dead *Bordetella pertussis* bacteria or acellular components (pertussis toxoid plus pertactin or another adhesin)

DtxR see **diphtheria toxin regulation protein**

Dual-use agent biological agents that can be used as bioweapons or in nonmalicious ways

DUS see **DNA uptake sequence**

Dysentery a type of diarrhea in which stools contain blood and mucus; associated with infection by Shiga toxin- or Shiga-like toxin-producing bacteria

EAE see **enterocyte attachment and effacement**

ECM see **extracellular matrix**

Edema excessive fluid in the tissues; associated with the edema factor component of anthrax toxin

Edible vaccines plants that have been genetically modified to produce a bacterial protein antigen for oral vaccination

EF-2 see **elongation factor 2**

Effector proteins see **exoenzymes**

Efflux mechanism (efflux pump) cytoplasmic membrane proteins that mediate resistance to tetracycline, macrolides, quinolones, and other classes of antibiotics by pumping them out of the bacterial cytoplasm

Efflux pump see **efflux mechanism**

EHEC see **enterohemorrhagic *Escherichia coli***

Ehrlichia chaffeensis an arthropod-borne pathogen that causes human monocytotrophic ehrlichiosis (HME) disease

EIEC see **enteroinvasive *Escherichia coli***

El Tor the O1 strain of *Vibrio cholerae* associated with more recent pandemics

Elastase an enzyme that degrades the elastin component of the extracellular matrix; may be important in causing lung damage in *Pseudomonas aeruginosa* infections

Electron beams streams of high-energy electrons produced by an electron gun and used to irradiate food

Elongation factor 2 (EF-2) a protein that plays an essential role in host cell protein synthesis; target of diphtheria toxin and *Pseudomonas* ExoA toxin

EMBL NDB European Molecular Biology Laboratory Nucleotide Database; Europe's library of nucleotide sequence data

Emerging infectious diseases new diseases that appear or become known, often due to increased human contact with a microbe that has been around for years

Endemic continually present at low levels in the community (describing a disease)

Endocarditis inflammation of heart valves

Endocytosis engulfment (usually receptor mediated) of extracellular material into a vacuole by a eukaryotic cell; bacteria and bacterial toxins often trigger their uptake into cells via this mechanism

Endospore a survival form of some bacteria that protects against environmental extremes

Endosymbiont a microbe that lives inside the cells of its symbiotic partner

Endothelial cells cells constituting the endothelium

Endothelium a thin layer of cells lining the interior surfaces of blood vessels, lymphatic vessels, and the heart

Endotoxic shock see **septic shock**

Endotoxin see **lipopolysaccharide**

ent **genes** genes encoding the enterotoxins of *Staphylococcus aureus*

Enteric relating to the gastrointestinal tract

Enterococci gram-positive cocci once considered members of group D streptococci but now part of the genus *Enterococcus*; commensal species that have emerged as opportunistic pathogens

Enterocyte attachment and effacement (EAE) altered ultrastructure of the apical surfaces of mucosal cells caused by proteins produced by EPEC

Enterohemorrhagic *Escherichia coli* (EHEC; *Escherichia coli* O157:H7) strains of *E. coli* that cause dysentery-like disease but rarely invade host cells; produce Shiga toxin; may cause hemolytic uremic syndrome

Enteroinvasive *Escherichia coli* (EIEC) strains of *E. coli* that cause invasive disease similar to that caused by *Shigella* but do not produce Shiga toxin and do not cause hemolytic uremic syndrome

Enteropathogenic *Escherichia coli* (EPEC) strains of *E. coli* that produce ultrastructural changes in mucosal cells of the small intestine and cause infant diarrhea

Enterotoxigenic *Escherichia coli* (ETEC) strains of *E. coli* that produce two toxins, one a cholera-like toxin (heat-labile toxin [HLT]) and the other a peptide hormone-like toxin (heat-stable toxin [HST])

Enterotoxin an exotoxin that acts specifically on the intestinal mucosa

Entrez an integrated search and retrieval system used by the National Center for Biotechnology Information (NCBI) for assembling data from major life science databases, including literature sources, sequences, and many other databases; the combined information is available at http://www.ncbi.nlm.nih.gov/sites/gquery

EPEC see **enteropathogenic *Escherichia coli***

Epidemic a disease that appears sporadically and affects many individuals in a community

Epidermis the outermost layer of the skin

Epithelia the layers of cells that cover the surfaces of the body and body cavities

Epitope a portion of an antigen recognized by an antibody-binding site; usually 5 to 9 amino acids in length for protein antigens

Epitope-based targeting signals a technique used to enhance dendritic cell binding properties to enhance immune responses and antigen presentation

Epsilon toxin a diarrhea-inducing exotoxin produced by *Clostridium perfringens*

erm genes regulating resistance by means of attenuation in gram-positive bacteria

ermA, *-B*, *-F*, and *-G* antibiotic resistance genes conferring resistance to macrolides, streptogramins, and lincosamides

Erythema redness or rash of the skin due to dilation of blood vessels

Erythromycin a member of the macrolide family of antibiotics

Eschar a black necrotic lesion surrounded by edema; characteristic of cutaneous anthrax

Escherichia coli O157:H7 see **enterohemorrhagic *Escherichia coli***

ETA exfoliative toxin of *Staphylococcus aureus* that cleaves desmoglein-1 of epidermal cells

ETEC see **enterotoxigenic *Escherichia coli***

ETs see **exfoliative toxins**

Eukaryotes organisms in which DNA is enclosed in a nuclear membrane

Exfoliation sloughing of cells

Exfoliative toxins (ETs) dermolytic exotoxins produced by *Staphylococcus aureus*; cause symptoms of scalded-skin syndrome and bullous impetigo; one type cleaves desmoglein-1, which holds keratinocytes together

ExoA see **exotoxin A**

Exoenzymes toxins introduced directly into host cells by specialized secretion systems; also called effector proteins

Exotoxin a protein toxin produced by bacteria; usually secreted into the extracellular medium

Exotoxin A (ExoA) an AB protein toxin produced by *Pseudomonas aeruginosa*; it has the same mechanism of action as diphtheria toxin

ExPortal a membrane microdomain in *Streptococcus pyogenes*; may coordinate interactions between proteins secreted by the Sec system and membrane-associated chaperones

Extensively drug resistant (XDR) strains of bacteria, e.g., *Mycobacterium tuberculosis*, that are resistant to multiple types of antibiotics

Extracellular matrix (ECM) extracellular protein-polysaccharide material in which mammalian cells are embedded; contains collagen, hyaluronic acid, elastin, etc.

Extravasation see **transmigration**

Exudate fluid and cells that have escaped from blood vessels

Fab the portion of an antibody that contains a light chain and the amino-terminal part of the heavy chain; the portion that contains the antigen-binding sites

Factor B a serum complement component protein that is cleaved into Ba and Bb by factor D

Factor D a serum complement component protein that cleaves factor B into Ba and Bb

Factor H a serum complement component protein that binds with sialic acid residues on host cells and activates C3; C3b is formed but remains bound to H; targets degradation of C3b

Factor I a serum complement component protein that degrades C3b bound to factor H

Facultative bacteria that can use either fermentation or respiration to obtain energy, depending on whether oxygen is present

Fc the portion of an antibody that binds complement and attaches the antibody-antigen complex to phagocytes; mediates opsonization; region of antibody bound by some streptococcal and staphylococcal surface proteins

Fe utilization regulator see **ferric uptake regulator**

Febrile having a fever (elevated temperature)

Fenton reaction the nonenzymatic formation of hydroxyl radical from iron and hydrogen peroxide

Ferric uptake regulator (Fur) a repressor protein that controls iron-regulated genes in many gram-negative bacteria

Ferritin intracellular iron storage protein of mammalian cells

fha see **filamentous hemagglutinin**

Fha see **filamentous hemagglutinin**

Fibrin a fibrous protein involved in the formation of blood clots

Fibronectin a high-molecular-mass (~440-kDa) extracellular matrix glycoprotein found on the surfaces of many host cells that binds to membrane-spanning receptor proteins (integrins)

Fibronectin-binding proteins (FnBP) staphylococcal and streptococcal surface proteins that bind fibronectin

Filamentous hemagglutinin (Fha) an adhesin produced by *Bordetella pertussis*; often a component of acellular pertussis vaccines

Fim see **fimbriae**

Fimbriae (Fim) short thin fibrils on the surfaces of bacteria; pili

FITC (fluorescein isothiocyanate) a green fluorescent dye

Flagella rod-like protein structures projecting from bacterial cells; responsible for motility

Flagellin the main structural protein constituting flagella; a TLR agonist

Flavohemoglobin an enzyme that converts nitric oxide (NO) to NO_3^- using heme-bound O_2; resistance mechanism found in *Escherichia coli*

Flora see **microbiota**

Fluorescein isothiocyanate see **FITC**

Fluoroquinolone a member of the quinolone family of antibiotics

FnBP see **fibronectin-binding proteins**

Follicles patches of tissue found in intestine; contain cells of GALT

Food irradiation technology ionizing radiation techniques for irradiating food at the last stages of processing to eliminate microorganisms

Food-borne bacterial disease a disease acquired by ingesting food contaminated with bacteria

Food-borne infection an infection acquired by ingesting food contaminated with microbes

FoodNet a surveillance program to monitor cases of infection by food-borne pathogens, e.g., *Salmonella* and *Campylobacter*

Fosfomycin an antibiotic that blocks conversion of UDP-NAG to UDP-NAM during peptidoglycan synthesis

Freund's complete adjuvant a mixture of oil and other components, including mycobacterial cell walls; used to stimulate antibody response to protein antigens

Freund's incomplete adjuvant the same as Freund's complete adjuvant minus the mycobacterial components

Fur see **ferric uptake regulator**

Fura-2 a fluorescent dye reagent used to detect calcium levels inside tissue culture cells

Fv see **variable region**

GALT see **gastrointestinal-associated lymphoid tissue**

Gamma interferon (IFN-γ) a cytokine that stimulates monocytes and PMNs to leave the bloodstream and stimulates endothelial cells to produce selectins

Gamma rays high-energy photons produced by radioactive material (usually cobalt-60 or cesium-137); used to irradiate food and sterilize medical equipment and supplies

γ-δ T cells immune cells that may play a role in limiting the growth of intracellular pathogens

γ-Phage a lysogenic phage closely related to β- and ω-phages; does not encode diphtheria toxin

γ-Toxin a membrane-damaging exotoxin produced by *Clostridium perfringens*

Gangrene death of body tissue (necrosis) usually associated with loss of blood supply caused by bacterial invasion, bacterial toxins, putrefaction, or other diseases that cause cell death

GAS group A streptococci (e.g., *Streptococcus pyogenes*)

Gastritis inflammation of the stomach lining; can be caused by *Helicobacter pylori* infection in some individuals

Gastrointestinal-associated lymphoid tissue (GALT) the intestinal mucosal immune system characterized by production of secretory IgA (sIgA); a component of mucosa-associated lymphoid tissue (MALT)

GBS group B streptococci (e.g., *Streptococcus agalactiae*)

GenBank (National Center for Biotechnology Information [NCBI]) U.S. centralized library of various biological data, including nucleotide sequences

Gene amplification a process by which a single copy of a gene is replaced by multiple copies (e.g., genes imparting antibiotic resistance); can be through homologous recombination or through PCR

Gene chip see **DNA chip**

Gene conversion a process by which genetic information is transferred by a nonreciprocal mechanism; a process of pilin gene recombination responsible for antigenic variation in *Neisseria*

General secretory pathway (Sec system) a common secretion pathway shared by both gram-negative and gram-positive bacteria through which many proteins are exported from the cytoplasm; requires an N-terminal signal sequence on the protein being secreted

Generalized transducing phage a type of phage that occasionally packages bacterial genomic DNA into the viral capsid and, following infection, transfers the DNA fragments into a new bacterial cell, where the new DNA is integrated into the bacterial DNA of the recipient cell by homologous recombination

Genome the complete set of genes of an organism

Genome sequence the order of bases in the genome

Genomic plasticity variability seen in the genome; caused by numerous mechanisms

Genomic subtractive hybridization (GSH) a PCR-supported method for isolating genomic DNA sequences unique to specific strains of closely related bacteria

Genomics-based discovery identification of new targets for antibacterial drug discovery by comparing the genomes of a variety of pathogens for unique genes

Genotype the genetic constitution of an organism

Gentamicin an antibiotic in the aminoglycoside family

Gentamicin assay a cell-based infection assay that permits distinction between bacteria that attach to the surfaces of host cells and those that invade the cytoplasm of the host cell

Germ-free animals see **gnotobiotic animals**

Germ theory of infectious disease the idea that there is a connection between a microbe and a particular set of disease symptoms

Germination the conversion of a bacterium from a dormant spore to a metabolically active, vegetative state

GFP see **green fluorescent protein**

Glassification a procedure used to stabilize a vaccine by lyophilization of the protein antigen in the presence of sugar-containing stabilizers, such as trehalose

Global regulator a single regulator that controls the expression of a large number and many types of genes involved in a broad range of physiological processes

Glycopeptide antibiotics a group of antibiotics that contain carbohydrate groups (glycans) covalently attached to the side chains of amino acid residues and that inhibit peptidoglycan synthesis; includes vancomycin and teicoplanin

Glycopeptides a group of antibiotics that inhibit peptidoglycan synthesis; includes vancomycin and teicoplanin

Glycyl-glycine tetracycline see **tigecycline**

GM-CSF see **granulocyte-macrophage colony-stimulating factor**

Gnotobiotic animals animals raised in sterile environments that have no bacteria in or on them; used for infection models of diseases

Goblet cells glandular simple columnar epithelial cells that secrete mucin

Golgi system an intracellular organelle that processes proteins destined for excretion from mammalian cells and determines what route they take to their ultimate destination

Granulocyte-macrophage colony-stimulating factor (GM-CSF) a cytokine that triggers release of monocytes and PMNs from bone marrow into the bloodstream

Granuloma a special type of inflammatory lesion or mass of immune cells containing actively growing fibroblasts, macrophages, and lymphocytes that forms when the immune system tries to wall off foreign substances that it cannot remove by other means

Granulosome a cytolytic granule found in phagocytes, cytotoxic T cells (CTLs), and natural killer (NK) cells that shares properties with lysosomes; contains myeloperoxidase, perforin (pore-forming cytolysin), and granzymes

Granulysin a cytolytic pore-forming protein released by CD8$^+$ cytotoxic T cells (CTLs) and natural killer (NK) cells that are attached to host cells infected with intracellular pathogens and presenting antigens through the MHC-I complex

Granzymes a set of serine proteases produced and released by CD8$^+$ cytotoxic T cells (CTLs) and natural killer (NK) cells that enter host cells infected with intracellular pathogens through pores created by perforin and then initiate apoptosis (programmed cell death) by cleaving caspases

Green fluorescent protein (GFP) a green fluorescent protein (encoded by modifications to the *gfp* gene originally from a jellyfish) used as a reporter for gene transcription and when fused with a target protein as a marker for intracellular localization using fluorescence microscopy

Group A streptococci a designation of *Streptococcus pyogenes* based on surface carbohydrate antigen, according to the Lancefield grouping system

Group B streptococci a designation of *Streptococcus agalactiae* based on surface carbohydrate antigen, according to the Lancefield grouping system

GSH see **genomic subtractive hybridization**

GUS see **β-glucuronidase**

H antigens bacterial flagellar antigens recognized by antibodies; used to identify bacterial serotypes, especially in *Escherichia coli* and related species

HA see **hospital acquired**

HACCP see **hazard analysis and critical control point**

Halides disinfectants, such as chlorine and iodine, that oxidize and inactivate bacterial proteins

Hamamelitannin a small molecule isolated from witch hazel bark that inhibits the quorum-sensing regulator of methicillin-resistant *Staphylococcus aureus* (MRSA)

Hazard analysis and critical control point (HACCP) a food and drug safety management system based on a surveillance program to identify steps in the food- and drug-processing stream where contamination might occur

HB-EGF see **heparin-binding epidermal growth factor**

Heat shock proteins a class of proteins whose expression is increased by stress; can enhance immune response

Heat-labile toxin (HLT) a multisubunit AB_5 exotoxin produced by ETEC; similar to cholera toxin in mechanism of action and sequence

Heat-stable toxin (HST) a peptide hormone-like diarrheal toxin produced by ETEC; STIa mimics guanylin, an intestinal hormone that increases intracellular cyclic GMP levels

Heavy chain the larger of two types of proteins making up an antibody molecule; the receptor-binding and translocation domains of botulinum neurotoxins

Hematogenous produced by, derived from, or spread by blood

Heme a group of porphyrin-like proteins that tightly bind to iron; includes metalloproteins, such as peroxidases and electron transfer proteins, and hemoglobin

Hemoglobin a heme-containing protein in the human body that binds iron

Hemolysin a pore-forming protein toxin that causes lysis of erythrocytes

Hemolytic-uremic syndrome (HUS) acute irreversible kidney failure caused by Shiga toxin, produced by *Shigella* strains, and Shiga-like toxins, produced by STEC (e.g., *Escherichia coli* O157:H7) and EHEC (enterohemorrhagic *E. coli*) strains; characterized by he-

molytic anemia, acute renal (kidney) failure, low blood platelet count, and possibly death

Heparin-binding epidermal growth factor (HB-EGF) a protein that is involved in wound healing and cardiac hypertrophy; the mammalian cell-bound precursor serves as the cellular receptor for diphtheria toxin

Hepatotoxin a bacterial toxin that targets liver cells

Herd immunity protection of unvaccinated people in a population where most people are vaccinated

Hfq-dependent sRNAs in gram-negative bacteria, the largest class of small RNAs (sRNAs) that use Hfq protein as a chaperone to facilitate base pairing between sRNA and mRNA targets

HGT see **horizontal gene transfer**

Hib a conjugate vaccine against *Haemophilus influenzae* type b; consists of a polysaccharide capsule attached covalently to a protein (tetanus toxoid, diphtheria toxoid, or meningococcal group B outer membrane protein)

High-performance liquid chromatography (HPLC) a form of column chromatography in which a pump is used to move material through a densely packed column, resulting in high resolution of the compounds being separated

High-throughput screening (HTS) an approach for scientific discovery that uses automated processing (robotics) or combinatorial libraries to screen massive numbers of compounds or to perform large numbers of assays or reactions (usually in 96-well or 384-well microplates) to identify active compounds, drugs, mutants, or other targeted products with the desired properties

High-throughput sequencing an approach used to sequence genomes that parallelizes the sequencing process to produce thousands to millions of sequence reads at once; methods include shotgun (Sanger) sequencing of large numbers of clones derived from DNA isolated from a mixed microbial population; 454 pyrosequencing, which uses oil emulsion PCR amplification on beads, followed by pyrosequencing with luciferase-based light generation as a readout; and Solexa/Illumina sequencing, which uses bridge amplification by PCR of DNA fragments attached to primers on a slide, followed by sequential incorporation of nucleotides reversibly labeled with fluorescent dye terminators

Histamine a vasoactive compound released by basophils and mast cells to trigger an inflammatory response by increasing capillary permeability to white blood cells to allow them access to the site of infection

Histatins antimicrobial peptides, produced by host cells, that are secreted in saliva

Histidine kinase the signal-transducing protein in bacterial two-component regulatory systems that autophosphorylates on a histidine residue in response to an environmental condition or signal and transfers the phosphoryl group to an aspartate residue on its cognate response regulator; also called sensor kinase

HLT see **heat-labile toxin**

HME see **human monocytotrophic ehrlichiosis**

H-NS a histone-like (nucleoid-associated DNA-binding) protein that is a global repressor of transcription in gram-negative bacteria; represses transcription of foreign genes acquired by horizontal gene transfer

HO· see **hydroxyl radical**

Holins phage proteins that form pores in bacterial membranes

Horizontal gene transfer (HGT) lateral movement of a group of genes from one bacterium to another through various mechanisms, including conjugation, transduction, and transposition

Hospital acquired (HA) refers to an infection acquired in a hospital setting

Hospital-acquired infection see **nosocomial infection**

Host (i) a human or animal body in which a resident microbiota lives; (ii) a human or animal body colonized and attacked by bacterial pathogens; (iii) in the context of bacterial viruses, the bacterium that is attacked by bacteriophage

Host-parasite interaction the relationship between the human or animal body (host) and the invading bacterium (parasite)

HPLC see **high-performance liquid chromatography**

HST see **heat-stable toxin**

HTS see **high-throughput screening**

Human monocytotrophic ehrlichiosis (HME) potentially fatal febrile illness due to *Ehrlichia* species infecting monocytes

Human umbilical vein endothelial cells (HUVEC) primary human endothelial cells, obtained from the umbilical vein, used to study bacterial transit across the endothelium and to study macromolecule transport, blood coagulation, fibrinolysis, immune cytokine responses, and cell adhesion

Humoral immunity adaptive, MHC-II-dependent, CD4$^+$-dependent, antibody-mediated immunity

HUS see **hemolytic-uremic syndrome**

HUVEC see **human umbilical vein endothelial cells**

Hyaluronate lyase (Hyl) an enzyme produced by *Streptococcus pneumoniae* and other bacterial pathogens that dissolves hyaluronic acid, thereby promoting bacterial spread and providing food for the bacteria

Hyaluronic acid an anionic, nonsulfated glycosaminoglycan (mucopolysaccharide) widely distributed throughout connective, epithelial, and neural tissues; a major component of the extracellular matrix; also a component of the group A streptococcal extracellular capsule that is important for immune evasion

Hyaluronidase an enzyme that degrades hyaluronic acid

Hydrogen peroxide (H$_2$O$_2$) a reactive oxygen compound produced by metabolism that acts by oxidizing proteins and damaging DNA; used as an external antiseptic

Hydroxamate a class of siderophores

Hydroxyl radical (HO·) a reactive form of oxygen that can kill bacteria

Hyl see **hyaluronate lyase**

Hypermutable having higher-than-normal mutation rates

Hypochlorite a reactive form of chlorine toxic to bacteria

Hypothiocyanite (OSCN$^-$) a reactive form of thiocyanate that is highly toxic to bacteria; oxidation of thiocyanate by hydrogen peroxide into hypothiocyanite is catalyzed by lactoperoxidase

icaA, -B, -C, -D genes encoding enzymes that produce PNSG (a polysaccharide adhesin) on the surfaces of *Staphylococcus aureus* and *Staphylococcus epidermidis* cells

ICAM see **intercellular adhesion molecule**

iC3b factor I-catalyzed proteolytic digestion product of C3b that opsonizes but does not form C3 convertase

ID$_{50}$ see **50% infectious dose**

IEL see **intestinal epithelial lymphocytes**

IFN-γ see **gamma interferon**

IgA see **immunoglobulin A**

IgE see **immunoglobulin E**

IgG see **immunoglobulin G**

IgG1 to IgG4 see **immunoglobulin G1 to immunoglobulin G4**

IgM see **immunoglobulin M**

IL-1 and IL-6 see **interleukins 1 and 6**

IL-1 receptor antagonist (IL-1ra) an anti-inflammatory cytokine

IL-2 see **interleukin 2**

IL-3 see **interleukin 3**

IL-4 see **interleukin 4**

IL-8 see **interleukin 8**

IL-10 see **interleukin 10**

IL-13 see **interleukin 13**

Illumina sequencing see **high-throughput sequencing**

ILs see **interleukins**

Imipenem a β-lactam antibiotic

Immune system a complex collection of phagocytic, cytotoxic, and antibody-producing cells and protein components that protect the body from infection

Immunogen an antigen that induces an immune response

Immunogenic the ability to elicit a robust antibody or T-cell response

Immunoglobulin A (IgA) the major class of antibody found in mucosal tissues; produced as a dimer and secreted into the lumen as sIgA

Immunoglobulin E (IgE) a class of monomeric antibody thought to play a role in control of metazoan and protozoan parasites; involved in allergic responses

Immunoglobulin G (IgG) the major monomeric antibody class present in the serum and tissue fluids; produced by plasma B cells and binds to many kinds of pathogens to cause agglutination, immobilization, activation of the classical complement pathway, opsonization for phagocytosis, and neutralization of bacterial toxins; important for antibody-dependent cell-mediated cytotoxicity (ADCC)

Immunoglobulin G1 to immunoglobulin G4 (IgG1 to IgG4) subtypes of human IgG

Immunoglobulin M (IgM) the first class of multimeric (usually pentameric) antibody produced by activated B cells in response to an antigen; possesses high avidity for antigen and is very effective at complement activation

Immunoglobulins antibodies

Immunohistochemistry a histological technique used to detect bacteria in tissue samples that uses labeled antibodies against the bacteria to visualize the bacteria through microscopy

Immunoprecipitation a technique to precipitate a protein antigen out of solution by using an antibody specific for that protein; the antibody is usually coupled to a solid substrate, such as a bead

Immunoproteomics a proteomic technique for identifying bacterial antigens that elicit distinct host responses

Immunosuppression suppression (dampening) of the normal immune response

Immunotoxins hybrid proteins containing the catalytic cytotoxic domain of a toxin (such as the A fragment of DT) and a cell-targeting antibody or receptor ligand that recognizes a specific host cell type; used as a basis for new therapeutic approaches against cancer cells or virus-infected cells

In vitro an environment outside the body (usually a test tube)

In vivo inside the body, as in an animal model of infection

In vivo expression technology (IVET) a high-throughput screening approach for identifying bacterial genes that are expressed only when the bacteria are in the host

In vivo infection model an animal model used to study colonization and invasive diseases caused by bacterial pathogens

In vivo-induced antigen technology (IVIAT) an antibody-based genomic method used to identify genes induced during human infections that uses antibodies present in the sera of convalescent patients to identify antigens of the pathogen that were expressed during infection and that elicited an immune response

IND see **investigational new drug**

Indels spontaneous DNA mutations resulting from insertions or deletions

Indolyl-galactoside see **X-Gal**

Induced tolerance see **anergy**

Infant botulism flaccid paralysis that results when botulinum neurotoxin is produced by *Clostridium botulinum* colonizing the colon of an infant

Infection the successful colonization of the body by a microorganism capable of causing damage to the body (i.e., a pathogen)

Infectious capable of causing disease

50% Infectious dose (ID$_{50}$) the number of microorganisms required to cause infection (as evidenced by a clearly defined symptom) in 50% of experimentally infected animals at a given time following infection; a measure of infectivity

Inflammation an immune response of the body to irritants, such as infection by a pathogen; characterized by redness, swelling, pain, and heat

Inflammatory response see **inflammation**

Injectosome the delivery apparatus of the type 3 and type 4 secretory systems, consisting of a complex of many proteins, that directly injects the effector proteins into the host cell

InlA see **internalin A**

InlB see **internalin B**

Innate immune system the host defenses that are always present and effective against most bacteria; includes physical barriers, complement, phagocytic cells, and the washing action of fluids

Inoculum the number of bacteria (in CFU) introduced into a culture medium, tissue culture, or animal model of infection

INSD see **International Nucleotide Sequence Database**

Insertion sequences sites flanking a transposon and consisting of inverted repeats that direct transposition (insertion) of the transposon via a transposase into the new regions of DNA

Institutional review board (IRB) a board made up of a diverse group of professionals at a given institution that must approve human clinical trials or research experiments involving human subjects

Integrase an enzyme encoded by a plasmid gene that mediates insertion of DNA (such as a plasmid) into the DNA of a recipient cell

Integrins heterodimeric protein receptors (comprised of α and β subunits) on the surfaces of cells that mediate attachment between the cell and the surrounding tissues (cells or extracellular matrix); integrins on PMNs bind to ICAMs on endothelial cells; binding stops the movement of PMNs through the bloodstream

Integron an integrating element probably responsible for the evolution of plasmids carrying multiple antibiotic resistance genes

Intellectual property (IP) a term referring to inventions (such as antibiotics and drugs) for which patents are issued to protect the owner's property rights

Intercellular adhesion molecule (ICAM) an adhesion molecule of endothelial cells that binds to integrins on PMNs; aids PMN migration out of blood vessels

Intergenic see **intergenic recombination**

Intergenic recombination homologous recombination of genes from different bacterial cells

Interleukin 2 (IL-2) a cytokine produced by T cells; stimulates T-cell proliferation; at high levels, causes nausea, vomiting, malaise, and fever

Interleukin 3 (IL-3) a cytokine that stimulates release of monocytes and PMNs from bone marrow into the bloodstream

Interleukin 4 (IL-4) an anti-inflammatory cytokine

Interleukin 8 (IL-8) a cytokine that stimulates monocytes and granulocytes to leave the bloodstream and move to a site of infection

Interleukin 10 (IL-10) an anti-inflammatory cytokine

Interleukin 13 (IL-13) an anti-inflammatory cytokine

Interleukins (ILs) protein cytokines, produced and secreted by leukocytes, monocytes, dendritic cells, macrophages, T cells, B cells, and other immune cells, that mediate inflammatory responses (inflammation, septic shock, and fever); maturation, differentiation, and/ or proliferation of immune cells; chemotaxis; angiogenesis; and production and release of cytokines and chemokines

Interleukins 1 and 6 (IL-1 and IL-6) cytokines produced by monocytes and macrophages; may contribute to septic shock symptoms

Internalin A (InlA) an adhesin protein of *Listeria monocytogenes* that binds E-cadherin on host cells and facilitates adhesion and invasion

Internalin B (InlB) an adhesin protein of *Listeria monocytogenes* that binds the receptor tyrosine kinase Met on host cells and facilitates adhesion and invasion

International Nucleotide Sequence Database (INSD) large primary sequence databases consisting of GenBank, EMBL NDB, DDBJ, UniProtKB, SwissProtKB, and PDB RCSB

Intestinal epithelial lymphocytes (IELs) cytotoxic T cells of the mucosal immune system; CD8 composed of γ and δ proteins

Intragenic recombination homologous recombination of genes within a single bacterium

Intranasal inoculation (i) instilling a vaccine in the nasal passages; activates the BALT; (ii) inoculating bacteria by inhalation into the nose in animal models; depending on the model, the bacteria colonize the nasopharynx or progress from the nasopharynx and cause invasive disease

Invasin a bacterial surface protein that provokes endocytic uptake by host cells

Invasion frequency the ratio of gentamicin-resistant CFUs to cell-associated CFUs at the end of a gentamicin protection assay

Invasion success curve a method for determining the maximum number of internalized bacteria per host cell; used to compare pathogenic bacterial strains

Invasive capable of penetrating the host's defenses; capable of entering host cells or passing through mucosal surfaces and spreading in the body

Inversion of sequences a process by which bacteria invert the sequence upstream of a particular gene, such as a promoter; one mechanism leading to phase variation

Investigational new drug (IND) a promising new drug that must be tested in two animal models of infection to evaluate efficacy and safety; if the FDA approves the IND application for a particular drug candidate, the drug can enter a phase I clinical trial

IP see **intellectual property**

IRB see **institutional review board**

IS elements see **insertion sequences**

IVET see **in vivo expression technology**

IVIAT see **in vivo-induced antigen technology**

J chain a peptide linking IgA monomers or IgM monomers

JGI Genomics Department of Energy Joint Genome Institute; database for many eukaryotic and microbial genomes

K antigen the capsular antigen used for classification of *Escherichia coli*

K1 antigen a component of the antiphagocytic capsule of *Escherichia coli* strains that cause septicemia and meningitis

Kallidin a vasoactive peptide that enhances extravasation

Kanamycin an antibiotic in the aminoglycoside family

Keratinocyte a stratified squamous cell constituting up to 95% of the epidermis; maintains the acidic environment of the skin; forms the keratin layer that protects the skin and the underlying tissues from the external environment; produces keratin and cytokines; produces cytokines that modulate the immune response of the skin

Ketolides new derivatives of macrolide antibiotics

Knock-in mice transgenic mice that have had human genes introduced into their genomes and express certain human proteins in infection models

Knockout mice transgenic mice that have disruptions in specific genes

Koch's postulates a set of postulates that must be met to prove a particular bacterial pathogen causes a particular disease; there are four postulates that must be satisfied, but today there are additional modern molecular methods for satisfying these postulates

Kupffer cell a resident (fixed) macrophage in the liver

L forms bacteria that lack a cell wall

Lactoferrin a mammalian host protein found in secretory fluids, such as milk, saliva, tears, and nasal secretions, that binds iron with high affinity

Lactoperoxidase a heme-iron-containing enzyme found in mucus that catalyzes the oxidation of inorganic and organic compounds (chloride, bromide, iodide, and thiocyanate) by hydrogen peroxide to form reactive oxygen species (hypochlorite, hypobromite, hypoiodite, and hypothiocyanite)

LacZ see **β-galactosidase**

lacZ **gene** a gene that encodes β-galactosidase

Lamina propria a thin layer of connective tissue and fibrillar material embedded in a mucopolysaccharide matrix that is located beneath the epithelial layer of mucosa

Lancefield grouping system immunology-based classification of surface antigens used to identify and characterize *Streptococcus* species; devised by prominent microbiologist Rebecca Lancefield

Langerhans cells a type of antigen-presenting cell (dendritic cells) found in the epidermis (skin) and lymph nodes

Laser capture microscopic dissection (LCM) a procedure for attaching isolated host cells to transparent parafilm by exposure to a low-energy near-infrared laser pulse, so that the nucleic acids or proteins can then be isolated from the attached cells

Latency a period of inactivity

Lateral surface the side portion of polarized epithelial cells

LBP see **LPS-binding protein**

LCM see **laser capture microscopic dissection**

LD$_{50}$ see **50% lethal dose**

Leader peptide a short regulatory peptide involved in transcriptional or translational attenuation mechanisms of gene control; encoded by transcripts from leader regions of operons controlled by attenuation (e.g., 100 bp upstream of the start codon for erythromycin resistance protein); leader peptides are not signal sequences

Leader region the upstream transcript region between the 5' end of a transcript and the translation start site of the first structural gene; often encodes regulatory leader peptides

Leader sequence see **signal sequence**

Lecithin a phospholipid (e.g., phosphatidylcholine, phosphatidylethanolamine, phosphatidylserine, and phosphatidylinositol) component of host cell membranes

Lecithinase a phospholipase protein toxin produced by *Clostridium perfringens* or *Listeria monocytogenes* that disrupts host cell membranes

LEE see **locus of enterocyte effacement**

50% Lethal dose (LD$_{50}$) the number of bacteria (in CFU) or the amount of toxin required to kill 50% of the animals experimentally inoculated within a given period of time

Leukocidin an extracellular pore-forming toxin produced by *Staphylococcus aureus* that kills leukocytes

Leukocyte a white blood cell; cells of the immune system

Leukotoxin a pore-forming protein toxin produced by some bacteria that kills leukocytes

Light chain the smaller of two proteins making up the basic antibody structure; the catalytic-domain-containing portion of botulinum neurotoxins

Lincomycin a member of the lincosamide family of antibiotics

Lincosamides a family of antibiotics that binds to the 23S rRNA of the 50S ribosomal subunit

Linezolid a member of the oxazolidinone family of synthetic antibiotics

Lipid A (endotoxin) the toxic portion of lipopolysaccharide (LPS) from gram-negative bacteria that is embedded in the outer membrane of gram-negative bacteria

Lipinski's "Rule of Five" a "rule of thumb" used by medicinal chemists to evaluate whether a potential new lead compound for a drug candidate is orally active without being toxic in humans

Lipooligosaccharide (LOS) a component of the outer membrane of some gram-negative bacteria (e.g., *Neisseria* and *Yersinia* species) that is similar to lipopolysaccharide (LPS) but with a shorter O antigen

Lipopeptides molecules consisting of a lipid covalently connected to a peptide; bacterial components recognized by TLR1

Lipopolysaccharide (LPS) a component of the gram-negative outer membrane; consists of lipid A (the toxic portion), a core made up of a series of sugars, and the O antigen, a long carbohydrate chain

Liposomes see **microspheres**

Lipoteichoic acid (LTA) a major component of the gram-positive bacterial cell surface consisting of a lipid-linked teichoic acid; the lipid is embedded in the cytoplasmic membrane, and the teichoic acid is exposed on the bacterial surface

Listeriolysin O (LLO) a pore-forming cytotoxin produced by *Listeria monocytogenes*; similar to enzymes produced by streptococci (streptolysin O and pneumolysin) and clostridia (perfringolysin O)

LLO see **listeriolysin O**

Locus of enterocyte effacement (LEE) a 35.6-kb pathogenicity island in the chromosome of EPEC that contains all EPEC genes necessary for the EAE phenotype

Loose connective tissue regions located under layers of epithelial cells; consists of extracellular matrix containing fibrous proteins, blood vessels, and other components

LOS see **lipooligosaccharide**

LPS see **lipopolysaccharide**; endotoxin

LPS O antigen see **O antigen**

LPS-binding protein (LBP) a plasma protein that binds lipopolysaccharide (LPS) and complexes with CD14

and TLRs on monocytes and macrophages to stimulate cytokine production through the TLR system

LPXTG　　an amino acid sequence motif of proteins cleaved by sortase of gram-positive bacteria and used to covalently link the cleaved protein to the peptidoglycan

LTA　　see **lipoteichoic acid**

luc **gene**　　the gene for firefly luciferase

Luciferase　　an enzyme commonly used as a reporter of transcription; catalyzes the production of bioluminescence (light) through the oxidation of a pigment (luciferin)

LukF　　a component subunit of the leukocidin produced by *Staphylococcus aureus*

LukS　　a component subunit of the leukocidin produced by *Staphylococcus aureus*

Lumen　　the cavity in an organ not inside the body

lux　　the gene for bacterial luciferase

Lux　　see **bacterial luciferase**

Lyme disease　　a systemic disease caused by *Borrelia burgdorferi* that is transmitted to humans by deer ticks

Lymph　　the fluid moving through the lymphatic system of the body; monitored by phagocytes

Lymph node　　an accumulation of lymphoid tissue positioned along lymphatic vessels that contain most of the lymphocytes of peripheral blood; the site where bacteria and toxins are removed from the circulation

Lymphadenopathy　　swollen lymph nodes; a diagnostic sign of infection

Lymphatic system　　a network of vessels that collects tissue fluid (lymph) and returns it to circulation; prevents excess buildup of fluid in tissues

Lymphocyte　　a leukocyte involved in immune response and chronic inflammation

Lymphokine　　a cytokine produced by lymphocytes

Lyophilization　　a procedure used for freeze-drying a sample that is carried out in a vacuum

Lysis　　disruption of a cell membrane to release its cytoplasmic contents

Lysogenic corynebacteriophages (β-phage and ω-phage)　　lysogenic phages that carry genes for diphtheria toxin

Lysogenic phages　　bacteriophages whose genomes integrate into the bacterial chromosome; may encode toxin genes; can excise from the chromosome and become lytic in response to stress

Lysogenic phase　　the phase in a bacteriophage reproductive cycle in which the bacteriophage genome is integrated into and maintained stably in the bacterial genome

Lysosomal granules　　see **lysosome**

Lysosome　　a mammalian cell organelle containing hydrolytic enzymes and other compounds toxic to bacteria

Lysozyme　　an enzyme that degrades glycan chains of bacterial peptidoglycan

Lytic phase　　the stage in a bacteriophage reproductive cycle in which the bacteriophage replicates and lyses the bacterial host

M (microfold) cells　　cells in Peyer's patches that engulf bacteria from the lumen of the gut and deliver them to antigen-presenting cells (macrophages) in the gastrointestinal mucosa

M protein　　a *Streptococcus pyogenes* surface protein that binds to extracellular matrix proteins and the Fc portion of IgG; plays an antiphagocytic role

MAC　　see **membrane attack complex**

Macrolactone　　a small polyketide-derived lipid-like bacterial toxin

Macrolides　　a family of antibiotics that inhibit protein synthesis by binding to 23S rRNA in the 50S ribosomal subunit

Macrophage　　a large mononuclear antigen-presenting cell of the immune system having phagocytic activity; develops from activation of a monocyte during an immune response

Magnetotaxis　　the movement of bacteria toward a magnetic field

Major histocompatibility complex (MHC)　　a protein complex on the surfaces of macrophages that binds foreign (e.g., bacterial) peptides and displays them on the macrophage surface where they are recognized by the cognate receptors on T cells; there are two classes of MHCs: MHC-I and MHC-II

Malaise　　not feeling well; a general feeling of discomfort or uneasiness

MALT　　see **mucosa-associated lymphoid tissue**

Mannose-binding lectin (MBL)　　a calcium-dependent serum protein of the collectin protein family that is produced by the liver, binds to mannose found on bacterial surfaces, and activates complement

Margination　　the flattening of PMNs against a blood vessel wall prior to transmigration

Mast cells　　resident cells of several tissue types that contain granules rich in histamine, heparin, and other substances that when released can attract phagocytes to the site of bacterial invasion; also can produce cytokines

Matrix　　an artificial substrate used for binding and growing tissue culture cells

MBC　　see **minimal bactericidal concentration**

MBL　　see **mannose-binding lectin**

MD2　　an accessory protein for TLR4

MDR　　see **multidrug resistant**

mecA **gene**　　the gene encoding a penicillin-binding protein (PBP2a) in *Staphylococcus aureus* that is not readily inhibited by methicillin

Mechanism-based inhibitor (suicide substrate)　　a compound that irreversibly blocks the active site of an enzyme

Membrane attack complex (MAC)　　a multimeric complex of complement components C5b to C9 that binds to LPS on the cell surfaces of gram-negative bacteria and inserts into the membranes to form pores, thereby killing the bacteria

Memory B cells see **memory cells**

Memory cells T or B cells or their descendents that persist for long periods in the body; they allow the body to respond rapidly to a second encounter with a microbe

Memory T cells see **memory cells**

Meninges the system of membranes covering the brain and spinal cord

Meningitis inflammation of the meninges

Metabolite-sensing riboswitches RNA elements that act as switches to control expression of genes involved in the biosynthesis of small ligands that bind to the switch; present in untranslated regions of the transcripts that they regulate

Metagenome all of the genetic material present in a sample; consists of the genomes of many different organisms

Metagenomic analysis a means of determining the genetic, proteomic, and metabolic potential of the microbiota

Methicillin-resistant *Staphylococcus aureus* **(MRSA)** strains of *Staphylococcus aureus* resistant to methicillin (as well as other antibiotics); currently can be treated only with vancomycin

Metronidazole a synthetic antibiotic that makes breaks in DNA; must be activated by bacterial cell flavodoxin not present in host cells

MF59 a squalene adjuvant

MHC see **major histocompatibility complex**

MHC-I an MHC type that when complexed with an epitope triggers activation and proliferation of CD8 cytotoxic T cells; found on most cells in the body

MHC-II an MHC type that when complexed with an epitope leads to activation and proliferation of helper T cells; found on only a few cell types of the immune system (e.g., B cells and APCs)

MIC see **minimal inhibitory concentration**

MICA and MICB protein complexes displayed on the surfaces of infected human cells; stimulate gamma-delta (γ-δ) T cells

Microarray see **microarray technology**

Microarray technology a multiplex biotechnique used in molecular biology to assess gene expression. The technique involves binding thousands of DNA oligonucleotide segments of genes (such as specific sequences of each gene in a genome) to a chip (made of a chemical, glass, or silicon matrix). Labeled probes of cDNA or cRNA from a sample (such as that obtained from RNA isolated from bacteria grown under different conditions) are then hybridized with the DNA on the chip to determine the relative abundances of the different target genes

Microbial surface components recognizing adhesive matrix molecules (MSCRAMMs) surface adhesins of *Staphylococcus aureus* and *Streptococcus* species that bind host extracellular matrix proteins

Microbiome the entire microbial populations, especially their genomes, in the human body and their interactions with their environment

Microbiota the microbial communities of the body

Microbiota shift disease a shift in the population of microbes that normally resides in a particular site, resulting in disease at that site

Microcapsules see **microspheres**

Microspheres beads made of resorbable inert material in which vaccine proteins are encapsulated; they serve as a type of adjuvant

Microvilli finger-like projections on the apical surfaces of mucosal absorptive cells

mIg membrane immunoglobulin expressed on the surface of a resting B cell

Minimal bactericidal concentration (MBC) the lowest concentration of antibiotic that will kill bacteria

Minimal inhibitory concentration (MIC) the lowest concentration of antibiotic that will prevent the growth of bacteria

MLST see **multilocus sequence typing**

mob **genes** genes on mobilizable plasmids that permit them to take advantage of transfer machinery of other, self-transmissible plasmids in the same cell

Mobilizable see **mobilizable plasmid**

Mobilizable plasmid a plasmid that is not self-transmissible but can be cotransferred with a self-transmissible plasmid

MOI see **multiplicity of infection**

Monobactams a class of antibiotics in the β-lactam family

Monocyte a mononuclear phagocyte circulating in blood; differentiates into a macrophage upon activation

Monophosphoryl lipid A (MPL) a TLR4-specific agonist; used as a vaccine adjuvant

Morbidity sickness

Moribund in a dying state; near death

Mortality fatality; lethality

MPL see **monophosphoryl lipid A**

MRSA see **methicillin-resistant** *Staphylococcus aureus*

MSCRAMMs see **microbial surface components recognizing adhesive matrix molecules**

Mucin the high-molecular-weight, heavily glycosylated protein component of mucus; a complex, viscous, sticky mixture of glycoproteins covering mucosal membranes; produced and secreted by goblet cells onto mucosal surfaces

Mucoid a wet, glistening appearance of colonies of bacteria that produce capsules on agar plates

Mucosa-associated lymphoid tissue (MALT) the specialized immune system that protects all mucosal surfaces; includes GALT, BALT, SALT, and NALT

Mucosal epithelia   the epithelial layer that lines respiratory, urogenital, and intestinal tracts; covered with a layer of mucus

Mucus   see **mucin**

Multidrug efflux pumps   integral membrane protein transporters that use energy to actively pump or excrete antibiotics and other small molecules out of the cell

Multidrug efflux system   see **multidrug efflux pumps**

Multidrug resistant (MDR)   resistant to more than one antibiotic

Multidrug-resistant bacteria   bacteria that are resistant to more than one antibiotic

Multidrug-resistant TB   tuberculosis (TB) caused by strains of *Mycobacterium tuberculosis* that are resistant to most commonly used antibiotics

Multilocus sequence typing (MLST)   a multiplex method for sequencing DNA regions of several (seven or more) housekeeping or virulence genes at a time; allows comparison of different isolates of a single bacterial species

Multiple organ system failure   the cause of fatality in septic shock

Multiplicity of infection (MOI)   a unit of measure used in an inoculation, defined as the ratio of infectious agent to target cell

Mupirocin   an antibiotic effective against most MRSA strains that targets tRNA synthetases; cannot be used systemically because of rapid degradation but is used to clear nasal colonization by MRSA

Mycolactone   a polyketide-derived lipid-like toxin produced by *Mycobacterium ulcerans* that has both cytotoxic and immunosuppressive properties; the causative agent of Buruli ulcers

Myeloperoxidase   a lysosomal heme-containing peroxidase enzyme that forms reactive oxygen species from reaction of hydrogen peroxide with halides (such as chloride ion to form hypochlorous acid), which is highly toxic to bacteria during a respiratory burst

N-acyl homoserine lactones (AHLs)   a class of intraspecies quorum-sensing signals used by gram-negative bacteria

NADPH oxidase   a membrane-bound enzyme located in phagosomes that produces reactive oxygen intermediates when the phagosome fuses with the lysosome containing myeloperoxidase

Nalidixic acid   a member of the quinolone family of antibiotics that inhibits DNA gyrase

NALT   see **nasal-associated lymphoid tissue**

Nasal inoculation   introduction of a vaccine by inhalation; stimulates the NALT

Nasal-associated lymphoid tissue (NALT)   the MALT of nasal passages

Natural killer (NK) cell   a cytotoxic lymphocyte of innate immune defense cells that attacks infected host cells expressing antigen-bound MHC-I molecules on their surfaces

Natural products   antibiotics obtained from bacteria, fungi, plants, or animals isolated from soil or other environmental sources; usually products of secondary metabolism

Natural transformation   the process by which DNA released from a donor cell into the environment is taken up by a recipient cell

Naturally competent   see **naturally transformable**

Naturally transformable   cells that are able to take up DNA from the environment without chemical treatment ("assistance")

Necrosis   death of tissue; characterized by chromatin flocculation and disappearance of organelles

Necrotizing fasciitis   a destructive wound infection caused by *Streptococcus pyogenes*; same as streptococcal gangrene

Neonatal-meningitis-causing *E. coli* (NMEC)   strains of *Escherichia coli* that cross the intestinal mucosa of newborns and enter the bloodstream and then the cerebral spinal fluid to cause meningitis

Neurotoxin   a toxin specific for nerve cells

Neutralization of microbe or toxin   blocking the action of a microbe or toxin by binding of specific antibodies to the surface of the microbe or to the toxin

Neutropenia   a decrease in the number of neutrophils (PMNs) in the blood

Neutrophil   also called polymorphonuclear leukocytes (PMNs); a leukocyte containing granules in which the granules do not stain dark using hematoxylin and eosin stains; one of the first inflammatory cells to migrate to the site of infection

Nitric oxide (NO)   a reactive nitrogen intermediate found in phagocytes; toxic to microbes; used as a mammalian signaling molecule

NK cell   see **natural killer cell**

NMEC   see **neonatal-meningitis-causing *E. coli***

NMPDR   National Microbial Pathogen Database Resource; a curated database of annotated genome data for a number of bacterial pathogens

NO   see **nitric oxide**

NOD1 and NOD2   intracellular pattern recognition proteins that recognize peptidoglycan inside host cells; analogous to TLRs

NOD17   see **NOD1 and NOD2**

Nonoxidative killing   a mechanism by which phagocytes kill bacteria in which toxic reactive oxygen species are not involved

Norfloxacin   an antibiotic in the fluoroquinolone family

Normal microbiota   see **resident microbiota**

Nosocomial infection   an infection acquired in the hospital

O antigen   the polysaccharide side chains on lipopolysaccharide (LPS) of gram-negative bacteria; used as the basis for determining the serogroup of a bacterial strain

O157:H7   the major pathogenic group of EHEC that causes bloody diarrhea and HUS

O$_2^-$ see **superoxide radical**

Obligate   required

ω-Phage   a temperate bacteriophage that carries the gene for diphtheria toxin

ONOO$^-$ see **peroxynitrite**

Open reading frame (ORF)   a segment of DNA sequence that encodes a putative polypeptide of unknown function; also see **URF**

Operator   a sequence of DNA to which a regulatory protein binds; usually used to denote the binding site of a repressor protein that blocks transcription

Operon   a genetic unit of expression in bacteria consisting of a transcriptional start site; one or more genes, usually with related functions; and a transcriptional terminator

Operon fusion   see **transcriptional fusion**

Opportunist   see **opportunistic pathogen**

Opportunistic pathogen   a microorganism capable of infecting and causing disease, but only when host defenses are compromised

Opsonin   an antibody or complement component C3b that attaches to the bacterial surface and enhances the ability of phagocytes to engulf the bacterium

Opsonization   the enhancement of phagocytosis by attachment of an antibody or complement component C3b to the bacteria

Opsonizing antibodies   antibodies that attach to bacteria and enhance phagocytosis

Oral vaccines   vaccines administered by the mouth; they stimulate MALT

ORF   see **open reading frame**

Organ cultures   model systems comprising portions of organs kept viable in vitro; usually include multiple cell types

OSCN$^-$ see **hypothiocyanite**

Otitis media   inflammation or infection of the middle ear

Oxazolidinones   a class of antibiotics that bind to the 50S ribosomal subunit

Oxidative burst (respiratory burst)   the production of reactive oxygen intermediates by phagocytes in the phagolysosomes

Oxidative killing   the killing of bacteria by toxic reactive oxygen or nitrogen species inside phagocytes

P   the whole-cell pertussis component of the DTP vaccine

PAF   see **platelet-activating factor**

PAI   see **pathogenicity island**

PaLoc   see **pathogenicity locus**

***p*-Aminobenzoic acid**   a substrate of the first enzyme in the tetrahydrofolic acid pathway; inhibited by the sulfonamide family of antibiotics

PAMPs   see **pathogen-associated molecular patterns**

Pandemic   an epidemic of an infectious disease that is spreading through many different countries or even worldwide

Pangenome   see **supragenome**

Panresistant   resistant to most types of antibiotics

Parasite   an organism that lives at the expense of another

Paroxysm   a severe attack of symptoms (associated with coughing in whooping cough)

Partner-switching mechanism   a regulatory signaling mechanism that occurs under stress conditions, whereby an anti-sigma factor forms alternative complexes with a sigma factor, keeping it inactive, or with an anti-anti-sigma factor, releasing the sigma factor so it can interact with RNA polymerase

Passive immunization   the transfer of active humoral immunity by injecting antibodies against a particular pathogen or toxin obtained from an immunized person or animal into an unimmunized individual

***Pasteurella multocida* toxin (PMT)**   a single-polypeptide AB-type toxin that interacts with and deamidates the α subunit of heterotrimeric G proteins (G_q or G_i), thereby disrupting host cell regulatory networks; causes cellular proliferation, bone destruction, and weight loss

Pasteurellosis   a disease caused by *Pasteurella multocida*

Patch   a small piece of material containing many tiny prongs used to deliver vaccines via the skin

Pathogen   a microorganism capable of colonizing a host and causing disease

Pathogen-associated molecular patterns (PAMPs)   particular molecular features or motifs of bacteria and other pathogens that are recognized by special receptors of innate immune cells (TLRs and Nods) and play an important role in triggering innate immune responses

Pathogen evolution   changes in genetic makeup that allow microbes to acquire new virulence factors or resist antibiotics

Pathogenicity   the ability of a bacterium to cause disease

Pathogenicity island (PAI)   a collection of genes involved in pathogenesis clustered together on DNA; usually acquired by horizontal gene transfer (HGT); often have a different G+C content than the rest of the bacterial chromosome

Pathogenicity locus (PaLoc)   a pathogenicity island of *Clostridium difficile* containing genes encoding toxins A and B (TcdA and TcdB) and regulatory proteins

PBP2a   a penicillin-binding protein encoded by *mecA* in *Staphylococcus aureus*

PBPs   see **penicillin-binding proteins**

PCR   see **polymerase chain reaction**

PCV-7   a seven-valent conjugate pneumococcal vaccine

PDB   see **protein sequence database, protein structure database**

PDB RCSB Protein Data Bank, Research Collaboratory for Structural Bioinformatics; a protein structure model database

PECAM see **platelet-endothelial cell adhesion molecule**

Penicillin an antibiotic in the β-lactam family

Penicillin-binding proteins (PBPs) proteins (normally located on the outer surface of the cytoplasmic membrane) that bind penicillins; include enzymes for peptidoglycan synthesis and turnover, especially transpeptidases

Peptidoglycan the polysaccharide backbone with peptide cross-links that covers the surface of the cytoplasmic membrane and gives bacteria their shape

Perforin a pore-forming cytolysin-like protein found in granules of CD8$^+$ cytotoxic T cells (CTLs) and natural killer (NK) cells; forms holes in membranes of host cells infected with intracellular pathogens

Peroxynitrite (ONOO$^-$) a reactive nitrogen compound formed from reaction of nitrous oxide with superoxide radical; toxic for bacterial and human cells

Persister a nongrowing cell in a bacterial population that survives antibiotic treatment

Pertactin an adhesin on the surfaces of *Bordetella pertussis* cells

Pertussis (whooping cough) an acute respiratory disease caused by *Bordetella pertussis*; characterized by paroxysmal cough (whooping cough), excess mucus production, and vomiting

Pertussis toxin (PT, also PTx) a multisubunit AB-type protein toxin from *Bordetella pertussis*; consists of subunits S1 to S5, where S1 ADP-ribosylates G$_i$, a G protein that regulates host cell adenylyl cyclase activity, and the S2 to S5 subunits form a pentameric receptor-binding complex

Pertussis toxoid the inactive form of pertussis toxin used in the DTaP vaccine

Petechiae rash; skin lesions

Peyer's patches follicles (organized lymphoid tissue) in the small intestine that contain cells of the mucosa-associated immune system (MALT)

PFGE see **pulsed-field gel electrophoresis**

PG see **phosphatidylglycerol**

Phage display regions from pathogen proteins incorporated into the coats of bacteriophage; used for screening potential vaccine targets

Phagocyte a host immune cell (white blood cell) adapted specifically to engulf (phagocytose) and destroy bacteria, dead or dying cells, and other foreign particulate matter

Phagocytic cell see **phagocyte**

Phagocytosis the ingestion (engulfment) of foreign particles by a phagocytic cell

Phagolysosome a vacuole resulting from the fusion of a phagosome and a lysosome

Phagosome a vacuole resulting from ingestion of particulate material, including bacteria, by phagocytes

Phalloidin a fungal toxin that binds tightly to F-actin but not G-actin; can be cross-linked to fluorescent dyes, such as rhodamine or FITC, for visualization of actin filaments in cells by fluorescence microscopy

Pharmacokinetics (PK) the study of the distribution of a compound (a drug, such as an antimicrobial compound) in the body

Pharyngitis inflammation of the pharynx; a sore throat

Phase variation a regulatory mechanism involving on-off control of some bacterial genes, such as those encoding surface proteins, that allow response to rapidly varying environmental conditions

Phenol-soluble modulin (PSM) a short amphipathic molecule; a weak toxin produced by *Staphylococcus epidermidis* and other *Staphylococcus* species

Phenotype a characteristic property or trait of an organism, such as morphology, growth rate, or ability to grow on selective media

PhoA see **alkaline phosphatase**

phoA **gene** the gene encoding alkaline phosphatase; used as a reporter for expression of genes that are secreted

Phosphatidylglycerol (PG) a glycerophospholipid that is an integral part of the bacterial membrane

Phosphatidylinositol a negatively charged phospholipid found as a minor constituent of the cytosolic side of eukaryotic cell membranes that may act as a second-messenger molecule in eukaryotes

Phospholipase an enzyme that hydrolyzes phospholipids that comprise the lipid bilayer of a host cell, thereby disrupting the cell membrane

Phototaxis the movement of bacteria toward light

Phylochip a 16S rRNA gene microarray chip comprised of thousands of oligonucleotide spots, each corresponding to 16S rRNA genes from various microbial species in a sample; used for profiling microbial communities

Phylotypes biological types that classify organisms on the basis of their phylogenetic relationships

Pili (also fimbriae) long, thick protein structures on the surfaces of bacteria that mediate adherence through a special protein(s) at the tip

Pilicide a new class of potential antibiotics designed to block pilus biogenesis

Pilin the major protein subunits packed in a helical array that form the shaft of a pilus

PK see **pharmacokinetics**

Plaque (i) the clear area formed in a monolayer of tissue culture cells when some of the cells are lysed by infecting bacteria, such as the clear area formed on fibroblast monolayers by *Listeria*; (ii) the clear area formed by lytic bacteriophage on an agar plate confluent with bacterial culture; (iii) a biofilm formed on the surfaces of teeth

Plaque assay a technique using tissue culture cells to assess cell-to-cell spread of intracellular bacteria

Plasma the noncellular portion of blood that contains proteins, sugars, hormones, and elements necessary for clot formation

Plasma cell a mature form of a B cell that produces massive amounts of antibodies

Plasmid an autonomously replicating extrachromosomal DNA segment; most are circular, but *Borrelia* and a few other bacterial species have linear (self-complementary) plasmids

Platelet-activating factor (PAF) a cytokine produced by macrophages that contributes to dilation of blood vessels

Platelet-endothelial cell adhesion molecule (PECAM) a protein that assists PMNs to move between endothelial cells

PLO see **pneumolysin**

PMN see **polymorphonuclear leukocyte**

PMT see *Pasteurella multocida* **toxin**

Pneumolysin (PLO) a cytotoxic, pore-forming protein produced by *Streptococcus pneumoniae*; similar to SLO, PFO, and LLO; binds to cholesterol in host cell membranes

PNSG see **poly-*N*-succinyl-β-1,6 glucosamine**

Polar a condition where the two ends of a structure differ from each other; for example, genetic elements or cells may be polar

Polarized cells mammalian cells that have different surfaces and different components on the different faces of the cell; most mucosal epithelial cells are polar

Poly(IC) double-stranded RNA a synthetic double-stranded RNA that acts as a mucosal adjuvant

Poly-Ig receptor a receptor on the basal surface of mucosal epithelial cells to which IgA binds; responsible for transcytosis of IgA, a portion of which then becomes the secretory piece of sIgA

Polyketide synthases a family of large enzyme complexes that produce polyketides, secondary metabolites that are precursors to mycolactones, some antibiotics, and other toxins

Polymerase chain reaction (PCR) a method for amplifying DNA in vitro; involves priming with oligonucleotide primers complementary to nucleotide sequences flanking a target sequence with subsequent replication of the target sequence using a thermostable DNA polymerase

Polymorphonuclear leukocyte (PMN) a short-lived professional phagocyte (white blood cell) that circulates in the body; characterized by the presence of granules in the cytoplasm and varying shapes of the nucleus

Poly-*N*-succinyl-β-1,6 glucosamine (PNSG) a polysaccharide adhesin on the surfaces of *Staphylococcus aureus* and *Staphylococcus epidermidis* cells; adheres to plastic

Polysaccharide capsule one type of capsule made of exopolysaccharide molecules; imparts antiphagocytic properties to a pathogen

Polyvalent antigen an antigen with multiple antibody-binding sites

Pore-forming cytotoxin a type of bacterial toxin that destroys the integrity of cell membranes

Porin a protein constituent of pores in the outer membrane of gram-negative bacteria that allows diffusion of nutrients

Postsurgical infection an infection that occurs after surgery; can cause serious complications

(p)ppGpp a bacterial compound that acts as a signal to change transcriptional patterns in response to amino acid limitation and other stresses; part of the stringent response system in bacteria

Prebiotics compounds that are supposed to foster growth of "good" bacteria in the gut

prfA **gene** a gene that encodes a positive regulatory protein that activates transcription of itself, listeriolysin O (LLO), and a number of other virulence genes in *Listeria monocytogenes*

Probiotic a preparation of live bacteria that is taken intentionally to bolster or alter the microbiota of the intestine or vagina

Progenitor toxin a toxin complex containing botulinum neurotoxin and other chaperone-like proteins that protect the neurotoxin from digestion in the stomach

Programmed cell death see **apoptosis**

Proinflammatory cytokines cytokines that aid in the process of inflammation

Promoter a site on DNA where RNA polymerase binds and initiates transcription

Prophage a lysogenic bacteriophage that has integrated into the bacterial chromosome

Prophylaxis an action taken for protection against disease, such as giving antibiotics to people exposed to certain bacterial pathogens

Prospective study a study of disease or treatment in humans that begins in the present and moves into the future

Proteases enzymes that degrade proteins by hydrolyzing the peptide bonds

Protein A a surface protein of *Staphylococcus aureus* that binds the Fc portion of antibodies

Protein activity-modifying sRNAs a small group of small RNAs (sRNAs) that function by altering the activities of target proteins to which they bind

Protein-conducting channel an oligomeric assembly of SecYEG complexes through which secreted proteins are initially inserted

Protein microarray an orderly arrangement of proteins corresponding to all the genes in a genome that are embedded or covalently attached to a matrix, such as a membrane or chip

Protein sequence database (PDB) a bioinformatics resource containing sequences of amino acids in specific proteins

Protein structure database (PDB) a bioinformatics resource containing structures of specific proteins

Proteoarray see **protein microarray**

Proteomic profiling a high-throughput microarray technique for determining patterns of gene and protein expression in bacterial or host cells; can be used as a diagnostic tool

Protozoa single-celled eukaryotes

Pseudomembrane a sheet-like layer of debris (fibrin, mucin, and dead host cells) that covers a large area of the colon (pseudomembranous colitis) or the throat (diphtheria)

Pseudomembranous colitis infection of the colon characterized by a pseudomembrane

PSM see **phenol-soluble modulin**

Psoralen a furocoumarin compound that upon UV irradiation cross-links DNA, thereby blocking DNA synthesis, replication, and repair in bacteria, but not protein synthesis or metabolism; used to produce live vaccines with bacteria that do not proliferate but do produce antigens

PT see **pertussis toxin**

ptsAB genes required for secretion of pertussis toxin

PTx see **pertussis toxin**

ptxAB genes encoding pertussis toxin subunits

PubMed a bioinformatics database for life science literature sources

Pulldown assay see **immunoprecipitation**

Pulsed-field gel electrophoresis (PFGE) an electrophoretic separation technique in which large pieces of DNA are subjected to bursts of electrical charge in different directions while migrating through a gel; used to compare bacterial strains in disease outbreaks

Purulent associated with the formation of pus

Pus whitish to brownish exudates produced by vertebrates during an inflammatory response to infection involving the accumulation of fibrin, PMNs, fragments of host cells, and released DNA

Pyogenic pus forming

Pyrogenic fever inducing

Pyrosequencing see **high-throughput sequencing**

Pyruvate oxidase (SpxB) an enzyme of *Streptococcus pneumoniae* and some other bacterial species that produces endogenous hydrogen peroxide

QACs see **quaternary ammonium compounds**

qPCR see **quantitative real-time PCR**

QS-21 a derivative of QuilA that is used as an adjuvant

Quantitative real-time PCR (qPCR) a PCR-based method for determining the relative numbers of different microbial species in a population; based on calculating the relative concentrations of 16S rRNA genes of each species; can be combined with reverse transcription to determine relative transcript amounts (RT-qPCR)

Quaternary ammonium compounds (QACs) disinfectants that intercalate into phospholipid bilayer membranes causing bacteria to lose essential ions and other small molecules

QuilA a saponin vaccine adjuvant

Quinolones a family of antibiotics that inhibit DNA gyrase; includes nalidixic acid

Quinupristin a streptogramin antibiotic

Quorum the minimum number of a group of bacteria that act together to generate a particular response

Quorum-sensing system a regulatory system that recognizes a particular bacterial signal (autoinducer), thereby sensing the density of bacteria in the area; controls either repressors or activators of specific sets of genes in response to bacterial population density

R domain the portion of diphtheria toxin that binds to a protein receptor on a host cell

Rabbit ileal loop model a procedure for testing the impacts of toxins and other diarrheal agents on the small intestine, whereby a segment of the rabbit intestinal ileum is isolated and the toxin, bacterium, or other potential diarrheal agent is injected into the lumen and after a period of time is measured for distension or swelling to indicate whether the agent has diarrhea-inducing properties

Rac 18 a protein targeted by toxins A and B of *Clostridium difficile*

Rational drug design an approach to drug development based on the synthesis of chemicals designed to bind to and inactivate a host target molecule based on the structures of the compound and target

RDB see **Ribosomal Database**

Recipient cell a cell that takes up DNA (i) from the environment or (ii) directly from another cell

Recombinase-based IVET (RIVET) a modified version of IVET involving the use of recombination that increases the ability to isolate genes only weakly expressed in vivo; uses site-specific DNA resolvase

Reemerging infectious diseases the appearance of older diseases once thought to be under control

Regulator a protein that controls the expression or activity of another target protein (e.g., a protein that controls transcription of a specific gene or set of genes)

Regulon genes located at different locations that have promoter/operator or promoter/activator binding regions that all recognize the same regulatory protein(s)

Reporter fusion see **transcriptional fusion**

Reporter gene a gene encoding an easily assayable enzyme or a readily observable fluorescent protein that is used in transcriptional fusions or translational fusions to measure gene expression

Reservoir a group of microorganisms that serve as a source for new genetic material

Resident microbiota a population of normally nonvirulent bacteria found routinely in a site in the bodies of most healthy adults or children

Respiratory burst see **oxidative burst**

Response regulator one component of a two-component regulatory system that mediates the cellular response; usually a DNA-binding protein that changes transcription in response to phosphorylation

Retrospective study a study of disease outbreaks after the fact that provides information on disease transmission or progression in humans by analyzing previously accumulated data

RGD tags Arg-Gly-Asp peptide motifs that act as ligands for cell surface integrins

Rheumatic fever a febrile illness that occurs several weeks after infection (sore throat) with *Streptococcus pyogenes*; can be accompanied by damage to heart valves

Rheumatic heart disease heart valve damage following rheumatic fever

Rho mammalian small G proteins (GTPases) targeted by a number of bacterial protein toxins, such as the cytotoxic necrotizing factors of *Escherichia coli* and toxins A and B of *Clostridium difficile*

Rhodamine a red (fluorone) fluorescent dye

Rho-dependent transcriptional terminators RNA structures that require bacterial Rho (ρ) factor protein to cause transcription termination

Ribosomal Database (RDB) a bioinformatics database that provides online data analysis, alignment, and annotation of bacterial and archaeal small-subunit 16S rRNA gene sequences

Ribosome protection protection of ribosomes from tetracycline by a bacterial cytoplasmic protein

Rifampin an antibiotic that inhibits bacterial RNA polymerase; also called rifampicin

RIVET see **recombinase-based IVET**

RNA interference (RNAi) the posttranscriptional silencing of specific genes by double-stranded RNA; involves two types of small RNA, micro-RNA (miRNA) and short-interfering RNA (siRNA), that bind to mRNA and induce the Dicer complex that degrades double-stranded RNA and prevents expression of proteins

RNAi see **RNA interference**

16S rRNA gene microarray chips see **phylochips**

SAg see **superantigen**

Sak see **staphylokinase**

Salmonella enterica **serovar Typhimurium DT104** a multidrug-resistant strain responsible for outbreaks among humans in Europe and the United States

SALT see **skin-associated lymphoid tissue**

Sanger sequencing see **high-throughput sequencing**

Saponins plant-derived triterpene glycosides used as vaccine adjuvants

SAR see **structure-activity relationship**

Scalded-skin syndrome a disease in infants caused by exfoliative toxins produced by *Staphylococcus aureus* where the upper layers of skin peel off

SCC*mec* see **staphylococcal cassette chromosome, *mec***

SCOTS see **selective capture of transcribed sequences**

SEA, SEB, SEC1, SEC2, SEC3, SED, and SEE different antigenic types of staphylococcal enterotoxins (SEs) produced by *Staphylococcus aureus*; SEA is the most common cause of food-borne disease

Sec system see **general secretory pathway**

SecA a molecular motor whose ATPase activity provides part of the energy for the protein translocation process of the general secretory pathway

SecA1 a protein that is part of the accessory Sec system; closely related to SecA proteins of other systems

SecA2 a protein that is part of the accessory Sec system

SecB a protein chaperone that binds to proteins targeted for secretion through the general secretory pathway

Secondary metabolites compounds that appear to have no direct role in energy metabolism or essential biosynthetic reactions; examples are pigments and antibiotics

Secretory IgA (sIgA) the dimeric form of IgA with two IgA molecules bound by a piece of the receptor protein (secretory piece) that is found in luminal secretions; provides local immune protection for mucosal membranes by binding both mucin and bacteria, which are then sloughed off

Secretory piece a portion of the IgA receptor of mucosal cells that becomes attached to IgA as it passes through the mucosal cells; the secretory piece plus IgA becomes sIgA

SecYEG subunits of the protein-conducting channel of the general secretory pathway

Select agents a list of bacteria, viruses, and toxins that are considered to be potential bioterror agents (biothreats)

Selectins a family of cell adhesion molecules (CAMs); surface transmembrane glycoproteins of endothelial cells that bind to PMNs

Selective capture of transcribed sequences (SCOTS) a reverse transcriptase-PCR-based method that can be used to identify genes transcribed only under in vivo or in vitro conditions

Self-transmissible plasmid see **conjugative plasmid**

Semisynthetic antibiotics antibiotics originally isolated from nature but then chemically modified to improve their characteristics

Sensor kinase see **histidine kinase**

Sepsis (i) a condition resulting from microbes or microbial products in the blood; (ii) systemic inflammatory response syndrome plus culture-documented infection

Septic shock　a systemic reaction caused when bacterial cell wall components (LPS, LTA, and peptidoglycan fragments) trigger the release of cytokines that have a variety of effects on control of body temperature and blood pressure; symptoms include fever, hypotension, disseminated intravascular coagulation, acute respiratory disease, and multiple organ system failure

Septicemia　a systemic disease in which microorganisms multiply in the blood or are continuously seeded into the bloodstream

Serogroup　a classification of bacterial strains based on a particular group of bacteria containing a common, shared antigen, such as the O antigen of LPS for classification of *Escherichia coli* strains

Serological classification (serogroup; serotype)　a scheme based on reactivity of bacterial surface antigens with antibodies (e.g., O antigen of LPS, H antigen of flagella, and C antigen of streptococci); see **Lancefield grouping system**

Serotype　the serological classification of a bacterial subgroup within a species (e.g., classification of *Escherichia coli* based on H antigen or of *Streptococcus pneumoniae* based on capsule antigens)

Serotyping　a classification of bacteria based on the reactivities of their surface molecules with different antibodies

Serovar　a subgroup classification of botulinum neurotoxins based on the non-cross-reactivity of antibodies against one group of the toxin proteins with those of another group

Serum　the fluid portion of the blood without clotting factors; the component of blood that is collected after coagulation; contains antibodies

Serum resistant　describes bacteria that can resist the killing action of serum; in gram-negative bacteria, serum resistance results from alteration in the O antigen so that MAC cannot nucleate around LPS

Serum sensitive　describes bacteria that cannot resist the killing action of serum; gram-negative bacteria that are susceptible to MAC

SEs　see **staphylococcal enterotoxins**

Severe sepsis　the third stage of septic shock characterized by organ dysfunction and low blood pressure

Sexually transmitted disease　disease transmitted by sexual activity

Shiga-like toxin (SLT)　an AB-type exotoxin produced by some species of intestinal pathogens (e.g., EHEC); has an activity like that of Shiga toxin

Shiga toxin (ST)　an AB-type protein toxin produced by *Shigella* that cleaves rRNA and stops protein synthesis in host cells; responsible for HUS during *Shigella* or EHEC infection

Shigella dysenteriae　a gram-negative bacterium that causes dysentery; produces Shiga toxin

Short interfering RNA (siRNA)　double-stranded RNA molecules 21 to 23 nucleotides long that bind to homologous mRNA and that induce the Dicer complex to cause selective degradation of that mRNA

Shotgun sequencing　see **high-throughput sequencing**

Sialic acid　N- or O-substituted derivatives of neuraminic acid, a nine-carbon sugar found commonly on mammalian cell glycolipids and glycoproteins; used by some pathogens to coat their surfaces and evade the host immune response by mimicking host cell surfaces

Siderophores　low-molecular-weight compounds produced by bacteria that chelate iron; involved in iron acquisition by bacteria

sIgA　see **secretory IgA**

sIgA protease　a bacterial enzyme that cleaves human sIgA at the hinge region

Sigma factor　a subunit of bacterial RNA polymerase that allows RNA polymerase to recognize a particular class of promoter; binds to the core RNA polymerase subunits to form a holoenzyme

Signal recognition particle (SRP)　a ribonucleoprotein (protein-RNA) complex to which signal peptides of integral membrane proteins bind and are targeted to the endoplasmic reticulum in eukaryotes or the plasma membrane in prokaryotes

Signal sequence　a sequence of 15 to 26 amino acids at the N termini of proteins that target them for secretion via the general secretory system

Signal transduction　a cascade of signaling proteins that mediate messages within a cell through external stimulation of a bacterial or mammalian cell surface receptor, which triggers a set of protein modifications, such as phosphorylation/dephosphorylation, that affect gene expression and metabolism

Signature-tagged mutagenesis (STM)　a procedure combining in vitro-barcoded transposon mutagenesis with in vivo selection to screen for mutants that do not grow in an animal host

Simple epithelium　a monolayer of epithelial cells covering body surfaces where absorption or secretion takes place

Single-nucleotide polymorphism (SNP)　a spontaneous mutation in which single nucleotides are changed; can result in amino acid substitution or protein chain termination

siRNA　see **short interfering RNA**

SIRS　see **systemic inflammatory response syndrome**

Site-specific DNA resolvase　a protein that causes cleavage of DNA at a specific recognition sequence; used in RIVET

Skin-associated lymphoid tissue (SALT)　a specialized set of cells that confront bacterial invaders in the area immediately underlying the skin and attempt to prevent their access to the bloodstream

Slipped-strand misrepair (also slipped-strand synthesis)　a process by which phase variation can occur; results from a frameshift during DNA replication in regions of highly repetitive sequence; a frequent basis of phase variation in *Neisseria gonorrhoeae*

Slipped-strand synthesis　see **slipped-strand misrepair**

SLO　see **streptolysin O**

SLT　see **Shiga-like toxin**

Small RNA (sRNA)　transcripts that are complementary to mRNAs of target genes; can modulate levels of virulence proteins

SNAP-25　synaptosomal-associated protein 25; a SNARE complex protein associated with the plasma membrane in neurons that facilitates synaptic vesicle-plasma membrane fusion during neurotransmission; complexed with the synaptic vesicle protein synaptobrevin (also called VAMP) and another plasma membrane protein, syntaxin; serves as a substrate for cleavage by botulinum neurotoxins (BoNTs)

SNARE proteins　a set of neuronal proteins (comprised of SNAP-25, synaptobrevin, and syntaxin) that form a SNARE complex involved in synaptic vesicle fusion with the plasma membrane during neurotransmission; intracellular targets of botulinum neurotoxins (BoNTs)

SNP　see **single-nucleotide polymorphism**

Sod　see **superoxide dismutase**

Sortase　a membrane transpeptidase of streptococci and other gram-positive bacteria that cleaves polypeptides between the threonine and glycine of the LPXTG motif and catalyzes covalent peptide bond formation between surface proteins, including pilin subunits, and amino acids in the peptidoglycan of the cell wall, thereby anchoring the proteins to the bacterial cell surface

Specialized transducing phages　lysogenic phages that can transduce bacterial sequences close to the phage attachment site, thereby packaging some of the host DNA along with the phage DNA in the phage particle; they undergo both lytic and lysogenic phases

Specific-pathogen-free (SPF) animals　animals raised in an environment free of a particular pathogen or pathogens, yet naturally colonized by other microbes

SPF　see **specific-pathogen-free animals**

Spleen macrophages　resident (fixed) macrophages of the spleen

Spore　a metabolically dormant form of some bacteria, derived from vegetative cells, that is highly resistant to extreme environmental conditions

Sputum　the material coughed up from an infected lung

Squalene　a precursor of cholesterol and other steroids; a component of a new adjuvant approved for use in humans

Squamous cells　flattened scale-like cells on the outer layer of the skin and other surfaces, e.g., the stomach lining and respiratory and digestive tracts

sRNA　see **small RNA**

SRP　see **signal recognition particle**

SSH　see **suppressive subtractive hybridization**

ST　see **Shiga toxin**

Staphylococcal cassette chromosome, *mec* (SCC*mec*)　a mobile genetic element containing the *mecA* gene and the genes regulating its expression in *Staphylococcus aureus*

Staphylococcal enterotoxins (SEs)　superantigens produced by *Staphylococcus aureus*

Staphylokinase (Sak)　an exoprotein toxin produced by *Staphylococcus aureus* that inactivates plasminogen, leading to dissolution of clots

Stem-loop structure　an internal paired structure that forms in mRNA that can be used in regulatory mechanisms, such as erythromycin resistance

STM　see **signature-tagged mutagenesis**

Stochastic processes　the random noise (chance events) in regulatory circuits that can lead to subpopulations with different phenotypes and environmental responses

Stratified epithelium　multiple layers of cells that make up the epithelium

Strep throat　pharyngitis due to *Streptococcus pyogenes*

Streptococcal gangrene　see **necrotizing fasciitis**

Streptogramins　a group of antibiotics that act by binding the ribosome and preventing protein translocation

Streptokinase　a blood clot-dissolving enzyme produced by *Streptococcus pyogenes* that binds and activates plasminogen to produce plasmin

Streptolysin O (SLO)　an oxygen-labile pore-forming toxin produced by *Streptococcus pyogenes*

Structure-activity relationship (SAR)　a process used in drug discovery to modify chemicals and determine the effects of the changes on biological drug properties, such as antibiotic efficacy

Substrate　see **matrix**

Subunit vaccine　a vaccine that consists of a few purified proteins, such as detoxified toxins, adhesins, or other bacterial surface proteins

Suicide substrate　see **mechanism-based inhibitor**

Sulbactam　a suicide-substrate-based inhibitor of β-lactamase

Sulfonamides　a family of antibiotics that inhibit an enzyme in the pathway that leads to synthesis of tetrahydrofolic acid; sulfa drug antibiotics

Superantigen (SAg)　peptide toxins that bind to MCH-TCR complexes with or without specific antigen bound and thereby stimulate large populations of T cells to produce cytokines

Superbugs　microorganisms that are resistant to most antibiotics

Supercoiling　a topological state of closed, circular, double-stranded DNA where the DNA strands are momentarily broken, twisted around their axis, and rejoined

Superoxide dismutase (Sod)　an enzyme that catalyzes conversion of superoxide radical to hydrogen peroxide

Superoxide radical (O₂⁻) a reactive oxygen species produced during phagocytosis

Suppressive subtractive hybridization (SSH) a PCR-based technique that involves amplification of only cDNA fragments that differ between a control and an experimental transcriptome; a combination of GSH and SCOTS

Supragenome (also pangenome) a large gene pool comprised of the full complement of all core and distributed genes in a population of bacteria of the same species; because there can be large variation in gene content among closely related strains, this increases the opportunity for bacteria to easily acquire new genes

Surface adhesin a molecule found on the surface of bacteria that recognizes and binds to a host cell receptor molecule

Surveillance programs programs developed to monitor the appearance of new diseases, increased incidence of known diseases, antibiotic-resistant bacteria, and other emerging infectious disease or epidemiological patterns

Swiss-ProtKB Swiss Institute of Bioinformatics protein sequence database

SYBR green a dye that binds to double-stranded PCR products; enhances fluorescence

Symptoms the physiological effects of bacterial colonization or invasive disease that are apparent to the infected host

Synaptobrevin (also VAMP) a SNARE complex protein associated with the synaptic vesicle membrane in neurons that facilitates synaptic-vesicle–plasma membrane fusion during neurotransmission; complexed with the plasma membrane proteins SNAP-25 and syntaxin; serves as a substrate for cleavage by botulinum neurotoxins (BoNTs)

Syndrome the symptoms that characterize a specific disease

Synercid an antibiotic consisting of two streptogramin antibiotics

Syntaxin a SNARE complex protein associated with the plasma membrane in neurons that facilitates synaptic-vesicle–plasma membrane fusion during neurotransmission; complexed with the synaptic vesicle protein syntaxin and another plasma membrane protein, SNAP-25; serves as a substrate for cleavage by botulinum neurotoxins (BoNTs)

Systemic affecting the whole organism rather than a specific organ or tissue

Systemic inflammatory response syndrome (SIRS) the first stage of septic shock

T the tetanus toxoid component of the DTP or DTaP vaccine

T cell a thymus-derived lymphocyte of the adaptive immune system; T helper cells activate macrophages (Th1) or stimulate antibody production by B cells (Th2); cytotoxic T cells (CTLs) kill host cells infected by specific intracellular pathogens

T helper (Th) cells T cells that activate macrophages or stimulate antibody production by B cells; they have CD4 on their surfaces

TA teichoic acid

TA modules see **toxin-antitoxin modules**

TAT see **twin-arginine transport system**

T-box riboswitch a tRNA-mediated mechanism that controls amino acid biosynthesis and aminoacyl tRNA synthetase expression in gram-positive bacteria

TcdA one of the large AB-type protein toxins produced by *Clostridium difficile* that catalyzes the monoglucosylation of small Rho GTPases (also referred to as *C. difficile* toxin A); responsible for symptoms of pseudomembranous colitis

TcdB one of the large AB-type protein toxins produced by *Clostridium difficile* that catalyzes the monoglucosylation of small Rho GTPases (also referred to as *C. difficile* toxin B); responsible for symptoms of pseudomembranous colitis

TcdC a regulatory protein that acts as an anti-sigma factor that inhibits synthesis of *Clostridium difficile* toxins A and B

TcdR an alternative sigma factor that mediates transcription of toxin genes of *Clostridium difficile*

T-cell-independent antibody response the activation of B cells by nonpeptide antigens, such as polysaccharides, lipids, or nucleic acids, without interaction with T cells that leads to production of antibodies (without memory); absent in infants

T-cell-independent antigen an antigen, such as a polysaccharide, lipid, or nucleic acid, that interacts directly with B cells, bypassing APCs and T cells and directly stimulating an antibody response; no memory cells are produced

T-cell receptor (TCR) a protein complex on the surface of T cells that recognizes a specific epitope bound to an MHC complex

TcpH one component of a two-component regulatory system that along with TcpP is involved in regulating the expression of the *toxT* and *tcp* genes of *Vibrio cholerae*

TcpP one component of a two-component regulatory system that along with TcpH is involved in regulating the expression of the *toxT* and *tcp* genes of *Vibrio cholerae*

TCR see **T-cell receptor**

TCS see **two-component regulatory system**

TCT see *Bordetella* **tracheal cytotoxin**

Teichoic acids polysaccharides of glycerol phosphate or ribitol phosphate molecules linked via phosphodiester bonds that are found interwoven with and covalently attached to the peptidoglycan of gram-positive bacteria; they represent as much as 50% of cell wall material

Teicoplanin a glycopeptide antibiotic that inhibits cell wall biosynthesis

Telithromycin a member of the ketolide family of antibiotics; the only member on the market

Temperate bacteriophages bacteriophages that can integrate into the bacterial chromosome (lysogeny) or enter the lytic cycle and kill the bacteria; temperate phages can encode toxin genes (e.g., diphtheria toxin)

TeNT see **tetanus toxin**

Terminal restriction fragment length polymorphism (T-RFLP) a PCR-based method for profiling microbial communities in which the 16S rRNA gene is PCR amplified using primers to each end that are labeled with a different fluorescent dye; the resulting PCR product is cleaved with restriction enzymes, and the fragments from each end are separated by size using a DNA sequencer

Tetanus spastic paralysis caused by tetanus toxin

Tetanus toxin (TeNT) an AB-type neurotoxin produced by *Clostridium tetani* that acts by cleaving synaptobrevins to cause spastic paralysis

Tetracyclines a family of antibiotics that bind to 16S rRNA in the 30S ribosomal subunit and distort the A site

Tetrahydrofolic acid an essential cofactor in the biosynthetic pathway leading to formyl-methionine and nucleic acid precursors; synthesis is inhibited by trimethoprim and sulfonamide antibiotics

tetX a gene encoding an enzyme that modifies tetracycline under aerobic conditions

TGGE see **denaturing/thermal gradient gel electrophoresis**

Th cells see **T helper cells**

Thermostable riboswitches riboswitches that regulate translation by masking ribosomal binding sites at low temperature and thereby preventing protein translation

Thymus-dependent antigens antigens that require T cells, which mature in the thymus to activate B cells

Tigecycline a tetracycline derivative with a bulky glycyl-glycine side chain

Tight junctions (or zonula occludens) closely associated areas where epithelial cell membranes join together to form a tight barrier that prevents fluids from moving between the lumen and substratum

Tissue plasminogen activator (tPA) a protein that activates plasminogen to produce plasmin, which results in an increase in fibrinolytic activity and dissolving of blood clots

TLR3 a toll-like receptor that recognizes poly(IC) double-stranded RNA

TLR4 a transmembrane signaling receptor that recognizes LPS bound to LBP and MD-2 to trigger innate immune responses

TLR5 a toll-like receptor that recognizes flagellin

TLR9 a toll-like receptor that recognizes CpG oligonucleotides

TLRs see **Toll-like receptors**

TNF-α see **tumor necrosis factor alpha**

Tolerance the dormant (nongrowing) condition where bacteria are able to avoid being killed by antibiotics; bacteria begin growing again when the antibiotic is removed

Toll a transmembrane receptor in *Drosophila* required for resistance to fungal infections

Toll-like receptor 4 see **TLR4**

Toll-like receptors (TLRs) a group of mammalian surface proteins that interact with various PAMPs (LPS, peptidoglycan fragments, and other microbial components) and signal innate immune responses, such as changes in cytokine production

Toxic shock syndrome (TSS) a potentially lethal disease caused by strains of *Staphylococcus aureus* that produce the superantigen toxic shock syndrome toxin 1 (TSST-1); symptoms include fever, rash, exfoliation of the palms and soles of the feet, and toxic shock

Toxic shock syndrome toxin (TSST) a superantigen exotoxin produced by some strains of *Staphylococcus aureus*

Toxic shock syndrome toxin 1 (TSST-1) an exotoxin produced by some strains of *Staphylococcus aureus*; a superantigen

Toxin A see **TcdA**

Toxin B see **TcdB**

Toxin neutralization the binding of antibodies to toxins, thereby preventing toxin action, such as binding of the toxin to host target cells

Toxin-antitoxin (TA) modules two-component protein complexes consisting of a toxin that, when expressed by itself, shuts down various cellular processes, including those susceptible to antibiotics, and an antitoxin that binds tightly to the toxin and prevents toxin action; may allow persisters to survive in environments containing antibiotics

Toxin-antitoxin systems see **toxin-antitoxin modules**

Toxinoses diseases in which symptoms are due entirely to the action of toxins

Toxoid a protein toxin that has been chemically or heat treated to destroy its toxicity but retain its immunogenicity; the inactive fragment or recombinant form of a protein toxin

Toxoid vaccine a vaccine that consists of toxoids (inactivated toxins); examples are diphtheria vaccine and tetanus vaccine

ToxR a component of the regulatory system that, along with ToxS, controls the expression of at least 17 genes in *Vibrio cholerae*; regulated genes include those for toxin, pili, serum resistance, and outer membrane proteins

ToxS a component of the regulatory system that, along with ToxR, controls the expression of at least 17 genes in *V. cholerae*; regulated genes include those for toxin, pili, serum resistance and outer membrane proteins

ToxT   a global regulator that activates the expression of *tcp*, *ctxAB*, and other virulence genes in *Vibrio cholerae*; its synthesis is regulated by ToxR, ToxS, TcpP, and TcpH

ToxT regulon   a regulon containing *tcp*, *ctxAB*, and other genes located in different places on the *Vibrio cholerae* chromosome that are regulated similarly by ToxT

tPA   see **tissue plasminogen activator**

Transconjugants   the progeny that result from conjugation

Transcription factor   a regulatory protein that binds to specific bacterial gene promoters and changes transcript levels

Transcriptional activator   a regulatory protein that facilitates binding of RNA polymerase to promoters and increases initiation of transcription; a positive regulator

Transcriptional antiterminators   factors that allow RNA polymerase to resume transcription following premature transcription termination

Transcriptional fusion   a hybrid gene with the promoter/operator of one gene (such as a virulence gene) fused to a promoterless reporter gene encoding an assayable enzyme or fluorescent protein

Transcriptional repressor   a protein that prevents transcription of genes by binding to an operator

Transcriptional response   the global changes in transcription, in terms of relative amounts determined by microarrays, in bacterial or host cells in response to adherence, invasion, or other stimuli

Transcriptional terminators   Rho factor-independent RNA secondary structures (hairpin loops) followed by a run of uridine residues that cause RNA polymerase to terminate transcription

Transcriptome   the full complement of RNA transcripts determined by relative transcript amounts for each gene in an organism

Transduction   the transfer of genetic information from one bacterial cell to another by a bacteriophage

Transfection   a process by which nucleic acids are introduced into mammalian cells

Transferrin   a blood and tissue glycoprotein synthesized by the liver that binds tightly to and sequesters iron

Transformant   the progeny resulting from DNA transformation

Transformation   a process in which competent bacteria take up free DNA from the environment

Transgenic mice   mouse strains with specific genetic alterations; examples are knockout mice and knock-in mice

Transglycosylation   a process by which *N*-acetylmuramic acid and *N*-acetylglucosamine residues of peptidoglycan are linked to form the polysaccharide glycan backbone

Translational attenuation   a type of regulation that controls the expression of genes involved in imparting resistance to translation-inhibiting antibiotics, such as erythromycin

Translocase complex   a part of the general secretory pathway; part of a protein-conducting channel through the cell membrane

Translocation   the movement of the catalytic A subunit of an AB-type toxin into the cytoplasm of a host cell

Transmigration   the movement of natural killer cells and PMNs across the blood vessel wall into tissues

Transpeptidation   the formation of peptide bonds between peptidoglycan peptides that link different peptidoglycan chains

Transposase   an enzyme that catalyzes the transposition of transposons and insertion sequences

Transposon   a segment of DNA containing insertion sequences flanking one or more genes encoding a transposase and virulence genes or antibiotic resistance genes

Transposon mutagenesis   the process of generating mutations by inserting a transposon carrying a selectable marker randomly into genes

T-RFLP   see **terminal restriction fragment length polymorphism**

Triclosan   an antiseptic/disinfectant incorporated into many plastic products

Trimethoprim   a bacteriostatic antibiotic that inhibits an enzyme in the tetrahydrofolic acid biosynthetic pathway

Tropism   a biological phenomenon indicating the growth, orientation, or movement of a microbe toward an external stimulus

TSS   see **toxic shock syndrome**

TSST   see **toxic shock syndrome toxin**

Tumor necrosis factor alpha (TNF-α)   a proinflammatory cytokine produced by monocytes and macrophages in response to LPS

Twin-arginine transport (TAT) system   a transmembrane protein export system dedicated to the export of fully folded proteins out of the cell; found in both gram-positive and gram-negative bacteria

Two-component regulatory system (TCS)   a bacterial signal transduction system in which one protein senses the signal and then phosphorylates the second protein to produce the form that responds to the signal, often regulation of transcription

Type I toxin   a bacterial toxin that is not translocated into a host cell but binds to the target cell surface; an example is superantigens

Type II toxin   a bacterial toxin that acts on the host cell membrane; examples are phospholipase and pore-forming cytotoxins

Type III toxin   a classical AB-type toxin that has a binding region that recognizes a specific host cell receptor, a translocation region that mediates translocation across the cell membrane, and a catalytic toxic domain that enters the host cell cytoplasm and modifies a host protein, rRNA, signaling molecule, metabolite, or other host cell process

uidA   a gene encoding *Escherichia coli* β-glucuronidase (GUS); a reporter gene

Uni ProtKB   Universal Protein Resource Knowledge-base; a database that provides protein translations of nucleotide sequences from the nucleotide sequence databases

Unidentified reading frame (URF)   a gene sequence encoding a putative polypeptide of unknown function that has no homology to known proteins; see **open reading frame**

UPEC   see **uropathogenic *Escherichia coli* strains**

URF   see **unidentified reading frame**

Urinary tract infection (UTI)   an infection of any portion of the urinary tract

Uroepithelium   the epithelial cells lining the urinary tract

Uropathogenic *Escherichia coli* (UPEC) strains   strains of *E. coli* that cause urinary tract infections

Uroplakins   mannose-containing glycoproteins located on the luminal surfaces of bladder epithelial cells that are the sites of attachment of UPEC strains

Usher   a protein dimer in the outer membrane that aids in the assembly and export of pilin

UTI   see **urinary tract infection**

Vaccination   the stimulation of an adaptive immune response to a specific pathogen or antigen by administering a vaccine

Vaccine   a suspension of microorganisms (usually killed or attenuated) and/or their products (toxoids, adhesins, and surface proteins), usually including an adjuvant, that is used for immunization

Valence   the number of antigen- or ligand-binding sites available for binding epitopes on an antigen or ligands on a receptor

VAMP   see **synaptobrevin**

vanA and *vanB* **genes**   vancomycin resistance genes that encode an enzyme that makes D-ala-D-lactate

Vancomycin   a glycopeptide antibiotic that inhibits bacterial cell wall biosynthesis

Vancomycin-intermediate susceptibility *Staphylococcus aureus* (VISSA)   strains of *S. aureus* that are approaching the point of being resistant to vancomycin

Vancomycin-resistant enterococci (VRE)   strains of *Enterococcus* (usually *Enterococcus faecium*) that are resistant to vancomycin; a common cause of nosocomial infections

vanH **gene**   a vancomycin resistance gene that encodes lactate dehydrogenase, which catalyzes the conversion of pyruvate to lactate

vanX **gene**   a vancomycin resistance gene that encodes an enzyme that cleaves D-ala from D-ala-D-ala

Variable region (Fv)   the N-terminal portion of the Fab domain of an antibody molecule that binds to a specific antigen

Vascular   related to blood vessels of the circulatory system

Vasoactive compounds   compounds released by mast cells that dilate blood vessels

Vasodilation   the dilation (widening) of blood vessels in response to hormone-induced relaxation of smooth muscle cells within the vessel walls

Vector   (i) a transmitter of infectious microorganisms, such as an arthopod; (ii) a virus or plasmid used in cloning

Verotoxin   an AB-type protein toxin related to Shiga-like toxin (SLT) that is produced by some strains of *Escherichia coli*

Virosome   a vesicle composed of a lipid membrane with membrane-bound viral proteins (often hemagglutinin or neuraminidase) that serves as an efficient antigen carrier and adjuvant in vaccine designs

Virotypes   the different strains of a particular bacterial species that are grouped based on their virulence factors or virulence properties, such as the different strains of *Escherichia coli*

Virotyping   the classification of various strains of a particular species of bacteria based on the types of virulence factors or different disease strategies exhibited

Virstatin   a small molecule that inhibits ToxT; discovered by high-throughput phenotypic screening

Virulence   the ability of a microorganism to cause disease

Virulence factor   a bacterial product or strategy that contributes to the ability of the bacterium to survive in the host and/or cause infection

VISSA   see **vancomycin-intermediate susceptibility *Staphylococcus aureus***

VRE   see **vancomycin-resistant enterococci**

Warhead   the reactive portion of a drug or antibiotic that often forms a covalent bond with its target to inactivate it (e.g., the β-lactam ring found in β-lactam antibiotics)

Water-borne infections   bacterial infections acquired from contaminated water

Weapon of mass destruction (WMD)   a bioweapon used as an implement in a state-sponsored war conducted on a grand scale

Weapon of mass disruption (WMD)   a bioweapon used on a small scale by groups or individuals to spread terror and disrupt normal routine operations

Whole-body biophotonic imaging   a noninvasive technique for monitoring disease progression in live animals whereby the bacteria (or sometimes animal hosts) are engineered with reporter genes that express bioluminescence when activated during colonization or invasive disease

Whooping cough   see **pertussis**

WMD see **weapon of mass destruction** or **weapon of mass disruption**

Wound botulism a paralytic disease caused by botulinum neurotoxin produced by *Clostridium botulinum* growing in a wound

XDR see **extensively drug resistant**

X-Gal 5-bromo-4-chloro-3-indolyl-β-D-galactopyranoside, a chromogenic substrate for β-galactosidase that produces a blue color when cleaved

X rays radiation generated when an electron beam is directed at a thin plate of gold or other metal; used to irradiate food

Yops the outer membrane or secreted proteins of *Yersinia;* some are part of a type 3 secretion system (T3SS), and others are T3SS effector proteins that are delivered directly to eukaryotic cells and cause toxic effects on the host cells

Zinc β-lactamase a type of β-lactamase that can cleave many classes of β-lactams and is not inhibited by clavulanic acid; does not contain a serine residue in its active site

Zonula occludens see **tight junctions**

Zoonosis an animal disease that can be transmitted to humans

Index